European Mineralogical Union
Notes in Mineralogy

Commissioning Editor: R. Oberti

Volume 16

MINERAL REACTION KINETICS: MICROSTRUCTURES, TEXTURES, CHEMICAL AND ISOTOPIC SIGNATURES

UNIVERSITY TEXTBOOK

Edited by

W. HEINRICH and R. ABART

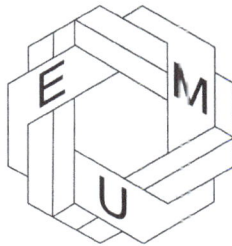

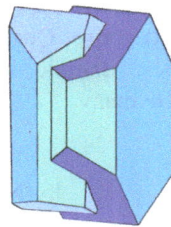

Published by the European Mineralogical Union and the Mineralogical Society of Great Britain & Ireland, London, 2017

The publication of this textbook is supported by the European Mineralogical Union

EMU Notes in Mineralogy
A series published under the auspices of the European Mineralogical Union (EMU).

Initiator of the EMU Schools and the EMU Notes in Mineralogy
Giovanni Ferraris, Torino, President of the EMU 1992−1996

Commissioning Editor: R. Oberti, Pavia (previous editors: Giovanni Ferraris, Torino,
Tamás G. Weiszburg and Gábor Papp, Budapest)

Editors of this Volume
W. Heinrich, Helmholtz-Zentrum Potsdam, Germany and
R. Abart, University of Vienna, Austria

Managing Editor and Indexer: Kevin Murphy, London

Front cover design: Michel H. Guay

Executive Committee of the EMU (2016−2020)
Michael Carpenter, UK, (President), Wilhelm Heinrich, Germany (Past-President),
Ewa Słaby, Poland (Vice President), Isabelle Daniel, France (Vice President),
Juraj Majzlan, Germany (Secretary) and Wim van Westrenen, The Netherlands
(Treasurer)

On the front cover: Transmitted-light image of coronitic metagabbro from the Precambrian crystalline basement complex of the Eastern Ghats in Orissa, India. During granulite-facies overprint of a gabbro protolith, (pinkish) garnet rims ~20 μm wide formed along the interfaces between pre-existing hedenbergite (green) and plagioclase (transparent) grains. Quartz is present as a discontinuous rim along garnet/hedenbergite interfaces. Sample and image courtesy of Prof. Michael Raith, University of Bonn, Germany.

ISSN: 1417 2917
ISBN: 978-0903056-63-2

The EMU Notes in Mineralogy Series

Published volumes

Volume	Year	Editors	Title
1	1997	S. Merlino	*Modular aspects of minerals*
2	2000	D.J. Vaughan R. Wogelius	*Environmental mineralogy*
3	2001	C.A. Geiger	*Solid solutions in silicate and oxide systems*
4	2002	C. Gramaccioli	*Energy modelling in minerals*
5	2003	D.A. Carswell R. Compagnoni	*Ultrahigh pressure metamorphism*
6	2004	A. Beran E. Libowitzky	*Spectroscopic methods in mineralogy*
7	2005	R. Miletich	*Mineral behaviour at extreme conditions*
8	2010	F.E. Brenker G. Jordan	*Nanoscopic approaches in earth and planetary sciences*
9	2010	G. Christidis	*Advances in the characterization of industrial minerals*
10	2010	M. Prieto	*Ion-partitioning in ambient-temperature aqueous systems*
11	2011	M.F. Brigatti A. Mottana	*Layered mineral structures and their application in advanced technologies*
12	2012	J. Dubessy M.-C. Caumon F. Rull	*Applications of Raman Spectroscopy to earth sciences and cultural heritage*
13	2013	D.J. Vaughan R.A. Wogelius	*Environmental mineralogy II*
14	2013	F. Nieto K.J.T. Livi	*Minerals at the nanoscale*
15	2015	M.R. Lee H. Leroux	*Planetary mineralogy*
16	2017	W. Heinrich R. Abart	*Mineral reaction kinetics: microstructures, textures, chemical and isotopic signatures*

Copies of the EMU Notes (volumes 1–7) are distributed in Europe by the larger member societies of the European Mineralogical Union:

Società Italiana di Mineralogia e Petrologia:
www.socminpet.it

Mineralogical Society of Great Britain & Ireland:
www.minersoc.org

Société Française de Minéralogie et de Cristallographie:
www.sfmc-fr.org

and by the Secretary of the EMU, Prof. Juraj Majzlan:
juraj.majzlan@uni-jena.de

in America by the Mineralogical Society of America:
www.minsocam.org

Institutional orders as well as individual requests from outside Europe and America
should be sent to the Secretary of EMU, Prof. Juraj Majzlan:
juraj.majzlan@uni-jena.de

For volumes 8 onwards, the Mineralogical Society of Great Britain acts as co-
publisher and copies may be ordered from www.minersoc.org

or from:
Mineralogical Society
12 Baylis Mews,
Amyand Park Road,
Twickenham TW1 3HQ
UK

E-mail: admin@minersoc.org
Tel. + 44 (0)20 8891 6600
Fax: + 44 (0)20 8891 6599

Contents

Introduction

RAINER ABART[1] and WILHELM HEINRICH[2]

[1] *University of Vienna, Department of Lithospheric Research, Althanstrasse 14, A-1090 Vienna, Austria, e-mail: rainer.abart@univie.ac.at*
[2] *Helmholtz-Zentrum Potsdam, Deutsches GeoForschungsZentrum (GFZ) Telegrafenberg, D-14473 Potsdam, Germany, e-mail: whsati@gfz-potsdam.de*

In the late 20th century advances in experimentation and in material characterization greatly fostered the development of internally consistent thermodynamic data. Together with the development of thermodynamic modelling tools this enhanced our ability to analyse phase equilibria in rocks and to obtain accurate quantitative information on the conditions of magmatic and metamorphic crystallization. This gave an unprecedented boost to mineralogy, petrology and geochemistry and helped illuminate long-standing questions in geodynamics as well as in geo- and cosmochemistry. Attainment of thermodynamic equilibrium among the phases constituting a rock and metastable preservation of equilibrium phase relations, which are indispensible pre-requisites for application of equilibrium thermodynamics, could be demonstrated or, in many cases, were tacitly assumed.

With the ever increasing spatial resolution in the characterization of microstructures, textures and chemical as well as isotopic patterns and the continuously improving sensitivity and precision in mineral chemical and isotopic analysis, increasing evidence has emerged indicating that phase relations in rocks from high-temperature environments may be modified during exhumation and cooling and that the original equilibrium phase relations may not have been preserved or that equilibrium may never have been attained fully ever at high temperatures. Such situations are indicated by the chemical zoning of minerals, non-equilibrium element partitioning, and by reaction microstructures comprising reactant and product phases in specific geometric arrangements such as in corona microstructures. The isotopic and chemical patterns and reaction microstructures observed in minerals and rocks reflect the processes involved in their genesis. The mechanisms of chemical mass transport, the breaking and making of chemical bonds at reaction interfaces, nucleation and interface migration and the coupling among these processes determine the microstructures, textures and isotopic and chemical patterns of minerals and rocks. If the associated kinetics is understood, this allows the dynamics of crystallization, including the evolution of the physical and chemical conditions, to be reconstructed. This provides crucial information complementing the insights that may be obtained from the thermodynamic analysis of phase equilibria and from radiometric age dating.

©Copyright 2017 the European Mineralogical Union and the Mineralogical Society of Great Britain & Ireland
DOI: 10.1180/EMU-notes.16.1

This volume accompanies an EMU School that is meant to bring contemporary research on mineral reaction kinetics to the attention of young researchers and to put it into the context of recent developments in related disciplines. The school and the accompanying volume cannot give a comprehensive review of the current state of geomaterials research. Rather a selection of topics, methods and concepts, which the contributors deem currently most relevant and instructive, is presented.

The aim is to provide a methodologically sound insight into the theoretical foundations of mineral reaction kinetics, to help students to become acquainted with contemporary methods in experimentation and analytical techniques, and to give worked examples that illustrate recent advances in geoscience based on an improved characterization and understanding of mineral and rock systems. The volume addresses three broad topical fields.

In the first part recent developments in mineralogical and petrological experimentation and instrumental analyses are reviewed. In particular, scanning and transmission electron microscopy, secondary ion mass spectrometry and synchrotron-based material characterization are addressed. In addition, theoretical approaches including selected topics in irreversible thermodynamics and atomistic modelling are presented.

In the second part the processes underlying mineral reactions, including diffusion, interfaces and interface motion as well as nucleation and growth, are investigated and put into the context of geomaterials. Both theoretical foundations and geoscience applications are discussed.

The third part is devoted to specific fields in geomaterials research. Knowledge from experimentation, theory and modelling is integrated with contemporary analytics to address problems in different fields of geological and planetary research. The various aspects addressed include crystallization from the melt, reactions of minerals with aqueous fluids, metamorphic mineral reactions, crystal orientation relations, reaction-induced deformation and chemical–mechanical feedback, as well as the links between mineral reaction and (isotope) geochemical signatures.

This volume emerged from the joint initiative of an international team of researchers with their backgrounds in mineralogy, petrology, geochemistry, geology, materials science, physics and chemistry, who were brought together through the fascinating topic of geomaterials research as their common denominator. We thank all the contributors for their enthusiasm and for the effort they spent on writing great contributions. We do hope that the choice of topics and presentations collected in this volume will be received well in the community and in particular by the scientists of the young generation. With this volume we intend to strengthen the links among a range of diverse research directions and to stimulate new thoughts and initiatives in geomaterials research that can improve our understanding of how Earth and other planets work.

Rainer Abart
Wilhelm Heinrich
Vienna and Potsdam
September 2016

Acknowledgements for financial support

We would like to thank the following institutions and societies for their generous financial support and sponsorship of the EMU School on "Mineral Reaction Kinetics: Microstructures, Textures, Chemical and Isotopic Signatures".

European Mineralogical Union (EMU)

Deutsche Mineralogische Gesellschaft (DMG)

Deutsche Forschungsgemeinschaft (DFG)

Società Italiana di Mineralogia e Petrologia (SIMP)

Der Wissenschaftsfonds (FWF)

Universität Wien (UNIVIE)

Fakultät für Geowissenschaften, Geographie und Astronomie

Österreichische Mineralogische Gesellschaft (ÖMG)

Helmholtz-Zentrum Potsdam - Deutsches GeoforschungsZentrum (GFZ)

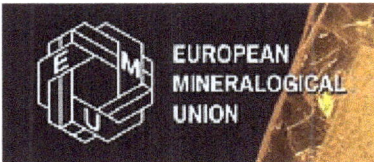

New avenues in experimentation on diffusion-controlled mineral reactions

RALF MILKE[1],* WILHELM HEINRICH[2], LUTZ GÖTZE[1] and SUSAN SCHORR[3]

[1] Freie Universität Berlin, Institut für Geologische Wissenschaften, Malteserstr. 74-100, 12249 Berlin, Germany, e-mail: milke@zedat.fu-berlin.de
[2] GeoForschungsZentrum Potsdam, Telegrafenberg, 14773 Potsdam, Germany
[3] Helmholtz-Zentrum Berlin, Hahn-Meitner-Platz 1, 14109 Berlin, Germany
* corresponding author

Mineral coronas and reaction rims are frequent features in many metamorphic rocks commonly interpreted as having been controlled by solid-state diffusion or by diffusion in an undefined medium. In material science the term 'interlayer growth' is often used for such processes. However, the terms 'reaction bands', 'corona structures', or 'rim structures' are commonly used as descriptive terms in petrology and we will use these below.

The formation of reaction bands, coronas and reaction rims between incompatible phases requires that one or more chemical components are mobile. Knowledge of the transport mechanism and relative mobility of the distinct chemical components is of prime interest for the interpretation of reaction band sequences, growth rates, and subsequently forming fabrics and textures. Recent experimentation at high pressures and temperatures has shown that even minute traces of water are decisive in changing reaction mechanisms in silicate systems. In this chapter we present avenues of experimentation classified as water-rich, water-poor and water-absent. All of them are important in order to gain deeper understanding of mineral-reaction kinetics in the Earth in different environments, and of the formation of texture and zoning patterns in rock assemblages of the Earth or Earth-like planets.

Experimental simulation and parameterization of variables relevant for mineral-reaction kinetics and the development of microstructures and textures requires a specific experimental approach. This consists of (1) application of time (or temperature) series; (2) miniaturization of experimental setups; and (3) experiments with perfectly defined geometries of the phases involved. This chapter specifies experimental setups for investigating mineral-reaction kinetics at high *P-T*, including setups for time-resolved, real-time monitoring of mineral nucleation and growth between incompatible phases at high *T*. We address mainly the experimental strategies in elucidating the controls of mineral-reaction kinetics rather than responding to results of particular studies, which can be found in many other chapters in this volume and in previous review articles (*e.g.* Dohmen and Milke, 2010; Watson and Dohmen, 2010). This holds also for the applied analytical methods as many experimental setups are specifically designed for subsequent analytical procedures.

1. Conventional experiments at high *P* and *T*

1.1. Early work on kinetics of hydrothermal reactions

Over the decades, high-temperature, high-pressure experimentation has provided tremendous insight into our understanding of rock formation and the intensive and extensive variables that control phase assemblages and phase compositions. The vast majority of countless studies aimed at attaining equilibrium in the investigated systems, and applied equilibrium thermodynamics to derive phase properties and phase stabilities in *P-T-X* space. With regard to reaction kinetics, much of the early high-pressure, high-temperature experimental work focused on determining rates of dehydration, devolatilization and hydration reactions under metamorphic conditions. This was done by simply measuring the time- and temperature-dependent turnover of univariant reactions at conditions far from equilibrium using conventional hydrothermal equipment. Usually, fine-grained reactant powders were used at high fluid-solid ratios (using H_2O or CO_2 or both). The reaction turnover was quantified by determining the shift in relative amounts of solid reactants and products by X-ray methods (*e.g.* Schramke *et al.*, 1987), by determining the amount of fluid released during a devolatilization reaction (*e.g.* Heinrich *et al.*, 1989; Lüttge and Metz, 1993a; Milke and Metz, 2002) or by measuring the weight loss of a single reactant crystal after experiment (*e.g.* Holdaway, 1966; Kerrick, 1968). A few experimentalists have moved towards studying mineral reactions in natural rock cylinders that underwent *P-T-X* conditions outside their previous geological context; in these investigations there was always water present to mediate the mineral reactions (Lüttge and Metz, 1993b; Shvedenko *et al.*, 2006; Metz and Milke, 2012; Jolis *et al.*, 2013). The controlling mechanism in such experiments at high fluid-solid ratios is dissolution-reprecipitation. In a seminal paper Walther and Wood (1984) concluded that the rates of such reactions basically follow zero-order kinetics controlled by the reacting surface area with the rate constant given by $\log K = -2900/T - 6.85$, where T is the absolute temperature and K has the units gram-atoms of oxygen per square centimetre per second. They argued that this would hold for all silicates and for reactions involving dissolution, fluid production, or solid-solid reactions in the presence of a fluid of moderate to high pH over a wide temperature range. This holds true only as long as sufficiently large surface areas of the reactants are available to react with water, which may be prevented by progressive encasing of one the reactants by product phases. Many more implications can be studied by a view of mineral pairs brought directly into contact and out of equilibrium at well controlled conditions and, most significantly, well defined geometry.

1.2. Growth kinetics of reaction layers, coronas and reaction rims

Appropriate experimental setups give information about the sequence and relative width of mineral layers in a reaction band, allow for extraction of quantitative relations among the effective diffusivities of the mobile components and may give insight into the controlling mechanism in transport-controlled reactions, *e.g.* volume *vs.* grain boundary diffusion. In this section we refer to the study of reaction rims that form according to transport-controlled mechanisms with growth rates $\Delta x \sim t^{1/2}$, *i.e.* diffusion-controlled processes. This includes water-absent experiments as well as rim growth experiments where transport occurs along wetted grain boundaries which, in extreme cases, is diffusive transport through a pore fluid (*e.g.* Jonas *et al.*, 2015). The formation of replacement rims by interface-controlled mechanisms with zero-order growth rates $\Delta x \sim t$ (*e.g.* Putnis, 2009) is not addressed here.

In the geoscience context two different experimental approaches have been used frequently to investigate the transport-controlled kinetics of reaction rims: (1) growth of mono- or bimineralic polycrystalline reaction products between two incompatible single crystals; and (2) growth of polycrystalline products between reactants consisting of large seed crystals embedded within a fine-grained matrix. Such studies aim to determine the transport properties of polycrystals (where grain boundaries may provide the most efficient pathways) in terms of diffusivities and/or diffusion coefficients of the mobile components. Those values are derived from the measurement of chemical fluxes through reaction rims in samples recovered after the experiment. For both experimental approaches the reconstruction of chemical fluxes across evolving reaction rims requires the identification of the position of the initial contact between the reactants after the experiment. To identify the initial interface several methods have been applied: (1) isotopically doped reactants that leave a step in isotope composition in the reaction rim at the sites of the initial interface (^{18}O and ^{29}Si in quartz for quartz-consuming reactions; Milke *et al.*, 2001); (2) chemically variable reactants (*e.g.* variable Mg/Fe ratios in olivine-consuming reactions) that leave a signal of the initial interface on BSE images (*e.g.* Milke *et al.*, 2009a, 2011); (3) imaging of textural changes at the initial contact, in part with the help of EBSD methods (*e.g.* Abart *et al.*, 2004); and (4) setting inert Pt markers on the reactant interface before the reaction (*e.g.* Fisler *et al.*, 1997; Gardés *et al.*, 2011; Joachim *et al.*, 2011; Nishi *et al.*, 2013; see also Gaidies *et al.*, 2017, this volume, on the Kirkendall effect). The use of Pt markers is a straightforward method to keep track of the initial interface, but some care is required in the interpretations of fluid-bearing systems (see below). It is possible that counter-diffusion of chemical components and/or vacancies may produce porosity along an interface between reactant and product. Consequently, the markers might be pinned to pores by mechanical forces and are not spatially inert anymore. The inert marker method works well in dry systems but might be spoiled when interface mobility is enhanced by the presence of water.

1.3. Formation of reaction layers in the solid state at virtually dry conditions

In high-pressure, high-temperature experimental setups virtually dry conditions are achieved only by using the two-single-crystal approach because employing fine-grained powders as reactants always introduces traces of water into the system, which is adsorbed on the powder surfaces, and which has a dramatic effect on the reaction kinetics (see below, and Gaidies *et al.*, 2017, this volume). Furthermore, high-*P-T* experiments using the sandwich approach can be performed at essentially dry conditions in internally heated pressure vessels, piston cylinder and multi-anvil-presses only if appropriate pressure media are used. Apparently dry reactions between two single crystals without adding water in solid media presses are not necessarily dry because sealed noble metal capsules at high P and T are prone to diffusion of minute amounts of water from the surrounding pressure-transmitting solid medium, *e.g.* pyrophyllite or graphite (Yund, 1997), fluorite (Milke *et al.*, 2001; Joachim *et al.*, 2012), and boron nitride (Milke *et al.*, 2009a). For assemblies containing parts made from natural fluorite, entrance of water into the capsule during heating to 900°C at 1.2 GPa is revealed by the presence of OH defects in previously dry reactants after the experiments (Joachim *et al.*, 2012). In solid-media presses virtually dry conditions (or conditions with a well defined small amount of added water) are achieved only by using hygroscopic pressure media such as polycrystalline crushable Al_2O_3 where the water fugacity was low enough not to affect the reaction rates (Gardés *et al.*, 2011). Internally heated pressure vessels (IHPV) and Paterson-type gas deformation devices with dry argon as pressure medium always ensure dry conditions. IHPVs have an upper pressure limit of 1.0 GPa, however, which is even lower for Paterson-type equipment.

The two-single-crystal (or sandwich) approach has been applied recently to several reactions at various P-T and nominally dry conditions by using sundry experimental devices. By performing time series, overall rim growth rates, transport mechanisms, and diffusivities of the respective components through the product layers have been determined. Examples are: (1) the reaction between forsterite and quartz to produce a polycrystalline enstatite single layer of the form [Fo|En|Qtz], and between periclase and quartz to produce a polycrystalline double layer of [Per|En|Fo|Qtz] at 1.5 GPa, 1100−1400°C, 5 min to 72 h in a piston-cylinder apparatus (Gardés and Heinrich, 2011; Gardés *et al.*, 2012); (2) the reaction between monticellite and wollastonite to polycrystalline åkermanite [Mtc|Åk|Wo] at 0.5 GPa, 1000−1200°C, 5 min to 60 h in an internally heated pressure vessel using dry Ar pressure (Joachim *et al.*, 2011); (3) the reaction between monticellite and wollastonite to a symplectitic lamellar intergrowth of diopside and merwinite [Mtc|alternating Di + Mer lamellae| Wo] at 1.2 GPa, 900°C, 5−65 h in a piston-cylinder apparatus (Joachim *et al.*, 2012); (4) the reaction beween periclase and stishovite to polycrystalline perovskite [Per|Pv|Sti] at 24−50 GPa, 1800−2000°C, 15 min to 25 h in a Kawai-type multi-anvil press (Nishi *et al.*, 2013); (5) the reaction between calcite and magnesite to polycrystalline dolomite [Cal|Dol|Mgs] at 0.4 GPa, 750−850°C, 3−146 h annealing time in a Paterson-type gas deformation apparatus (Helpa *et al.*, 2014); and (6) the reaction between periclase and corundum to spinel [Per|Spi|Cor] at 1.0−4.0 GPa, 1200−2000°C, 3 min to 240 h in a piston-cylinder apparatus (Watson and Price, 2002). Dimensions of the various setups vary with experimental device; however, the geometry of the reaction assemblies is basically the same. An example for a piston-cylinder starting assembly that uses a double sandwich for simultaneous growth of [Fo|En] double layers between Per and Qtz, and an [En] single layer between Fo and Qtz is shown in Fig. 1.

While overall growth rates are obtained simply by measuring the reaction rim width after different run durations, the net component fluxes can be determined only if the

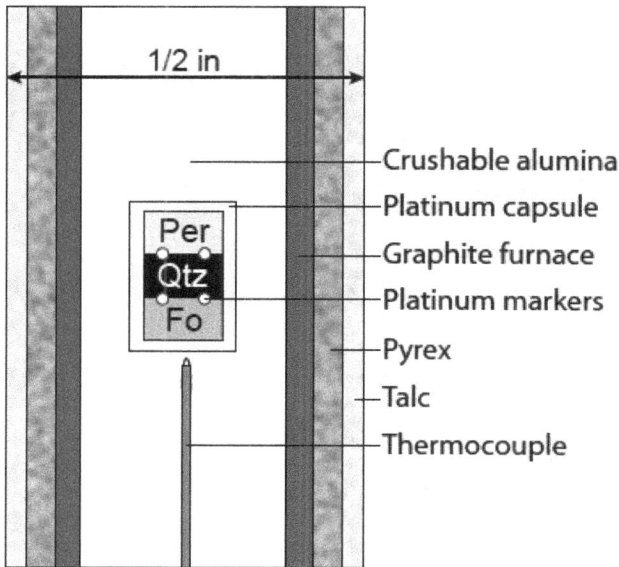

Figure 1. Sketch of a piston-cylinder assembly. The sample is a sandwich consisting of polished slices of periclase, quartz and forsterite single crystals. A forsterite–enstatite double layer grows between periclase and quartz, and an enstatite single layer between forsterite and quartz. The use of dry crystal slices and crushable alumina as pressure medium minimizes water uptake from the cell. From Gardés *et al.* (2011), reprinted by permission of Springer.

position of the original interface between two reactant phases is known. In all studies cited above, the position of the original interface after the experiment was determined through the use of inert markers consisting of small Pt particles considered to be immobile (Fig. 2a,b; see also Fisler *et al.*, 1997; Yund, 1997). We illustrate the principle of this strategy by referring to the experimental results on [Per|Fo|En|Qtz] double rims and [Fo|En|Qtz] single rims (Gardés *et al.*, 2011, see above). Electron microprobe (EMP) back-scattered images revealed that in forsterite–enstatite double rims the Pt markers sit at the Per|Fo interface (Fig. 3a), and in enstatite single rims, exactly in the centre of the enstatite rim (Fig. 3b). Mass-balance considerations taking into account the partial reactions at the respective interfaces indicated unequivocally that MgO was the only mobile component in both the polycrystalline forsterite–enstatite double and in the enstatite single layer, and that its mobility controls overall growth. In the case of double rims, any significant SiO_2 mobility relative to MgO would have shifted the Per|Fo interface away from the markers. Similarly, in enstatite single layers the En|Qtz interface would have been shifted farther away from the Pt markers than the En|Fo interface if mobility of the SiO_2 component had occurred, which was not the case. This is illustrated schematically in Fig. 2a,b.

That rim growth in virtually dry systems is controlled by MgO as the only mobile component was demonstrated for a number of reactions, *e.g.* for [Fo|En|Qtz],

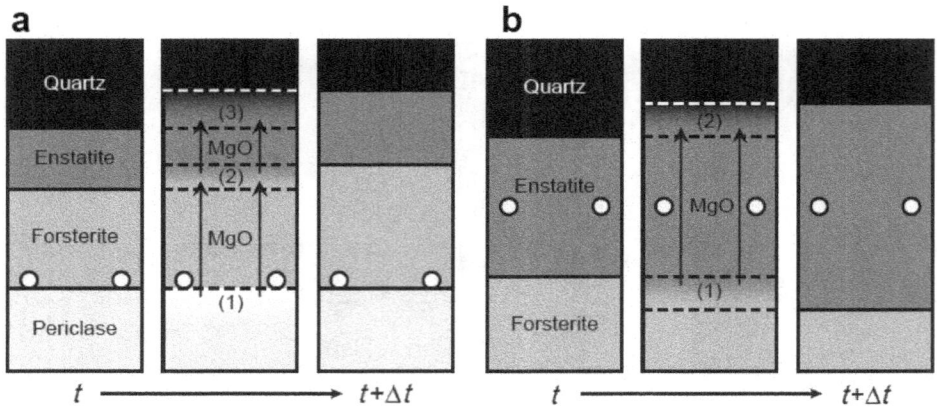

Figure 2. Schematic growth mechanisms of (a) the forsterite–enstatite double layer and (b) the enstatite single layer. Arrows indicate the direction of MgO flux. (a) (1) Periclase decomposes and the released MgO diffuses through the forsterite layer. (2) Part of the MgO component reacts at the forsterite–enstatite interface to produce forsterite. The remaining MgO diffuses through the enstatite layer before (3) reacting at the quartz surface producing enstatite. Quartz is consumed by enstatite and enstatite by forsterite. The platinum markers (bright dots) remain at the reverse side of the growing fronts, *i.e.* at the periclase–forsterite interface. (b) (1) Forsterite decomposes, and new enstatite is produced. The MgO excess diffuses through the enstatite layer and (2) reacts at the quartz surface to produce an equal amount of enstatite. The platinum markers remain in the centre of the layer. From Gardés *et al.* (2011), reprinted by permission of Springer.

[Per|En|Fo|Qtz], [Per|Pv|Sti] and [Mtc|Åk|Wo] (*cf.* above). There are, however, significant differences in overall rim growth rates. Time series showed that the

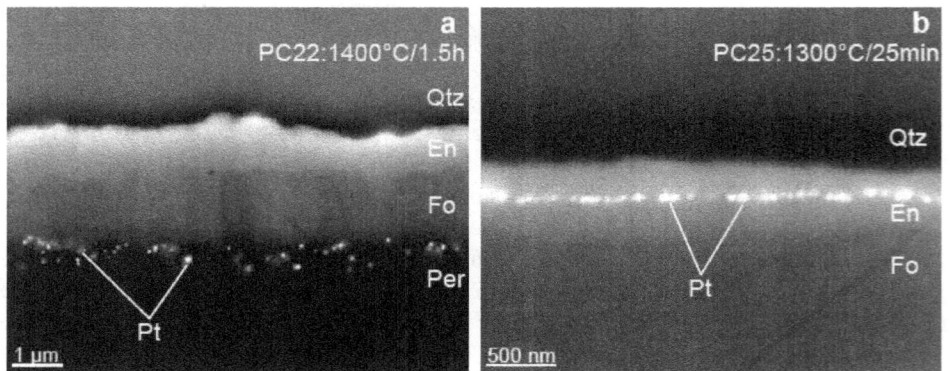

Figure 3. EMP back-scattered electron micrographs of reaction layers. (a) Polycrystalline forsterite–enstatite double rim produced between periclase and quartz at 1.5 GPa, 1400°C, 1.5 h. Inert Pt markers sit at the Per|Fo interface. (b) Polycrystalline enstatite single rim produced between forsterite and quartz at 1.5 GPa, 1300°C, 25 min. Inert Pt markers are located in the centre of the enstaite layer. From Gardés *et al.* (2011), reprinted by permission of Springer.

formation of polycrystalline åkermanite between monticellite and wollastonite follows strictly a parabolic rate law, whereas growth of enstatite between forsterite and quartz, forsterite|enstatite between periclase and quartz, and perovskite between periclase and stishovite does not, but is slower. Textural observations by transmission electron microscopy (TEM) revealed that in all of these systems ongoing reaction leads to pronounced grain coarsening within the reaction layer, which reduces the grain boundary density perpendicular to the MgO flux, indicative of grain boundary diffusion to control the latter reactions. Combination of textural observations, *i.e.* measuring the grain boundary densities of recovered samples with rim-growth kinetic data, allowed us to derive MgO grain boundary diffusivity in forsterite, enstatite (Gardés and Heinrich, 2011); and perovskite (Nishi *et al.*, 2013) at the respective conditions. Regarding the åkermanite-forming reaction from monticellite + wollastonite single crystals, the successive decrease in the grain boundary area fraction did not affect overall rim growth. Consequently, grain boundary diffusion in this case could account for only a minor fraction of the overall mass transfer, and rim growth is essentially controlled by volume diffusion, for which effective MgO volume diffusion coefficients in åkermanite have been derived (Joachim *et al.*, 2011). These examples highlight the fact that detailed textural characterization of reaction products is indispensable for the correct interpretation of kinetic data from reaction time series.

The single-crystal sandwich approach can be modified and extended in several ways. For example, multiple sandwich piles can be used as initial setups (Watson and Price, 2002; Milke *et al.*, 2009a). By using and calibrating the temperature gradient in a piston-cylinder cell this allows us to determine rim development at different temperatures in a single experiment (Fig. 4). Also, reactant crystals can be arranged with respect to distinct crystallographic orientations, which then allows identification of potential topotactic relationships between reactants and products (Götze *et al.*, 2014).

1.4. Dry or wet? Presence of minute amounts of water controls growth kinetics

Recent experimental work on the effect of minute amounts of water on reaction rim kinetics in the systems $MgO-SiO_2-H_2O$ and $CaO-MgO-SiO_2-H_2O$ specifically addressed the 'dry *vs.* wet" problem (Fisler *et al.*, 1997; Yund *et al.*, 1997; Milke *et al.*, 2001, 2007, 2009a, 2013; Gardés *et al.*, 2011, 2012; Joachim *et al.*, 2012). For example, in the CMS(H) system, growth of enstatite rims between forsterite and quartz in the presence of traces of water is about four orders of magnitude faster than the dry system. Water in this respect is not a chemical component in the sense of thermodynamics, but acts as a catalyst that speeds up reaction kinetics. Even though the effect is well established, experimentalists struggled with the dry *vs.* wet concept, particularly with the problem: how much water is required to make a system a wet one?

Piston-cylinder experiments of Joachim *et al.* (2012) on [Mtc|alternating Di + Mer lamellae|Wo] using monticellite + wollastonite sandwiches as reactants have shown that continuous entrance of minute amounts of water into the reaction cell is possible if

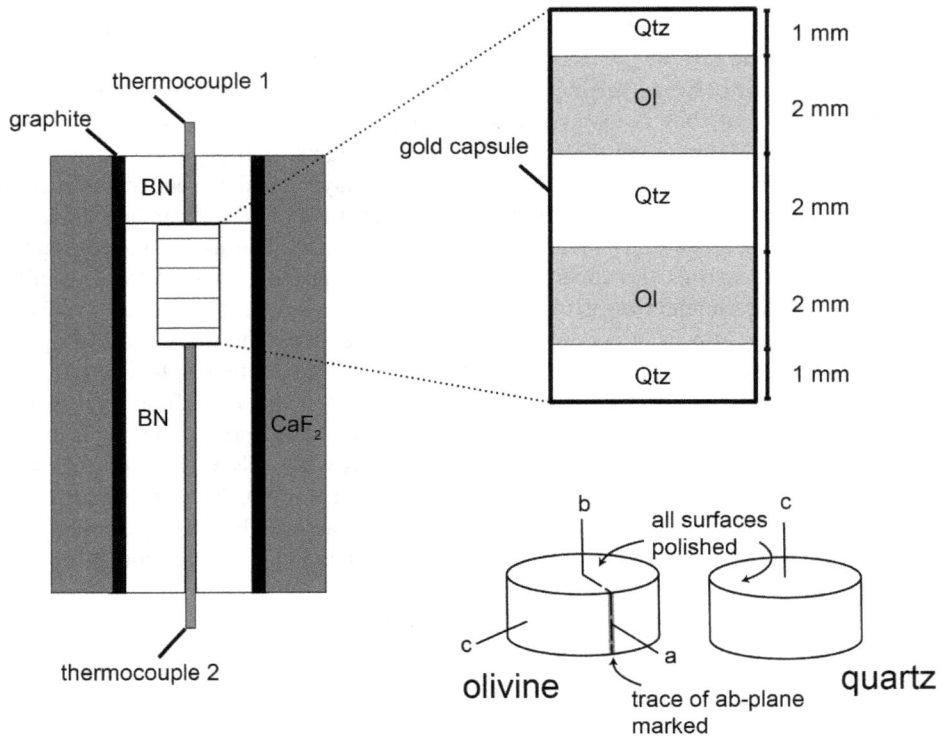

Figure 4. Multiple sandwich setup in rim growth experiments in a temperature gradient. A stack of olivine and quartz single crystal discs in a gold capsule is placed off-centre in the graphite heater such that each run contains four olivine-quartz reactant pairs at four different temperatures. Temperature is measured with two thermocouples placed at both ends of the capsule. Quartz and olivine are cut and piled with known crystallographic orientation. Modified after Milke *et al.* (2009).

the pressure-transmitting solid media are not completely dry. This is revealed by the presence of OH defects in the reactants after the experiments, and particularly by significant changes in the reaction kinetics. Not only do overall reaction rates speed up but the resulting microstructures may also change completely due to changes in relative component mobility, in this case by mobilization of both the CaO and SiO_2 components in addition to MgO (Abart *et al.*, 2012; Joachim *et al.*, 2012).

   The influence of the amount of water on transport mechanisms and rates could not be evaluated quantitatively in these experiments due to continuous water ingression. Uncontrolled water ingression into the reaction cell in piston-cylinder experiments was also demonstrated by Milke *et al.* (2013) for orthopyroxene growth between olivine and quartz (see below).

   The effect of water on intergranular mass transport on growth of the [Fo|En|Qtz] and the [Per|En|Fo|Qtz] rims using the same setup was determined by Gardés *et al.* (2012) by adding distinct amounts of powdered $Mg(OH)_2$ to the starting sandwich, which quickly releases water during heating. Their results are in line with the large set of kinetic data on [Fo|En|Qtz], that had been interpreted as representing either "dry" or "wet" conditions depending on experimental set-up (Dohmen and Milke, 2010; Gardés *et al.*, 2012). All experiments are clearly divided into these two groups following rather straight lines in the Arrhenius diagram with a separation of 2 to 4 orders of magnitude in diffusivity (Fig. 5). In order to explain this kinetic dichotomy, Gardés *et al.* (2012) proposed a classification scheme of intergranular mass transport of MgO as a function of the water-solid ratio (Fig. 6). In their experiments, the jump from anhydrous to hydrous-saturated grain boundary transport was interpreted as occurring at ~500 to 1000 ppm water in the bulk, which was accompanied by an increase in intergranular MgO (and possibly SiO_2) diffusivity of four to five orders of magnitude (regime 2; Fig. 6). This dominates as long as excess water is distributed into isolated pores, and allows diffusivity in hydrous-saturated grain boundaries of polycrystalline forsterite and enstatite to be interpreted as MgO diffusivity. Water contents of >2 wt.% of the bulk produced an interconnected network of water-filled channels, also mobilized the SiO_2 component and further increased overall rim growth rates by one to two orders of magnitude (regime 4; Fig. 6). As water-filled porosity increases, transport properties tend towards those of bulk water, thus inducing a continuous enhancement of mass transport as a function of the water–solid fraction (regime 4; Fig. 6).

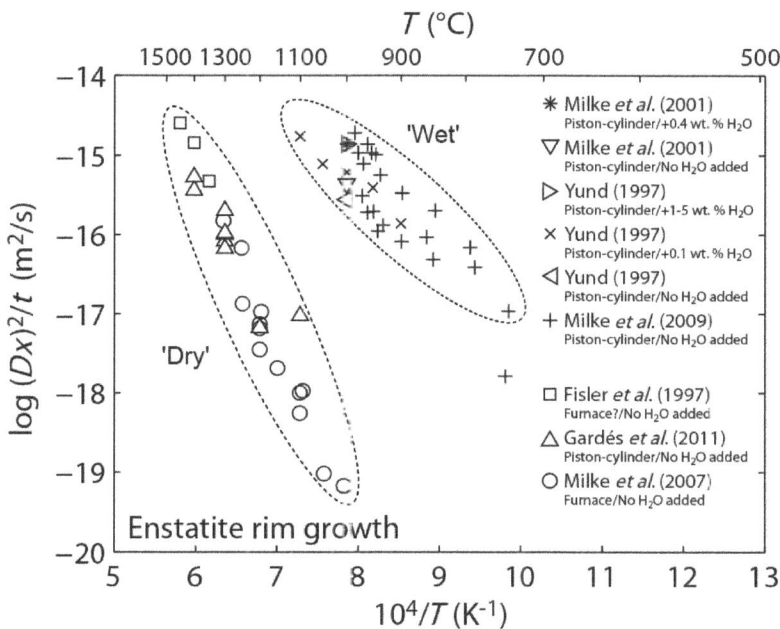

Figure 5. Synopsis of all rim-growth kinetic data for Opx rims between Ol and Qtz in various studies and setups. Two rather linear temperature series emerge in the Arrhenius diagram that are interpreted as "wet" and "dry". From Gardés *et al.* (2012), reprinted by permission of Springer.

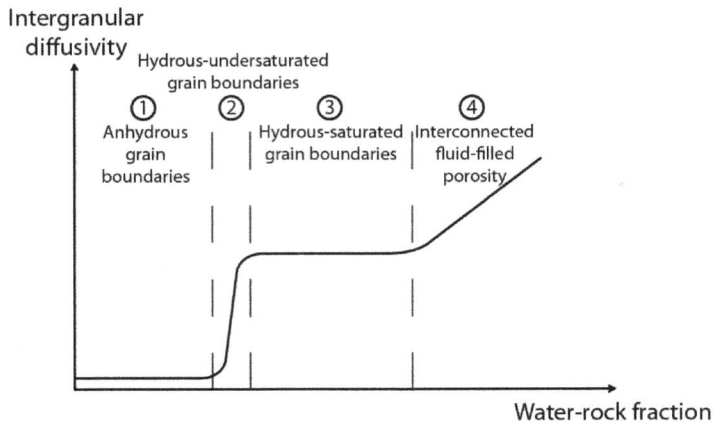

Figure 6. Classification scheme of intergranular mass transport as a function of water-rock fraction derived from [Fo|En] and [En] rim growth experiments. Regime (1): Diffusion in anhydrous grain boundaries, which holds as long as water can be accommodated in the volume of the minerals. (2) Diffusion in hydrous-undersaturated grain boundaries is a transition regime that is most likely to be rarely observed as it occurs in a very narrow range in this system. (3) Diffusion in hydrous-saturated grain boundaries, which is the first regime occurring in water-saturated conditions and corresponds to acceleration of reaction kinetics of several orders of magnitude. (4) Diffusion in interconnected fluid-filled porosity, where continuous enhancement of mass transport delevelops due to the concomitant increases in porosity, interconnectivity and diffusivity with high water-rock fraction. From Gardés *et al.* (2012), reprinted by permission of Springer.

The jump from the virtually dry stage 1 to conditions at hydrous-saturated grain boundaries (stage 3) as suggested by Gardés *et al.* (2012) is possibly very narrow. Milke *et al.* (2013) investigated that regime for the same reaction and suggested that it occurs if only few ppm of free water are present in the bulk, which may, moreover, concentrate at the reaction interface during the course of the experiment (Fig. 7). Reactions allowed by diffusion along wetted grain boundaries or fluid-filled pores may lead to specific fabrics (Milke *et al.*, 2009a). By combination of 3D imaging and FTIR analysis, Milke *et al.* (2013) determined that a few tens of ppm of water are sufficient to make a fundamental difference in growth textures. In line with Gardés *et al.* (2012) they still found diffusion-controlled kinetics in their experiments. The mechanism acting at the propagating interfaces, though, that defines the boundary conditions for diffusion through the reaction rims, is fluid-aided dissolution-precipitation where growth of the reaction products follows the migration of a μm- or sub-μm-sized pore (Fig. 7). With regard to this specific reaction at given P and T conditions, their answer is: "very little water is necessary to make a dry solid system wet" (where "wet" is a synonym for regime 3 of Gardés *et al.*, 2012). The threshold between dry and wet is not defined by water fugacity but rather by the amount of water present as a free fluid per interface area, and was estimated at ~1 wt-ppm water per 10 mm^2 silicate interface area (Milke *et al.*, 2013). These experiments highlight the tremendous effect that minute amounts of water may exert on rim-forming mineral reactions in the solid state. Obviously, much

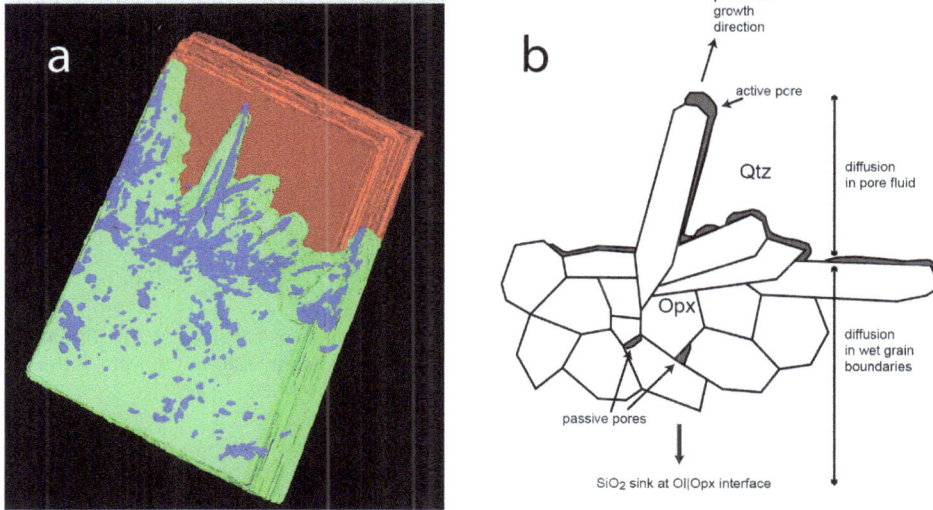

Figure. 7. Left: 3D tomography of an Opx reaction rim and the Opx|Qtz interface. Qtz is red, Opx is green, pore space filled with aqueous fluid is blue. *Right*: Schematic image of the growth structures and controlling mechanisms at the Opx|Qtz reaction front. The propagation of the reaction front is led by the migration of "active pores". Isolated porosity left behind in the Opx rim is termed "passive pores". Modified after Milke *et al.* (2013).

more experimental work for different reactions in different systems is required. Correct kinetic interpretation of reaction rims and coronas from natural rocks becomes extremely difficult if information about water contents, *i.e.* dry *vs.* wet, and "how wet?" is not available.

A different experimental approach is the growth of product rims or coronas between reactants consisting of seed crystals embedded within a fine-grained matrix. It has been found that in high-pressure experiments fine-grained matrix material used as a reactant *cf.* above) is not completely dry, and adsorbed water at the matrix surfaces definitely makes the system wet. Rim growth using this approach probably represents a transport regime that corresponds to hydrous saturated grain boundary transport of the respective components. This accelerates overall rim growth rates by several orders of magnitude and may, in some cases, mobilize additional components that are virtually immobile in dry systems.

This has been demonstrated for the formation of polycrystalline enstatite rims, [Fo|En|Qtz], with seed crystals of olivine in powdered quartz matrix, and *vice versa* (Milke *et al.*, 2001, 2009a; Abart *et al.*, 2004). Experiments were performed at various reaction conditions in conventional hydrothermal and piston cylinder apparatuses. Overall reaction rates are in line with that of Gardés *et al.* (2012) for wet sandwich experiments. Inert Pt-markers were not applied here; however, growth textures clearly reveal two different portions of the rim (Fig. 8) From the original interface ~70% of enstatite palisades grew towards forsterite and 30% of granular enstatite towards quartz,

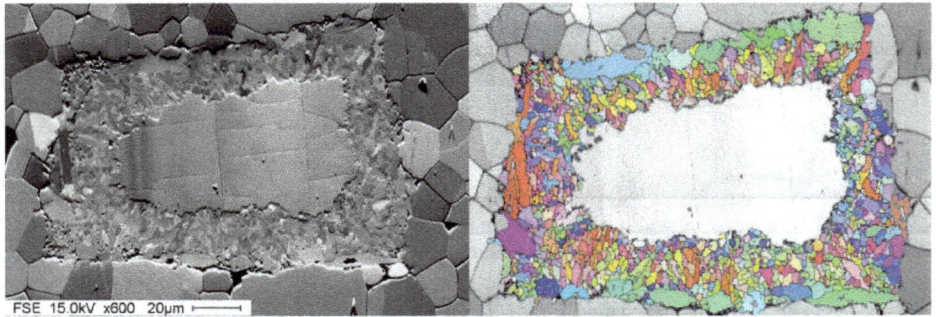

Figure. 8. Left: Forward scattered electron image of an enstatite rim around forsterite in a polycrystalline quartz matrix. Image is 200 μm long. Enstatite in the inner portion of the reaction rim forms a palisade microstructure, whereas towards the quartz matrix a granular microstructure develops. *Right*: Crystal orientation map of the enstatite polycrystals obtained from EBSD scans. Enstatite palisades are inherited from the forsterite crystal orientation indicating that growth at the Fo|En interface is directed inwards normal to the interface. In the outer portion, enstatites are much coarser and their crystal orientations are related to the orientation of the interface. The outer part covers ~30% of the total rim volume. Modified after Abart *et al.* (2004).

as shown by crystal orientation maps obtained by EBSD methods (Abart *et al.*, 2004). In these experiments, isotopically-doped quartz enriched in ^{18}O and ^{29}Si was used to identify the relative fluxes of the components allowing introduction of a model which accounts for simultaneous layer growth and superimposed silicon and oxygen self-diffusion under wet conditions (Abart *et al.*, 2004).

The experimental strategy of applying seed crystals in a "wet" powder matrix allows further investigation of the effect of the matrix rheology on rim growth rates, demonstrated by means of the [Ol|Opx|Qtz] system (Milke *et al.*, 2009b; Schmid *et al.*, 2009). Orthopyroxene rim growth around olivine grains in a quartz matrix was compared to rim growth around quartz grains in an olivine matrix. At constant P and T, within a single capsule in a piston-cylinder apparatus, orthopyroxene rims grow faster around quartz clasts in an olivine matrix than around olivine clasts in a quartz matrix (Fig. 9). This behaviour was interpreted in terms of matrix rheology, where in the two different matrix-inclusion arrangements the olivine matrix behaves softer and the quartz matrix more rigid. It was concluded that the strain energy associated with matrix deformation reduces the free energy that drives orthopyroxene rim growth, and that the quartz matrix behaves more strongly than the olivine matrix. Textural observations showed that in both cases SiO_2 is also mobile, in line with all observations in wet systems, and that the diffusivity of MgO exceeds that of SiO_2. It was further argued that the relative mobility of MgO and SiO_2 at given conditions appears to be controlled by energy minimization due to volume-conserving replacement at the compressive Ol|Opx interface. Such experiments demonstrate that diffusivity along chemical potential gradients to reaction sites is dependent on rheology, and that the relative diffusivity of components during reaction rim growth is a function of local volume changes at the rim's interfaces.

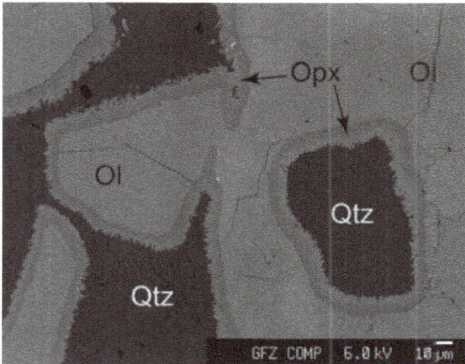

Figure 9. Orthopyroxene rims around olivine seeds in a quartz matrix (*left*), and around quartz seeds in olivine matrix (*right side*). Rims around quartz seeds in the olivine matrix are wider than the rims around olivine seeds in the quartz matrix. Sectioning effects are largely cancelled out by the measuring routine, which gives average rim widths of 6.1 μm around olivine seeds, and 10.3 μm around quartz seeds. From Milke *et al.* (2009b), reprinted by permission of Wiley.

Natural olivine with X_{Fo} of 0.901 and 400 ppm of Ni has been used to investigate the transport behaviour of Mg, Fe and Ni during orthopyroxene rim growth in dry sandwich assemblies as well as using seeds in a "wet" powdered quartz matrix (Milke *et al.*, 2011). Textural observations showed that under dry conditions, orthopyroxene grew in equal amounts on both sides of the reactions, indicating growth control by diffusion of the MgO plus FeO components only. In the wet system, the proportion of orthopyroxene that grew into olivine exceeded that growing into quartz, indicating additional mobilization of SiO_2 in line with arguments presented above. Irrespective of dry or wet, a distinct zonation in X_{Fe} across the orthopyroxene rim developed. At the Ol|Opx interface the Opx composition is fixed at $X_{En} = 0.91$ due to equilibrium fractionation between Ol and Opx. The resulting local iron excess is compensated by formation of Opx with a higher X_{Fe} at the Opx|Qtz interface, which is out of equilibrium with the reactant olivine (Fig. 10). Nickel is forced to diffuse back into Ol producing a Ni-enriched zone ahead of the Ol|Opx reaction front, due to the much lower capacity of Opx to accommodate Ni (Fig. 11). This leads to greater Ni concentrations in the Opx rim than obtained by equilibrium partitioning between the original Ol and Opx produced. Relative mobilities of the various components during this process have been

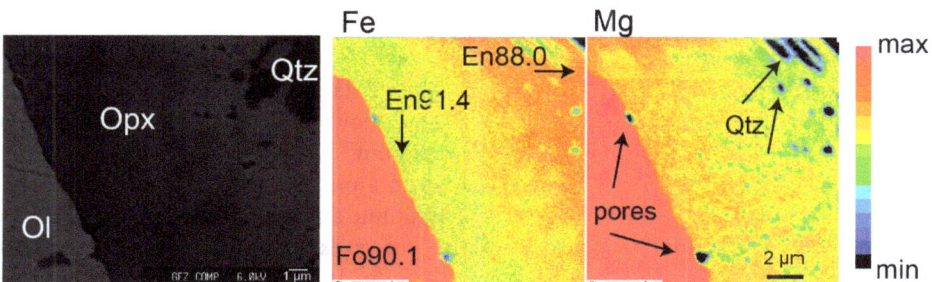

Figure 10. BSE image and X-ray distribution maps for Fe and Mg from an experimentally synthesized orthopyroxene rim between olivine (Fo$_{91.0}$) and quartz. Orthopyroxene is En$_{91.4}$ at the Ol|Opx interface, and En$_{88.0}$ at the Opx|Qtz interface. From Milke *et al.* (2011), reprinted by permission of Springer.

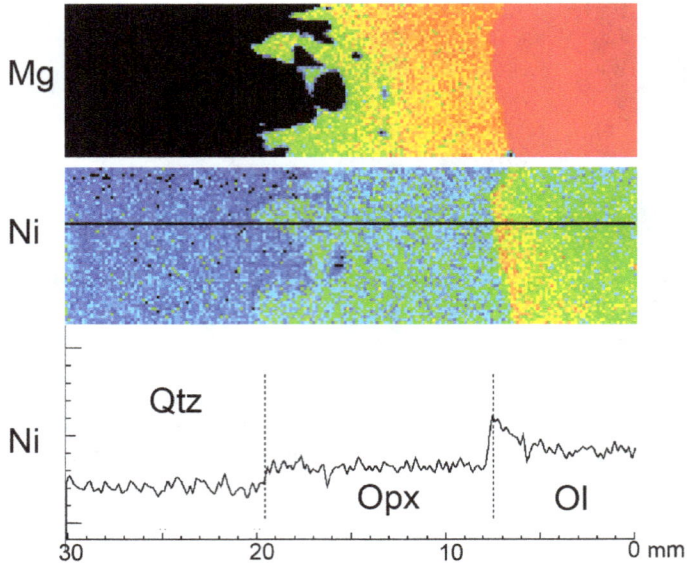

Figure 11. Ni and Mg distribution across an orthopyroxene rim between quartz and olivine. The trace of the Ni profile is depicted as a line of the Ni map. Ni in Opx is enriched by about twice the original concentration in a 2 μm wide zone in olivine. From Milke *et al.* (2011), reprinted by permission of Springer.

modelled (Milke *et al.*, 2011), highlighting the importance of kinetic element-fractionation to our understanding of mantle rocks.

1.5. Reaction rims, fluid production and fluid release

Diffusion-controlled formation of wollastonite rims between calcite and quartz is a phenomenon frequently observed in many contact metamorphic aureoles worldwide (*e.g.* Joesten and Fisher, 1988; Heinrich, 1993). This implies that mineral reaction rims may also form during devolatilization reactions, where fluid is produced by one of the reactants. This was confirmed experimentally by the reaction quartz + calcite → wollastonite + CO_2, resulting in [Qtz|Wo|Cc] layers (Milke and Heinrich, 2002; Milke and Wirth, 2003). Again, contrasting replacement features at dry (*i.e.* CO_2-present, water-absent) and wet conditions at various P and T have been demonstrated using the seed-in-matrix procedure. Dry conditions were achieved by pressurizing open-sample capsules, with pure CO_2 as a pressure medium, in conventional hydrothermal equipment. Wet experiments were carried out in a piston-cylinder device with CaF_2 as a solid pressure medium, which introduced minute amounts of water into the sealed capsule. Under dry conditions, wollastonite grows exclusively on quartz surfaces, whereas under wet conditions it grows on calcite. In both cases the released CO_2 produces open pores (Fig. 12), and rim growth follows a parabolic rate law indicating that the reaction is diffusion-controlled. Textural observations show clearly that wollastonite growth on quartz is controlled by grain-boundary diffusion of the SiO_2

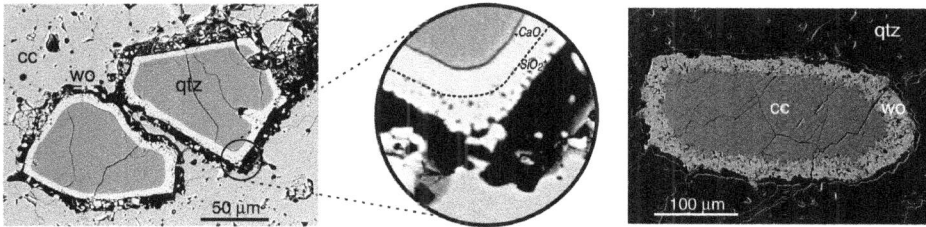

Figure 12. BSE images of experimentally produced wollastonite rims between quartz and calcite. *Left*: Dry conditions. Wollastonite rims grown on quartz grains in hot isostatically-pressed calcite matrix. The rims are separated from calcite by CO_2-filled voids. The outer part of the rims contains CO_2-filled pores (detail). The boundary between the dense and the porous part of the rims is interpreted as the original quartz–calcite interface. *Right*: Wet conditions. Wollastonite rims grown on calcite. No closed rims formed. The rims are porous due to CO_2 release at the calcite surface. From Milke and Heinrich (2002), reprinted by permission of Wiley.

component, whereas the formation of porous wollastonite rims on calcite is controlled by a complex diffusion-advection mechanism, where the relative contributions of the mobility of the CaO and SiO_2 components remain unclear. Overall, rim growth rates are at least one order of magnitude higher in the wet compared to the dry system at the respective temperatures. This again illustrates the importance of careful characterization of the starting materials and setups before interpreting reaction kinetics in experimental systems.

The Paterson-type gas pressure vessel is suitable for investigating the interplay between the kinetics of devolatilization reactions and fluid expulsion (Fig. 13), demonstrated by the same wollastonite-producing reaction under dry conditions (Milsch *et al.*, 2003). Hot-pressed synthetic rock cylinders consisting of quartz seeds in a calcite matrix with low initial porosity react to form wollastonite rims around quartz

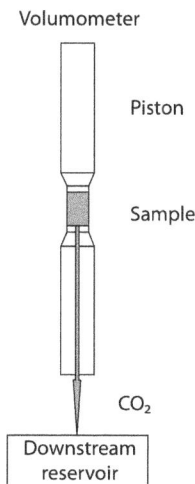

Figure 13. Synthetic rock assembly setup inside a Paterson-type gas medium apparatus used for hot isostatic pressurizing and mineral reaction experiments. Samples are placed between spacers with a central bore connected to a pore fluid system. Pore pressure is generated by a volumometer and monitored upstream and downstream. During ongoing reaction pore-pressure changes in the isolated downstream reservoir are used to monitor reaction-induced CO_2 release from the sample chamber. From Milsch *et al.* (2003), reprinted by permission of Springer.

at 900°C, 150 MPa, and 950°C, 300 MPa temperature and confining pressure. Measured widths of wollastonite rims *vs.* time show a parabolic growth law, very similar to that determined by Milke and Heinrich (2002). Release of CO_2 was measured continuously in a downstream reservoir. While the chemical reaction proceeds continuously, it is most striking that the fluid release is not continuous but occurs episodically and in pulses. Ongoing reaction produces an increase in pore pressure, which in this system, once having attained a critical value of ~20 MPa, is released spontaneously until a new cycle is built up. These experiments show that properties such as pore pressure, fluid-filled porosity and permeability are transient and mobility of fluid in a reacting rock volume is a short-lived process.

2. Experiments with elaborate setups

2.1. Reactions in thin films

Deposition of thin films onto crystal substrates and subsequent reaction between film and crystal at high temperature is an elegant method for investigating mineral reaction kinetics, particularly because large variations of chemical compositions of the film can be adjusted to the particular problem at hand. There is a wide variety of chemical (CVD) or physical (PVD) deposition methods that are used for production of thin films onto reaction substrates that all have severe drawbacks when applied to silicates (for overviews see, *e.g.* Wasa *et al.*, 2004; Martin, 2010). By far the most versatile method with respect to silicate phases is deposition of films after ablation of material by a pulsed laser (pulsed laser deposition, PLD; Dohmen *et al.*, 2002; Eason, 2007; Watson and Dohmen, 2010). Due to the high energy of the laser pulses and the resulting high temperatures at the target surface, we can ensure that all the target constituents are evaporated simultaneously (Shaginyan, 2001). The high degree of stoichiometric transfer from target to substrate is presumably the main reason for the widespread application of PLD (Schou, 2009). Deposition is ensured to occur at extremely dry conditions. Still there are disadvantages, such as formation of a limited area of uniform film thickness and ejection of micro-sized globules or particles (Wasa *et al.*, 2004; see also Dohmen *et al.*, 2002). Another practical problem consists of maintaining a constant material flux through the plasma from target to specimen because evolving target surface morphology during the ablation process leads to a shift in the angle of the plasma stream towards the laser beam, which might lead to serious problems regarding the stoichiometry of the deposited thin films (Shaginyan, 2001). Still, this method has proven to be the best choice so far for producing thin films of materials mimicking widespread geomaterials. In geological sciences this technique has been applied widely to produce source layers in diffusion couples (*e.g.* Dohmen *et al.*, 2007). Here, we briefly address the fabrication of multilayered reaction couples in thin film geometry using PLD for miniaturized reactive diffusion experiments, subsequent experimentation and methods of evaluation. The use of thin-film samples yields manifold advantages over conventional rim-growth experiments. Synthetic thin films can be tailored, thus offering flexibility in chemical composition and providing the ideal topology for subsequent analytical methods such as secondary ion mass spectrometry (SIMS) and Rutherford back-scattering (RBS) depth scanning. Under these conditions very small growth rates between film and substrate or simple element diffusion from film into substrate can be measured accurately. The experimentally accessible temperature range is thus extended by several hundred degrees down to lower temperatures. The temperature range of diffusion-controlled reactions in the Earth's crust is entered, thereby, even for virtually dry conditions, where silicate reactions are slow and chemical or isotopic profiles formed by reaction-diffusion are very short.

As a model system to introduce this method for geologically relevant mineral reaction pairs, Milke *et al.* (2007) used orthopyroxene rim-growth between an olivine reactant layer and a quartz crystal substrate. Two-layer setups had been prepared consisting of a thin amorphous seed layer of Opx composition applied directly onto the ultrapolished quartz surface and an olivine reaction layer of ~1 μm thickness on top. Seed layers are generally recommended for reactive systems with rather sluggish nucleation at low *T* such as (Ol|Opx|Qtz). Where nucleation is fast, *e.g.* for spinel production between Al_2O_3 and MgO (Cor|Spl|Per; see section 4) they are not necessary. Rim thickness before and after the experiments was determined independently by RBS and combined FIB-TEM imaging (Fig. 14). Opx layer growth follows a parabolic rate law and extrapolation to higher

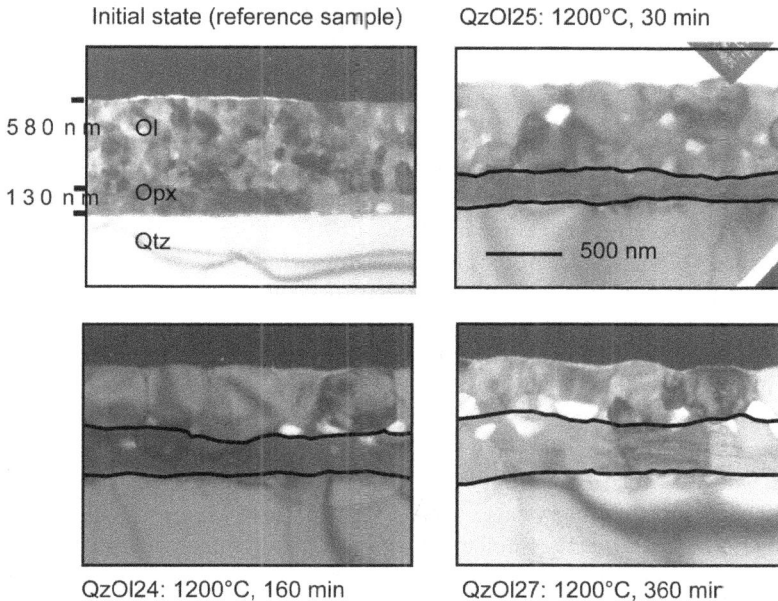

Figure 14. FIB-TEM images of Ol | Opx thin films on a Qtz substrate before the experiment and after Opx layer growth. Starting assembly (*above, left-hand side*) consisting of a 130 nm-thick enstatite seed layer and a 580 nm-thick olivine reactant layer. With subsequent reaction the interfaces deviate progressively from their initial planar morphology. Increase in layer thickness must be measured by averaging series of measurements according to defined methods. Modified after Milke *et al.* (2007).

temperatures matches exactly earlier high-T data at ambient pressure (Fisler *et al.*, 1997); these results were confirmed by Gardés *et al.* (2011) in their virtually dry piston-cylinder sandwich experiments on the same reaction.

2.2. Synthesis and use of bicrystals

Volume diffusion in silicate crystals is much slower than grain boundary diffusion but can still be more efficient in the long run. There is, however, overall agreement that grain-boundary diffusion prevails over volume diffusion during most small-scale transport processes in large portions of solid Earth where reactions take place within $\leqslant 10^6$ my (Joesten, 1991; Dohmen and Milke, 2010). Grain boundary diffusion is, among other parameters, controlled by the structure of the grain boundary itself. It is well known from studies on metals and alloys that their diffusivity may vary by several orders of magnitude with varying crystallographic orientations of the adjacent crystals (*e.g.* Mishin and Herzig, 1999) as a consequence of differing grain-boundary energies. In general, it is to be expected that specific well fitting grain boundaries have lower excess energies and therefore lower transport rates. In the context of geomaterials, experimental determination of grain-boundary diffusivities requires starting materials in which the structure of the grain boundary as well as the crystallographic misorientation is known exactly. This is achieved by synthesis of bicrystals with a well defined single grain boundary. The method of choice is the direct wafer bonding technique (*e.g.* Tong and Gösele, 1999).

This technique has been adopted for use with geomaterials by using bicrystals of K-feldspar (Heinemann *et al.*, 2001), olivine (Heinemann *et al.*, 2005) and yttrium-aluminium-garnet (YAG) (Hartmann et al., 2010). The procedure requires several steps. Single-crystal plates oriented using X-ray diffraction are cut to dimensions of, for example, 4 mm × 4 mm × 1 mm, and are subsequently chemo-mechanically polished with silica slurry. Surface roughness measurements by using interferometry reveals that the final roughness of the surface of the chips is as low as 0.3 to 0.4 nm. Thereafter, the chips are wet-chemically cleaned using a five-step cleaning procedure in a clean-room environment (Hartmann *et al.*, 2010). In step 1, a hydrogen peroxide cleaning solution at 80°C is applied, followed by step 2, a per iodic treatment in order to decompose organic adsorbants. Steps 3 and 4 involve cleaning of the surface with acetone and isopropanol, respectively. Finally, in step 5, the surface is saturated with pure adsorbed water. Subsequently, an initial contact between the two chips is established by applying a slight pressure with tweezers. Final bonding is achieved by heating in a vacuum furnace for several days. Annealing bicrystals of YAG (Hartmann *et al.*, 2010), silicon (Tong and Gösele, 1999) and forsterite (Heinemann *et al.*, 2005) by this method revealed that the maximum stability of the grain boundary is attained at annealing temperatures (in K) of ~60% of the melting temperature in the respective system. Bicrystals produced by this method are mechanically inseparable. A cartoon illustrating individual stages in direct bonding of K-feldspar crystals is shown in Fig. 15 (Heinemann *et al.*, 2001).

The structure and property of the grain boundary is revealed by HRTEM investigations after producing thin foils by focused ion beam (FIB) (Wirth, 2017;

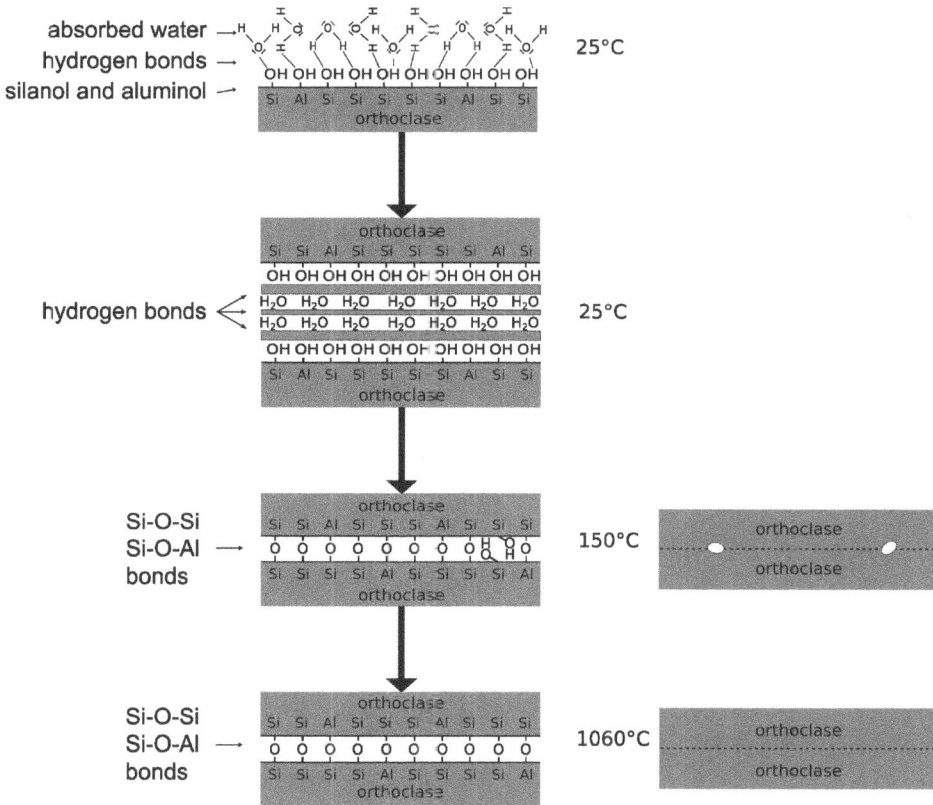

Figure 15. Individual stages in direct bonding of K-feldspar bicrystals. At room temperature, hydrogen bonds form between opposing surfaces initially covered with silanol and aluminol groups and adsorbed water molecules. Water loss and polymerization of silanol and aluminol groups occur at annealing temperatures of ~150°C. Rearrangement of Al and Si along the interface and closure of voids require ~1000°C. From Heinemann *et al.* (2001), reprinted by permission of Springer.

this volume). Figure 16a shows an energy-filtered HRTEM image of a synthesized K-feldspar bicrystal with the interface (010) perpendicular to the image. The grain boundary is that of a Manebach twin, shows no defects and is free of amorphous material. Figure 16b displays a near $\Sigma 5$ (210)/[100] grain boundary of a YAG bicrystal. For details of the Σ notation in terms of the coincidence site lattice (CSL) model see Petrishcheva and Abart, (2017; this volume). Again, the lattice fringes of the opposing crystals touch each other at the interface and no amorphous material or material with different structures is observed.

For single-crystal diffusion experiments, perfect bicrystals with different misorientations are covered by the desired diffusant source, using PLD. Thin films are several hundreds of nm thick and applied perpendicular to the bicrystal grain boundary. This setup is then used for high-temperature annealing experiments

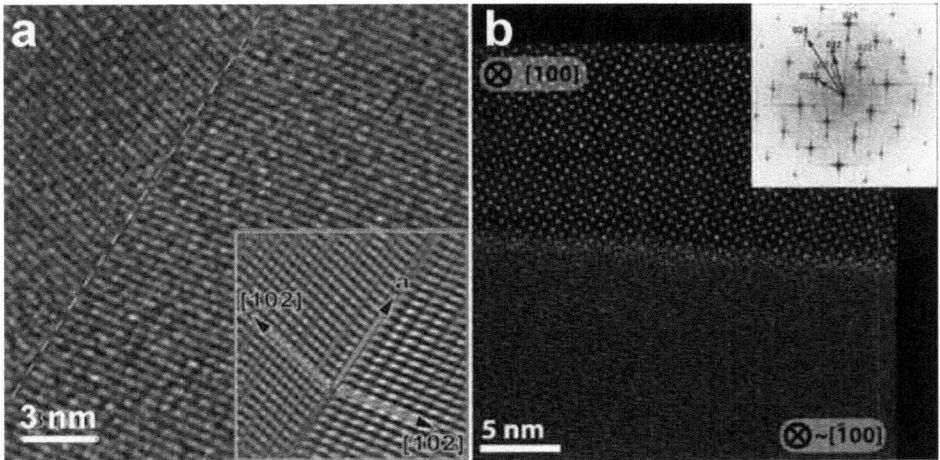

Figure 16. (a) Energy-filtered HRTEM image of a synthesized K-feldspar interface (interface perpendicular to the image. There are no defects along the coherent Manebach twin boundary. Insert is a Fourier-filtered image. The trace of the interface is parallel to the *a* direction. From Heinemann *et al.* (2001). (b) HRTEM image of a near Σ5 (210)/[100] YAG grain boundary. The [100] direction of the upper crystal is parallel to the incident beam. [100] is the common axis of a perfect Σ5 twin boundary. As the misorientation with respect to this orientation is small, both crystals show the diffraction pattern of the [100] zone axis rotated by 36.9°. Nevertheless, the lower crystal is not perfectly aligned in [100]. The TEM lamella is very thin and the diffraction spots are still visible in the diffraction pattern, but with reduced intensities. In the lattice fringe image only the (240) planes of the lower crystal are visible. From Hartmann *et al.* (2010), reprinted by permission of Springer.

(Fig. 17). After annealing, site-specific sample preparation is performed by cutting thin foils using the FIB technique. Concentration profiles of the diffusant into the bicrystals and along the grain boundary are measured by ATEM. The simple diffusion geometry allows for appropriate TEM analysis because it is possible to orient the grain boundary parallel to the axis of the incident electron beam and the EDX detector, whereby artefacts from planar defects are excluded (Marquardt *et al.*, 2010, 2011). Volume and grain boundary diffusion are easily distinguished and measured on the same sample from one experiment using identical experimental and analytical settings.

As an example, Fig. 18 shows a TEM bright field (BF) image of a FIB foil from an annealing experiment modified after Marquardt et *al.* (2011). A YAG bicrystal is covered by a 380 nm thin film of Yb-doped YAG and annealed at 1723 K for 2 h. The thin film recrystallized, grew epitactically on the bicrystal and adopted the preset grain boundary. The ATEM window for Yb analysis is set at 20 nm × 40 nm and rasterizing the foil with this window reveals the two-dimensional volume and grain boundary diffusion pattern of Yb in YAG. Results including numerical modelling of volume and grain-boundary diffusion for this system are presented in Petrishcheva and Abart (2017, this volume). For this particular sample, Yb grain-boundary diffusion is about five orders of magnitude faster than volume diffusion. Qualitative Yb maps of the diffusion zone in the YAG

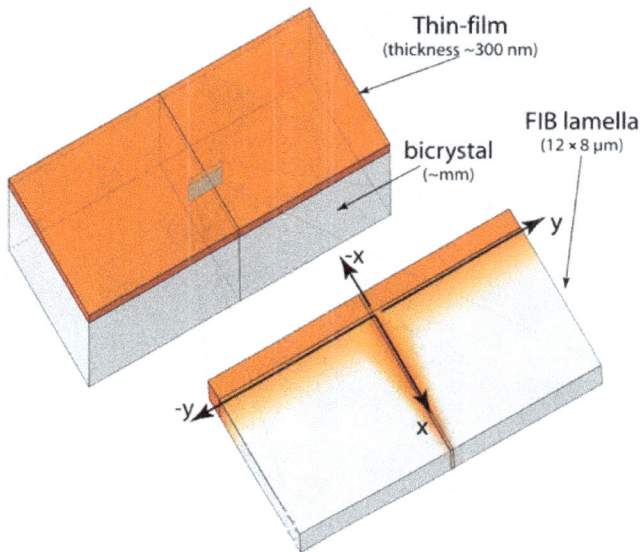

Figure 17. Schematic sketch of the sample geometry for volume- and grain-boundary-diffusion studies. The bicrystal is grey, the thin-film diffusant source red, and the grain boundary is perpendicular to the thin film. Enlarged sketch of a TEM lamella prepared using the focused ion beam (FIB) technique after the experiment. The foils have dimensions of about 15 μm × 10 μm and are ~100 nm thick. The size of the bicrystal is in the mm range. Fast diffusion from the source into the crystals and along the grain boundary produces concentration gradients as indicated by colour intensities. Modified after Marquardt *et al.* (2011).

bicrystal with sub-micrometre resolution can also be obtained by nano-X-ray fluorescence analysis using synchrotron radiation (Schroer, 2017, this volume), using FIB-cut lamellae of 10 μm × 8 μm × 0.25 μm in size. An example is shown in Fig. 19, where V-shaped Yb diffusion (compare Fig. 17) into the grain boundary is indicated clearly (Marquardt *et al.*, 2011). In summary, this new methodical combination of using bicrystals, pulsed laser deposition and ATEM now allows exact quantification of grain boundary diffusion coefficients of distinct elements along distinct grain boundaries over a wide range of temperatures. This well controlled characterization of single grain boundaries might potentially also allow for detailed insight into changes of grain boundary configurations in the presence of very few ppm of water. A wide field of unresolved problems can be tackled by this experimental method; not only with respect to grain boundary transport properties but also regarding reactivity and mechanical behaviour in the dry regions of the Earth's crust, mantle and in extraterrestrial bodies such as Mars, because grain boundaries play an important role in these processes.

3. Experiments at non-hydrostatic stress

Mineral reactions under hypogene conditions occur commonly in mechanical stress fields in the Earth's crust and mantle. Recent experimentation addressed the interplay

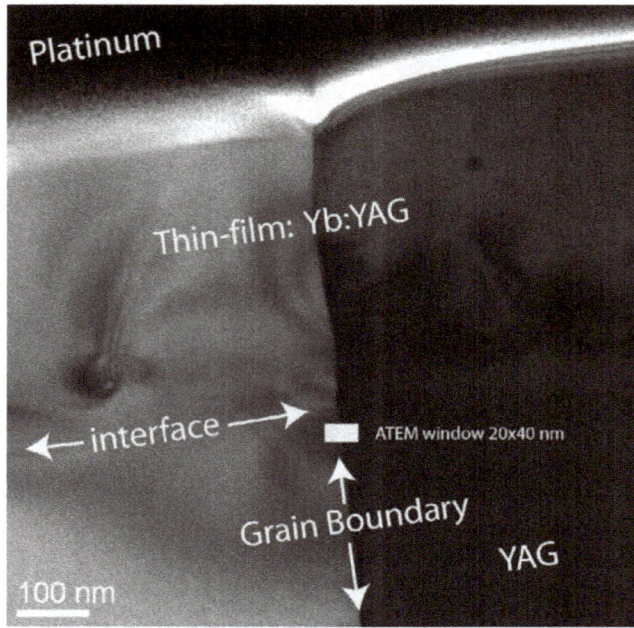

Figure 18. Bright field image of a FIB foil showing a YAG bicrystal with a Yb-YAG thin film after 2 h of annealing at 1723 K. The boundary of the interface between bicrystal and substrate and the bicrystal grain boundary are indicated by arrows. The thin film recrystallized, grew epitaxially on the bicrystal and adopted the preset grain boundary. Note the size of the window set for ATEM analysis. Modified after Marquardt *et al.* (2011).

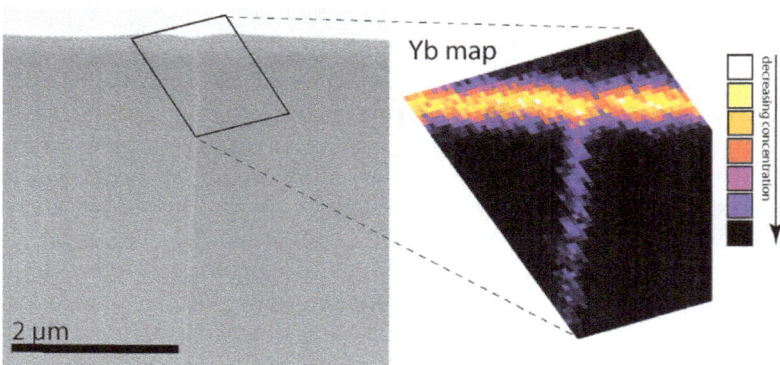

Figure 19. Left: STEM image of a FIB foil showing a YAG bicrystal with a Yb YAG thin film after 24 h of annealing at 1723 K. The camera length is chosen to probe a mixture of Z contrast and diffraction contrast; the thin-film is fully epitactic. The platinum layer is the brightest grey in the image, the thin film appears in light grey and the two parts of the bicrystal are dark grey. The region mapped using the YbLα peak from synchrotron X-ray fluorescence at the ESRF is shown in the enlargement. From Marquardt *et al.* (2011), reprinted by permission of Springer.

between minimization of chemical and mechanical energy in systems brought to out-of-equilibrium conditions. Following experimental traditions, the small number of studies available can be divided into two types of experimental setups: (1) reactant grains in a reactant matrix, and (2) single crystalline or polycrystalline reactant stacks with well defined interfaces.

The first approach was pioneered by de Ronde *et al.* (2004, 2005) in a series of papers on the mineral reaction plagioclase + olivine → orthopyroxene + clinopyroxene + spinel ± garnet (see also de Ronde and Stünitz, 2007). They studied this reaction under shear deformation, using isostatically hot-pressed plagioclase-olivine aggregates as a starting material in a Griggs apparatus, where, in addition to the confining pressure *via* a solid media cell, a deviatoric stress is applied to the sample. At experimental conditions of 900°C, 1000−1600 MPa and shear strain rates of 5×10^{-5} s^{-1} in a rock salt cell it was found that both the onset of reaction as well as the reaction rate is catalysed strongly by deformation. At static conditions thermodynamically driven reactions proceed very sluggishly. Enhancement of reaction rates in deformed samples was attributed largely to much higher nucleation rates, even in the absence of water (de Ronde and Stünitz, 2007).

Experimentation in artificial mineral aggregates bears similarities to reactions in natural rocks but lacks control on deformation and interface directions. Recent studies of the effect of deviatoric stress used simpler geometry and reactions with minimized degrees of freedom. The experimental approach consists of producing reaction layers between oriented single crystals in a dead load uniaxial creep apparatus with dry argon ambient pressure plus exerting a constant load by means of a piston (Fig. 20). By imposing vertical stresses of 1.2 to 29 MPa at temperatures ranging from 1150 to 1350°C, Götze *et al.* (2010) produced polycrystalline layers of enstatite between forsterite and quartz crystals [Fo|En|Qtz], enstatite−forsterite double layers between periclase and quartz [Per|En|Fo|Qtz], and spinel, MgAl$_2$O$_4$, between periclase and corundum [Per|Spi|Cor]. They found that at a given temperature the effect of stress variation is larger than anticipated from the modification of the thermodynamic driving force for reaction due to the storage of elastic strain energy in the reactant phases. By using the same device the spinel-forming reaction, *i.e.* [Per|Spi|Cor] triggered at 1350°C, 157 h run duration and varying uniaxial load, was investigated in detail by Keller *et al.* (2010). The rate of spinel growth increased fourfold when stress normal to the reaction interface increases from 3 to 30 MPa due to stress-induced changes in the grain-boundary structure. At low applied stress, low-index "coincidence site lattice" (Petrishcheva and Abart, 2017, this volume) spinel grain boundaries with slow diffusion coefficients dominated, related to epitactic growth of spinel on corundum. Increasing stress triggered epitactic growth of spinel on periclase, and caused corundum-grown spinel grains to rotate out of epitaxy, and grain boundaries with high diffusivity became dominant This effect was concluded to outweigh the hitherto emphasized influence of grain size on the bulk transport properties of polycrystals.

The Paterson-type gas deformation apparatus enables triaxal and torsion experiments at high temperature and high confining gas pressure (Paterson and

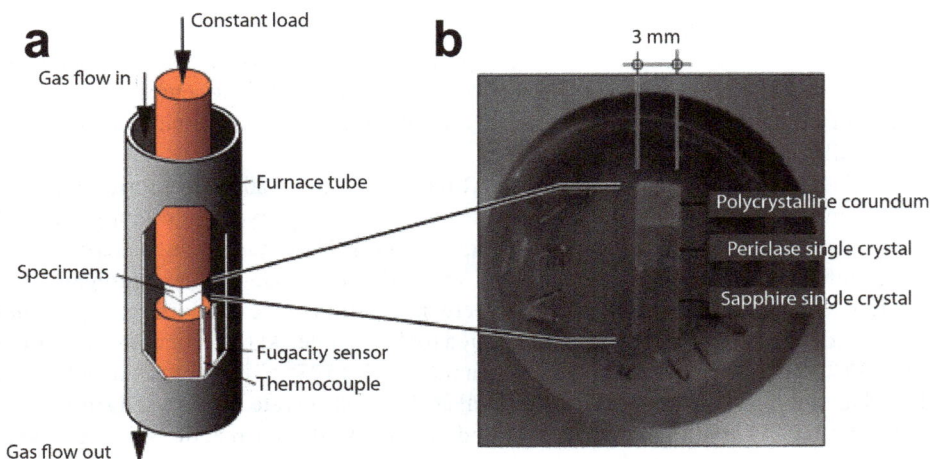

Figure 20. (a) Schematic drawing of a uniaxial creep apparatus with constant dry argon flow at ambient pressure. Differential stresses are applied dead-weight on the top of the assembly. (b) Setup of the reactant couples for producing spinel layers between corundum (termed sapphire here) and periclase. Cubes consisting of single-crystal corundum and of polycrystalline corundum are positioned above and below the periclase reactant crystal to unravel the effects of polycrystalline versus single crystal nature of reactants on spinel growth. From Keller *et al.* (2010), reproduced with the permission of the Mineralogical Society of America.

Olgaard, 2000). This device was used to investigate the effect of non-isostatic stress and strain on growth of dolomite layers between calcite and magnesite single crystals, [Cal|Dol|Mgs] (Helpa *et al.*, 2014, 2015; Fig. 21). Experimental conditions were 750–850°C, 400 MPa confining pressure, stresses between 7 and 38 MPa, and run durations of up to 171 h. Dolomite grew in both directions from the original interface. Palisade-shaped dolomite grew towards magnesite and granular dolomite towards calcite (Fig. 22). In addition to the layer-forming reaction, diffusion of magnesium into the calcite reactant caused a wide zone of Mg-calcite. Microstructural observations demonstrated that inelastic deformation was partitioned into calcite and reaction products, while magnesite remained undeformed. At high strain, the Mg-calcite zone was wider, suggesting faster growth kinetics, interpreted to be related to additional diffusion pathways provided by enhanced dislocation activity. At the experimental conditions, the dolomite-layer growth kinetics are diffusion-controlled and independent of the applied axial stress and strain in triaxial compression, in agreement with thermodynamic calculations indicating that the stress-strain-induced change in the driving force for dolomite growth is negligible (Helpa *et al.*, 2015).

Independence of layer growth kinetics of experimentally applied stress-strain-induced changes does not necessarily hold for silicate systems at high P and T conditions. It is quite conceivable that deformation-induced structural defects in reactants as well as in product layers would have significant effects not only on diffusion behaviour, reaction mechanisms and reaction rates but also on the mobility of

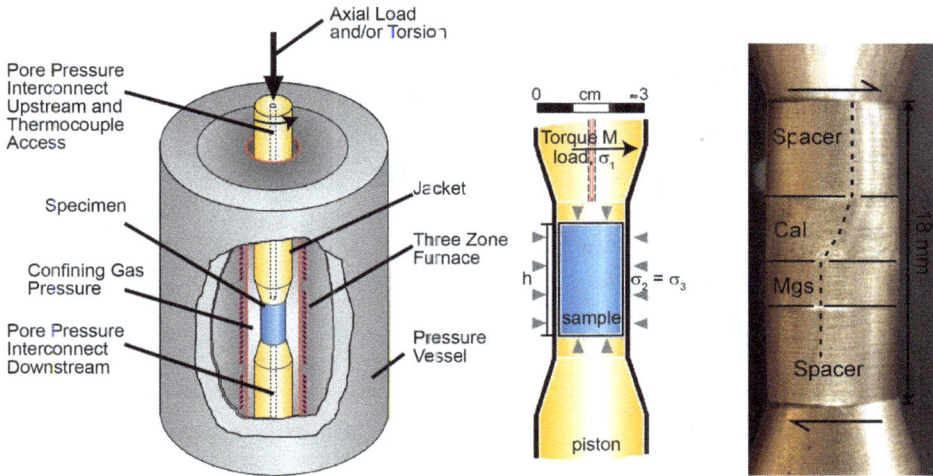

Figure 21. Schematic drawing of the pressure vessel (*left*) and sample assembly (*centre*) used for high-pressure, high-temperature axial compression and torsion experiments with the Paterson-type deformation apparatus. The whole assembly is jacketed by a metal sleeve to prevent intrusion of the confining pressure medium, *i.e.* argon gas. *Right:* Jacketed calcite-magnesite pair after torsion with dextral shear sense. Dashed black line shows the distribution of strain indicated by wrinkles in the copper sleeve. The sample was twisted for 29 h with a maximum equivalent stress of 23 MPa at 750°C. The resulting dolomite layer between calcite and magnesite is not visible here. Modified after Helpa *et al.* (2015).

reaction interfaces, among others. It is clear that much more experimental work is required in order to gain more insight into the interplay between mineral reaction kinetics and deformation at geologically relevant conditions.

4. Time-resolved X-ray diffraction and *in situ* monitoring of mineral reactions

Recent developments in experimental and analytical methodology now allow direct determination of layer-forming reactions by time-resolved *in situ* monitoring, making use of high-energy synchrotron radiation sources. The experimental strategy here is to grow reaction layers between a single crystal substrate and a thin-film reactant deposited on the substrate by PLD (see above), and monitoring growth using

Figure 22. Internal microstructure of a dolomite rim that formed in the sequence Cal|Do|Mgs in triaxial compression for 76 h, 750°C. Palisade-shaped dolomite develops towards the Do|Mgs interface (lower), granular dolomite towards Mg-calcite (upper). Modified after Helpa *et al.* (2015).

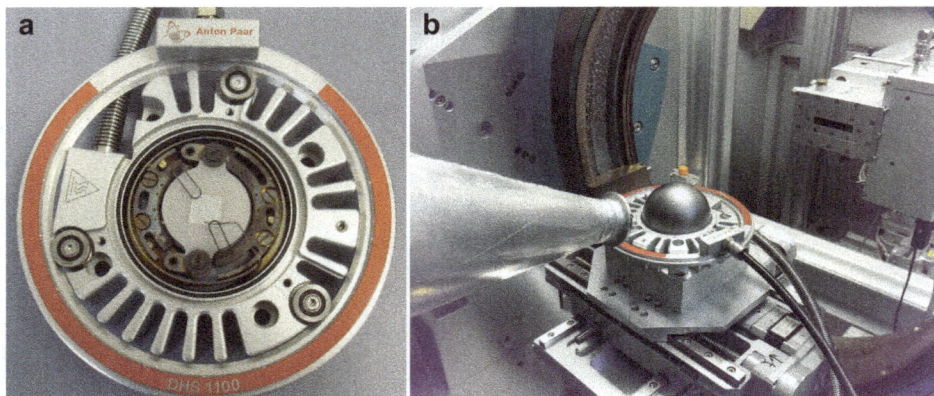

Figure 23. (a) View into the heating attachment for *in situ* synchrotron layer growth experiments in thin films; the sample in the centre is clipped to the heating plate; (b) heating attachment fully mounted for energy dispersive XRD analysis under high temperature. Modified after Götze *et al.* (2014).

synchrotron X-ray diffraction spectra at very short time-scales. Figure 23 shows a high-temperature cell attached to the KMC-2 beamline of BESSY II, Berlin, Germany. At the EDDI beamline, complete diffraction spectra of growing layers in energy-dispersive mode can be obtained on time-scales in the range of tens of seconds per pattern (Genzel *et al.*, 2007).

This new method has been applied recently to the reaction periclase + corundum → spinel (Götze *et al.*, 2014; Götze, 2015). Per|Spi|Cor reaction layers were produced at 800–1000°C, ambient pressure, and reaction durations of 5 to 180 min. Experimental starting setups used both PLD-deposited periclase reactant films on (0001)-oriented corundum single crystals, as well as Al_2O_3 films on (111)-oriented periclase single

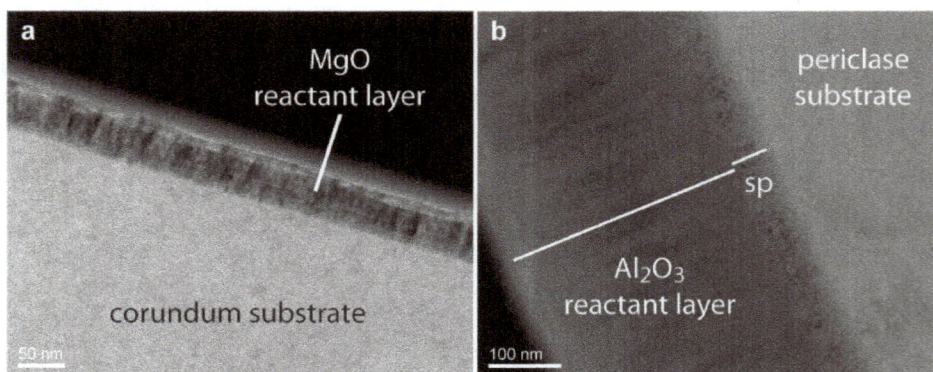

Figure 24. TEM bright field micrographs of FIB foils used as starting setups for spinel layer growth between periclase and corundum: (a) An ~60 nm-thick periclase thin film applied directly to an ultra-polished corundum substrate; (b) corundum thin film on periclase substrate with intermediate spinel seed layer. From Götze (2015).

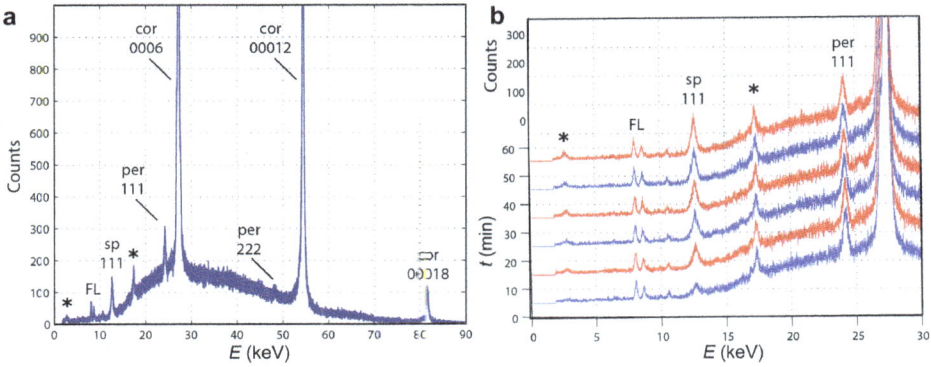

Figure 25. (a) Typical energy-dispersive XRD spectrum (EDDI, Bessy II) of spinel layer growth on corundum substrate; the initially amorphous MgO reactant layer as well as the spinel reaction layer crystallized {111}-oriented. (b) Evolution of Bragg reflections with time at a target temperature of 900°C. The development of the spinel 111 Bragg peak between 5 and 56 min is a direct measure of steady spinel layer growth. Modified after Götze *et al.* (2014). Per =periclase; cor = corundum.

crystals, and sometimes with a thin PLD-deposited seed film of spinel to prevent inhibited layer growth by delayed nucleation (Fig. 24).

Typical energy-dispersive X-ray diffraction patterns are shown in Fig. 25. The strongest signal comes from the substrate, but the integral intensity of the spinel 111 peak can be used with high accuracy as a measure of spinel rim thickness. The absolute thickness is subsequently parameterized by FIB-TEM analysis of the reaction product (Fig. 26). This experimental approach is ideally suited to detecting the initial stages of

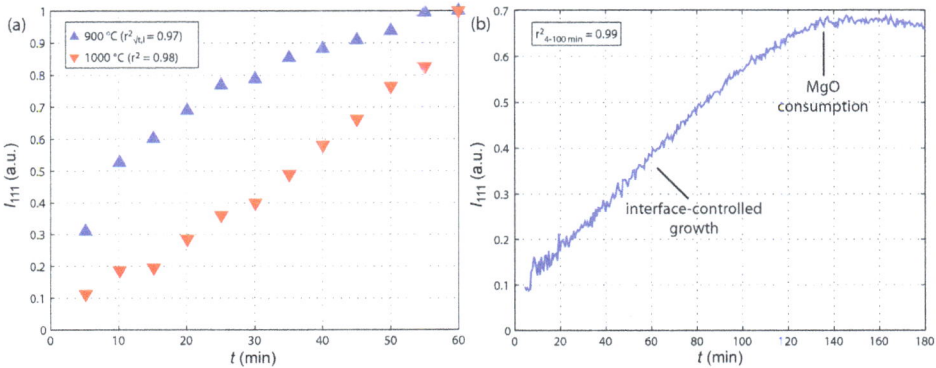

Figure 26. (a) Increase of the integral intensities of the 111 spinel reflections during 1 h experiments perfomed at EDDI at 900 and 1000°C. At 900°C growth kinetics shows a transition from linear growth at initial stages (interface-controlled reaction) to transport-controlled parabolic growth, whereas at 1000°C the rate law is linear, showing only the interface-controlled reaction at higher temperature. (b) Increase of spinel 111 integral intensity measured at the KMC-2 beamline at BESSY II by angle-dispersive XRD. The acquisition time per measurement was reduced to 30 s by using an area detector. Modified after Götze *et al.* (2014).

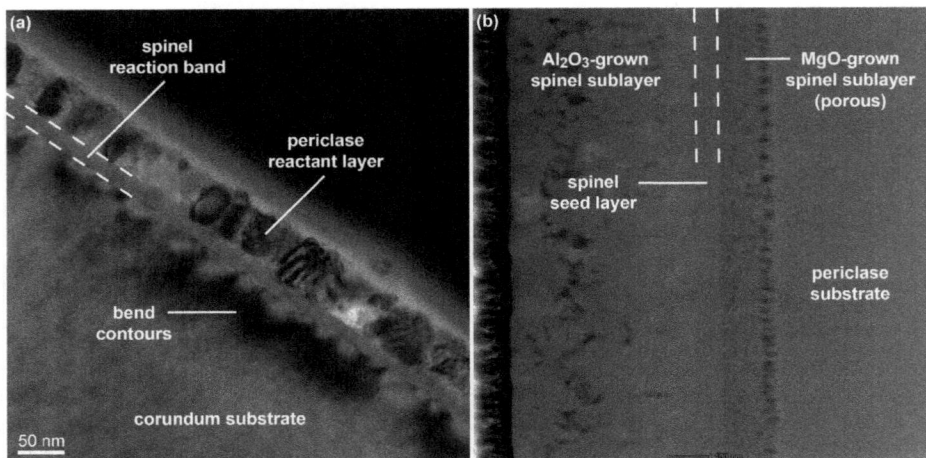

Figure 27. TEM bright field micrographs of FIB sections after thin film layer growth experiments. Local volume changes (compressive or dilative) can be assessed directly from imaging. (a) Growth of a 15 nm thick spinel layer between periclase reactant film and corundum substrate after 1 h at 900°C; growth-induced deformation of the substrate is indicated by bend contours; (b) spinel growth between corundum film and periclase substrate with a spinel seed film. The spinel seed film is still visible between a sublayer that grew topotaxially into the periclase substrate and a sublayer that consumed the Al_2O_3 reactant film. The spinel layer is ~400 nm thick. Porosity in the reaction rim growing towards the periclase substrate indicates local negative reaction volume. Modified after Götze *et al.* (2014).

the kinetics of reaction-layer formation. For example, at 900°C spinel growth shows a transition from linear growth, corresponding to interface reaction control, to parabolic growth, corresponding to diffusion control, which occurs at a layer thickness of <10 nm. At 1000°C, growth is largely linear up to a layer thickness in excess of 300 nm (Götze *et al.*, 2014; Fig. 27). The evolution of interface-controlled to diffusion-controlled kinetics in reaction rims in the initial stage of rim growth formation when the product layers are typically far below 1 μm thick has been predicted by theory (*e.g.* Abart and Petrishcheva, 2011, and references therein). The method used by Götze *et al.* (2014) and Götze (2015) is able to resolve growth rates starting from rim thicknesses of a few lattice planes (Figs. 25, 26). In addition, texture formation is studied by this new method as well as crystallization-induced deformation at reaction fronts (Fig. 26), both of which are factors relevant to the energy state of potentially reactive interfaces and therefore change their reactivity. The method of *in situ* monitoring of reactions in thin films can now be transferred to more intricate silicate systems. It allows unique new insight into mechanical-chemical energy feedback and energy minimization at reactive interfaces in the early stage of nucleation and growth processes.

Acknowledgements

The authors thank Bastian Joachim and Ralf Dohmen for reviewing this chapter and for all their helpful comments.

References

Abart, R. and Petrishcheva, E. (2011) Thermodynamic model for reaction rim growth: Interface reaction and diffusion control. *American Journal of Science*, **311**, 517–527.

Abart, R., Kunze, K., Milke, R., Sperb, R. and Heinrich, W. (2004) Silicon and oxygen self-diffusion in enstatite polycrystals: the Milke *et al.* (2001) rim growth experiments revisited. *Contributions to Mineralogy and Petrology*, **147**, 633–646.

Abart, R., Petrishcheva, E. and Joachim, B. (2012) Thermodynamic model for growth of cellular reaction rims. *American Mineralogist*, **97**, 231–240.

de Ronde, A.A. and Stünitz, H. (2007) Deformation-enhanced reaction in experimentally deformed plagioclase-olivine aggregates. *Contributions to Mineralogy and Petrology*, **153**, 699–717.

de Ronde, A.A., Heilbronner, R., Stünitz, H. and Tullis, J. (2004) Spatial correction of deformation and mineral reaction in experimentally deformed plagioclase-olivine aggregates. *Tectonophysics*, **389**, 93–109.

de Ronde, A.A., Stünitz, H., Tullis, J. and Heilbronner, R. (2005) Reaction-induced weakening of plagioclase-olivine composites. *Tectonophysics*, **409**, 85–106.

Dohmen, R. and Milke, R. (2010) Diffusion in polycrystalline materials: Grain boundaries, mathematical models, and experimental data. Pp. 921–970 in: *Diffusion in Minerals and Melts* (Y. Zhang and D.J. Cherniak, editors). Reviews in Mineralogy and Geochemistry, **72**. Mineralogical Society of America and the Geochemical Society, Chantilly, Virginia, USA.

Dohmen, R., Becker, H.W., Meißner, E., Etzel, T. and Chakraborty, S. (2002) Production of silicate thin films using pulsed laser deposition (PLD) and applications to studies in mineral kinetics. *European Journal of Mineralogy*, **14**, 1155–1168.

Dohmen, R., Becker, H.W. and Chakraborty, S. (2007) Fe-Mg diffusion in olivine I: experimental determination between 700 and 1,200 degrees C as a function of composition, crystal orientation and oxygen fugacity. *Physics and Chemistry of Minerals*, **34**, 389–407.

Eason R., editor (2007) *Pulsed Laser Deposition of Thin Films: Applications-Led Growth of Functional Materials*. John Wiley & Sons, Inc.., Hoboker, New Jersey, USA.

Fisler, D.K., Mackwell, S.J. and Petsch, S. (1997) Grain boundary diffusion in enstatite. *Physics and Chemistry of Minerals*, **24**, 264–273.

Gaidies, F., Milke, R., Heinrich, W. and Abart, R. (2017) Metamorphic mineral reactions: Porphyroblast, corona and symplectite growth. Pp. 469–540 in: *Mineral Reaction Kinetics: Microstructures, Textures and Chemical Compositions* (R. Abart and W. Heinrich, editors). EMU Notes in Mineralogy, **16**. European Mineralogical Union and Mineralogical Society of Great Britain & Ireland, London.

Gardés, E. and Heinrich, W. (2011) Growth of multilayered polycrystalline reaction rims in the MgO-SiO$_2$ system, part II: modelling. *Contributions to Mineralogy and Petrology*, **162**, 37–49.

Gardés, E., Wunder, B., Wirth, R. and Heinrich, W. (2011) Growth of multilayered polycrystalline reaction rims in the MgO-SiO$_2$ system, part I: experiments *Contributions to Mineralogy and Petrology*, **161**, 1–12.

Gardés, E., Wunder, B., Marquardt, K. and Heinrich, W. (2012) The effect of water on intergranular mass transport: new insights from diffusion-controlled reaction rims in the MgO-SiO$_2$ system. *Contributions to Mineralogy and Petrology*, **164**, 1–16.

Genzel, C., Denks, I., Gibmeier, J., Klaus, M. and Wagener, G. (2007) The materials science synchrotron beamline EDDI for energy-dispersive diffraction analysis. *Nuclear Instruments and Methods in Physics Research A*, **578**, 23–33.

Götze, L.C. (2015) *Growth of magnesio-aluminate spinel in thin film geometry: In situ monitoring using synchrotron X-ray diffraction*. Dissertation, Freie Universität, Berlin.

Götze, L.C., Abart, R., Rybacki, E., Keller, L.M., Petrishcheva, E. and Dresen, G. (2010) Reaction rim growth in the system MgO-Al$_2$O$_3$-SiO$_2$ under uniaxial stress. *Mineralogy and Petrology*, **99**, 263–277.

Götze, L.C., Abart, R., Milke. R., Schorr, S., Zizak, I., Dohmen, R. and Wirth, R. (2014) Growth of magnesio-aluminate spinel in thin-film geometry: *in situ* monitoring using synchrotron X-ray diffraction and thermodynamic model. *Physics and Chemistry of Minerals*, **41**, 681–693.

Hartmann, K., Wirth, R. and Heinrich, W. (2010) Synthetic near Σ5 (210)/[100] grain boundary in YAG fabricated by direct bonding: structure and stability. *Physics and Chemistry of Minerals*, **37**, 291–300.

Heinemann, S., Wirth, R. and Dresen, G. (2001) Synthesis of feldspar bicrystals by direct bonding. *Physics and*

Chemistry of Minerals, **28**, 685−692.

Heinemann, S., Wirth, R., Gottschalk, M. and Dresen, G. (2005) Synthetic [100] tilt grain boundaries in forsterite: 9.9 to 21.5°. *Physics and Chemistry of Minerals*, **32**, 229−240.

Heinrich, W. (1993) Fluid infiltration through metachert layers at the contact aureole of the Bufa del Diente intrusion, NE-Mexico: Implications for wollastonite formation and fluid immiscibility. *American Mineralogist*, **78**, 804−818.

Heinrich, W., Metz, P. and Gottschalk, M. (1989) Experimental investigation of the kinetics of the reaction 1 tremolite + 11 dolomite = 8 forsterite + 13 calcite + 9 CO_2 + 1 H_2O. *Contributions to Mineralogy and Petrology*, **102**, 163−173.

Helpa, V., Rybacki, E., Abart, R., Morales, L.F.G., Rhede, D., Jerábek, P. and Dresen, G. (2014) Reaction kinetics of dolomite rim growth. *Contributions to Mineralogy and Petrology*, **167**, 1001.

Helpa, V., Rybacki, E., Morales, L.F.G. and Dresen, G. (2015) Influence of stress and strain on dolomite rim growth: a comparative study *Contributions to Mineralogy and Petrology*, **170**, 16.

Holdaway, M.J. (1966) Hydrothermal stability of clinozoisite plus quartz. *American Journal of Science*, **264**, 643−667.

Joachim, B., Gardés, E., Abart, R. and Heinrich, W. (2011) Experimental growth of åkermanite reaction rims between wollastonite and monticellite: evidence for volume diffusion control. *Contributions to Mineralogy and Petrology*, **161**, 389−399.

Joachim, B., Gardés, E., Velickov, B., Abart, R. and Heinrich, W. (2012) Experimental growth of diopside + merwinite reaction rims: The effect of water on microstructure development. *American Mineralogist*, **97**, 220−230.

Joesten, R. (1991) Kinetics of coarsening and diffusion-controlled mineral growth. Grain boundary diffusion kinetics in silicate and oxide minerals. Pp. 507−582 in: *Contact Metamorphism* (D.M. Kerrick, editor) Reviews in Mineralogy and Geochemistry, **26**. Mineralogical Societry of America, Chantilly, Virginia, USA.

Joesten, R. and Fisher, G. (1988) Kinetics of diffusion-controlled mineral growth in the Christmas Mountains (Texas) contact aureole. *Geological Society of America Bulletin*, **100**, 714−732.

Jolis, E.M., Freda, C., Troll, V.R., Deegan, F.M., Blythe, L.S., McLeod, C.L. and Davidson, J.P. (2013) Experimental simulation of magma-carbonate interaction beneath Mt. Vesuvius, Italy. *Contributions to Mineralogy and Petrology*, **166**, 1335−1353.

Jonas, L., Müller, T., Dohmen, R., Baumgartner, L. and Putlitz, B. (2015) Transport-controlled hydrothermal replacement of calcite by Mg-carbonates. *Geology*, **43**, 779−782.

Keller, L.M., Götze, L.C, Rybacki, E., Dresen, G. and Abart, R. (2010) Enhancement of solid-state reaction rates by non-hydrostatic effects on polycrystalline diffusion kinetics. *American Mineralogist*, **95**, 1399−1407.

Kerrick, D.M. (1968) Experiments on the upper stability limit of pyrophyllite at 1.8 and 3.8 kb water pressure. *American Journal of Science*, **266**, 204−214.

Lüttge, A. and Metz, P. (1993a) Mechanism and kinetics of the reaction: 1 dolomite + 2 quartz = 1 diopside + 2 CO_2 investigated by powder experiments. *The Canadian Mineralogist*, **29**, 803−821.

Lüttge, A. and Metz, P. (1993b) Mechanism and kinetics of the reaction 1 dolomite + 2 quartz = 1 diopside + 2 CO_2: a comparison of rock-sample and powder experiments. *Contributions to Mineralogy and Petrology*, **115**, 155−164.

Marquardt, K., Petrishcheva, E., Abart, R., Gardés, E., Wirth, R., Dohmen, R., Becker, H.-W. and Heinrich, W. (2010) Volume diffusion of ytterbium in YAG: thin-film experiments and combined TEM-RBS analysis. *Physics and Chemistry of Minerals*, **37**, 751−760.

Marquardt, K., Petrishcheva, E., Gardes, E., Wirth, R., Abart, R. and Heinrich, W. (2011) Grain boundary and volume diffusion experiments in yttrium aluminium garnet bicrystals at 1,723 K: a miniaturized study. *Contributions to Mineralogy and Petrology*, **162**, 739−749.

Martin P.M. (editor) (2010) *Handbook of Deposition Technologies for Films and Coatings − Science, Applications and Technology*. Elsevier Inc., Amsterdam.

Metz, P. and Milke, R. (2012) Mechanism and kinetics of forsterite formation in metamorphic siliceous dolomites: Findings from a rock-sample experiment. *European Journal of Mineralogy*, **24**, 59−72.

Milke, R. and Heinrich, W. (2002) Diffusion-controlled growth of wollastonite rims between quartz and calcite: comparison between nature and experiment. *Journal of Metamorphic Geology*, **20**, 467−480.

Milke, R. and Metz, P. (2002) Experimental investigation on the kinetics of the reaction wollastonite + calcite + anorthite = grossular + CO_2. *American Journal of Science*, **302**, 312–345.

Milke, R. and Wirth, R. (2003) The formation of columnar fiber texture in wollastonite rims by induced stress and implications for diffusion-controlled corona growth. *Physics and Chemistry of Minerals*, **30**, 230–242.

Milke, R., Wiedenbeck, M. and Heinrich, W. (2001) Grain boundary diffusion of Si, Mg, and O in enstatite reaction rims:a SIMS study with isotopically doped reactands. *Contributions to Mineralogy and Petrology*, **142**, 15–26.

Milke, R., Dohmen, R., Becker, H.-W. and Wirth, R. (2007) Growth kinetics of enstatite reaction rims studied on nano-scale, Part I: Methodology, microscopic observations and the role of water. *Contributions to Mineralogy and Petrology*, **154**, 519–533.

Milke, R., Kolzer, K., Koch-Müller, M. and Wunder, B. (2009a) Orthopyroxene rim growth between olivine and quartz at low temperatures (750–950°C) and low water concentration. *Mineralogy and Petrology*, **97**, 223–232.

Milke, R., Abart, R., Kunze, R., Koch-Müller, M., Schmid, D. and Ulmer, P. (2009b) Matrix rheology effects on reaction rim growth I: evidence from orthopyroxene rim growth experiments. *Journal of Metamorphic Geology*, **27**, 71–82.

Milke, R., Abart, R., Keller, L. and Rhede, D. (2011) The behavior of Mg, Fe, and Ni during the replacement of olivine by orthopyroxene: experiments relevant to mantle metasomatism. *Mineralogy and Petrology*, **103**, 1–8.

Milke, R., Neusser, G., Kolzer, K. and Wunder, B. (2013) Very little water is necessary to make a dry solid silicate system wet. *Geology*, **41**, 247–250.

Milsch, H., Heinrich, W. and Dresen, G. (2003) Reaction-induced fluid flow in synthetic quartz-bearing marbles. *Contributions to Mineralogy and Petrology*, **146**, 286–296.

Mishin, Y. and Herzig, C. (1999) Grain boundary diffusion: recent progress and future research. *Materials Science and Engineering: A*, **260**, 55–71.

Nishi, M., Nishihara, Y. and Irifune, T. (2013) Growth kinetics of $MgSiO_3$ perovskite reaction rim between stishovite and periclase up to 50 GPa and its implication for grain boundary diffusivity in the lower mantle. *Earth and Planetary Science Letters*, **377–378**, 191–198.

Petrishcheva, E. and Abart, R. (2017) Interfaces. Pp. 295–345 in: *Mineral Reaction Kinetics: Microstructures, Textures and Chemical Compositions* (R. Abart and W. Heinrich, editors). EMU Notes in Mineralogy, **16**. European Mineralogical Union and Mineralogical Society of Great Britain & Ireland, London.

Paterson, M.S. and Olgaard, D.L. (2000) Rock deformation tests to large shear strains in torsion. *Journal of Structural Geology*, **22**, 1341–1358.

Putnis, A. (2009) Mineral replacement reactions. Pp. 87–124 in: *Thermodynamics and Kinetics of Water–Rock Interaction* (E.H. Oelkers and J. Schott, editors). Reviews in Mineralogy and Geochemistry, **70**. Mineralogical Society of America and the Geochemical Society, Chantilly, Virginia, USA.

Schmid, D.W., Abart, R., Podladchikov, Y.Y. and Milke, R. (2009) Matrix rheology effects on reaction rim growth II: coupled diffusion and creep model. *Journal of Metamorphic Geology*, **27**, 71–82.

Schou, J. (2009) Physical aspects of the pulsed laser deposition technique: The stoichiometric transfer of material from target to film. *Applied Surface Science*, **255**, 5191–5198.

Schramke, J.A., Kerrick, D.M. and Lasaga A.C. (1987) The reaction muscovite + quartz = andalusite + K-feldspar + water. Part I. Growth kinetics and mechanism. *American Journal of Science* **287**, 517–559.

Schroer, C.G. (2017) Spatially resolved materials characterization with synchrotron radiation. Pp. 165–180 in: *Mineral Reaction Kinetics: Microstructures Textures and Chemical Compositions* (R. Abart and W. Heinrich, editors). EMU Notes in Mineralogy, **16**. European Mineralogical Union and Mineralogical Society of Great Britain & Ireland, London.

Shaginyan, L.R. (2001) Pulsed laser deposition of thin films: Expectations and reality. In *Handbook of Thin Films*, Five-Volume Set (H.S. Nalwa, editor). Academic Press, New York.

Shvedenko, G.Y., Reverdatto, V.V., Bul'bak, T.A. and Bryksina, N.A. (2006) Bimetasomatic zoning in the CaO-MgO-SiO_2-H_2O-CO_2 system: experiments with the use of natural rock samples. *Petrology*, **14**, 515–527.

Tong, Q.Y. and Gösele, U. (1999) *Semiconductor Wafer Bonding: Science and Technology*. VCH-Wiley, New York.

Walther, J.V. and Wood, B.J. (1984) Rate and mechanism in prograde metamorphism. *Contributions to Mineralogy and Petrology*, **88**, 246–259.

Wasa, K., Kitabatake, M. and Adachi, H. (2004) *Thin Film Materials Technology – Sputtering of Compound Materials*. Springer.

Watson, E.B. and Dohmen, R. (2010) Non-traditional and emerging methods for characterizing diffusion in minerals and mineral aggregates. Pp. 61–105 in: *Diffusion in Minerals and Melts* (Y. Zhang and D.J. Cherniak, editors). Reviews in Mineralogy and Geochemistry, **72**. Mineralogical Society of America and the Geochemical Society, Chantilly, Virginia, USA.

Watson, E.B. and Price, J.D. (2002) Kinetics of the reaction $MgO + Al_2O_3 = MgAl_2O_4$ and Al-Mg interdiffusion in spinel at 1200-2000°C and 1.0-4.0 GPa. *Geochimica et Cosmochimica Acta*, **66**, 2123–2138.

Wirth, R. (2017) Spatially resolved materials characterization with TEM. Pp. 97–130 in: *Mineral Reaction Kinetics: Microstructures, Textures and Chemical Compositions* (R. Abart and W. Heinrich, editors). EMU Notes in Mineralogy, **16**. European Mineralogical Union and Mineralogical Society of Great Britain & Ireland, London.

Yund, R.A. (1997) Rates of grain boundary diffusion through enstatite and forsterite reaction rims. *Contributions to Mineralogy and Petrology*, **126**, 224–236.

EMU Notes in Mineralogy, Vol. 16 (2017), Chapter 3, 37–95

Scanning electron microscopy and electron backscatter diffraction

STEFAN ZAEFFERER[1] and GERLINDE HABLER[2]

[1] *Max-Planck-Institut für Eisenforschung, Max-Planck-Str. 1, D-40237 Düsseldorf, Germany, e-mail: s.zaefferer@mpie.de*
[2] *Department of Lithospheric Research, Vienna University, Althanstraße 14, 1090 Vienna, Austria, e-mail: gerlinde.habler@univie.ac.at*

Scanning electron microscopy (SEM) is one of the most frequently used techniques for near- surface characterization of most solid (and some liquid or liquid-containing) materials with a lateral resolution ranging between 1 nm (for surface-morphology observations) and 1 μm (for certain elemental-composition measurements). It works by scanning a finely focused electron beam over a surface and recording a variety of signals obtained from the electron-beam illuminated volume. Besides observation of the morphology of the surface of materials by recording of secondary electrons, SEM allows the measurement of elemental composition (*e.g.* by X-ray spectroscopy) and crystallographic nature (by backscattered electrons) of microscopic volumes beneath the surface. Even information on the chemical bonding (by Auger electrons) and on the electronic state (*e.g.* by cathodoluminescence) of microscopically small-volume elements can be obtained. Furthermore direct observation of crystal lattice defects is possible. Many of the observations and measurements can be performed simultaneously, thus referring to the same position on a sample surface. Furthermore, materials may be modified during the measurements, *e.g* by heating or mechanical loading, and changes can be observed either *in situ* or by means of interrupted tests.

Particularly powerful techniques for characterization of crystalline materials in SEM are electron backscatter diffraction (EBSD) and EBSD-based orientation microscopy. These techniques allow quantitative characterization of microstructures (*i.e.* defect arrangements) of bulk crystalline materials down to a lateral resolution of ~50 to 200 nm (depending on material and microscope conditions). The largely automated analysis of electron backscatter diffraction patterns (EBSPs) yields the crystallographic phase, orientation, defect density, and, potentially, elastic stress state of the illuminated crystal volume. Crystal orientation mapping (COM) is performed by scanning the electron beam over the sample and recording and analyzing a diffraction pattern from every point of the scan grid. The data obtained can then be plotted, for example, in the form of orientation, misorientation or phase maps of the scanned area. These maps reveal all kinds of morphological data such as grain size, grain shape, spatial distribution of phases and defects and much more. Besides this, the orientation data represent the texture of the investigated area. Together with EBSD scanning, further signals can be recorded, *e.g.* the elemental composition via energy-dispersive X-ray spectroscopy (EDX) or opto-electronic properties *via* cathodoluminescence (CL). This enhances the strength of the technique even further.

DOI: 10.1180/EMU-notes.16.3

1. Principles of scanning electron microscopy

1.1. Introduction

Because scanning electron microscopy is so versatile and frequently used, a large number of text books and monographs exist on the subject (*e.g.* Reimer, 1998; Reed, 2005; Goldstein *et al.*, 2007; Echlin, 2009). Without detailing the underlying physical principles the following text intends to give a brief overview on the possibilities, limitations and recent developments of the technique.

1.2. The scanning principle

Unlike transmission electron microscopy or light microscopy, the SEM does not use lenses to create an image of the sample. Instead, the basic principle of SEM is the measurement of a particular signal arising from the interaction of a finely focused electron probe with the sample. Lenses are only required to demagnify the electron source point into a finely focused electron beam. By scanning the electron probe over the sample surface and recording the measured signal for every scan point a bitmap image of the scanned surface can be reconstructed. Scanning is performed by a set of crossed deflection coils which are controlled by a scan generator which itself is computer-controlled. The scan extension, number of scan points, dwell time per scan position and the scan mode can all be selected from wide ranges through the operation software. The magnification of the scan image is given by the ratio of the size of the image on the screen to that of the scanned area on the sample. The spatial resolution of the technique is given by the size of the interaction volume of the electron probe with the sample for the respective observation signal chosen. The size of the interaction volume depends on material properties (*e.g.* bulk Z-number, density), beam properties (*e.g.* acceleration voltage, probe diameter, electron source type) and the measured signal type. Figure 1 demonstrates the scanning principle.

1.3. The basic set-up of an SEM

The most important elements of an SEM are shown in Fig. 2 and are described briefly below.

The electron probe, also called the primary electron beam, is created in the SEM using an electron source and a series of electron lenses that demagnify (focus) the originally relatively broad electron beam into a small circular spot on the sample.

Three different types of electron sources are commonly in use, namely the thermionic electron emission from a tungsten filament or from a LaB_6 crystal, the field-assisted thermionic emission from a thin ZrO-coated tungsten tip ('Shottky' or 'thermal' field emission) and the field emission from a thin tungsten tip ('cold' field emission). These three emitter types have quite different characteristics; a few of them are summarized in the Table 1.

The construction details of the electron source are beyond the scope of this chapter but it is important to note that the electrons emitted are accelerated in an electric field and focused to a small spot, the first cross-over. Usually the electrons are accelerated to a kinetic energy between 1 and 40 keV. Lower electron energies in the range from 0 to 1 keV,

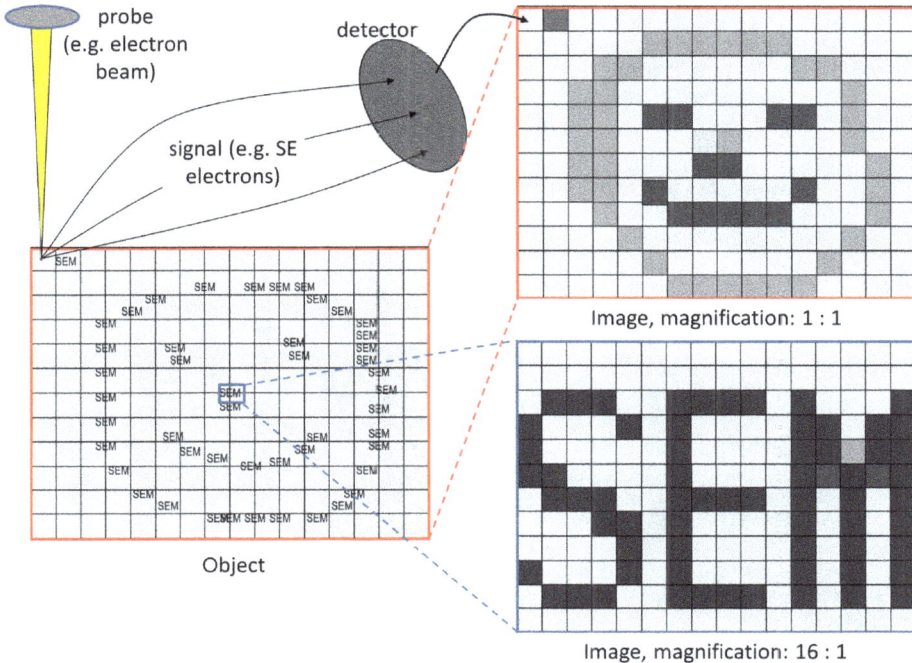

Fig. 1. The scanning principle of an SEM. An object displayed on the right side is scanned by a finely focussed probe which moves on a regular grid across the object. At every position of the grid the emitted signal is recorded by a detector and the signal is saved as one image pixel at the respective position. A lens is only required to create the focused probe but not to create the image. The magnification of the image is given by the size of the scan grid on the sample with respect to the size of the grid image seen by the user. The image resolution is given by the size of the probe and its interaction volume within the sample.

which are desired for many low-voltage applications (Bell and Erdman, 2013), are produced by deceleration of the electron beam in a retarding electrical field either inside of the electron column ('booster' technology) or towards the sample ('cathode lens' technology) (Frank *et al.*, 2007).

Employing a set of electrostatic and/or electromagnetic condenser lenses the accelerated electrons leaving the electron gun are changed into an almost parallel beam of defined total beam current and convergence. The convergence defines the final spot size of the beam on the sample. The beam then enters a set of crossed deflection coils which deflect the beam in the x and y directions in order to enable beam scanning. Low-magnification imaging/high probe deflection requires relatively strong deflection coil voltages. Finally the beam is focused onto the sample by the objective lens. Inside of the objective lens there is a further set of coils that are required to correct for probe astigmatism.

The signals that arise from the interaction of the electron beam with the sample at a given scan position can be recorded for that position using a number of dedicated

electron gun
creates the electron
beam by thermal
and/or field emission

condenser
demagnifies the cross-
over, determines
achievable spot size

objective system
focuses the beam
corrects astigmatism
determines magnification

vacuum system
creates the required
vacuum conditions

cathode
Wehnelt cylinder
anode
first cross-over

condenser lens 1

condenser lens
aperture

condenser lens 2

objective lens
scan coils
stigmator

objective lens
aperture

detector

sample

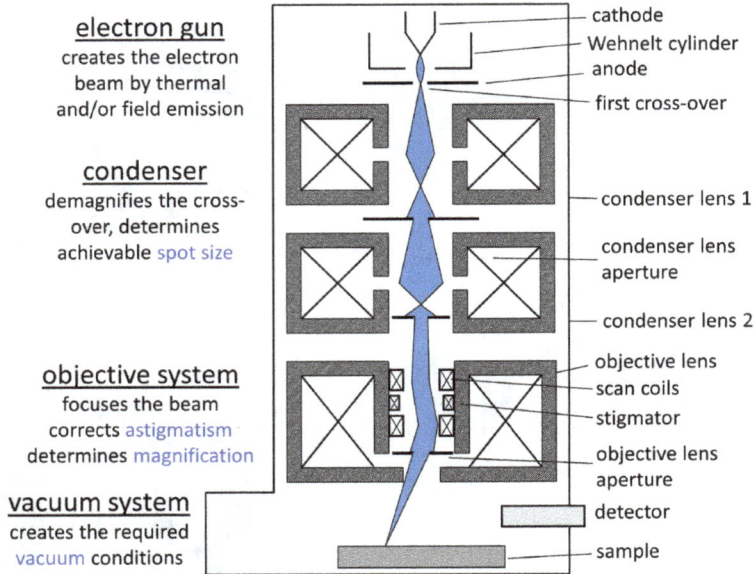

Fig. 2. Principle of construction of an SEM column. The controlling electronics and the scan generator are not displayed.

detectors arranged around the specimen chamber and are used to build an image or a point-dataset. As electrons interact strongly with any kind of matter, including gas molecules, the electron beam column must be well evacuated, usually to a level of 10^{-8} Pa in the electron gun part of field emission SEMs. The specimen chamber is usually at a higher pressure, of the order of 10^{-4} Pa. If the beam path in the chamber is short enough and suitable detectors are used, as is the case in so-called environmental SEMs, the pressure in the chamber may also be raised to 10^5 Pa in order to observe samples that contain liquid water (Stokes, 2008). Moreover, elevated chamber pressure by insertion of water vapour enables investigation of samples with low electrical conductivity, as ionization of the gas by the electron beam leads to conductivity of the atmosphere around the sample.

Besides the main microscope column an SEM also requires a dedicated scan generator and an electronic control system that synchronizes scanning and signal recording. A dedicated sample stage with piezo or step motors on up to six axes (x, y, z translation, sample tilt and rotation and a tilted z axis for eucentric positioning) allows flexible positioning of the sample. Nowadays all of this is controlled by computer with dedicated DA (digital-analogue) converters.

1.4. Signals from electron-matter interaction

There are a number of signals that arise from the interaction of an electron beam with matter. Each of these signals carries different information about the material at the position of exposure to the electron beam. The signals can be classified into electron

Table 1. Characteristics of common electron source types. The beam brightness measures the electron density in the beam as current per areas and solid angle. The diameter of the virtual source is the beam diameter at the first beam cross-over.

	Thermionic emission W/LaB6	Field-assisted thermionic emission	Field emission
Emission temperature	W: 2500 to 3000 K LaB6: 1400–2000 K	1800 K	RT to 1500 K
Electron energy spread	W: 1 to 3 eV LaB6: 0.5 to 2 eV	0.5 eV	0.2 to 0.5 eV
Max. beam current, brightness	W: 10^6 A/cm^2 sr LaB6: 10^7 A/cm^2 sr	10^8 A/cm^2 sr	10^9 A/cm^2 sr
Diameter of virtual source	W: 30 μm LaB6: 10 μm	15 nm	3 to 5 nm
Time of stable beam emission	several hours, up to days	weeks or longer	few hours
Life time	~20–200 h	1–5 years	5–10 years
Operation mode	regular filament exchange and re-adjustment required	works service-free over many years	needs to be flashed regularly
Vacuum conditions	nothing specific	HV	UHV
Typical application/ resolution	general low spatial resolution imaging and analytics with high current requirements; SE imaging with ~5 nm resolution; ideal for EDX/WDX, CL	high spatial resolution imaging and analytics with medium current requirements; SE imaging with ~1 nm resolution; ideal for EBSD and EDX at all length scales	ultra-high spatial resolution imaging, analytics is limited by low beam current and stability; SE imaging with <1 nm resolution; ideal for high-resolution imaging

and photon signals. Both electrons and photons will be emitted with very different characteristics depending on the particular interaction mechanism of the primary beam electrons with the sample. Figure 3 shows the distribution of energy of the electrons emitted from a sample surface during high-energy electron illumination. The different electrons carry very different information about the sample. Photons also show very different characteristics, with energies from the visible-light spectrum (cathodoluminescence) all the way to energies in the area of soft and hard X-rays for X-ray spectrometry. In the following we will discuss the most frequently used signals in some more detail.

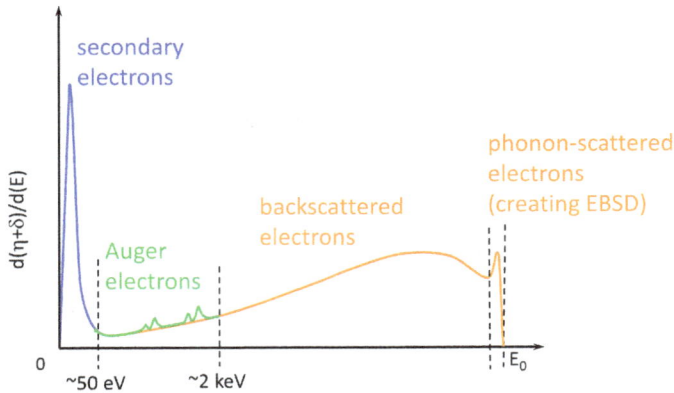

Fig. 3. Electron energy spectrum emitted from a sample illuminated with electrons of energy E_0. The relative heights of the peaks may change according to sample and illumination conditions.

1.4.1. Secondary electrons

The most commonly used signal in an SEM is secondary electron (SE) emission. SEs arise in large numbers and are collected and detected effectively by an Everhart-Thornley (ET) scintillator detector. SEs are generated as a result of different inelastic scattering processes of the primary beam electrons. By definition, all electrons emitted from the sample surface with energies of <50 eV are called SE. Due to their low energy SEs cannot travel far through the material, thus they leave the sample from only a very shallow depth below the sample surface and very close to the location of the primary electron beam. As a consequence the spatial resolution of SE is the highest of all signals. The low energy of SEs also allows collection of these electrons by a positive electrical field that attracts them towards the ET detector. SEs mainly carry information about the sample morphology because the SE intensity depends strongly on the inclination angle of the sample surface relative to the primary beam and relative to the ET position. Edges and corners emit more SEs than flat surfaces, giving rise to an edge-contrast effect of the SE signal-intensity distribution. Due to the shallow depth of SE emission, SE images are very sensitive to surface contamination, sample preparation artefacts and charging effects. A typical SE image is shown in Fig. 4. As SE emission is comparable to the emission of visible light from a surface, SE images give a familiar and natural impression.

Besides SE being created directly by primary beam electrons, SE can also be created from backscattered electrons, which leave the sample at an appreciable distance from the footprint of the primary beam. These SEs are called SE2 and they may reduce the contrast of SE images because they are emitted from a comparatively large area around the footprint of the primary beam. The use of an in-lens SE detector (positioned inside the objective lens and not in the chamber as is the case for an ET detector) at small working distance (<3 mm) reduces the amount of SE2 arriving at the detector and may, therefore, deliver SE images with greater contrast and spatial resolution.

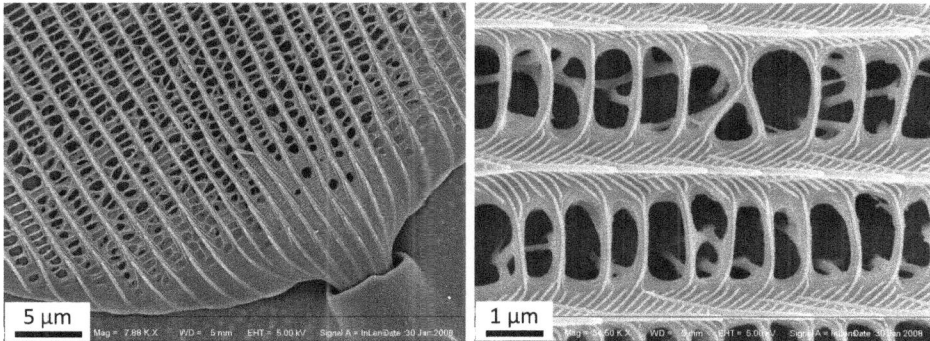

Fig. 4. Secondary electron image of a detail of a butterfly wing at two different magnifications. Images have been recorded with an in-lens SE detector. Note the familiar appearance of the image although the features are magnified 30,000 times. Courtesy of Helge Fabritius, MPIE.

1.4.2. Backscattered electrons

Another commonly used signal is backscattered electrons (BSE). These electrons can be considered to be primary beam electrons that have been changed in energy and propagation direction by elastic or inelastic scattering events at the sample matter. Most detected BSE have conserved relatively high energy, of the order of 80% of the primary electron energy or higher, which is why they may travel relatively far in the sample and show straight travelling trajectories once they have left the sample. BSEs are, therefore, best detected by a solid-state photo diode detector or a special scintillator detector (Robinson detector) positioned directly above the sample. The amount of electrons backscattered from a sample varies with three important material properties, which are responsible for the contrast in BSE images: first, the BSE yield varies with the mass density of the material: the higher the material density, the higher the BSE signal intensity. This contrast is referred to as the mass or atomic-number or Z-contrast (Fig. 5). The brighter is a pixel in a BSE image, the higher is the average atomic number of the corresponding volume on the sample. Due to the larger travelling distance of BSEs in the sample material, the lateral and depth resolution of Z-contrast images is relatively poor compared to the SE signal. Depending on primary electron energy and sample density it may vary between 50 nm for gold at 10 kV and 4 µm for aluminium at 30 kV. A second contrast obtained with BSE is due to the topography of the sample. In fact, the angular distribution of BSE varies with the inclination of the surface at the illuminated spot. This dependency can be recorded with a directionally sensitive detector, *e.g.* a 4-quadrant BSE detector which records the BSE intensity in four different directions above the sample. A sample image for topography contrast is given in Fig. 6b. Finally, when crystalline material is investigated, the BSE yield also depends on the orientation and perfection of the atomic lattice. This contrast is also called electron channelling contrast (ECC) or orientation contrast (OC). ECC is, in principle, similar to diffraction contrast in TEM and, therefore, can only be explained using the wave nature of the electrons. The physics of channelling contrast will be described in more detail in conjunction with electron

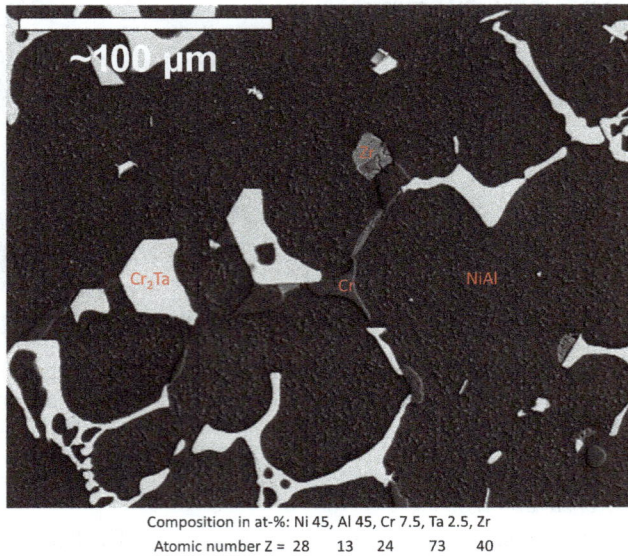

Composition in at-%: Ni 45, Al 45, Cr 7.5, Ta 2.5, Zr
Atomic number Z =   28   13   24   73   40

Fig. 5. Backscattered electron image with atomic number (Z) contrast of an NiAl base alloy with additions of Cr, Ta and Zr. The brightness of the image corresponds directly to the average atomic number of the phases. Alloy composition and atomic numbers are marked in the table.

backscatter diffraction in section 2.2. ECC enables direct observation of lattice defects such as dislocations, stacking faults or nanotwins (Fig. 7). The lateral resolution of ECC is ~10 nm which is significantly better than that of Z-contrast. The probing depth below the sample surface is of the order of 50 to 100 nm, depending on beam energy and scattering power of the material.

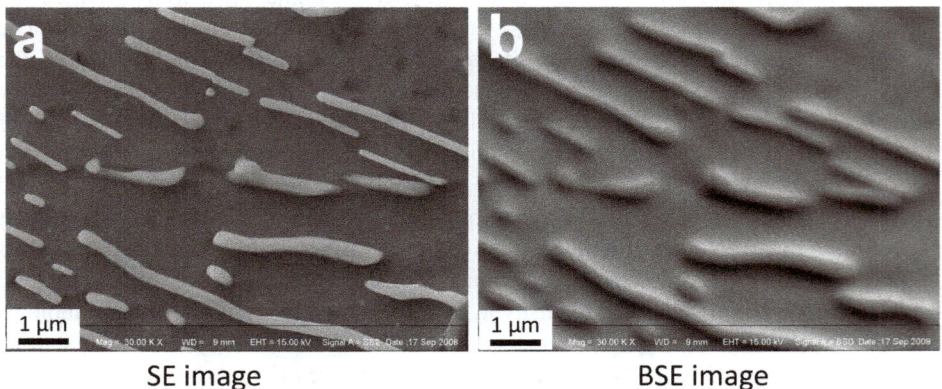

SE image   BSE image

Fig. 6. Comparison of topographic contrast obtained with secondary electrons (a) and backscattered electrons (b) from pearlite, a lamellar eutectoid structure of bcc iron with Fe_3C cementite. The BSE image is obtained from a 4-quadrant detector by subtraction of the signal from one half from that of the second half.

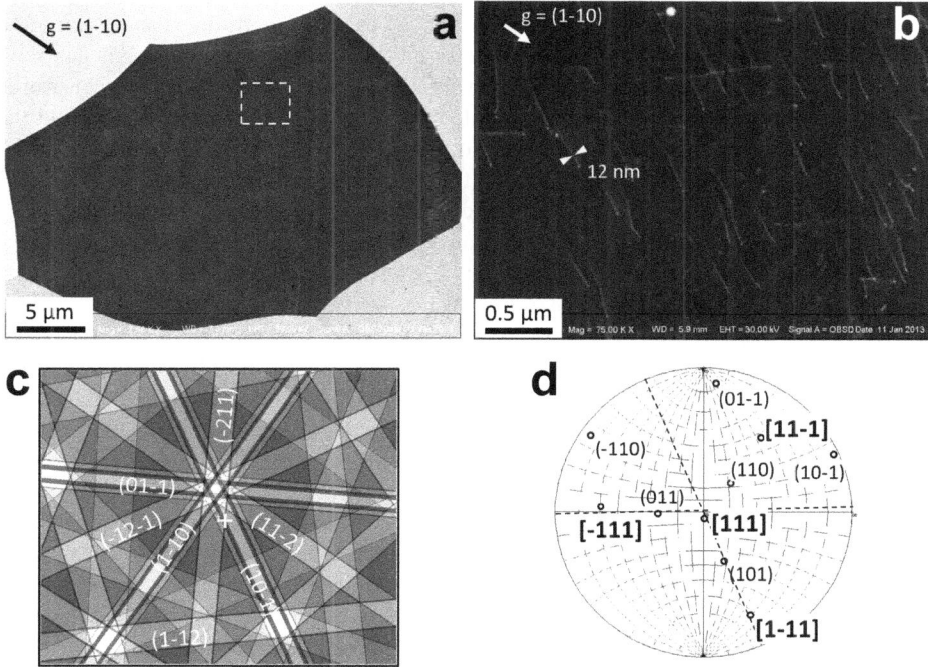

Fig. 7. BSE image showing electron channelling contrast (ECC) on an Fe 3 wt.% Si electrical steel with bcc crystal structure. (a) An individual grain in channelling mode (*i.e.* very few electrons are backscattered, most are channelling into the crystal) for ($1\bar{1}0$) two-beam diffraction conditions. (b) Enlargement of the area marked in a. The fine white lines are dislocation lines. Four different slip systems are visible. (c) Simulated electron channelling pattern for the grain orientation in a. The primary beam position is marked by a white cross. It is on the ($1\bar{1}0$) Kikuchi line. (d) Stereographic projection of the {110} planes and <111> directions for the orientation of the crystal. The <111> directions correspond exactly to the dislocation line directions in b, indicating that all these dislocations have screw character. See Zaefferer and Elhami (2014) for more detailed explanations.

Closely related to BSE and ECC is the electron backscatter diffraction (EBSD) signal which is frequently used to obtain information on crystal orientation, crystal lattice perfection, crystal lattice symmetry and, to a certain extent, lattice parameters of crystalline matter. EBSD is the basis of orientation contrast microscopy which is a very powerful and versatile technique for observation of many aspects of crystalline matter. A detailed treatment of these important techniques will be presented in section 2.

1.4.3. X-ray emission

Another frequently used signal is X-ray emission. Interaction of high-energy electrons with matter may lead to emission of electromagnetic radiation which can have, at most, the energy of the exciting electrons (0.1 to some 10 keV). X-rays are created through two processes of electron-matter interaction. The continuous deceleration of electrons in the electric field of the target material electrons causes a continuous spectrum of X-rays, also called "Bremsstrahlung". At the same time, the primary electrons can have sufficient energy to ionize the target atoms by kicking out electrons from inner orbital shells. The subsequent re-filling of these electron holes by electrons from outer orbital shells leads to the emission of characteristic X-rays, the energy of which corresponds exactly to the energy difference between the different orbitals involved in ionization and re-filling. The radiation energy is, therefore, specific for each element referred to as characteristic X-ray lines. X-ray emission is recorded and analysed by dedicated detectors based either on energy dispersion (EDX or EDS for energy dispersive (X-ray) spectrometry) or wavelength dispersion (WDX or WDS for wavelength dispersive (X-ray) spectrometry). X-ray spectrometry is mainly used for obtaining qualitative and quantitative information about the elemental composition of the illuminated volume. The wavelength of the characteristic X-ray lines emitted and their particular combinations give information about which elements are contained in the volume. The intensity of characteristic radiation energies allows quantification of the element concentrations. The spatial resolution of X-ray spectrometry depends on the primary electron energy and the characteristic X-ray energy of the particular element. It varies between ~50 nm to ~0.5 μm; for many minerals such as silicates, carbonates, phosphates, oxides and hydroxides the spatial resolution may even be poorer, of the order of 1−2 μm. EDS and WDS detection have quite different characteristics with respect to detectable elements, energy resolution, mass detection limit, and time needed for measurement. EDS is comparatively insensitive for light elements (below C) or very low-energy radiation of other elements; it has a rather poor energy resolution of the order of tens of eV, and a rather high detection limit of 0.1 at.%. On the positive side, the measurement rate is very high, because the entire X-ray energy spectrum is measured in parallel. In contrast, with WDS, very low-energy radiation can be analysed with good accuracy, WDS has a comparatively high energy resolution of a few eV, a detection limit of 0.001 at.%, but the time needed for a measurement is much longer than for an EDS analysis, because different wavelengths are measured sequentially and the detector efficiency is significantly lower than that of an EDS detector. Nowadays almost all SEM are equipped with an EDS detector. WDS detectors are less common and are more frequently found on dedicated SEMs called electron probe microanalysers (EPMA). Examples of the use of EDS and WDS will be found throughout this paper, most of them in conjunction with observations of microstructures using EBSD. For detailed information the reader is referred to the book by Goldstein *et al.* (2007).

1.4.4. Auger electrons

Another signal commonly used in SEM is auger electrons (AE), which are created, similarly to characteristic X-rays, by the ionization of the target atoms and subsequent re-filling of the missing shell electrons. In the case of AE the de-excitation energy is not released as a photon (*i.e.* as X-rays), but it is released as an electron that leaves the atom from a specific orbital. The AE energy corresponds to the energy difference between the three electron shells which are involved in the AE formation. The probability for a non-radiative emission of an AE increases with decreasing energy of the shells involved. Therefore, the AE yield is more pronounced for light elements (Z < ~20) while X-ray emission is more pronounced for heavier elements. Furthermore, the energy of individual Auger lines is highly sensitive to the bonding state of the atom. This is why AE can not only be used for elemental but also for bonding analysis. The energy of most AE is in the range of 10 to 2000 eV (see Fig. 3), which means that they cannot cover large distances within the material; the typical escape depth is of the order of ten atomic layers only. The AE signal is, therefore, extremely surface-composition sensitive. The lateral resolution is of the order of some 10 nm. The high surface sensitivity is one of the major reasons why AE spectroscopy and imaging is relatively complicated compared to other SEM techniques. The sample has to be extremely clean and the vacuum level of the SEM very good, at the ultra-high vacuum level. Furthermore, electron energy spectrometry is technically more complicated than, for example, X-ray spectrometry which is why usually only dedicated instruments are equipped with an AE spectrometer. A comprehensive overview of this technique was given by Hofmann (2013).

One consequence of the high surface-sensitivity of AE spectroscopy is that non-conductive samples, *e.g.* minerals, cannot be coated to increase their surface conductivity. Nevertheless, AE spectrometry can be successfully applied for the investigation of mineral samples when the sample is steeply tilted to an angle larger than ~30° to 60°. This leads to a high yield of BSE leaving the sample. If, at the same time, the acceleration voltage is carefully adjusted, the BSE may create a sufficiently large amount of SE to compensate the charge carried into the sample by the primary electrons. AE can then be collected also from geological-samples, like silicate (*e.g.* amphiboles and feldspars), carbonates, and salts. A good account of the application of AE to geological samples can be found in Nesbitt and Pratt (1995). Very valuable information on AES on non-conductive materials is also available in Hofmann (2013).

1.4.5. Cathodoluminescence

Cathodoluminescence (CL), as AE and X-ray emission, is an analytical and imaging technique which is particularly useful for investigation of minerals and semiconductors. The term CL refers to the emission of photons in the visible, infrared or ultraviolet range of the electromagnetic spectrum after excitation with high-energy electrons. The energy of the emitted photons is in the range 1 to 6 eV. A CL signal may be recorded from all materials and defects which allow electronic transitions in the appropriate energy range, which is the case for many semiconductors and insulators. The transitions usually occur between the occupied conduction band and the unoccupied valence band and may proceed via deep or shallow energy levels inside the band gap. The light-emitting radiative transition competes with non-radiative transitions which create phonons instead. One distinguishes intrinsic and extrinsic transitions. Intrinsic transitions are those between energy levels of the basic crystal, *e.g.* across the band gap of a solar cell material. Extrinsic transitions, in contrast, are transitions caused by defects, *i.e.* point defects or extended defects. Point defects may be trace-element ions which create energy levels in the band gap of minerals, in particular transition element or rare earth element ions. Commonly, different element ions emit very different colours (which was used for decades in colour cathode ray tubes, for example). Extended defects may be, for example, dislocations or grain boundaries in semiconductors, which create dangling or false bonds related to energy levels in the band gap. In semiconductors yielding intrinsic radiation, extended defects may also be points of non-radiative charge-carrier recombinations. These defects may then appear dark in an otherwise emitting surrounding.

Unfortunately, the theory of CL emission is much less straight-forward than that of X-ray emission (see Sieber (2013) and Yacobi and Holt (1990) for an introduction). There are some elements, in particular the rare earth elements, which emit sharp lines which correspond to inner-atomic transitions (f–f or d–f orbital transitions) similar to X-ray lines. Rare earth elements and, for example, Mn^{2+} ions, give very intense signals, such that trace amounts of these cations at concentrations of the order of 10 ppm, can be detected. With dedicated spectral analysis methods even concentrations in the ppb range can be observed. CL is, therefore, a sensitive analytical technique for investigation of minerals (see Pagel et al., 2000). Intrinsic transitions usually give less sharp spectral peaks; often even only very broad, band-like peaks are visible. One reason for this is that the thermal energy of electrons may interfere with the valence-conduction band transition. Cooling the sample may result in much sharper peaks that can be evaluated more easily.

The lateral resolution of CL corresponds approximately to the size of the electron range of the primary electrons inside of the sample and, therefore, depends heavily on the acceleration voltage. However, the voltage cannot be tuned to an arbitrarily small value because it must exceed a threshold value to obtain a reasonable signal to noise ratio. In most practical situations the lateral resolution will therefore be of the order of 1 μm.

Detection of CL requires a particular detector set-up. An ellipsoidal or parabolic mirror is placed upon the sample, which collects as much of the emitted light as possible from the sample and focuses it onto a light-sensitive detector, either a photomultiplier or a highly sensitive CCD with or without a spectrometer in front of it. The signal can be recorded either in pan-chromatic, i.e. light of all wavelengths is collected simultaneously, or in wavelength dispersive mode, using a prism and a CCD detector. The latter allows discrimination of different kinds of defects, although deciphering the nature of the defects is not straightforward.

Figure 8 displays an example of CL measurements on a polycrystalline CdTe thin-film solar cell absorber material. From comparison with the orientation microscopy image, measured via EBSD, it is clear that the large-angle grain boundaries do not show radiative emission while the grain interior and areas of twin boundaries show intense emission at different wavelengths.

1.4.6. Electron beam-induced current

Finally, the electron beam-induced current (EBIC) technique will be mentioned briefly. This technique is used primarily for investigation of semi-conductors with an intrinsic

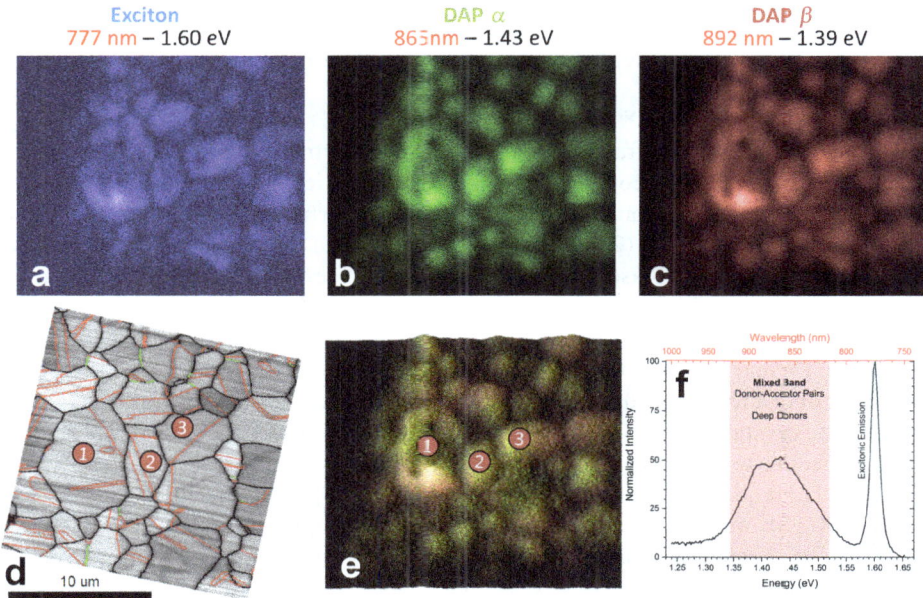

Exciton
777 nm – 1.60 eV

DAP α
865nm – 1.43 eV

DAP β
892 nm – 1.39 eV

Fig. 8. An example of wavelength-resolved CL imaging on a polycrystalline thin-film CdTe solar cell obtained at liquid nitrogen temperature. (a–c) Images obtained with different characteristic wavelengths. The emission spectrum is displayed in f with important emission lines (excitons, donor-acceptor pairs, DAP) being marked. (e) A colour-mixed map of the three individual maps. (d) An EBSD pattern quality map (grey shaded) with grain boundary traces coloured by misorientation angle. Image courtesy of G. Stechmann, MPIE.

pn junction or a pn junction applied to a sample specifically to perform EBIC investigations. A comprehensive overview on the technique can be found in Reimer (1998). The principle of the technique is that minority charge carriers created within the semiconductor by the high-energy electron beam will be separated in a pn junction and then lead to a measurable current flowing from the sample. The current is of an order of 10^{-9} to 10^{-7} A. EBIC is used to observe morphological features of pn junctions, but also to measure the lifetime and diffusion lengths of minority charge carriers. It can be used further to observe dislocations or grain boundaries inside or close to the pn junction by the fact that these extended defects lead to effective recombination of charge carriers and, therefore, to a significant drop in the EBIC signal. In the case that a semiconductor without an intrinsic pn junction is to be observed, a thin metallization layer can be applied to the surface which leads to a Shottky barrier with a similar function as an intrinsic pn layer. The spatial resolution of the technique is of the order of the diffusion length of the charge carriers, *i.e.* 1 to several micrometres and improves with decreasing voltage.

2. Electron backscatter diffraction (EBSD)-based orientation microscopy of minerals

2.1. Introduction: An overview of EBSD

Electron backscatter diffraction (EBSD) is an electron diffraction technique in scanning electron microscopy and, as such, only applicable to crystalline matter. A comprehensive overview on many aspects of this powerful and versatile technique can be found in Schwartz *et al.* (2009). From individual EBSD patterns (EBSP) information about crystallographic phase, crystal orientation and defect densities of small sample volumes can be obtained. More advanced applications of EBSP evaluation also allow the determination of residual stresses (Wilkinson *et al.*, 2006) or of details of the crystal structure (Winkelmann and Nolze, 2015a). In combination with electron beam scanning over a regular grid on a sample, EBSPs constitute the basis of a powerful SEM technique, crystal orientation mapping (or microscopy), which will be abbreviated throughout this text as COM. Note that other names and acronyms are frequently in use for this technique, namely orientation imaging microscopy (OIM) which is a trade mark of EDAX/TSL and automated crystal orientation mapping (ACOM). COM enables quantitative characterization of microstructures (covering the material defect structure, the grain/phase boundary geometry, crystal shapes, shape preferred orientations and the spatial phase arrangements) and textures (*i.e.* the distribution of crystallographic orientations) with a lateral resolution in the $10-100$ nm range on sampling areas ranging from a few square micrometres to several square centimetres in size.

EBSPs are, in principle, Kikuchi diffraction patterns observed in reflection rather than transmission. It was Nishikawa and Kikuchi (1928) who observed these patterns for the first time. For a long time, these patterns were considered as rather exotic until Venables and Harland (1973) and then Dingley (1984) came up with an experimental set up and software that allowed recording of EBSPs and extracting crystallographic orientation information from them (see, for example, Schwarzer, www.ebsd.info/history.htm for a good account on this early history). This stimulated further research resulting in the development of EBSD-based orientation mapping by Adams *et al.* (1993), who also coined the name orientation imaging microscopy (OIM)[1]. For this technique an electron beam scans over a suitable sample and an EBSD pattern is obtained and analysed at every scan point. Fully automated pattern analysis became possible mainly due to the work of Krieger-Lassen *et al.* (1992) and Krieger-Lassen (1994) who introduced the application of a Hough transform for the geometrical analysis of these patterns. Based on this work several commercial and non-commercial EBSD-based COM systems were developed and EBSD evolved into one of the most versatile tools for characterization of microstructures and crystallographic textures.

---

[1] Note again, the acronym 'OIM', though used widely in the community, is not accepted by all groups because it is a commercial trademark of EDAX/TSL. We therefore will refer to it here as COM for crystal orientation microscopy/mapping.

Several text books and review articles have given an exhaustive overview of theoretical and practical aspects of the technique (*e.g.* Dingley, 2004; Pouchou, 2004; Randle and Engler, 2009; Schwartz *et al.*, 2009; Randle, 2013). Perhaps the greatest attribute of EBSD-based COM is the fact that the technique yields a significant amount of numerical information for every measurement position, *e.g.* the space coordinates, the orientation matrix (a rotation matrix that rotates an external sample coordinate system into the crystal lattice coordinate system), the best-fitting crystallographic phase (selected from a list of predetermined phases), the EBSP contrast and a variety of confidence measures. In combination with other techniques, particularly with X-ray spectrometry (EDX) further data can be collected simultaneously. These datasets are used to produce, for example, crystal orientation or phase maps of the scanned area on the sample. Determination of the change of orientation from point to point enables misorientation and grain-boundary maps to be produced as well. Besides construction of a large variety of different maps it is also possible to extract statistical data sets from the sample, *e.g.* orientation distributions (textures), charts about stereological data of microstructures and many others. Some characteristic data of the technique are compiled in Table 2.

One of the latest important achievements in EBSD research and development is the improved understanding of the physics of EBSP formation and the consequent creation

Table 2. Some characteristic values for EBSD-based orientation microscopy.

Feature	Value	Remarks
Lateral resolution	30–200 nm on bulk samples 5–50 nm on thin foils	depending on sample material and acceleration voltage in transmission below the sample
Depth resolution	10–40 nm	depending on sample material and acceleration voltage
Smallest measurable volume	$\sim 10^4 - 10^3$ nm^3	depending on sample material and acceleration voltage, the smallest value is obtained for t-EBSD on thin samples.
Angular resolution	0.5–0.01°	large value obtained with standard pattern analysis (Hough transform), small value by pattern cross-correlation
Measurable area	$\mu m^2 - cm^2$	
Measurement rate	1–1000 patterns s^{-1}	slow for high-quality patterns, rapid for large maps with high-noise and low-resolution EBSPs

of computer software for simulation of realistic EBSPs by Winkelmann et al. (2007). This work enabled new means of pattern analysis and new information on crystallographic structures (Winkelmann and Nolze, 2015a). Other recent significant activities have been triggered by the introduction of cross correlation techniques for analysis of EBSD patterns (Wilkinson et al., 2006; Maurice and Fortunier, 2008). These techniques allow a substantially improved angular resolution of pattern analysis and, therefore, enable the determination of very small lattice rotations and, to a certain extent, determination of elastic lattice strains (section 2.6.2).

EBSD-based COM has become a standard tool for characterization of micro-structures of crystalline materials, including metals (Humphreys, 2001; Zaefferer et al., 2012), intermetallics (Zambaldi et al., 2009), semiconductors (Wilkinson, 2000), ceramics and minerals (Prior et al., 1999), water ice (Prior et al., 2015), crystallizing glasses (Wisniewski et al.,2011) or biominerals (Griesshaber et al., 2010), down to grain sizes in the range 100−200 nm diameter, with meaningful step-sizes down to ~30−200 nm (note that these values depend heavily on the material and the acquisition parameters, the values at the lower end can be expected for heavy metals, those at the larger end for beam-sensitive materials, as for example some carbonate and silicate minerals). It enables, on the other end of the size scale, quantitative observation of samples as large as several square-centimetres. With the new transmission-EBSD technique, the spatial resolution of the technique may reach or exceed TEM-based COM methods (see section 2.4.1).

Although COM is, by nature, a technique for characterization of near-surface properties of materials it can be extended into a powerful 3-dimensional characterization technique through combination with some sort of serial sectioning, such as sputtering with a focused ion beam (FIB) (e.g. Zaefferer et al., 2008; Uchic et al., 2004), or with mechanical polishing (Pirgazi et al., 2014). The 3D characterization of microstructures allows, in particular, a detailed analysis of grain and phase boundaries (Rohrer, 2011; Konijnenberg et al., 2012; Groeber and Jackson, 2014). In addition this EBSD-based COM is used for in situ observations of deformation (Ojima et al. 2011; Yan et al., 2015; Tasan et al., 2014), recrystallization and grain growth (Humphreys and Ferry, 1996; Seward et al., 2002; Field et al., 2007; Nakamichi et al., 2008) or phase-transformation processes (Seward et al., 2004). The ability of orientation mapping to determine crystal orientation relationships has contributed significantly to the understanding of displacive (and in some cases also diffusive) phase transformations, e.g. in steels or titanium alloys (Seward et al., 2004; Gourgues-Lorenzon, 2009; Takayama et al., 2012), but also in minerals (Abart et al., 2004). The orientation mapping data obtained from EBSD measurements of the low-temperature phase can be used to reconstruct the high-temperature microstructure using the known crystallographic orientation relation between the high- and low-temperature phase, e.g. (Cayron, 2007; Germain et al., 2007; Miyamoto et al., 2010). The fact that EBSD-based COM is, by nature, a quantitative technique predestines it for combination with numerical simulations, e.g. of mechanical properties or evolution during mechanical loading (Tasan et al., 2014; Zhang et al., 2014, 2015), microstructure evolution during annealing (Choi and Cho, 2005; Raabe and Hantcherli, 2005; Wang *et al.*, 2011) or during phase transformation (Kundin *et al.*, 2011). Finally, the combination of COM with several other techniques, *e.g.* EDX (Chen and Thomson, 2010), EBIC and CL (Abou-Ras *et al.*, 2010), or nano-indentation (Zaafarani *et al.*, 2006), yields additional information on functional and structural properties of materials. Although all applications mentioned here yield important information on material properties and processes, they are not without shortcomings. The most significant limitations will be summarized in the conclusions at the end of this text.

In the following, an overview of important aspects of the technique will be presented, including the physical basis of EBSD (section 2.2), the analysis of patterns (section 2.3) and the spatial and angular resolution (section 2.4). Methodological particularities with respect to geomaterials applications will be pointed out. In the subsequent sections various applications and emerging techniques will be illustrated, including an example on intergrowth of feldspar and felspathoid phases in the section on textures and microstructures (section 2.5), an example, on the chirality of quartz, in the section on pattern simulation (section 2.6.1) as well as an example on diffusion zones in feldspar (section 2.6.2) and on dislocation density determination in garnet (section 2.6.3) to illustrate the use of the cross-correlation EBSD technique. The following two sections are intended to give the reader a general idea on the versatility of EBSD and COM by means of two examples, one from metallurgy and one from geological materials.

2.1.1. An example from metallurgy: Thermal barrier coating on a single crystal Ni-based superalloy

Ni-based single-crystal superalloys are used for the turbine blades in jet engines where they run at high temperatures, high stresses and a strongly corrosive atmosphere. In order to increase the operation temperature of these alloys they are, nowadays, coated with thermal barrier coatings. The formation mechanisms of such a thermal barrier coating were investigated using EBSD together with EDS measurements (Wöllmer *et al.*, 2003). The coating was produced by so-called pack cementation, where a single-crystal superalloy was packed into a bed of an Al-rich salt mixture and heated at 1330 K for ~10 h. During this process the Al-salts decompose and aluminium diffuses from the single-crystal surface into the interior where, eventually, the initial base material composition Ni_3Al is changed into $NiAl$ with higher corrosion resistance and higher melting point. The phase transformation goes along with precipitation of most of the refractory elements which are soluble in the Ni_3Al (fcc-based) but not in the $NiAl$ (bcc-based) phase. Figure 9a shows a backscatter electron image of a cross section through the coating. From top to bottom the cross section shows different microstructures and textures: the top is represented by a single crystal with typical γ-γ' microstructure, labelled (i). Here the crystal structures correspond to an fcc matrix (γ) with a large amount of $L1_2$-ordered cuboidal precipitates (γ'). The latter is an ordered version of the fcc structure and cannot be distinguished from the fcc phase on the basis of the EBSD pattern geometry alone. The γ-γ' area is followed by a number of precipitation zones (ii) with precipitates of different size, grain shape and crystal structure. At the bottom of

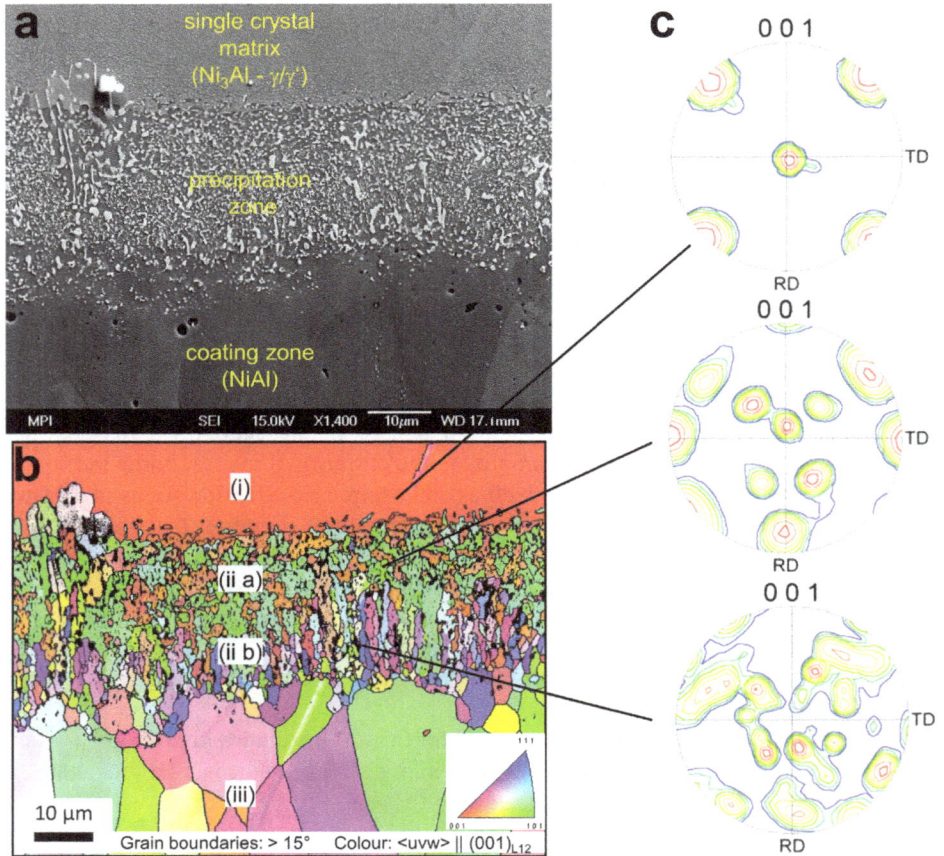

Fig. 9. Microstructure of a thermal barrier coating on a single crystal Ni-based superalloy. (a) BSE image. (b) IPF map for a direction perpendicular to the observation plane. (c) Normal pole figures in stereographic projection of the (001) planes for three different sample zones. See Wöllmer *et al.* (2003) for details.

Fig. 9b a precipitate-free zone (iii) consists of coarser grains with B2 crystal structure, which is an ordered bcc crystal lattice. The latter constitutes the desired heat and oxidation-resistant coating layer. EBSD-based COM was performed on this microstructure. Note that, in order to perform proper orientation assignment, the crystal structures expected in the microstructure need to be known beforehand or have to be determined upon the measured diffraction patterns before COM maps can be constructed. Figure 9b shows the EBSD-measured COM map of this microstructure, displayed in so-called inverse pole figure (IPF) colouring.

Inverse pole figure coding is a very common colour code for presentation of COM results. It displays the crystallographic direction parallel to one predefined sample direction, which must be mentioned together with the map. Every direction in a crystal of given symmetry can be described uniquely in the standard orientation triangle for this symmetry. Therefore, by overlaying the standard orientation triangle with a colour code consisting of red, green and blue for the corners of this triangle, every crystallographic direction is assigned a unique colour. Note, that this form of display describes only crystallographic *directions*, not *orientations*. In the present case, the colour code in Fig. 9b displays a direction perpendicular to the image plane which here is parallel to one of the [001] directions of the single crystal matrix.

The IPF map reveals many more details than the BSE image. In particular it shows that the first precipitation zone occurs within the single-crystal matrix, followed by two distinct domains within the precipitation zone, (iia) and (iib), in which the B2 matrix material shows equiaxial grains (iia) or columnar grains (iib). The crystallographic textures of these two layers are distinctively different (Fig. 9c). The grains in layer (iia) show a crystallographic orientation relationship (COR) with the single-crystal matrix of type $(110)_{B2} \parallel (111)_{L12}$ and $[001]_{B2} \parallel [1\bar{1}0]_{L12}$ which is the well known Bain orientation relationship. In contrast, the columnar grains do not show any clear COR with the single crystal. Finally, the pure B2 layer, (iii), shows a random texture, as far as that can be concluded from the limited number of grains measured (Fig. 9c).

In addition to the crystallographic and morphological information, the combination of the EBSD measurements with EDX measurements allows conclusions on the diffusion and phase formation mechanisms as suggested by Wöllmer *et al.* (2003). The inspection of the electron diffraction patterns from individual precipitate crystals may allow the identification of the crystallographic phase using a search in crystallographic data-bases (Fig. 10). Figure 10a shows the BSE image with one precipitate crystal being marked. Figure 10b displays the EDX spectrum obtained from this area and Fig. 10c the EBSD pattern. The composition displayed below Fig. 10b is obtained from a quantitative, standard-less analysis of the spectrum. Using this analysis, 34 phases which match this analysis approximately are identified from the ICDD X-ray powder diffraction database. Indexing the EBSP for these different phases reveals 22 matches, with one being particularly good in terms of interplanar angles and lattice plane spacing fit. This match is shown in Fig. 10d. A visual inspection reveals, indeed, a good fit of the indexed bands but, at the same time, a large amount of bands that remain non-indexed. This is due to the fact that the powder diffraction files (PDF) list the intensities of the lattice plane reflections as measured by X-ray diffraction, which differ significantly, however, from those obtained through electron diffraction. In a second step, it is important, therefore, to compare the reflections occurring in the EBSP with those proposed in the PDF. With the help of dedicated software (*e.g.* the program Delphi by EDAX/TSL) further reflectors may be added to the reflector list until a consistent and reasonable simulation of the diffraction pattern has been achieved. The solution obtained in the present case is displayed in Fig. 10e, where the (4 0 $\bar{4}$ 0) and (2 0 $\bar{2}$ 5) reflectors have been added. As a result the marked precipitate crystal is identified as C36 hexagonal Laves phase $TaNi_2$.

2.1.2. An example from geosciences: Quartz–forsterite reactive diffusion
A second example typical for an application of EBSD in mineralogy is presented in Fig. 11, taken from an investigation of phase formation and diffusion mechanisms in

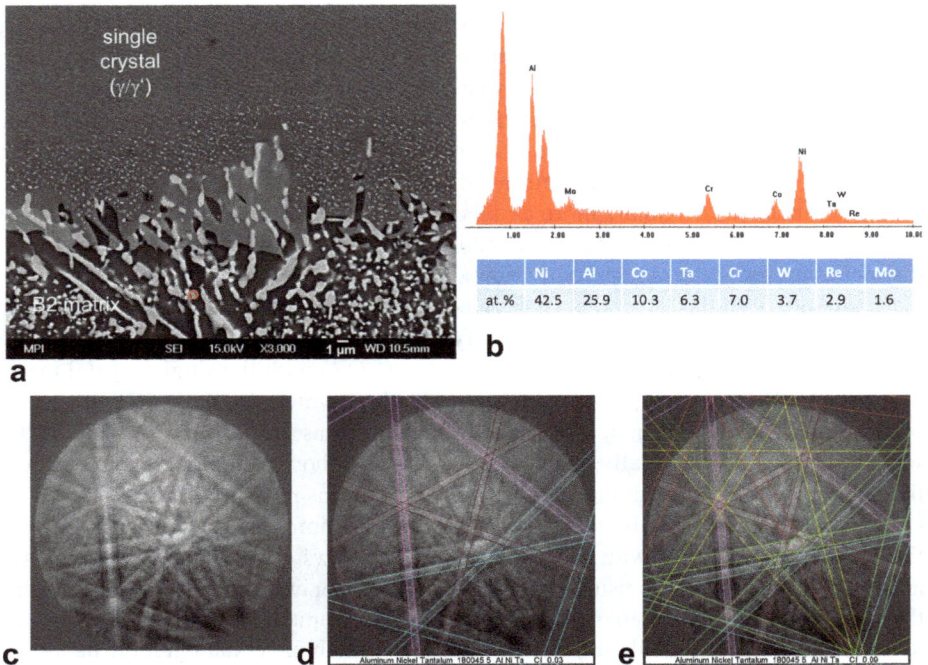

Fig. 10. Combined use of EBSD and EDX for the determination of the phase of a precipitation in the same material as that in Fig. 8. (a) BSE image with one precipitate marked; (b) EDX spectrum from that area together with the calculated composition obtained by standard-less ZAF correction; (c) EBSD pattern from the marked area; (d) EBSD pattern indexed for a C36 phase PDF #180045 5; (e) same indexing as in d but with reflectors being adapted to electron diffraction intensities.

minerals (Abart *et al.*, 2004): powders of forsterite (Mg-endmember of the olivine group with the basic composition $Mg_2[SiO_4]$) single crystals and quartz (SiO_2) were mixed, compacted and then annealed at 1000°C at a hydrostatic pressure of 1 GPa for 24 h. The quartz was highly enriched with ^{29}Si and ^{18}O isotopes in order to trace the diffusion of the individual elements. Annealing led to formation of polycrystalline reaction rims of enstatite (a chain silicate of the orthopyroxene group with the composition $Mg_2[Si_2O_6]$) between the forsterite single crystals and the surrounding polycrystalline quartz matrix. The material was subsequently investigated using EBSD-based COM and additional secondary ion mass spectrometry (SIMS). Figure 11a displays an orientation contrast (electron channelling contrast) image of the investigated area collected by a forescattered electron detector (FSD), Fig. 11b displays a phase map (colour overlay) combined with EBSD pattern contrast information (grey shading), and Fig. 11c shows an IPF map of enstatite combined with an EBSD pattern contrast map for quartz and forsterite (grey shading). Diffusion profiles of oxygen and silicon across the reaction rim (along the line marked in Fig. 11a) are shown in Fig. 11d. The enstatite reaction rim shows two easily

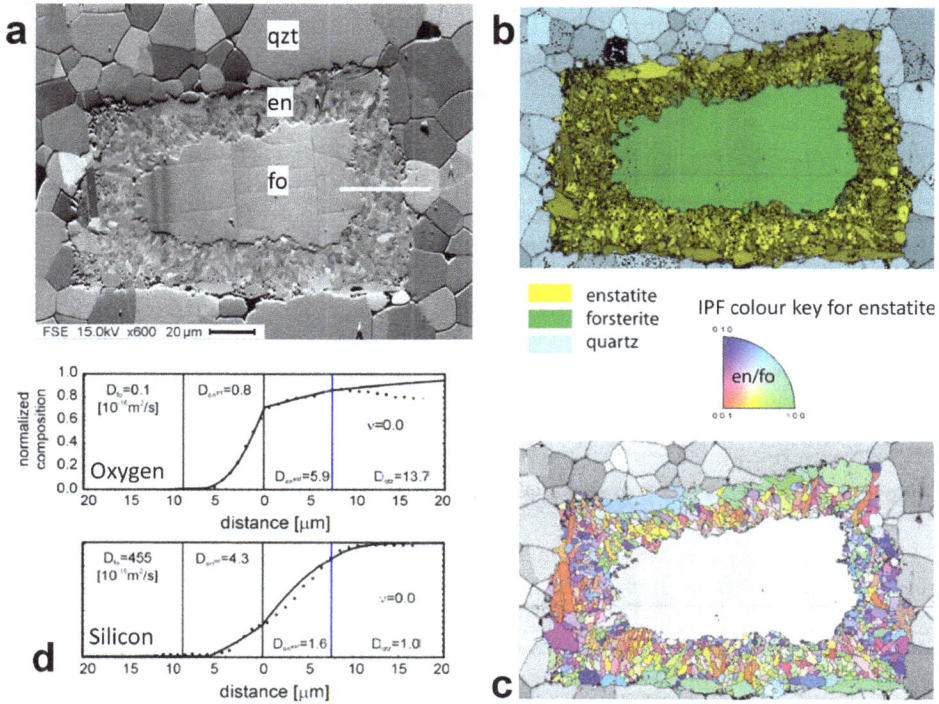

Fig. 11. Microstructure and elemental and phase distribution obtained from a diffusion experiment between quartz and forsterite. (a) FSD (forward-scatter detector) orientation contrast (electron channelling contrast) image. (b) EBSD-determined phase distribution (colour coded overlay) and combined EBSD pattern contrast map (grey-shaded overlay). (c) IPF-coloured map of the enstatite reaction rim indicating the crystal direction normal to the observation plane. (d) Oxygen and silicon diffusion profiles measured with secondary ion mass spectroscopy (solid lines) and calculated from a diffusion model (dotted lines). The white line in a indicates the principal position of the profiles in d. See Abart *et al.* (2004) for more details.

distinguished zones, an inner zone adjacent to forsterite with small columnar grains and an outer zone adjacent to quartz with coarser crystals elongated parallel to the reaction front. The position where both enstatite microstructural domains meet is interpreted as corresponding to the original quartz–forsterite interface, indicating that both reaction fronts propagated, one towards quartz and the other towards forsterite. This information is important for the correct interpretation of diffusion profiles.

The authors of that study subsequently discussed possible diffusion scenarios in line with the measured diffusion profiles. Three possible processes were taken into account: (1) diffusion of MgO from the forsterite–enstatite interface to the enstatite–quartz interface, (2) diffusion of SiO_2 from the enstatite–quartz interface to the forsterite–enstatite interface, and (3) interdiffusion of ionic species such as 2 Mg^{2+}/Si^{4+}.

Important additional information derived from EBSD analyses was the observation that the reaction rims were completely closed, which led to the conclusion that the volumetric changes associated with the phase transformation must have caused a heterogeneous stress distribution. It was shown that the diffusion occurred by two different mechanisms in the two different domains of the enstatite rim. Silicon self-diffusion was approximately equal in both domains, but oxygen self-diffusion was significantly enhanced in the outer, coarser-grained rim zone compared to the inner rim zone. These findings were used to simulate diffusion profiles (dotted lines in Fig. 11d) for comparison with the measured profiles. It was suggested that the different diffusion coefficients were related to different local stresses in the two different rim zones. Dilation at the outer reaction front was supposed to induce a greater water fugacity at the enstatite–quartz interface, which in turn enhanced the oxygen diffusion in the outer enstatite rim zone compared to the inner zone.

2.2. Geometry and physics of EBSD-pattern formation

In order to obtain EBSD patterns a well polished sample surface is illuminated with a stationary electron beam at shallow incidence angle (usually of the order of 20°), as displayed in Fig. 12a. The incoming primary beam experiences multiple scattering events by quasi-elastic but incoherent interaction with the material, creating a point source of electrons within the material emitting electrons into all possible trajectories. In a second step these electrons may undergo (coherent and elastic) Bragg diffraction on any of the lattice planes of the crystal. As the electrons are emitted from a point-like source inside the crystal, the locus of Bragg diffracted electrons is on large pairs of so-called Kossel cones, each of which with an opening angle of $180° - 2\theta$ (θ being the Bragg angle), as indicated in Fig. 12c. These Kossel cones become visible on a detector which is positioned close to the sample (at a distance of few centimetres) in the direction of optimum diffraction contrast as is indicated in Fig. 12a. A typical detector consists of a phosphor screen of a diameter of a few centimetres observed from the rear side by a highly light-sensitive camera (see section 2.3.1).

A characteristic EBSD pattern obtained from a CdTe thin film solar cell absorber layer is displayed in Fig. 12b. The pattern shows approximately straight bright bands, so called Kikuchi (or K-) bands, which usually have relatively sharp edges, called Kikuchi (or K-) lines. The latter can be imagined, as displayed in Fig. 12c, as the intersection of the Kossel cones of diffracted electrons with the flat detector. Because the Bragg angle is small (of the order of 1 to 5°) for typical electron energies in scanning electron microscopy (~10 to 30 keV) the intersections of the Kossel cones with the detector are almost straight K-lines (in fact, they are slightly curved hyperbolae). The two Kossel cones for one set of lattice planes enclose the K-band, the width of which is approximately proportional to the double Bragg angle for the corresponding set of lattice planes. As indicated in Fig. 12c, the centre plane between the two cones corresponds to the diffracting lattice plane set. Note that, due to the gnomonic projection, in most cases the lattice plane projection does not correspond exactly to the middle of the K-band; also, there is no visible trace of the position of this plane in the EBSD pattern.

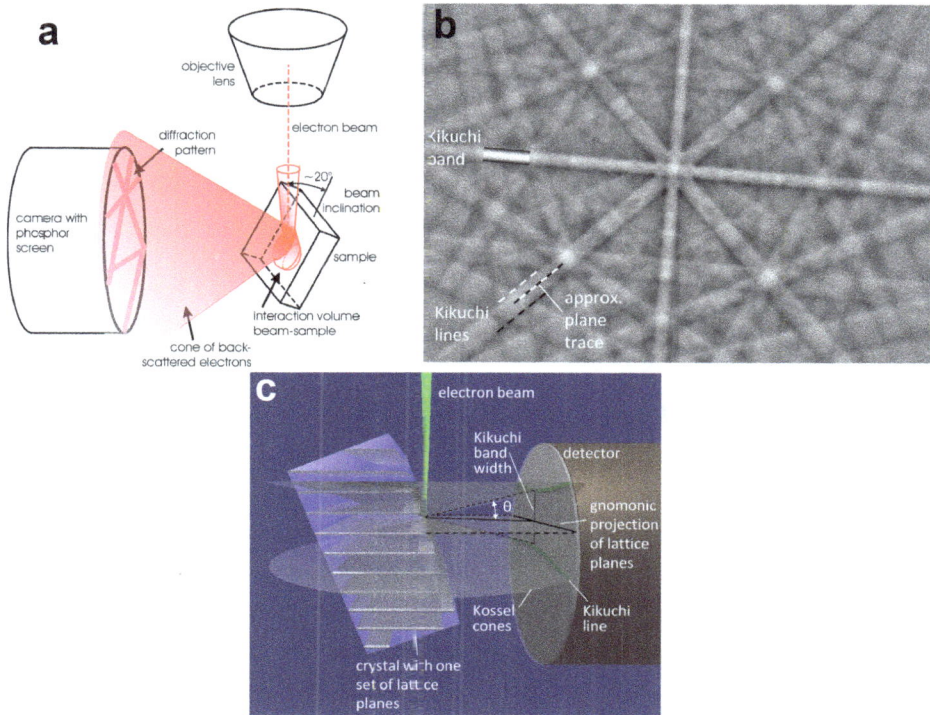

Fig. 12. (a) Experimental set-up for measurement of EBSD patterns in a scanning electron microscope. (b) An EBSD pattern of CdTe, obtained at 15 kV acceleration voltage. (c) Schematic drawing explaining the formation of Kikuchi bands and Kikuchi lines from Kossel cones.

The simple geometric construction of the K-bands described above would yield an intensity distribution consisting of bright Kikuchi lines (lines of Bragg diffraction) enveloping dark and homogeneous Kikuchi bands. However, in reality the intensity distribution is exactly the opposite: dark Kikuchi lines envelope a bright Kikuchi band with various internal fine structures. This band intensity profile can be explained only by taking into account dynamical electron diffraction theory (Winkelmann *et al.*, 2007). The physics of electron backscatter diffraction may be best understood as an inversion of the process of electron channelling as illustrated in Fig. 13. In electron channelling (also called 'channelling-in'), indicated in Fig. 13a, a plane electron wave (corresponding to a primary beam electron) enters a crystal and forms an electron density field coherent with the translational symmetry of the atomic lattice (indicated by the blue wave pattern in Fig. 13a). The electron density distribution changes in a characteristic manner with the incidence angle of the primary electron wave onto the crystal and this distribution may be calculated using dynamical electron diffraction theory. Depending on the distribution of this density field at the position of the atom nuclei, more or fewer electrons are scattered incoherently by interaction with the atoms

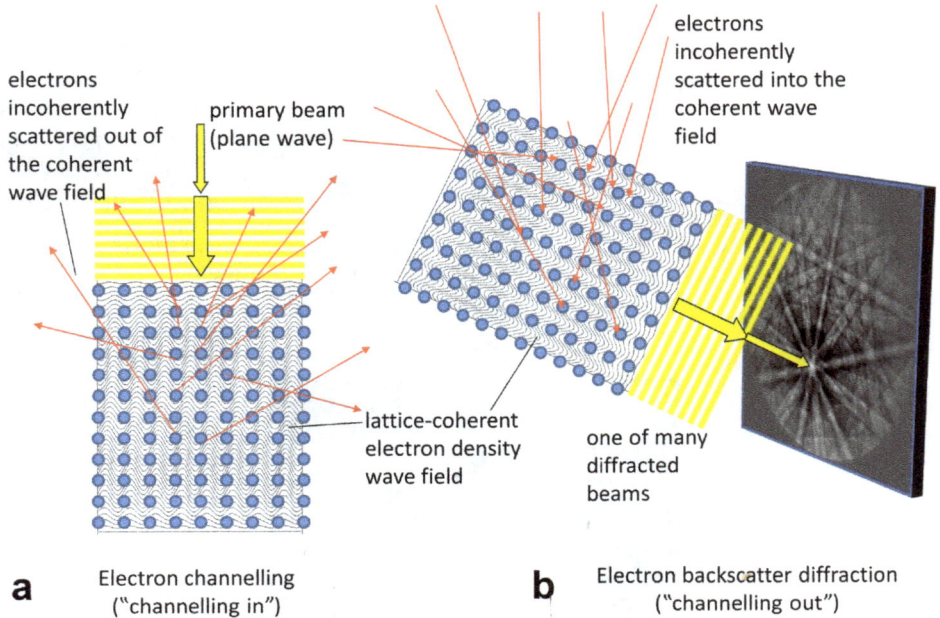

Fig. 13. Comparison of electron channelling (a) and electron backscatter diffraction (b). Yellow features indicate plane electron waves travelling in vacuum. The blue wave features indicate the electron density wave field inside of the crystal. The red arrows indicate incoherently scattered electrons.

and are eventually recorded as backscattered electrons on a detector placed somewhere out of the sample. The probability of an electron being backscattered out of the wave field depends on the scattering power of the atoms and the electron density at the position of scattering. The total backscatter intensity is obtained by integration of these local backscattering events over the total penetration depth of the primary electron wave into the crystal. In the case that the primary beam travels at an angle very close to the Bragg angle of one set of lattice planes one finds that the electron density distribution is at a minimum close to atomic nuclei and at a maximum between them. As a consequence the electrons travel almost unaffected through the crystal (they 'channel') and the amount of backscattered electrons is small. In contrast, if the primary electron wave is travelling at an angle smaller than the Bragg angle (e.g. exactly parallel to the lattice plane) the electron density is maximum at the atom positions, while it is low in between them. In this case the resulting amount of backscattered electrons is large. The total intensity profile obtained for one set of lattice planes is displayed in Fig. 14a and b (Zaefferer and Elhami, 2014).

For EBSD (displayed in Fig. 13b), the situation described above is reversed, as if all electron movements occurred in the opposite direction, as is suggested by the reciprocity theorem of electron diffraction. The electrons are actually 'channelling out': for every position on the EBSD detector a plane wave is leaving the crystal. Each

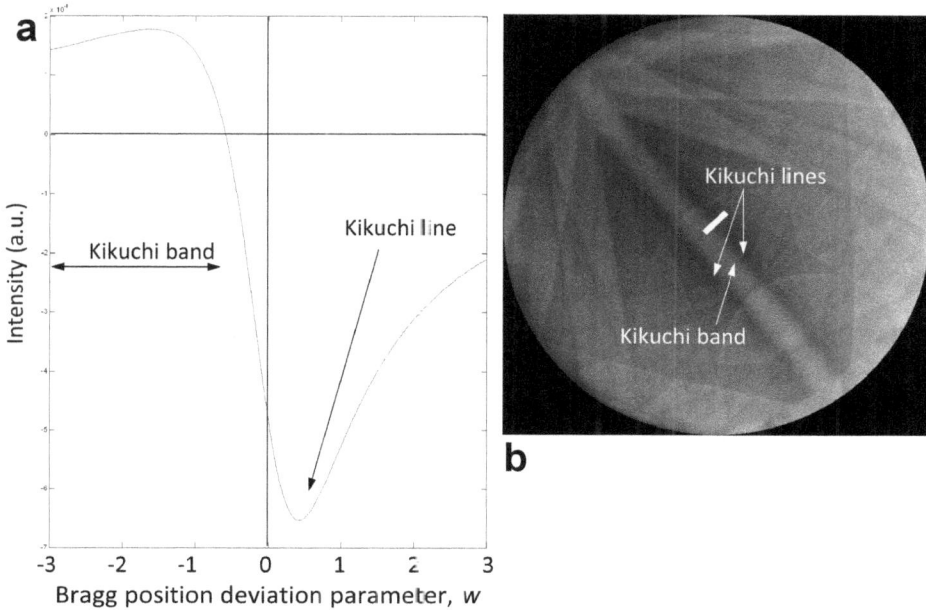

Fig. 14. (a) Principal shape of a channelling intensity profile (in arbitrary units, a.u.) across one set of lattice planes. The Bragg angle corresponds to $w = 0$, a value of $w = 3$ corresponds approximately to an angular deviation of 0.5°. (b) A typical electron channelling pattern created by rocking the beam over a certain angular range and recording the backscattered electron intensity. The white line indicates a potential position of the profile displayed in a.

of these plane waves is associated with a lattice-coherent electron density wave field formed inside of the crystal. These wave fields are 'supplied' with electrons, which are incoherently and randomly scattered (mainly by low-energy loss phonon scattering (Wang, 2003; Zaefferer, 2007)) out of the primary electron beam into the wave fields. The probability that an electron is scattered into one of the individual coherent wave fields depends on the scattering power of the atoms which scatter the primary electron wave and on the probability (or intensity) of the outgoing wave field at the respective scattering position. The total intensity obtained in a given direction is obtained by integration over the total depth of material from which electrons still contribute to the pattern. The intensity profile that is created by this channelling-out process is, in principle, identical to that created by the channelling-in process, displayed in Fig. 14a: for angles close to the Bragg angle of a lattice plane one obtains low intensities, the Kikuchi lines, while for angles smaller than the Bragg angle high intensities are obtained. These latter constitute the bright Kikuchi bands.

The quantitative description for this process has been worked out by Rossouw *et al.* (1994) and related studies. The work of Rossouw was taken further and applied to the backscattering process (Winkelmann, 2009). This author developed a computer program for simulation of EBSPs resulting in very realistic calculations. Figure 15 shows a simulated EBSD pattern and compares it with a real pattern; the similarities are striking. Note, however, that some characteristic differences exist. One is band asymmetry, *i.e.* the appearance of bright and dark K-lines along one K-band which is missing in the simulation, and which is due to anisotropic illumination in the experimental case. Nevertheless, the realistic pattern simulations have triggered a number of very interesting applications, particularly on materials with more complicated crystal structures like minerals. These applications will be discussed in section 2.6.1.

2.3. Practice of recording and analysis of EBSD patterns

2.3.1. Pattern recording

EBSD patterns are generated at relatively low signal intensity, compared to, for example, TEM diffraction patterns. Modern EBSD detectors, therefore, consist of a highly sensitive CCD camera, which observes an Al-coated phosphor screen facing the sample. The phosphor screen converts the electron signal into visible light. The sensitivity and acquisition rate of modern cameras is ever increasing (some cameras deliver up to about 1000 frames per second), and it is frequently the sample or the microscope which limits the rate of acquisition. With respect to acquisition of patterns from minerals, increasing sensitivity is of particular interest because the patterns from minerals are often weak caused by their low scattering power and by the presence of a conductive coating applied to the surface. Also, certain minerals are rather beam sensitive[2], in which case it is important to keep the exposure time as short as possible.

The raw patterns usually need to be processed further before they can be analysed. First, the recorded pattern is, in most cases, binned to speed up the image transfer process and to increase sensitivity. Note that the greater the binning the lower is the angular resolution. Therefore, measurements at high data collection rates (which also usually use a low-resolution Hough transform, see section 2.4.2) show less orientation details (*e.g.* small-angle grain boundaries) than measurements at low rates and high angular resolution.

For most materials, the signal-to-background ratio in the raw patterns is small, *i.e.* the K-bands display only a faint modulation on an intense background. Good patterns may have a signal-to-background ratio as strong as 5 to 20% (*e.g.* EBSPs from high-quality crystals of metals). Minerals and defected crystals, in contrast, may yield a significantly lower signal-to-background ratio. It is important, therefore, to remove the background portion of the total signal from the pattern. This is usually done either by subtracting (or dividing by) a static background image, obtained by integration over many different patterns of the same material. Alternatively, the background may be calculated by a low-pass filter. Good low-pass filtering may require more processing

---

[2] Beam-sensitive minerals are, *e.g.* feldspars, quartz, carbonates, apatite, mica; robust minerals are *e.g.* garnet, olivine, pyroxene, rutile, zircon.

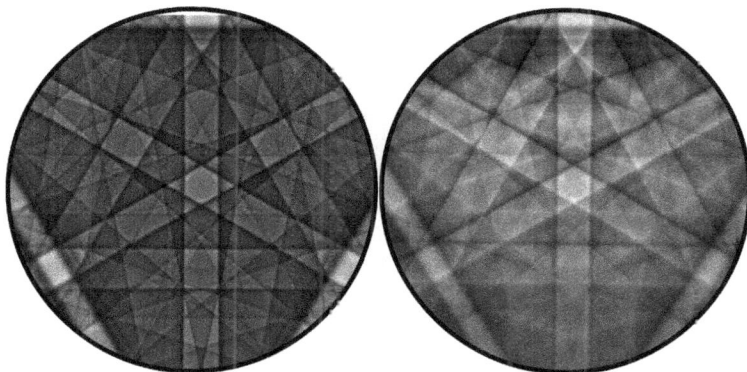

Fig. 15. Simulated (left) and experimental EBSD pattern of Al at 15 kV acceleration voltage. The centre pole is the [111] zone axis.

time than subtraction of a static background. In contrast, high-speed filtering may also remove crystallographic features from the diffraction pattern and may, therefore, result in less contrast in the final pattern image than the static background subtraction. On the other hand, the calculated filter has great advantages when dealing with multi-phase materials of very different atomic number: in this case diffraction patterns may have very different average brightness, which leads to strongly over- or under-saturated patterns when using static background subtraction.

2.3.2. Pattern analysis

Following pattern collection and processing a band-detection algorithm is invoked which detects the straight K-bands with the ultimate aim of determining the position of the corresponding lattice planes. This algorithm is, in all cases, a so-called Hough transformation (HT) (Krieger-Lassen, 1994). In principle the algorithm simply sums up the intensities of all pixels along straight lines in the image, characterized by their normal parameters θ (angle of line normal with respect to one axis of a reference system) and ρ (distance of the line from the reference system centre measured along the line normal), and places this intensity sum in a (θ, ρ)-collector space (the 'Hough space'). As indicated in Fig. 16, a straight line in the pattern is described by:

$$\rho = x \sin \theta + y \cos \theta \Leftrightarrow y(\rho, \theta, x) = \frac{\rho - x \sin \theta}{\cos \theta} \tag{1}$$

where x and y are Cartesian coordinates of the line in the pattern frame and ρ and θ are its polar coordinates. The summation is then:

$$HT(\rho, \theta) = \sum_x I(y(\rho, \theta, x)) \tag{2}$$

where I is the intensity of a pixel at position x,y. Note that all parameters, *i.e.* x, y, ρ and θ, are discrete coordinates of bitmap images of the pattern or the Hough space.

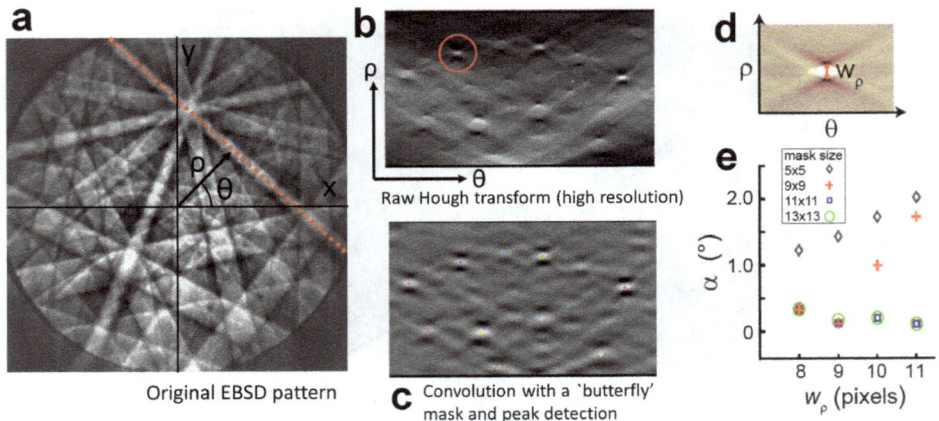

Fig. 16. Principles of band determination by the Hough transformation. (a) EBSD pattern of Cu at 15 kV. The Cartesian (x,y) and polar (ρ,θ) coordinate systems are indicated. (b) Raw Hough transform calculated with a resolution of $\theta = 0.5°$ and $\rho = d/160$, d: diameter of the EBSD pattern. The red circle indicates the peak belonging to the band marked in red in a. (c) Same as b after convolution with a 13×13 'butterfly'-shaped convolution mask and with detected maxima marked. (d) Details of a maximum in the HT space indicating the peak width w_ρ. (e) Plane trace position detection error from a peak in HT depending on the mask size and peak (i.e. band) width. (d) and (e) are taken from Ram et al. (2015) – see that work for more details. Reproduced here with the permission of the International Union of Crystallography (http://journals.iucr.org).

The pixel resolutions of the pattern bitmap and of the Hough transform are important parameters determining the angular resolution of orientation determination (see, e.g. Ram et al., 2015).

An original diffraction pattern and the resulting Hough transform are displayed in Fig. 16. It can be seen that bright, straight lines in the EBSD pattern are transformed into double-winged maxima in the Hough space (see Fig. 16d). The detection of these maxima is usually done by convolution of the image with a mask of constant size. As a result, for every peak, one position is found where the mask fits best and this is selected as the characteristic band position. From this position the lattice plane intersection can be estimated with an accuracy which finally allows calculation of the orientation with an accuracy of ~0.5° to 1°. A detailed treatment of this subject including further literature indications can be found in Ram et al. (2015) and Ram (2015) (see also section 2.4.2). From Fig. 16d it becomes clear that the size of the convolution mask has to be selected carefully for optimum detection accuracy. It is generally recommended to choose a mask which is too large rather than too small, even though larger masks require longer calculation times. It should also be noted that not all commercially available software packages allow user selection of convolution mask sizes.

2.3.3. Pattern centre determination

From the detected lattice-plane traces on the detector plane the 3-dimensional positions of the lattice planes or their normal vectors need to be deduced. This task is

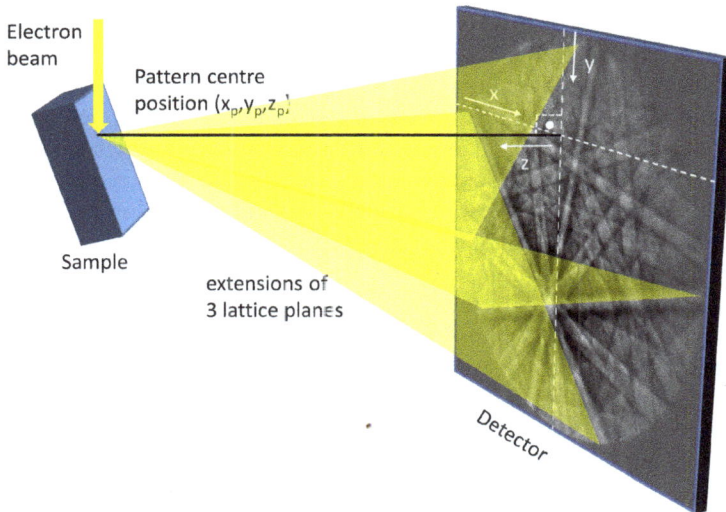

Fig. 17. Schematic display of the position of the pattern projection centre together with three lattice plane projections. The PC is the position of the electron beam on the sample, measured in detector (or pattern) coordinates.

straightforward if the projection centre (PC) of the EBSP is known. As shown in Fig. 17 the PC is the position of the electron beam on the sample, measured in detector coordinates. The PC position is highly sensitive to changes in working distance, and therefore has to be calibrated for each particular sample domain of interest. All commercial EBSD-analysis software contains algorithms to determine the PC approximately. This position is then stored and can be used for reasonably accurate indexing of the patterns of a map. In contrast, the exact PC determination, which is required for precise orientation and lattice distortions determination in particular, turns out to be one of the most challenging tasks in EBSD analysis. Conventionally, the PC is detected by a systematic search for optimum fit between recalculated and measured lattice plane positions (Krieger Lassen, 1999). For greater accuracy it is necessary to invoke significantly more complicated approaches, *e.g.* ray-tracing by moving the detector towards or away from the sample (Maurice *et al.*, 2011), or by placing spheres into the ray path which cast characteristic shadows as proposed by Dingley (see the contribution of A. Day in Schwartz *et al.* (2009)). Ram (2015) has also proposed a method for accurate PC detection. So far, none of these techniques provides a PC position accuracy sufficient to determine lattice plane distortions caused by elastic stresses, *i.e.* to determine the absolute value of residual stresses or strains.

2.3.4. Indexing and orientation determination
Finally, the lattice planes determined from the Kikuchi bands are assigned Miller indices ('indexing') by comparison of the measured interplanar angles with those pre-calculated for potential crystal structures. In addition, interplanar distances calculated from the

detected band width may be used as a further indexing criterion. Note that the use of band width for indexing is usually only necessary for complicated crystal structures or when different, closely related structures are to be distinguished. From a triplet of consistently indexed lattice planes, it is straightforward to calculate the crystal orientation, provided the lattice planes do not belong to the same zone axis. As all detected lattice plane positions have errors, there are usually, however, several potential crystal orientations that can be calculated from one triplet of lattice planes. One powerful approach to obtain the most confident orientation solution consists of calculating orientations for all possible triplets of bands extracted from a given pattern. Following this, the most probable orientation is that which obtains the largest amount of votes, V. The 'confidence index' (CI) of this solution is calculated as CI $= V_1 - V_2/\Sigma V_i$, where 1 and 2 indicate the solution with the highest and second highest votes and i runs over all votes. Thus, CI is 1 when there is only one solution possible, and it is 0 when there are two equally possible solutions. The higher the CI value, the more probable is the detected solution. A value of -1 may be selected for cases when no solution is possible. The above-mentioned extremely robust indexing and confidence calculation algorithm has been developed by the former TSL team (Adams *et al.*, 1993) and is implemented in the OIM DC software by TSL/EDAX. Other computer programs use serial indexing approaches where all or a selection of all detected Kikuchi bands need to be indexed consistently. The confidence is then usually given by the number of consistently indexed bands over all detected bands. To the best knowledge of the author, all these approaches are less robust than the voting scheme and potentially lead to higher numbers of non-indexed points in orientation maps.

   The different approaches to the indexing of diffraction patterns differ in the amount of information that is used in parallel. An algorithm which matches all bands at the same time has been proposed by Callahan and De Graef (2013) and Chen *et al.* (2015): it consists of a comparison of measured EBSD patterns with stored simulated patterns, pre-calculated for all orientations distributed regularly over orientation space. Using image-correlation techniques the best-fitting pre-calculated pattern is determined and selected as a solution. A similar technique was realized successfully by Rauch and Dupuis (2005) for indexing of TEM diffraction patterns. Although the transition of this approach to EBSD patterns is less straightforward than it might appear, this technique is supposed to result in very robust indexing, though at lower angular resolution or lower speed than the conventional HT-based approach. The big advantage of this approach is that it matches both geometry and intensity distribution of EBSD patterns, in contrast to the classical HT approach which is mainly based on geometrical information. The image correlation technique is, therefore, potentially better suited for indexing weak and/or complicated patterns than the HT-based algorithms. Examples of this approach will be discussed in section 2.6.1.

2.4. Spatial and angular resolution

2.4.1. Spatial resolution

For a microscopy technique, spatial resolution is one of the most important properties. Frequently, only the lateral resolution is reported, *i.e.* the ability of a technique to resolve two closely spaced points on a surface. In the case of COM techniques this means the ability to separate two points of different orientation and it is important to note also that the orientational difference between the points influences this ability. The ability to distinguish large misorientations (>~5°) is relevant for the resolution of grain boundaries, while the ability to distinguish small misorientations has relevance, *e.g.* for investigations of geometrically necessary dislocations (GND) (see section 2.6.3). For COM techniques it is furthermore important to realize that the lateral resolution is not sufficient to describe the resolving power of these techniques because they are, in fact, probing a 3-dimensional volume, called interaction volume or spatial resolution.

For illustration and comparison, Fig. 18 shows a sketch of the interaction volume of EBSD and of a TEM diffraction experiment. For EBSD the interaction volume has a highly anisotropic shape, which is on the sample surface extended by ~3 times as much in the direction perpendicular to the tilt axis than in the direction parallel to it. At the same time the penetration depth below the surface is only ~1/5 of the longest direction of the interaction volume. Note that EBSD, therefore, is a rather surface-sensitive

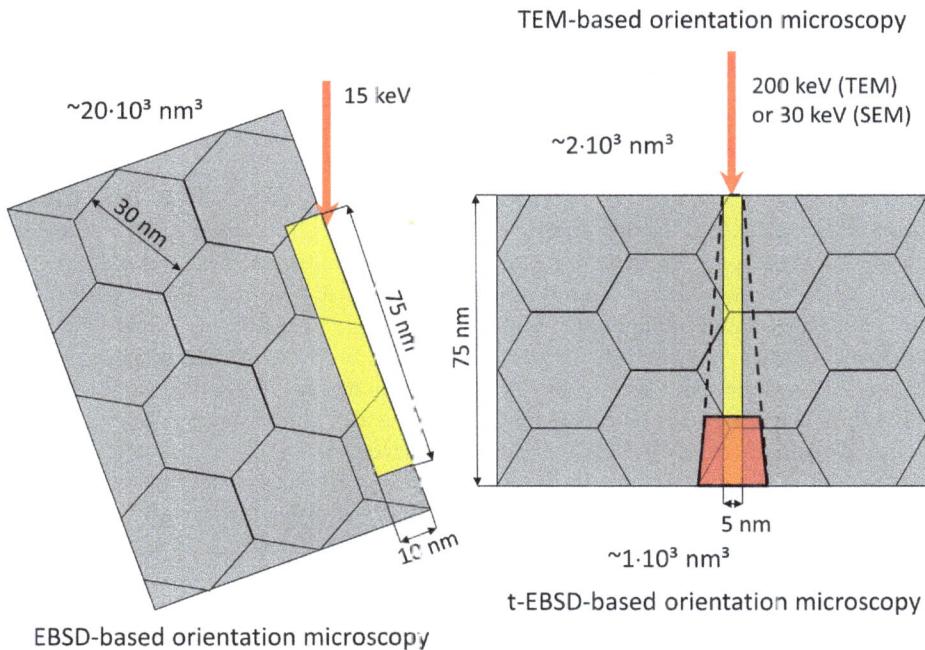

Fig. 18. Schematic illustration of the spatial resolution of different diffraction techniques. The left side shows a bulk specimen observed at its surface, the right side a thin specimen (thickness = 75 nm) observed in transmission. The hexagons correspond to nano-crystals with 30 nm grain size; the yellow areas indicate the respective diffraction volume. The area shaded in red corresponds to the interaction volume observed by t-EBSD. It is obvious that EBSD and TEM-based techniques both have interaction volumes which are too large to resolve individual grains. t-EBSD offers the best chance to measure these.

technique, probing only 10−20 nm below the surface. A typical volume measured on an iron sample (Z = 26) at an acceleration voltage of 15 kV may have the dimensions 75 nm × 25 nm × 15 nm. This volume increases with increasing acceleration voltage and decreasing atomic number of the sample (Steinmetz and Zaefferer, 2010). For many minerals the average atomic number is rather low and, consequently, the lateral resolution is also relatively low[3] (of the order of 100−200 nm for the largest extension in the interaction volume). Additionally, the non-conductive nature of most minerals requires a thin carbon film to be deposited onto the surface, which leads to a further increase of the lateral resolution value. In contrast, for TEM the volume is less anisotropic and the spatial resolution is mainly determined by the thickness of the sample. In total, the diffraction volume of TEM may be about 10 times smaller than that of EBSD. Humphreys *et al.* (1999) and Zaefferer (2007) pointed out that the physical resolution, *i.e.* the size of the interaction volume, has to be distinguished from the effective resolution. The latter corresponds to the smallest volume in which two well distinguished orientations (with a misorientation of >~5°) can still be distinguished by means of deconvolution software. The effective resolution value may be significantly smaller than the physical one. Dingley (2004), for example, reported the observation of grains with 9 nm diameter in platinum. In contrast, when very similar orientations are to be distinguished, *e.g.* to determine the density of GNDs from local lattice rotations, the resolution value becomes as large as the physical resolution value because very similar overlapping patterns cannot be distinguished by the software. Note that the step size during measurements, which is frequently reported, is not a measure for the spatial resolution capability of a method. The latter can be larger or smaller than the step size used.

Nowadays, the advent of 'transmission-(t-)EBSD', or transmission Kikuchi diffraction, TKD, has further improved the spatial resolution capability of EBSD-based COM, however, by sacrificing the important advantage of conventional EBSD of allowing observation of bulk samples. t-EBSD works on thin foils and the patterns are recorded as transmitted electron signal below the sample (Keller and Geiss, 2012), in most cases at the conventional EBSD detector position (note that the company Bruker Nano offers a detector which observes the patterns directly below the sample). It has been shown that the t-EBSD signal stems from a small volume at the exit surface of the primary beam (Suzuki, 2013). This volume, indicated in Fig. 18 by red colouring, may be as small as 10 nm × 10 nm × 10 nm and, therefore, be even smaller than that of TEM-based COM. t-EBSD, for the first time, makes truly nanocrystalline materials accessible to observation with EBSD-based equipment (Trimby, 2012).

An alternative approach to decrease the spatial resolution value of COM, while not giving up the bulk nature of samples, is to use classical EBSD at reduced acceleration

---

[3] In order to avoid confusion, a comment on the use of the term lateral 'resolution' is required: resolution is the smallest length distance at which two distinguishable features can still be observed separately. A 'large' resolution value, therefore, means a large distance, and is in literature called 'low' resolution. Inversely, a 'small' resolution value indicates a 'high' resolution. We will thus use the adjectives 'large' and 'small' for values when talking about 'low' or 'high' resolution, respectively.

voltage. It has been shown (Humphreys *et al.*, 1999; Steinmetz and Zaefferer, 2010) that a reduced acceleration voltage leads to a reduced interaction volume. Simple calculations indicate that the lateral extention of the interaction volume may be reduced by a factor of four when the accelerating voltage is dropped from 15 to 5 kV. This would lead to a reduction in the total interaction volume by a factor of 64. Unfortunately, when reducing the beam energy many problems become more severe, *e.g.* contamination, beam drift, or camera sensitivity, which finally leads to the fact that in practice spatial resolution does not improve, but rather deteriorates (Humphreys *et al.*, 1999).

An example, where reduced acceleration voltage was employed to improve spatial resolution on minerals, is the work of Goetz *et al.* (2011) on the primary layer of the modern brachiopod *Gryphus vitreus*, a mollusc-like ocean animal. Here, COM was used to observe the microstructure and texture of fine-grained calcium carbonate constituting the primary layer of their shell. A reduction of the acceleration voltage to 10 kV and sufficiently long camera exposure times allowed users to observe structures as small as 100 nm which revealed important details of the growth mechanism. Figure 19 reproduces one result from this study which shows that a fan-like, primary layer is followed by a fibrous secondary layer. Both structures, although morphologically very different, show almost identical textures as visible in the pole figures (Fig. 19b). The authors concluded that this indicates a common nucleation mechanism based on amorphous calcium carbonate for both layers.

2.4.2. Angular resolution

Besides spatial resolution, angular resolution is an important criterion in assessing a COM technique. Angular resolution is the smallest misorientation between two measurement positions that can still be resolved reliably. In the literature, the angular resolution of the classical EBSD technique has been reported as 0.5 to 1° (Wright *et al.*, 2012). There are two main reasons for this relatively poor resolution (Ram *et al.*, 2015): one is the insufficient accuracy of band detection using the Hough transform. As discussed previously and displayed in Fig. 16d and e, the band-detection accuracy can be improved by correct selection of the size of the HT convolution filter. A second significant source of error is the inaccurate position of the pattern centre, which itself is inaccurate because of inaccurate band detection. A new way to improve angular resolution by improving the band detection accuracy has recently been proposed by Ram *et al.* (2014), who used a mathematical model of the general shape of Kikuchi bands to obtain the position with a resolution value 10 times smaller. Nevertheless, the angular resolution value is usually largely sufficient when microstructures with grains and subgrains are to be observed. It may, however, not allow determination of the density of geometrically-necessary dislocations (see section 2.6.3) or study of subgrain structures in materials that have undergone deformation recovery leading to very small misorientations (Humphreys *et al.*, 2001). In this study a Kuwahara filter was used to display the subgrain structure. This filter sharpens subgrain boundaries which otherwise would appear blurred.

One way to increase angular resolution significantly is to use EBSP cross-correlation as developed by Wilkinson *et al.* (2006), originally with the intention to determine

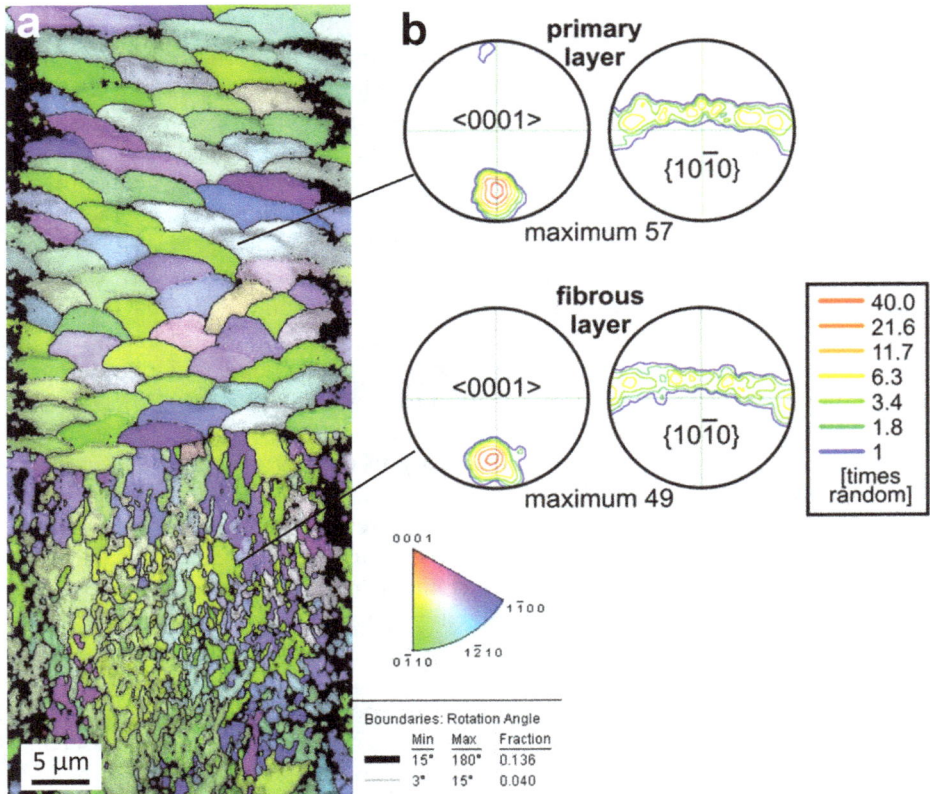

Fig. 19. High-resolution EBSD map of the primary and the fibrous shell layers of the modern brachiopod G. vitreus, measured at 10 kV and a step size of 100 nm. (a) EBSD maps coloured in an IPF colour scheme. The black regions correspond to measured points, with a confidence index of <0.1. Small-angle grain boundaries are marked with grey lines, grain boundaries with black lines. The lower part of the map corresponds to the fibrous secondary shell layer, the upper part to the primary shell layer. (b) Pole figures of the two different microstructural areas displayed in a. Figures modified from Goetz et al. (2011).

pattern distortion caused by elastic strain. In fact, lattice distortion, as well as small misorientation, causes characteristic shifts of different areas of an EBSD pattern. These shifts can be determined by comparing the pattern with a reference pattern of almost similar orientation and without distortion. In this way, lattice rotations down to 0.01° and elastic strains down to 10^{-4} can be determined and mapped. Cross-correlation EBSD will be described in some more detail in section 2.6.2.

2.5. Microstructure and texture investigations based on EBSD

EBSD-based orientation microscopy has been developed and is used mainly to describe microstructures and textures in a quantitative manner. The term microstructure denotes the geometry and arrangement of all kinds of lattice defects that disturb the perfect atomic ordering of crystalline matter, including dislocations, grain and phase boundaries, inclusions and precipitations, cracks and elastic and plastic strain fields. Lattice defects significantly alter the properties of defect-free material, which, again, is very important for the mechanical and chemical behaviour of materials. At the same time the existence of microstructures and their particular character allows conclusions on the mechanisms and the progressive material evolution. In natural rocks the microstructure evolution is mainly controlled by either non-hydrostatic stress ('deformation microstructures') and/or by changes in hydrostatic stress and/or temperature conditions inducing mineral reactions ('reaction microstructures') and thermally controlled microstructural changes.

The crystallographic texture indicates the entity of crystallographic orientations of polycrystalline materials (Bunge, 1987) as reflected by the crystallographic orientation distribution. Any non-random orientation distribution of crystalline material is termed a crystallographic preferred orientation (CPO), which is reflected by the preference of some crystallographic orientations with respect to others. Because every single crystal has anisotropic properties, a textured material will also show anisotropic properties provided the texture is not random. CPOs are, therefore, of interest for many engineering materials as well as for natural rocks, *e.g.* when dealing with the propagation velocity of seismic waves. The determination of anisotropic elastic properties of rocks and minerals from EBSD measurements has been dealt with in great detail by Mainprice and others (Mainprice *et al.*, 2011) and is part of the *MTEX* software package, a *Matlab*-based model kit for description of textures and microstructures (Bachmann *et al.*, 2010). An even more important aspect of texture in the geosciences is, however, that CPOs of polycrystalline material may give very valuable information on the processes that occurred during petrogenesis (*e.g.* direction of shear, temperature).

2.5.1. An example of the use of EBSD for studying microstructure formation: Intergrowth structures of feldspar and feldspathoid minerals.
An example where EBSD was employed to obtain information on the microstructure development in geomaterials is the work of Ageeva *et al.* (2012) on the genetic implications of particular feldspar–feldspathoid intergrowths observed in rocks of the Khibiny massif. The authors investigated coarse-grained, inclusion-bearing alkali feldspar single crystals in a fine-grained matrix. The alkali feldspar crystals contain polymineralic inclusions of feldspar and nepheline and feldspar and kalsilite, which not only show a characteristic morphology (micrographic or lamellar intergrowths) but also a characteristic orientation relationship with each other and with the host crystal. As a first investigative approach, light-optical polarization microscopy was applied. This method, however, only reveals the full orientation relationship for optically biaxial crystal structures, *i.e.* in the present case only for K-feldspar. For nepheline and kalsilite only the direction of the optical axis can be determined, which coincides with the hexagonal [0001] axis. Therefore, EBSD was employed to obtain a more accurate and complete orientation relationship between the different phases at significantly higher spatial resolution.

Crystallographic orientation maps and corresponding pole figures (Fig. 20) indicate that intergrown K-feldspar and kalsilite phases show predominantly the specific crystallographic orientation relationship [0001]Ks // [010]Or and each of the three [10$\bar{1}$0] Ks // [100]Or, [101]Or and [001]Or respectively, as schematically indicated in Fig. 20b and c. In contrast, isometric nepheline inclusions do not show a COR with the orthoclase host. Therefrom the authors concluded that nepheline represented initial inclusions in the hosting alkali feldspar, and was later replaced by the kalsilite–orthoclase intergrowths. Combined microstructural and COR data led to the further conclusion that metasomatic alteration of nepheline formed kalsilite and orthoclase, which involved introduction of silica and potassium, supposedly derived from magmatic fluids expelled during crystallization of the nepheline-syenite magmas of the Khibiny massif.

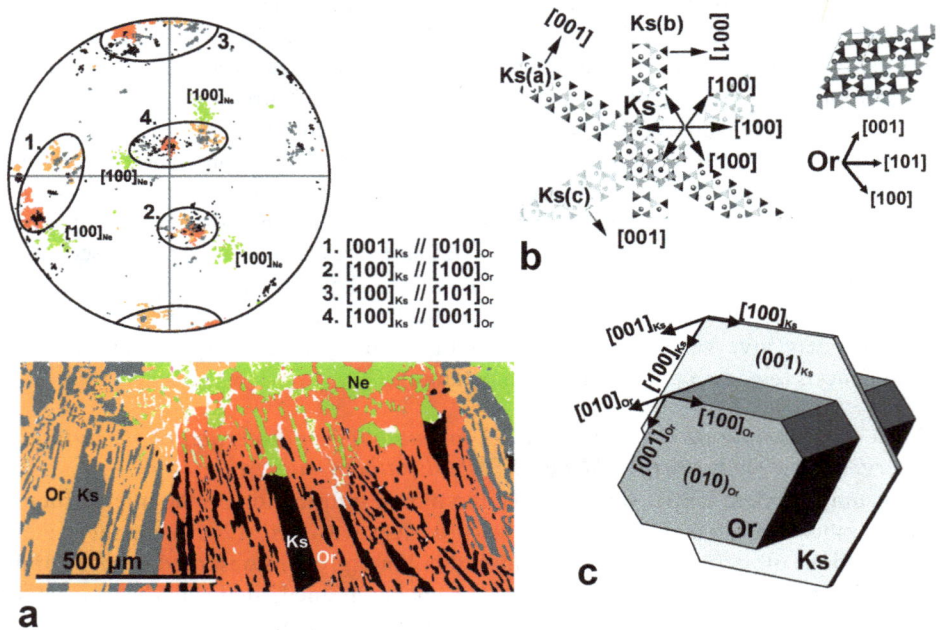

Fig. 20. Microstructural and structural description of feldspar and feldspathoid mineral intergrowths. (a) Orientation relation among nepheline (NE, green), kalsilite (ks, black, grey) and potassium feldspar (OR, red, orange) next to an isometric nepheline grain. The zone in immediate contact with the isometric nepheline grain consists of a lamellar nepheline–K-feldspar zone, which is followed by a kalsilite-feldspar zone further away from the isometric nepheline grain; the pole figure (upper hemisphere) shows the main crystallographic directions of feldspar, nepheline of both the isometric grain as well as of nepheline and kalsilite. (b) Projection of four different shape orientations of kalsilite lamellae with respect to feldspar viewed down the [010] direction of feldspar (Or). (c) Crystal model showing the predominant COR between kalsilite (ks) and K-feldspar (Or) in the intergrowths (feldspar crystal is cut along the (010) plane). The kalsilite lamellae 'Ks', 'Ks(a,b,c)' show a shape-preferred orientation perpendicular to the [001]$_{Ks}$ and all [100]$_{Ks}$ directions of a hexagonal cell according to the main crystallographic planes. Figures modified after Ageeva *et al.* (2012).

In the study described above a monoclinic and a triclinic crystal-structure description for feldspar and a hexagonal crystal structure description for nepheline and kalsilite have been used. For the latter the authors did not use the actual Laue group 6/*m* but the higher symmetric one 6/*m* 2/*m* 2/*m*. Using the 6/*m* reference structure led to two solutions with antipodal orientations, identical vote numbers and, consequently, a confidence index of 0. Using the higher symmetry obviously solved this problem. It should be mentioned here, however, that the approach of using the higher symmetry structure must not be considered a general solution to the problem, because, in this way, two orientations, which are actually distinguished by the absence of mirror symmetry about the $(10\bar{1}0)$ plane, may erroneously be identified to be the same. Rather, the problem has to be solved by identification of those Kikuchi bands which are actually breaking the assumed mirror symmetry in the 6/*m* 2/*m* 2/*m* group and their inclusion into the indexing procedure. These bands may, however, be of low intensity and therefore difficult to detect by the classical HT approach. A possible solution to this problem may be the use of direct EBSD pattern comparison as will be presented with an example in section 2.6.1.

2.5.2. Crystallographic texture

The presence of any crystallographic texture influences virtually all properties of polycrystalline material. In the geosciences, texture is of great importance for the understanding of deformation mechanisms as well as temperature conditions and/or strain rates during rock deformation. Besides deformation, recrystallization or grain growth can also lead to non-random orientation distributions. In addition, phase transformations may preserve texture information when they are associated with the formation of crystallographic orientation relationships between deformed reactant and product phases. Traditionally, crystallographic preferred orientations are determined by X-ray diffraction (XRD) pole figure measurement and subsequent pole figure inversion, see, *e.g.* (Bunge, 1987). EBSD, however, developed into a very important competitor for XRD for a number of reasons: EBSD directly measures orientations, rather than plane distributions. No corrections for defocusing or absorption are to be made, no peak overlaps occur when measuring multiphase materials, texture fields are spatially resolved and, additionally, microstructural information such as size, phase distributions or correlated misorientations are obtained. One disadvantage is the limited spatial resolution of EBSD, which impedes the measurement of nano-crystalline (less than a few 10 to a few 100 nm grain size) or highly strained material. A second disadvantage can be the lack of statistical representativeness of the EBSD-measured data. This point will be addressed below by means of an example from metallurgy.

A crystallographic texture is called strong when all crystal orientations group around a defined range of preferred orientations with a deviation of few degrees only; it is called weak when only some of the crystals follow the preferred orientation rules with a wide spread around the preferred orientations, or random when there is no preferred crystallographic orientation developed, respectively. This classification is, however, only reasonable when a sufficiently large number of crystals has been considered in the measurement; this is usually the case for classical XRD but it becomes crucial when applying EBSD. According to Baudin *et al.* (1995), depending on the texture strength, at least 10^2 (for strong textures) to 10^3 grains have to be measured. Measurements on a TRIP (transformation induced plasticity) steel with a relatively weak texture (Davut, 2012), however, suggest that ~10^4 grains are required. In this case, the number of grains sufficient to describe the texture was measured by means of the perfection of sample symmetry reflected in pole figures. A number of 10^4 grains required for satisfactory texture description was also confirmed by Wright *et al.* (2007). For all EBSD-based texture measurements it is, thus, very important to carefully select the probed sample volume considering the material grain size.

Davut and Zaefferer (2010) developed a new orientation mapping scheme which determines many small-step size maps for local information distributed at some mutual distance to cover large sample areas for statistically representative measurements, as displayed in Fig. 21. This scheme requires approximately the same measurement duration to deliver texture information with the same reliability as XRD-based pole figure data, however, with all above-mentioned advantages of EBSD-based analyses. Figure 22 demonstrates the power of EBSD-based texture measurements of large

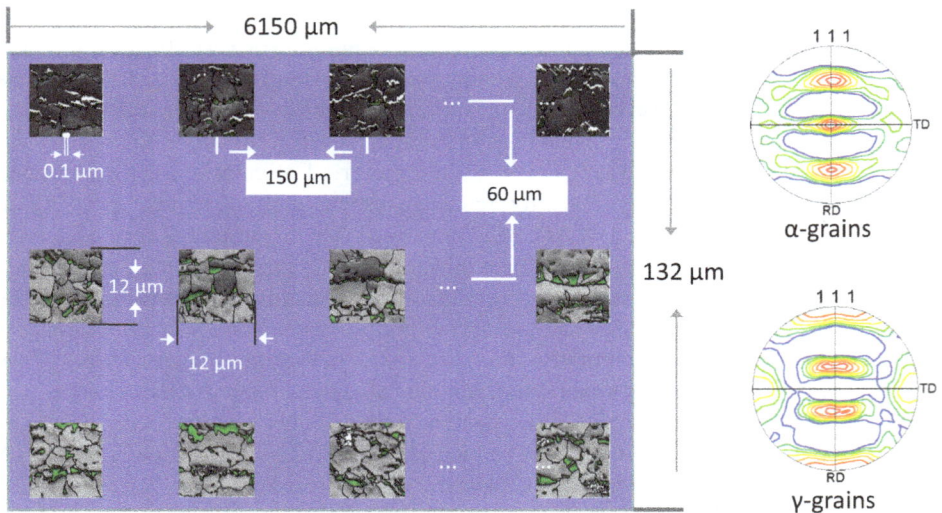

Fig. 21. A new EBSD-based texture-measurement scheme that allows acquisition of representative data set sizes by combination of measurement of small maps, measured with small step size, distributed over a large sample surface (modified from Davut and Zaefferer, 2010). Example: cold rolled and heat-treated low-alloyed TRIP steel consisting of a microstructure of ferrite (α-phase), displayed in grey values, and ~10 vol.% of residual austenite (γ-phase), displayed in green. The well developed orthotropic symmetry of the orientation distribution displayed in pole figures (caused by the production process consisting of hot and cold rolling) indicates that a sufficient number of crystal orientations has been measured.

sample domains applied to an example of a hot-rolled Fe-Si sheet with heterogeneous texture distribution over the cross section: the surfaces show a strong shear texture, caused by friction of the rolls with the material, while the texture in the centre indicates orthotropic plane-strain deformation. The EBSD data collection rate was very high (analysis duration of <30 min) whereas collection of an XRD dataset with the same information content would require a time-consuming serial sectioning procedure.

2.6. Advanced techniques

2.6.1. Simulation of diffraction patterns and new indexing methods

In section 2.2 the physical model used to describe EBSD patterns was introduced. This model has been employed by Winkelmann in order to generate realistic EBSP simulations (Winkelmann *et al.*, 2007; Schwartz *et al.*, 2009). This approach has significant potential for advanced EBSD applications, in particular for the distinction of complicated crystal structures or orientation determination for structures with pseudo symmetries (Nolze *et al.*, 2016). The dynamic nature of Kikuchi patterns leads to the fact that Friedel's law, which states that the electron intensity diffracted from planes (hkl) is identical to that from planes $(\bar{h}\bar{k}\bar{l})$, does not apply. This means that Kikuchi diffraction can distinguish all point groups of crystals, in contrast to spot diffraction in TEM or X-ray diffraction which can only distinguish the Laue groups,

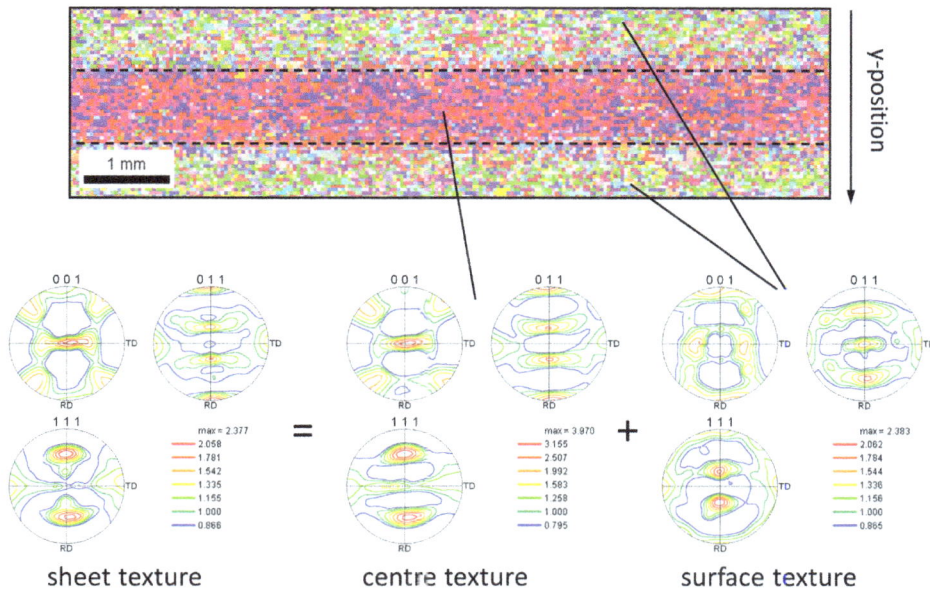

Fig. 22. EBSD-based measurement of a texture field on a large cross section of a hot rolled Fe-Si steel sheet. The different texture types in different domains are clearly visible: shear textures close to the surface and plane-strain texture in the centre.

which contain a centre of symmetry. Thus, EBSD patterns (and other Kikuchi patterns) may show the $(hkl)-(\bar{h}\bar{k}\bar{l})$ asymmetry caused by the absence of a centre of symmetry. Winkelmann and Nolze (2015a) demonstrated that the use of realistic diffraction-pattern simulations allows determination of this asymmetry: for the cubic but non-centrosymmetric GaP with Zincblende structure (space group $F\bar{4}3m$) they demonstrated that the direction of the polar [111] axis can be distinguished, *i.e.* that EBSD allows the distinction of the [111] and the [$\bar{1}\bar{1}\bar{1}$] crystal direction. The major results from this work are summarized in Fig. 23.

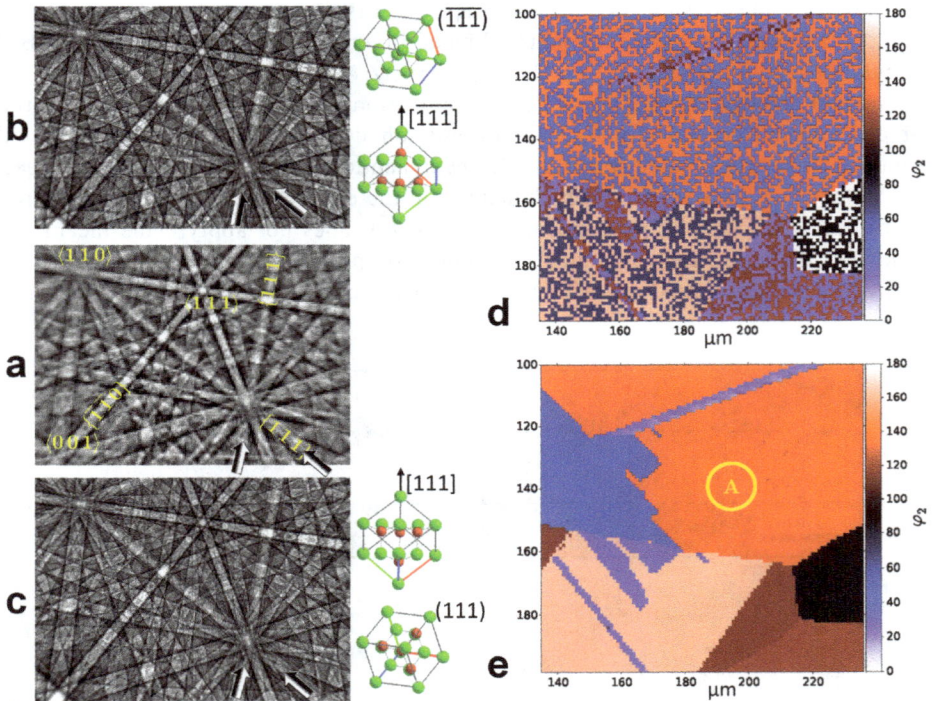

Fig. 23. Use of pattern-matching for indexing of non-centrosymmetric GaP crystals. (a) An experimental EBSP obtained from crystal A in e. (b,c) Two possible simulated diffraction pattern solutions for a. Their orientations just differ by the sense of the polar [111] axis as indicated by the lattice unit cells. Note that the (111)-plane views (at the very top and very bottom, respectively) correspond to the orientation of the pattern whereas the additional side views of the unit cells reveal the different order of the Ga and P atom layers more clearly. The arrows in b and c indicate those Kikuchi bands which show a pronounced intensity asymmetry that can be used to distinguish the two orientations. The asymmetry of the arrows corresponds to the asymmetry of the bands. It is clear that c fits better to a than b. (d) Orientation map from a polycrystal of GaP indexed taking into account only the geometry of the pattern: the two possible polar directions cannot be distinguished and lead to random scatter. (e) Same map as d but indexed taking into account the intensity distribution. The polar directions are correctly distinguished. Figures modified from Winkelmann and Nolze (2015a,b).

In a different study, Winkelmann and Nolze (2015b) showed that it is, in principle, possible to determine the chirality of trigonal α-quartz. α-quartz crystallizes in the truly chiral point group 32, which means that it can form two enantiomorphic crystals with space groups $P3_121$ and $P3_221$ (containing a left-rotating or right-rotating 3-fold screw axis). As displayed in Fig. 24a and b these crystal structure varieties are mirror-symmetric to each other but they cannot be superimposed by any sort of rotation, similar to the left- and right-hand pair. The enantiomorphism of α-quartz has important consequences for its piezoelectric properties, as the two enantiomorphic crystals show opposite polarity. Note that properties which are, by themselves, centrosymmetric, *e.g.* thermal conductivity or elasticity, are not affected by the enantiomorphism. Using dynamically simulated patterns the authors could show that the diffraction patterns of left-handed and right-handed quartz crystals show, indeed, weak but characteristic differences. Their results are reproduced in Fig. 24c and d which show the two

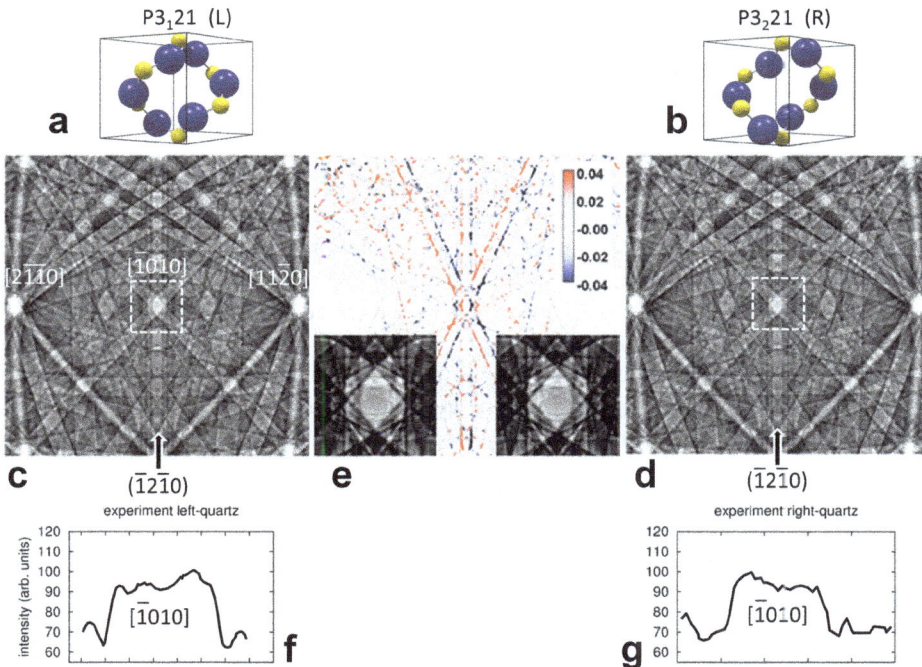

Fig. 24. Use of EBSD to distinguish left-handed (left side) and right-handed (right side) quartz. (a, b) Sketches of the atomic configuration and their space groups. (c, d) EBSD patterns simulated for both enantiomorphic crystals centred to the [$\bar{1}$010] pole. (e) Differences between the patterns in c and d calculated as A_c (see text). The insets in this figure display the details of the two different [$\bar{1}$010] poles marked by a white square in c and d. (f, g) Intensity profiles determined from experimental diffraction patterns (not shown), measured through the centre of the [$\bar{1}$010] poles. Graphics modified from Winkelmann and Nolze (2015a,b). Note that the original graphs also contained data from Dauphiné twins which are omitted here for simplification.

simulated diffraction patterns, whereas Fig. 24e displays the normalized difference between both which is calculated by $A_c = \frac{I_R - I_L}{I_R + I_L}$ where I_R and I_L correspond to the signal intensities in the patterns of right- and left-handed quartz, respectively. The insets in these figures zoom into the [$\bar{1}$010] zone axes which show clear left- and right-hand intensity differences. When analysing true EBSPs from left- and right-handed crystals these differences are, however, for two reasons very difficult to spot. The first reason is that experimental patterns usually show a certain amount of blur which is due to the detection system and due to the energy variation of the electrons, caused by the inelastic scattering events. The second reason is the existence of excess and deficiency lines in true patterns which are due to the anisotropic intensity distribution in the illumination source inside the crystal. Of the two Kikuchi lines enveloping a Kikuchi band, the line which is closer to the direction of highest backscatter intensity (which is approximately the direction of reflection of the primary beam) appears darker while the other line is brighter than their average intensities. This illumination contrast superimposes onto the crystallographic contrast and needs to be separated carefully from the latter. The excess-deficiency contrast is zero for those bands which run through the pattern centre. Therefore such bands are most reasonably selected for the investigation. Furthermore, although R-L differences may be found all over the pattern as shown in Fig. 24e, it seems that the differences are strongest in the ($\bar{1}2\bar{1}0$) band which corresponds to the pseudo mirror plane. The authors selected a ($\bar{1}2\bar{1}0$) band running exactly vertically across the screen and showed that it is, indeed, possible to determine sufficient details to distinguish the two crystal-structure varieties. Figure 24f and g reproduces the intensity profiles obtained at a cross section through the [$\bar{1}$010] zone axis across the ($\bar{1}2\bar{1}0$) band. The contrast is clearly asymmetric which characterizes the right- or left-handed crystal. Note that it is important to observe the profile into a well defined direction, for example the positive c-axis direction, in order not to confuse the left and right sides.

The investigation described above shows that EBSD has great potential, not only for microstructure but also for crystal structure analysis, including crystallographically challenging structures. In combination with dynamic simulations it may become possible to determine even unknown crystal structures, although this certainly still has a way to go.

2.6.2. Cross-correlation EBSD for the measurement of elastic strain fields

As discussed in section 2.4.2, the angular resolution of Hough Transform (HT)-based EBSD analysis is in the range of ~0.5 to 1°. The reasons for this relatively large value are manifold: the Hough transform itself is of limited angular resolution; the HT can detect only straight features though K-lines are actually curved; finally, the pattern centre is known only approximately. Another related shortcoming of HT-based EBSD analysis is the very inaccurate determination of lattice constants. The reasons for this are, first, the rather small values of typical Bragg angles, which correspond to band widths and cannot be measured with good accuracy. Second, the dynamic nature of the diffraction process leads to rather blurred diffraction patterns as compared to XRD patterns. As a consequence and in contrast to XRD, an elastic expansion or compression of the lattice cannot be measured accurately. Nevertheless, the analysis of EBSPs can be improved significantly, so that very small rotations and even elastic lattice distortions can be measured. This analysis is based on the careful comparison of an EBSP from a measurement position with a very similar EBSP from an assumingly strain- and rotation-free reference position using image cross-correlation (ICC) techniques. This technique has been developed mainly by Wilkinson *et al.* (2006), originally with the intention to determine pattern distortions caused by elastic strains. This paper also references important preliminary papers. A recent overview was presented in Wilkinson *et al.* (2014). The technique is based on the fact that lattice distortions as well as small misorientations cause characteristic shifts of different areas within an EBSP, as illustrated schematically in Fig. 25. These shifts can be determined by comparing the measured pattern with a reference pattern of a crystal of almost identical orientation and without lattice distortions. To this end the pattern is segmented into reasonably sized regions of interest (ROI). For each of these ROIs the shift is determined by ICC. Because elastic strains and small lattice rotations lead to image shifts significantly smaller than 1 pixel, sub-pixel resolution needs to be achieved in the ICC by fitting of the cross correlation peaks with a smoothing function, *e.g.* a cubic spline. In this way shifts as small as a few hundreds of a pixel can be detected from high quality EBSPs. Figure 26 shows an example of the determination of lattice distortion: for each ROI, marked by white squares in the unstrained image, a shift vector is determined, which is displayed as a green arrow on the strained image. From the derived shift vector field the deformation gradient tensor can be calculated and from this the rotation and distortion tensors inferred using the assumption of infinitesimal

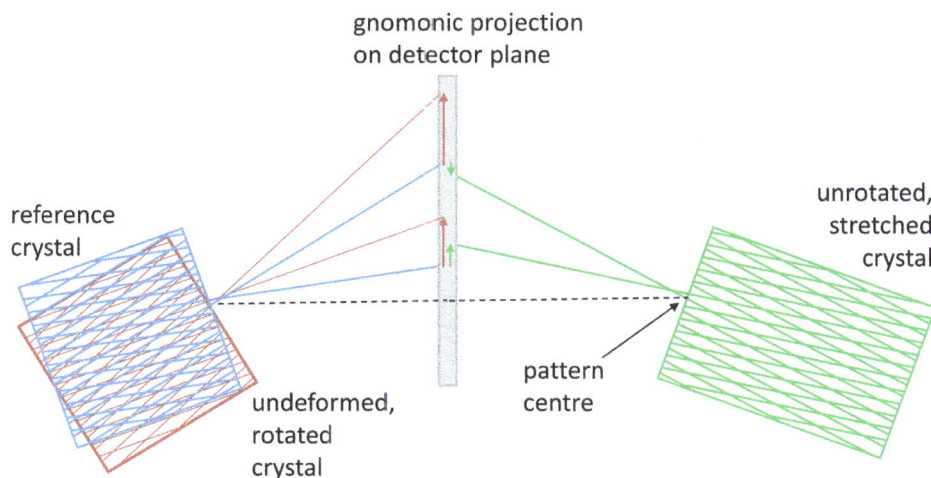

Fig. 25. Schematic display of the shift of Kikuchi bands or their respective intersections (crystallographic poles) by a rotation or stretch of the crystal lattice. The arrows on the detector plane indicate the observed shifts with respect to the reference position. The blue crystal represents the strain-free reference crystal, the red crystal shows the effect of a small rotation, the green crystal that of a small stretch. For all crystals the pattern centre relative to the detector plane, indicated by the black dotted line, is the same.

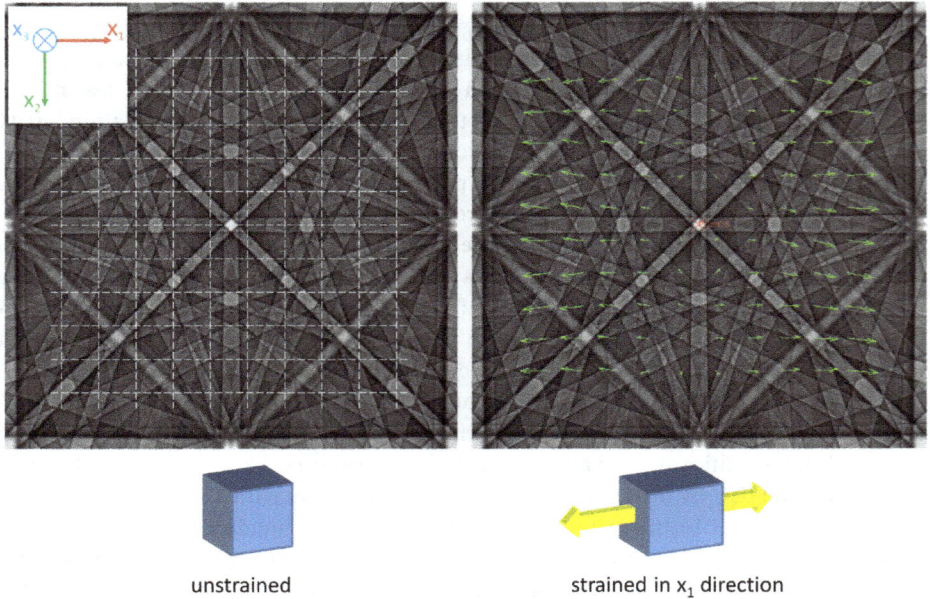

Fig. 26. Illustration of pattern cross-correlation for determination of small crystal lattice distortions or rotations using simulated diffraction patterns from a crystal with known lattice distortion. The white squares indicate the regions of interest in the unstrained (simulated) pattern. The green arrows indicate the shifts of the ROIs after an extension of the lattice in the x_1 direction. Figure modified from Jaepel (2015).

strains. Currently a strain accuracy in the order of 10^{-4}, and of rotations of the order of 0.01° can be reached. The shifts of local EBSP domains determined correspond to changes of the angles between crystallographic directions. This means that only lattice rotations and the deviatoric part of the strain tensor can be determined directly. The missing three normal, or hydrostatic components can also be determined presuming that the measurement is performed at the surface of the material and that there is no stress component perpendicular to this surface. It is important to realize that elastic strains (or residual stresses) can, currently, only be measured relative to a reference point in the same crystal, *i.e.* only 3rd kind residual stresses are accessible. For the measurement of absolute lattice constants the pattern centre would have to be determined very accurately (to within ~1 µm), which is currently impossible.

One important precondition for reliable application of the cross-correlation technique is that the features compared in the two images are only shifted with respect to each other, but not distorted. When comparing EBSD patterns, this condition is fulfilled as long as the lattice rotations stay small. Larger lattice rotations exceeding ~3° lead, however, to severe distortions caused by the gnomonic projection of the pattern and lead to increasingly large errors in stress and strain determination. The cross-correlation technique was, therefore, originally employed to measure residual stresses in semiconductor materials (Wilkinson, 2000; Wilkinson *et al.*, 2014) which contain very few dislocations and, consequently, very small lattice rotations. In contrast, the application to plastically deformed metals (*e.g.* Vaudin *et al.*, 2008; Dingley *et al.*, 2010; Britton and Wilkinson, 2011; Wilkinson and Karamched, 2011; Jaepel, 2015) leads to less reliable results because they create stronger rotations caused by geometrically necessary dislocations. In the meantime the problem posed by large lattice rotations has been solved largely by employing so-called pattern remapping techniques (Maurice *et al.*, 2012; Britton and Wilkinson, 2013; Britton *et al.*, 2013). In this technique the lattice rotation between measured and reference orientation is determined using a standard HT-based approach. Subsequently, one of the patterns is remapped onto the orientation of the other one by a pixel-by-pixel operation. Finally, the remaining differences are determined using the cross-correlation approach. With this approach misorientation up to ~10° can be compensated satisfactorily and the accuracy of strain measurement be improved significantly.

Similar to semiconductor materials, minerals usually develop only small plastic rotations and seem, therefore, suitable candidates for application of the CC technique. However, they pose other difficulties, as discussed further below. One of the first applications of CC EBSD to the determination of lattice parameter changes and residual stresses in minerals was performed by Schäffer *et al.* (2014) investigating the kinetics of phase transformation by interdiffusion of K and Na in alkali feldspar. A Na-rich sanidine single crystal of gem quality was immersed in a melt of KCl at 850°C for 32 days. The monoclinic crystal was cut and polished along the (001) and (010) planes as indicated in Fig. 27A. During immersion, K-Na diffusion profiles developed perpendicular to the (001) and (010) planes of the crystal which showed significantly different compositional zoning characteristics. Figures 27B and C reproduce these results: While the diffusion profile parallel to the *a* axis [100] is wide and smooth, the profile along the *b* axis, [010] is very short and sharp. At the same time the crystal shows the development of characteristic cracks which indicate the presence of residual stresses. In order to measure these residual stresses CC EBSD was employed. One problem that arises with this measurement is that the EBSPs from feldspar crystals have rather low contrast and the crystal lattice deteriorates rapidly when exposed to an intense electron beam. Therefore, the illumination was spread over narrow windows of 2 µm × 25 µm, positioned exactly perpendicular to the diffusion gradient. This resulted in well integrated patterns, however, with a light blur of ~0.5 pixels (corresponding to 25 µm) in horizontal direction caused by the varying pattern centre positions across the illumination window. After recording the diffraction patterns along lines parallel to the *a* and *b* directions, the crystal distortion was analysed using the software *CrossCourt 3* by BLG Productions. A pattern from the original, undistorted sanidine crystal was taken for reference. The results obtained on the sharp compositional boundary along the [010] direction are displayed in Fig. 28 as a strain tensor matrix. Schematic drawings of the lattice parameters for both crystal directions are shown in Fig. 29. It was found that the lattice strain in the *b* direction (corresponding to ε_{11}) shows a peculiar change over the narrow diffusion profile: the *b* lattice-parameter goes through a maximum in the centre of the diffusion zone, while the *a* parameter shows a monotonic decrease and the

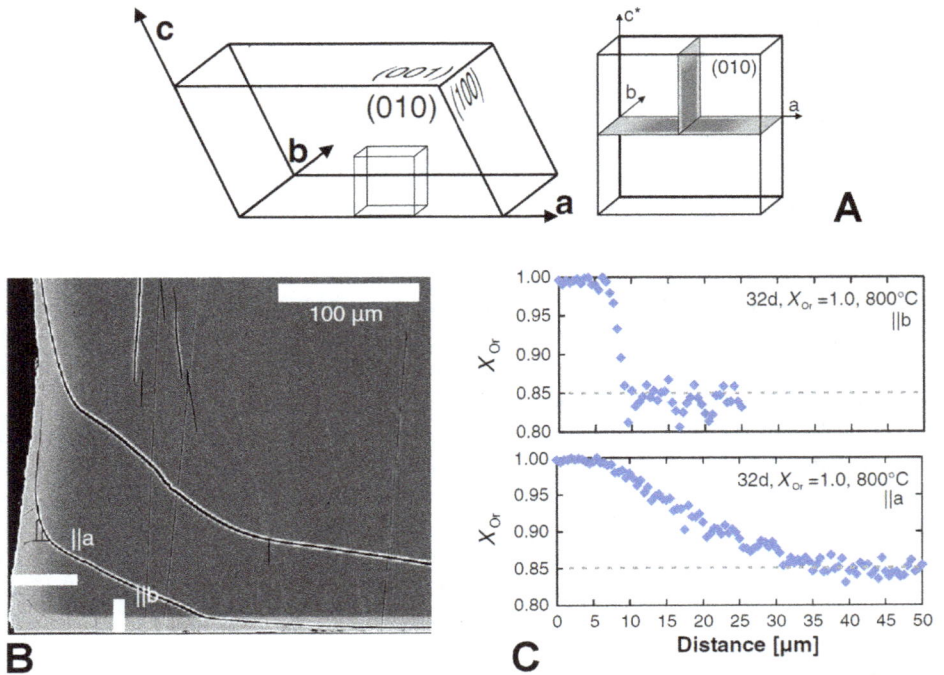

Fig. 27. Investigation of the lattice constant changes in a Na-K alkali feldspar interdiffusion experiment using EBSD pattern cross correlation. (A) Sample geometry of the prepared plates; (*left*) orientation of the plate within the unit cell of feldspar, (*right*) after the experiment the plate was cut in two mutually perpendicular directions (grey planes) to allow documentation of interdiffusion fronts in three different crystallographic directions. (B) BSE image viewed on the (001) plane of feldspar; bright areas along the grain surfaces indicate the K-rich domains, darker areas correspond to unexchanged Na-rich feldspar. (C) FEG-EMP (electron microprobe) measurements of the cation ratio (K^+/Na^+) X_{Or} along the profiles indicated in the BSE image. Experimental conditions of diffusion experiments are given in the profile plots. Figures modified from Schäffer *et al.* (2014).

c parameter remains unchanged. This systematic change of lattice strain does not conform to the change of lattice strain in a stress-free crystal of alkali feldspar when subject to a similar composition change. The changes can, instead, be explained by a confinement of the *a* lattice-parameter by the large substrate crystal. These constraints, however, do not reach far into the diffusion front which leads to a short-distance overshoot of the lattice parameter. On the *a* side, in contrast, the diffusion gradient is so widespread and *b* lattice parameter changes so little that the confinement does not become visible. Finally, the authors hypothesized that the measured lattice-parameter changes may significantly influence the diffusion rate, either by a change of driving force with additional mechanical stresses or by a related change of the atomic structure and energy barriers for atomic jumps.

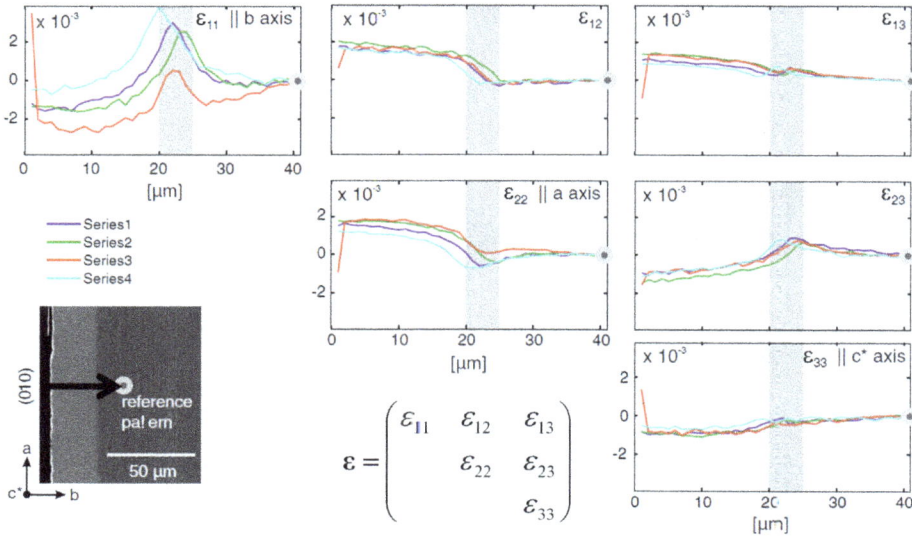

Fig. 28. Line profiles of the different components of the strain tensor measured across a sharp diffusion front (sample derived from an 850°C, 32 days diffusion experiment); the differently coloured profiles represent different measurements executed to test reproducibility, the reference pattern is located at the right hand side as represented by the grey dot and the edge of the crystal is on the left hand side. The vertical grey bar gives the width of the compositional transition zone between the exchanged rim (K-rich sanidine) and the un-exchanged core domain (Na-rich sanidine). Figure modified from Schäffer *et al.* (2014).

2.6.3. Quantification of plastic strain fields (geometrically necessary dislocations, GND) by cross-correlation EBSD

The cross correlation analysis of EBSD patterns not only reveals elastic distortions of the crystal unit cell but also small rotations caused by dislocations. Dislocations display a plastic, *i.e.* non reversible, relaxation of macroscopic elastic strains into linear, strongly localized lattice defects. Nevertheless, when a material is plastically deformed, many dislocations invade the volume and this distribution of dislocations can be considered, as proposed by Nye (1953), as a homogeneous strain field. The dislocations stored in the strain field may be separated into geometrically necessary dislocations (GND) and statistically stored dislocations (SSD). This distinction of dislocations is rather artificial and more based on the way of measuring them than on a physical argument: for GNDs the long-range stress field associated with a dislocation line has relaxed in favour of a long-range lattice rotation. All those dislocations which are not associated with a long-range rotation field are classified as SSDs. This includes relaxed dislocation dipoles which consist of two dislocation segments of opposite sign (*e.g.* created by dislocation loop), as well as unrelaxed individual dislocations. Figure 30 gives a schematic overview on the two 'types' of dislocations. While SSDs do not create any measurable misorientation at a significant distance from the

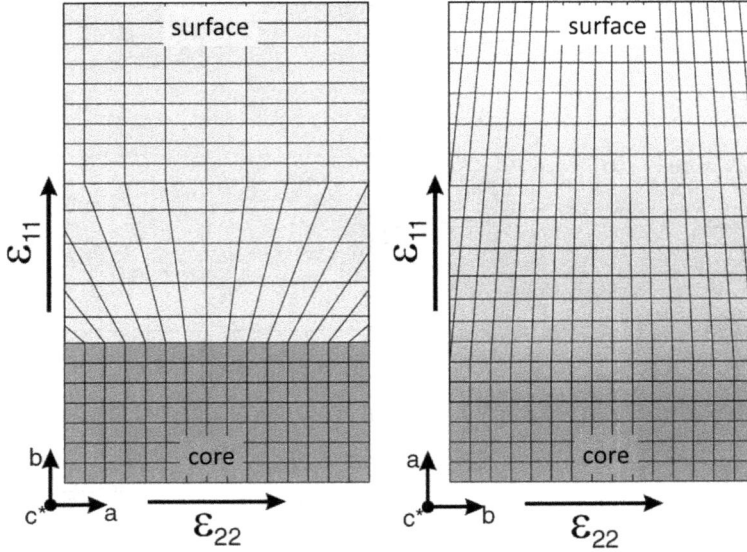

Fig. 29. Schematic diagrams of the lattice distortion across the sharp (*left*) and broad (*right*) diffusion fronts propagating parallel to the *b* and *a* directions, respectively; the dark grey colour represents part of the unexchanged core, and the light grey represents part of the exchanged rim. Modified from Schäffer *et al.* (2014).

dislocation core they lead, however, to a blurring of EBSPs acquired from this location which becomes stronger with increasing amount of dislocations. This blur becomes visible in the pattern quality value although this is not a quantitative measure.

In contrast to SSDs, GNDs build up lattice rotations and form, for example, small-angle grain boundaries. In fact, as shown by Nye (1953), Kröner (1959) and others, the curl of the lattice rotation field is directly related to the total sum of the Burgers vectors of all GNDs in a given volume:

$$\alpha_{ij} = \text{curl}(\mathbf{g}) = -e_{jkl}\mathbf{g}_{il,k} \tag{3a}$$

$$\alpha_{ij} = \sum_{k=1}^{K} \rho^k b_i^k \otimes t_i^j \tag{3b}$$

Here, α_{ij} is the dislocation (or Nye) tensor, $\mathbf{g}$ is the 3-dimensional orientation field, and e_{jkl} is the Levi-Civita permuation symbol, *e.g.* $\alpha_{12} = \frac{\partial g_{13}}{\partial x_2} - \frac{\partial g_{12}}{\partial x_3}$. The summation is carried out over all K different dislocation systems with their Burgers vector length b^k, their line element t^k and their density ρ^k. The symbol $\otimes$ indicates the dyadic product. If the orientation field is known, for example, from EBSD measurements, equation 3 can be used to calculate the dislocation tensor α, which has nine components allowing determination of the densities of nine independent slip systems. In most materials with crystal-plastic behaviour, however, more than nine slip systems are available. In the fcc

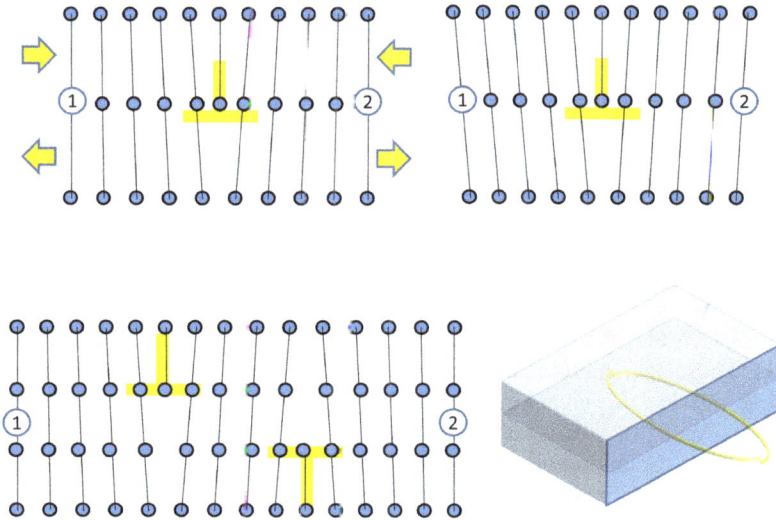

Fig. 30. Illustration of different types of dislocations in strain fields. The top left dislocation is an unrelaxed dislocation which does not create long-range lattice rotations (considered to be a statistically stored dislocation, SSD). In the top right the same dislocation has relaxed its strain field into a rotation field and changed into a geometrically necessary dislocation (GND). The bottom left figure illustrates a dipole without long-range rotation field (SSD). The bottom right figure illustrates the 3D arrangement of the dipole.

lattice, for example, with <110> Burgers vectors and {111} slip planes a total of 12 edge dislocations and six screw dislocations are possible. As a result the equation system 3 is heavily under-determined and one may find in this case $\binom{9}{18}$ = 48620 possible solutions to 3. One way to select a unique solution from amongst these is to select the one which corresponds to the smallest total line energy. For a more detailed mathematical treatment and further references see, *e.g.* Konijnenberg *et al.* (2015).

Full appreciation of these calculations requires knowledge of the full 3-dimensional orientation field as measured, for example by 3D-EBSD (Konijnenberg *et al.*, 2015). Wheeler *et al.* (2009) showed, however, that with 2D-EBSD reasonable dislocation information can also be obtained. It should be mentioned that it is also possible to estimate the total GND densities directly from local misorientation information using the so-called kernel average misorientation (*KAM*) (Konijnenberg *et al.*, 2015):

$$\rho_{GND} = \frac{c}{ban}KAM \qquad (4)$$

where *b* is the average length of the Burgers vector, *a* is the measurement step size, *n* is the radius of the kernel in pixels and c is a geometric constant of the order of 3, the *KAM* value is given in radians. Note that this quantification only returns the total dislocation density but does not allow conclusions about the type of dislocations. An overview on potential EBSD-based techniques for dislocation density analysis can be found in Wright *et al.* (2011).

One example for the application of CC-EBSD measurements to the determination of GNDs in a geological sample is the investigation by Griffiths *et al.* (2014) on the origin of bleaching zones in garnet crystals from the Koralpe region in the Eastern Alps. The Permian metapegmatite garnet crystals investigated show abundant inclusions of different sizes and distributions. The coarse-grained (up to 1 cm in size) almandine-spessartine garnet contains numerous micro-inclusions of several different phases, *e.g.* corundum, rutile, ilmenite, zircon and wyllieite-group phosphate with grain sizes between a few nm and up to 500 nm. In the transmitted light microscope the garnet crystals showed bleached zones with significantly lower density of micro-inclusions (Fig. 31). These sparse inclusions were, however, at up to 10 μm, significantly larger than those outside the bleaching zones. The present study addressed the causes of bleaching along the trails of coarsened micro-inclusions. The geometry and arrangement of the inclusion trails indicated that they correspond to healed cracks. Element distribution maps measured by WDX showed that the cracks were not necessarily associated with compositional changes; thus a chemical bleaching could be excluded. Finally, CC-EBSD maps were performed across the bleaching zones flanking the inclusion trails and very small lattice orientation changes with a maximum of 0.4° were found (Fig. 32). These orientation changes are too large to be created by elastic stresses, and it was concluded that they are due to GNDs. Figure 32 shows the orientation changes with respect to the centre of the bleaching zone as misorientation angle profiles over distance to the centre (Fig. 32a–b), and as pole figures showing the distribution of the rotation axes on either side of the trail centre (Fig. 32 c–d). Note that the axes plots are reasonable only if the orientation noise is small enough, which is, in

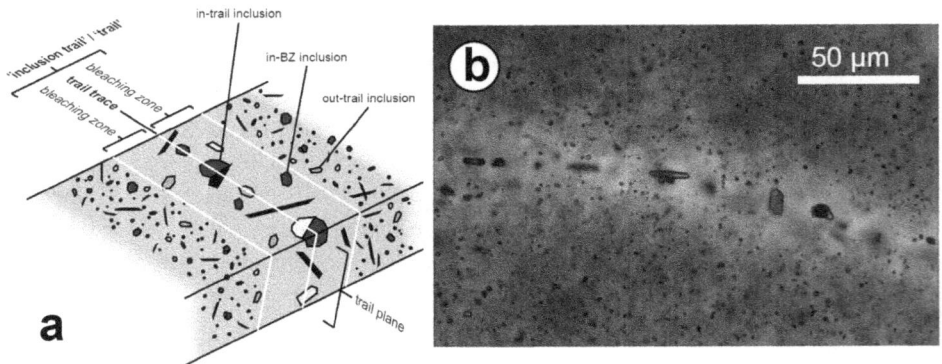

Fig. 31. Investigations on inclusions in healed fractures in garnets. (a) Idealized sketch of the inclusion trail microstructure in metapegmatite garnet labelled with the definitions used in the original paper (Griffith *et al., 2014*). (b) Photomicrograph of an inclusion trail microstructure consisting of a trail of coarsened inclusions and an adjacent bleaching zone where the finer-grained inclusion fraction is almost absent. Reproduced from Griffiths *et al.* (2014) with the permission of Springer.

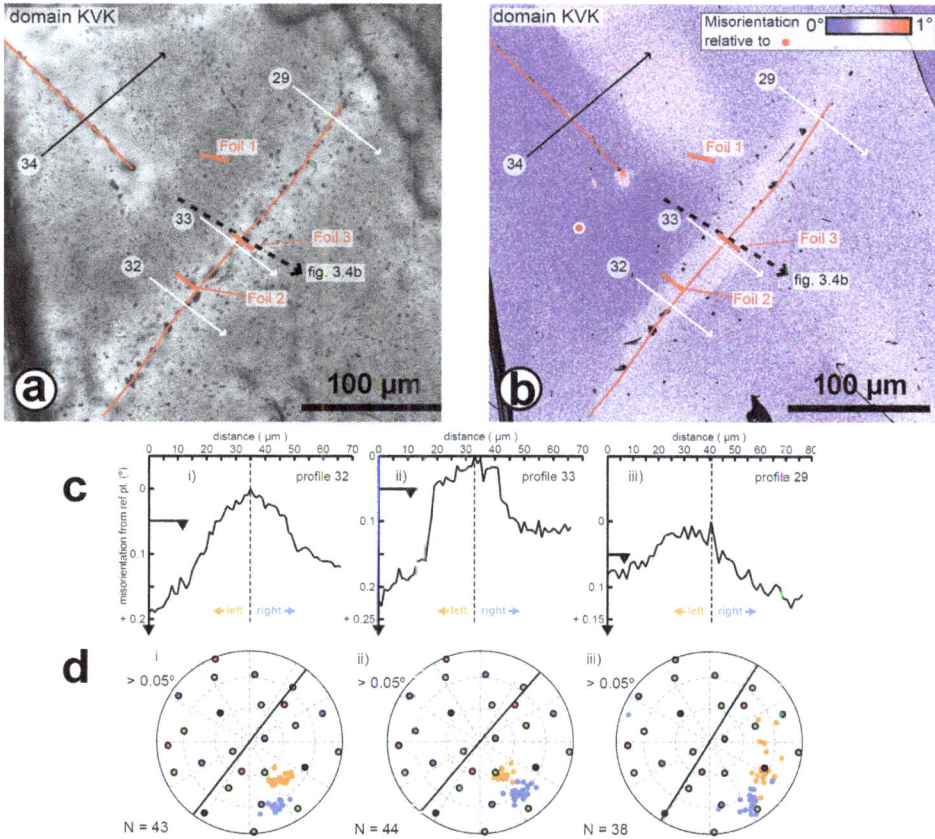

Fig. 32. Continued from Fig. 31. (a) Thin-section photograph of domain KVK with measurement locations marked, not all of which are discussed here. (b) Hough transform-based EBSD map of this domain, colour coded for relative misorientation angle to the marked reference point. Locations of the EBSD cross correlation profiles (part c) are marked by solid arrows. Red lines give location of trail centre (crack plane). (c) Profiles of misorientation angle to a reference point (marked by a vertical dotted line) with distance across the microstructure, obtained using cross correlation-based EBSD. Positive misorientation angles are plotted downward. Arrows at the left indicate the threshold misorientation angle of 0.05° below which misorientation axes are not plotted. Coloured labels 'left' and 'right' indicate the colour assigned to misorientation axes in d that belong to points to the left and right of the reference point, respectively. (d) Antipodal, upper hemisphere, equal area pole figure plots of misorientation axes for misorientation angles >0.05° in c above, colour coded according to their location to the left or right of the reference point. The thick line indicates the trace orientation of the trail where it cuts each profile. Garnet directions are plotted as coloured circles for <100> (dark grey), <110> (blue), <111> (red) and <112> (green). Reproduced from Griffiths *et al.* (2014) with the permission of Springer.

the present case, established by the use of the CC EBSD technique. Here the orientation noise is of the order of 0.05° to 0.02°, while a classical HT analysis would result in values which are about ten times larger. Bate *et al.* (2005) developed an analytical relationship between the angular inaccuracy of the misorientation axis, ϕ, the angular resolution, δ, and the misorientation, ω:

$$\phi = \arctan\left(\frac{\delta}{\omega}\right) \tag{5}$$

According to this equation, the misorientation axis inaccuracy increases with increasing orientation noise and decreasing misorientation angle. For the present case an orientation resolution of 0.04° resulted in an accuracy of the misorientation axis of $\pm39°$ for a misorientation angle of 0.05° and of $\pm5°$ for a misorientation angle of 0.45°. Thus, the authors used only those misorientations larger than 0.05° to plot the misorientation axis in Fig. 32c. It was found that the rotation axes were different on the two different sides of the centre line of the bleaching zone, which was interpreted to be due to two different sets of dislocations being activated on either side of the crack tip.

From all these observations the authors concluded that the lattice rotations are due to dislocations that had been created at the propagating crack tip during the fracturing process. The density of dislocations was found to be of the order of 10^{11} m^{-2}, which is ~10 times higher than in regions outside the bleaching zone. These dislocations led to enhanced diffusion rates due to pipe diffusion which was the reason for Ostwald ripening of the particles and consequent bleaching of the dislocation-populated zones. The authors estimated that the diffusion rate was enhanced by a factor of at least 10^2 by the presence of the dislocations.

2.7. Conclusions: Possibilities and challenges

As shown above EBSD and EBSD-based orientation microscopy are extremely powerful and versatile techniques for quantification of many aspects of crystal structures, microstructures and textures of crystalline materials on bulk samples.

With respect to crystal structures, many space groups can be distinguished and, in the future, the atomic occupation of lattices might become accessible by realistic simulation of diffraction patterns. The dimensions of the crystal unit cell can be determined only with relatively low accuracy (10^{-2}), while the angles between the unit-cell axes may be measured quite accurately (10^{-4}).

Concerning microstructures, many defects can be quantified in number and kind by EBSD-based COM, including dislocation densities, grain and phase boundaries, and residual stresses. The latter are determined by EBSD pattern cross-correlation which allows, however, only measurement of residual stresses of the 3rd kind. For the 1st and 2nd kinds of residual stresses, the accuracy of band detection and pattern centre positioning is still too low.

EBSD-based COM is also a very powerful tool to measure micro and macro textures of materials both at high (nm-scale) resolution and across large (several cm-scale) sample domains. For the latter, correct probing of the sample is very important to obtain representative data. In this case texture measurements of higher quality than those obtained with XRD can be made, with additional information on location and microstructure of the measured volume.

A serious limitation of EBSD with respect to many modern engineering materials is its spatial resolution: the smallest measurable volume is ~10^4 nm^3, which is 10 to 100 times larger than that measured by TEM diffraction. Truly nanocrystalline materials (with grain volumes of the order of 10^3 nm^3) are, therefore, not yet accessible using EBSD on bulk samples. Their microstructures may, however, be measured with transmission (t)-EBSD, which requires thin foil samples and, therefore, gives up one of the largest advantages of EBSD, the site-specific measurement on bulk materials. A valuable alternative approach appears to be the reduction of acceleration voltage coupled with direct detection and new indexing methods for low-intensity patterns.

A number of aspects of EBSD and EBSD-based COM have not been discussed in detail in this text, but should, nevertheless be mentioned and accessed here again (see the introductory section for literature). One is the possibility of performing 3-dimensional COM by combination of COM with some sort of serial sectioning, like sputtering with a focused ion beam (FIB), milling with a broad ion beam, or mechanical polishing. These techniques allow more comprehensive description of grain and phase boundaries, boundary networks and other 3-dimensional features like cracks. Despite the potential of these techniques they are comparatively slow, and, of more concern, they are destructive. This means that the investigated material volume is destroyed during the measurement and cannot be investigated further, *e.g.* to observe microstructural changes over time or to measure physical or chemical properties. Nevertheless, work is currently going on to compensate partly for these shortcomings, for example by interrupted 3D EBSD or measurement of 3D data at an edge of a sample (Mandal *et al.*, 2014). A second group of applications not discussed in detail comprises all kinds of *in situ* EBSD techniques, which allow observation of, for example, deformation processes during *in situ* straining or microstructural changes during heating or cooling. Despite the interesting results that can be obtained from these observations, it should be kept in mind that these *in situ* experiments always allow observation only of processes at the free surface of a sample, which may deviate significantly from those proceeding inside the sample, either because of different stress states or because of different surrounding materials in terms of, for example, grain boundary or dislocation networks. Also the free surface itself may, by its surface energy, contribute to the process.

Finally, EBSD measurements are increasingly used in combination with materials-simulation tools, *e.g.* crystal plasticity modelling, modelling of recrystallization and grain growth during annealing or modelling of phase transformations (see the references in the introduction). The combination with modelling, on the one hand, enables us to follow processes which are not directly observable by EBSD. On the other hand, however, the input is usually a measurement from a sample surface – with all the limitations mentioned above – or from a 3D material volume, with the consequence that the original material volume no longer exists after serial sectioning. The 'holy grail' for all these observations and related modelling approaches are, therefore, non-destructive 3D COM methods as they are provided by synchrotron radiation techniques (*e.g.* Jakobsen *et al.*, 2007; King *et al.*, 2011; Wang *et al.*, 2013). These techniques nowadays reach spatial resolutions in the order of that of EBSD but are, of course, expensive and have limited access.

References

Abart, R., Kunze, K., Milke, R., Sperb, R. and Heinrich, W. (2004) Silicon and oxygen self diffusion in enstatite polycrystals: the Milke *et al.* (2001) rim growth experiments revisited. *Contributions to Mineralogy and Petrology*, **147**, 633–646.

Abou-Ras, D., Caballero, R., Kavalakkatt, J., Nichterwitz, M., Unold, T., Schock, H.-W., Bücheler, S. and Tiwan, A.N. (2010) Electron backscatter diffraction: Exploring the microstructure in Cu(InGa)(SSe)$_2$ and CdTe thin-film solar cells. *Photovoltaic Specialists Conference (PVSC)*, **35th IEEE**, 418–423.

Adams, B.L., Wright, S.I. and Kunze, K. (1993) Orientation imaging: the emergence of a new microscopy. *Metallurgical Transactions*, **24A**, 819–831.

Ageeva, O.A., Abart, R., Habler, G., Borutzky, B.Y. and Trubkin, N.V. (2012) Oriented feldspar-feldspathoid intergrowths in rocks of the Khibiny massif: genetic implications. *Mineralogy and Petrology*, **106**, 1–17.

Bachmann, F., Hielscher, R. and Schaeben, H. (2010) Texture analysis with MTEX – free and open source software toolbox. *Solid State Phenomena Series*, **160**, 63–68.

Bate, P.S., Knutsen, R.D., Brough, I. and Humphreys, F.J. (2005) The characterization of low-angle boundaries by EBSD. *Journal of Microscopy*, **220**, 36–46.

Baudin, T., Jura, J., Penelle, R. and Pospiech, J. (1995) Estimation of the minimum grain number for the orientation distribution function calculation from individual orientation measurements on Fe–3%Si and Ti–4Al–6V alloys. *Journal of Applied Crystallography*, **28**, 582–589.

Bell, D.C. and Erdman, N. (editors) (2013) *Low Voltage Electron Microscopy: Principles and Applications.* Wiley, Weinheim, Germany.

Britton, T.B. and Wilkinson, A.J. (2011) Measurement of residual elastic strain and lattice rotations with high resolution electron backscatter diffraction. *Ultramicroscopy*, **111**, 1395–1404.

Britton, T.B. and Wilkinson, A.J. (2013) High resolution electron backscatter diffraction measurements of elastic strain variations in the presence of larger lattice rotations. *Ultramicroscopy*, **114**, 82–95.

Bunge, H.J. (1987) Three-dimensional texture analysis. *International Materials Review*, **32**, 265–291.

Callahan, P.G. and De Graef, M. (2013) Dynamical electron backscatter diffraction patterns. Part I: Pattern simulations. *Microscopy and Microanalysis*, **19**, 1255–1265.

Cayron, C. (2007) ARPGE: a computer program to automatically reconstruct the parent grains from electron backscatter diffraction data. *Journal of Applied Crystallography*, **40**, 1183–1188.

Chen, C.-L. and Thomson, R.C. (2010) The combined use of EBSD and EDX analyses for the identification of complex intermetallic phases in multicomponent Al–Si piston alloys. *Journal of Alloys and Compounds*, **490**, 293–300.

Chen, Y.-H., Park, S.U., Wei, D., Newstadt, G., Jackson, M., Simmons, J.P., De Graef, M. and Hero, A.O. (2015) A dictionary approach to EBSD indexing. *Journal of Microscopy and Microanalysis*, **21**, 739–752.

Choi, S.H. and Cho, J.H. (2005) Primary recrystallization modelling for interstitial free steels. *Materials Science Engineering A*, **405**, 86–101.

Davut, K. (2012) Relation between microstructure and mechanical properties in a low-alloyed TRIP steel. PhD thesis, RWTH Aachen, Germany.

Davut, K. and Zaefferer, S. (2010) Statistical reliability of phase fraction determination based on electron backscatter diffraction (EBSD) investigations on the example of an Al-TRIP steel. *Metallic Materials Transactions A*, **41**, 2187–2196.

Dingley, D.J. (1984) On-line determination of crystal orientation and texture determination in an SEM. *Proceedings of the Royal Microscopical Society*, **19**, 74–75.

Dingley, D.J. (2004) Progressive steps in the development of electron backscatter diffraction and orientation imaging microscopy. *Journal of Microscopy*, **213**, 214–224.

Dingley, D.J., Wilkinson, A.J., Meaden, G. and Karamched, P.S. (2010) Elastic strain tensor measurement using electron backscatter diffraction in the SEM. *Journal of Electron Microscopy*, **59 (Supplement)**, S155–S163.

Echlin, P. (2009) *Handbook of Sample Preparation for Scanning Electron Microscopy and X-Ray Microanalysis*. Springer, Heidelberg, Germany.

Field, D.P., Bradford, L.T., Nowell, M.M. and Lillo, T.M. (2007) The role of annealing twins during recrystallization of Cu. *Acta Materialia*, **55**, 4233−4241.

Frank, L., Müllerova, I., Matsuda, K. and Ikeno, S. (2007) Cathode lens mode of the SEM in materials science applications. *Materials Transactions*, **48**, 944 − 948.

Germain, L., Dey, S.R., Humbert, M. and Gey, N. (2007) Determination of parent orientation maps in advanced titanium-based alloys. *Journal of Microscopy*, **227**, 284−291.

Goetz, A.J., Steinmetz, D.R., Griesshaber, E., Zaefferer, S., Raabe, D., Kelm, K., Irsen, S., Sehrbrock, A. and Schmahl, W.W. (2011) Interdigitating biocalcite dendrites form a 3-D jigsaw structure in brachiopod shells. *Acta Biomaterialia*, **7**, 2237−2243.

Goldstein, J., Newbury, D.E., Joy, D.C., Lymar, C.E., Echlin, P., Lifshin, E., Sawyer, L. and Michael, J.R. (2007) *Scanning Electron Microscopy and X-ray Microanalysis*, 3rd edition. Springer, New York.

Gourgues-Lorenzon, A.-F. (2009) Application of electron backscatter diffraction to the study of phase transformations: present and possible future. *Journal of Microscopy*, **233**, 460−473.

Griesshaber, E., Neuser, R. and Schmahl, W.W. (2010) The application of EBSD analysis to biomaterials: microstructural and crystallographic texture variations in marina carbonate shells. Pp. 23−34 in: *Biominerals and Biomineralization Processes* (L. Fernández Díaz and J.M. Astilleros García-Monge, editors). Seminarios de la Sociedad Española de Mineralogía, **7**. Sociedad Española de Mineralogía, Madrid.

Griffiths, T.A., Habler, G., Rhede, D., Wirth, R., Ram, F. and Abart, R. (2014) Localization of submicron inclusion re-equilibration at healed fractures in host garnet. *Contributions to Mineralogy and Petrology*, **168**, 1077.

Groeber, M.A. and Jackson, M.A. (2014) DREAM.3D: A digital representation environment for the analysis of microstructure in 3D. *Integrating Materials and Manufacturing Innovation*, **3**, 5.

Hofmann, S. (2013) *Auger- and X-Ray Photoelectron Spectroscopy in Materials Science*. Springer Series in Surface Science, Springer, Heidelberg, Germany.

Humphreys, F.J. (2001) Review grain and subgrain characterisation by electron backscatter diffraction. *Journal of Materials Science*, **36**, 3833−3854.

Humphreys, F.J. and Ferry, M. (1996) Combined in-situ annealing and EBSD of deformed aluminium alloys. *Materials Science Forum*, **217-222**, 529−534.

Humphreys, F.J., Huang, Y., Brough, I. and Harris, C. (1999) Electron backscatter diffraction of grain and subgrain structures - resolution considerations. *Journal of Microscopy*, **195**, 212−216.

Humphreys, F.J., Bate, P.S. and Hurley, P.J. (2001) Orientation averaging of electron backscattered diffraction data. *Journal of Microscopy*, **201**, 50−58.

Jaepel, T. (2015) *Feasibility study on local elastic strain measurements with an EBSD pattern cross correlation method in elastic-plastically deforming materials*. PhD thesis, RWTH Aachen, Germany.

Jakobsen, B., Poulsen, H.F., Lienert, U. and Pantleon, W. (2007) Direct determination of elastic strains and dislocation densities in individual subgrains in deformation structures. *Acta Materialia*, **55**, 3421−3430.

Keller, R.R. and Geiss, R.H. (2012) Transmission EBSD from 10 nm domains in a scanning electron microscope. *Journal of Microscopy*, **245**, 245−251.

King, A., Ludwig, W., Herbig, M., Buffière, J.-Y., Khan, A.A., Stevens, N. and Marrow, T.J. (2011) Three-dimensional in situ observations of short fatigue crack growth in magnesium. *Acta Materialia*, **59**, 6761−6771.

Konijnenberg, P.J., Zaefferer, S., Lee, S.B., Rollett, A.D., Rohrer, G.S. and Raabe, D. (2012) Advanced tomographic tools for reconstruction and analysis of 3D-EBSD datasets. *Materials Science Forum*, **702-703**, 475−478.

Konijnenberg, P.J., Zaefferer, S. and Raabe, D. (2015) Assessment of geometrically necessary dislocation levels derived by 3D-EBSD. *Acta Materialia*, **99**, 402−414.

Krieger-Lassen, N.C. (1994) *Automated determination of crystal orientations from electron backscattering patterns*. Ph.D. Thesis, Danmarks Tekniske Universitet, Lyngby, Denmark.

Krieger-Lassen, N.C. (1999) Source point calibration from an arbitrary electron backscattering pattern. *Journal of Microscopy*, **195**, 204–211.

Krieger-Lassen, N.C., Conradsen, K. and Juul-Jensen, D. (1992) Image processing procedures for analysis of electron diffraction patterns. *Scanning Microscopy*, **6**, 115–121.

Kröner, E. (1959) Allgemeine Kontinuumstheorie der Versetzungen und Eigenspannungen. *Archive for Rational Mechanics and Analysis*, **4**, 273–334.

Kundin, J., Raabe, D. and Emmerich, H. (2011) A phase-field model for incoherent martensitic transformations including plastic accommodation processes in the austenite. *Journal of Mechanics and Physics of Solids*, **59**, 2082–2102.

Mainprice, D., Hielscher, R. and Schaeben, H. (2011) Calculating anisotropic physical properties from texture data using the MTEX open-source package. Pp. 175–192 in: *Deformation Mechanisms, Rheology and Tectonics: Microstructures, Mechanics and Anisotropy* (D.J. Prior, E.H. Rutter and D.J. Tatham, editors). Geological Society Special Publications, **360**. Geological Society, London.

Mandal, S., Pradeep, K.G., Zaefferer, S. and Raabe, D. (2014) A novel approach to measure grain boundary segregation in bulk polycrystalline materials in dependence of the boundaries' five rotational degrees of freedom. *Scripta Materialia*, **81**, 16–19.

Maurice, C. and Fortunier, R. (2008) A 3D Hough transform for indexing EBSD and Kossel patterns. *Journal of Microscopy*, **230**, 520–52.

Maurice, C., Dzieciol, K. and Fortunier, R. (2011) A method for accurate localisation of EBSD pattern centres. *Ultramicroscopy*, **111**, 140–148.

Maurice, C., Driver, H.J. and Fortunier, R. (2012) On solving the orientation gradient dependency of high angular resolution EBSD. *Ultramicroscopy*, **113**, 171–181.

Miyamoto, G., Naomichi, I., Naoki, T. and Furuhara, T. (2010) Mapping the parent austenite orientation reconstructed from the orientation of martensite by EBSD and its application to ausformed martensite. *Acta Materialia*, **58**, 6393–6403.

Nakamichi, H., Humphreys, F.J. and Brough, I. (2008) Recrystallization phenomena in an IF steel observed by in situ EBSD experiments. *Journal of Microscopy*, **230**, 464–471.

Nesbitt, H.W. and Pratt, A.R. (1995) Application of Auger-Electron Spectroscopy to Geochemistry. *The Canadian Mineralogist*, **33**, 243–259.

Nishikawa, S. and Kikuchi, S. (1928) The Diffraction of Cathode Rays by Calcite. *Proceedings of the Imperial Academy (of Japan)*, **4**, 475–477.

Nolze, G., Winkelmann, A. and Boyle, A.P. (2016) Pattern matching approach to pseudosymmetry problems in electron backscatter diffraction. *Ultramicroscopy*, **160**, 146–154.

Nye, J.F. (1953) Some geometrical relations in dislocated crystals. *Acta Materialia*, **1**, 153–162.

Ojima, M., Adachi, Y., Suzuki, S. and Tomota, Y. (2011) Stress partitioning behavior in an fcc alloy evaluated by the in situ/ex situ EBSD-Wilkinson method. *Acta Materialia*, **59**, 4177–4185.

Pagel, M., Barbin, V., Blanc, P. and Ohnenstetter, D. (editors) (2000) *Cathodoluminescence in Geosciences*. Springer, Berlin.

Pirgazi, H., Ghodrat, S. and Kestens, L.A.I. (2014) Three-dimensional EBSD characterization of thermo-mechanical fatigue crack morphology in compacted graphite iron. *Materials Characterization*, **90**, 13–20.

Pouchou, J.L. (editor) (2004) *L'analyse EBSD Principes et applications*. EDP Sciences Les Ulis, France, 44927.

Prior, D.J., Boyle, A.P., Brenker, F., Cheadle, M.C., Day, A., Lopez, G., Peruzzo, L., Potts, G.J., Reddy, S., Spiess, R.,. Timms, N.E., Trimby, P., Wheeler, J. and Zetterström, L. (1999) The application of electron backscatter diffraction and orientation contrast imaging in the SEM to textural problems in rocks. *American Mineralogist*, **84**, 1741–1759.

Prior, D.J., Lilly, K., Seidmann, M., Vaughan, M., Becroft, L., Easingwood, R., Diebold, S., Obbard, R., Daghlian, C., Baker, I., Caswell, T., Golding, N., Goldsby, D., Durham, W.B., Piazolo, S. and Wilson, C.J. (2015) Making EBSD on water ice routine. *Journal of Microscopy*, **259**, 237–256.

Raabe, D. and Hantcherli, L. (2005) 2D cellular automaton simulation of the recrystallization texture of an IF sheet steel under consideration of Zener pinning. *Computational Materials Science*, **34**, 299–313.

Ram, F. (2015) *The Kikuchi bandlet method for the intensity analysis of the Electron Backscatter Kikuchi Diffraction Patterns*. PhD thesis, RWTH Aachen, Germany.

Ram, F., Zaefferer, S. and Raabe, D. (2014) Kikuchi bandlet method for the accurate deconvolution and localization of Kikuchi bands in Kikuchi diffraction patterns. *Journal of Applied Crystallography*, **47**, 264–275.

Ram, F., Zaefferer, S., Jaepel, T. and Raabe, D. (2015) The fidelity analysis of the crystal orientations and misorientations obtained by the classical electron backscatter diffraction technique. *Journal of Applied Crystallography*, **48**, 797–813.

Randle, V. (2013) *Microtexture Determination and its Applications*. Maney Publishing, Leeds, UK.

Randle, V. and Engler, O. (2009) *Introduction to Texture Analysis: Macrotexture Microtexture and Orientation Mapping*, 2nd. edition. CRC Press, Boca Raton, Florida, USA.

Rauch, E.F. and Dupuy, L. (2005) Rapid spot diffraction patterns identification through template matching. *Archives of Metallurgy and Materials*, **50**, 87–99.

Reed, S.J.B. (2005) *Electron Microprobe Analysis and Scanning Electron Microscopy in Geology* 2nd edition, Cambridge University Press, Cambridge, UK.

Reimer, L. (1998) *Scanning Electron Microscopy: Physics of Image Formation and Microanalysis* 2nd edition, Springer, New York.

Rohrer, G. (2011) Measuring and interpreting the structure of grain-boundary networks. *Journal of the American Ceramics Society*, **94**, 633–646.

Rossouw, C.J., Miller, P.R., Josefsson, T.W. and Allen, L.J. (1994) Zone-axis back-scattered electron contrast for fast electrons. *Philosophical Magazine A* **70**, 985–998.

Schäffer, A.K., Jäpel, T., Zaefferer, S., Abart, R. and Rhede, D. (2014) Lattice strain across Na–K interdiffusion fronts in alkali feldspar: an electron back-scatter diffraction study. *Physics and Chemistry of Minerals*, **41**, 795–804.

Schwartz, A.J., Kumar, M., Adams, B.L. and Field, D.P. (2009) *Electron Backscatter Diffraction in Materials Science*. Springer, New York

Schwarzer, R. website www.ebsd.info/history.htm.

Seward, G.G.E., Prior, D.J., Wheeler, J., Celotto, S., Halliday, D.J.M., Paden, R.S. and Tye, M.R. (2002) High-temperature electron backscatter diffraction and scanning electron microscopy imaging techniques: In-situ investigations of dynamic processes. *Scanning*, **24**, 232–240.

Seward, G.G.E., Celotto, S., Prior, D.J., Wheeler, J. and Pond, R.C. (2004) In situ SEM-EBSD observations of the hcp to bcc phase transformation in commercially pure titanium. *Acta Materialia*, **52**, 821–832.

Sieber, B. (2013) Cathodoluminescence – Principes physiques et systèmes de détection. *Les Sélections, Techniques de l'Ingénieur, Paris*, **3**, 792–807.

Steinmetz, D. and Zaefferer, S. (2010) Towards ultra-high resolution EBSD by low accelerating voltage. *Materials Science and Technology*, **26**, 640–645.

Stokes, D. (2008) *Principles and Practice of Variable Pressure: Environmental Scanning Electron Microscopy (VP-ESEM)*. Wiley, Weinheim, Germany.

Suzuki, S. (2013) Features of transmission EBSD and its application. *JOM*, **65**, 1254–1263.

Takayama, N., Miyamoto, G. and Furuhara, T. (2012) Effects of transformation temperature on variant pairing of bainitic ferrite in low carbon steel. *Acta Materialia*, **60**, 2387–2396.

Tasan, C.C., Hoefnagels, J.P.M., Diehl, M., Yan, D., Roters, F. and Raabe, D. (2014) Strain localization and damage in dual phase steels investigated by coupled in-situ deformation experiments and crystal plasticity simulations. *International Journal of Plasticity*, **63**, 198–210.

Trimby, P.W. (2012) Orientation mapping of nanostructured materials using transmission Kikuchi diffraction in the scanning electron microscope. *Ultramicroscopy*, **120**, 16–24.

Uchic, M.D., Groeber, M., Wheeler, R., Scheltens, F. and Dimiduk, D.M. (2004) Augmenting the 3D characterization capability of the dual beam FIB-SEM. *Microscopy and Microanalysis*, **10**, 1136–1137.

Vaudin, M.D., Gerbig, Y.B., Stranick, S.J. and Cook, R.F. (2008) Comparison of nanoscale measurements of strain and stress using electron back scattered diffraction and confocal Raman microscopy. *Applied Physics Letters*, **93**, 193116.

Venables, J.A and Harland, C.J. (1973) Electron backscattering patterns – a new technique for obtaining crystallographic information in the scanning electron microscope. *Philosophical Magazine A*, **27**, 1193–1200.

Wang, L., Li, M., Almer, J., Bieler, T. and Barabash, R. (2013) Microstructural characterization of polycrystalline materials by synchrotron X-rays. *Frontiers of Materials Science*, **7**, 156 − 169.

Wang, S., Holm, E.A., Suni, J., Alvi, M.H., Kalu, P.N. and Rollett, A.D. (2011) Modeling the recrystallized grain size in single phase materials. *Acta Materialia*, **59**, 3872−3882.

Wang, Z.L. (2003) Thermal diffuse scattering in sub-angstrom quantitative electron microscopy - phenomenon effects and approaches. *Micron*, **34**, 141.

Wheeler, J., Mariani, E., Piazolo, S., Prior, D.J., Trimby, P. and Drury, M.R. (2009) The weighted Burgers vector: a new quantity for constraining dislocation densities and types using electron backscatter diffraction on 2D sections through crystalline materials. *Journal of Microscopy*, **233**, 482−494.

Wilkinson, A.J. (2000) Advances in SEM-based diffraction studies of defects and strains in semiconductors. *Journal of Electron Microscopy (Tokyo)*, **49**, 299−310.

Wilkinson, A.J. and Karamched, P.S. (2011) High resolution electron back-scatter diffraction analysis of thermally and mechanically induced strains near carbide inclusions in a superalloy. *Acta Materialia*, **59**, 263−272.

Wilkinson, A.J., Meaden, G. and Dingley, D.J. (2006) High-resolution elastic strain measurement from electron backscatter diffraction patterns: new levels of sensitivity. *Ultramicroscopy*, **106**, 307−320.

Wilkinson, A.J., Britton, T.B., Jiang, J. and Karamched, P.S. (2014) A review of advances and challenges in EBSD strain mapping. *IOP Conference Series: Materials Science and Engineering*, **55**, 012020.

Winkelmann, A. (2009) Dynamical simulation of electron backscatter diffraction patterns. Pp. 21−32 in: *Electron Backscatter Diffraction in Materials Science* (A.J. Schwartz, M. Kumar, B.L. Adams and D.P. Field, editors). Springer, Berlin.

Winkelmann, A. and Nolze, G. (2015a) Point-group sensitive orientation mapping of non-centrosymmetric crystals. *Applied Physics Letters*, **106**, 72101.

Winkelmann, A. and Nolze, G. (2015b) Chirality determination of quartz crystals using electron backscatter diffraction. *Ultramicroscopy*, **149**, 58−63.

Winkelmann, A., Trager-Cowan, C., Sweeney, F., Day, A.P. and Parbrook, P. (2007) Many-beam dynamical simulation of electron backscatter diffraction patterns. *Ultramicroscopy*, **107**, 414−421.

Wisniewski, W., Völksch, G. and Rüssel, C. (2011) The degradation of EBSD-patterns as a tool to investigate surface crystallized glasses and to identify glassy surface layers. *Ultramicroscopy*, **111**, 1712−1719.

Wöllmer, S., Zaefferer, S., Göken, M., Mack, T. and Glatzel, U. (2003) Characterization of phases of aluminized nickel base superalloys. *Surface and Coating Technology*, **167**, 83−96.

Wright, S.I., Nowell, M.M. and Bingert, J.F. (2007) A comparison of textures measured using X-ray and electron backscatter diffraction. *Metallurgical and Materials Transactions A*, **38**, 1845−1855.

Wright, S.I., Nowell, M.M. and Field, D.P. (2011) A review of strain analysis using electron backscatter diffraction. *Microscopy and Microanalysis*, **17**, 136−329.

Wright, S.I., Basinger, J.A. and Nowell, M.M. (2012) Angular precision of automated electron backscatter diffraction measurements. *Materials Science Forum*, **702-703**, 548−553.

Yacobi, B.G. and Holt, D.B. (1990) *Cathodoluminescence Microscopy of Inorganic Solids*. Springer, New York.

Yan, D., Tasan, C.C. and Raabe, D. (2015) High resolution in situ mapping of microstrain and microstructure evolution reveals damage resistance criteria in dual phase steels. *Acta Materialia*, **96**, 399−409.

Zaafarani, N., Raabe, D., Singh, R.N., Roters, F. and Zaefferer, S. (2006) Three-dimensional investigation of the texture and microstructure below a nanoindent in a Cu single crystal using 3D EBSD and crystal plasticity finite element simulations. *Acta Materialia*, **54**, 1863−1876.

Zaefferer, S. (2007) On the formation mechanisms spatial resolution and intensity of backscatter Kikuchi patterns. *Ultramicroscopy*, **107**, 254−266.

Zaefferer, S. and Elhami, N.-N. (2014) Theory and application of electron channelling contrast imaging under controlled diffraction conditions. *Acta Materialia*, **75**, 20−50.

Zaefferer, S., Wright, S.I. and Raabe, D. (2008) 3D orientation microscopy − a new dimension of microstructure characterisation. *Metallurgical and Materials Transactions A*, **39**, 374−389.

Zaefferer, S., Elhami, N.N. and Konijnenberg, P. (2012) Electron backscatter diffraction (EBSD) techniques for studying phase transformations in steels. Pp. 557−587 in: *Phase Transformations in Steels* (E. Pereloma and D.V. Edmonds, editors). Woodhead Publishing, Sawston, Cambridgeshire, UK.

Zambaldi, C., Zaefferer, S. and Wright, S.I. (2009) Characterization of order domains in gamma-TiAl by orientation microscopy based on electron backscatter diffraction. *Journal of Applied Crystallography*, **42**, 1092−1101.

Zhang, C., Li, H., Eisenlohr, P., Liu, W., Boehlert, C.J., Crimp, M.A. and Bieler, T.R. (2015) Effect of realistic 3D microstructure in crystal plasticity finite element analysis of polycrystalline Ti-5Al-2.5Sn. *International Journal of Plasticity*, **69**, 21−35.

Zhang, T., Collins, D.M., Dunne, F.P.E. and Shollock, B.A. (2014) Crystal plasticity and high-resolution electron backscatter diffraction analysis of full-field polycrystal Ni superalloy strains and rotations under thermal loading. *Acta Materialia*, **80**, 25−38.

EMU Notes in Mineralogy, Vol. 16 (2017), Chapter 4, 97–130

Spatially resolved materials characterization with TEM

Richard WIRTH

Helmholtz Centre Potsdam, GFZ German Research Centre For Geosciences, 4.3, Telegrafenberg, 14473 Potsdam, Germany, e-mail: wirth@gfz-potsdam.de

Transmission electron microscopy (TEM) is an ideal tool for characterizing materials in terms of chemical composition, micro- and nanostructure as well as crystal structure up to atomic level. The major advantage of TEM is superior spatial resolution. It is possible to obtain chemical composition and structure information from even nm-sized crystals and that allows unambiguous identification of the phases present in a TEM sample. Chemical information is supplied by EDX analysis, electron energy-loss spectroscopy (EELS) and to a small extent by Z-contrast imaging (high-angle annular dark-field imaging – HAADF). Micro- and nanostructure information is provided applying bright-field and dark-field imaging based on diffraction contrast and high-resolution imaging (HREM). Structure information is obtained from selected area electron diffraction (SAED), convergent beam electron diffraction on nm-sized crystals, precession electron diffraction which allows crystal-structure determination and fast Fourier transforms (calculated diffraction-pattern FFT) from high-resolution images.

1. Introduction

This chapter on spatially resolved materials characterization with transmission electron microscopy (TEM) is not to be considered as another contribution to the already existing excellent textbooks on electron microscopy (Hirsch *et al.*, 1977; Putnis, 1992; Williams and Carter, 1996; Joy *et al.*, 1986; Fultz and Howe, 2001). Rather, it is more focused on applications of TEM methods in the geosciences. Detailed theoretical contributions to the different chapters can be found in the textbooks referenced in the text.

TEM is an ideal tool for identifying unambiguously micrometre- or nanometre-sized crystalline or non-crystalline objects based on their crystal structure and their chemical composition. Phase identification is important but can be demanding, especially the investigation of nanometre-sized inclusions in minerals. For example, nano-inclusions discovered in diamond established novel ideas on the formation of diamonds. The chemical composition of the nano-inclusions found in diamond allowed reconstruction of the fluid or melt composition from which diamond has nucleated and grown (Klein BenDavid *et al.*, 2006; Logvinova *et al.*, 2008; Kaminsky *et al.*, 2009, 2015). It needs to be emphasized, however, that this kind of investigation was feasible only when the focused ion beam (FIB) sample preparation technique was made available for TEM sample preparation. FIB is a site-specific sample preparation technique that allows for sputtering electron transparent foils out of almost any material (Wirth, 2004, 2009).

The basic principle of TEM is that accelerated electrons with a typical kinetic energy of 200 keV penetrate through an electron transparent slice of matter. With geomaterials, FIB-prepared foils of minerals or rock slices are usually 150 nm thick. However, they can also be fabricated thicker as in the case of diamond (up to 250 nm) or substantially thinner (as small as 30 nm) if required. The electron beam penetrating through the sample interacts with the atoms constituting the specimen resulting in diverse scattering events. Recording the different interactions of the primary electrons with electrons or nuclei of the specimen applying suitable detectors allows extraction of structural and chemical information from the sample (Fig. 1). Principally, there are elastically scattered electrons (no momentum transfer from the primary electron onto the target electron) and inelastically scattered electrons (momentum transfer from the primary electron to the target electron or nucleus. Bragg scattered electrons are elastically scattered primary electrons scattered at the outer shell electrons of target atoms thus providing information about the crystal structure and symmetry of the target. Electrons that are elastically scattered at inner shell electrons can be used for Z-contrast imaging using high-angle annular dark field detectors (HAADF). The scattering angle and scattering intensity depend on the atomic number of the scattering atom. Inelastic scattering occurs when a primary electron transfers enough momentum onto an inner shell target electron to excite it into unoccupied states or into the vacuum. Instantaneously, the empty site will be filled with an outer shell electron and the energy difference is emitted as an X-ray photon. Chemical information about the sample can be derived from the excited X-ray photons that are collected with an energy dispersive X-ray detector (EDX). However, excited X-ray photons contain enough energy to excite outer-shell electrons of the same atom, which can then be recorded as Auger electrons (Auger electron spectroscopy). Auger electrons probe the surface chemistry of a specimen. Because of their comparatively low energy they will be absorbed very rapidly if they are excited at a greater depth of the sample. The energy-loss of primary electrons

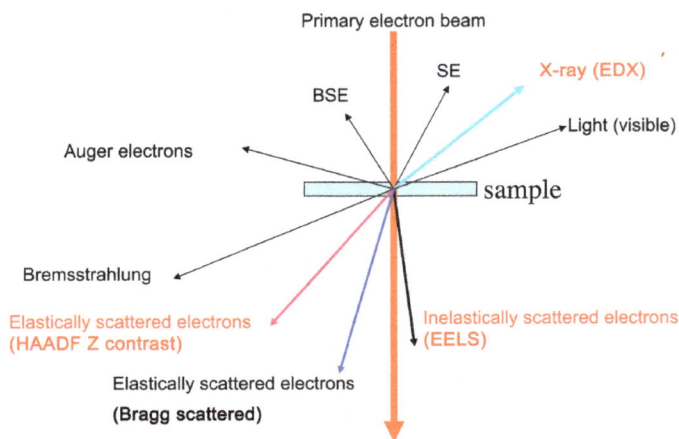

Fig. 1. The interaction of high-energy electrons with matter causes several scattering events that can be used in TEM.

during inelastic scattering can be measured with electron energy-loss spectroscopy (EELS). Inelastic scattering is predominantly forward scattering under low scattering angles (<1°). Secondary electrons (SE), and back-scattered electrons (BSE) usually are not used in TEM applications. Details of the interactions of accelerated electrons with matter are given elsewhere (Williams and Carter, 1996; Fultz and Howe, 2001).

Full characterization of nano-inclusions in minerals requires knowledge of chemical composition and structural properties. Chemical information from the specimen can be derived from EDX analysis, EELS and Z-contrast imaging (HAADF). Diffraction contrast imaging (grain boundaries, dislocations, stacking faults, antiphase domain boundaries), electron diffraction and high-resolution imaging (HRTEM) provide structural information.

2. Chemical composition of materials

2.1. EDX analysis

X-ray photons are generated when an incident primary electron (*e.g.* 200 keV) transfers enough energy to a core electron to eject it from its energy level into an unoccupied state or into the vacuum. If the electron is ejected into the vacuum the atom is ionized. For example, an electron from the innermost copper K-shell requires the energy of 8.98 keV to be ionized (Williams and Carter, 1996). An outer shell electron of higher energy level falls into this empty state thus emitting the energy difference as an X-ray photon. In the case of a $CuK\alpha_1$ X-ray photon, which is emitted during the transition of an electron from the L_3 level to the K-shell of a copper atom, the energy released is 8.04778 keV (Fultz and Howe, 2001). The X-ray photon that is emitted can be detected using a Li-doped silicon detector. The X-ray photon creates an electron−hole pair in the Si detector. A high voltage applied to the detector causes the electrons moving towards the positive end and the hole that can be considered as positive charge towards the negatively charged side of the silicon crystal thus creating an electrical pulse that can be registered by the detector. More details on Si-Li detectors and their working principles are presented elsewhere (Joy *et al.*, 1986; Williams and Carter, 1996).

The X-ray intensities of the different K or L or M lines of the elements constituting a sample can be measured qualitatively or quantitatively using standard material. The X-ray intensity of a particular element concentration depends on the X-ray fluorescence yield of the element. The X-ray fluorescence yield, ω, is a measure for the number of electrons needed to excite one single X-ray photon (Fig. 2). For example, for carbon ($Z = 6$) $\omega = 10^{-3}$, *i.e.* 1000 electrons are needed to excite 1 X-ray photon. An element with a higher atomic number such as Germanium (Ge) requires only 2 electrons to excite 1 X-ray photon ($\omega = 0.5$). For elements with an atomic number <30, Auger electrons will predominate over X-ray photons thus resulting in a low X-ray fluorescence yield of the element.

2.1.1. Quantitative analysis in analytical electron microscopy (AEM)
Quantitative analysis in TEM is based on the Cliff-Lorimer equations (Cliff and Lorimer, 1975):

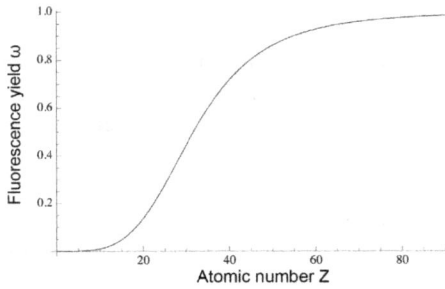

Fig. 2. X-ray fluorescence yield depending on the atomic number Z of an element.

$$\frac{C_A}{C_B} = k_{AB}\frac{I_A}{I_B} \tag{1}$$

$$C_A + C_B = 1 \tag{2}$$

The Cliff-Lorimer equations relate the ratio of the concentration of two elements A and B in the sample to the measured X-ray intensities, I, of the elements A and B in the sample. The factor k_{AB} needs to be determined from a standard reference material. It considers the ionization cross-section σ, the fraction of the total line intensity measured, the absorption of the X-ray photons depending on sample thickness and some detector properties. Quantitative X-ray analysis with TEM is described in great detail in the textbook by Joy, Romig and Goldstein (1986). Standard reference materials for calibration of the k_{AB}-factor can be a mineral grain that was analysed by the electron probe microanalysis (EPMA), or a part of the sample was analysed by some other chemical method (*e.g.* wet chemistry, inductively coupled plasma mass spectrometry [ICPMS], *etc.*) to determine its chemical composition. After that, a TEM foil can be cut with FIB technique from the same area that was measured with the microprobe. That TEM foil is then used as a standard reference material. Because of the strong influence of matrix effects on the detected X-ray intensity, it is crucial to use standard materials with chemical composition and structure at least close to that of the unknown sample. The number of the k_{AB}-factor is valid only if the sample foil has approximately the same thickness as the standard foil, because the value of the k_{AB}-factor increases with increasing foil thickness (Fig. 3). In geomaterials k_{AB} factors are usually based on Si (B), and in materials sciences very often Fe is used as reference (B).

Focused ion beam-prepared TEM foils are ideal for line scans and element mapping because of their homogeneous thickness. The measured intensities of individual X-ray lines in element maps or line scans are interpretable only if the sample thickness does not change and the X-ray intensity is related only to the concentration of the element in the sample and not to varying sample thickness.

2.1.2. Detection limit of elements analysed and total error calculation of the results
The detection limit of an element in energy dispersive X-ray analysis (EDX) is an important factor when measuring low element concentrations. In general, the detection

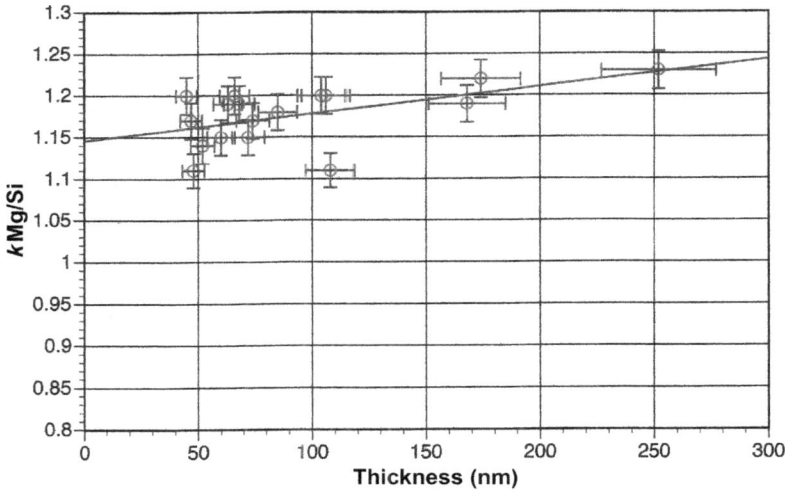

Fig. 3. Dependence of the k_{AB} factor — here $k_{Mg/Si}$ — on the thickness of the TEM foil.

limit is much higher (100s of ppm) than that of electron microprobe analysis (10s of ppm) using wavelength dispersive spectroscopy (WDS). Before the concentration detection limit can be determined it is mandatory to decide if the observed peak intensity is statistically valid peak intensity. A peak is statistically valid if the following criterion is fulfilled:

$$I_A - I_b^A \geq 3\sqrt{2I_b^A} \tag{3}$$

where I_A is the measured intensity of the element A in the EDX spectrum and I_b^A is the background intensity of that peak. The concentration detection limit C_{DL} of an element with the known concentration C_A in a sample is given by the equation:

$$C_{DL} = C_A \frac{3\sqrt{2I_b^A}}{I_A - I_b^A} \quad \text{(Goldstein and Williams, 1989)} \tag{4}$$

For example, the detection limit of Pt-group elements in sulfide matrix using EDX is ~700 ppm (Wirth *et al.*, 2013).

2.1.3. Total error in X-ray analysis analytical electron microscopy (AEM)
Another very important criterion to evaluate data from EDX analyses is to evaluate the statistical error and the absolute error of the measurement. Quantitative EDX analysis requires calculation of the statistical and the absolute errors for both the k_{AB} factor and the concentration of the elements in the sample and addition of both errors. Low X-ray intensities resulting in poor counting statistics are a major source of significant statistical error. Gaussian behaviour describes the counting statistics. That is $\sigma = \sqrt{N}$

with N the number of accumulated counts. At the 3σ confidence level (99.73%) the error in the number of accumulated counts is $3\sigma = \sqrt{N}$. Then, the relative error in the number of counts is $3\sqrt{N}/N \times 100$. A relative error in the intensity of a particular element of $\pm 3\%$ is achieved by 10,000 accumulated counts above background. The error in the measured concentration of an element in a mineral is larger in a single measurement than in n individual measurements. The same holds true for the determination of the k_{AB} factor. Therefore, n measurements should be done to minimize the error. The relative error of k_{AB} based on a 99.73% confidence level is determined as follows:

$$\% \text{ error} = \left[\frac{t_{99}^{n-1}}{\sqrt{n}} \frac{s}{\bar{k}_{AB}} \right] \times 100 \tag{5}$$

In that equation $t_{99}{}^{n-1}$ is the Student t value calculated from n analyses at the 99.73 % confidence level (3σ) (cf. Brownlee, 1960), s is the standard deviation from n measurements, and $\bar{k}_{AB}$ is the mean of k_{AB} from n spectra. That calculation accounts for the relative error of the k_{AB} factor. Error calculation of the concentration of an element in a mineral with n measurements follows the same equation.

$$\% \text{ error} = \left[\frac{t_{99}^{n-1}}{\sqrt{n}} \frac{s}{\bar{C}_A} \right] \times 100 \tag{6}$$

Here again s is the standard deviation of n individual measurements and C_A is the mean of the concentration of the element A from each of the n spectra.

Errors add up, that means the relative error of k_{AB} and the relative error of C_A must be added up to a total error. Table 1 shows an example of the error calculation of MgO, Al_2O_3, CaO and TiO_2 measured from orthopyroxene using an orthopyroxene standard. Details and examples of error calculation are presented elsewhere (Brownlee, 1960; Goldstein and Williams, 1989; Williams and Carter, 1996).

2.2. High-angle annular dark-field imaging (HAADF)

The interaction of primary electrons with matter results in scattering of the primary electrons at the inner and outer shell (K) electrons of an atom. The scattering angle of the inner shell electrons is larger ($>>10$ mrad) than those scattered at the outer shell electrons (<10 mrad). A high-angle annular dark field detector (HAADF), a semiconductor device, collects the scattered electrons. The intensity of the primary electrons scattered at the inner shell electrons of an atom and detected by HAADF depends on the atomic number (Z). A detector that collects electrons scattered at the inner shell electrons of an atom produces a Z-contrast image. A typical HAADF detector is a semiconductor device designed as a disk with a hole in the centre. Because of the much smaller scattering angle of the electrons that were scattered at the outer shell electrons of an atom (Bragg scattered electrons causing diffraction contrast) the Bragg scattered electrons might fall through the hole and will not contribute to the image contrast. It depends on the selected camera length of the TEM and which kind of electrons is collected using the HAADF detector. Using a short camera length (e.g.

Table 1. Error calculation of an EDX analysis from OPX.

Error of the k_{AB} factor from n measurements

$n = 17$	$k_{Mg/Si}$	$k_{Al/Si}$	$k_{Ca/Si}$	$k_{Ti/Si}$
$\bar{x}$	1.18	1.05	0.88	0.78
s	0.035	0.103	0.173	0.692
Error (%)	1.85	6.11	12.2	55.2
$n = 17$	MgO	Al_2O_3	CaO	TiO_2
$\bar{x}$	33.16	4.38	0.89	0.12
s	0.56	0.40	0.20	0.09
Error (%)	1.05	5.7	14.5	48.4
Total rel. error (%)	2.9	11.81	26.7	103.6

$\bar{x}$ = mean of n measurements
s = standard deviation

75 mm) the detector collects the high-angle scattered electrons only, thus providing a Z-contrast image (Fig. 4). A camera length in the range of 220–330 mm collects both high-angle scattered electrons (Z contrast) and Bragg scattered electrons thus displaying Z-contrast plus diffraction contrast (dislocations, grain boundaries) (Fig. 4).

The resolution of the HAADF imaging depends on the sample thickness and should not exceed 40 nm for high-resolution imaging. Recorded scattering intensity depends also on sample thickness. Pores in a sample are visible in the image as slightly darker contrast because at the location of the pore the sample thickness is reduced and thus fewer electrons are scattered. Note that in TEM bright-field (BF) imaging the contrast is reversed and pores will appear brighter than the matrix. In that case, the reduced thickness will cause less absorption of electrons resulting in a brighter image contrast. HAADF imaging is very sensitive for nano-inclusions of high atomic number elements in a silicate or sulfide matrix. For example, Pb spheres of only a few nanometres in diameter as inclusions in a zircon matrix can be identified easily in a HAADF image even at lower magnification due to the high brightness (Z-contrast) in the HAADF image (Fig. 5) (Kusiak *et al.*, 2015).

2.3. Electron energy-loss spectroscopy (EELS)

Primary electrons with a typical energy of 200 keV pass through the sample interacting with the electrons of the atoms that constitute the sample. Along with these interactions inelastic scattering events occur. During inelastic scattering the primary electron loses energy because it transfers momentum onto a target atom or electron. The energy loss of the primary electrons is measured with electron energy-loss spectroscopy (EELS).

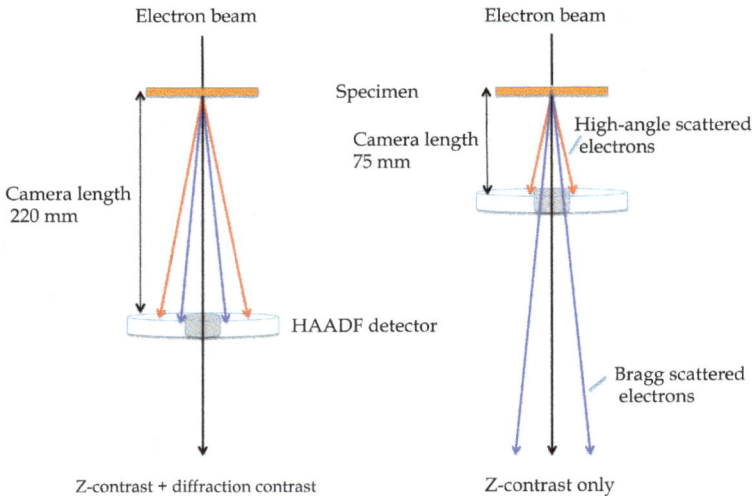

Fig. 4. The position of the HAADF detector defined by the camera length determines which kind of scattered electrons are detected. At a small camera length (*e.g.* 75 mm) only those electrons that were scattered at the inner shell electrons of the sample are collected displaying a Z-contrast image. At a larger camera length (*e.g.* 220 mm) additionally, the Bragg scattered electrons are collected thus producing a Z-contrast plus diffraction contrast image.

A typical EEL spectrum as shown in Fig. 6 can be subdivided in three different energy regions:

1. The *zero-loss peak* representing elastically forward scattered electrons or non-scattered electrons without energy-loss.
2. The *low-loss region* with the plasmon peak extending up to an energy region of ~50 eV representing electrons that have interacted with the weakly bonded outer-shell electrons or the delocalized electrons in a sample.
3. The energy region above 50 eV, the so-called *core-loss region*, represents those primary electrons that have interacted with the more tightly bonded electrons of the inner shells (K, L, M) of an atom.

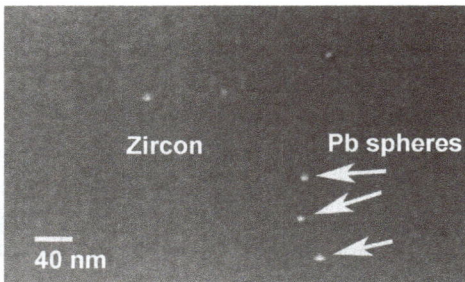

Fig. 5. HAADF Z-contrast image displaying Pb-nanospheres in zircon.

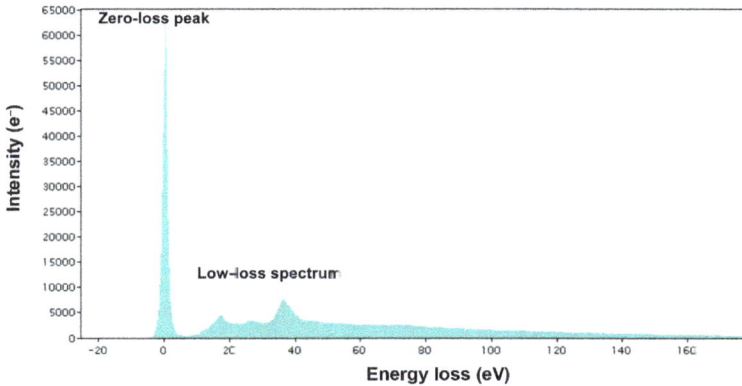

Fig. 6. Electron energy-loss (EEL) spectrum. Zero-loss peak that represents the unscattered and elastically scattered electrons plus low-loss spectrum.

The different energy-loss regions provide special and characteristic information about the atoms constituting the specimen. The zero-loss peak can be used to determine the energy resolution of the spectrometer. For example, electrons emitted from a field emission gun (FEG) electron source usually have an energy resolution of ~0.9 eV (full width at half maximum of the peak (FWHM)).

The plasmon peak characterizes the low-loss spectrum. In that case the primary electrons transfer energy to the weakly bonded outer shell electrons or the free electron gas in metals thus causing longitudinal collective oscillations of these electrons. The typical energy loss of these electrons is in the range 5–40 eV (Fig. 6).

An electron can be ejected into the vacuum, thus ionizing the atom, when primary electrons transfer sufficient energy to an electron of the K, L, M or an outer shell. The energy loss due to ionization or just to excite an inner shell electron to an unoccupied state in an outer shell level defines the high-loss spectrum or core-loss spectrum (Fig. 7). In Fig. 7 the core loss spectrum of NiO is displayed in the energy range of the Ni-$L_{3,2}$ edge. The edge onset of the Ni-L_3 edge is at 855 eV. One of the 2s or the 2p electrons of the L-shell of Ni has been excited and it could fill an unoccupied 3d state (Ni has two unoccupied 3d states). To excite a 1s electron from the oxygen K-shell 532 eV must have been transferred from the primary electron to the excited electron. In the case of oxygen 532 eV is the critical ionization energy, E_c, to overcome the ionization threshold. The 1s electron at the K-shell can be ejected into the vacuum or into an unoccupied 2p state (oxygen has 2 unoccupied p states on the L shell). The scattering angle

$$\Theta_E = \frac{E}{2E_0} \qquad (7)$$

of the scattered primary electron is in general very small, and in the case of oxygen it is 1.3 mrad. E is the energy-loss of the primary electron at the onset of the edge (*e.g.* O-K edge 532 eV) after momentum transfer, and E_0 is the electron energy of the incident

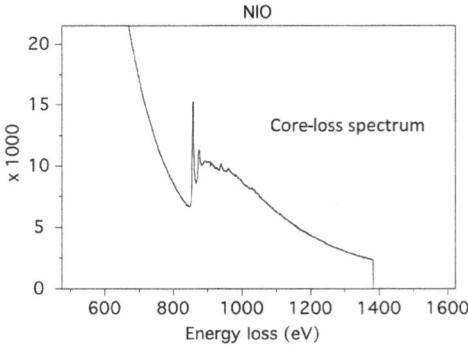

Fig. 7. EEL spectrum. Core-loss spectrum showing the Ni-L3,2 edge.

electron beam (200,000 eV). The critical ionization energy, E_c, defined by the edge onset has the greatest probability of ionizing an atom. That means, with increasing transferred energy the probability of ionizing an atom or exciting a 1s electron into an unoccupied state decreases. The scattering cross section σ for such a scattering event decreases with increasing energy-loss thus reducing the probability of ionization. In the case of a hydrogen atom the energy-loss spectrum would resemble a saw tooth curve with a sharp onset and a gradual decrease with increasing energy-loss (Fig. 8a). However, in a sample with finite thickness the primary electron will experience multiple scattering resulting in a plural scattering background (Fig. 8b). The core-loss edge is superimposed onto the background. Consequently, good quality EEL spectra require thin samples (<<100nm) for good peak/background ratios. Usually, the total edge shape, which is termed the energy-loss near edge structure (ELNES), extends ~50 eV from the edge onset to higher energy losses. The ELNES of a particular atom in a crystal or any solid-state material is more complicated and contains information about the electronic structure of the atom. Iron is a good example to demonstrate that. The edge onset can vary when changing the valence state of an ion, which is then called 'chemical shift'. For example, a valence change of iron from Fe^{2+} to Fe^{3+} will alter the

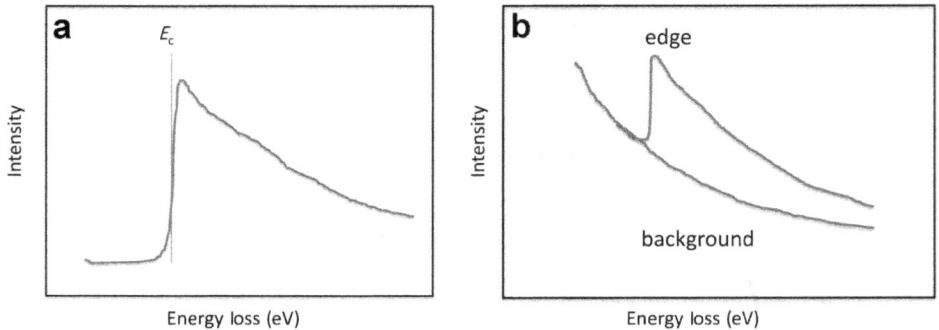

Fig. 8. EEL spectrum. (a) Idealized hydrogen K-edge spectrum with the critical energy E_c to remove the 1s electron from its K-shell position. (b) Idealized core-loss edge superimposed on the background.

screening of the positive charge of the nucleus by the electrons. If there is one electron less present (Fe^{3+}) screening the charge of the nucleus, the inner shell electrons from the K, L and M shells will be bonded slightly more strongly to the nucleus, thus higher energy is required to excite an electron from the L shell to an empty state on the 3d level of the M shell. In the EEL spectrum of Fe^{3+} the onset of the $L_{3,2}$ edge is shifted ~1.7 eV to higher energies (707.8 eV Fe^{2+} and 709.5 eV Fe^{3+}; van Aken *et al.*, 1998). The valence state of an ion can be determined from an EEL spectrum.

ELNES can also be used as a fingerprint to identify individual phases. This has been proposed and demonstrated previously (Taftø and Zhu, 1982; Rask *et al.*, 1987; Kurata *et al.*, 1993; Garvie *et al.*, 1994). ELNES of the carbon C-K edge is a very good example of how the edge structure can be used as a fingerprint to identify a polymorph. The C-K edges of amorphous carbon, graphite and diamond are significantly different in edge-onset and near edge structure and can be used to discriminate between the carbon phases without using electron diffraction. The basic component of graphite and amorphous carbon are planar rings with 6 carbon atoms. Graphite and amorphous carbon are characterized by strong σ-bonds or sp^2 hybrid orbitals within the hexagonal ring structure. Additionally, there are non-hybridized p_z orbitals forming the π-bonds normal to the hexagonal sheets. The EEL spectrum of graphite and amorphous carbon contains and displays the information about the two different bond types σ- and π-bonds. In the EEL spectrum, there is a pre-peak with an edge-onset at 284 eV, which represents the π-bonds, followed by a stronger peak with an onset at ~290 eV representing the sp^2-bonds or σ-bonding (Fig. 9a). The only difference between the EEL spectra of graphite and amorphous carbon is that the sp^2-related peak is more structured showing 3 peaks, whereas the sp^2-related peak in amorphous carbon is smooth and rounded (Fig. 9b). The π-bond related peak at 284 eV appears in both the spectra of graphite and amorphous carbon. Pure sp3 hybrid bonds without π-bonding characterize diamond. Therefore, in diamond, the pre-peak due to π-bonds is absent and the edge-onset of the C-K edge representing the sp^3 bonds is at ~289 eV followed by a second peak with its maximum at 299.8 eV and a third peak with a maximum at ~305 eV (Fig. 9c). The carbon C-K edge allows identification of amorphous carbon, graphite or diamond using the ELNES of the carbon C-K edge as a fingerprint.

The individual core-loss edges can be used for high-resolution element mapping. There are two methods of element mapping: a three-windows technique and a two-windows jump ratio technique. The basic procedure of the three-windows technique is as follows (using the O-K edge as an example). An image is acquired applying an energy window of 30 eV to the O-K edge with an offset of 15 eV towards higher energy loss from the edge onset (532 eV). Additionally, two windows each 30 eV wide are placed at lower energy loss than the edge onset with an offset of 18 eV and 48 eV, respectively, in regard to the onset of the O-K edge. The two pre-edge windows are used to calculate the background intensity and to subtract the background intensity from the intensity at the O-K edge (*Digital Micrograph* software package®). The residual O-K edge intensity represents the oxygen concentration in the image.

The two-windows technique is different. In that case, only two images are acquired with a defined width of the energy windows and offset from the edge onset. The post-

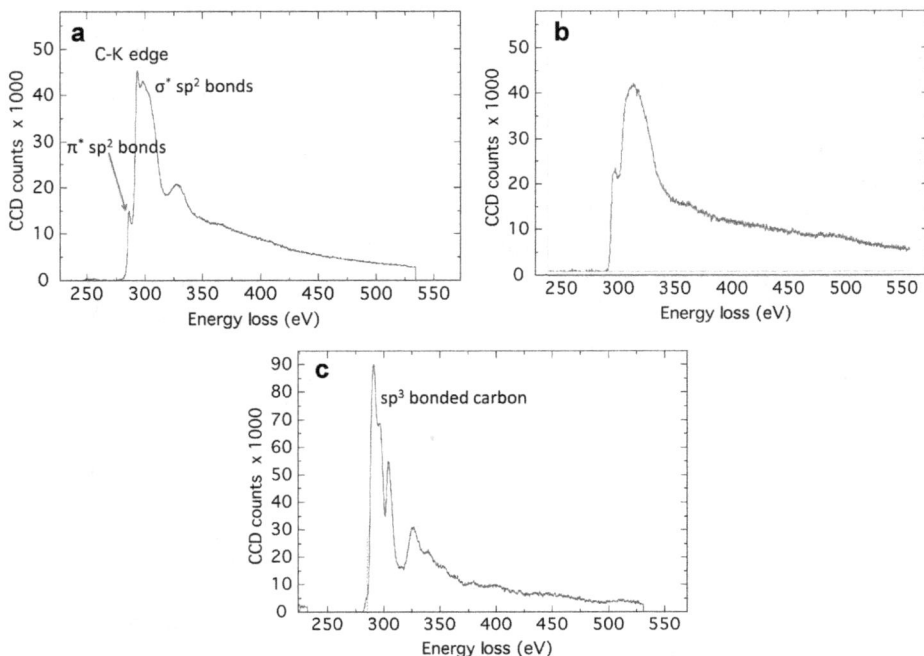

Fig. 9. EEL spectrum of carbon C-K edge. (a) Carbon C-K edge of graphite. Note the pre-edge peak (arrow) that indicates the π bonds in graphite. (b) Carbon C-K edge of amorphous carbon. Both bond types π and σ bonds are visible; the pronounced edge structure of the σ bonds is converted into a rounded peak, however. (c) Carbon C-K edge structure of diamond. The signal from the π electrons is almost absent and the σ bonds are replaced by sp^3 bonding. The signature of the carbon C-K edge can be used as a fingerprint of the carbon phase present.

edge image is divided by the pre-edge image creating a jump ratio image, which is very sensitive to thickness variations. An example of element mapping with EELS using the three windows technique is shown in Fig. 10a,b. A narrow platelet in diamond, ~300 nm wide, is composed of graphite and Fe-carbide (Fe_3C). Element maps of the C-K edge (Fig. 10b) and the Fe-$L_{3,2}$ edge (Fig. 10a) show the distribution of graphite and Fe-carbide within that narrow platelet on a nanometre scale (Kaminsky and Wirth, 2011).

Another useful application of EELS is the measurement of the local sample thickness of a TEM foil. The knowledge of the foil thickness is required for quantitative EDX analysis incorporating absorption correction. The amount of inelastic scattering in a sample increases with increasing specimen thickness. That means that the ratio between the intensity of the primary unscattered electrons and/or elastically scattered electrons and the intensity of the inelastically scattered electrons contains information about the sample thickness. The sample thickness, t

$$t/\lambda = \ln(I_{tot}/I_0) \tag{8}$$

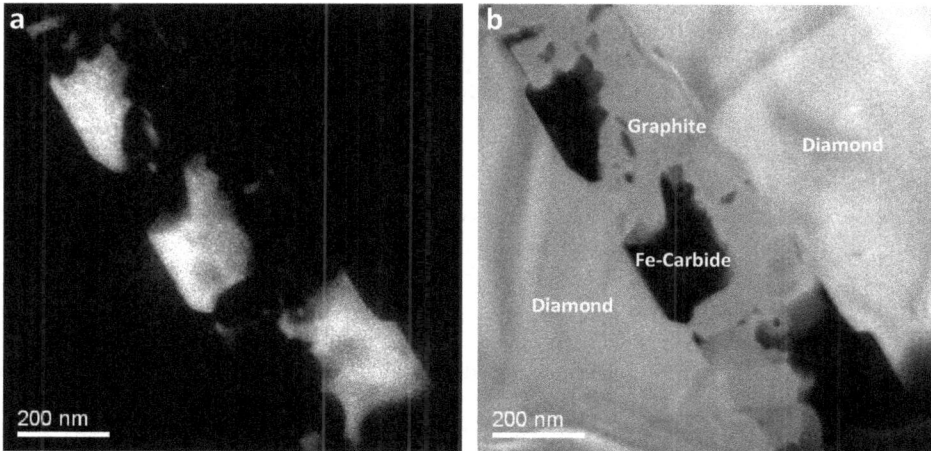

Fig. 10. EELS element mapping applying the three-window technique. Fe carbide precipitated in diamond. (a) Element map of Fe using the Fe-$L_{3,2}$ edge. (b) Element map of carbon using the C-K edge.

is the ln of the ratio of the total intensity of the spectrum, I_{tot}, and the intensity of the zero loss peak, I_0, multiplied by the mean free path, λ, of the electrons in the sample. Details of the calculation of λ are presented elsewhere (Egerton, 1996; Williams and Carter, 1996). There are several other applications of EELS, which are beyond the scope of this chapter and can be found elsewhere together with many fundamental details of EEL spectroscopy and technical descriptions (Joy *et al.*, 1986; Egerton, 1996; Williams and Carter, 1996).

3. Structural information

3.1. Electron diffraction

Primary electrons penetrating through a specimen are scattered elastically at the outer shell electrons of the atoms constituting the sample. This kind of electron scattering is called Bragg scattering, and if the conditions of the Bragg equation

$$n\lambda = 2d\sin\theta \qquad (9)$$

are fulfilled, diffraction intensity can be recorded (film, image plate, CCD camera). In the Bragg equation n is the order of a reflection (*e.g.* (100), (200)....), λ is the wavelength of the electrons at a given accelerating voltage (*e.g.* 0.00251 nm relativistic wavelength at 200 keV), d is the spacing of the reflecting lattice plane (*hkl*), and θ is the scattering angle. In conventional electron diffraction patterns the specimen is illuminated with a parallel electron beam using the condenser lenses to adjust illumination. The objective lens focuses the diffracted electrons in the back focal plane forming the first diffraction pattern, which is finally magnified and imaged onto the fluorescence screen by the subsequent lens systems. There are two techniques to generate electron diffraction patterns. The most commonly used technique is the selected area electron diffraction pattern (SAED) with parallel illumination of the sample. SAED contains structural information from a scattering volume that is defined by a diffraction aperture (10 up to 800 μm in diameter). The SAED pattern consists of discrete and symmetrically arranged scattering intensity spots (Fig. 11a). The convergent beam electron diffraction (CBED) technique uses a focused electron beam thus allowing electron diffraction from even nanometre-sized crystals. The scattering volume is defined by the size of the electron beam and not by an aperture. The electron diffraction pattern is composed of diffraction discs in the back focal plane (Fig. 11b). The diameter of the discs depends on the electron beam size and the camera length used. The CBED patterns are evaluated using the same procedure as for the SAED pattern. Measuring the distances between the diffraction spots or between the discs provides the length of a diffraction vector $\vec{g}$ in reciprocal space. The d spacing d_{hkl} (real space) is calculated by applying the equation

$$\lambda L = \vec{g} d_{hkl} \tag{10}$$

with L the camera length (mm), λ the relativistic wavelength of the accelerated electron (200 kV; 0.00251 nm), $\vec{g}$ (mm) and d_{hkl} (nm). Usually, electron diffraction patterns do not represent the true scattering intensity because of dynamical scattering conditions (thick sample); therefore, crystal symmetry and crystal structure determination is almost impossible. That problem has been overcome by introducing precession electron diffraction (Vincent and Midgley, 1994). The precession technique is equivalent to the Buerger precession technique applied in X-ray crystallography where the specimen is precessed with respect to the incident X-ray beam (Buerger, 1970). In a TEM, the sample is static and the electron beam is precessing around the optical axis on a conical surface. As a result of that precession more reflections in the reciprocal space appear, the scattered intensity is closer to the real scattering intensity and the final diffraction pattern is much closer to true kinematic diffraction

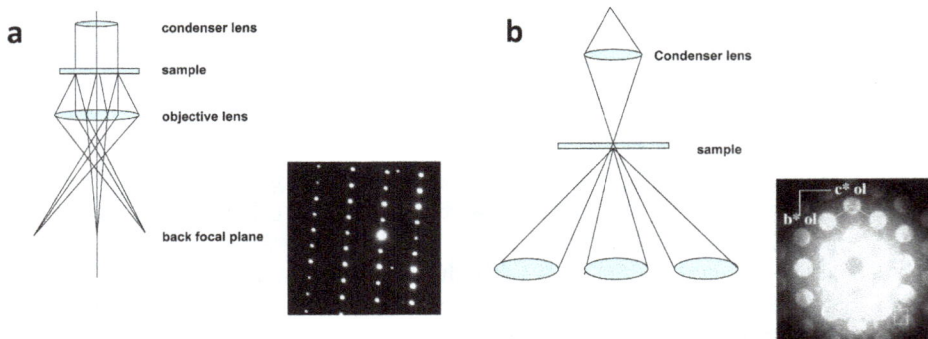

Fig. 11. Electron diffraction. (a) Microscope setting for selected area electron diffraction (SAED). Note the diffracted electrons form sharp reflection spots on the viewing screen. (b) Microscope setting for convergent beam electron diffraction (CBED). Note the diffraction discs.

conditions. With precession electron diffraction the occurrence of forbidden reflections due to multiple scattering is reduced significantly thus allowing determination of the space group. With the precession electron diffraction technique it is possible to perform structure analysis even from nanometre-sized crystals (Koch-Müller *et al.*, 2014; and ref. therein). The major problem with that technique is that electron irradiation sensitive material might become amorphous during acquisition of the diffraction pattern due to irradiation damage.

With nanocrystalline and electron irradiation sensitive material there is only one option to get structural data. High-resolution lattice fringe images can be acquired from different low-indexed zone-axis orientations with very short (0.1 s) acquisition time. From the lattice fringe image a diffraction pattern can be calculated using fast Fourier Transform (FFT) which provides the same information as an electron diffraction pattern.

3.1.1. Diffraction contrast in TEM

Without the use of an objective aperture, the contrast of TEM bright-field image is very poor because only mass absorption contrast is relevant for the image contrast. The objective aperture is located at the back focal plane where the diffraction pattern is displayed. There are objective apertures available in the range 10–800 μm in diameter. Placing the objective aperture with the smallest diameter (10 μm) around the central spot representing the non-scattered electrons from the primary beam excludes all other diffracted electrons from image formation (Fig. 12a). That kind of image is called a bright-field (BF) image with strong image contrast. The second option to place an objective aperture is to position it over one of the diffracted intensities (Fig. 12b). This imaging mode is called dark-field (DF) imaging. The diffracted intensity only is used for image formation. For example, dark-field imaging is useful for imaging a particular crystal in fine-grained material, exsolution lamellae, twins, dislocation lines, and grain- or phase boundary planes.

Another option is to place a large objective aperture such that it includes the undiffracted intensity together with different scattered intensity with the objective aperture centred at the undiffracted primary beam. Under these conditions all diffracted intensity within the aperture plus the undiffracted intensity contribute to image formation. That kind of imaging is called phase contrast imaging. It is used for high-resolution imaging. More details about phase contrast imaging will be presented in the section on high-resolution imaging.

Diffraction contrast and the basic physics of this method are described in more detail elsewhere (Hirsch *et al.*, 1977; Edington, 1975a,b; Williams and Carter, 1996; Fultz and Howe, 2001).

3.2. Information derived from electron diffraction patterns

Electron diffraction patterns contain the following important structural information:
    Crystalline or non-crystalline state
    Single crystal or polycrystal (*e.g.* nano-diamond)
    Lattice parameters

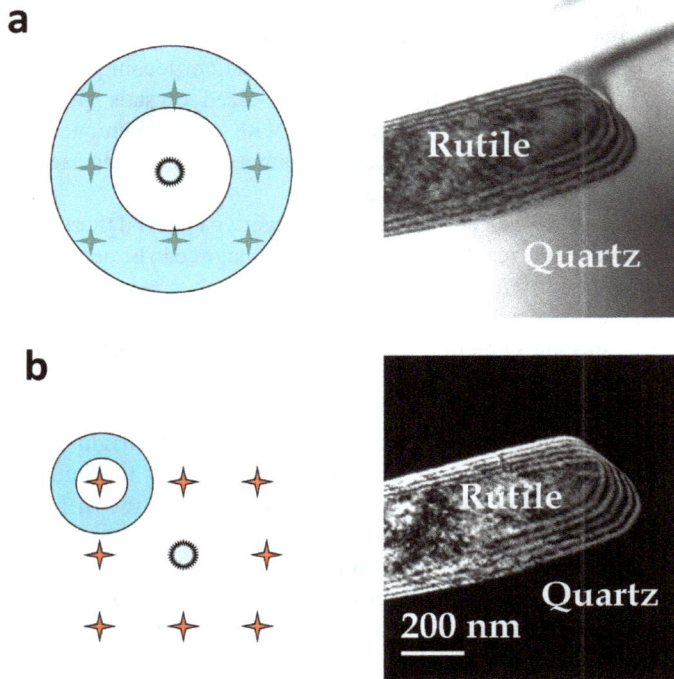

Fig. 12. Bright-field (BF) and dark-field (DF) imaging in TEM. (a) In bright-field imaging the objective aperture excludes all diffracted intensity from image formation. Only the intensity of the primary electron beam passes through the objective aperture. The corresponding BF image of a rutile needle in quartz is shown on the right hand side of the image. The sequence of dark and bright contrasted lines along the interfaces represents thickness fringes. The sequence of contrast lines starts with bright contrast followed by a dark contrast. (b) Using one of the diffracted intensities for image formation only, excluding all other intensities with the objective aperture is called DF imaging. The corresponding DF image on the right hand side of the image shows again the alternating bright and dark contrast at the interface rutile/quartz. In the dark-field image the sequence of thickness contrast starts with dark contrast followed by bright contrast lines.

   Crystal symmetry and space group
   Ordering of atoms in the crystal lattice (*e.g.* hydrous olivine)
   Orientation relationship of two crystals (*e.g.* Mg-ferrite and ferro-periclase)
   Exsolution by spinodal decomposition (*e.g.* satellite reflections feldspar)
   Mosaic crystals

3.2.1. Crystalline or non-crystalline state

Diffraction patterns display the crystalline or non-crystalline state of the material investigated. Crystals reveal regular and symmetric discreet scattering intensities whereas non-crystalline matter shows only broad and diffuse scattering intensity rings. However, nanocrystalline materials with randomly oriented crystals and average grain size of <5 nm also display only broad scattering intensity along Debye-Scherrer rings. There is substantial peak broadening in X-ray diffraction analysis with decreasing grain size from 100 to 1 nm (Glocker, 1971). However, peak broadening can also be caused by elastic strain in the crystals. If there is only broad scattering intensity in the diffraction pattern the amorphous or nanocrystalline state of the material can be verified with high-resolution lattice fringe imaging searching for the presence or absence of lattice fringes. Only if there are no lattice fringes visible is it possible to identify unambiguously the non-crystalline state of a material (Toy *et al.*, 2015). More detailed information about diffraction line broadening can be found elsewhere (Glocker, 1971; Fultz and Howe, 2001).

3.2.2. Single crystal or polycrystalline material
The diffraction pattern can also indicate the presence of a single crystal or polycrystalline material. Randomly oriented crystals create Debye-Scherrer rings in the electron diffraction pattern. Each radius of the individual rings corresponds to a particular $\vec{g}$ vector in reciprocal space and corresponding d spacing in real space can be calculated from equation 10. Usually, the Debye-Scherrer rings are not completely occupied with diffraction intensity because the number of scattering crystals in a typical TEM foil is too small to produce enough diffraction spots to form a complete ring pattern. Only in the case of nanocrystalline material is there a chance to observe the Debye-Scherrer rings completely occupied. If the individual diffraction spots on the ring cluster it suggests preferred orientation of the individual crystals. This has been demonstrated using the electron diffraction pattern (SAED) from nanocrystalline diamond included in orthopyroxene (Wirth and Rocholl, 2003). There is scattering intensity on rings with different radii. These rings can be assigned to (111), (022) and (113) diamond reflections. The clustered intensity on the individual rings suggests a weak preferred orientation relationship between the individual nanodiamonds.

A special case of oriented nanocrystals in a host mineral is presented in Fig. 13a,b. A ferro-periclase inclusion in diamond hosts oriented nanocrystalline precipitations of Mg-ferrite along ancient dislocation lines (Fig. 13a) (Wirth *et al.*, 2014). The diffraction spots from the ferro-periclase host are bright with much weaker spots in between that represent the nanocrystalline Mg-ferrite (Fig. 13b). Nanocrystalline Mg-ferrite behaves like a single crystal. The unit-cell parameter of cubic Mg-ferrite is 0.858 nm and that of the host cubic ferro-periclase is 0.422 nm. Therefore, in reciprocal space the reflections of Mg-ferrite are located half way between the host ferro-periclase reflections. Both phases have the same zone axis orientation [$\bar{1}01$].

3.2.3. Unit-cell parameters
Unit-cell parameters can be derived from electron diffraction patterns acquired in suitable low-indexed zone-axis orientation. In those cases where there are only nanocrystals (<50 nm), which are sensitive to electron irradiation damage, diffraction patterns can be obtained only from high-resolution lattice fringe images in low-indexed zone axis orientation. Fast Fourier Transform processing converts the lattice fringe image into a diffraction pattern. The evaluation of several diffraction patterns from

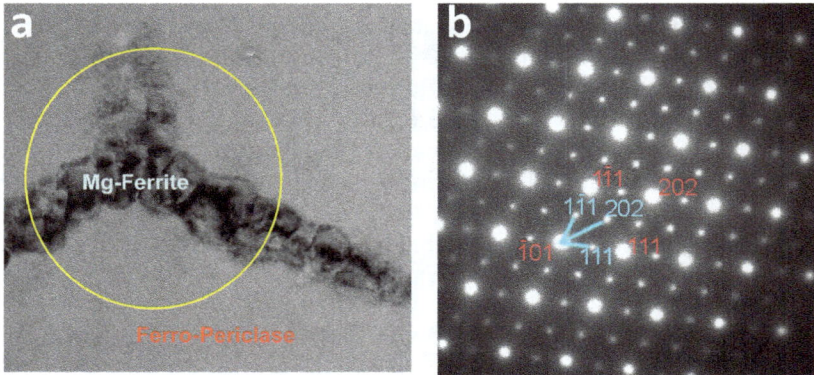

Fig. 13. Nanocrystals of Mg-ferrite precipitated in a ferro-periclase inclusion in deep-seated diamond from Juina, Brazil. (a) BF image of Mg-ferrite with dark diffraction contrast patches in a homogeneous ferro-periclase matrix. Circle indicates the volume used for the electron diffraction pattern. (b) SAED pattern with strong scattering intensity from the ferro-periclase matrix and the weaker scattering intensities from the Mg-ferrite inclusions. Note the rigorous orientation relationship between inclusion and host.

different zone axis orientation can provide the unit-cell parameters of a crystal. This technique has been applied successfully to nanocrystalline inclusions (*e.g.* phase Egg) in diamond and oriented nanocrystalline inclusions of $(Mg,Fe,Cr)TiO_3$ perovskite in olivine (Wirth and Matsyuk, 2005; Wirth *et al.*, 2007). If the nanocrystals are randomly oriented that procedure can be very time consuming as in the case of phase Egg. However, when the nanocrystals are strictly oriented with respect to the host (*e.g.* $(Mg,Fe,Cr)TiO_3$ perovskite in olivine) then there is a chance to observe the very weak scattering intensity of the nanocrystals in a conventional SAED pattern. In that case the scattering intensities from the individual scattering nanocrystals add up to more intensive and finally visible scattering intensity in the diffraction pattern. Image plates are very useful as the recording medium because of the very high dynamic range of the image plates. That means, it is possible to observe very weak scattering intensity from the nanocrystals beside the very strong scattering intensity from the host.

If the nanocrystals are not electron irradiation sensitive the precession technique described above provides the best results in resolving the crystal structure (*e.g.* Koch-Müller *et al.*, 2014; and references therein). The precession technique is also most valuable for the determination of space group and crystal symmetry. Usually, the TEM samples are not thin enough ($\ll 50$ nm) to apply the kinematic scattering theory, which calculates the scattering amplitude scattered by a crystal under conditions in which the amplitude of the scattered wave is a small fraction of the incident wave (Hirsch *et al.*, 1977). If the conditions of a thin foil and single scattering of the primary electron are not fulfilled, the square of the scattering amplitude (which is the scattering intensity),

$$\text{Amplitude} = \Phi g = \frac{\exp(2\pi i k r)}{r} F(\Theta \tag{11}$$

$$F(\Theta) = \sum_j f_j \times e^{i\pi(hx_j + k_y + lz_j)} \tag{12}$$

does not represent the true scattering intensity. Measuring the real scattering intensity is the basic requirement for structure calculations. In equation 11 $F(\Theta)$ is the scattering amplitude from the unit cell, Φ_g is the scattering amplitude of the crystal with r representing the periodic distance in a lattice and k is the wave vector. The indices (hkl) represent the reflecting lattice planes and (x, y, z) are the coordinates of the individual atoms in the unit cell. The atomic scattering factor for an individual atom j is f_j in equation 12. Electron scattering in the usually thicker TEM foils (100–150 nm) is theoretically treated by the dynamical scattering theory. The dynamical theory considers the interaction of the electron wave with the periodic potential of the atoms in the crystal, which results in an attenuation of the scattered intensity with increasing sample thickness and an interaction of the primary electron wave with the scattered electron wave. A detailed treatment of the kinematic and dynamic theory of electron scattering can be found elsewhere (Edington, 1975a,b; Hirsch et al., 1977; Williams and Carter, 1996).

3.2.4. Crystal symmetry and space group
Because we usually use relatively thick samples (100–150 nm) we have to be aware that due to multiple scattering, the extinction rules, which are defined by the crystal symmetry, cannot be applied. That means that, in electron diffraction patterns, forbidden reflections might occur. Olivine is a good example to demonstrate that. The space group of olivine with an orthorhombic structure is *Pbnm*. That space group does not allow reflections such as (100), (010) and (001). However, in real diffraction patterns they might be present. The fact that forbidden reflections might occur in electron diffraction patterns and in FFT from high-resolution lattice fringe images is important when evaluating diffraction pattern and comparing the observed *d* spacing with the calculated *d* spacing. Therefore, it is necessary to calculate also the forbidden *d* spacing otherwise it is very often impossible to index the observed diffraction pattern correctly and thus identify a phase.

3.2.5. Ordering
Hydrous olivine is an example to demonstrate ordering processes occurring in a crystal causing a change in crystal symmetry. Mg vacancies in olivine charge-compensated by hydrogen (hydrous olivine) occur at random when they form in the Earth's mantle. Due to the high temperature in the mantle, the Mg vacancies become mobile and start ordering in the crystal structure along the (100) plane, thus forming a hydrous olivine super structure Hy-2a (Khisina and Wirth, 2002). Hydrous olivine was observed as oriented thin lamellae (<50 nm wide) in olivine. In high-resolution lattice fringe images the lattice fringes of hydrous olivine show a completely different pattern from that of olivine (fig. 5a in Khisina and Wirth, 2002; Churakov et al., 2003)). Hydrous olivine is very sensitive to electron irradiation damage and therefore the contrast related to hydrous olivine disappears rapidly. Because there were many of the hydrous olivine

lamellae present, and all of them had the same orientation with respect to the olivine host SAED pattern could be acquired successfully. SAED diffraction patterns from such olivine display weak and streaked diffraction intensity additional to the strong spots of the host olivine. The streaks extend normal to the lamellae of hydrous olivine. It is well known in TEM investigations that thin lamellae cause streaking in the electron diffraction normal to the lamellae (Hirsch *et al.*, 1977). The strong olivine reflections in that sample oriented with the [010] zone axis of olivine parallel to the electron beam are the allowed reflections based on the space group *Pbnm* ((200); (002); fig. 7a in Khisina and Wirth, 2002). However, in between there are forbidden reflections of (100) and (001) visible with much less intensity. In between the (001) reflection and the primary beam spot appears an additional spot thus indicating a new reflecting plane with the d spacing of $2 \times d_{001}$. Ordering of the Mg vacancies creates a change in crystal symmetry thus producing additional reflections in electron diffraction.

3.2.6. Satellite reflections from spinodal decomposition in alkali-feldspar

If hypersolvus alkali feldspar of intermediate composition is brought into the two-phase equilibrium domain by cooling, it tends to exsolve forming perthite, micro- or cryptoperthite, depending on the characteristic dimensions of the exsolved phases (Abart *et al.*, 2009). Exsolution of a new stable phase from a host mineral may occur by nucleation and growth or by spinodal decomposition (Cahn, 1968). In the final state of the exsolution it is not possible to discriminate from the microstructure whether the exsolution was a spinodal or a nucleation and growth process. However, spinodal decomposition is an uphill diffusion process that in the initial state displays a peculiar microstructure in TEM. The most striking feature of spinodal decomposition microstructure is the presence of very narrow, lens-shaped contrast in bright- and dark-field images (Fig. 14a). The electron diffraction pattern of such a crystal displays a single crystal pattern. However, the reflections show satellites normal to the lens-shaped contrast in the bright-field image indicating the spinodal decomposition lamellae (Fig. 14a). The distance between two satellite spots in reciprocal space of the diffraction pattern can be converted into the mean width of the lamellae in real space using the same procedure of transforming the imaging vector in a diffraction pattern into d spacing.

The same information about the mean width of the lamellae can be obtained from intensity profiles across the lamellae in dark-field images (Fig. 14b).

3.2.7. Mosaic crystals

A mosaic crystal is basically a single crystal that is composed of very small (<<100 nm) individual crystalline blocks that are slightly misoriented in relation to each other. In nature, such mosaic crystals commonly form when strongly metamict zircon is hydrothermally treated. The newly formed zircon is composed of individual zircon nanocrystals with almost identical crystallographic orientation. Electron diffraction patterns of mosaic crystals show diffraction intensity that is smeared and distorted (Geisler *et al.*, 2004; Anderson *et al.*, 2008). Mosaic crystals of phase Egg have been reported as inclusions in diamond (Wirth *et al.*, 2007). Figure 15 shows an example of a

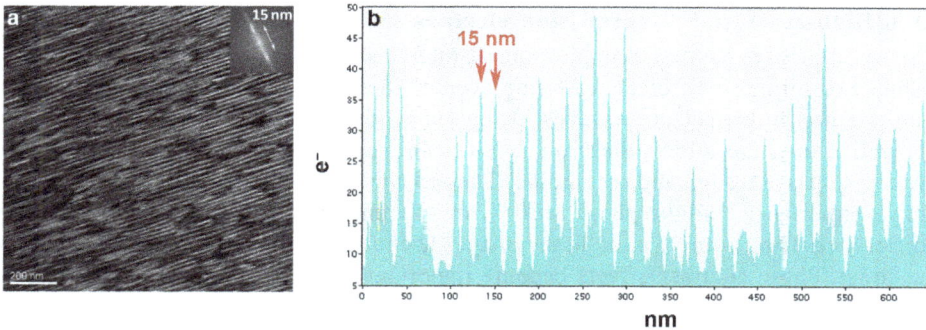

Fig. 14. Spinodal decomposition in alkali feldspar. (a) TEM BF image of spinodal exsolution lamella in alkali feldspar. The inset in the upper right part shows a Fast Fourier Transform (FFT) of the BF image. This is not a high-resolution lattice fringe image. The diffraction spots in the FFT are smeared thus indicating a slightly variable width of the lamellae. The distance between the diffraction spots (maxima from the intensity profile along the FFT) indicates an average lamella thickness of 15 nm. (b) Intensity profile across the image in 14a showing the same average distance between the spinodal decomposition lamellae.

mosaic crystal of phase Egg. In that case several individuals, each <50 nm in size and with almost the same crystallographic orientation are joined thus forming a larger single crystal. Nanometre-sized mosaic crystals are more common in nature. They are usually observed when crystal formation did not occur by nucleation and growth but by assembling precursors to stable crystals. This is referred to as non-classical crystallization (Teng, 2013). Another example of mosaic crystal are nanometre-sized anatase particles forming chains of much larger crystals (Penn and Banfield, 1999). In a recent paper by Schäbitz et al. (2015) the formation of nanometre-sized crystals of Calcite III, a high-pressure form of calcite, was also related to the formation of mosaic crystals from nanometre-sized precursor phases.

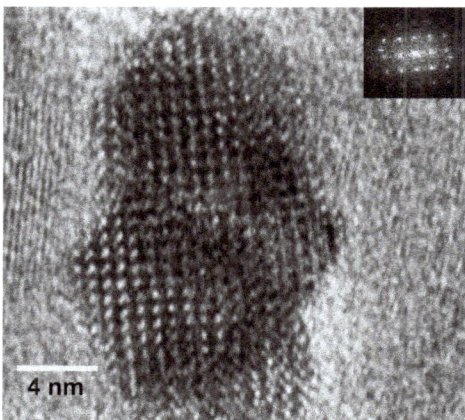

Fig. 15. High-resolution lattice fringe image of nano-inclusions of phase Egg observed as inclusions in diamond from Juina, Brazil. Nanocrystals of phase Egg are attached to each other thus forming a larger mosaic crystal with the individual crystals slightly misoriented. The diffraction patterns inserted (FFT) show that the crystals represent a single crystal.

3.3. High-resolution transmission electron microscopy (HRTEM)

An electron wave passing through a sample interacts with the atoms and electrons of the sample resulting in different scattering events. These interactions induce slight phase shift of the scattered electron waves. Using a high-resolution mode of the TEM, a larger objective aperture allows the primary electron wave plus several scattered electron waves to pass through and recombine. The recombination of the scattered waves that have experienced slight phase shift with the unscattered wave creates intensity differences from point to point in the image. The intensity differences are called phase contrast. Phase contrast effects become important when image details <1.5 nm are resolved, and that applies for high-resolution imaging.

More sophisticated theoretical treatments of phase shifts of the scattered waves caused by the atoms and electrons of the sample are found elsewhere (Cowley and Iijima, 1976). Consider a diffraction pattern from a crystal in which only the primary beam and two strongly excited diffraction spots are visible. If an objective aperture be placed in such a way that it allows only the primary beam and the two diffracted beams to pass through, then, after recombination, a high-resolution image with periodic changes in intensity (fringes) results. The periodicity of the fringes is inversely proportional to the distance between the two diffraction spots in the diffraction pattern thus corresponding to the periodicity d_{hkl} of the lattice planes in the crystal structure (Fig. 16). The position of the maxima and minima (intensity) in the lattice fringe image depends on the phase shift of the two diffracted beams. The phase shift depends on the sample thickness and the crystallographic orientation of the sample. The contrast – dark or bright – depends on sample thickness z and deviation from the Bragg position (deviation parameter, s). It is important to note that the location of the reflecting lattice planes in the crystal does not coincide with the lattice fringes. If the sample thickness z changes, dark fringes might be replaced by bright fringes. The reason for that behaviour is the dependence of the intensity of the primary electron beam and the diffracted beam having passed through the sample from the sample thickness z. That effect is commonly observed in wedge-shaped samples that have been prepared by argon ion milling. Additional image shift is caused by spherical aberration of the objective lens c_s, astigmatism of the objective lens, deviation parameter s and defocus, Δf. High-resolution imaging requires thin foils and perfect alignment of the crystal with a low-indexed zone axis parallel to the optical axis of the TEM. Perfect alignment of the crystal is accomplished when the diffracted intensity in the electron diffraction pattern is almost equal (Fig. 17).

The transfer of information from the sample into the diffraction pattern and into the phase contrast image is described by the contrast transfer function (CTF).

$$H(\mathbf{u}) = A(\mathbf{u}) \times E(\mathbf{u}) \times B(\mathbf{u}) \qquad (13)$$

In equation 13, $\mathbf{u}$ is a reciprocal lattice vector or the spatial frequency ($1/d_{hkl}$) for a particular direction in the crystal structure. High frequencies correspond to short distances in the lattice. In high-resolution imaging we are looking for high spatial frequencies, and therefore, we need to use a large objective aperture to include these

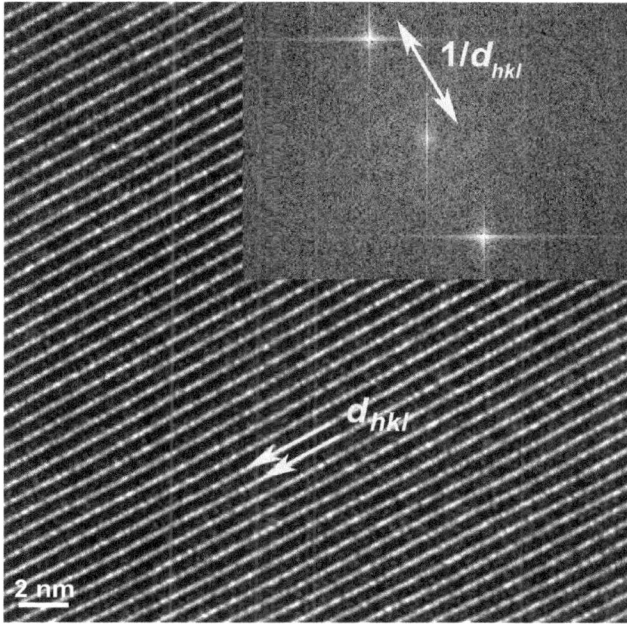

Fig. 16. High-resolution lattice fringe image. The distance between the bright and dark contrasts represents the d spacing of the reflecting lattice planes. The corresponding diffraction pattern (FFT) is given in the inset. Note that the distance between the primary electron beam and the diffracted beam is $1/d_{hkl}$ (reciprocal space). The d spacing in the image is real space d_{hkl}.

high spatial frequencies in the image formation. $A(\mathbf{u})$ is the aperture function. The objective aperture cuts off the high spatial frequencies greater than the diameter of the aperture. $E(\mathbf{u})$ is the envelope function describing properties of the lens itself. $E(\mathbf{u})$ represents the attenuation of the electron wave (envelope function) caused by

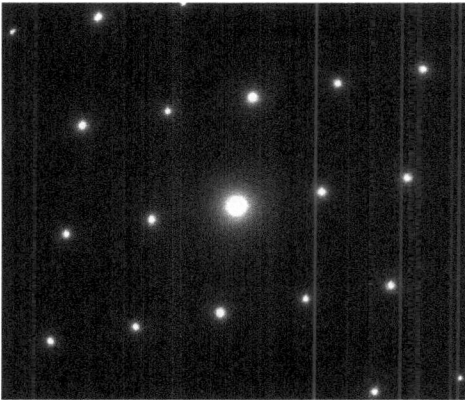

Fig. 17. Electron diffraction pattern with a low-indexed zone axis aligned almost parallel to the electron beam. Note that under that conditions the scattered intensity of the different spots is approximately the same.

instabilities of the electron current (I) and instabilities in the accelerating voltage (V). The most important term in the contrast transfer function $H(\mathbf{u})$ is the term $B(\mathbf{u})$, which considers the aberration of the objective lens. $B(\mathbf{u})$ is expressed as:

$$B(\mathbf{u}) = \exp i\chi(\mathbf{u}) \tag{14}$$

$$\chi(\mathbf{u}) = \pi \times \Delta f \times \lambda \times \mathbf{u}^2 + 1/2\pi \times c_s \times \lambda^3 \times \mathbf{u}^4 \tag{15}$$

In equation 15, Δf is the defocus used, c_s is the spherical aberration coefficient, λ is the wavelength of the electrons and $\mathbf{u}$ is the spatial frequency (reciprocal lattice vector $1/d_{hkl}$). To solve equation 14 the Euler formalism (equation 16) is applied:

$$B(\mathbf{u}) = \cos \chi(\mathbf{u}) + i\sin\chi(\mathbf{u}) \tag{16}$$

The term $\cos \chi(\mathbf{u})$ describes the amplitude contrast and $i\sin \chi(\mathbf{u})$ represents phase contrast.

Figure 18 is the contrast transfer function (CTF) calculated for defocus $\Delta f = -68$ nm, spherical aberration coefficient $c_s = 1.3$ mm and electron wavelength $= 0.00251$ nm. For $\sin \chi(\mathbf{u}) = 0$ no information will be transferred from the sample to the image. With increasing spatial frequency the graph of $\sin \chi(\mathbf{u})$ gets negative approaching the value -1. Between the spatial frequencies 1.5 (nm^{-1}) and 3.5 (nm^{-1}) that function $\sin \chi(\mathbf{u})$ shows high negative values approaching -1 and thus provides a broad band of good transmittance. That is the range of spatial frequencies where the most information is transferred from the sample to the image and it results in positive phase contrast. At $\mathbf{u} = 4.2$ (nm^{-1}) the CTF in Fig. 18 is 0, and then changes into positive phase contrast. When the graph of the function $\sin \chi(\mathbf{u})$ crosses the spatial frequency axis $\mathbf{u}$ (nm^{-1}) for the first time the value of the function is 0 and that value of $\mathbf{u}$ (nm^{-1}) defines the point resolution of the TEM. For $\Delta f = -68$ nm, spherical aberration coefficient, $c_s = 1.3$ mm and electron

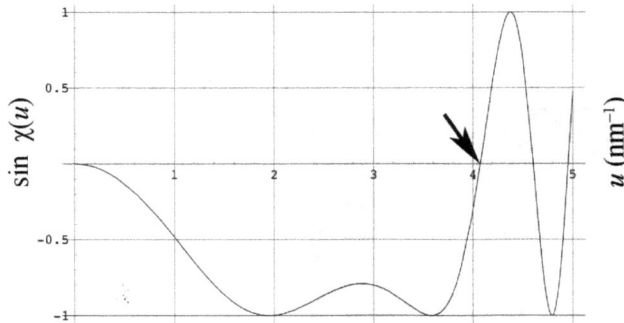

Fig. 18. Contrast Transfer Function (CTF) as a function of spatial frequency. Note the plateau region of the CTF between 2 and 3.5 nm^{-1} with the value of $\sin\chi(\mathbf{u})$ approximately -1. This is the region of CTF where most of the information is transferred from the object to the image. The first transect of the graph of CTF with the spatial frequency axis (arrow) defines the point resolution of the TEM. The CTF presented here is defined for a spherical aberration coefficient $c_s = 1.3$ mm and Scherzer defocus is -68 nm. The point resolution is 0.25 nm.

wavelength = 0.00251 nm; the point resolution is 0.25 nm (*e.g.* FEI TECNAI F20 X-Twin). In a TEM with spherical aberration coefficient c_s = 2 mm (*e.g.* Philips CM 200) the point resolution will decrease to 0.303 nm (Fig. 19). The CTF also changes dramatically with changing defocus Δf. The crucial question is now at which defocus the CTF is maximized? The optimal defocus has come to be known as 'Scherzer' defocus Δf_{Sch}.

$$\Delta f_{Sch} = -1.2(c_s\lambda)^{1/2} \qquad (17)$$

The contrast transfer function $H(\mathbf{u}) = A(\mathbf{u}) \times E(\mathbf{u}) \times B(\mathbf{u})$ is modified if we add the contribution from the envelope function $E(\mathbf{u})$ to the term $B(\mathbf{u})$. $E(\mathbf{u})$ can be considered as a damping function. $E(\mathbf{u})$ considers chromatic aberration, which is caused by instabilities in the high voltage and the objective lens current. In the CTF without envelope function the oscillations of the function $\sin \chi(\mathbf{u})$ principally could extend to unlimited values of $\mathbf{u}$ (nm^{-1}). In reality, if the envelope function is applied, the function $\sin \chi(\mathbf{u})$ dies off at a certain value of $\mathbf{u}$ (nm^{-1}). Beyond that value there is no more information in the image (Fig. 20). The value of $\mathbf{u}$ (nm^{-1}) where CTF approaches 0 is defined as the information limit of the TEM. Beyond the information limit there is no meaningful interpretation of any feature in the image possible.

There are two important imaging limits represented by the CTF. The first crossover of the graph of the function $\sin \chi(\mathbf{u})$ with the $\mathbf{u}$ (nm^{-1}) axis defines the point resolution and the damping of the function $\sin \chi(\mathbf{u})$ to almost 0 by the envelope function, which defines the information limit. If the information limit is beyond the Scherzer resolution limit (point resolution), image simulation software is needed to interpret any detail beyond the Scherzer limit (Williams and Carter, 1996).

From high-resolution lattice-fringe images we can easily derive the spatial frequencies $1/d_{hkl}$ by applying a FFT. The FFT is a calculated diffraction pattern from the lattice fringe image. The diffraction spots with a vector length of $1/d_{hkl}$ are calculated from the information stored in the whole image. Therefore, they provide more precise *d*-spacing determinations than measuring the *d* spacing directly from

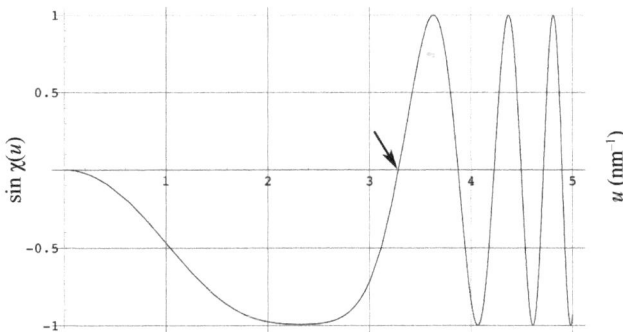

Fig. 19. Dependence of the contrast transfer function CTF on the spherical aberration coefficient c_s ($c_s = 2$ mm). The CTF at the same Scherzer defocus results in a smaller plateau region and a point resolution of 0.303 nm.

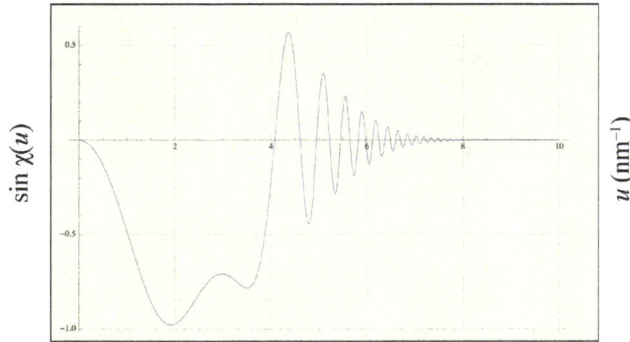

Fig. 20. CTF with the envelope function $E(\mathbf{u})$ applied, which considers chromatic aberration due to instabilities in high-voltage and electron current. Without the envelope function the CTF would oscillate and extend to unlimited values of $\mathbf{u}$ (nm^{-1}). With the envelope function applied the CTF dies off at a certain value of $\mathbf{u}$ (nm^{-1}). That value defines the information limit of the TEM.

some locations in the lattice fringe image. The d spacing calculated from the FFT and the angles between the spots (diffraction vectors) in the diffraction pattern can be used to identify a particular phase unambiguously. It is only necessary to compare the observed d_{hkl} and the adjacent angles between the diffraction vectors with calculated d spacings and angles. At that point it needs to be emphasized that only in the case of a very thin sample (<30 nm), where we can apply kinematic theory of electron diffraction (Hirsch *et al.*, 1977) do the extinction rules based on the space group of a crystal apply. In thicker samples it is expected that forbidden reflections occur and contribute to the diffraction pattern and to the lattice fringe image. Consequently, a diffraction pattern or FFT can only be indexed and attributed to a particular crystal if we include forbidden reflections.

However, if we want to interpret contrast details in the high-resolution image with respect to individual atom columns or crystal structural details, image simulation is required. Such an image simulation requires the knowledge of the crystal structure, the experimental conditions such as sample thickness z, defocus Δf, deviation parameter s, spherical aberration of the objective lens c_s and the wavelength of the primary electron λ. Only if the simulated image based on the imaging parameters matches the experimental image is a meaningful interpretation of the contrast possible (*e.g.* Marquardt *et al.*, 2011; their fig. 2). More detailed information about high-resolution imaging in TEM is provided elsewhere (Buseck *et al.*, 1988; Spence, 1988; Carter and Williams, 1996, Fultz and Howe, 2001).

High-resolution lattice-fringe imaging is a powerful tool in geosciences to identify minerals, even nanometre-sized crystals. It is not, however, a tool for crystal structure analysis. Usually, silicate minerals are sensitive to electron irradiation damage and therefore become amorphous very quickly under the electron beam. Quartz, a most common silicate is especially sensitive and transforms into a non-crystalline state very rapidly. To acquire a high-resolution lattice-fringe image from quartz a very short exposure time (<0.5 s) is required; after that the imaged volume is amorphous. With nanometre-sized crystals convergent beam electron diffraction (CBED) would be the right technique to acquire an electron diffraction pattern. However, focusing the electron beam for CBED would immediately render the structure amorphous. CBED is not an option for most silicate nanocrystals. The only technique to get structural information from nanoinclusions in minerals (*e.g.* inclusions in diamond) is high-resolution lattice fringe imaging and subsequent calculation of FFTs. This technique has been applied effectively in numerous cases (Wirth *et al.*, 2005, 2007, 2013; Khisina and Wirth, 2002; Kaminsky *et al.*, 2009; Schäbitz *et al.*, 2015).

4. Microstructural features observed in TEM

4.1. Grain shape

The grain shape of micrometre- or nanometre-sized grains can be displayed very well in bright-field or dark-field images. In the dark-field image in Fig. 21 a diamond crystal included in zircon is displayed. The imaging vector for the dark-field image was the (111) reflection of diamond. The alternating bright and dark contrasted layers along the zircon–diamond interface are thickness fringes. It is evident that in some locations the

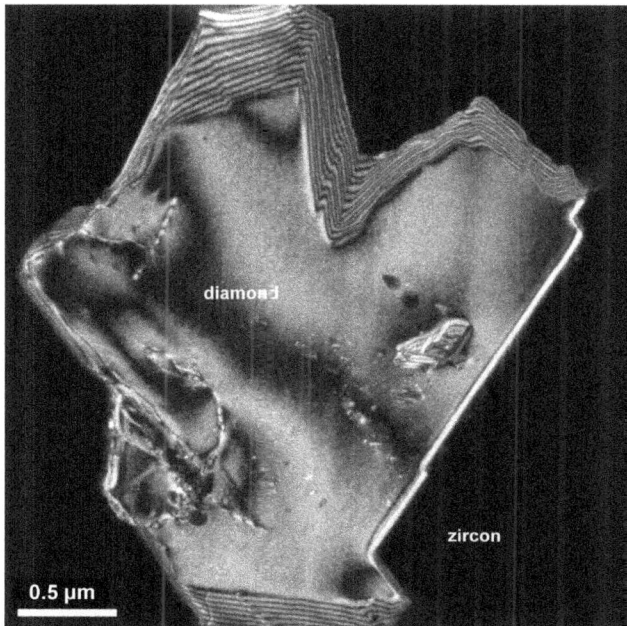

Fig. 21. Dark-field image (DF) of a diamond inclusion in zircon. The interfaces of the diamond crystal are characterized by the alternating dark and bright contrast representing thickness fringes. Note the roughness of the diamond surface and the locally different inclination of the interface that gives the image an almost three-dimensional appearance. A few dislocation lines are also visible in the left part of the diamond.

interface is parallel to the electron beam with no thickness fringes visible whereas in other locations the interface is inclined differently and rather rough. The image appears to be almost three dimensional. The broad dark contrasts on the diamond surface are diffraction contrasts. A few dislocation lines are also visible in the left part of the diamond crystal. The alternating bright and dark contrast of the dislocation in the upper left part of diamond near the interface suggests a steep inclination of the dislocation line with the surface. Zircon is completely dark because the (111) reflection of diamond was used exclusively for the dark-field imaging.

4.2. Twins

Twinning is a common feature in minerals, and TEM is a tool to identify the twin law using electron diffraction and diffraction contrast imaging. In diamond grown by chemical vapour deposition or gas phase condensation {111} nanotwins are very common. These twins are a characteristic microstructural feature of such diamond formation under low-pressure conditions (Shechtman, 1994; Tsuno et al., 1994). Recently diamond aggregates found as inclusions in lava from Kamchatka volcanoes have been investigated with TEM. One of the striking microstructural features of these diamonds is intensive twinning along the {111} planes (Fig. 22). Aside from other observations, the presence of nanotwins suggested that these diamonds were formed in local microenvironments in lavas by chemical vapour deposition (Kaminsky et al., 2016).

4.3. Symplectites

Symplectites comprise a vermicular intergrowth of usually submicrometre-sized crystals that replace a host mineral at a sharp reaction front. Garnet is a typical example for symplectite formation with anorthite + orthopyroxene and spinel replacing garnet (Degi et al., 2010). Another example is majoritic garnet from the Snap Lake kimberlite

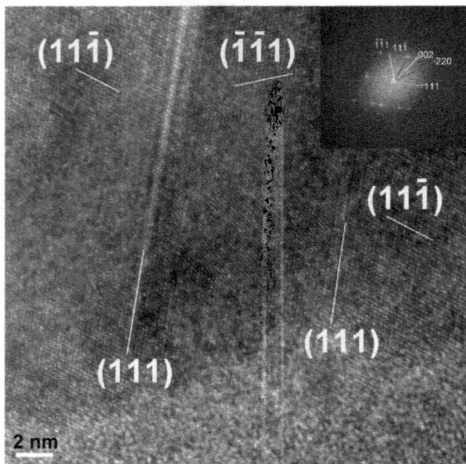

Fig. 22. High-resolution lattice fringe image of nanometre-sized twins in diamond. The twinning plane is the (111) plane in diamond. The twinning is also visible in the diffraction pattern (FFT).

(North West Territories, Canada) that was partially transformed into orthopyroxene + clinopyroxene + Cr-spinel + coesite. The majoritic garnet is separated from the lamellar intergrowths of very narrow lamellae (<50 nm) of Opx + Cpx + Cr-spinel + coesite by a sharp reaction boundary (Fig. 23). Such small lamellae can be identified only using high-resolution lattice fringe imaging combined with EDX analyses. The FFTs from the lattice-fringe images provide structural data (*d* spacing) that can be combined with the chemical composition thus identifying the phases. For example, high-resolution element mapping of the X-ray intensities revealed the presence of a SiO_2 phase, which subsequently was identified from lattice-fringe images as coesite. That majoritic garnet occurs as an inclusion in diamond. During uplift of the diamond to the Earth's surface pressure and temperature decreased so that majorite became unstable and the transformation started *via* a moving reaction front. Equilibrium of majoritic garnet was achieved during mantle residence time at 1380°C at 11 GPa (Sobolev *et al.*, 2008).

4.4. Dislocations

Dislocations are crystal defects that can be incorporated in the crystal during crystal growth or during later plastic deformation. The dislocation line and the Burgers vector define a dislocation. There are three different types of dislocations: There are edge dislocations with the Burgers vector normal to the dislocation line; in case of a screw dislocation the Burgers vector is parallel to the dislocation line; and finally, there are dislocation loops with a screw component (Burgers vector parallel to the dislocation line) and an edge component (Burgers vector normal to the dislocation line). Screw dislocations and edge dislocations are basic elements of plastic deformation of materials. Dislocation loops are created by vacancy condensation or by condensation of atoms or molecules on particular lattice planes. More details about dislocations can be found elsewhere (Hull and Bacon, 1984). Dislocations can move along a crystallographic direction in a crystal by a glide mechanism. Usually, plastic deformation

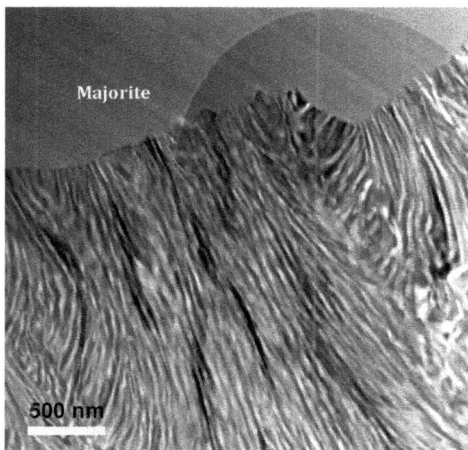

Fig. 23. HAADF image of a symplectite composed of Opx, Cpx, Cr-spinel and coesite replacing a majoritic garnet inclusion in diamond from Slave Craton. The round feature in the background is due to the perforated carbon film on which the TEM lamella rests.

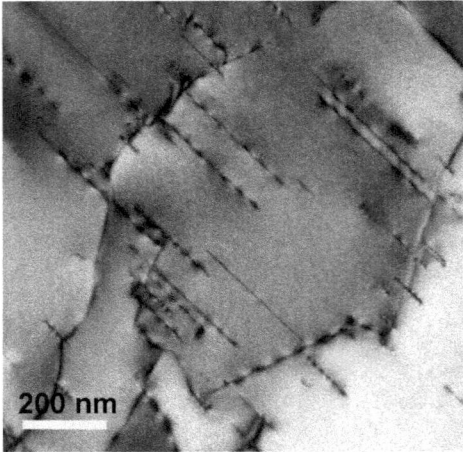

Fig. 24. TEM bright-field image of diamond with dislocation lines. Note that most of the dislocation lines are straight thus indicating deformation of the diamond by dislocation glide.

occurs by dislocation glide. The presence of straight dislocation line contrast in BF- or DF-imaging indicates dislocation glide in TEM (Fig. 24). Very often, in plastically deformed materials parallel dislocation lines are visible simultaneously. The dislocation line visualizes the end of the inserted half-plane in the crystal structure. If atoms are added to the end of the half plane or removed from the half plane the dislocation line contrast starts curving finally resulting in a strongly curved dislocation line. That process is called dislocation climb (Fig. 25). Adding atoms to the half plane or removing atoms from the half plane is a diffusion process that is always thermally activated. A certain threshold energy (*i.e.* the activation energy) that is characteristic of a particular mineral must be overstepped to start the process. That means, if we observe

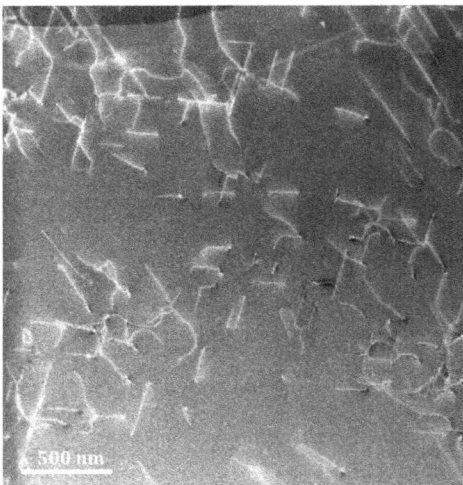

Fig. 25. HAADF diffraction contrast image of diamond. The bright contrasted lines are dislocation lines that are mostly curved and twisted. Such curved dislocations indicate dislocation climb, a thermally activated process suggesting a longer residence time under high temperature after deformation.

predominantly straight dislocation lines (*e.g.* in diamond in Fig. 24) we might conclude that the crystal has been deformed plastically. In the case of diamond that deformation occurred under high-temperature conditions most likely in the upper mantle. However, if only straight dislocation lines are visible we can conclude that the residence time of the diamond under high-temperature conditions was not long. If the diamond had been deformed and had resided for a longer period under high-temperature conditions, the dislocations would have started to climb thus forming strongly curved dislocation lines. In such a particular case we can assume that the diamond has been transported rapidly to the surface after deformation. If the threshold temperature for thermally activated processes such as dislocation climb by diffusion of atoms is known − in the case of quartz it is ~300°C (Voll, 1976) − we can deduce a temperature estimate for that particular mineral. If the dislocations start moving by glide and climb they can rearrange in the crystal finally reducing the stored energy of the crystal. The dislocations rearrange thus forming walls or cells (Fig. 26). The volume that is defined by the cell wall is free of dislocations. It is a process that is called recovery. The formation of dislocation walls is an early stage of recovery. Continuous movement of dislocations ends up in an arrangement of the dislocations in low-angle grain boundaries, which is shown in Fig. 27

The misorientation between two grains determines the number of dislocations in a low-angle grain boundary (Heinemann *et al.*, 2005). The transition between small-angle and large-angle grain boundaries was defined by the invisibility of individual dislocation lines. If the number of dislocation lines along a low-angle grain boundary increases repeatedly, at a certain number the dislocation cores will overlap and individual dislocation lines are no longer visible (Gleiter and Chalmers, 1972). Heinemann *et al.* (2005) demonstrated that in olivine there are individual dislocation lines visible even at 21° misorientation. That contradicts the observations in metals and alloys where the transition between low-angle and high-angle grain boundaries is in the range of 10−15° misorientation. The

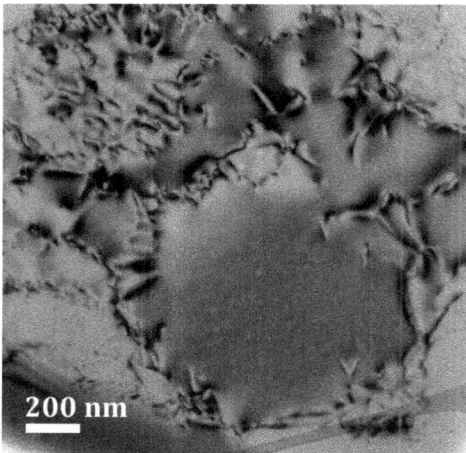

Fig. 26. TEM bright-field image of a calcite crystal with dislocations forming a dislocation-free cell.

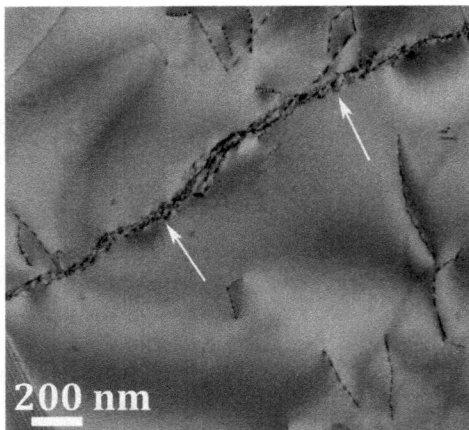

Fig. 27. TEM bright-field image of olivine with dislocations and an array of dislocations (arrows) forming a low-angle grain boundary in olivine. Note the thickness fringes along the surface of olivine.

conclusion from that observation is that the structure of grain boundaries in silicates is different from the structure of grain boundaries in metals. This is quite reasonable because in metals there is no need for charge compensation and in silicates there are rigid units in the form of the SiO_4 tetrahedra.

References

Abart, R., Petrishcheva, E., Wirth, R. and Rhede, D. (2009) Exsolution by spinodal decomposition II: Perthite formation during slow cooling of anatexites from Ngoronghoro, Tanzania. *American Journal of Science*, **309**, 450–475.

Anderson, A.J., Wirth, R. and Thomas, R. (2008) The alteration of metamict zircon and its role in the remobilization of High-Field-Strength elements in the Georgeville granite, Nova Scotia. *The Canadian Mineralogist*, **46**, 1–18.

Aken van, P.A., Liebscher, B. and Styrsa, V.J. (1998) Quantitative determination of iron oxidation states in minerals using Fe $L_{2,3}$-edge electron energy-loss near-edge structure spectroscopy. *Physics and Chemistry of Minerals*, **25**, 323–327.

Brownlee, K.O. (1960) *Statistical Theory and Methodology in Science Engineering*. Wiley, Table II, p 548.

Buerger, M.J. (1970) *Contemporary Crystallography*. Mc-Graw-Hill Book Company, New York.

Buseck, P.R., Cowley, J.M. and Eyring, L. (1988) *High-resolution Transmission Electron Microscopy and Associated Techniques*. Oxford University Press, New York.

Cahn, J.W. (1968) Spinodal decomposition. *Transactions of the Metallurgical Society of AIME*, **242**, 166–180.

Cliff, G. and Lorimer, G.W. (1975) The quantitative analysis of thin specimens. *Journal of Microscopy*, **103**, 203–207.

Churakov, S.V., Khisina, N.R., Urusov, V.S. and Wirth, R. (2003) First principle study of $(MgH_2SiO_4).n(Mg_2SiO_4)$ hydrous olivine structures. I. Crystal structure modeling of hydrous olivine Hy-$2a(MgH_2SiO_4).3(Mg_2SiO_4)$. *Physics and Chemistry of Minerals*, **30**, 1–11.

Cowley, J.M. and Iijima, S. (1976) The direct imaging of crystal structures. Chapter 2.5, pp. 123–136 in: *Electron Microscopy in Mineralogy* (H. Wenk, editor). Springer. Berlin, Heidelberg, New York.

Dégi, J., Abart, R., Török, K., Bali, E., Wirth, R. and Rhede, D. (2010) Symplectite formation during decompression induced garnet breakdown in lower crustal mafic granulite xenoliths: mechanisms and rates. *Contributions to Mineralogy and Petrology*, **159**, 293–314.

Edington, J.W. (1975a) *Electron Diffraction in the Electron Microscope*. Philips Technical Library. Monographs in Practical Electron Microscopy in Materials Science, vol. **2**.

Edington, J.W. (1975b) *Interpretation of Transmission Electron Micrographs*. Philips Technical Library.

Monographs in Practical Electron Microscopy in Materials Science, vol. **3**.

Egerton, R.F. (1996) *Electron Energy-loss Spectroscopy in the Electron Microscope*. Plenum Press, New York.

Fultz, B. and Howe, J.M. (2001) *Transmission Electron Microscopy and Diffractometry of Materials*. Springer, Berlin, Heidelberg, New York.

Garvie, L.A.J., Craven, A.J. and Brydson, R. (1994) Use of electron energy-loss near-edge fine structure in the study of minerals. *American Mineralogist*, **79**, 411−425.

Geisler, T., Seydoux-Guillaume, A-M., Wiedenbeck, M., Wirth, R., Berndt, J., Ming Thang, Mihailova, B., Putnis, A., Salje, E. and Schlüter, J. (2004) Periodic precipitation pattern formation in hydrothermally treated metamict zircon. *American Mineralogist*, **89**, 1341−1347.

Gleiter, H. and Chalmers, B. (1972) *High-angle Grain Boundaries*. Pergamon Press, Oxford, UK.

Glocker, R. (1971) *Materialprüfung mit Röngenstrahlung*. Springer. Berlin Heidelberg New York.

Goldstein, J.I. and Williams, D.B. (1989) Quantitative X-ray analysis. Pp. 155−217 in: *Principles of Analytical Electron Microscopy* (D.C. Joy, A.D. Romig and J.I. Goldstein, editors). Plenum. New York and London.

Heinemann, S., Wirth, R., Gottschalk, M. and Dresen, G. (2005) Synthetic [100] tilt grain boundaries in forsterite: 9.9 to 21.5°. *Physics and Chemistry of Minerals*, **32**, 229−240.

Hirsch, P., Howie, A., Nicholson, R., Pashley, D.W. and Whelan, M.J. (1977) *Electron Microscopy of Thin Crystals*. Reprint of the 1967 issue of the edition published by Butterworths, London. Robert E. Krieger Publishing Co., Inc.

Hull, D. and Bacon, D.J. (1984) *Introduction to Dislocations*. International Series on Materials Science and Technology, Pergamon Press, Oxford, UK.

Joy, D.C., Romig, A.D. and Goldstein, J.I. (1986) *Principles of Analytical Electron Microscopy*. Plenum, New York and London.

Kaminsky, F.V. and Wirth, R. (2011) Iron carbide inclusions in lower-mantle diamond from Juina, Brazil. *The Canadian Mineralogist*, **49**, 555−572.

Kaminsky, F., Wirth, R., Matsyuk, S., Schreiber, A. and Thomas, R. (2009) Nyerereite and nahcolite inclusions in diamond: evidence for lower-mantle carbonatitic magmas. *Mineralogical Magazine*, **73**, 797−816.

Kaminsky, F.V., Wirth, R. and Schreiber, A. (2015) A microinclusion of Lower-Mantle rock and other minerals and nitrogen Lower-Mantle inclusions in a diamond. *The Canadian Mineralogist*, **53**, 83−104.

Kaminsky, F.V., Wirth, R., Anikin, L.P., Morales, L. and Schreiber, A. (2016) Carbonado-like diamond from the Avacha active volcano in Kamchatka, Russia. *Lithos*, doi.org/10.1016/j.lithos.2016.02.021

Khisina, N.R. and Wirth, R. (2002) Hydrous olivine $(Mg_{1-y}Fe_y^{2+})_{2-x}v_xSiO_4H_{2x}$ – a new DHMS phase of variable composition observed as nanometer sized precipitations in mantle olivine. *Physics and Chemistry of Minerals*, **29**, 98−111.

Klein-BenDavid, O., Wirth, R. and Navon, O. (2006) TEM imaging and analysis of microinclusions in diamonds: A close look at diamond-growing fluids. *American Mineralogist*, **91**, 353−365.

Koch-Müller, M., Mugnaioli, E., Rhede, D., Spezzale, S., Kolb, U. and Wirth, R. (2014) Synthesis of quenchable high-pressure form of magnetite (h-Fe_3O_4) with composition $^{Fe1}(Fe^{2+}_{0.75}Mg_{0.26})$ $^{Fe2}(Fe^{3+}_{0.70}Cr_{0.15}Al_{0.11}Si_{0.04})_2O_4$. *American Mineralogist*, **99**, 2405−2415.

Kurata, H., Lefevre, E., Colliex, C. and Brydson, R. (1993) Electron energy-loss near-edge structures in the oxygen-edge spectra of transition-metal oxides. *Physical Review B*, **47**, 13763−13768.

Kusiak, M.A., Dunkley, D.J., Wirth, R., Whitehouse, M.J., Wilde, S.A. and Marquardt, K. (2015) Metallic lead nanospheres discovered in ancient zircons. *Proceedings of the National Academy of Sciences of the United States of America (PNAS)*, **112**, 4958−4963.

Logvinova, A.M., Wirth, R., Fedorova, E.N. and Sobolev, N.V. (2008) Nanometre-sized mineral and fluid inclusions in cloudy Siberian diamonds: new insights on diamond formation. *European Journal of Mineralogy*, **20**, 317−331.

Marquardt [Hartmann], K., Ramasse, Q.M., Kisielowski, C. and Wirth, R. (2011) Diffusion in yttrium aluminium garnet at the nanometer-scale. Insight into the effective grain boundary width. *American Mineralogist*, **96**, 1521−1529.

Penn, L.R. and Banfield, J.F. (1999) Morphology development and crystal growth in nanocrystalline aggregates under hydrothermal conditions: insights from titania. *Geochimica et Cosmochimica Acta*, **63**, 1549−1557.

Putnis, A. (1992) *Introduction to Mineral Sciences*. Cambridge University Press, Cambridge, UK.

Rask, J.H., Miner, B.A. and Buseck, P.R. (1987) Determination of manganese oxidation states in solids by electron energy-loss spectroscopy. *Ultramicroscopy*, **21**, 321–326.

Schäbitz, M., Wirth, R., Janssen, C. and Dresen, G. (2015) First evidence of $CaCO_3$-III and $CaCO_3$-IIIb high-pressure polymorphs of calcite: Authigenically formed in near surface sediments. *American Mineralogist*, **100**, 1230–1235.

Shechtman, D. (1994) Twin-determined growth of diamond films. *Materials Science and Engineering*, **A184**, 113–118.

Sobolev, N., Wirth, R., Logvinova, A. and Pokhilenko, N. (2008) An unusual partial retrograde phase transition of majoritic garnet from Snap Lake diamond. In: *Jahrestagung der Deutschen Mineralogischen Gesellschaft* (Berlin 2008), Abstracts.

Spence, J.C.H. (1988) *Experimental High-Resolution Electron Microscopy*. Oxford University Press, Oxford, UK.

Taftø, J. and Zhu, J. (1982) Electron energy-loss near-edge structure (ELNES), a potential technique in the studies of local atomic arrangements. *Ultramicroscopy*, **9**, 349–354.

Teng, H.H. (2013) How ions and molecules organize to form crystals. *Elements*, **9**, 189–194.

Toy, V.G., Mitchell, T.M., Driuventak, A. and Wirth, R. (2015) Crystallographic preferred orientations may develop in nanocrystalline materials on fault planes due to surface energy interactions. *Geochemistry Geophysics Geosystems*, **16**, 2549–2563.

Tsuno, T., Imai, T. and Fujimori, N. (1994) Twinning structure and growth hillock on diamond (001) epitaxial film. *Journal of Applied Physics* , **33**, 4039–4043.

Vincent, R. and Midgley, P. (1994) Double conical beam-rocking system for measurement of integrated electron diffraction intensities. *Ultramicroscopy*, **53**, 271–282.

Voll, G. (1976) Recrystallization of quartz, biotite and feldspars from Erstfeld to the Leventina Nappe, Swiss Alps, and its geological significance. *Schweizer Mineralogisch Petrographische Mitteilungen*, **56**, 641–647.

Williams, D.B. and Carter, C.B. (1996) *Transmission Electron Microscopy*. Plenum Press, New York.

Wirth, R. (2004) A novel technology for advanced application of micro- and nanoanalysis in geosciences and applied mineralogy. *European Journal of Mineralogy*, **16**, 863–876.

Wirth, R. (2009) Focused Ion Beam (FIB) combined with SEM and TEM: Advanced analytical tools for studies of chemical composition, microstructure and crystal structure in geomaterials on a nanometre scale. *Chemical Geology*, **261**, 217–229.

Wirth, R. and Rocholl, A. (2003) Nanocrystalline diamond from the Earth's mantle underneath Hawaii. *Earth and Planetary Science Letters*, **211**, 357–369.

Wirth, R. and Matsyuk, S. (2005) Nanocrystalline $(Mg,Fe,Cr)TiO_3$ perovskite inclusions in olivine from a mantle xenolith, kimberlite pipe Udachnaya, Siberia. *Earth and Planetary Science Letters*, **233**, 325–336.

Wirth, R., Vollmer, C., Brenker, F., Matsyuk, S. and Kaminsky, F. (2007) Inclusions of nanocrystalline hydrous aluminium silicate "Phase Egg" in superdeep diamonds from Juina (Mato Grosso State, Brazil). *Earth and Planetary Science Letters*, **259**, 384–399.

Wirth, R., Reid, D. and Schreiber, A. (2013) Nanometer-sized Platinum-Group Minerals (PGM) in base metal sulfides: new evidence for an orthomagmatic origin of the Merensky Reef PGE ore deposit, Bushveld Complex, South Africa. *The Canadian Mineralogist*, **51**, 143–155.

Wirth, R., Dobrzhinetskaya, L., Harte, B., Schreiber, A. and Green, H.W. (2014) High-Fe (Mg, Fe)O inclusion in diamond apparently from the lowermost mantle. *Earth and Planetary Science Letters*, **404**, 365–375.

EMU Notes in Mineralogy, Vol. 16 (2017), Chapter 5, 131–164

Trace element and isotope analysis using Secondary Ion Mass Spectrometry

ETIENNE DELOULE[1] and NATHALIE VALLE[2]

[1] Centre de Recherches Pétrographiques et Géochimiques, UMR 5873 CNRS - Université de Lorraine, BP20, 54501 Vandœuvre-lès-Nancy, France, email: deloule@crpg.cnrs-nancy.fr
[2] Luxembourg Institute of Science and Technology, Materials Research and Technology Department, 41 Rue du Brill, L-4422 Belvaux, Luxembourg, email: nathalie.valle@list.lu

The determinations of trace element and isotopic ratio distributions in rocks and in minerals are key tools for the understanding of geological processes, and they have undergone continuous development for more than 50 years. However, these geochemical tools are not accessible by conventional mineralogical and petrographic *in situ* methods based on electron or light beams. Therefore, secondary ion mass spectrometry (or ion microprobe) techniques have been used more and more for high-sensitivity *in situ* analyses at scales down to tens of nanometers. The aim of this paper is to present in the first part the principle of ion probe analysis and the characteristics of the most commonly used ion microprobes today. Then two applications will be presented. The first focuses on the study of alteration processes during hydrothermal alteration of glass. The combination of experiments with isotopically marked fluids and high-resolution *in situ* isotopic measurements of the marked species allows definition of the main chemical reaction occurring during the alteration. The second example summarizes the contribution of ion microprobe measurements to the understanding of Li and Li-isotope distributions and behaviours in geological processes, and to the determination of Li isotope fractionation during diffusion processes.

1. Introduction to SIMS and ion microprobe technique

Fifty years ago, the ion microprobe or Secondary Ion Mass Spectrometry (SIMS) was first conceived as a surface-imaging technique, secondary ion microscopy, in order to determine the distribution of chemical species at a solid surface (Slodzian, 1964). For that, primary ions accelerated by an electrostatic field of a few kV are used to sputter the sample surface. The secondary ions emitted from the sample surface are then analysed in a mass spectrometer. The development of ion microprobe technology has allowed rapid measurement of the abundance of trace and ultra-light elements with high accuracy and spatial resolution (*e.g.* H, Li, Be, B) and of isotopes (Shimizu *et al.*, 1978; Shimizu and Hart, 1982; Zinner and Crozaz, 1986; Reed, 1989). Moreover, another field of investigation was opened with the determination of elements in a depth profile below the sample surface and with the 3D imaging of sample compositions.

The ion microprobe permits *in situ* analysis of any kind of solid material prepared as polished or thin sections. Nearly all elements from H to U can be detected, and many can be analysed quantitatively down to parts per million (ppm) levels, and below. However, this method suffers from limitations due to the dependence of the formation of secondary ions on the sample chemistry (matrix effects), and on the complexity of the secondary ions mass spectra, due to the interferences of numerous polyatomic ions. Different families of ion microprobes have been developed during the last decade, including the classical magnetic sector small-size instrument (CAMECA IMS 7f and 7f Geo), the magnetic sector large-size high mass resolution instrument (ASU SHRIMP and CAMECA IMS 1270-1280), the magnetic sector high spatial resolution instrument (CAMECA NanoSIMS), and finally the time-of-flight SIMS instrument such as IonTof instruments, which are dedicated mostly to surface analysis.

This chapter will be focused on the magnetic sector ion microprobe, the most frequently used instrument in Earth Sciences. The different possibilities for obtaining *in situ* 3D high-sensitivity and high-resolution measurements will be compared. Finally, two representative mineralogical applications will be presented.

2. Instrumental set up for *in situ* 3D chemical and isotopic measurements

2.1. Physical basis

An ion microprobe uses a focused primary ion beam to bombard the sample surface in a vacuum chamber (Fig. 1). The impact of the primary ions on the sample atoms produces collision cascades and the ejection or sputtering of particles. Some of these particles are secondary ions, which can be accelerated by an electrostatic field for mass spectrometry analysis. Thus, the ion probe consists of a source of primary ions, a sample chamber, a secondary ions optic, which transfers the secondary ions from the sample surface to the

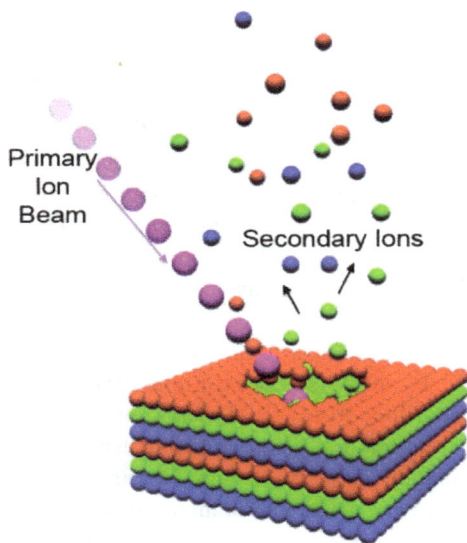

Primary
Ion
Beam

Secondary Ions

Fig. 1. Ion microprobe schematic view: atoms and molecules are sputtered from the sample surface and subsurface atom layers.

mass spectrometer entrance, and the mass spectrometer itself. The first generation of ion microprobes was conceived as ion microscope (Slodzian, 1964), the transfer optic allowing the transmission of the ionic image of the sample, with a spatial resolution of ~0.5 µm and a lens coverage of several hundred µm. With the development of the IMS 4f, the direct imaging of the sample was complemented by the acquisition of scanned images, realized by scanning the primary beam over a square area and the acquisition of digital images with a lateral resolution down to 70 nm (Migeon *et al.*, 1989). Thanks to the destructive nature of the primary-ion bombardment, a SIMS analysis gives progressive access to the inner regions of the sample. Therefore, storing the elemental maps as a function of depth allows three dimensional analysis and representation of the samples.

2.2. Primary and secondary ion optics

2.2.1. The primary ion sources

The primary ion beam is composed of atoms or molecules that transfer their energy momentum to the dislodged target atoms through a collision cascade and induce the formation of the secondary ion beam from the sample.

The ionization of the species of the primary beam allows them to be accelerated by an electrostatic field. Then they are focused through apertures by a succession of electrostatic lenses and deflectors, before reaching the target. The final beam diameter ranges from a few tens of micrometers down to a few tens of nanometers. Many elements can be ionized and used as primary ions to sputter the target, but a primary beam of electropositive elements will enhance the emission of electronegative elements and *vice versa*. Therefore, the two most common primary beam species are Cs^+, used to induce emission in non-metals as negative secondary ions, and O^- (or O_2^-), used to emit metals as positive secondary ion species. The use of these two species can increase secondary ionization yields by several orders of magnitude in comparison with chemically inert species such as Ar.

Low-melting-point metals such as Ga and Cs, are first vapourized by source heating and then ionized on a high-temperature frit (effectively a mesh), producing positive ions by electron loss. The ion beam is extracted with a potential of 3 to 10 kV and focused into the primary column. Metal sources are extremely bright and can produce focused beams of diameters of 50 nm or less.

Negative primary beams are produced in gas sources. The common type of gas source for SIMS is a duoplasmatron, which operates with an arc discharge through which a gas is fluxed. The plasma is magnetically confined by a coil and extracted through an anode extraction aperture to define the ion beam. The duoplasmatron is renowned for its high beam density and low angular divergence. It can also be used to generate a positive ion beam (O or Ar). The two modes require slight modifications of the extraction geometry. For negative ion production, the duoplasmatron also generates an electron beam, which can be excluded by setting the extraction apertures out of the central axis. In the positive mode, the electrons are heading in the opposite direction, and so the duoplasmatron and apertures can be aligned on the central axis.

More recently, radiofrequency (RF) sources have been adapted as SIMS primary ion sources. Induced coupling produces the primary ion plasma. RF sources are expected to be extremely bright and more stable through time than a duoplasmatron.

2.2.2. Sample chamber and ion optic

The Cameca IMS 3f to Cameca IMS 7f instruments are a family of small radius ion microprobes developed by Cameca during the last 40 years that can be considered as the reference instrument for materials analysis. It was designed as an ion microscope with stigmatic refocusing that allows direct imaging of the target. The IMS XF has proven to be a highly versatile instrument.

The primary beam has an incidence angle of 60° to the target (Fig. 2). The secondary ion beam is extracted normal to the target. The sample surface is 5 mm away from the extraction plate, producing a large field gradient that facilitates imaging. The secondary extraction potential has been increased from 4.5 to 10 kV from the IMS 3f to the IMS 7f. Secondary ions are analysed *via* a double-focusing mass spectrometer combining a spherical electrostatic sector analyser (ESA) with a radius of 17.3 cm, and

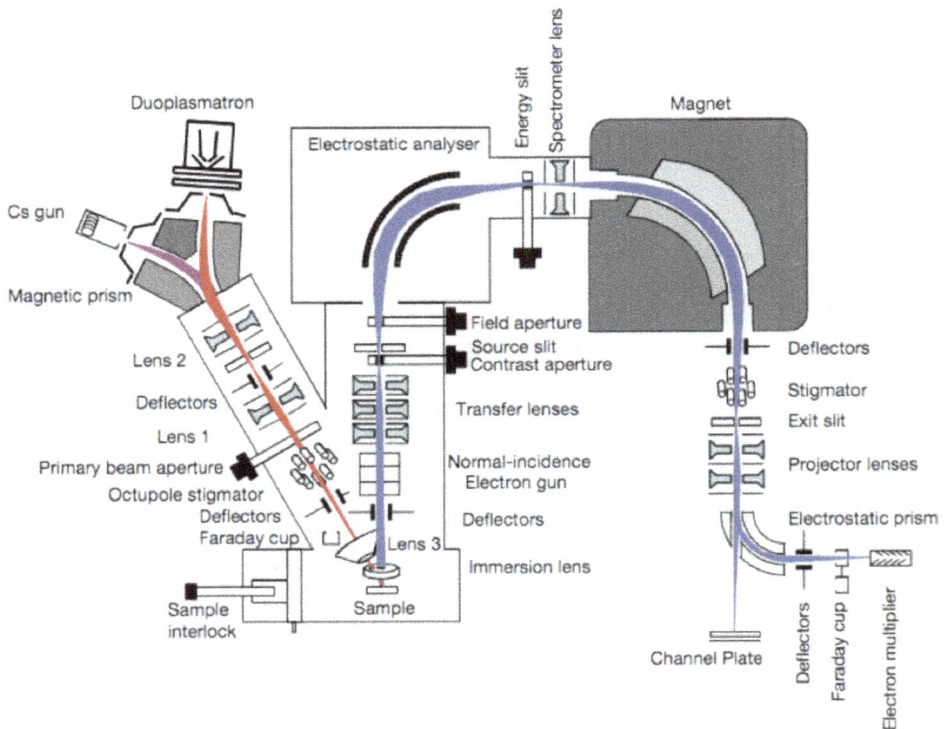

Fig. 2. Schematic layout of the Cameca IMS-3f, an ion microscope capable of direct ion imaging of the target.

a magnetic sector with a radius of 12.7 cm. An energy slit is located at the energy crossover between the two sectors (Fig. 2). This energy slit is usually closed at a few tens of eV width to limit chromatic aberration, and can be used to select the high-energy secondary ions for the measurements in energy filtering mode. Entrance and exit slits are used to define the mass resolution of the mass spectrometer (*cf.* section 1.2). Apertures are used to limit the secondary ion beam. The contrast aperture and the field aperture, set on an image crossover, allow limiting the angular dispersion of the beam and the selection of ions from a particular region of the sample surface, respectively. The detection system consists of a channel plate for the direct acquisition of ion microscope images adjusted with the projector lenses, and a counting system with a Faraday cup (FC) electrometer and an ion counting detector or electron multiplier (EM). The ultimate mass resolution for those instruments is ~5000, but the transmission starts to decrease above 600.

For the 4f and following spectrometers, two ion sources (duoplasmatron and Cs gun) are mounted with switching using a magnetic prism. Usually, the beam is refocused directly from the source and can provide an illuminated area of >100 mm and down to ~200 nm.

The Cameca IMS 1270 series (1280 and 1280 HR) has kept the same ion optic as the IMS 3f series, using similar primary column and source chamber, but incorporating a large mass analyser to provide a greater transmission and sensitivity at high mass resolution. Both the ESA and magnet have 58.5 cm turning radii through 90°. These instruments can be equipped with a multicollection detection system, with five FC or EM fixed on mobile chariots. The 1270 series retains direct ion imaging ion microscopes, but it has also a high transmission mode where the sensitivity is increased at the expense of the spatial resolution.

The SHRIMP I and II (Fig. 3) were developed as geological ion microprobes with the purpose of maintaining high sensitivity at high mass resolution (Ireland *et al.*, 2008). The primary column utilizes either a Cs gun or a duoplasmatron that can be switched into the column. Primary ions from the duoplasmatron are accelerated to 10 keV, those from the Cs gun to 5 kV. Primary ions are mass-separated in a Wien filter. The primary beam is tilted at 45° to the sample. The secondary ion beam is extracted normal to the target. A bias voltage of ~700 V is applied to the extraction electrode before the secondary beam sees the full extraction voltage. This produces a low-extraction field. The SHRIMP operates solely as an ion microprobe with no ion microscope capability. The mass analyser is a Matsuda (1974) double-focusing mass spectrometer. The 72° cylindrical ESA with a radius of 127 cm is associated with a 46° magnet with a radius of 100 cm. A quadrupole lens and an energy slit are placed between the ESA and the magnet. This large-radius instrument was designed to run at high mass resolution (6000) with a high transmission. It is therefore mostly devoted to isotopic ratio measurements.

The NanoSIMS was built as an ion microprobe with extremely high lateral resolution, down to 50 nm. The design is based on coaxial delivery of the primary ion beam and extraction of secondary ions (Fig. 4). In this configuration, aberrations can be minimized allowing the formation of an ion beam with an extremely small diameter.

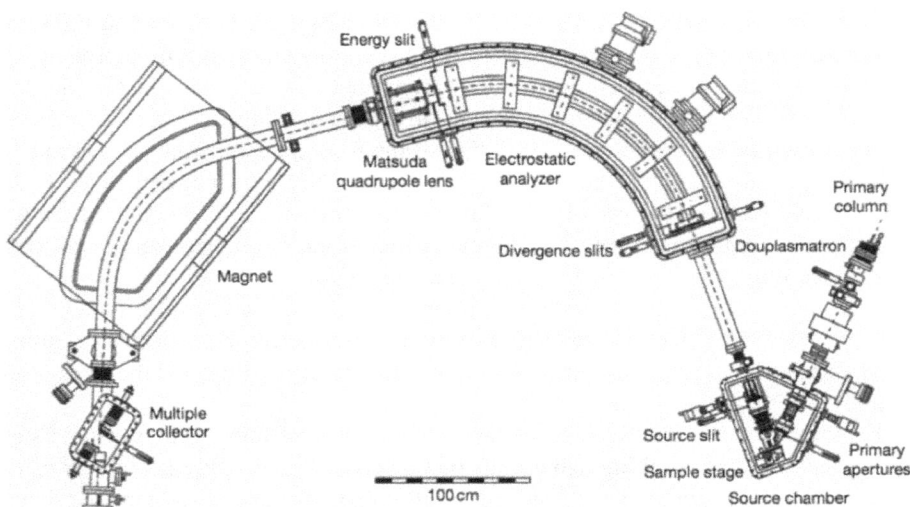

Fig. 3. Schematic layout of the SHRIMP IIe/MC.

Primary ions are generated in either a duoplasmatron (O^-) or a metal source (Cs^+). Unlike the other SIMS instruments, the ion sources are mounted vertically above the secondary extraction column. The primary ion beams are accelerated, focused and then redirected by 90° through an electrostatic prism to align along the secondary extraction axis. The electron gun is also fitted at 90° to the secondary extraction column, and the electron beam is deflected into the secondary extraction axis with a magnetic prism. The primary ion beam is focused down to the sample surface with only 400 µm between the sample surface and the extraction window. This short distance allows a smaller spot size for a given beam current. The minimum spot size is ~50 nm for the Cs^+ beam (for 0.3 pA) and ~200 nm for O^- (for 0.3 pA). The charge compensation when working with Cs is produced by matching the electron potential to that of the sample. Optical viewing of the sample is limited to an off-axis position but enables the positioning of the sample into the geometry of the sample stage (*i.e.* a fixed offset can be applied from the viewing location to the analysis position).

The short working distance also provides high collection efficiency in the secondary beam extraction. The secondary beam requires deflection to compensate for the magnetic prism responsible for electron insertion, and the electrostatic insertion of the primary ion beam. The secondary beam is focused to the source slit to enter the mass analyser, a double-focusing ion optic based on a Mattauch–Herzog design. A 90° deflection through the ESA produces an energy crossover at the energy slit. The magnet structure is very open, allowing a large mass fraction of the secondary beam to be refocused to the focal plane, and such to measure simultaneously several different chemical elements. The detection system consists of one fixed detector and a set of mobile detectors (5 or 7). Unlike the other Cameca SIMS instruments, the NanoSIMS is

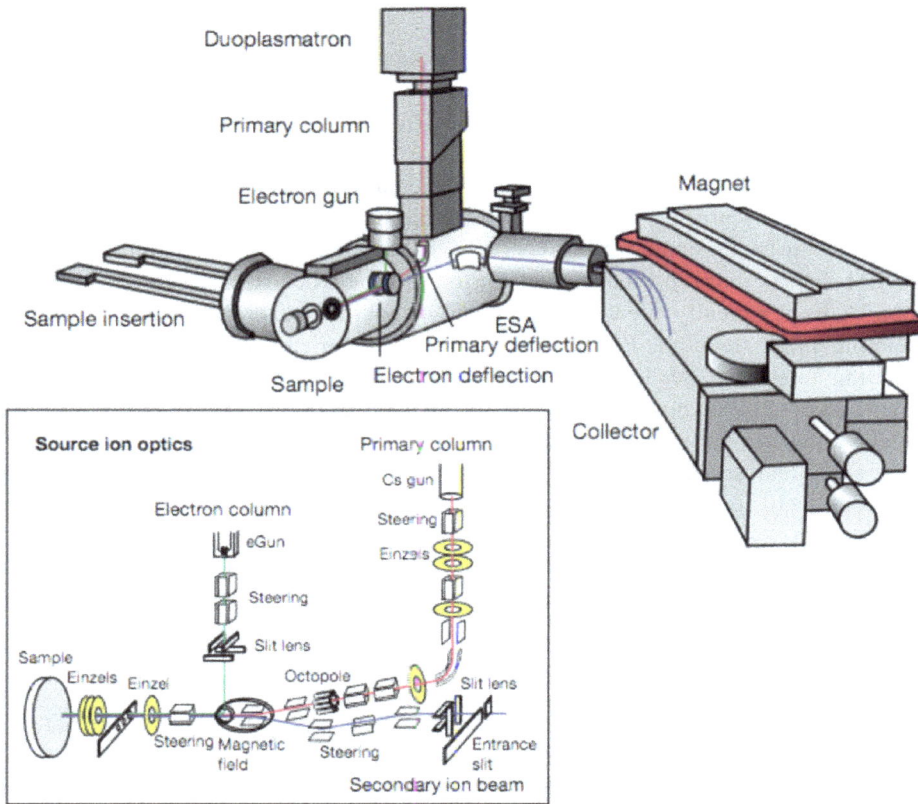

Fig. 4. Schematic layout of the Cameca NanoSIMS. Inset shows the ion optical beam lines for the primary, electron and secondary beams in the source region.

an ion microprobe without direct imaging of the sample. Ion images are only obtained by scanning the primary beam over the surface.

2.3. Secondary ion detection

The secondary ion intensity obtained during SIMS analysis depends on multiple parameters, such as (1) the primary beam intensity that ranges from pA (10^{-12} A) for NanoSIMS measurements up to tens of nA (10^{-9} A) for microprobe; (2) the nature of the analysed element, that can be present as a trace or as a major element in the sample; and (3) on the analytical conditions. Depending on the sample and on the goal of the measurement, ion beam intensities can vary from less than 0.1 c.p.s. (counts per second) up to >10^9 c.p.s. Therefore, high-sensitivity ion counters (EM) have to be used for weak signals ($\leqslant 2 \times 10^5$ c.p.s.$^{-1}$), and Faraday cups (FC) for stronger signals ($\geqslant 10^6$ c.p.s.). The developments of higher-sensitivity FC is expected to extend their use towards lower intensities (~2×10^5 c.p.s.) in the coming years.

2.3.1. High-sensitivity ion counter (EM)

Ion counters are based on electron multipliers. Incoming ions strike the first dynode (or conversion electrode) to generate several electrons. On a discrete electrode electron multiplier (Fig. 5), the electrons are accelerated towards successive dynodes, and the electron cascade increases exponentially, with a total current amplification in the range of 10^6 or more. Each incoming ion produces a current pulse at the exit of the EM. A discriminator discards output pulses below a certain voltage, as they are considered as the electronic background, while other pulses are counted. The pulse width at the exit of the EM defines its dead time, *i.e.* the period of time when it cannot process another count. This is typically a few tens of ns, and it should be taken into account when the counting rates are high. For accurate isotope measurements, particularly at high counting rates, the dead time of the system must be known accurately and be independent of the count rate. As the dead time depends on the EM aging, the measured species and their energy and polarity (Zinner et al., 1986), it should be determined for each analytical session. The EM aging is linked to at least two points: the first electrode's erosion by the incoming ions, the last electrode's damaging by the high electron flux.

Therefore, they cannot be used for high secondary ion currents, but are usually limited to secondary ion beam currents below 1×10^6 c.p.s. Consequently as to the increasing amount of counted ions, their yield decreases with time.

2.3.2. Faraday cups

To measure ion-beam intensities greater than 10^6 c.p.s., all instruments use Faraday cups. The arrival of ions on a metal detector effectively creates a current. The Faraday cup is a simple assembly used to trap the secondary ions. Ions pass through the collector slit and enter the cup. As the ions impact in the cup, secondary electrons can be produced and, to avoid any current loss, an electron repellent is set in front of the cup with a moderate negative voltage. Incoming charges pass through a resistor, and the voltage across the resistor is measured (Fig. 6). The resistor value can be adjusted to optimize the sensibility of the collection, with high currents ($\sim 5 \times 10^9$ c.p.s.) being measured through low resistor values (10^9 to 10^{10} Ohm) and low currents ($\sim 1 \times 10^6$ c.p.s.) through higher resistor values (10^{11} to 10^{12} Ohm). The measurement

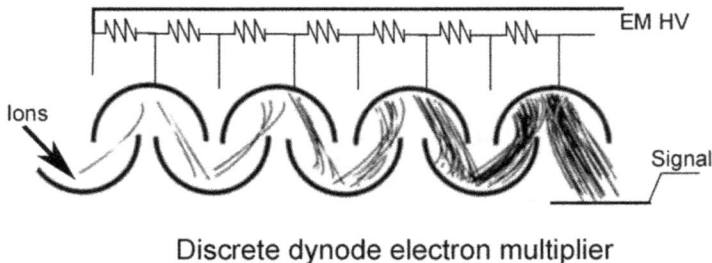

Discrete dynode electron multiplier

Fig. 5. Schematic representation of a discrete electrode electron multiplier.

Faraday cup - FC

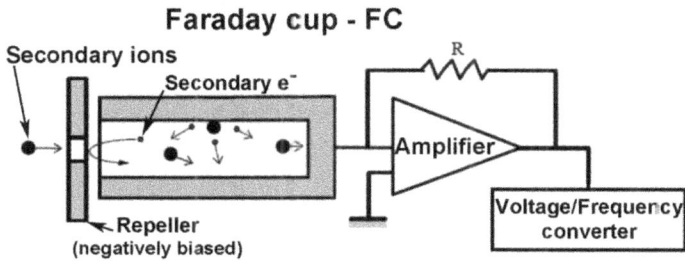

Fig. 6. Schematic representation of a Faraday Cup detection system.

of low currents on a Faraday cup is limited by the intrinsic noise of the system. In a resistor, there is a background current simply due to the thermal noise of electrons.

2.3.3. 2D detectors

In ion microscope mode, the secondary ion images can be viewed by using 2D detectors such as micro channel plates. A micro channel plate (Fig. 7) consists of 10 μm-long oblique cylindrical holes, separated by 20 μm from each other, in a resistive plate. Each ion impact generates an electron cascade in the channel, due to the strong electric field gradient. At the exit of the channel, electrons impact a phosphor glass that is imaged using a camera. An alternative method is the collection of electrons on a resistive anode encoder that allows a digital output to be produced.

2.4. Monocollection, multicollection

The first generations of ion probes (CAMECA IMS 3f, 4f, 6f, IMS 1270, SHRIMP I and II) were built with a single detection system, combining one EM and one FC at the same position. Therefore, to measure different elements or different isotopes of the same element, the value of the magnetic field is changed according to their mass-to-charge ratio, following:

$$B = \frac{\sqrt{2V\frac{m}{z}}}{R}$$

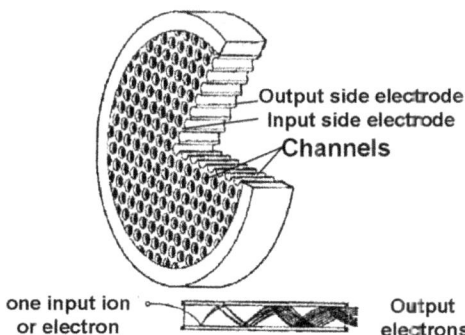

Fig. 7. Schematic representation of a micro channel plate.

where B is the magnetic field, V the acceleration voltage, m the mass and z the charge of the secondary ion, and R the magnetic sector radius.

In monocollection mode, the different species are measured successively with a waiting time between each species to allow the magnetic field to stabilize, and eventually for the FC measurement system to reach a steady state. To obtain good precision, the measurements of the different species are repeated over several measurement cycles, and are interpolated in time to correct any time-dependent increase or decrease of the signals. Usually, for the measurements of trace- or light-element concentrations, a precision of a few % can be achieved by accumulating a few tens of cycles, with counting times for each species ranging from a few to tens of seconds, depending on the counting rate. But to obtain a good precision (‰ or less) for isotopic measurements, the number of cycles needed increases significantly.

In order to allow high-precision isotopic measurements, the Cameca IMS 1270 and then the SHRIMP SI have been equipped with multiple collection measurement systems, in which up to five detectors can be used simultaneously to measure the different isotopes of an element or different elements, with a mass dispersion up to 1.125 (the ratio between the highest and lowest mass measured simultaneously). The NanoSIMS detection system was designed to measure simultaneously several elements on a wide mass range, with mass dispersion of 13.2 for the NS 50 and 21 NS 50L. In multicollection mode, the different isotopes or the different elements are measured simultaneously on detectors previously located at the relevant radial position for a given magnetic field value. The major benefits are the elimination of the beam noise associated with the primary beam, and the much shorter measurement time needed to reach a high precision, as all isotopes are measured simultaneously. Typically, C, O, S, isotopic measurements can be achieved in a few minutes with precisions of 0.1‰ when the counting rates are high enough to use Faraday cups. Multiple ion counting or a combination of ion counting and FC can also be used, on the NanoSIMS or on the other ion probe, but since the ion counters are prone to drift and are particularly sensitive to the count rate, the precision is more limited.

2.5. Mass resolution, sensitivity and detection limits

2.5.1. Mass resolution

The mass resolution of a mass spectrometer (Mass Resolving Power: MRP) represents its capacity to separate ions of different masses, calculated as

$$\text{MRP} = \frac{M}{\Delta M}$$

where M is the mass of the ions, and ΔM the mass difference between them.

To separate ^{206}Pb from ^{207}Pb, the mass resolution needed is 207, whilst to separate ^{56}Fe (55.9349 a.m.u. (atomic mass units)) from $^{40}\text{Ca}^{16}\text{O}$ (39.9626 +15.9949 = 55.9575 a.m.u.), the mass resolution needed is ~2500. A main difficulty in SIMS analysis of natural samples is that natural samples contain numerous chemical elements. Therefore, the secondary ion mass spectrum includes many molecular ions and a high mass resolution is required to separate them. For light elements, from H to Fe, most molecular isobars (hydride, oxide, *etc.*) can be separated from atomic species with a mass resolution ⩽5000, but other compounds may require higher resolutions (Fig. 8). For the elements in the mass range from Fe to the *REE*, the mass resolution needed becomes >10,000, and for the heavier elements it decreases to ~6000 (Pb isotopes, U-Th decay series).

The optimal peak shape is trapezoidal (Figs 8 and 9). A flat top is essential to get high-precision measurements because small variations in the magnetic field will not change the secondary ion beam intensity. The peak shape is determined mainly by the source slit image and the exit slit (or collector slit) widths (Fig. 9). The mass resolution is defined by the sum of the source slit image width and the exit slit width, and the width of the flat top peak by their difference. Therefore, the source slit image width has to be kept at less than half the exit slit width. When the mass resolution becomes high (>300 on a small sector ion probe, >3000 for a large sector ion probe), the source slit width controls the transmission directly, which decreases as a function of it. At high mass resolution, the transmission will also depend on the beam size at the source, *i.e.* the size of the primary beam and of the sputtered area, and of the magnification of the beam from the source to the collector.

The dispersion of the mass spectrometer controls the physical distance between peaks, and the mass resolution increases with the mass dispersion. This is the reason for the large sector ion microprobe (SHRIMP and Cameca IMS 1270 series). Note that the NanoSIMS can also reach high mass resolution due to the smaller size of the secondary beam source. The effective mass resolution depends on the instrument and on the analytical condition for each analysis. The mass resolution required for a specific

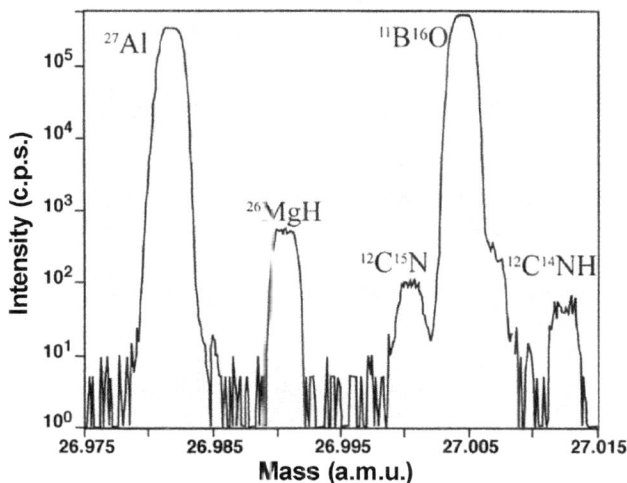

Fig. 8. High resolution (MRP = 8000) mass spectrum for CN analysis. Measured on the CRPG Cameca IMS 1270 ion microprobe, Cs^+ primary beam and negative secondary ions.

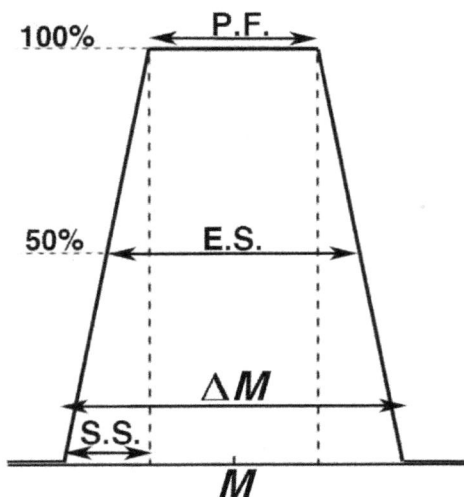

Fig. 9. Schematic representation of secondary ions peak-shape formation. The peak is formed by the passage of the source slit image (S.S.) across the collector slit (E.S.). A flat top peak is obtained when the source slit image (S.S.) is narrower than the exit slit (E.S). The peak width (ΔM) is equal to S.S. + E.S, the peak top flat width (P.F.) to E.S. − S.S. The mass resolution is $M/\Delta M$.

analysis can be calculated from the nuclide masses, but when the mass of the analysed element increases, the number of isobaric polyatomic interferences increases quickly. As an example, the possible isobaric interferences for Sr isotopes include Ca_2, Mg_2O_2, Si_2O_2, FeO_2, Al_2O_2, $AlSiO_2$ and others, which are all present as major elements in silicates, so that a mass resolution >30,000 is required.

2.5.2. Lateral and depth spatial resolution

The spatial resolution of SIMS is determined mainly by the primary ion beam diameter for the lateral resolution, and by the primary ion stopping depth for the depth resolution. The highest lateral resolution is provided by the NanoSIMS instrument, which is designed especially with a primary beam perpendicular to the sample surface. The primary beam can then be focused on an area as small as 50 nm in diameter. However, such a small primary beam has a limited intensity, down to a few pA, and produces only weak secondary ion intensities. Thus its use is limited to secondary ion imaging for major or minor elements. High precision measurements or trace-element measurements require higher primary beam intensities larger than ~1 nA and up to several tens of nA. Therefore, the beam diameter becomes larger than 1 μm, and is even currently up to 10 or 20 μm.

The depth reached by the primary ions depends on the density and the crystallographic orientation of the target, but mainly on the nature and the energy of the primary ions. O^- primary ions will reach a depth of ~2 nm in a silicate target with a 100 V acceleration voltage, 50 nm at 10 kV; and Cs^+ ions will reach a depth of 3 nm at 1 kV, and 20 nm at 10 kV (Schiott, 1970). To perform analyses with a high depth resolution of a few nm to evaluate processes such as fluid–rock interaction, surface alteration or contamination, low-energy beams (a few hundred eV) are used, with heavy ions or molecular primary ions. The primary beam is often defocused or scanned over a large surface to decelerate the sputtering. For conventional analyses, the primary ion energy is typically from 10 to 13 kV, and a mixing layer of a few tens of nm is formed under the beam. With an intensity of several nA, a focused primary beam will sputter the target from a few tens to 100s of r.m per minute.

2.6. Calibration and matrix effects

2.6.1. Chemical fractionation

The SIMS analysis of natural samples produces a mass spectrum including a large variety of atomic and molecular species formed by the major, minor and trace elements. Each element belongs to several species of the mass spectrum, as mono- or multi-charged single atomic ion, as oxide, hydride and multi-atomic molecular species.

Furthermore only a small fraction of the material sputtered from the target is ionized, most of the material is ejected as neutral species. At the beginning of ion probe development (1970–1980) it was expected that the theoretical understanding of the secondary ion emission process would allow quantification of element and ion abundances from SIMS spectra. Various types of models have been proposed based on the existence of plasma in local thermodynamic equilibrium (LTE) at the sample surface during ion bombardment (Anderson and Hinthorne, 1972; Simonds *et al.*, 1976; Rudenauer and Steiger, 1976). However, the model predictions were often far from the true values (Rudat and Morrison, 1979) and the measurements are now usually calibrated following empirical models based on the comparison between the signal intensity for a certain element or isotope in the phase under examination with the corresponding signal intensity in a reference material. Note that the signal measured can be either atomic or molecular species. For instance, the water content in a glass or mineral may be determined by measuring H^-, H^+ or OH^- secondary ions. But the measured species should be free of any isobaric contribution. H^- or H^+ can be measured at low mass resolution, OH^- should be measured at a mass resolution ($M/\Delta M$) >6000 to be separated from $^{17}O^-$. To measure trace elements such as *REE*, the mass resolving power needed to separate the oxides is >10,000, which is not possible with small magnetic field radius ion probes such as the Cameca xf.

The energy-filtering technique was first proposed by Shimizu (1978) to analyse a series of elements ranging in mass from Na to Co in plagioclase, and was extended by Shimizu *et al.* (1978) to the analysis of trace elements, including *REE*. This technique is based on measuring only the ions which are emitted from the sample with a high energy, typically >80 or 100 eV. As the energy spread of molecules is much more restricted than that of single ions (Fig. 10), this allows the ratio between single ions and molecules to be increased by several orders of magnitude, and to minimize molecular isobaric interferences. Zinner and Crozaz (1986) improved this filtering technique for *REE* measurements: a range of masses between mass 133 and 191 was measured and deconvoluted into atomic and oxide species. The measurement of multiple isotopes of elements allows the deconvolution quality to be checked, and the ion counting rates from individual elements are related to the signal from a major element the concentration of which can be determined by electron microprobe (usually Si or Ca). For each element, a relative sensitivity factor (measured ratio divided by the true ratio)

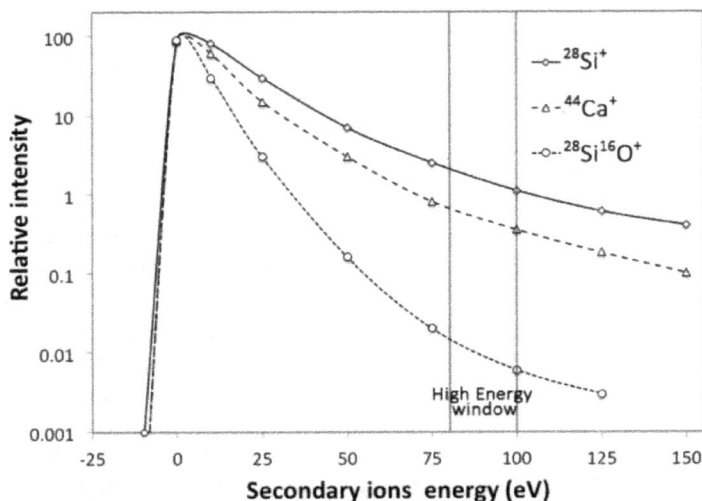

Fig. 10. Energy distribution of secondary ions for Si, Ca and SiO, measured on plagioclase with an O-primary beam and an energy aperture of 10 eV. In measuring the high-energy ion (90 ±10 eV) the contribution of $^{28}Si^{16}O$ on ^{44}Ca will be about two orders of magnitude lower than in measuring the low-energy ions (0 ±10 eV).

is defined on reference materials, and used to obtain the concentration of unknown samples. One of the benefits of energy filtering is that the relative sensitivity factors for high-energy ions show much less matrix sensitivity than the low-energy ions. As such, standards can have a wider range of application where they are suitable.

2.6.2. Isotopic fractionation

As the secondary ion emission results from the collision cascade following the impact of the primary ions with sample atoms, the emission yield will depend on the secondary ion mass. Therefore an isotope mass fractionation will occur during sputtering and secondary ionization, and depend on the sample composition and structure (Slodzian *et al.*, 1980; Shimizu and Hart, 1982b). Moreover, instrumental factors can also influence the observed isotopic mass fractionation, such as secondary ion energy and field aperture (Deloule *et al.*, 1992).

A significant difficulty in getting significant *in situ* isotopic ratios using an ion probe is to have appropriate reference materials. Indeed, it is not easy to get reference minerals with homogeneous isotopic composition at the micrometer scale in sufficient amounts to measure them by conventional methods and to share them on a large scale. To be able to measure natural samples of variable composition, different empirical models for elemental isotopic fractionation have been developed. Deloule *et al.* (1991) showed that hydrogen isotopic fractionation during amphibole and mica measurements can be calculated as a function of the total electronegativity of the cations on the octahedral sites (Fig. 11). Chaussidon and Albarede (1992) use a similar model for

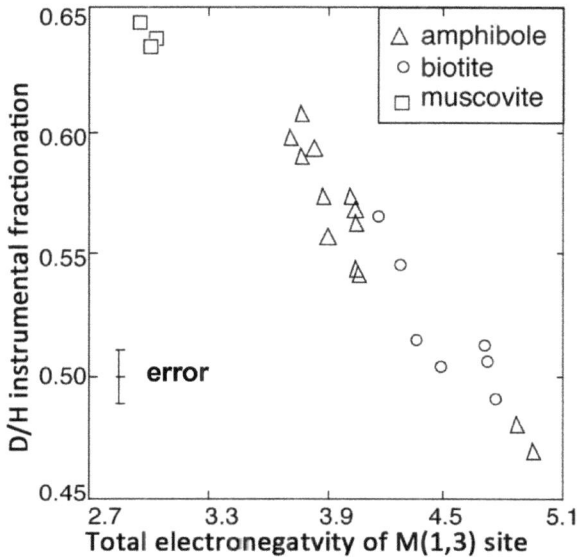

Fig. 11. D/H isotopic fractionation *vs*. the total electronegativity of the octahedral cations for amphibole, biotite and muscovite (Deloule *et al.*, 1991).

Boron isotopic fractionation from tourmaline. Eiler *et al.* (1997) developed a two-collision kinetic calculation to model the instrumental isotopic fractionation of high-energy (350 eV) oxygen secondary ions. Vielzeuf *et al.* (2005) showed that for the O isotope measurement in Fe–Mg–Ca garnet a three-pole linear mixing model allowed the instrumental fractionation to be corrected accurately. Also for O isotopes measurement, Valley and Kita (2009) reported a non-linear mixing model between the Mg and Fe poles for olivine and pyroxene isotopic fractionation. Bell *et al.* (2009) reported a linear relationship between the instrumental isotopic fractionation and the Mg/(Mg+Fe) ratio for the Li isotope measurement by SIMS in olivine. Similar calibration models have been published for many other elements (*e.g.* C, N, Mg, Si, S, Ca, Fe) during the last decade. Therefore the *in situ* determination of isotopic ratios using SIMS is strongly dependent on the comparison of samples with standards of the same chemistry and mineralogy. If the matrix is poorly defined or heterogeneous at the analytical scale, it may be extremely difficult to standardize and resulting data should be treated with caution. Furthermore, in some cases, such as O analysis in magnetite or hematite, the crystal orientation may affect the instrumental mass fractionation (Huberty *et al.*, 2010). But for well defined matrices the analytical precision can reach 0.1% for the *in situ* analysis of major or minor elements.

2.7. Detection limits and precision

The detection limits in SIMS analysis are related mainly to the secondary ion yields. The sputtering yield is high with 5 to 15 atoms extracted from the target for each incoming primary ion. However, only a small proportion of these atoms extracted are ions, most of them being neutral. The ionization yield relies largely on the ionization potential for positive secondary ions and on the electron affinity for negative ions of the analysed elements. Elemental ionization yields vary over several orders of magnitude between different elements (Hinton, 1990). The ionization yield relative to the primary ion intensity can reach several percent. For instance, this is the case for oxygen in silicates, which reaches a relative ionization yield for $^{16}O^-$ intensities close to 10^9 c.p.s./nA (1 nA $\approx 6 \times 10^9$ Cs^+ ions/s) by using a Cs^+ primary beam. The yield is even more important for more refractory elements. Thus, the SIMS instruments allow their detection at levels ranging from ppm to ppb (10^{15} to 10^{12} atoms/cm^3). The very low detection limit of SIMS instruments relies on (1) the ionization yields of the elements, (2) a very low background, due to the ultra-high vacuum condition of the analysis and very limited instrumental contamination, and (3) the high sensitivity of the ion counting device with an electronic noise of a few counts per minute.

When measuring trace element contents, the analytical precision may depend largely on the number of counted ions, *i.e.* the statistical error. But for minor or major element contents or isotopic ratios, the precision relies mostly on the quality of the measurement calibration. Indeed, as SIMS analyses are *in situ* measurements, *i.e.* without any chemical preparation, they are strongly affected by matrix effects. These matrix effects affect the elemental relative ionization yield, and the instrumental isotopic fractionation. A precision of a few percent for chemical contents and of a few hundred ppm and even lower for isotopic ratios can be reached, provided that appropriate reference material is used for the calibration. Over recent decades, many reference materials have been developed, allowing trace-element or isotopic-ratio measurements in various matrixes, but many remain to be defined.

3. Trace-element and isotope analysis in altered glasses

A glass in contact with water is not inert. Different reactions occur and lead to the formation of an alteration layer on the surface of the pristine glass. This layer is depleted in some elements and has the capability to retain others. It is essential to know the efficacy of this altered layer as a containment for hazardous elements. This is true, for instance, for lead in a crystal. Added to glass in order to increase its brilliance, lead is released by the crystal when it comes into contact with liquid (*e.g.* alcohol in a carafe). Because its presence bears health risks, public health laws have regulated its release. Thus, test procedures have been defined to assess whether the Pb release exceeds the permissible limits. In the case of nuclear waste glass, the quantification of the release of radionuclides over periods as long as several millions of years (physical radioactive period of the most radioactive elements) is of great importance, but very challenging. Throughout the world, glass has been chosen to confine highly radioactive wastes, because it can accommodate a large number of elements in its microstructure, and has great durability (Ojovan and Lee, 2011). But, how is it possible to ensure a durable confinement of radioactive elements over such a long period of time? Such a long-term prediction requires the understanding of alteration mechanisms in order to establish a solid basis for geochemical modelling of the long-term behaviour of nuclear waste glass. With this aim, experimental alterations of nuclear waste glass have been performed for >30 years. In spite of the crucial insights gained through these numerous short-term studies, the long-term behaviour of glass is still the subject of debate. With the time factor being limited in laboratory experiments, other studies have focused on natural glasses such as several thousand year old basaltic glass or obsidian, or several centuries old historical glass, such as stained glass windows, all being considered as analogues of nuclear glasses (Lutze *et al.*, 1985; Sterpenich and Libourel, 2001; Libourel *et al.*, 2011; Parruzot *et al.*, 2015).

3.1. Experimental alteration of glasses

Far away from the geological scale, the laboratory scale provides access to the kinetics of initial alteration stages. In glass alteration experiments, glass (monoliths or powder) is brought into contact with a leaching solution in order to study the exchange between glass and solution. Each experiment can be configured to evaluate independently the effect of different parameters on fluid–glass interaction. The duration of the majority of these experiments has been no more than several months. The longest experiment described in the literature lasted 26 years (Gin *et al.*, 2011). These experimental alterations have allowed scientists to study how factors such as pH, temperature, renewal or composition of the altered solution affects the glass alteration process.

Borosilicate glasses are currently used to confine nuclear wastes. Consisting of ~30 elements, their compositions are very complex and are tailored to ensure that the fission products are immobilized. For example, the French R7T7 nuclear glass can accommodate up to 18.5 wt.% of fission product oxides the individual concentrations of which are <1 at.% (trace elements). This borosilicate glass contains large amounts (>1 wt.% oxide) of sodium, aluminium, calcium, iron, lithium, molybdenum, zinc, zirconium, caesium and rare earth elements. For most studies, a non-radioactive glass analogous to the radioactive glass (called SON68) is used to simplify its manipulation. Other experiments have also been conducted on glass with simplified compositions in order to understand the behaviour of one particular element or to ease the interpretation of the results.

Despite the precautions taken, water will inevitably come into contact with nuclear glass during its lengthy storage. Thus, water is the main factor responsible for glass alteration and the spreading of radioactive elements. The complexity of nuclear glass alteration lies in the complexity of glass composition and the existence of simultaneous reactions. Each reaction influences the kinetics and mechanisms of the other reactions (Bunker, 1994).

   (1) A preferential exchange (also known as interdiffusion) between the network modifiers of the glass (mainly alkali metals) and hydrogenated species (H^+, H_3O^+, H_2O, OH^-) in solution. This is predominant during the first step of the alteration.
   $$\equiv Si\text{-}O\text{-}Na^+ + H_3O^+ \rightarrow \ \equiv Si\text{-}OH + Na^+ + H_2O$$
   $$\equiv Si\text{-}O\text{-}Na^+ + H_2O \rightarrow \ \equiv Si\text{-}OH + Na^+ + OH^-$$

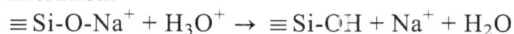

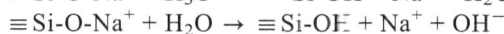

(2) Hydrolysis reactions of the glass network through the covalent bonds Si$-$O$-$M (M is a network former) that can lead to its dissolution at a later stage.

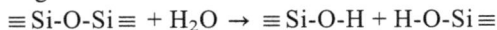

$\equiv$ Si-O-Si$\equiv$ + H$_2$O $\rightarrow$ $\equiv$ Si-O-H + H-O-Si$\equiv$

(3) Condensation reactions

$\equiv$ Si-O-H + H-O-Si$\equiv$ $\rightarrow$ $\equiv$ Si-O-Si$\equiv$ + H$_2$O

(4) Secondary phase precipitation resulting from a thermodynamic equilibrium between a solid and a solution in contact.

During the experimental alterations of glass, the leachates are commonly analysed by inductively coupled plasma-atomic emission spectroscopy, absorptiometry or inductively coupled plasma-mass spectrometry, electron microscopy, Raman spectroscopy, X-ray and neutron reflectometry, extended X-ray absorption fine structure, nuclear magnetic resonance, SIMS, and more recently, atom probe to characterize the altered layers. Therefore, it is possible to follow the exchanges between glass and the solution in contact.

Among the different techniques already used, the SIMS technique has contributed to acquiring new insights in predicting the durability of nuclear waste glasses mainly through the acquisition of elemental depth profiles, but also through isotopic measurements and elemental mapping.

3.2. Characterization of altered layers

3.2.1. Partitioning of elements: example of rare earth elements and actinides

The SIMS elemental depth profiling provides important information about the distribution of elements in the different layers that formed consecutively during the glass alteration process (Lodding *et al.*, 1985; Lodding and Engström, 1992; Koenderink *et al.*, 2000; Riciputi *et al.*, 2002). For depth profiling, the use of a negative oxygen primary ion beam is beneficial. Firstly, oxygen enhances the ionization of the elements of interest confined in the glass, because these elements are electropositive (excellent sensitivity favourable for trace elements). Secondly, during analysis, the low-charge effect induced by the accumulation of positive charges on the surface of the glass can be corrected easily by applying an offset voltage to the surface of the glass. This can be done by limiting the energy window to ~20 eV and by regularly monitoring the position of the distribution of energy. This distribution is obtained by recording the intensity of one of the minor isotopes of silicon (^{29}Si or ^{30}Si). The shift in energy evolves only during the first stages of the analysis (pre-equilibrium state) and then remains roughly constant (a few tens of volts). The use of Cs$^+$ or O$_2^+$ primary ions instead of O$^-$ ions is also possible but requires the use of an electron gun to compensate the positive charges that accumulate over the scanned surface. Nevertheless, the existence of an electric field can induce migration of the most mobile elements such as alkalis (Migeon *et al.*, 1990) and distort their true distribution. In any case, the surface has to be coated with a metallic layer (preferably gold).

The application of an additional offset (between 80 and 100 V) allows polyatomic isobaric interferences to be eliminated. The mass-filtering based on the difference in energy distributions between mono- and polyatomic ions has already been applied successfully to silicates for the measurements of *REE*s with detection limits below <50 ppb and 200 ppb for the light *REE*s and the heavy *REE*s, respectively (Zinner and Crozaz, 1986; MacRae *et al.*, 1993). More recently it was used to follow the behaviour of these elements during the experimental alteration of borosilicate glass SON68 (Valle *et al.*, 2010). It enabled us to acquire elemental depth profiles of La, Ce, Pr and Nd by suppressing their respective isobaric interferences (for instance ^{138}BaH and ^{94}Zr^{29}Si^{16}O for ^{139}La). In this way, it was highlighted that the secondary phases, phyllosilicates and phosphates, play a crucial role by trapping *REE*s (Fig. 12). These elements are also partly retained within the hydrated glass (also called gel). These alteration products were able to confine almost 100% of *REE*s in the alteration conditions described by Valle *et al.* (2010). During a five-year alteration of the SON68 glass in interaction with Boom clay, *REE*s and U were found to be mobile whereas Th was found to be "inert" (retention above 80%; Lodding and Van Iseghen, 2001). Retention above 60% and above 40% was observed for *REE*s (Ce and Nd) and U, respectively. These elements, considered to be analogues of the tetra- (Np-U) and tri- (Pu and Am) valent actinides, bring crucial data related to the active glass. A few similar SIMS analyses carried out on radioactive glass have shown a retention of Th, Np, Pu and Am in the alteration layer and a greater mobility for uranium (Valcke *et al.*, 2007). The energy filtering leads to a transmission loss for monoatomic ions by a factor of 50. This transmission loss is similar to that encountered at a $M/\Delta M_{(50\%)}$ of 5000 on a small-radius sector instrument (Sangely *et al.*, 2015).

3.2.2. Hydration, hydrolysis and interdiffusion

SIMS is one of the techniques of choice to study the penetration of water into glass. Water can penetrate into glass by diffusion through its void space between oxygen

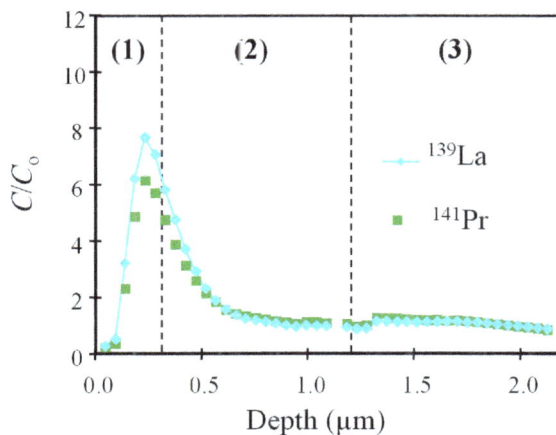

Fig. 12. Example of the partitioning of *REE*s in the three layers which make up the altered glass SON68, obtained by SIMS: (1) phyllosilicates; (2) gel; and (3) pristine glass. Some phosphates are also present at the interface between (1) and (2).

atoms in the structure or *via* hydrolysis and condensation reactions with metal–oxygen bonds. Generally, the rate at which water penetrates the glass structure controls the kinetics of the other reactions occurring between glass and water (Bunker, 1994).

Provided that the vacuum inside the SIMS machine is low enough to decrease the H background, the detection limit of H_2O measured by SIMS can be as low as 5 ppm (Rhede and Wiedenbeck, 2006). By applying the SIMS analytical protocol defined by Deloule *et al.* (1995) and Fick's law of diffusion, Sterpenich and Libourel (2006) determined apparent diffusion coefficients of H in different buried medieval stained glasses altered over a period of 1000 y. They demonstrated a dependence of the apparent diffusion coefficient of H on the degree of polymerization of the glass (defined as the non-bridging oxygens per tetrahedrally coordinated cation). The higher the degree of polymerization of the glass (more open in its structure), the lower the permeation of water in the structure. Such evolution cannot be explained by the simple permeation of water through the porosity of the glass. This correlation reflects the preponderant role of the hydrogen–cation exchange during glass hydration.

Alkali metals and boron are known to be the most mobile elements during the alteration of glass. Thus, boron that is not trapped in secondary phases is usually used to determine the thickness of the alteration layer (from its concentration in solution). The shapes of alkali metals or boron and hydrogen SIMS profiles (sigmoidal profiles) are anti-correlated and mark out the zone of glass hydration. Because interdiffusion is faster than the hydrolysis of the glass network, a zone of interdiffusion is present at the alteration front. It has been impossible to determine the thickness of this zone from SIMS depth profiles carried out on conventional dynamic SIMS instruments (CAMECA IMS Xf). Indeed, the majority of measurements were obtained with an O^- bombardment of high-impact energy (14.5 keV), which is detrimental for depth resolution. Moreover, the depth resolution is primarily reduced by the roughness of the altered glass that results from the precipitation of secondary phases (size of the analysed area which is several tens of micrometers in diameter). This roughness is further intensified during SIMS analysis. So, the thicker the altered layer, the greater the depth resolution. Although the depth resolution will inevitably be reduced by the intrinsic roughness of the interface pristine glass/altered glass, a very low-energy SIMS bombardment (<500 eV) should give better results, at least for short alteration times. Indeed, the broadening of this interface is <30 nm in the study of Fearn *et al.* (2006) with a 1 keV energy caesium beam.

To circumvent the problem of roughness, SIMS imaging (NanoSIMS instrument) has been used recently to study the interdiffusion zone (Gin *et al.*, 2011). Both O^- primary ions favourable for the ionization of boron and alkali metals and Cs^+ primary ions favourable for optimizing the lateral resolution were tested. The thickness of the diffusion front was determined from the evolution of normalized boron intensities through the interface between pristine glass and altered glass (from linescans on boron NanoSIMS images; Fig. 13). The most precise value was obtained with the Cs^+ bombardment. It was found to be ~170 nm. More recent studies by atom probe have demonstrated that the thickness can be as small as 5 nm and support the hypothesis of a boron release controlled by interface reaction (Gin *et al.*, 2013).

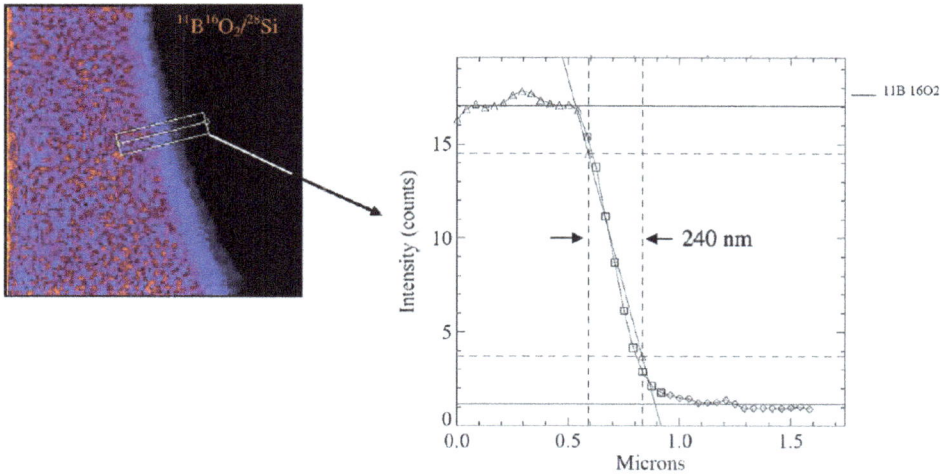

Fig. 13. Investigation of the interface between altered and pristine glass by SIMS imaging. Reprinted with permission from Gin *et al.* (2011). Copyright (2011) the American Chemical Society.

3.3. Isotopic markers of alteration

The alteration of glass has long been a subject of debate. Some researchers have considered altered glass as a residual skeleton of the pristine glass resulting from selective leaching including interdiffusion between alkalis and hydrogen species. This pathway was observed mainly for soda-lime silicate glasses (Rana and Douglas, 1961a,b; Doremus, 1975). Others support the hypothesis that the altered glass is the result of dissolution-reprecipitation reactions, which implies thermodynamic equilibrium between the solution and the altered glass. Alteration by *in situ* condensation reactions was also reported for nuclear waste glass (Pederson, 1987; Jégou *et al.*, 2000; Gin, 2001; Tsomaia *et al.*, 2003; Frugier *et al.*, 2008).

Isotopic tracing can be used to discern between these hypotheses. The idea is simple. Glass is composed mainly of silicon and oxygen. Silicon has three stable isotopes ^{28}Si, ^{29}Si, ^{30}Si with different abundances, 92.21%, 4.70% and 3.09%, respectively. Oxygen has three stable isotopes ^{16}O, ^{17}O, ^{18}O with abundances, 99.759%, 0.037% and 0.204%, respectively. By modifying the proportion of Si and O isotopes in the leaching solution in contact with the glass (or the opposite; Geisler *et al.*, 2015), it is possible to observe the inward and outward migration of the two main components of the glass (Valle *et al.*, 2010; Gin *et al.*, 2015). Valle *et al.* (2010) have studied the experimental alteration of the SON68 glass by water labelled with ^{29}Si and ^{18}O under dynamic conditions (Fig. 14).

The SIMS measurements of the O and Si isotopes were carried out at a mass resolution of 1600 and 3500, respectively, to eliminate isobaric interferences ($H_2{}^{16}O$ at mass 18 and ^{28}SiH at mass 29). From the variation of the Si and O isotopic ratios along SIMS depth profiles three different zones were distinguished: the phyllosilicate layer with the largest shifts in O and Si isotopic ratios, the gel with intermediate values, and the pristine glass identified by its natural isotopic ratios (Fig. 15a). Comparison of the

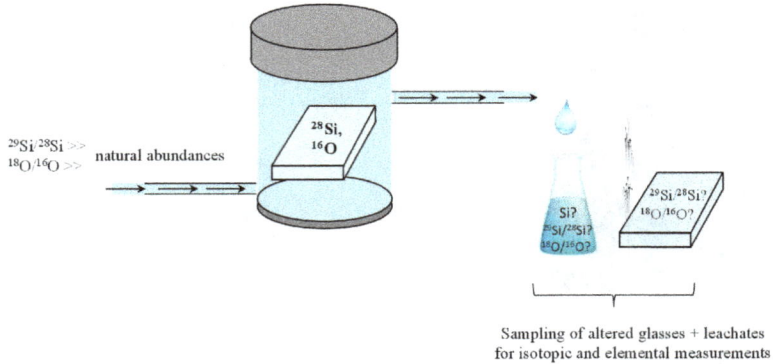

Sampling of altered glasses + leachates
for isotopic and elemental measurements

Fig. 14. Principle of isotopic tracing of glass-alteration mechanisms under dynamic conditions.

isotopic ratios in the leachates and in the altered glass in the course of the experimental alteration enabled us to highlight inward and outward diffusion of silicon and different mechanisms (Fig. 15b). The isotopic signature of the phyllosilicates corresponds at any time to that of the leachate in contact with the altered glass indicating that these silicate minerals grow by precipitation reactions on the surface of the glass. The isotopic signature of the gel takes a value lower than that of the leachate in contact. It was then concluded that the gel is formed by successive hydrolysis/condensation reactions. The incorporation of ^{18}O in the altered glass occurred *via* recondensation of hydrolysed siloxane as already proposed by Pederson *et al.* (1986) and Westrich *et al.* (1989):

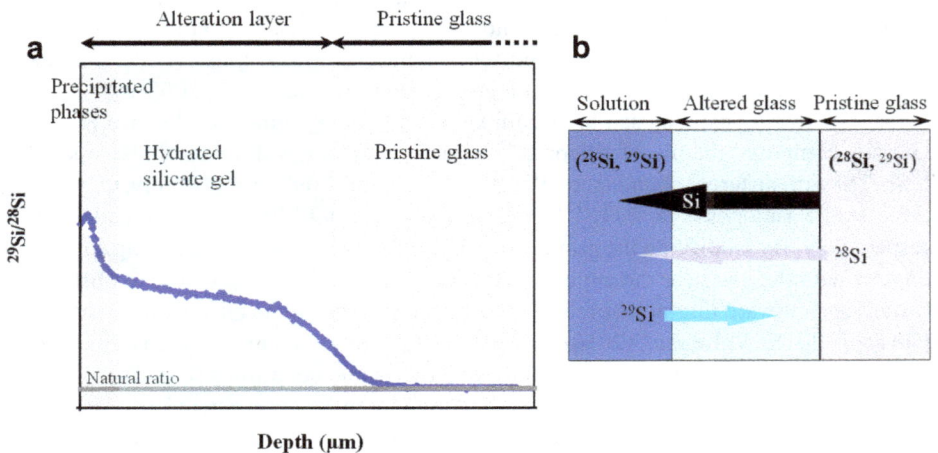

Fig. 15. Left: example of Si isotopic variations observed in the SON 68 sample altered in ^{29}Si-labelled water. Similar profiles were acquired for oxygen (see Valle *et al.*, 2010). Right: schematic representation of inward and outward diffusion of silicon (elemental and isotopes) during isotope tracing of the alteration of glass.

$$\equiv Si\text{-}^{16}O\text{-}Si \equiv\ +\ H_2{}^{18}O\ \rightarrow\ \equiv Si\text{-}^{16}O\text{-}H + H\text{-}^{18}O\text{-}Si \equiv \qquad \text{(hydrolysis reaction)}$$
$$\equiv Si\text{-}^{16}O\text{-}H + H\text{-}^{18}O\text{-}Si \equiv\ \rightarrow$$
$$(50\%)\ \equiv Si\text{-}^{16}O\text{-}Si \equiv\ +\ H_2{}^{18}O \qquad \text{(condensation reaction)}$$
$$(50\%)\ \equiv Si\text{-}^{18}O\text{-}Si \equiv\ +\ H_2{}^{16}O \qquad \text{(condensation reaction)}$$

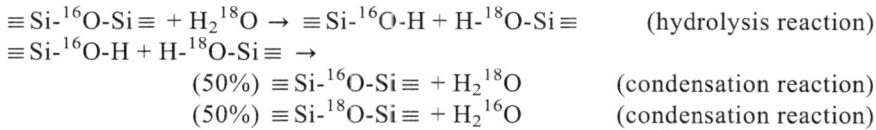

Similarly, Valle *et al.* (2010) proposed an incorporation of ^{29}Si in the altered glass *via* successive hydrolysis and condensation reactions:

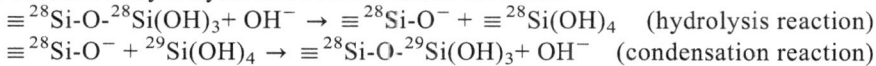

$$\equiv {}^{28}Si\text{-}O\text{-}{}^{28}Si(OH)_3 + OH^- \rightarrow\ \equiv {}^{28}Si\text{-}O^- + \equiv {}^{28}Si(OH)_4 \quad \text{(hydrolysis reaction)}$$
$$\equiv {}^{28}Si\text{-}O^- + {}^{29}Si(OH)_4 \rightarrow\ \equiv {}^{28}Si\text{-}O\text{-}{}^{29}Si(OH)_3 + OH^- \quad \text{(condensation reaction)}$$

Deuterium combined or not with a minor isotope of oxygen (preferentially ^{18}O which is less expensive) can also be used to trace the circulation of water in the alteration layer (Anovitz *et al.*, 2009; Verney-Carron *et al.*, 2015) or to understand better the ion-exchange reaction (Pederson *et al.*, 1986; McGrail *et al.*, 2001). In this way, for instance, Mc Grail's isotopic study demonstrated that the rupture of the O−H bond is the rate-limiting reaction during ion exchange. This conclusion was supported by a decrease of the Na-release rate by 30% observed in D_2O (compared with H_2O) without any modification of the matrix dissolution rate.

The examples described above show how the SIMS technique has provided new insights and improved our understanding of glass alteration. The great potential of the SIMS technique is certainly due to the fact that it allows very efficient acquisition of crucial *in situ* data for major, minor as well as trace elements in a single analytical run and that it can track isotopes.

In spite of the numerous investigations focused on glass alteration during the last 30 years, the mechanisms involved have not yet been clearly understood as evidenced by recent works (Geisler *et al.*, 2015; Gin *et al.*, 2015; Hellmann *et al.*, 2015). Published in 2015, these papers proposed new interpretations of glass-alteration mechanisms partly based on SIMS results (NanoSIMS instrument or Time-Of-Flight SIMS instrument operating in dynamic SIMS mode). Thus, they also confirm the technique's potential applied to glass alteration today. Furthermore, we can expect to take advantage of the latest developments in SIMS in terms of lateral and depth resolution to refine our interpretation of the experimental glass alteration in the coming years.

4. Isotopic signature of Li and Li diffusion

Lithium is the lightest alkali element (atomic number 3). It is a moderately incompatible trace element in the silicate Earth, enriched $10-15$ fold in crustal rocks (~24 ppm for the upper continental crust, 16 ppm for the total crust (Rudnik and Gao, 2003) compared to mantle rocks, where concentrations are typically $1-2$ ppm (Jagoutz *et al.*, 1979; Ottolini *et al.*, 2004). Lithium is highly mobile in fluid-related processes such as seafloor alteration (Stoffyn-Egli and Mackenzie, 1984; Seyfried *et al.*, 1984; Chan and Edmond, 1988; Chan *et al.*, 1992) or continental brine formation. Even though the Li may initially have been provided by Li-bearing aluminosilicates, nowadays most Li production relies on continental brines with large Li content. Lithium has low ionization energy and a small covalent radius (128 ± 7 pm), compared to other alkalis (Na: 166 ± 9 pm, K: 203 ± 12 pm) and is highly mobile in minerals. As its radius is smaller than that of Mg (141 ± 7 pm), it can substitue for Mg in the major minerals of peridotite, pyroxene and olivine (Zjang and Wright, 2012). The first ion microprobe study dedicated to Li effectively pointed out that Li diffusion in obsidian, albite and orthoclase glasses was fast even at low temperature (300°C; Jambon and Semet, 1978).

4.1. Li isotope systematics

Lithium has two isotopes, ^6Li (7.59%) and ^7Li (92.41 %). Due to the high relative mass difference between these two isotopes (~14%), significant isotopic fractionation could be expected in geological processes. The Li isotopic composition is currently expressed in δ notation:

$$\partial^7Li = \left(\frac{\frac{^7Li}{^6Li}\,sample}{\frac{^7Li}{^6Li}\,reference} - 1 \right) \times 100$$

where the reference value is defined by the NIST L-SVEC Li_2CO_3 standard.

Lithium isotopes display extensive fractionation during processes associated with fluid–rock interaction (Chan and Edmond, 1988; Chan et al, 1992, 1993, 1994; You et al., 1995). The Li isotopic signature of seawater ($\delta^7Li = +32.3‰$) is different from that of basalts (+4.7‰, Chan and Edmond, 1988) and hydrothermal fluids (+6 to +11‰; Chan et al., 1993). The Li isotope composition of altered basalts mainly depends on temperature, water–rock ratios and secondary mineral composition (Chan and Edmond, 1988; Chan et al., 1992, 1994; You et al., 1996; Moriguti and Nakamura, 1998).

Decitre et al. (2001) presented the first Li isotopic composition measurement by SIMS, by using a Cameca IMS3f ion microprobe. Petrographic thin sections were gold-coated. An O-primary beam accelerated at 10 kV was focused onto the sample surface with a diameter of 25 μm with 10−20 nA current intensity. Positive secondary ions were accelerated through 4.5 kV. For isotopic ratio determination, a mass resolution of 1100 was used for separating ^6LiH from ^7Li, and no energy filtering was applied. For concentration determination, ^7Li and ^{30}Si were measured at a mass resolution of 500 with an energy offset of 80 ± 10 eV. Secondary ions were counted on an electron multiplier in pulse counting mode. Counting times were 6 s for ^6Li and 3 s for ^7Li over 120 peak jumping cycles for isotopic analyses, 4 s for ^7Li and ^{30}Si over 30 cycles for concentration analyses. The chemical and isotopic instrumental fractionation, which is mainly dependent on instrumental parameters and on variations in the sample composition and structure (Deloule et al., 1992), was monitored in measuring standards before and after each sample. The standards used were natural minerals containing 1.5 to 1200 ppm Li. They cover the compositional and structural range encountered in the minerals of the samples.

In order to determine the Li concentration, the apparent Li/Si ratios (($Li/Si)_{SIMS}$) were calculated from ^7Li and ^{30}Si intensities using their atomic abundances. These apparent Li/Si ratios were then corrected for the relative useful yield of Li and Si determined on standards during the same analytical session (Fig. 16a). The measurement precision (2σ) was better than 10% for concentrations >5 ppm, and better than 30% for concentrations <1 ppm. The instrumental isotopic mass

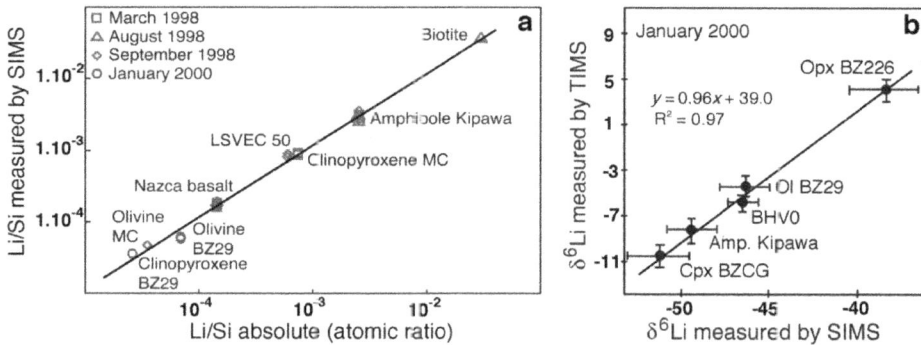

Fig. 16. (a) Ionization yield calibration curves of standards for four analytical sessions: Li/Si measured by ion microprobe *vs*. Li/Si measured by conventional methods. For description of the standards see Decitre *et al.* (2001). (b) δ^6Li measured by TIMS (Bristol University) *vs*. δ^6Li measured by ion microprobe for the same standards as part a.

fractionation for Li was determined on a set of reference minerals measured by thermo ionization mass spectrometry (TIMS) and by ion microprobe (Fig. 16b; with the isotopic ratio expressed in δ^6Li). Standards with δ^6Li ranging from -10 to $+4‰$ plot on a single line with a slope of 1, demonstrating the absence of a matrix effect on the instrumental Li isotopic fractionation. This allowed the same correction to be applied regardless of the mineral analysed. The instrumental mass fractionation expressed in δ^6Li units was determined at $-40.5‰$ (Fig. 16b), with a reproducibility (2σ) better than $2‰$. Note that the constant instrumental mass fractionation observed on different minerals during this study is due mainly to their small compositional variation due to only a weak change of the Mg number (Mg# = Mg/(Mg+Fe)). Indeed, Bell *et al.* (2009) pointed out that for the measurement of Li isotopes in olivine, the instrumental mass fractionation changed by $1.3‰$ per Mg# unit, inducing the generation of more reference materials (Bell *et al.*, 2009; Su *et al.*, 2015).

Decitre *et al.* (2001) presented Li contents and isotopic compositions, which were measured on a suite of peridotites and gabbros from the South West Indian ridge. Altered minerals are enriched in Li relative to fresh minerals. In harzburgites, fresh olivine and pyroxene present Li contents ranging from 0.6 to 0.7 ppm and 0.5 to 1.5 ppm respectively, while serpentines – their alteration products – contain 1 to 20 ppm Li, inversely correlated with the temperature. In gabbro, pyroxene is altered to clay minerals with Li contents in clay up to 35 ppm, while plagioclase is altered to chlorite with Li contents up to 23 ppm. In dunites, the Li content of serpentine is low: 1 ppm. In all the analysed samples, the lithium isotopic composition is heavier in altered minerals than in fresh minerals, as expected. The lithium isotopic ratios measured on serpentine vary by $5‰$ in each sample and by 8 to $9‰$ between samples. A main result from this study was to recognize that during the alteration of the oceanic crust by hydrothermal fluid, the secondary minerals were enriched in Li, with increasing δ^7Li along the hydrothermal flow (Fig. 17). Therefore, the altered oceanic crust is enriched in Li with δ^7Li values ranging from -2 to $+25$, while δ^7Li of unaltered MORB is typically $\sim+4$.

During the following years, much work was done to characterize better the Li chemical and isotopic distribution in various geodynamic contexts, showing large Li isotopic compositional variations during subduction of oceanic crust with δ^7Li values ranging from -30 to $+25$‰, which is ascribed to fluid–rock interactions (Tomascak et al., 2002; Seitz et al., 2003, 2004; Kobayashi et al., 2004; Ottolini et al., 2004; Rudnick et al., 2004; Teng et al., 2004).

4.2. Li isotope fractionation by diffusion

Richter et al. (2003) published an experimental work entitled "Isotope fractionation by chemical diffusion between molten basalt and rhyolite". In that paper, they showed that Li diffuses faster than the other major or trace elements ($100 \times$ to $1000 \times$ faster than Ca in basaltic melt and rhyolitic melt, respectively), and that an isotopic fractionation of ~40‰ resulted from the Li diffusion with a concentration ratio of 15. The isotopic fractionation during diffusion results from the relationship between the diffusivity and the mass of components, which is expressed as

$$\frac{D_{i,1}}{D_{i,2}} = \left(\frac{m_{i,2}}{m_{i,1}}\right)^{\beta}$$

where $D_{i,1}$ and $D_{i,2}$ are the diffusivities of isotopes 1 and 2 of element i, and $m_{i,1}$ and $m_{i,2}$ are the atomic masses of the isotopes (Richter et al., 1999). Lithium isotopes are strongly fractionated by diffusion because of (1) their large mass ratio ($7/6 \approx 1.166$),

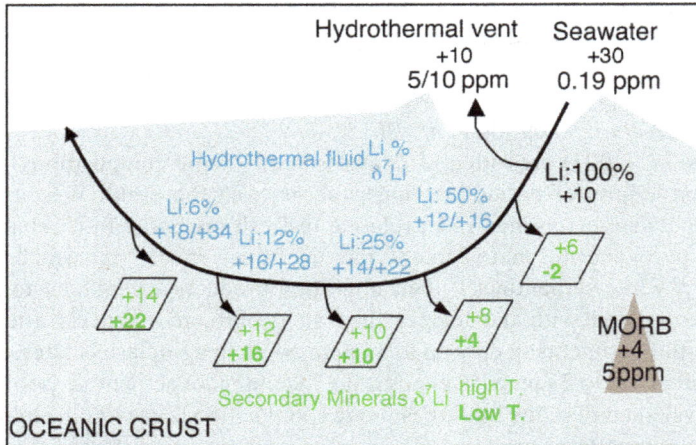

Fig. 17. Schematic representation of the evolution of the Li content and δ^7Li values of hydrothermal fluid and secondary minerals formed during the interaction of the fluid with the peridotite. The initial fluid is considered to be similar to high-temperature vent fluids. For each step, 50% of the Li of the fluid is taken up by the secondary minerals. The isotopic fractionation factor between fluid and minerals ranges from 1.004 to 1.012, depending on the temperature of the fluid (Chan et al., 1992). Redrawn from Decitre et al. (2001).

and (2) a large kinetic fractionation parameter (for Li β ≈ 0.215, ~3 times higher than for Ca).

The paper by Richter *et al.* (2003) caused significant interest in the measurement of Li diffusion profiles at various scales. To determine the cooling rate of the oceanic crust at different depths, Coogan *et al.* (2005) developed a new geospeedometer based on the extent of diffusion of lithium from plagioclase into clinopyroxene during cooling. They performed high-temperature experiments to determine both the temperature dependence of the partitioning of Li between plagioclase and clinopyroxene and the diffusion coefficient for Li in clinopyroxene. Then, they determined the cooling rate of the lavas and dykes from ODP Hole 504B by measuring the Li content and isotopic composition in pyroxene and plagioclase, and inferred extremely rapid cooling rates, up to 450°C h^{-1} in the upper part of the sheeted dyke complex.

In analysing the lunar meteorite NWA 479, Barrat *et al.* (2005) reported Li contents of 3.2 to 11.8 ppm and δ^{7}Li values ranging from +2.4 to +15.1‰ in olivine and Li contents of 2.8 to 18.4 ppm and δ^{7}Li values ranging from −0.2 to +16.1‰ in pyroxene phenocrysts. They proposed that the wide range of δ^{7}Li values observed in the phenocrysts results from its isotopic fractionation by diffusion. In using a model similar to that reported in Fig. 18, they showed that this process was able to produce large isotopic heterogeneities in a very short time over a distance of a few hundreds of μm, in agreement with the cooling rate of the meteorite after the extraction from the Moon.

Lundstrom *et al.* (2005) used Li to study the interaction between melt and rock around conduits for melt ascent in dunite from the Trinity Ophiolite. They measured Li isotope compositions of clinopyroxene and some olivine within these sequences using ion probe techniques. Their results show large variations in ^{7}Li/^{6}Li occurring in a consistent pattern across three transects from dunite to plagioclase lherzolite.

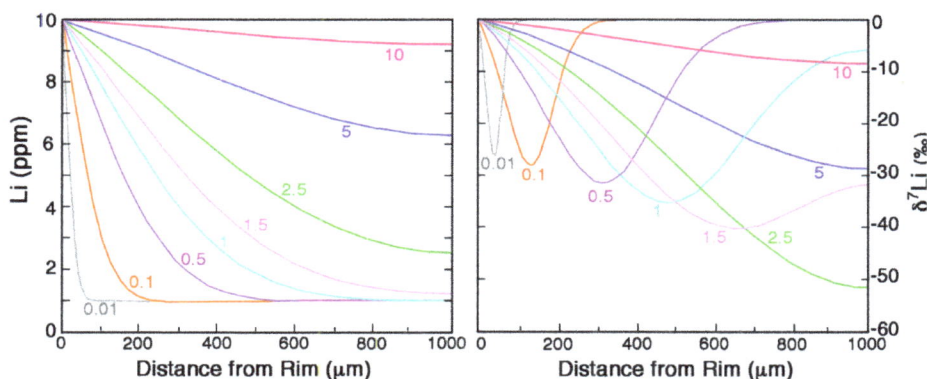

Fig. 18. Numerical modelling of Li diffusion in clinopyroxene. On the left, the evolution of Li contents with time, on the right the evolution of δ^{7}Li with time. The numbers on the curves are duration in years. The Li content at the mineral surface is ten times that of the initial mineral interior, and the initial δ^{7}Li value for both the mineral surface and interior are set to zero for clarity. The Li concentration and isotopic composition at the surface are kept constant over time.

Specifically, measurements along the traverse reveal low δ^7Li values in the harzburgite adjacent to the dunite returning to higher values farther from the dunite, with a typical offset of $-10‰$ in the low δ^7Li trough. This pattern is consistent with a process whereby Li isotopes are fractionated during diffusion through a melt either from the dunite conduit to the surrounding peridotite, or from the surrounding peridotite into the dunite conduit. Numerical modelling showed that large changes in δ^7Li as a function of distance can be created in year-to-decade timescales on a few metres distance.

Parkinson et al. (2007) and Gallagher and Elliot (2009) presented detailed numerical modelling of Li isotopic fractionation during diffusion at high temperature in magmatic systems. In the following years, the interpretation of low δ^7Li values has developed according to two different models: (1) diffusion-driven kinetic fractionation of Li isotopes (Lundstrom et al., 2005; Teng et al., 2006; Jeffcoate et al., 2007; Marschall et al., 2007; Rudnick and Ionov, 2007; Ionov and Seitz, 2008; Kaliwoda et al., 2008; Gao et al., 2011; Tang et al., 2012), and (2) metasomatism of lithospheric mantle by low-δ^7Li melt/fluid (e.g. Nishio et al., 2004; Tang et al., 2007, 2012; Agostini et al., 2008; Gu et al., 2016).

In the first model, the isotope signature is generated in the course of local Li exchange, whereas in the second model input of a low-δ^7Li melt/fluid is implied. Based on experimental results, Wunder et al. (2006, 2007) suggested that subduction and concomitant dehydration of altered oceanic crust, containing chlorite and clinopyroxene with Li in six-fold coordination, releases fluids enriched in ^{7}Li into the fore-arc mantle (Chan et al., 2002; Elliott et al., 2003) and consequently introduce a low-δ^7Li component into the deeper mantle. This suggestion supports the model that low δ^7Li in eclogites were derived from Li isotopic fractionation during progressive metamorphism of the subducting slab (Zack et al., 2003). This is supported by the low δ^7Li values measured later in eclogites (Halama et al., 2011; Marschall et al., 2007). Taking into account the only moderate incompatibility of Li (Brenan et al., 1998a,b), Agostini et al. (2008) suggested that only the very dehydrated slab will have a low δ^7Li, which will be preserved in the local mantle. This is in agreement with the observation that the Li isotopic composition of the deep mantle as recorded by carbonatites since the Archaean (2.7 Ga) up to recent time does change significantly, with a δ^7Li value of 4.1 ± 1.3‰ (Halama et al., 2008).

It seems clear in this review that our understanding of the Li isotope behaviour in high-temperature magmatic processes has changed completely over the last two decades, and that this understanding was largely obtained by in situ ion microprobe measurements of both Li content and isotopic composition. Measurements of potential zonations in Li concentrations and isotope patterns in individual grains are also of key importance for the interpretation of the isotopic signature of natural samples.

5. Concluding remarks

Over the last 50 years, secondary ion mass spectrometry has benefitted from great advances in the development of new instruments with ever increasing performance and the SIMS technique has experienced increasing usage in many fields of research in Earth Sciences. Ion microprobes are now able to provide 3D secondary ion images with a lateral resolution of ~50 nm and a depth resolution of a few nm for very different disciplines including mineralogy, geochemistry, environmental sciences, material sciences, biology, medicine and many more. They are also able to provide high precision and high sensitivity measurements of light-element and trace-element contents, as well as the isotopic composition of many elements with a spatial resolution of a few μm. They seem to be the most appropriate instruments to study processes during solid–solid or solid–fluid interactions, and to characterize the exchange through the interface. They are also powerful tools for measuring the isotopic composition of geological or extraterrestrial samples at high resolution.

Developments expected in the coming years rely mainly on the development of larger collections of reference materials and on the improvement of analytical procedures to provide more accurate results adapted to the large variety of geological samples. This is a necessary prerequisite for exploiting the instrumental improvement toward higher sensitivity and precision.

References

Agostini, S., Ryan, J.G., Tonarini, S. and Innocenti, F. (2008) Drying and dying of a subducted slab: coupled Li and B isotope variations in Western Anatolia Cenozoic Volcanism. *Earth and Planetary Science Letters*, **272**, 139–147.

Anderson, C.A. and Hinthorne, J.R. (1972) U, Th, Pb and REE abundances and $^{207}Pb/^{206}Pb$ ages of individual minerals in returned lunar material by ion microprobe mass analysis. *Earth and Planetary Science Letters*, **14**, 195–200.

Anovitz, L.M., Cole, D.R. and Riciputi, L.R. (2009) Low-temperature isotopic exchange in obsidian: Implications for diffusive mechanisms. *Geochimica Cosmochimica Acta*, **73**, 3795–3806.

Barrat, J.A., Chaussidon, M., Bohn, M., Gillet, P., Gopel, C. and Lesourd, M. (2005) Lithium behavior during cooling of a dry basalt: An ion-microprobe study of the lunar meteorite Northwest Africa 479 (NWA 479). *Geochimica et Cosmochimica Acta*, **69**, 5597–5609.

Bell, D.R., Hervig, R.L., Buseck, P.R. and Aulbach, S. (2009) Lithium isotope analysis of olivine by SIMS: Calibration of a matrix effect and application to magmatic phenocrysts. *Chemical Geology*, **258**, 5–16.

Brenan, J.M., Neroda, E., Lundstrom, C.C., Shaw, H.F., Ryerson, F.J. and Phinney, D.L. (1998a) Behaviour of boron, beryllium and lithium during melting and crystallization: constraints from mineral–melt partitioning experiments. *Geochimica et Cosmochimica Acta*, **62**, 2129–2141.

Brenan, J.M., Ryerson, F.J. and Shaw, H.F. (1998b) The role of aqueous fluids in the slab-to-mantle transfer of boron, beryllium, and lithium during subduction: experiments and models. *Geochimica et Cosmochimica Acta*, **62**, 3337–3347.

Bunker, B.C. (1994) Molecular mechanisms for corrosion of silica and silicate glasses. *Journal of Non-Crystalline Solids*, **179**, 300–308.

Chan, L.H. and Edmond, J.M. (1988) Variation of lithium isotope composition in the marine environment: a preliminary report. *Geochimica Cosmochimica Acta*, **52**, 1711–1717.

Chan, L.H., Edmond, J.M., Thompson, G. and Gillis, K. (1992) Lithium isotopic composition of submarine basalts: implications for the lithium cycle in the ocean. *Earth and Planetary Science Letters*, **108**, 151–160.

Chan, L.H., Edmond, J.M. and Thompson, G. (1993) A lithium isotope study of hot springs and metabasalts from mid-ocean ridge hydrothermal systems. *Journal of Geophysical. Research*, **98 (B6)**, 9653–9659.

Chan, L.H., Zhang, L. and Hein, J.R. (1994) Lithium isotope characteristics of marine sediments. *EOS Transactions, American Geophysical Union*, **75**, 314.

Chan, L.H., Leeman, W.P. and You, C.F. (2002) Lithium isotopic composition of Central American volcanic arc lavas: implications for modification of subarc mantle by slab-derived fluids: correction. *Chemical Geology*, **182**, 293–300.

Chaussidon, M. and Albarède, F. (1992) Secular boron isotope variations in the continental crust: an ion microprobe study. *Earth and Planetary Science Letters*, **108**, 229–241.

Coogan, L.A., Kasemann, S.A. and Chakraborty, S. (2005) Rates of hydrothermal cooling of new oceanic upper crust derived from lithium-geospeedometry. *Earth and Planetary Science Letters*, **240**, 415–424.

Decitre, S., Deloule, E., Reisberg, L., James, R., Agrinier, P. and Mével, C. (2001) Behavior of Li and its isotopes during serpentinization of oceanic peridotites. *Geochemistry, Geophysics, Geosystems*, **3**, 1–20.

Deloule, E., France-Lanord, C. and Albarède, F. (1991) D/H analysis of mineral by ion probe. Pp. 53–62 in: *Stable Isotope Geochemistry: A Tribute to Samuel Epstein*, (H.P. Tayor, J.R. O'Neill and I.R. Kaplan, editors). Special Publication N°3. The Geochemical Society.

Deloule, E., Allé, P. and Chaussidon, M. (1992) Instrumental limitations for isotope ratios measurements with a Cameca IMS 3f ion microprobe: the example of H, B, S, Sr. *Chemical Geology*, **101**, 187–192.

Deloule, E., Paillat, O., Pichavant, M. and Scaillet, B. (1995) Ion microprobe determination of water in silicate glasses: methods and applications. *Chemical Geology*, **125**, 19–28

Doremus, R.H. (1975) Interdiffusion of hydrogen and alkali ions in a glass surface. *Journal of Non-Crystalline Solids*, **19**, 137–144.

Eiler, J.M., Graham, C.M. and Valley, J.W. (1997) SIMS analysis of oxygen isotopes; matrix effects in complex minerals and glasses. *Chemical Geology*, **138**, 221–244.

Elliott, T., Thomas, A., Jeffcoate, A.B. and Niu, Y. (2003) Li isotope composition of the upper mantle. *EOS Transactions, American Geophysical Union*, **84**, 1608.

Fearn, S., McPhail, D.S., Morris, R.J.H. and Dowsett, M.G. (2006) Sodium and hydrogen analysis of room temperature glass corrosion using low energy Cs SIMS. *Applied Surface Science*, **252**, 7070–7073.

Frugier, P., Gin, S., Minet, Y., Chave, T., Bonin, B., Godon, N., Lartigue, J.E., Jollivet P., Ayral A., De Windt L. and Santarini G. (2008) SON68 nuclear glass dissolution kinetics: current state of knowledge and basis of the new GRAAL model. *Journal of Nuclear Materials*, **380**, 8–21.

Gallagher, K. and Elliott, T. (2009) Fractionation of lithium isotopes in magmatic systems as a natural consequence of cooling. *Earth and Planetary Science Letters*, **278**, 286–296.

Gao, Y., Snow, J.E., Casey, J.F. and Yu, J. (2011) Cooling-induced fractionation of mantle Li isotopes from the ultraslow-spreading Gakkel Ridge. *Earth and Planetary Science Letters*, **301**, 231–240.

Geisler, T., Nagel, T., Kilburn, M.R., Janssen, A., Icenhower, J.P., Fonseca, R.O.C., Grange, M. and Nemchin, A.A. (2015) The mechanism of borosilicate glass corrosion revisited. *Geochimica et Cosmochimica Acta*, **158**,112–129.

Gin, S. (2001) Protective effect of the alteration gel: a key mechanism in the long-term behavior of nuclear waste glass. *Material Research Society Proceedings*, **663**, 207–215.

Gin, S., Guittonneau, C., Godon, N., Neff, D., Rebiscoul, D., Cabi, M. and Mostefaoui, S. (2011) Nuclear glass durability: New insight into alteration layer properties. *Journal of Physical Chemistry C*, **115**, 18696–18706.

Gin, S., Ryan, J.V., Schreiber, D.K., Neeway, J. and Cabié, M. (2013) Contribution of atom-probe tomography to a better understanding of glass alteration mechanisms: application to a nuclear glass specimen altered 25 years in a granitic environment. *Chemical Geology*, **349–350**, 99–109.

Gin, S., Jollivet, P., Fournier, M., Angeli, F., Frugier, P. and Charpentier, T. (2015) Origin and consequences of silicate glass passivation by surface layers. *Nature Communications*, **6**, 6360, 1–8.

Gu, X., Deloule, E., France, L. and Ingrin, J. (2016) Multi-stage metasomatism revealed by trace element and Li isotope distributions in minerals of peridotite xenoliths from Allègre volcano (French Massif Central). *Lithos*, **264**, 158–174.

Halama, R., McDonough, W.F., Rudnick, R.L. and Bell, K. (2008) Tracking the lithium isotopic evolution of the mantle using carbonatites. *Earth and Planetary Science Letters*, **265**, 726–742.

Halama, R., John, T., Herms, P., Hauff, F. and Schenk, V. (2011) A stable (Li, O) and radiogenic (Sr, Nd) isotope perspective on metasomatic processes in a subducting slab. *Chemical Geology*, **281**, 151–166.

Hellmann, R., Cotte, S., Cadel, E., Malladi, S., Karlsson, L.S., Lozano-Perez, S., Martiane Cabié, M. and Seyeux, A. (2015) Nanometre-scale evidence for interfacial dissolution–reprecipitation control of silicate glass corrosion. *Nature Materials*, **14**, 307–311.

Hinton, R.W. (1990) Ion microprobe trace-element analysis of silicates: Measurement of multi-element glasses. *Chemical Geology*, **83**, 11–25.

Huberty, J.M., Kita, N.T., Kozdon, R., Heck, P.R., Fournelle, J.H., Spicuzza, M.J., Xu, H. and Valley J.W. (2010) Crystal orientation effects in $\delta^{18}O$ for magnetite and hematite by SIMS. *Chemical Geology*, **276**, 269–283.

Ireland T.R., Clement S., Compston W., Foster, J.J., Holden, P., Jenkins, B., Lanc, P., Schram, M. and Williams, S. (2008) The development of SHRIMP. *Australian Journal of Earth Sciences*, **55**, 937–954.

Ionov, D.A. and Seitz, H.M. (2008) Lithium abundances and isotopic compositions in mantle xenoliths from subduction and intra-plate settings: mantle sources vs. Eruption histories. *Earth and Planetary Science Letters*, **266**, 316–331.

Jagoutz, E., Palme, H., Baddenhausen, H., Blum, K., Cendales, M., Dreibus, G., Spettel, B., Lorenz, V. and Wänke, H. (1979) The abundance of major, minor and trace elements in the earth's mantle as derived from primitive ultramafic nodules. *Lunar and Planetary Science Conference Proceedings*, **10**, 2031–2050.

Jambon, A. and Semet, M.P. (1978) Lithium diffusion in silicate glasses of albite, orthoclase and obsidian composition: an ion microprobe determination. *Earth and Planetary Science Letters*, **37**, 445–450.

Jeffcoate, A.B., Elliott, T., Kasemann, S.A., Ionov, D., Cooper, K. and Brooker, R. (2007) Li isotope fractionation in peridotites and mafic melts. *Geochimica et Cosmochimica Acta*, **71**, 202–218.

Jégou, C., Gin, S. and Larche, F. (2000) Alteration kinetics of a simplified nuclear glass in an aqueous medium: effects of solution chemistry and of protective gel properties on diminishing the alteration rate. *Journal of Nuclear Materials*, **280**, 216–229.

Kaliwoda, M., Ludwig, T. and Altherr, R. (2008) A new SIMS study of Li, Be, B and $\delta7Li$ in mantle xenoliths from Harrat Uwayrid (Saudi Arabia). *Lithos*. **106**, 261–279.

Kobayashi, K., Tanaka, R., Moriguti, T., Shimizu, K. and Nakamura, E. (2004) Lithium, boron, and lead isotope systematics of glass inclusions in olivines from Hawaiian lavas: evidence for recycled components in the Hawaiian plume. *Chemical Geology*, **212**, 143–161.

Koenderink, G.H., Brzesowsky, R.H. and Balkenende, A.R. (2000) Effect of the initial stages of leaching on the surface of alkaline earth sodium silicate glasses. *Journal of Non-Crystalline Solids*, **262**, 80–98.

Libourel, G., Verney-Carron, A., Morlok, A., Gin, S., Sterpenich, J., Michelin, A., Neff, D. and Dillmann, P. (2011) The use of natural and archeological analogues for understanding the long-term behavior of nuclear glasses. *Comptes Rendus Geosciences*, **343**, 237–245.

Lodding, A.R. and Engström, E.U. (1992) Elemental depth profiling of nuclear waste glasses after two-years burial in a salt geology. *Journal of the American Ceramic Society*, **75**, 2702–2706.

Lodding, A. and Van Iseghen, P. (2001) In-depth distributions of elements in leached layers on two HLW waste glasses after burial in clay; step-scan by SIMS. *Journal of Nuclear Materials*, **298**, 197–202.

Lodding, A., Odelius, H., Clark, D.E. and Werme, L.O. (1985) Element profiling by Secondary Ion Mass Spectrometry of surface layers in glasses. *Mikrochimica Acta*, **11**, 145–161.

Lundstrom, C.C., Chaussidon, M., Hsui, A.T., Kelemen, P. and Zimmerman, M. (2005) Observations of Li isotopic variations in the Trinity Ophiolite: evidence for isotopic fractionation by diffusion during mantle melting. *Geochimica et Cosmochimica Acta*, **69(3)**, 735–751.

Lutze, W., Malow, G., Ewing, R.C., Jercinovic, M.J. and Keil K. (1985) Alteration of basalt glasses: implications for modeling the long-term stability of nuclear waste glasses. *Nature*, **314**, 252–255.

MacRae, N.D., Bottazzi, P., Ottolini, L. and Vannucci, R. (1993) Quantitative REE analysis of silicates by SIMS: Conventional energy filtering vs. specimen isolation mode. *Chemical Geology*, **103**, 45–54.

Marschall, H.R., von Strandmann, P.A.P., Seitz, H.M., Elliott, T. and Niu, Y. (2007) The lithium isotopic composition of orogenic eclogites and deep subducted slabs. *Earth and Planetary Science Letters*, **262**, 563–580.

Matsuda, H. (1974) Double focusing mass spectrometers of second order. *International Journal of Mass Spectrometry and Ion Physics*, **14**, 219–233.

McGrail, B.P., Icenhower, J.P., Shuh, D.K., Liu, P., Darab, J.G., Baer, D.R., Thevuthasen, S., Shutthanandan, V., Engelhard, M.H., Booth, C.H. and Nachimuthu, P. (2001) The structure of $Na_2O-Al_2O_3-SiO_2$ glass: Impact on sodium ion exchange in H_2O and D_2O. *Journal of Non-Crystalline Solids*, **296**, 10–26.

Migeon, H.N., Schuhmacher, M., Legoux, J.J. and Rasser B. (1989) 3-dimensional analysis of trace elements with the CAMECA IMS-4F. *Fresenius Zeitschrift für Analytische Chemie*, **333**, 333–334

Migeon, H.N., Schuhmacher, M. and Slodzian, G. (1990) Analysis of insulating specimens with the Cameca IMS4f. *Surface and Interface Analysis*, **16**, 9–13.

Moriguti, T. and Nakamura, E. (1998) Across-arc variation of Li isotopes in lavas and implications for crust/ mantle recycling at subduction zones. *Earth and Planetary Science Letters*, **167**, 167–174.

Nishio, Y., Nakai, S.I., Yamamoto, J., Sumino, H., Matsumoto, T., Prikhod'ko, V.S. and Arai, S. (2004) Lithium isotopic systematics of the mantle-derived ultramafic xenoliths: implications for EM1 origin. *Earth and Planetary Science Letters*, **217**, 245–261.

Ojovan, M.I. and Lee, W.E. (2011) Glassy wasteforms for nuclear waste immobilization. *Metallurgical and Materials Transactions A*, **42A**, 837–851.

Ottolini, L., Le Fèvre, B. and Vannucci, R. (2004) Direct assessment of mantle boron and lithium contents and distribution by SIMS analyses of peridotite minerals. *Earth and Planetary Science Letters*, **228**, 19–36.

Parkinson, I.J., Hammond, S.J., James, R.H. and Rogers, N.W. (2007) High-temperature lithium isotope fractionation: insights from lithium isotope diffusion in magmatic systems. *Earth and Planetary Science Letters*, **257**, 609–621.

Parruzot, B., Jollivet, P., Rébiscoul, D. and Gin, S. (2015) Long-term alteration of basaltic glass: Mechanisms and rates. *Geochimica et Cosmochimica Acta*, **154**, 28–48.

Pederson, L.R. (1987) Comparison of sodium leaching rates from a $Na_2O.3SiO_2$ glass in H_2O and D_2O. *Physics and Chemistry of Glasses*, **28**, 17–21.

Pederson, L.R., Baer, D.R., McVay, G.L. and Engelhard, M.H. (1986) Reaction of soda lime silicate glass in isotopically labeled water. *Journal of Non-Crystalline Solids*, **86**, 369–380.

Rana, M.A. and Douglas, R.W. (1961a) The reaction between glass and water. Part 1. Experimental methods and observations. *Physics and Chemistry of Glasses*, **2**, 179–195.

Rana, M.A. and Douglas, R.W. (1961b) The reaction between glass and water. Part 2. Discussion for the results, *Physics and Chemistry of Glasses*, **2**, 196–204.

Reed, S.J.B. (1989) Ion microprobe analysis – a review of geological applications. *Mineralogical Magazine*, **53**, 3–24.

Rhede, D. and Wiedenbeck, M. (2006) SIMS quantification of very low hydrogen contents. *Applied Surface Science*, **252**, 7152–7154.

Richter, F.M., Liang, Y. and Davis, A.M. (1999) Isotope fractionation by diffusion in molten oxides. *Geochimica et Cosmochimica Acta*, **63**, 2853–2861.

Richter, F.M., Davis, A.M., DePaolo, D.J. and Watson, E.B. (2003) Isotope fractionation by chemical diffusion between molten basalt and rhyolite. *Geochimica et Cosmochimica Acta*, **67**, 3905–3923.

Riciputi, L.R., Elam, J.M., Anovitz, L.M. and Cole, D.R. (2002) Obsidian diffusion dating by secondary ion mass spectrometry: a test using results from Mound-65, Chalco, Mexico. *Journal of Archaeological Sciences*, **29**, 1055–1075.

Rudat, M.A. and Morrison, G.H. (1979) Evaluation of several semi-theoretical methods for quantitative secondary ion mass spectrometric analysis after discrimination-correction of data. *Analytical Chemistry*, **51**, 1179–1187.

Rudenauer, F.G. and Steiger, W. (1976) Quantitative evaluation of SIMS spectra using Saha-Eggert type equations. *Vacuum*, **26**, 537–543.

Rudnik, R.L. and Gao, S. (2003) Composition of the continental crust Pp. 1–64 in: *Treatise on Geochemistry*, **3** (H.D. Holland and K.K. Turekian, editors). Elsevier.

Rudnick, R.L. and Ionov, D.A. (2007) Lithium elemental and isotopic disequilibrium in minerals from peridotite xenoliths from far-east Russia: product of recent melt/fluid–rock reaction. *Earth and Planetary Science Letters*, **256**, 278–293.

Rudnick, R.L., Tomascak, P.B., Njo, H.B. and Gardner, L.R. (2004) Extreme lithium isotopic fractionation during continental weathering revealed in saprolites from South Carolina. *Chemical Geology*, **212**, 45–57.

Sangely, L., Boyer, B., de Chambost, E, Valle, N., Audinot, J-N, Ireland, T., Wiedenbeck, M., Aléon, J., Jungnickel, H., Barnes, J.-P., Bienvenu, P. and Breuer, U. (2015) Secondary Ion Mass Spectrometry. Chapter 15 in: *Sector Field Mass Spectrometry for Elemental and Isotopic Analysis* (T. Prohaska, J. Irrgeher and N. Jakubowski, editors). New Developments in Mass Spectrometry (Book 3). Royal Society of Chemistry, London.

Schiott, H.E. (1970) Approximations and interpolation rules for ranges and range stragglings. *Radiation Effects*, **6**, 107–113.

Seitz, H.M., Brey, G.P., Stachel, T. and Harris, J.W. (2003) Li abundances in inclusions in diamonds from the upper and lower mantle. *Chemical Geology*, **201**, 307−318.

Seitz, H.M., Brey, G.P., Lahaye, Y., Durali, S. and Weyer, S. (2004) Lithium isotopic signatures of peridotite xenoliths and isotopic fractionation at high temperature between olivine and pyroxenes. *Chemical Geology*, **212**, 163−177.

Seyfried, W.E., Janecky, D.R. and Mottl, M.J. (1984) Alteration of the oceanic crust: implications for geochemical cycles of lithium and boron. *Geochimica et Cosmochimica Acta*, **48**, 557−569.

Shimizu, N. (1978) Analysis of the zoned plagioclase of different magmatic environments: A preliminary ion-microprobe study. *Earth and Planetary Science Letters*, **39**, 398−406.

Shimizu, N. and Hart, S.R. (1982a) Applications of the ion microprobe to geochemistry and cosmochemistry. *Annual Review in Earth and Planetary Sciences*, **10**, 483−526.

Shimizu, N. and Hart, S.R. (1982b) Isotope fractionation in secondary ion mass spectrometry. *Journal of Applied Physics*, **53**, 1303−1311.

Shimizu, N., Semet, M.P. and Allegre, C.J. (1978) Geochemical applications of quantitative ion-microprobe analysis. *Geochimica et Cosmochimica Acta*, **42**, 1321−1334.

Simonds, D.S., Baker, J.E. and Evans, C.A. (1976) Evaluation of the local thermal equilibrium model for quantitative secondary ion mass spectrometric analysis. *Analytical Chemistry*, **48**, 1341−1348.

Slodzian, G. (1964) Etude d'une méthode d'analyse locale chimique et isotopique utilisant l'émission ionique secondaire. *Annales de Physique*, **9**, 591−598.

Slodzian, G., Lorin, J.C. and Havette A. (1980) Isotopic effect on the ionisation probabilities in secondary ion emission. *Journal de Physique*, **41**, L555−L558

Sterpenich, J. and Libourel, G. (2001) using stained glass windows to understand the durability of toxic waste matrices. *Chemical Geology*, **174 (1−3)**, 181−193.

Sterpenich, J. and Libourel, G. (2006) Water diffusion in silicate glasses under natural weathering conditions: evidence from buried medieval stained glasses. *Journal of Non-Crystalline Solids*, **352**, 5446−5451.

Stoffyn-Egli, P. and Mackenzie, F.T. (1984) Mass balance of dissolved lithium in the oceans. *Geochimica et Cosmochimica Acta*, **48**, 859−872.

Su, B.-X., Gu, X.-Y., Deloule, E., Zhang, H.-F., Li, Q.-L., Li, X.-H., Vigier, N., Tang, Y.-J., Tang, G.-Q., Liu, Y., Brewer, A., Pang, K.-N., Mao, Q. and Ma, Y.-G. (2015) Orthopyroxene, clinopyroxene and olivine reference standards for in-situ lithium isotope measurements. *Geostandards and Geoanalytical Research*, **39**, 357−369.

Tang, Y.J., Zhang, H.F., Nakamura, E., Moriguti, T., Kobayashi, K. and Ying, J.F. (2007) Lithium isotopic systematics of peridotite xenoliths from Hannuoba, North China Craton: implications for melt−rock interaction in the considerably thinned lithospheric mantle. *Geochimica et Cosmochimica Acta*, **71**, 4327−4341.

Tang, Y.J., Zhang, H.F., Deloule, E., Su, B.X. Ying, J.F., Xiao, Y. and Hu, Y. (2012) Slab-derived lithium isotopic signatures in mantle xenoliths from northeastern North China Craton. *Lithos*, **149**, 79−90.

Teng, Z., McDonough, W.F., Rudnick, R.L., Dalpé, C., Tomascak, P.B., Chappell, B.W. and Gao, S. (2004) Lithium isotopic composition and concentration of the upper continental crust. *Geochimica et Cosmochimica Acta*, **68**, 4167−4178.

Teng, F.Z., McDonough, W.F., Rudnick, R.L. and Walker, R.J. (2006) Diffusion-driven extreme lithium isotopic fractionation in country rocks of the Tin Mountain pegmatite. *Earth and Planetary Science Letters*, **243**, 701−710.

Tomascak, P.B., Widom, E., Benton, L.D., Goldstein, S.L. and Ryan, J.G. (2002) The control of lithium budgets in island arcs. *Earth and Planetary Science Letters*, **196**, 227−238.

Tsomaia, N., Brantley, S.L., Hamilton, J.P., Pantano, C.G. and Mueller, K.T. (2003) NMR evidence for formation of octahedral Al and tetrahedral Si network during dissolution of aluminosilicate glass and crystal. *American Mineralogist*, **88**, 54−67.

Valcke, E., Smets, S., Labat, S., Lemmens, K., Van Iseghem, P., Gysemans, M., Thomas P , Van Bree, P., Vos, B. and Van den Berghe, S. (2007) An integrated in situ corrosion test on alpha-active HLW glass − phase II. Detailed final report of SCK·CEN for the EC project CORALUS II.

Valle, N., Verney-Carron, A., Sterpenich, J., Libourel, G., Deloule, E. and Jollivet, P. (2010) Elemental and isotopic (^{29}Si and ^{18}O) tracing of glass alteration mechanisms. *Geochimica Cosmochimica Acta*, **74**, 3412–3431.

Valley, J.W. and Kita, N.T. (2009) In situ oxygen isotope geochemistry by ion microprobe. Pp. 19–63 in: *Secondary Ion Mass Spectrometry in the Earth and Planetary Sciences* (M. Fayek, editor). GAC/MAC Short Course, **41**.

Verney-Carron, A., Saheb, M., Loisel, C., Duhamel, R. and Remusat, L. (2015) Use of hydrogen isotopes to understand stained glass weathering. *Procedia Earth and Planetary Science*, **13**, 64–67.

Vielzeuf, D., Champenois, M., Valley, J.W., Brunet, F. and Devidal, J.L. (2005) SIMS analyses of oxygen isotopes: matrix effects in Fe–Mg–Ca garnet. *Chemical Geology*, **223**, 208–226

Westrich, H.R., Casey, W.H. and Arnold, G.W. (1989) Oxygen isotope exchange in the leached layer of labradorite feldspar. *Geochimica et Cosmochimica Acta*, **53**, 1681–1685.

Wunder, B., Meixner, A., Romer, R.L. and Heinrich, W. (2006) Temperature-dependent isotopic fractionation of lithium between clinopyroxene and high-pressure hydrous fluids. *Contributions to Mineralogy and Petrology*, **151**, 112–120.

Wunder, B., Meixner, A., Romer, R.L., Feenstra, A., Schettler, G. and Heinrich, W. (2007) Lithium isotope fractionation between Li-bearing staurolite, Li-mica and aqueous fluids: an experimental study. *Chemical Geology*, **238**, 277–290.

You, C.F., Chan, L.H., Spivack, A.J. and Gieskes, J.M. (1995) Lithium, boron, and their isotopes in sediments and pore waters of ocean drilling program Site 808, Nankai Trough: implications for fluid expulsion in accretionary prisms. *Geology*, **23**, 37–40.

You, C.F., Castillo, P.R., Gieskes, J.M., Chan, L.H. and Spivack, A.J. (1996) Trace element behavior in hydrothermal experiments: Implications for fluid processes at shallow depths in subduction zones. *Earth and Planetary Science Letters*, **140**, 41–52.

Zack, T., Tomascak, P.B., Rudnick, R.L., Dalpé, C. and McDonough, W.F. (2003) Extremely light Li in orogenic eclogites: the role of isotope fractionation during dehydration in subducted oceanic crust. *Earth and Planetary Science Letters*, **208**, 279–290.

Zinner, E. and Crozaz, G. (1986) A method for the quantitative measurement of Rare Earth Elements in the ion microprobe. *International Journal of Mass Spectrometry and Ion Process*, **69**, 17-38.

Zinner, E., Fahey, A.J. and McKeegan, K.D. (1986) Characterization of electron multipliers by charge distributions. Pp. 170–172 in: *Secondary Ion Mass Spectrometry SIMS V* (A. Benninghoven, R.J. Colton, D.S. Simons and H.W. Werner, editors). Springer Series in Chemical Physics, **44**, Springer, Berlin.

Zjang, F. and Wright, K. (2012) Lithium defects and diffusivity in forsterite. *Geochimica et Cosmochimica Acta*, **91**, 32–39.

Spatially resolved materials characterization using synchrotron radiation

Christian G. SCHROER

DESY Photon Science, Notkestr. 85, 22607 Hamburg, Germany
Department Physik, Universität Hamburg, Luruper Chaussee 149, 22761 Hamburg,
Germany, e-mail: christian.schroer@desy.de

The high brilliance of synchrotron radiation has allowed us to combine microscopy with X-ray analytical techniques such as X-ray fluorescence, absorption spectroscopy, and diffraction. In this way, elemental, chemical and structural information can be obtained locally and with high spatial resolution. The large penetration depth of X-rays in matter allows one to probe the inner structure of an object or a specimen inside a special sample environment, such as a chemical reactor or a pressure cell. Using tomographic techniques, the 3D information from inside a sample can be obtained *in situ* and without destructive sample preparation. In this chapter, we give an overview over X-ray microscopy techniques with chemical and structural contrast.

1. Synchrotron radiation

In a synchrotron radiation source, charged particles, such as electrons, move inside a storage ring at relativistic speeds, *i.e.* close to the speed of light. They emit electromagnetic radiation whenever they are accelerated, *e.g.* when they are forced onto a curved path by bending magnets or onto an oscillatory path inside so-called wigglers or undulators. Synchrotron radiation can cover a large part of the electromagnetic spectrum, ranging from THz radiation to hard X-rays. One important characteristic of synchrotron radiation is its collimation, *i.e.* the beam is emitted into a very small cone around the direction of motion of the charged particle. For example, a modern undulator source emits X-rays into a narrow cone with an opening angle of a few tens of microradiants, only. As a result, the beam is highly intense and widens to just a few millimetres at ~100 m from the source and is ideally suited to investigating the structure of matter. In the hard X-ray range, the spectral brightness of a modern synchrotron radiation source exceeds that of a laboratory X-ray source by ~12 orders of magnitude, making very sophisticated X-ray techniques possible, such as X-ray microscopy. An introduction to synchrotron radiation and techniques based upon it can be found in Als-Nielsen (2002).

Synchrotron radiation sources are typically user facilities, giving access to synchrotron radiation techniques to scientists from universities, research institutions and industry. Beamtime at these facilities is granted through a peer-reviewed proposal system. There are many synchrotron radiation facilities around the world (see www.lightsources.org). Examples of synchrotron radiation facilities are the European Synchrotron Radiation Facility (ESRF) in Grenoble, France, the Advanced Photon

Source (APS) near Chicago, USA, SPring-8 in Japan, PETRA III at DESY in Hamburg, Germany, Diamond Light Source in Didcot, Oxfordshire, UK, SOLEIL near Paris, France, and SSRL at SLAC in Menlo Park, California, USA. All these sources provide X-ray microscopy techniques to users amongst many other X-ray analytical techniques.

2. X-ray microscopy and tomography

One of the key strengths of X-rays is the large penetration depth in matter, making it possible to image inner structures of an object without destructive sample preparation, study objects in special sample environments such as chemical reactors or pressure cells, and investigate physical or chemical processes under working (operando) conditions as well as study samples in their natural environment, e.g. under high pressure and temperature.

In addition, X-ray microscopy can make use of various X-ray analytical techniques, such as X-ray fluorescence, diffraction and absorption spectroscopy. In this way, elemental, chemical and structural information can be obtained locally from inside a sample and with high spatial resolution. When combined with tomography, X-ray microscopy yields three-dimensional information from inside an object or a sample environment.

The spatial resolution in X-ray microscopy is limited in the same way as in visible-light microscopy, where the smallest distance between distinguishable features in a micrograph is given by Abbe's resolution formula:

$$d = \gamma \frac{\lambda}{2NA}, \quad NA = n \sin \alpha \qquad (1)$$

where λ is the wavelength of the light, NA is the numerical aperture of the objective lens, and γ is a factor (close to one) that depends on the shape of the aperture. The wavelength of X-rays lies in the range of a few nanometers down to 0.1 Å and is two to four orders of magnitude smaller than that of visible light.

By examining equation 1, one might expect a significantly enhanced spatial resolution for microscopy with hard X-rays compared to that with visible light due to the much shorter wavelength. However, the gain in resolution obtained by the reduced wavelength is mostly compensated by a small numerical aperture NA for X-ray optics. The numerical aperture is the product of the sine of the aperture angle α and the refractive index n of the medium surrounding the object (cf. equation 1). For visible light, numerical apertures of microscope objectives can approach and even slightly exceed unity. In the hard X-ray range, however, the aperture angles of the best imaging optics lie in the range of milliradians, resulting in numerical apertures in the range of 10^{-3}. Therefore, spatial resolutions in conventional X-ray microscopy (based on optics) are typically limited to a few tens of nanometers, today.

Due to the weak interaction of X-rays with matter and their short wavelength, the fabrication of X-ray optics with large numerical aperture and free of aberrations is very challenging. A variety of different X-ray focusing and imaging optics have been devised over recent decades based on reflection (Kirkpatrick and Baez, 1948; Mimura et al., 2007), diffraction (Vila-Comamala et al., 2009; Mimura et al., 2010; Morgan et *al.*, 2015) and refraction (Snigirev *et al.*, 1996; Schroer *et al.*, 2005; Schropp *et al.*, 2013). Today, most of these optics are technology limited, but diffractive optics have the potential of focusing X-rays down to <1 nm (Schroer, 2006; Yan *et al.*, 2007).

2.1. X-ray full-field microscopy

In X-ray full-field microscopy the sample is illuminated by X-rays and its transmission image is projected onto an X-ray sensitive detector. In its simplest form, this is done by illuminating the sample by the synchrotron radiation beam and positioning a high-resolution X-ray camera directly behind it. The resulting radiomicrograph is limited in spatial resolution by that of the detector. In the hard X-ray range, high-resolution X-ray cameras can resolve features of slightly <1 μm in size (Koch *et al.*, 1998; Stampanoni *et al.*, 2002).

To achieve greater spatial resolution, more sophisticated imaging schemes are needed, involving magnified imaging with X-rays (Fig. 1). Figure 1a shows a scheme based on magnified projection. For that purpose, the X-rays from the synchrotron radiation source are strongly focused to create a secondary source. The object is placed a distance L_1 behind the focus and a magnified projection image is formed on a detector placed a distance L_2 from the focus The magnification is $M = L_2/L_1$. The spatial resolution is predominantly limited by the size of the focus point and the imaging geometry (*cf.* section 2.2 for a discussion of the size of the focus). While this method is very dose efficient, it has the drawback of more complicated image formation (*cf.* Mokso *et al.*, 2007; Salditt *et al.*, 2009).

Alternatively, an objective optic can be used to generate a magnified image on the detector as shown in Fig. 1b. In that case, the object that is illuminated from behind is placed a distance L_1 ($> f$) before the objective lens with focal length f and a magnified image (magnification $M = L_2/L_1$) of the object is formed on the detector at a distance $L_2 = (L_1 f)/(L_1 - f)$ behind the lens. Typically, the illuminating X-rays are condensed onto the sample by a condensor optic (not shown in Fig. 1b), illuminating the object

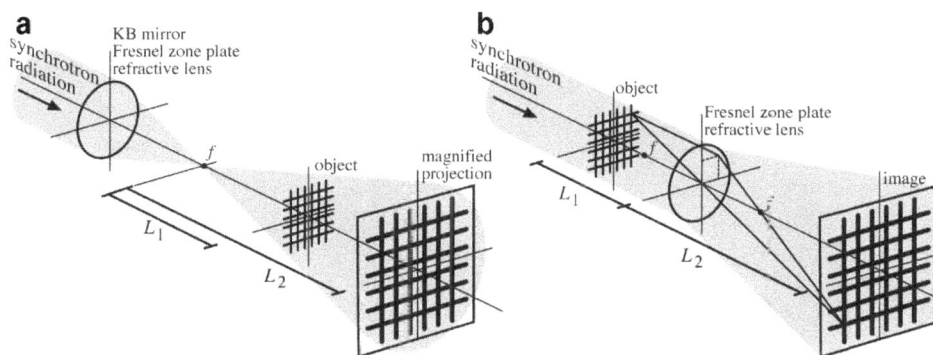

Fig. 1. Hard X-ray full-field imaging: (a) magnified projection imaging; and (b) X-ray imaging with a Fresnel zone plate or refractive lens as objective lens.

incoherently and matching the aperture of the objective lens. In that case, the spatial resolution is limited by equation 1. As the efficiency of the objective lens is typically significantly <1, the dose efficiency of this technique is low compared to that of magnified projection imaging (Fig. 1a).

2.2. X-ray scanning microscopy

In scanning microscopy the sample is scanned through a confined beam, recording at each position of the scan the X-ray analytical contrasts of interest. For example, by recording the X-ray fluorescence emitted by the sample, the element distribution inside the sample can be determined. By scanning the sample in the two directions perpendicular to the beam (cf. Fig. 2), two-dimensional elemental maps can be recorded. An important prerequisite for this is that the sample is thin enough not to exceed the depth of focus of the focused beam. In the X-ray range this is typically well fulfilled even for relatively thick samples, as the divergence angle of the focused beam is limited by the small numerical aperture of the focusing optic (cf. introductory paragraphs of section 2 above). In order to obtain three-dimensional element maps, tomographic scanning schemes are needed, involving in addition the rotation around an axis perpendicular to the beam (cf. Fig. 2; and section 2.3). The spatial resolution of the technique is limited by step size of the scan unless it is made smaller than the lateral focus size at the sample position. In the latter case, the resolution is given by the lateral beam size.

Figure 2 shows schematically the experimental setup for scanning microscopy. The X-rays from the synchrotron radiation source are focused onto the sample position in a strongly demagnifying geometry. In order to achieve a large demagnification and thus a small focus, the focal length f of the optic is typically chosen to be much smaller than

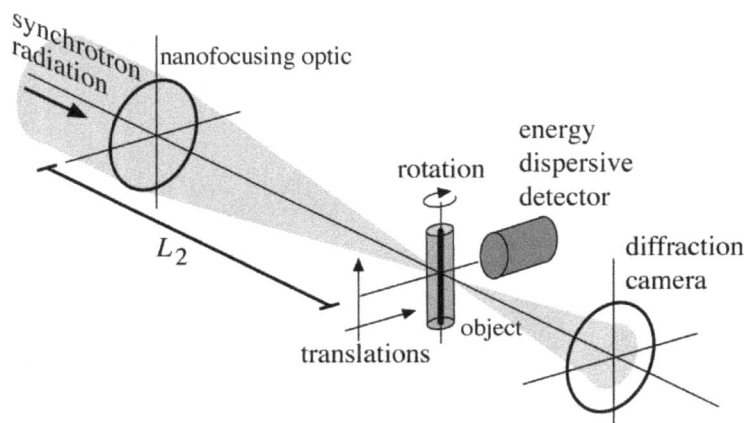

Fig. 2. Hard X-ray scanning microscopy: the X-ray source is imaged onto the sample in a strongly demagnifying geometry. The sample is scanned through the focused X-ray beam in the two translations perpendicular to the beam and, in addition, in rotation for tomography. At each position of the scan, the X-ray fluorescence, the transmission, the diffraction, or some other X-ray analytical contrast is recorded.

the source-to-optic distance L_1. In that case, the image of the source is formed at a distance $L_2 = (L_1 f)/(L_1 - f)$ behind the optic. The demagnified image of the source has a size of $b_{geo} = Mg$, where $M = L_2/L_1$ is the magnification and g is the lateral source size. The lateral size of the focus is thus at best b_{geo}. In addition, diffraction at the aperture of the focusing optic blurs the focus by an Airy disc of size d as given by equation 1. If aberrations of the optic can be neglected, the size of the focus is

$$b = \sqrt{b_{geo}^2 + d^2} \tag{2}$$

where it is assumed that the shape of the source as well as the aperture of the optic are Gaussian to good approximation. In case b_{geo} is smaller than the Airy disc size d, the beam is said to be diffraction limited. In that case, the size of focus and thus the resolution of the scanning microscope is limited to approximately d by the optic. The smallest diffraction-limited beam sizes reached so far are slightly <10 rm (Mimura *et al.*, 2010; Morgan *et al.*, 2015).

In principle, any X-ray analytical technique can be exploited as contrast, giving access to a wealth of information about the sample. For example, using absorption spectroscopy as contrast, the chemical state of a given element can be mapped (*cf.* section 3.2). Using small-angle X-ray scattering, the local nanostructure can be determined (Schroer *et al.*, 2006; Schaff *et al.*, 2015; Liebi *et al.*, 2015) or by using wide-angle X-ray scattering, the local crystalline phases can be imaged (Bleuet *et al.*, 2008).

2.3. Tomography

As X-rays penetrate through the object, X-ray micrographs (both in full-field and scanning mode) are typically two-dimensional projections of the three-dimensional object. The more complex the structure of the object, the more difficult are the interpretations of such individual projections. Using a series of such projections from different angles, the three-dimensional inner structure of the object can be reconstructed. An introduction to tomographic reconstruction can be found in Kak and Slaney (1988). In the following sections, tomographic imaging is illustrated for different X-ray microscopy techniques.

3. *In situ* imaging with chemical contrast

One way that X-rays can interact with matter is photo absorption. In this process, an X-ray photon with energy $E = h\nu$ is absorbed by an atom. Here, h is Planck's constant and ν is the frequency of the X-rays. The absorbed energy excites the atom, typically transferring a core electron into a free state. This excitation depends heavily on the energy levels of the given element and is thus element specific. As it also depends on the free states available in the vicinity of the atom, photo absorption is also sensitive to the local chemical environment of the absorbing atom. In this way, the chemical state of a given element influences the absorption of X-rays and can be probed by measuring the latter. This the basis of X-ray absorption spectroscopy (XAS).

Once the atom is excited with a vacancy in a core energy level, the atom relaxes into its ground state in a cascade of secondary processes, such as X-ray fluorescence or Auger emission. In X-ray fluorescence, an electron from a higher atomic level is transferred to the vacant core state, and a photon is emitted with an energy $E = h\nu$ that corresponds to the energy difference of the two levels. As the atomic levels are element specific, X-ray fluorescence can be used to determine the elemental composition of a sample. Synchrotron-radiation-induced X-ray fluorescence is very sensitive and can therefore be used for trace-element analysis.

Besides measuring the X-ray absorption or fluorescence of the sample, the electrons emitted in the photo absorption or Auger process can also be detected, measuring their energy spectrum. This is the basis for photo emission spectroscopy (PES) or Auger spectroscopy. Both of these techniques are surface sensitive and require a vacuum environment for the sample. They find wide application in solid state physics but will not be considered in this chapter.

3.1. Fluorescence mapping and tomography

X-ray fluorescence can be used in X-ray scanning microscopy in order to image the chemical composition of a specimen. For that purpose, the sample is scanned through a focused X-ray beam. Along the beam path through the sample, X-rays are absorbed (Fig. 3), resulting in part in the emission of X-ray fluorescence. The fluorescence radiation can be recorded by an energy dispersive detector (Fig. 3). In this way, the full spectrum of X-rays emitted from the sample can be measured, giving information about the element composition of the illuminated part of the sample.

To make a two-dimensional fluorescence map of the sample, it is scanned through the beam in the two translational directions perpendicular to the beam. At each position of the scan, a full fluorescence spectrum is recorded. The X-ray fluorescence is emitted from all the parts that are illuminated along the beam (Fig. 3b). If the sample is too thick and its structure varies too much along this direction, a three-dimensional fluorescence

a scanning fluorescence microscopy   **b** fluorescence tomography

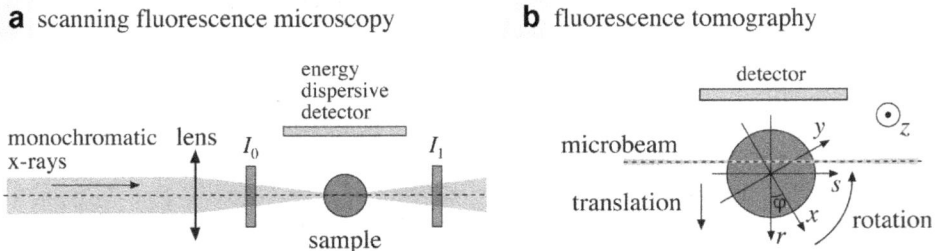

Fig. 3. X-ray microscopy combined with X-ray fluorescence analysis. (a) The sample is scanned through the focused beam. Incoming and transmitted radiation are measured by detectors I_0 and I_1 before and behind the sample, respectively. The energy dispersive detector records the X-rays emitted by the sample from the side. Fluorescence mapping and tomography only differ in how the sample is scanned through the beam. (b) Fluorescence tomography: the sample is scanned through the beam in translation perpendicular to the beam (along r and z axis) and in rotation φ around the z axis.

map is needed. It can be obtained by scanning the sample in tomographic mode (Simionovici *et al.*, 2000). After recording a two-dimensional map in r and z (Fig. 3), the sample is rotated by an integer fraction $\Delta\varphi$ of 360° and the next 2D map is recorded. This is repeated, until the sample has completed a half or full rotation (180° or 360°). The number M of projections needed to sample the tomogram properly corresponds to $M \approx (\pi/2)N$, where N is the number of scan points in the r direction (Kak and Slaney, 1988). If a two-dimensional slice through the sample is sufficient, it is scanned along the r direction, only, to record a projection. As the fluorescence emitted from inside the object can be absorbed by other parts of the sample on its way to the detector, special tomographic reconstruction schemes are needed to obtain quantitative reconstruction (Schroer, 2001).

As an example, a fluorescence tomogram (single tomographic slice) of a root of the mahogany tree is considered. Figure 4a shows fluorescence data recorded as function of translational position r and rotation angle φ of the root for different physiologically relevant ions. The tomogram was recorded at the ESRF, illuminating the sample with a focused hard X-ray beam ($E = 19.5$ keV) with a lateral size of 6 µm × 1.4 µm (Schroer *et al.*, 2002). 152 projections were recorded over 360°C, making 128 translational steps of 6 µm, each. At each scan point the X-ray fluorescence was recorded for 2 s.

The tomograms for the different elements were reconstructed using an algorithm that self-consistently corrects for self-attenuation effects (Schroer, 2001). The reconstructions are shown in Fig. 4b. The rubidium concentration is about three orders of magnitude smaller than that of the potassium. Nevertheless, a clear Rb distribution can be reconstructed, illustrating the high sensitivity of synchrotron radiation based X-ray fluorescence analysis.

X-ray fluorescence tomography has been applied in various fields of science, *e.g.* imaging the elemental content of cometary dust particles (Silversmit *et al.*, 2009) or the tissue-specific trace element level in the freshwater crustacean *Daphnia magna* that is a frequently used ecotoxicological model organism to investigate the mechanisms of toxicity of transition metals in the pelagic and benthic zones of freshwater bodies (De Samber *et al.*, 2008). A recent example of three-dimensional fluorescence tomography of aged fluid-catalytic-cracking catalyst particles can be found in Kalirai *et al.* (2015).

3.2. Chemical-state imaging and tomography using X-ray absorption spectroscopy

In order to understand chemical processes, such as catalytic reactions, it is not only necessary to know the element distribution inside the sample, but also the chemical state of certain elements of interest. The latter can be imaged by exploiting X-ray absorption spectroscopy (XAS). XAS can be combined with both X-ray full-field and scanning microscopy (Fig. 5). An overview over XAS microscopy techniques for chemical imaging is given in Grunwaldt and Schroer (2010).

X-ray absorption spectroscopy is a powerful tool for analysing the chemical state and local environment of a given element without the need for long-range order in the specimen. Experimentally, the sample is illuminated by monochromatic X-rays, the

a Cl-$K\alpha$ ($E = 2.62$ keV)   K-$K\alpha$ ($E = 3.31$ keV)   Ca-$K\alpha$ ($E = 3.69$ keV)   Rb-$K\alpha$ ($E = 13.38$ keV)

128 translations
(step size 6 μm)

0   50   100
intensity relative to maximum [%]

b Cl   K   Ca   Rb

200 μm

0   50   100
concentration relative [%]

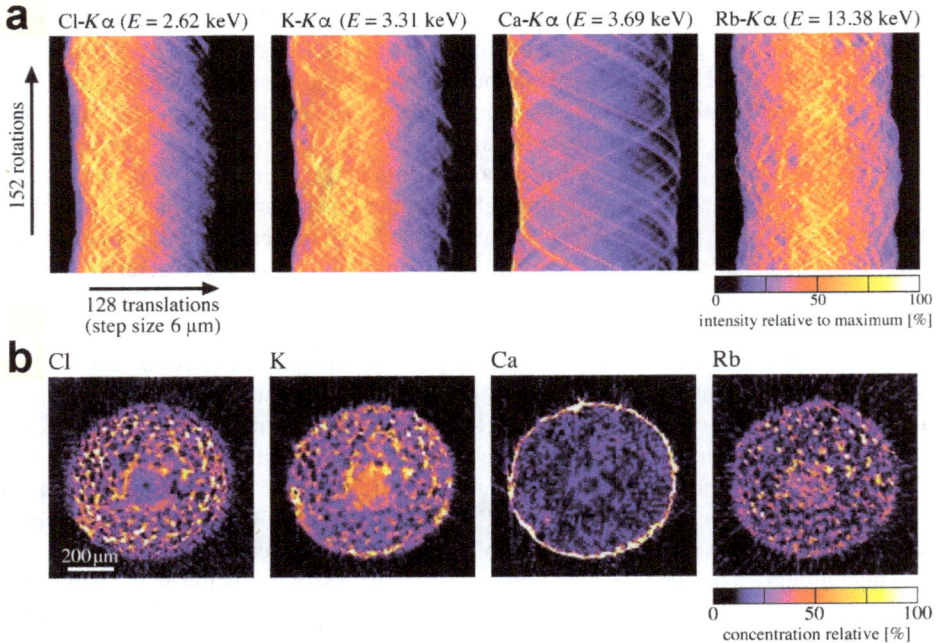

Fig. 4. (a) Tomographic data (sinograms) recorded from a root of the mahogany tree (Schroer 2001; Schroer *et al.*, 2002) for different physiologically relevant elements. If self-absorption of the fluorescence inside the sample can be ignred, the sinogram is symmetric (see the Rb-$K\alpha$ line). If self-absorption is relevant, the fluorescence data from parts facing the detector are brighter than those away from the detector (see for example Cl-$K\alpha$). (b) Tomographic reconstruction of the data in (a) using an algorithm correcting for self-absorption (Schroer, 2001).

energy of which is tunable around the absorption edge of the element of interest (Fig. 5) The intensity of the radiation before and behind the sample is measured. In the case of full-field imaging this is achieved by taking an image without and with the sample in the beam; in scanning microscopy the ionization chambers I_0 and I_1 measure these quantities.

From the ratio of I_0 and I_1 measured at a given energy E, the integral of the attenuation coefficient $\mu(x,y,z,E)$ can be extracted using Lambert-Beer's law:

$$I_1 = I_0 \cdot e^{-\int \mu(x,y,z,E)ds} \Rightarrow \int \mu(x,y,z,E)ds = \log\frac{I_0}{I_1} \qquad (3)$$

Here, the integral is taken along the X-ray beam for each position r and z as defined in Fig. 3b. An absorption spectrum is acquired by scanning the energy in an interval around the absorption edge. From these data, various pieces of information about the chemistry of the given element can be extracted. The exact position of the absorption edge yields the oxidation state of the element. The X-ray absorption near edge structure

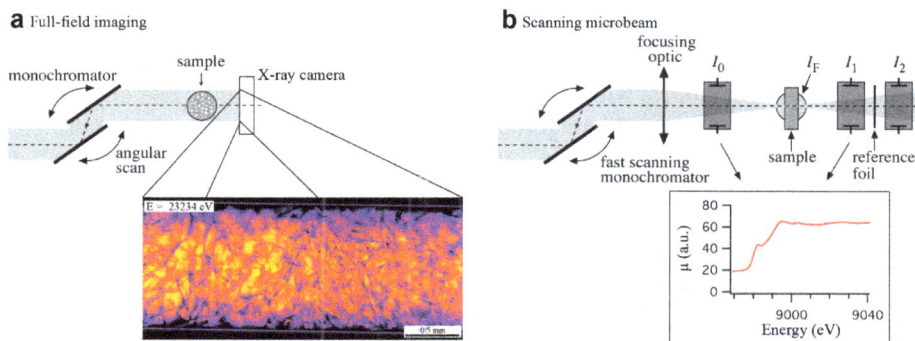

Fig. 5. X-ray microscopy combined with absorption spectroscopy: (a) full-field imaging using XAS contrast: as the energy of the X-rays is scanned over an absorption edge of a chemical element of interest, X-ray full-field transmission images are recorded for each energy. In this way, a full XAS spectrum is obtained for each pixel in the detector. (b) Scanning microscopy using XAS contrast: the sample is raster scanned through the focused beam, recording at each position of the scan a full absorption spectrum by varying the incident X-ray energy. Incident and transmitted intensity are measured by ion chambers I_0 and I_1, respectively. For energy calibration, the transmission through a reference sample is measured simultaneously with the ion chamber I_2.

(XANES) extending to ~50 eV above the edge probes the local projected density of states for the given element. The extended X-ray absorption fine structure (EXAFS) that extends to ~1 keV above the edge gives information about the local chemical neighbourhood of the element, yielding the interatomic distances with the neighbours and coordination numbers (Koningsberger and Prins, 1988). As an example, the XANES spectrum $\mu(E)$ of a CuO/ZnO catalyst is shown in Fig. 6.

The main advantage of full-field imaging with XAS contrast (*cf.* Fig. 5a) is the fast data acquisition, as only the energy of the X-rays incident on the sample needs to be scanned. For example, it is very suitable for *in situ* investigations of chemical processes and has been applied to study heterogeneous catalysts under working conditions inside a chemical reactor (Grunwaldt *et al.*, 2006; Hannemann *et al.*, 2007). During the partial oxidation of methane the oxidation state of the catalyst was shown to vary in a characteristic manner as a function of position inside the chemical reactor. This variation is consistent with the two-step chemical reaction of reforming rather than with the direct partial oxidation of methane, thus settling the question of the reaction mechanism under the given reaction conditions. These operando studies did not require a high temporal resolution due to the stationary reaction conditions.

In order to image the ignition of the catalytic reaction on the sub-second timescale, scanning the energy to acquire full absorption spectra is not practical. In some cases, a change in oxidation state of the catalyst changes its absorption coefficient $\mu(E)$ significantly at a given energy. For example, this is true for platinum just above the L_{III} absorption edge at $E = 11.596$ keV. By recording a series of X-ray images at this X-ray energy, it was possible to study the ignition of the catalytic partial oxidation reaction of methane in real time down to the micrometre scale (Kimmerle *et al.*, 2009).

The full-field imaging technique as depicted in Fig. 5a is limited in spatial resolution by the pixel size of the detector to the micrometre range. To reach greater spatial resolutions, XAS needs to be combined with full-field imaging techniques as discussed in section 2.1 and shown in Fig. 1. Full-field microscopy (*cf.* Fig. 1b), for example, was demonstrated imaging a Li-ion composite electrode containing Ni/NiO at the Ni K edge (Meirer *et al.*, 2011). By combining XAS microscopy with tomography, the Ni and NiO distribution in the electrode could be imaged with a spatial resolution of several tens of nanometers.

Alternatively, scanning microscopy can be combined with X-ray absorption spectroscopy. This was demonstrated both in the hard and soft X-ray regime (Kinney *et al.*, 1986; Ade *et al.*, 1992; Yun *et al.*, 1998; Youn *et al.*, 2001) and more recently to study *in situ* a Fischer Tropsch catalyst in the soft X-ray range (de Smit *et al.*, 2008), where a spatial resolution of a few tens of nanometers was reached. Scanning microscopy with absorption spectroscopic contrast needs not only to scan the sample through the beam, but also the energy. Therefore, two-dimensional XAS maps require three-dimensional scans (two lateral translations and the X-ray energy). Tomographic scanning schemes would even require four-dimensional scans resulting in very, if not prohibitively, long acquisition times. Fast scanning of either a spatial degree of freedom or the energy is mandatory. The technique has benefited greatly from the development of fast scanning monochromators (Frahm, 1989; Müller *et al.*, 2015) that allow the acquisition of full XAS spectra on the timescale of 10 ms.

Tomographic scanning XAS microscopy can be realized when restricting the scans to a few two-dimensional slices through the sample. This is illustrated here by imaging a CuO/ZnO catalyst inside a reactor capillary on a virtual slice through the reactor (Fig. 6). Similarly to fluorescence microtomography (*cf.* section 3.1) the capillary was scanned in the transverse direction through the microfocused X-ray beam in 90 steps of 10 μm each, recording at each scan position a full XANES spectrum around the Cu K edge using a quick scanning monochromator and data acquisition system (Richwin *et al.*, 2002; Schroer *et al.*, 2003). After the translational scan was completed, the sample was rotated by an integer fraction 360° and the transverse scan was repeated. This whole procedure was repeated until a full rotation was completed (101 projections).

From the transmission data, the integral of the attenuation coefficient along the microbeam was determined using Lambert-Beer's law (equation 3) for each energy in the spectrum. At fixed energy, the tomographic transmission data were reconstructed using standard tomographic techniques (Kak and Slaney, 1988). This reconstruction was done for each energy, yielding a full absorption spectrum at each location in the reconstruction (*cf.* Fig. 6). By fitting a set of reference spectra for metallic copper, $Cu(I)_2O$, and $Cu(II)O$ to the reconstructed spectra, the local concentration of metallic, monovalent and bivalent copper could be determined on the virtual slice through the capillary (Fig. 6). In this way, the chemical state of the copper sites in the CuO/ZnO catalyst were imaged non-destructively inside a reactor capillary. Details of this experiment were given by Schroer *et al.* (2003).

While most applications for absorption spectroscopic imaging, so far, are from chemistry, especially from catalysis and electro-chemistry, the methods can be applied to other fields, *e.g.* geo- and environmental science and biology.

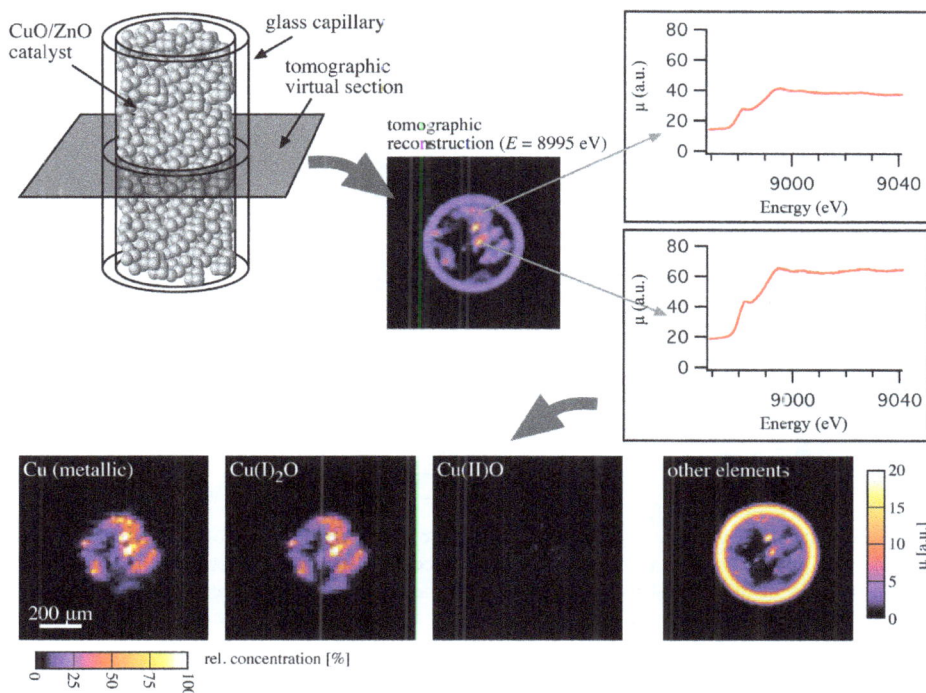

Fig. 6. XANES tomography of a CuO/ZnO catalyst inside a reactor capillary. A tomographic slice is recorded by scanning the sample through an X-ray microbeam, recording at each position of the scan a full XANES spectrum around the Cu K edge. The reconstructed tomographic data yield a full spectrum at each location in the tomographic slice. By fitting reference spectra for metallic copper, $Cu(I)_2O$ and $Cu(II)O$, to the reconstructed spectra, the distributions of the different oxidation states of copper and those of other elements can be obtained. While metallic and monovalent copper are present with very similar concentrations, the bivalent copper content is below the detection limit.

3.3. High-resolution imaging by scanning coherent diffraction microscopy

The highest spatial resolutions in X-ray microscopy are reached today using scanning coherent diffraction microscopy also known as ptychography (Schropp *et al.*, 2012). In this scanning microscopy method, the sample is scanned through the focused X-ray beam, recording at each position of the scan a far-field diffraction pattern as shown in Fig. 7a (Rodenburg and Faulkner, 2004). The step size of the scan is chosen such that the illuminated areas of neighbouring scan points partially overlap. From the set of diffraction patterns together with knowledge of their respective scan positions, the object can be reconstructed numerically (Thibault *et al.*, 2008, 2009; Maiden and Rodenburg, 2009; *cf.* Fig. 7b) with a spatial resolution that is significantly higher than in conventional scanning microscopy, *i.e.* the resolved features are smaller than the lateral size of the focused beam (Thibault *et al.*, 2008). This is shown in Fig. 8a where the ptychographic image of a test structure is shown at the same scale as the illuminating nanofocused X-ray beam (inset). Although the nanofocused X-ray beam

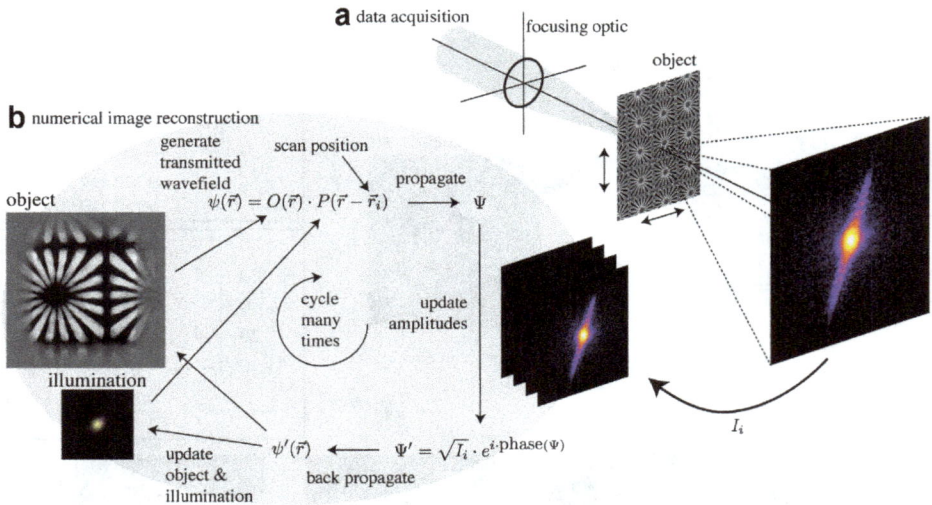

Fig. 7. Scanning coherent diffraction microscopy (ptychography). (a) For ptychographic data acquisition, the sample is scanned through a focused coherent X-ray beam, recording at each position of the scan a far-field diffraction pattern. (b) These data serve as input for an iterative numerical image-reconstruction scheme (Maiden and Rodenburg, 2009).

used for scanning the sample had a lateral size of 175 nm, the spatial resolution in the reconstructed image is of the scale of 10 nm. The contrast in the ptychogram is quantitative and describes the phase shift induced by refraction and the attenuation of the X-rays inside the object. For small objects the phase shift typically gives a stronger contrast than the attenuation.

As for the scanning microscopy techniques discussed in the previous sections, ptychography can also be combined with tomography (Dierolf *et al.*, 2010). The ptychogram (as shown in Fig. 8) is a projection of the index of refraction of the X-rays inside the object and the tomographic reconstruction yields the refractive index in three-dimensions as a function of position inside the object. In this way, high-resolution three-dimensional imaging down to 10 nm is possible for relatively large objects (up to 10 μm in size; Holler *et al.*, 2014). In the last few years, ptychographic tomography has quickly evolved into a standard technique that is being applied to many fields of science and engineering (Trtik *et al.*, 2013; Diaz *et al.*, 2014; Dam *et al.*, 2015; da Silva *et al.*, 2015; Pedersen *et al.*, 2015).

In the same way as the absorption of X-rays changes in a characteristic way as the energy is scanned across the absorption edge of a given element of interest (*cf.* section 3.2) so does the refraction and thus the phase contrast in ptychography. By scanning the energy across an absorption edge and recording ptychograms for each energy, high-resolution imaging with chemical contrast can be achieved (Beckers *et al.*, 2011; Hoppe *et al.*, 2013). When combined with tomography, 3D chemical imaging with high spatial resolution becomes possible (Donnelly *et al.*, 2015).

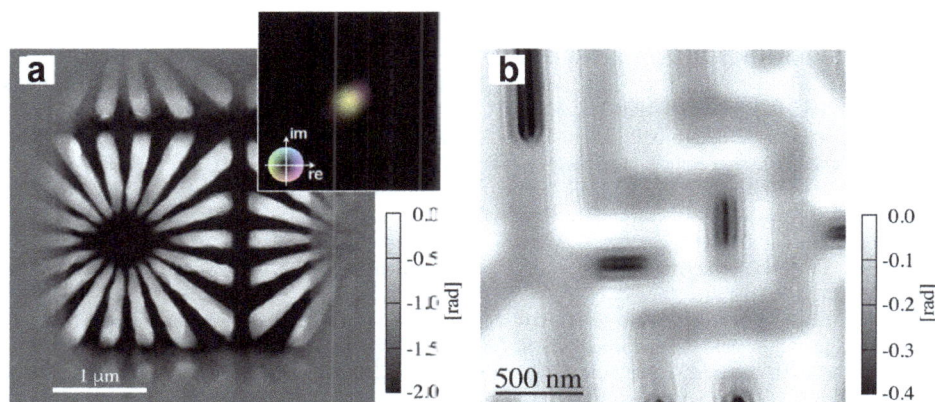

Fig. 8. (a) Ptychographic image of a test structure made of tungsten (1 μm thick) on a diamond substrate. The smallest features are ~50 nm in size (Schroer *et al.*, 2013). The focused X-ray beam used to record this ptychogram is shown in the inset (same scale), illustrating the gain in spatial resolution compared to the beam size. (b) Ptychographic reconstruction of a front-end processed and passivated microchip (Schropp *et al.*, 2011, 2012).

As for the other X-ray microscopy techniques, ptychography is also suited for investigating samples *in situ* or under operation conditions (Høydalsvik *et al.*, 2014; Baier *et al.*, 2016).

References

Ade, H., Zhang, X., Cameron, S., Costello, C., Kirz, J. and Williams, S. (1992) Chemical contrast in X-ray microscopy and spatially resolved XANES spectroscopy of organic specimens. *Science*, **258**, 972–975.

Als-Nielsen, J. (2002) *Elements of Modern X-Ray Physics*. John Wiley & Sons, Copenhagen.

Baier, S., Damsgaard, C.D., Scholz, M., Benzi, F., Rochet, A., Hoppe, R., Scherer, T., Shi, J., Wittstock, A., Weinhausen, B., Wagner, J. B., Schroer, C.G., and Grunwaldt, J.D. (2016) In situ ptychography of heterogeneous catalysts using hard X-rays: High resolution imaging at ambient pressure and elevated temperature. *Microscopy and Microanalysis*, **22**, 178–188.

Beckers, M., Senkbeil, T., Gorniak, T., Reese, M. Giewekemeyer, K., Gleber, S.C., Salditt, T. and Rosenhahn, A. (2011) Chemical contrast in soft X-ray ptychography. *Physical Review Letters*, **107**(20), 208101.

Bleuet, P., Welcomme, E., Dooryhée, E., Susini, J., Hodeau, J.L. and Walter, P. (2008) Probing the structure of heterogeneous diluted materials by diffraction tomography. *Nature Materials*, **7**, 468–472.

Da Silva, J.C., Mader, K., Holler, M., Huberthür, D., Diaz, A., Guizar-Sicairos, M., Cheng, W.C., Shu, Y., Raabe, J., Menzel, A. and van Bokhoven J.A. (2015) Assessment of the 3D pore structure and individual components of preshaped catalyst bodies by X-ray imaging. *CemCatChem*, **7**, 413–416.

Dam, H.F., Andersen, T.R., Pedersen, E.B.L., Thydén, K.T.S., Helgesen, M., Carlé, J.C., Jørgensen, P.S., Reinhardt, J., Søndergaard, R.R., Jørgensen, M., Bundgaard, E., Krebs, F.C. and Andreasen, J.W. (2015) Enabling flexible polymer tandem solar cells by 3D ptychographic imaging. *Advanced Energy Materials*, **2015**(5), 1400736.

De Samber, B., Evans, R., De Schamphelaere, K. Silversmit, G., Masschaele, B., Schoonjans, T., Vekemans, B., Janssen, C.R., Van Hoorebeke, L., Szalóki, I., Vanhaecke, F., Falkenberg, G. and Vincze, L. (2008) A combination of synchrotron and laboratory X-ray techniques for studying tissue-specific trace level metal distributions in *Daphnia magna*. *Journal of Analytical Atomic Spectrometry*, **23**, 829–839.

De Smit, E., Swart, I., Creemer, J.F., Hoveling, G.H., Gilles, M.K., Tyliszczak, T., Kooyman, P.J., Zandbergen,

H.W., Morin, C., Weckhuysen, B.M. and de Groot, F.M.F. (2008) Nanoscale chemical imaging of a working catalyst by scanning transmission X-ray microscopy. *Nature*, **456(7219)**, 222–225.

Diaz, A., Guizar-Sicairos, M., Poeppel, A., Menzel, A. and Bunk, O. (2014) Characterization of carbon fibers using X-ray phase nanotomography. *Carbon*, **67**, 98–103.

Dierolf, M., Menzel, A., Thibault, P., Schneider, P., Kewish, C.M., Wepf, R., Bunk, O. and Pfeiffer, F. (2010) Ptychographic X-ray computed tomography at the nanoscale. *Nature*, **467(7314)**, 436–440.

Donnelly, C., Guizar-Sicairos, M., Scagnoli, V., Holler, M., Huthwelker, T., Menzel, A., Vartiainen, I., Müller, E., Kirk, E., Gliga, S., Raabe, J. and Heyderman, L.J. (2015) Element-spectific X-ray phase tomography of 3D structures at the nanoscale. *Physical Review Letters*, **114(11)**, 115501.

Frahm, R. (1989) New method for time dependent X-ray absorption studies. *Review of Scientific Instruments*, **60**, 2515–2518.

Grunwaldt, J.D. and Schroer, C.G. (2010) Hard and soft X-ray microscopy and tomography in catalysis: Bridging the different time and length scales. *Chemical Society Reviews*, **39**, 4741.

Grunwaldt, J.D., Hannemann, S., Schroer, C.G. and Baiker, A. (2006) 2D - mapping of the catalyst structure inside a catalytic microreactor at work: Partial oxidation of methane over Rh/Al_2O_3. *Journal of Physical Chemistry B*, **110**, 8674–8680.

Hannemann, S., Grunwaldt, J.D., van Vegten, N., Baiker, A., Boye, P. and Schroer, C.G. (2007) Distinct spatial changes of the catalyst structure inside a fixed-bed microreactor during the partial oxidation of methane over Rh/Al_2O_3. *Catalysis Today*, **126**, 54–63.

Holler, M., Diaz, A., Guizar-Sicairos, M., Karvinen, P., Färm, E., Härkönen, E., Ritala, M., Menzel, A., Raabe, J. and Bunk, O. (2014) X-ray ptychographic computed tomography at 16 nm isotropic 3D resolution. *Scientific Reports*, **4**, 3857.

Hoppe, R., Reinhardt, J., Hofmann, G., Patommel, J., Grunwaldt, J.D., Damsgaard, C.D., Wellenreuther, G., Falkenberg, G. and Schroer, C.G. (2013) High-resolution chemical imaging of gold nanoparticles using hard X-ray ptychography. *Applied Physics Letters*, **102(20)**, 203104.

Høydalsvik, K., Fløystad, J.B., Zhao, T., Esmaeili, M., Diaz, A., Andreasen, J.W., Mathiesen, R.H., Rønning, M. and Breiby, D.W. (2014) *In situ* X-ray ptychography imaging of high-temperature CO_2 acceptor particle agglomerates. *Applied Physics Letters*, **104**, 241909.

Kak, A.C and Slaney, M. (1988) *Principles of Computerized Tomographic Imaging*. IEEE Press, New York.

Kalirai, S., Boesenberg, U., Falkenberg, G., Meirer, F. and Weckhuysen, B.M. (2015) X-ray fluorescence tomography of aged fluid-catalytic-cracking catalyst particles reveals insight into metal deposition processes. *ChemCatChem*, **7**, 3675–3682.

Kimmerle, B., Grunwaldt, J.D., Baiker, A., Glatzel, P., Boye, P., Stephan, S. and Schroer, C.G. (2009) Visualizing a catalyst at work during the ignition of the catalytic partial oxidation of methane. *Journal of Physical Chemistry C*, **113**, 3037–3040.

Kinney, J., Johnson, Q., Nichols, M., Bonse, U. and Nusshardt, R. (1986) Elemental and chemical-state imaging using synchrotron radiation. *Applied Optics*, **25**, 4583–4585.

Kirkpatrick, P. and Baez, A. (1948) Formation of optical images by X-rays. *Journal of the Optical Society of America*, **38**, 766–774.

Koch, A., Raven, C., Spanne, P. and Snigirev, A. (1998) X-ray imaging with submicrometer resolution employing transparent luminescent screens. *Journal of the Optical Society of America A*, **15**, 1940–1951.

Koningsberger, D.C. and Prins, R. (1988*) X-ray Absorption*. John Wiley & Sons, New York.

Liebi, M., Georgiadis, M., Menzel, A., Schneider, P., Kohlbrecher, J., Bunk, O. and Guizar-Sicairos, M. (2015) Nanostructure surveys of macroscopic specimens by small-angle scattering tensor tomography. *Nature*, **527(7578)**, 349–352.

Maiden, A.M. and Rodenburg, J.M. (2009) An improved ptychographical phase retrieval algorithm for diffractive imaging. *Ultramicroscopy*, **109**, 1256–1262.

Meirer, F., Cabana, J., Liu, Y., Mehta, A., Andrews, J.C. and Pianetta, P. (2011) Three-dimensional imaging of chemical phase transformations at the nanoscale with full-field transmission X-ray microscopy. *Journal of Synchrotron Radiation*, **18**, 773–781.

Mimura, H., Yumoto, H., Matsuyama, S., Sano, Y., Yamamura, K., Mori, Y., Yabashi, M., Nishino, Y., Tamasaku, K., Ishikawa, T. and Yamauchi, K. (2007) Efficient focusing of hard X-rays to 25 nm by a total reflection mirror. *Applied Physics Letters*, **9C**, 051503.

Mimura, H., Handa, S., Kimura, T., Yumoto, H., Yamakawa, D., Yokoyama, H., Matsuyarra, S., Inagaki, K., Yamamura, K., Sano, Y., Tamasaku, K., Nishino, Y., Yabashi, M., Ishikawa, T. and Yamauchi, K. (2010) Breaking the 10 nm barrier in hard X-ray focusing. *Nature Physics*, **6**, 122–125.

Mokso, R., Cloetens, P., Maire, E., Ludwig, W. anc Buffière, J.Y. (2007) Nanoscale zoom tomography with hard X-rays using Kirkpatrick-Baez optics. *Applied Physics Letters*, **90**, 144104.

Morgan, A.J., Prasciolu, M., Andrejczuk, A., Krzywinski, J., Meents, A., Pennicard, D., Graafsma, H., Barty, A., Bean, R.J., Barthelmess, M., Oberthuer, D., Yefanov, O., Aquila, A., Chapman, H.N. and Bajt, S. (2015) High numerical aperture multilayer Laue lenses. *Scientific Reports*, **5**, 09892.

Müller, O., Lützenkirchen-Hecht, D. and Frahm, R. (2015) Quick scanning monochromator for millisecond in situ and in operando X-ray absorption spectroscopy. *Review of Scientific Instruments*, **86**, 093905.

Pedersen, E.B.L., Angmo, D., Dam, H.F., Thydén, K.T.S., Andersen, T.R., Skjønsfjell, E.T.B., Krebs, F.C., Holler, M., Diaz, A., Guizar-Sicairos, M., Breiby, D.W. and Andreasen, J.W. (2015) Improving organic tandem solar cells based on water-processed nanoparticles, by quantitative 3D nanoimaging. *Nanoscale*, **7**, 13765–13774.

Richwin, M., Zaeper, R., Lützenkirchen-Hecht, D. and Frahm, R. (2002) Piezo-XAFS-time-resolved X-ray absorption spectroscopy. *Review of Scientific Instruments*, **73**, 1668–1670.

Rodenburg, J.M. and Faulkner, H.M.L. (2004) A phase retrieval algorithm for shifting illumination. *Applied Physics Letters*, **85**, 4795–4797.

Salditt, T., Giewekemeyer, K., Fuhse, C., Krüger S.P., Tucoulou, R. and Cloetens, P. (2009) Projection phase contrast microscopy with a hard X-ray nanofocused beam: Defocus and contrast transfer. *Physical Review B*, **79**, 184112.

Schaff, F., Bech, M., Zaslansky, P., Jud, C., Liebi, M., Guizar-Sicairos, M. and Pfeiffer, F. (2015) Six-dimensional real and reciprocal space small-angle X-ray scattering tomography. *Nature*, **527(7578)**, 353–356.

Schroer, C.G. (2001) Reconstructing X-ray fluorescence microtomograms. *Applied Physics Letters*, **79**, 1912–1914.

Schroer, C.G. (2006) Focusing hard X-rays to nanometer dimensions using Fresnel zone plates. *Physical Review B*, **74**, 033405.

Schroer, C.G., Benner, B., Günzler, T.F., Kuhlmann, M., Lengeler, B., Schröder, W.H., Kuhn, A.J., Simionovici, A.S., Snigirev, A. and Snigireva, I. (2002) High resolution element mapping inside biological samples using fluorescence microtomography. Pp. 230–239 in: *Developments in X-Ray Tomography III* (U. Bonse, editor). Volume 4503 of Proceedings of the SPIE.

Schroer, C.G., Kuhlmann, M., Günzler, T.F., Lengeler, B., Richwin, M., Griesebock, B., Lützenkirchen-Hecht, D., Frahm, R., Ziegler, E., Mashayekhi, A., Haeffner, D., Grunwaldt, J.D. and Baiker, A. (2003) Mapping the chemical states of an element inside a sample using tomographic X-ray absorption spectroscopy. *Applied Physics Letters*, **82**, 3360–3362.

Schroer, C.G., Kurapova, O., Patommel, J., Boye, P., Feldkamp, J., Lengeler, B., Burghammer, M., Riekel, C., Vincze, L., van der Hart, A. and Küchler, M (2005) Hard X-ray nanoprobe based on refractive X-ray lenses. *Applied Physics Letters*, **87**, 124103.

Schroer, C.G., Kuhlmann, M., Roth, S.V., Gehrke, R., Stribeck, N., Almendarez-Camarillo, A. and Lengeler, B. (2006) Mapping the local nanostructure inside a specimen by tomographic small angle X-ray scattering. *Applied Physics Letters*, **88**, 164102.

Schroer, C.G., Brack, F.E., Brendler, R., Hönig, S., Hoppe, R., Patommel, J., Ritter, S., Scholz, M., Schropp, A., Seiboth, F., Nilsson, D., Rahomäki, J., Uhlén, F., Vogt, U., Reinhardt, J. and Falkenberg, G. (2013) Hard X-ray nanofocusing with refractive X-ray optics: full beam characterization by ptychographic imaging. *Proceedings of the SPIE*, **8848**, 884807.

Schropp, A., Boye, P., Goldschmidt, A., Hönig, S., Hoppe, R., Patommel, J., Rakete, C., Samberg, D., Stephan, S., Schöder, S., Burghammer, M. and Schroer, C.G. (2011) Non-destructive and quantitative imaging of a nano-structured microchip by ptychographic hard X-ray scanning microscopy. *Journal of Microscopy*, **241**, 9–12.

Schropp, A., Hoppe, R., Patommel, J., Samberg, D., Seiboth, F., Stephan, S., Wellenreuther, G., Falkenberg, G.

and Schroer, C.G. (2012) Hard X-ray scanning microscopy with coherent radiation: Beyond the resolution of conventional X-ray microscopes. *Applied Physics Letters*, **100**, 253112.

Schropp, A., Hoppe, R., Meier, V., Patommel, J., Seiboth, F., Lee, H.J., Nagler, B., Galtier, E.C., Arnold, B., Zastrau, U., Hastings, J.B., Nilsson, D., Uhlen, F., Vogt, U., Hertz, H.M. and Schroer, C.G. (2013) Full spatial characterization of a nanofocused X-ray free-electron laser beam by ptychographic imaging. *Scientific Reports*, **3**, 1633.

Silversmit, G., Vekemans, B., Brenker, F.E., Schmitz, S., Burghammer, M., Riekel, C. and Vincze, L. (2009) X-ray fluorescence nanotomography on cometary matter from comet 81P/Wild2 returned by stardust. *Analytical Chemistry*, **81**, 6107–6112.

Simionovici, A.S., Chukalina, M., Schroer, C., Drakopoulos, M., Snigirev, A., Snigireva, I., Lengeler, B., Janssens, K. and Adams, F. (2000) High-resolution X-ray fluorescence microtomography of homogeneous samples. *IEEE Transactions on Nuclear Science*, **47**, 2736–2740.

Snigirev, A., Kohn, V., Snigireva, I. and Lengeler, B. (1996) A compound refractive lens for focusing high energy X-rays. *Nature*, **384**, 49–51.

Stampanoni, M., Borchert, G., Wyss, P., Abela, R., Patterson, B., Hunt, S., Vermeulen, D. and Rüegsegger, P. (2002) High resolution X-ray detector for synchrotron-based microtomography. *Nuclear Instruments and Methods in Physics Research A*, **491**, 291–301.

Thibault, P., Dierolf, M., Menzel, A., Bunk, O., David, C. and Pfeiffer, F. (2008) High-resolution scanning X-ray diffraction microscopy. *Science*, **321(5887)**, 379–382.

Thibault, P., Dierolf, M., Bunk, O., Menzel, A. and Pfeiffer, F. (2009) Probe retrieval in ptychographic coherent diffractive imaging. *Ultramicroscopy*, **109**, 338–343.

Trtik, P., Diaz, A., Guizar-Sicairos, M., Menzel, A. and Bunk, O. (2013) Density mapping of hardened cement paste using ptychographic X-ray computed tomography. *Cement and Concrete Composites*, **36(C)**, 71–77.

Vila-Comamala, J., Jefimovs, C., Raabe, J., Pilvi, T., Fink, R.H., Senoner, M., Maassdorf, A., Ritala, M. and David, C. (2009) Advanced thin film technology for ultrahigh resolution X-ray microscopy. *Ultramicroscopy*, **109**, 1360–1364.

Yan, H., Maser, J., Macrander, A., Shen, Q., Vogt, S., Stephenson, G.B. and Hyon Chol Kang (2007) Takagi-taupin description of X-ray dynamical diffraction from diffractive optics with large numerical aperture. *Physical Review B*, **76**, 115438.

Youn, H.S., Lee, D.H., Kim, K., Choi, H.J. and Koo, Y.M. (2001) μ-XAFS beam line at Pohang Light Source. *Nuclear Instruments and Methods in Physics Research A*, **467–468**, 1557–1559.

Yun, W., Pratt, S.T., Miller, R.M., Cai, Z., Hunter, D.B., Jarstfer, A.G., Kemner, K.M., Lai, B., Lee, H.R., Legnini, D.G., Rodrigues, W. and Smith, C.I. (1998) X-ray imaging and microspectroscopy of plants and fungi. *Journal of Synchrotron Radiation*, **5**, 1390–1395.

EMU Notes in Mineralogy, Vol. 16 (2017), Chapter 7, 181–214

Thermodynamic modelling of irreversible processes

J. SVOBODA[1], F. D. FISCHER[2] and E. KOZESCHNIK[3]

[1] Institute of Physics of Materials, Brno, Czech Republic, e-mail: svobj@ipm.cz
[2] Montanuniversität Leoben, Austria, e-mail: mechanic@unileoben.ac.at
[3] Technische Universität Wien, Austria, e-mail: ernst.kozeschnik@tuwien.ac.at

Non-equilibrium linear thermodynamics represents an effective tool for phenomenological description of processes in solids. It introduces local internal state variables as the mole fractions of individual components and provides evolution equations for them in the form of partial differential equations. Materials science has to treat rather complex systems and characterizes them by means of a limited number only of characteristic parameters (CPs) and their evolution, which can be extracted from the solution of partial differential equations. If, however, one utilizes Thermodynamic Extremal Principle (TEP) formulated in CPs, the task becomes much easier. Thus, the TEP can be considered a convenient tool for modelling which has been applied successfully to sintering, creep and rafting in superalloys, grain growth and precipitate coarsening or kinetics of precipitation, for example (in the program *MatCalc*).

In this chapter the TEP is presented in its general form as well as in discrete CPs. Its general form is used for the derivation of diffusion and creep equations in mechanically loaded multi-component systems with non-ideal sources and sinks for vacancies. The TEP formulated in discrete CPs is used for derivation of equations for grain growth and coarsening within the multi-object concept and distribution concept. Both concepts are compared successfully in an example for grain growth with initial bi-modal size distribution. The TEP is used in modelling of precipitation in solid multi-phase, multi-component systems. Comparison of an example with experiment indicates good applicability of the model in the prediction of and description of the evolution in rather complex systems.

1. Introduction to linear thermodynamics of irreversible processes

Thermodynamics allows a very effective phenomenological description of the state and processes in systems of many interacting particles. Although thermodynamics is based on a certain imagination of the atomic structure of matter, it deals with the properties of matter stemming from collective interactions of the atoms, which become apparent macroscopically and can be measured. Classical thermodynamics introduces state variables, which characterize the equilibrium state of the system. Examples of such state variables are, *e.g.* the temperature (being a measure of mean kinetic energy of particles in the system) or the mole fractions of individual components (characterizing the mean chemical composition of the system). Mutual interactions of the components are characterized by chemical potentials of the individual components and depend on the temperature, chemical composition and phases, characterizing the lattice structure.

Processes in a non-equilibrium system are treated by non-equilibrium thermo-dynamics. In this case the system is subdivided into a number of equilibrium sub-systems, called representative volume elements (RVEs), for which the equilibrium state variables are defined. The RVEs must be sufficiently large to allow a statistical approach but they must be small enough so that they still can be considered as homogeneous objects. The processes in the system are then treated as interactions between neighbouring RVEs exchanging matter and/or energy.

If the system is not too far from equilibrium, as is usually assumed for several kinds of materials at elevated temperatures, processes such as diffusion are slow and obey the laws of linear non-equilibrium thermodynamics; see the textbooks on thermodynamics by Callen (1966) or by de Groot and Mazur (2013) for example. The simplest well known phenomenological laws are first Fick's law for diffusion, representing a proportionality relation between the diffusive flux and the gradient of concentration of a component, Ohm's law relating the electric current to the intensity of the electric field or Fourier's law for heat conduction relating the heat flux to the temperature gradient. The proportionality coefficients (diffusion coefficient, electric conductivity or thermal conductivity) are kinetic material parameters depending on the chemical composition and microstructure of the material and usually also on the temperature.

If such phenomenological equations are supplemented with proper conservation laws as well as with boundary and contact conditions, they represent a set of partial differential equations for the evolution of the system. This set is usually complicated even for simple systems. Thus the models, based directly on the solution of the phenomenological equations, suffer from the need to simplify drastically the reality. However, materials science has to treat complicated systems and, on the other hand, only a limited number of most appropriate characteristic parameters (CPs) is required for the description of the system evolution. Thus, one solves the phenomenological equations under simplified conditions in a first step, and the time evolution of the CPs is extracted from the solution in a second step. This procedure is, however, often impossible, if complex interactions in complicated systems have to be described.

The above-mentioned problem of treatment of complicated systems can be avoided or simplified significantly by application of the Thermodynamic Extremal Principle (TEP), formulated directly in CPs. The first extremal formulation of a thermodynamic problem was provided by Onsager (1931), who showed that the phenomenological equations of nonequilibrium thermodynamics can be derived from the requirement of a constrained maximum of the total dissipation (potential) of the system. The principle was nearly forgotten for the next 60 years, although Ziegler (1963, 1977) published a very general variational approach ~30 years after Onsager. The TEP as formulated by Onsager and Ziegler, however, cannot help significantly in modelling the evolution of complex systems as it reproduces only the already known phenomenological equations.

The situation of applicability of TEP to the treatment of complicated systems changed drastically in 1991, when the TEP was formulated directly in CPs. Since then the TEP has provided a direct systematic way to derive the evolution equations of the CPs without needing to solve the partial differential equations (Svoboda and Turek,

1991). Up to now the TEP has been applied successfully to modelling of sintering, *e.g.* Svoboda and Riedel (1995), creep and rafting in superalloy single crystals, *e.g.* Svoboda and Lukáš (2000), grain growth and precipitate coarsening, *e.g.* Fischer *et al.* (2003), kinetics of precipitation (program package *MatCalc* http://matcalc.at), Svoboda *et al.* (2004b), Kozeschnik *et al.* (2004a), treatment of contact conditions at migrating interfaces, *e.g.* Svoboda *et al.* (2004a) and phase transformation kinetics *e.g.* Svoboda *et al.* (2011).

The microstructure in materials determines not only the material properties, it also bears important information about the history of the materials, *e.g.* of minerals and rocks. Thus, the models for the microstructure evolution kinetics provide an important tool for determination of the history of minerals and rocks. The TEP, formulated in discrete CPs, represents a handy tool also for this task by development of relevant models.

This chapter introduces the TEP and its application to derive evolution equations for diffusion and creep in multi-component systems with non-ideal sources and sinks for vacancies. Thus, the application of the TEP allows description of the diffusion-stress-vacancy interactions, rigorous derivation of driving forces for individual processes and their use in kinetic equations. The TEP, formulated in discrete CPs, is utilized in models treating the microstructure evolution as grain growth, characterizing the evolution of one-phase systems (minerals), and diffusional phase transformation characterizing the evolution of multi-component multi-phase systems (rocks).

2. The Thermodynamic Extremal Principle (TEP)

A general remark: Defining the setting we are working with, we introduce a representative volume element (RVE) with the volume, W. For the sake of simplicity we assume that all physical properties (*e.g.* material properties) and all physical quantities (*e.g.* the state variables $\mathbf{X},\mathbf{x}$) have (spatially) constant values in the RVE. If we consider a system consisting of a large ensemble of RVEs, then we replace the RVE by the volume element dV and integrate over the system. We work always in the actual configuration and, consequently, have to keep in mind that not only the above-mentioned physical properties and quantities but also W or dV depend on the time t.

2.1. Introduction of kinetic variables

First of all let us introduce thermodynamic quantities, such as the temperature T, the pressure p and the site fractions y_k of individual components and classify them as external or internal state variables. All these quantities may depend on their spatial position and time t. External state variables can be controlled from outside the RVE, *e.g.* by heating or cooling or by applying a loading stress to the body and consequently also to the RVE. In our case the external state variables are the temperature T, the chemical potentials μ_k of individual components and the stress state represented by the stress tensor $\boldsymbol{\sigma} = \sigma_H \boldsymbol{\delta} + \mathbf{s}$. The unity tensor is denoted as $\boldsymbol{\delta}$, σ_H is the hydrostatic stress defined as positive for tension (and sometimes denoted as $-p$ with p being the hydrostatic pressure) and $\mathbf{s}$ is the stress deviator. For the sake of simplicity, we collect all these quantities in a vector $\mathbf{X}$ and denote this vector as 'external variables'.

Internal state variables develop in the system and cannot be controlled directly from outside. We select as internal state variables the site fractions of the components, *i.e.* y_0 for vacancies, y_k, $k = 1,...,n$ for the substitutional components in the lattice positions and y_k, $k = n + 1,...n + m$ for the interstitial components, and the eigenstrain tensor ε_{gc} accounting for the volumetric mismatch of atoms of the individual components, shrinking or swelling due to vacancy annihilation or generation, and creep. For simplicity, we collect all internal state variables in a vector **x** and denote this vector as 'internal variables'. The time (or material) derivatives of the internal variables, $\dot{\mathbf{x}}$, can be selected as kinetic variables. However, it can be more convenient or necessary to substitute some internal variables $\dot{\mathbf{x}}$ by related quantities, *e.g.* the rates $\dot{y}_k$ by the divergences of the corresponding diffusive fluxes $\mathbf{j}_k$; for details see section 2.4. For the sake of simplicity, we formulate the TEP in terms of $\dot{\mathbf{x}}$.

2.2. Gibbs energy and its dissipation

We introduce now the Gibbs thermodynamic potential G in an RVE, the so-called 'Gibbs energy', which is assumed to be a smooth, generally nonlinear function of **X** and **x**. The rate G of the Gibbs energy follows in vector notation as

$$\dot{G} = \frac{\partial G}{\partial \mathbf{X}} \cdot \dot{\mathbf{X}} + \frac{\partial G}{\partial \mathbf{x}} \cdot \dot{\mathbf{x}} \tag{1}$$

Note that the dot between two vectors refers to the scalar product of these vectors. The dissipation $D(\mathbf{X},\mathbf{x},\dot{\mathbf{x}})$ in the system is defined as

$$D = \frac{\partial G}{\partial \mathbf{x}} \cdot \dot{\mathbf{x}} = W\mathbf{f}_{\dot{\mathbf{x}}} \cdot \dot{\mathbf{x}} \geq 0 \tag{2}$$

The interested reader is referred to the pioneering work by Coleman and Gurtin (1967) for further information.

The derivatives $\partial G/\partial \mathbf{x}$ determine the thermodynamic driving forces $\mathbf{f}_{\dot{\mathbf{x}}}$. The inequality 2 serves to fulfil the second law of thermodynamics.

Furthermore, we assign now a dissipation function Q to the evolution of the internal variables in an RVE as non-negative, smooth function, which can be interpreted as the 'constitutive' equation of the dissipation process. This type of function goes back to Lord Rayleigh in his 1871 paper (for the reference and more details see Fischer *et al.*, 2014), and was cited by Ziegler (1963) and in chapter 15 of his book from 1977 in his extremal principle as the constraint:

$$Q(\mathbf{X},\mathbf{x},\dot{\mathbf{x}}) = D(\mathbf{X},\mathbf{x},\dot{\mathbf{x}}) \tag{3}$$

The dissipation function Q is assumed as homogeneous function of order l of the kinetic variables $\dot{\mathbf{x}}$ such that

$$\frac{\partial Q}{\partial \dot{\mathbf{x}}} \cdot \dot{\mathbf{x}} = lQ \tag{4}$$

The dissipation function Q does not involve the driving forces $\mathbf{f}_{\dot{\mathbf{x}}}$, but it does involve kinetic material parameters such as diffusion coefficients. In the frame of linear nonequilibrium thermodynamics the homogeneity order parameter l gives the value 2, as is the case for diffusion or heat conduction. For time-independent plasticity l gives the value 1. For more general dissipation functions we refer to Hackl and Fischer (2008) and Hackl *et al.* (2011a,b).

2.3. The TEP in its original form

The TEP in its original form states that the dissipation D gives a maximum value, subject to the constraint $D = Q$. Further constraints with respect to the kinetic variables (*e.g.* mass balances, *etc.*) can be met by the TEP. When there are no further constraints, the Lagrangian L follows as

$$L = D - \lambda(Q - D) \tag{5}$$

with λ being a Lagrange multiplier. Maximizing L with respect to the rates of internal variables $\dot{\mathbf{x}}$ provides the relation between the thermodynamic forces $\mathbf{f}_{\dot{\mathbf{x}}}$ and the kinetic variables $\dot{\mathbf{x}}$ as

$$W\mathbf{f}_{\dot{\mathbf{x}}} = \left(\frac{Q}{(\partial Q/\partial \dot{\mathbf{x}}) \cdot \dot{\mathbf{x}}} \right) \frac{\partial Q}{\partial \dot{\mathbf{x}}} \tag{6}$$

Because Q is supposed to be an homogeneous function of order l (see equation 4), it follows that

$$W\mathbf{f}_{\dot{\mathbf{x}}} = \frac{1}{l} \frac{\partial Q}{\partial \dot{\mathbf{x}}} \tag{7}$$

Because one needs finally the evolution equations for $\dot{\mathbf{x}}$, it is necessary to invert equation 7. If Q is a quadratic function in $\dot{\mathbf{x}}$ ($l=2$), having the form $Q = \dot{\mathbf{x}}^T \cdot \mathbf{U}(\mathbf{X},\mathbf{x}) \cdot \dot{\mathbf{x}}$ with U being a positive-definite matrix involving material and structure parameters, then the evolution equations for $\dot{\mathbf{x}}$ follow as

$$\dot{\mathbf{x}} = W\mathbf{U}^{-1} \cdot \mathbf{f}_{\dot{\mathbf{x}}} \tag{8}$$

The following aspects are important:

- The TEP can be understood as a 'strong' form of the second law of thermodynamics. The second law of thermodynamics allows an infinite number of paths of the system with a positive dissipation D. The TEP selects only the path which corresponds with the constrained maximum of the dissipation D. Therefore, the TEP is not a 'Law of Physics' but a 'Selection Principle'. It provides evolution equations for the kinetic variables $\dot{\mathbf{x}}$.
- The maximization must be performed with respect to $\dot{\mathbf{x}}$, but must not be performed with respect to $\mathbf{X}$ or $\dot{\mathbf{X}}$ or $\mathbf{x}$! This is the most common mistake and has led in the past to the opinion that the TEP does not work or produces incorrect results.
- The TEP is ideally suited to incorporating further constraints such as mass balances, *etc.* by adding the constraints multiplied by corresponding Lagrange multipliers to the Lagrangian given by equation 5.
- No problem occurs from the mathematical point of view, if the order of homogeneity l is not equal to 2 but equal for all processes. If this is not the case, we refer to the papers by Hackl and Fischer (2008) and Hackl *et al.* (2011a,b).

2.4. Diffusion in a multicomponent system — an example of the application of TEP

2.4.1. Quantities involved

Let us assume n substitutional and m interstitial components in the system and collect the state variables and physical quantities, which are involved in a rather general, vacancy-mediated diffusion concept (see also section 2.1) as follows:

(i) External state variables T (assumed as constant in time, *i.e.* an isothermal state), the stress state $\boldsymbol{\sigma} = \sigma_H \boldsymbol{\delta} + \mathbf{s}$ and the chemical potentials μ_k, $k = 0,...,n+m$, of the vacancies ($k = 0$) and the components, expressed in J/mol.

(ii) Internal variables as site fractions as y_0 of the vacancies and y_k, $k = 1,...,n+m$, of the components, $\sum_{k=0}^{n} y_k = 1$, and the eigenstrain tensor $\boldsymbol{\varepsilon}_{gc}$, the rate of which is outlined below (see equation 21).

(iii) The partial molar volumes of the elements Ω_k, $k = 1,...,n+m$. Furthermore, referring to Svoboda *et al.* (2006), the following quantities are introduced: the partial molar volume $\bar{\Omega}_0$ of the vacancies as:

$$\bar{\Omega}_0 = \sum_{k=1}^{n} y_k \Omega_k / (1 - y_0) \qquad (9)$$

and the molar volume $\bar{\Omega}$ corresponding to one mole of lattice sites as:

$$\bar{\Omega} = \bar{\Omega}_0 + \sum_{k=n+1}^{n+m} y_k \Omega_k \qquad (10)$$

(iv) The specific elastic strain energy φ_{mech}, expressed in $\boldsymbol{\sigma}$ via Hooke's law. We assume small strain setting which allows additive decomposition of the strain tensor $\boldsymbol{\varepsilon}$ into an elastic contribution $\boldsymbol{\varepsilon}_e$ and an eigenstrain tensor $\boldsymbol{\varepsilon}_{gc}$. Then φ_{mech} follows as

$$\varphi_{\text{mech}} = \tfrac{1}{2}\boldsymbol{\sigma} : \boldsymbol{\varepsilon}_e \qquad (11)$$

Hooke's law follows as $\boldsymbol{\varepsilon}_e = \mathbf{C}\boldsymbol{\sigma}$, with $\mathbf{C}$ being the forth-order-compliance tensor, the components of which may depend on the site fractions $y_1,...,y_{n+m}$, yielding

$$\varphi_{\text{mech}} = \tfrac{1}{2}\boldsymbol{\sigma} : \mathbf{C}(y_1,...,y_{n+m})\boldsymbol{\sigma} \qquad (12)$$

(v) The molar chemical Gibbs energy g_{chem}, expressed as

$$g_{\text{chem}} = \mu_0 + \sum_{k=1}^{n} (\mu_k - \mu_0) y_k + \sum_{k=n+1}^{n+m} \mu_k y_k \qquad (13)$$

Note that the first sum on the r.h.s. of equation 13 stems from the replacement of y_0 by $1 - \sum_{k=1}^{n} y_k$; see (ii) above.

(vi) The potential energy P of the loading system acting on the RVE in the sense of standard continuum mechanics (applying the Principle of Virtual Work) as

$$P = -\sigma W : \varepsilon = -2\varphi_{mech} - \sigma W : \varepsilon_{gc} \tag{14}$$

which follows directly from (iv).

2.4.2. Rates of the quantities involved

Using equation 10 one finds after some algebra that:

$$\dot{\Omega} = \sum_{k=1}^{n} \dot{y}_k (\Omega_k - \bar{\Omega}_0)/(1 - y_0) + \sum_{k=n+1}^{n+m} \dot{y}_k \Omega_k \tag{15}$$

With N as number of moles of lattice sites in the RVE, the volume W of the RVE is given as $W = \bar{\Omega}N$. Its relative rate $\dot{W}/W$ follows as:

$$\frac{\dot{W}}{W} = \frac{\dot{\bar{\Omega}}}{\bar{\Omega}} + \frac{\dot{N}}{N}, \quad \alpha = \frac{\dot{N}}{N} \tag{16}$$

The quantity α is the rate at which lattice sites are generated (or, for a negative rate α, annihilated) and is introduced as a kinetic variable, see below. Finally, one can find with equation 16 the rate

$$\frac{d(W/\bar{\Omega})}{dt} = \alpha \frac{W}{\bar{\Omega}} = \dot{N} \tag{17}$$

Continuum mechanics teaches that the rate $\dot{W}/W$ of RVE follows as $div(\mathbf{v})$ with $\mathbf{v}$ being the material velocity. In case of small strain setting $div(\mathbf{v})$ is equivalent to the trace of the strain rate tensor $\dot{\varepsilon}$, *i.e.*

$$\text{div}(\mathbf{v}) = \dot{\varepsilon} : \delta \tag{18}$$

If one neglects the rate of the elastic strain component, then, combining equations. 16–18, one obtains

$$\text{div}(\mathbf{v}) = \varepsilon_{gc} : \delta = \frac{\dot{\bar{\Omega}}}{\bar{\Omega}} + \alpha \tag{19}$$

Consequently, the rates of W and dV can be expressed as

$$\dot{W} = (\dot{\varepsilon}_{gc} : \delta)W, \quad d\dot{V} = (\dot{\varepsilon}_{gc} : \delta)dV \tag{20}$$

Furthermore, one can formulate as simplest equation for the rate of $\dot{\varepsilon}_{gc}$

$$\dot{\varepsilon}_{gc} = \left(\frac{\dot{\bar{\Omega}}}{\bar{\Omega}} + \alpha\right)\frac{\delta}{3} + \dot{\varepsilon}_{cr} \tag{21}$$

that $\dot{\varepsilon}_{gc}$ includes *via* equation 15 the rates $\dot{y}_k$, $k = 1,..., n+m$, α and $\dot{\varepsilon}_{cr}$.

As next step let us calculate the rates of contributions to G given by equations 11–14 and repeat that the calculated rates concern only the internal variables. In other words, σW and μ_k, $k = 0,1,...,m+n$ are fixed with respect to time, which means that no rates with respect to σW and μ_k, $k = 0,1,...,m+n$ are involved.

Starting from equations 11−12 one finds with equations 15−20

$$\frac{\mathrm{d}}{\mathrm{d}t}(\varphi_{\mathrm{mech}}W) = \dot{\varphi}_{\mathrm{mech}}W + \varphi_{mech}\dot{W} =$$

$$\sum_{k=1}^{n+m}\frac{\partial\varphi_{\mathrm{mech}}}{\partial y_k}\dot{y}_k W - \varphi_{\mathrm{mech}}\left[\frac{\sum\limits_{k=1}^{n}\dot{y}_k(\Omega_k - \bar{\Omega}_0)/(1-y_0) + \sum\limits_{k=n+1}^{n+m}\dot{y}_k\Omega_k}{\bar{\Omega}} + \alpha\right]W \quad (22)$$

$$\frac{\mathrm{d}}{\mathrm{d}t}(g_{\mathrm{chem}}W/\bar{\Omega}) = \dot{g}_{\mathrm{chem}}\frac{W}{\bar{\Omega}} + g_{\mathrm{chem}}\left(-\frac{\dot{\bar{\Omega}}}{\bar{\Omega}^2}W + \frac{\dot{W}}{\bar{\Omega}}\right) =$$

$$\left[\sum_{k=1}^{n}(\mu_k - \mu_0)\dot{y}_k + \sum_{k=n+1}^{n+m}\mu_k\dot{y}_k\right]\frac{W}{\bar{\Omega}} + \left[\mu_0 + \sum_{k=1}^{n}(\mu_k - \mu_0)y_k + \sum_{k=n+1}^{n+m}\mu_k y_k\right]\frac{\alpha W}{\bar{\Omega}} \quad (23)$$

One must keep in mind that the rate $\dot{P}$ must be calculated from equation 14 as $-(\sigma W):\dot{\varepsilon}$. Then $-\mathrm{d}P$ can be considered as the incremental work done on the RVE during a time increment $\mathrm{d}t$, and $\dot{P}$ follows as

$$\dot{P} = -\sigma W : \dot{\varepsilon} = -\sigma W : (\dot{\varepsilon}_{gc} + \dot{\varepsilon}_e) = -(\sigma : \dot{\varepsilon}_{gc} + 2\dot{\varphi}_{\mathrm{mech}})W =$$

$$-\sigma_H\left[\frac{\sum\limits_{k=1}^{n}\dot{y}_k(\Omega_k - \bar{\Omega}_0)/(1-y_0) + \sum\limits_{k=n+1}^{n+m}\dot{y}_k\Omega_k}{\bar{\Omega}} + a\right]W - \sigma : \dot{\varepsilon}_{cr}W$$

$$-2\sum_{k=1}^{n+m}\frac{\partial\varphi_{\mathrm{mech}}}{\partial y_k}\dot{y}_k W \quad (24)$$

2.4.3. Dissipation D and dissipation function Q for diffusion

It is a well established fact that the dissipation functions of 'flow processes' (mass diffusion, heat conduction, electric current *etc.*) are quadratic functions of the fluxes. Therefore, we have to provide kinetic variables related to the fluxes. Here, we engage the conservation law for the atomic species as

$$\dot{y}_k = -\bar{\Omega}\mathrm{div}(\mathbf{j}_k) - \alpha y_k, \quad k = 1,...,n + m \quad (25)$$

for derivation see Svoboda *et al.* (2006), section 2. Therefore, we select as kinetic variables the divergences $\mathrm{div}(\mathbf{j}_k)$, α and $\dot{\varepsilon}_{cr}$. Insertion of equation 25 into equations 22−24 and summing up these three rates provides the rate $\dot{G}$ of the total Gibbs energy as:

$$\dot{G} = \sum_{k=1}^{n+m} \frac{\partial \varphi_{\text{mech}}}{\partial y_k} (\bar{\Omega}\text{div}(\mathbf{j}_k) + \alpha y_k) W -$$

$$\varphi_{\text{mech}} \left[\sum_{k=1}^{n} \text{div}(\mathbf{j}_k)(\Omega_k - \bar{\Omega}_0)/(1 - y_0) + \sum_{k=n+1}^{n+m} \text{div}(\mathbf{j}_k)\Omega_k - \frac{\bar{\Omega}_0}{\bar{\Omega}}\alpha \right] W -$$

$$\left[\sum_{k=1}^{n} (\mu_k - \mu_0)\text{div}(\mathbf{j}_k) + \sum_{k=n+1}^{n+m} \mu_k \text{div}(\mathbf{j}_k) \right] W + \mu_0 \frac{\alpha W}{\bar{\Omega}} +$$

$$\sigma_H \left[\sum_{k=1}^{n} \text{div}(\mathbf{j}_k)(\Omega_k - \bar{\Omega}_0)/(1 - y_0) + \sum_{k=n+1}^{n+m} \text{div}(\mathbf{j}_k)\Omega_k - \frac{\bar{\Omega}_0}{\bar{\Omega}}\alpha \right] W - \mathbf{s} : \dot{\boldsymbol{\varepsilon}}_{cr} \quad (26)$$

The according thermodynamic forces to the kinetic variables (*i.e.* the divergences $\text{div}(\mathbf{j}_k)$, α and $\dot{\boldsymbol{\varepsilon}}_{cr}$) follow as:

$$\frac{\partial \dot{G}}{\partial (\text{div}(\mathbf{j}_k))} = \left[\bar{\Omega}\frac{\partial \varphi_{\text{mech}}}{\partial y_k} + (\sigma_H - \varphi_{\text{mech}})\frac{(\Omega_k - \bar{\Omega}_0)}{(1 - y_0)} - (\mu_k - \mu_0) \right] W = -f_{j,k} W, \quad k = 1, ..., n$$

$$(27)$$

$$\frac{\partial \dot{G}}{\partial (\text{div}(\mathbf{j}_k))} = \left[\bar{\Omega}\frac{\partial \varphi_{\text{mech}}}{\partial y_k} + (\sigma_H - \varphi_{\text{mech}})\Omega_k - \mu_k \right] W = -f_{j,k} W, \quad k = n+1, ..., m$$

$$(28)$$

$$\frac{\partial \dot{G}}{\partial \alpha} = \left[\sum_{k=1}^{n+m} \left(\frac{\partial \varphi_{\text{mech}}}{\partial y_k} \right) y_k - (\sigma_H - \varphi_{\text{mech}})\frac{\bar{\Omega}_0}{\bar{\Omega}} + \frac{\mu_0}{\bar{\Omega}} \right] W = -f_\alpha W \quad (29)$$

$$\frac{\partial \dot{G}}{\partial \dot{\boldsymbol{\varepsilon}}_{cr}} = -\mathbf{s}W = -\mathbf{f}_{\dot{\boldsymbol{\varepsilon}}_{cr}} W \quad (30)$$

The dissipation D, expressed now in the a/m kinetic variables and the dissipation D_{system} for the whole system can now be formulated in an analogous way to equation 2 as:

$$D = \left[\sum_{k=1}^{n+m} f_{j,k}\text{div}(\mathbf{j}_k) + f_\alpha \alpha + \mathbf{f}_{\dot{\boldsymbol{\varepsilon}}_{cr}} : \dot{\boldsymbol{\varepsilon}}_{cr} \right] \quad (31)$$

$$D_{\text{system}} = \int_{\text{system}} \sum_{k=1}^{n+m} f_{j,k}\text{div}(\mathbf{j}_k)dV + \int_{\text{system}} f_\alpha \alpha dV + \int_{\text{system}} \mathbf{f}_{\dot{\boldsymbol{\varepsilon}}_{cr}} : \dot{\boldsymbol{\varepsilon}}_{cr}dV \quad (32)$$

The Gauss theorem is now applied to the first term on the r.h.s. of equation 32 for a closed system yielding

$$D_{\text{system}} = \int\limits_{\text{system}} \sum_{k=1}^{n+m} \text{grad}(f_{j,k}) \cdot \mathbf{j}_k \mathrm{d}V + \int\limits_{\text{system}} f_\alpha \alpha \mathrm{d}V + \int\limits_{\text{system}} \mathbf{f}_{\dot{\varepsilon}_{cr}} : \dot{\varepsilon}_{cr} \mathrm{d}V \quad (33)$$

The assumption of a closed system is acceptable because whether the system is open or closed must not influence the constitutive equations for the system evolution; it influences only the boundary conditions.

The application of the Gauss theorem allows transformation of the set of kinetic variables $\text{div}(\mathbf{j}_k)$, $k = 1,...n + m$, α and $\dot{\varepsilon}_{cr}$ to a new set of kinetic variables $\mathbf{j}_k$, $k = 1,...n + m$, α and ε_{cr} as a new set of kinetic variables. The driving forces $-\text{grad}(f_{j,k})$ are conjugated to $\mathbf{j}_k$, $k = 1,...n + m$, f_α to α and $\mathbf{f}_{\dot{\varepsilon}_{cr}}$ to $\dot{\varepsilon}_{cr}$.

As next step we formulate the dissipation function Q_{system} of the whole system as the simplest possible homogeneous function of order 2 in the new set of kinetic variables as:

$$Q_{\text{system}} = \int\limits_{\text{system}} \left\{ \left(\sum_{k=1}^{n+m} \mathbf{j}_k^2/A_k \right) + \left(\sum_{k=1}^{n} \mathbf{j}_k \right)^2 / A_0 \right\} \mathrm{d}V + \int\limits_{\text{system}} K\alpha^2 \mathrm{d}V + \int\limits_{\text{system}} S\dot{\varepsilon}_{cr} : \dot{\varepsilon}_{cr} \mathrm{d}V \quad (34)$$

Note that the term with A_0 represents the dissipation due to diffusion of vacancies taking into account that $\mathbf{j}_0 - \sum_{k=1}^{n} \mathbf{j}_k$. The coefficients $A_k(k = 1,...,n + m)$ can be taken from Svoboda *et al.* (2002) and read as $A_k = (y_0/y_0^{\text{eq}})y_k D_k/(\mathrm{R_g}T\bar\Omega)$ for $k = 1,...,n$ and as $A_k = y_k D_k/\mathrm{R_g}T\bar\Omega)$ for $k = n + 1,...,n + m$. The tracer diffusion coefficient for component k is denoted as D_k, and $\mathrm{R_g}$ is the gas constant. The quantity A_0 follows as $A_0 = -\sum_{k=1}^{n} A_k/(1 - f)$ with f being the geometric correlation factor dependent on the lattice type (*e.g.* fcc: $f = 0.7815$, bcc: $f = 0.7272$). The formulations of the A_k are based on Manning's theory of diffusion (Manning, 1970). The quantity K can be considered as bulk viscosity and S as shear viscosity. Derivations of K and S and their relations to the microstructure of the system can be found in Svoboda *et al.* (2006), section 5 and in Fischer and Svoboda (2011).

The structures of D_{system}, equation 33, and Q_{system}, equation 34, allow us to work with their specific values, *i.e.* with the integrands or local values D and Q in an RVE. The order of homogeneity of D with respect to kinetic variables is 1 and that of Q is 2. The TEP requires us to maximize D with respect to the kinetic variables, subjected to the constraint $D = Q$, which yields a set of leading equations, see section 2.3 above, with $\mathbf{p}$ standing for $\mathbf{j}_k$, α and $\dot{\varepsilon}_{cr}$ as

$$\frac{1}{2}\frac{\partial Q}{\partial \mathbf{p}} = W\mathbf{f_p} \quad (35)$$

and in a straightforward way to

$$-\text{grad}(f_{j,k}) = \frac{1}{A_k}\mathbf{j}_k + \frac{1}{A_0}\sum_{l=1}^{n} \mathbf{j}_l, \quad k = 1,...,n \quad (36)$$

$$-\text{grad}(f_{j,k}) = \frac{1}{A_k}\mathbf{j}_k, \quad k = n+1, ..., n+m \tag{37}$$

$$f_\alpha = K\alpha \tag{38}$$

$$\mathbf{f}_{\dot{\varepsilon}_{cr}} = S\dot{\varepsilon}_{cr} \tag{39}$$

The set of equations 36−39 can be inverted with the result

$$\mathbf{j}_k = \sum_{k=1}^{n} L_{ik}\text{grad}(f_{j,k}), \quad k = 1, ..., n \tag{40}$$

$$\mathbf{j}_k = -A_k\text{grad}(f_{j,k}), \quad k = n+1, ..., n+m \tag{41}$$

$$\alpha = f_\alpha/K \tag{42}$$

$$\dot{\varepsilon}_{cr} = \mathbf{f}_{\dot{\varepsilon}_{cr}}/S = \mathbf{s}/S \tag{43}$$

The coefficients L_{ik} are denoted in the literature as the Onsager coefficients.

2.5. Concluding remarks

Some comments may be helpful for the reader at this point:

- Cross terms
  As already emphasized by Manning (1970) the cross terms L_{ik} have their physical justification and disappear only for $f = 1$. The tacit assumption of $f = 1$ can be found very often in the literature, but, generally, is not *a priori* acceptable.

- Role of stress state
  Note that the role of stress state $\boldsymbol{\sigma}$ in the driving forces is also included in the specific elastic strain energy φ_{mech} and its derivatives with respect to y_k, but it is mostly pronounced by the hydrostatic stress σ_H. Because very often, φ_{mech} and its a/m derivatives can be ignored in comparison to σ_H, only σ_H needs to be considered. The stress deviator $\mathbf{s}$ neither appears in the constitutive equations for the fluxes $\mathbf{j}_k$ nor for the vacancy generation/annihilation rate α; however, it determines the creep rate $\dot{\varepsilon}_{cr}$, see equation 43.

- Evolution equations for the site fractions
  The evolution equations for the site fractions y_k, $k = 1, ..., n + m$ and $y_0 = 1 - \sum_{k=1}^{n} y_k$ follow after insertion of $\mathbf{j}_k$, equations 40−41 and α, equation 42 together with equation 10 into the conservation law for the atomic species, equation 25. The related partial differential equations (PDEs) are nonlinear in the y_k; think, *e.g.* about $\bar{\Omega}$ which depends on the individual y_k− values! Detailed formulations of the PDEs, together with the related boundary conditions, were given by Fischer and Svoboda (2014).

- Evolution equation for $\dot{\varepsilon}_{gc}$
  The rate of the eigenstrain tensor includes contributions from all kinetic variables $\mathbf{j}_k$, $k = 1, ..., n + m$, α and $\dot{\varepsilon}_{cr}$. Thus, the value of $\dot{\varepsilon}_{gc}$ can be

considered as known at any time, and ε_{gc} can be obtained by integration of $\dot{\varepsilon}_{cr}$ with respect to time. The stress state $\boldsymbol{\sigma}$ in the RVE is then given by standard stress-strain analysis coupling ε_{gc} with Hooke's law and accounting for respective mechanical boundary conditions.

- Role of vacancies
  The leading quantity is their generation/annihilation rate α. There are two limiting cases:
  (i) no sources and/or sinks for vacancies: $\alpha = 0$ and vacancies behave as a conserving substitutional component,
  (ii) ideal sources and/or sinks for vacancies: $\alpha = \mathrm{div}(\mathbf{j}_0)$ and $y_0 = y_0^{\mathrm{eq}}$, *i.e.* the equilibrium value is kept everywhere in the system.
  Note that the stress state can be influenced remarkably by the role of vacancies! There are some instructive examples available, which show the role of vacancies in a plane bilayer with cross-diffusion (see Svoboda and Fischer, 2009, 2011; Fischer and Svoboda, 2010, 2015).
- Rigorous derivation
  We refer the reader who is interested in details of the derivations in section 2.4 to the recent paper by Fischer *et al.* (2016).

3. Modelling of grain growth and precipitate coarsening

3.1. Introduction

Both grain growth and precipitate coarsening (Oswald ripening) are dissipative processes driven by the decrease in the total Gibbs energy of grain boundaries or in matrix/precipitate interfaces. The problem of grain growth (more precisely 'grain coarsening') concerns a wide range of materials and has been a favourite research topic for 50 years (Hillert, 1965, 1996; Atkinson, 1988; Fischer *et al.*, 2003; Saetre, 2004a,b). The term 'grain coarsening' reflects much better the fact that during the process only the large grains grow and the small grains shrink and disappear. Because the term 'grain growth' has been used widely for this phenomenon in the literature, we retain it here. The problem of bulk diffusion-controlled precipitate coarsening concerns a wide range of materials and has been under study for 50 years (Lifshitz and Slyozov, 1958, 1961; Wagner, 1961; Vorhees, 1985, 1992). Both grain growth and precipitate coarsening usually have a rather negative influence on mechanical properties and, thus, modelling of kinetics of these processes is of significant industrial relevance. Furthermore, grain sizes and their distributions are an important petrographic feature, which may shed light on rock formation.

Grain-boundary migration-controlled grain growth is often described by the classical Hillert (1965, 1996) theory, and the diffusion-controlled precipitate coarsening is usually treated by the classical Lifshitz-Slyozov-Wagner (LSW) theory (Lifshitz and Slyozov, 1958, 1961; Wagner; 1961). In both cases, each object is characterized by its effective radius R_i. The evolution equations for R_i contain the critical radius R_C as a parameter calculated from the radii distribution function and the requirement of conservation of the total volume of objects. The objects with radii $R_i < R_C$ shrink and the objects with radii $R_i > R_C$ grow. Both theories provide a steady-state radii distribution function with normalized radii $\rho = R/R_C$.

The thermodynamic models of grain growth have been refined, and the treatment has become successively complicated, see Rios *et al.* (2001, 2002, 2004). Non-steady-state distributions of grains are also a topic of current research. Such distributions have been obtained by the multi-phase-field (MPF) model, see Suwa *et al.* (2003, 2007) or by the Monte Carlo Potts (MCP) model, see *e g.* Zöllner and Streitenberger (2006).

A similar regime as in grain growth happens in precipitate coarsening. The pioneering work on precipitate coarsening was published by Greenwood (1956), Lifshitz and Slyozov (1958, 1961) and Wagner (1961), who incorporated the coupling of reaction at the interface and diffusion in the matrix. In this case, second-phase precipitates are distributed in the matrix and change their morphology by dissolving small precipitates employing transfer of mass *via* diffusion in the matrix to larger precipitates. For description of the kinetics of coarsening in a multicomponent system it is necessary to know the concentrations of all components in the precipitates and in the matrix and the diffusion coefficients of all components in the matrix (Fischer *et al.*, 2003).

In 1991, Onsager's and Ziegler's principle was first formulated in terms of discrete parameters characterizing the state of the system, the so-called 'characteristic parameters' (CPs), (Svoboda and Turek, 1991), and named as the 'Thermodynamic Extremal Principle' (TEP). The application of the TEP provides a systematic concept for the derivation of the evolution equations for these CPs for a wide range of problems in the context of linear non-equilibrium thermodynamics (Svoboda *et al.*, 2005).

The theoretical treatment of grain growth and precipitate coarsening is possible only by accepting some model assumptions, *e.g.* the approximation of each object by a volume-equivalent sphere. Then the system can be described by CPs, represented by the effective radii R_k of individual objects, and the evolution of R_k can be treated by application of the TEP (Fischer *et al.*, 2003). This treatment is called 'the multiobject concept', and the evolution equations for R_k obtained from the TEP are equivalent to those employed first by Hillert (1965, 1996) for grain growth and by Lifshitz and Slyozov (1958, 1961) and Wagner (1961) for precipitate coarsening.

The microstructure is usually characterized by the object radii distribution function $f(R) = dN_g/dR$, where dN_g is the number of objects in the interval $(R, R+dR)$. The TEP was also employed in distribution functions and referred to as 'the distribution concept' (Svoboda and Fischer, 2007), which provides the evolution equations for the time-dependent parameters of the distribution function.

The concepts based on the TEP require the proper formulation of the total dissipation function Q and of the total Gibbs energy G of the system (Fischer *et al.*, 2003). The Gibbs energy G contains only the total grain boundary (or interface) energy. The dissipation function Q is calculated as corresponding to the grain-boundary migration for grain growth and to diffusion in the matrix for precipitate coarsening.

3.2. Thermodynamic extremal principle formulated in discrete characteristic parameters

Let us assume a closed system at constant temperature and pressure as external variables. Then the total Gibbs energy G represents the characteristic potential, which is dissipated during system evolution. Let the CPs be q_i, $(i = 1,...,N)$, instead of the vector $\mathbf{x}$ in section 2.2, and their rates $\dot{q}_i$, $(i = 1,...,N)$. The time evolution of the system is assumed to be caused by diffusive fluxes and by motion of the interfaces, which lead to a change in the CPs expressed by the rates $\dot{q}_i$, $(i = 1,...,N)$. Let the total Gibbs energy G of the system be expressed as a function of CPs, $G = G(q_1,q_2,...,q_N)$. Its rate $\dot{G}$ is then given by the chain rule and is equal to the negative dissipation D, see also equations 1 and 2 as

$$\dot{G} = \sum_{i=1}^{N} \frac{\partial G}{\partial q_i} \dot{q}_i = -D \tag{44}$$

In order to calculate the total dissipation D and the dissipation function Q, one must perform a key step, which allows us (with reasonable simplifications and assumptions for the system geometry and by application of conservation laws) to express all diffusive fluxes $\mathbf{j}_k$ and the interface velocity v as functions of the CPs and their rates. Thus, one has to look for the expressions

$$\mathbf{j}_k = \sum_{i=1}^{N} \mathbf{g}_{ik}(q_1, q_2, ..., q_N)\dot{q}_i, \quad (k = 1, ..., n_C) \tag{45}$$

$$v = \sum_{i=1}^{N} h_i(q_1, q_2, ..., q_N)\dot{q}_i \tag{46}$$

where n_C is the number of components in the system. The quantities $\mathbf{j}_k$ and v are homogeneous functions of order 1 in $\dot{q}_i$, $(i = 1,...,N)$.

The dissipation function Q in an isotropic system can be expressed by Svoboda *et al.* (2005) as

$$Q = \sum_{k=1}^{n_C} \int_{\text{Volume}} \frac{R_g T}{c_k D_k} \mathbf{j}_k^2 dV + \int_{\text{Interfaces}} \frac{1}{M} v^2 dA \tag{47}$$

with R_g as gas constant, T the absolute temperature, c_k the concentration and D_k the tracer diffusion coefficient of component k, with M the grain boundary or interface mobility. Then one can insert equations 45 and 46 into equation 47 and perform the integrals with the result

$$Q = \sum_{i,j=1}^{N} U_{ij}(q_1, q_2, ..., q_N)\dot{q}_i\dot{q}_j \tag{48}$$

where U_{ij} is a symmetric and positive definite matrix of the CPs and takes into account material properties and the geometry of the system (compare with the text below equation 7). In many cases the chosen CPs q_i, $(i = 1,...,N)$ are not free, and some constraints exist amongst them, which can generally be expressed in the form

$$F_l(q_1, q_2, ..., q_N) \equiv 0, \quad (l = 1,...,m) \tag{49}$$

Differentiating equation 49 in time, one obtains the constraints in terms of the rates $\dot{q}_i$ as

$$\sum_{j=1}^{N} a_{jl}(q_1, q_2, ..., q_N)\dot{q}_j = 0, \quad (l = 1, ..., m), \quad a_{jl} = \frac{\partial F_l}{\partial q_j} \tag{50}$$

Now we repeat the Thermodynamic Extremal Principle (TEP) as outlined in detail in section 2 as: "A closed system under constant temperature and pressure evolves in a way which corresponds to the maximum dissipation D constrained by the balance equation $Q = D$ and some further constraints (equation 50)."

The necessary condition for the maximum dissipation D, constrained by $Q - D = 0$ and equation 50, is given according to section 2.3 as

$$\frac{\partial}{\partial \dot{q}_i}\left[D - \lambda(Q - D) + 2\sum_{i=1}^{r1}\beta_l\sum_{j=1}^{N}a_{jl}\dot{q}_j\right] = 0, \quad (i = 1, ..., N) \tag{51}$$

with λ and β_l, $(l = 1,...,m)$ being Lagrange multipliers. After inserting D and Q from equations 44 and 48, respectively, and evaluating the derivatives, one obtains

$$\frac{\partial G}{\partial q_i} - \lambda\left(2\sum_{j=1}^{N}U_{ij}\dot{q}_j + \frac{\partial G}{\partial q_i}\right) + 2\sum_{l=1}^{m}\beta_l a_{il} = 0, \quad (i = 1, ..., N) \tag{52}$$

If equations 52 are multiplied by $\dot{q}_i$, summed over i, $(i = 1,...,N)$, and equations 44 and 48 are used, one obtains

$$D - \lambda(2Q - D) = 0 \tag{53}$$

which, together with the constraint $Q - D = 0$, allows the evaluation of the value of λ as $\lambda = 1$. Then equations 52 can be rewritten in a final form as

$$\sum_{j=1}^{N}U_{ij}\dot{q}_j + \sum_{l=1}^{m}a_{il}\beta_l = -\frac{\partial G}{\partial q_i} = -\frac{\partial \dot{G}}{\partial \dot{q}_i}, \quad (i = 1, ..., N) \tag{54}$$

It is advantageous to note that the matrix U_{ij} can be calculated from the dissipation function Q as

$$U_{ij} = \frac{1}{2}\frac{\partial^2 Q}{\partial \dot{q}_i \, \partial \dot{q}_j}, \quad (i, j = 1, ..., N) \tag{55}$$

Equations 54, together with the constraints in equation 50, can always be solved with respect to $\dot{q}_i$, $(i = 1,...,N)$ and β_i $(i = 1,...,m)$. The rates $\dot{q}_i$ of the CPs can be integrated with respect to time to obtain $q_i = q_i(t)$, $(i = 1,...N)$, which is the aim of our efforts.

3.3. Thermodynamic treatment of grain growth and coarsening – multi-object concept

Let each object (grain or precipitate) be approximated by a volume-equivalent sphere and characterized by its effective radius R_i ($i = 1,...,N$).

In the case of precipitate coarsening, we assume that the system is chemically in equilibrium, and, thus, the total volume of precipitates remains constant. The only driving force in the system is the change in the total interface energy. Analogous assumptions are made for grain growth, assuming a constant total volume of grains. The only driving force in the system is the change in the total grain-boundary energy.

If the system consists of N objects of effective radii R_i ($i = 1,...,N$), and the specific grain-boundary energy (in the case of grain growth) or the interface energy (in the case of precipitate coarsening) is denoted by γ, then the variable contribution G to the Gibbs energy G of the system is given by

$$G = 4\pi\xi\gamma \sum_{i=1}^{N} R_i^2 \tag{56}$$

The factor ξ follows as $\xi = 1/2$ for grain growth (each grain boundary is shared by two grains) and $\xi = 1$ for precipitate coarsening.

To calculate the dissipation function Q in the system, one may start with equation 47. The grain-boundary velocity of grain i equals the rate $\dot{R}_i$ of the radius R_i.

In case of grain growth no diffusive fluxes occur in the system. The dissipation function Q_M due to migration of grain boundaries is then given by

$$Q_M = \frac{1}{2}\frac{4\pi}{M} \sum_{i=1}^{N} R_i^2 \dot{R}_i^2 \tag{57}$$

In the case of coarsening of precipitates, we assume that no dissipation occurs due to interface migration, and the total dissipation function is given exclusively by diffusion in the matrix. For application of the TEP, it is necessary to express the diffusive fluxes by means of the rates $\dot{R}_i$. Let each spherical precipitate with the radius R_i be surrounded by a diffusion zone of the radius Z_i. In each zone, only the radial diffusive fluxes j_{ki} of components k, $1 \leqslant k \leqslant n_C$, related to the rate $\dot{R}_i$, are considered. A uniform deposition or collection of all components k is supposed in the diffusion zone. The diffusion zone radii Z_i are assumed to be large enough such that the diffusion zones are overlapping. The matrix does not change its mean chemical composition given by concentrations c_k, and negligible concentration gradients are assumed in the matrix. The concentration of component k in the precipitates is denoted by $c_k{}^*$. No radial flux is assumed inside the precipitate, $j_{ki}(r) \equiv 0$ for $0 \leqslant r < R_i$. The radial flux $j_{ki}(r)$ of component k in the matrix outside of the precipitate i at the distance r from its centre is given by

$$j_{ki}(r) = \dot{R}_i(c_k - c_k^*)\frac{R_i^2}{r^2} \cdot \frac{Z_i^3 - r^3}{Z_i^3 - R_i^3}, \quad R_i \leq r \leq Z_i \tag{58}$$

The derivation of equation 58 is based on the contact condition at R_i (insert R_i for r in equation 58!) and a constant value of divergence of the flux in the diffusion zone.

Then the total dissipation function Q_D can be expressed as

$$Q_D = \sum_{i=1}^{N} \sum_{k=1}^{n_c} 4\pi \int_{R_i}^{Z_i} \frac{R_g T}{c_k D_k} j_{ki}^2 r^2 dr \tag{59}$$

Assuming $R_i \ll Z_i$, equation 59 can be formulated, after integration, as

$$Q_D = \sum_{i=1}^{N} 4\pi R_g T R_i^3 \dot{R}_i^2 \sum_{k=1}^{n_c} \frac{(c_k^* - c_k)^2}{c_k D_k} = 4\pi\varepsilon \sum_{i=1}^{N} R_i^3 \dot{R}_i^2, \quad \varepsilon = R_g T \sum_{k=1}^{n_c} \frac{(c_k^* - c_k)^2}{c_k D_k} \tag{60}$$

Thus, Q_D is, in first approximation, independent of the size of the diffusion zones, which are large compared to the dimension of the precipitate (note that most of the dissipation occurs nearest to the precipitates).

The assumption of constant total volume of all objects yields

$$\frac{4\pi}{3} \sum_{i=1}^{N} R_i^3 = \text{const} \Rightarrow \sum_{i=1}^{N} R_i^2 \dot{R}_i = 0 \tag{61}$$

The necessary condition for the constrained maximum of D is given, according to equation 51, as

$$\frac{\partial}{\partial \dot{R}_i} \left[D - \lambda(Q - D) + \beta 4\pi \sum_{j=1}^{N} R_j^2 \dot{R}_j \right] = 0, \quad i = 1, ..., N \tag{62}$$

with D following from equation 56 as

$$D = -8\pi\xi\gamma \sum_{i=1}^{N} R_i \dot{R}_i \tag{63}$$

The dissipation function Q is given by equation 57 for grain growth and by equation 60 for coarsening, and λ and β are Lagrange multipliers. Following the general procedure starting with equation 44 and leading to equation 54, equation 62 yields for grain growth

$$R_i^2 \dot{R}_i = 2M(-\gamma R_i + \beta R_i^2), \quad i = 1, ..., N \tag{64}$$

and for coarsening

$$R_i^2 \dot{R}_i = \frac{2(-\gamma + \beta R_i)}{\varepsilon}, \quad i = 1, ..., N \tag{65}$$

Insertion of equations 64 into equation 61 allows the evaluation of β. Consequently the evolution equations for individual grains during grain growth read

$$\dot{R}_i = 2\gamma M \left(\frac{1}{R_{CG}} - \frac{1}{R_i} \right), \quad i = 1, ..., N \tag{66}$$

with $R_{CG} = \sum_{j=1}^{N} R_j^2 / \sum_{j=1}^{N} R_j$ being the critical radius; see also section 3.1 and R_C there. Grains with radius R_{CG} (the second subscript G points to 'growth') neither grow nor shrink. One can observe that large (super-critical) grains grow, while small (sub-critical) grains shrink and must be eliminated from the system when they reach zero radius. As the total number N of grains decreases and the total volume of grains remains constant during the process, the mean volume as well as the mean radius $\bar{R}$ of the grains increase; a detailed treatment is presented in Fischer *et al.* (2003).

Analogously, insertion of equations 65 into equation 61 allows the evaluation of β and then the evolution equations for individual precipitates during coarsening read as

$$\dot{R}_i = \frac{2\gamma}{\varepsilon R_i} \left(\frac{1}{R_{CC}} - \frac{1}{R_i} \right), \quad i = 1, ..., N \tag{67}$$

with $R_{CC} = \sum_{j=1}^{N} R_j / N = \bar{R}$ being the critical radius (the second subscript C points to 'coarsening').

Here it should be mentioned that equations 66–67 are the outcome of a rational procedure based on application of the TEP.

3.4. Thermodynamic treatment of grain growth and coarsening – distribution concept

One can also apply the TEP directly to the distribution function itself (Svoboda and Fischer, 2007). For this, let the distribution function depend on n parameters $a_1(t),...,a_n(t)$, which determine the evolution of the distribution function

$$f(R,t) = f(R, a_1(t),...a_n(t)) \tag{68}$$

To avoid taking into account the condition of a fixed volume of the system as a constraint, we assume that all distribution functions considered f meet a condition analogous to equation 61 as

$$\int_0^\infty R^3 f \, \mathrm{d}R = 1 \tag{69}$$

at any time, *i.e.* the system has a fixed volume $4\pi/3$.

The well known continuity relation of the distribution function f in the radius space reads as

$$\frac{\partial f}{\partial t} + \frac{\partial (f \dot{R})}{\partial R} = 0 \tag{70}$$

for references see Svoboda and Fischer (2007).

Equation 70 can be integrated at a fixed time as

$$\int_R^\infty \frac{\partial f}{\partial t} dR' = (f\dot{R})|_R - (f\dot{R})|_\infty = (f\dot{R})|_R \qquad (71)$$

because $(f\dot{R})|_\infty = 0$, due to a finite value of $\dot{R}|_\infty$ and $f|_\infty = 0$.

Differentiation of equation 68 with respect to time yields for fixed R

$$\frac{\partial f}{\partial t} = \sum_{i=1}^n \frac{\partial f}{\partial a_i} \dot{a}_i \qquad (72)$$

Insertion of equation 72 into equation 71 enables us to calculate $\dot{R}$ as

$$\dot{R}(R) = \frac{1}{f} \sum_{i=1}^n \dot{a}_i \int_R^\infty \frac{\partial f}{\partial a_i} dR' \qquad (73)$$

Equation 73 represents a key equation for the treatment, as it allows us to express the total dissipation in the system by means of $\dot{a}_1,...,\dot{a}_n$.

Similarly to equation 56 the total Gibbs energy of the system is given by

$$G = 4\pi\xi\gamma \int_0^\infty R^2 f dR \qquad (74)$$

The dissipation function Q due to grain-boundary migration during grain growth follows, in analogy to equation 57, as

$$Q_M = \frac{2\pi}{M} \int_0^\infty R^2 f \dot{R}^2 dR = \frac{2\pi}{M} \int_0^\infty \frac{R^2}{f} \left(\sum_{i=1}^n \dot{a}_i \int_R^\infty \frac{\partial f}{\partial a_i} dR' \right)^2 dR \qquad (75)$$

and, due to diffusion during coarsening, in analogy to equation 60, as

$$Q_D = 4\pi\varepsilon \int_0^\infty R^3 f \dot{R}^2 dR = 4\pi\varepsilon \int_0^\infty \frac{R^3}{f} \left(\sum_{i=1}^n \dot{a}_i \int_R^\infty \frac{\partial f}{\partial a_i} dR' \right)^2 dR \qquad (76)$$

According to the TEP, the evolution of the system corresponds to the maximum of the total dissipation D constrained by $Q - D = 0$, Q stands for Q_M or Q_D (no further constraints exist for $\dot{a}_1,...,\dot{a}_n$). The treatment, analogous to that presented in the previous section, leads to evolution equations of the system given by a set of linear equations with respect to $\dot{a}_i$, $i = 1,...n$ (for details see Svoboda and Fischer, 2007),

$$\sum_{j=1}^n \frac{1}{2} \frac{\partial^2 Q}{\partial \dot{a}_i \partial \dot{a}_j} \dot{a}_j = -\frac{\partial G}{\partial a_i}, \quad i = 1,...,n \qquad (77)$$

with

$$\frac{\partial G}{\partial a_i} = 4\pi\xi\gamma \int_0^\infty R^2 \frac{\partial f}{\partial a_i} dR, \quad i = 1, ..., n \tag{78}$$

The second derivatives of the dissipation function Q follow for grain growth as

$$\frac{1}{2}\frac{\partial^2 Q_M}{\partial \dot{a}_i \partial \dot{a}_j} = \frac{2\pi}{M} \int_0^\infty \frac{R^2}{f} \int_R^\infty \frac{\partial f}{\partial a_i} dR' \int_R^\infty \frac{\partial f}{\partial a_j} dR' dR, \quad i,j = 1, ..., n \tag{79}$$

and for coarsening

$$\frac{1}{2}\frac{\partial^2 Q_D}{\partial \dot{a}_i \partial \dot{a}_j} = 4\pi\varepsilon \int_0^\infty \frac{R^2}{f} \int_R^\infty \frac{\partial f}{\partial a_i} dR' \int_R^\infty \frac{\partial f}{\partial a_j} dR' dR, \quad i,j = 1, ..., n \tag{80}$$

3.5. Comparison of the multi-object and distribution concept for grain growth in systems with initial bimodal distribution functions

For the treatment of systems with a bimodal grain radii distribution (e.g. Marthinsen et al., 1996), one can use a linear combination of two Rayleigh distribution functions as

$$f_B = KR[Ac_1^{5/2} \exp(-c_1 R^2) + (1 - A)c_2^{5/2} \exp(-c_2 R^2)] \tag{81}$$

meeting again the condition equation 69 of constant volume of the system for arbitrary values of parameters from the interval $0 \leqslant A \leqslant 1$, $c_1 > 0$, $c_2 > 0$. Note that A, c_1, c_2 replace the parameters a_1, a_2, a_3 in the previous treatment. Using equations 78, one can write

$$\frac{\partial G_B}{\partial A} = 2\pi\gamma K \int_0^\infty R^3 \left[c_1^{5/2} \exp(-c_1 R^2) - c_2^{5/2} \exp(-c_1 R^2)\right] dR = \pi\gamma K \left[c_1^{1/2} - c_2^{1/2}\right] \tag{82}$$

$$\frac{\partial G_B}{\partial c_1} = 2\pi\gamma A K c_1^{5/2} \int_0^\infty R^3 \left(\frac{5}{2c_1} - R^2\right) \exp(-c_1 R^2) dR = \frac{\pi\gamma A K}{2c_1^{1/2}} \tag{83}$$

$$\frac{\partial G_B}{\partial c_2} = 2\pi\gamma(1 - A) K c_2^{5/2} \int_0^\infty R^3 \left(\frac{5}{2c_2} - R^2\right) \exp(-c_2 R^2) dR = \frac{\pi\gamma(1 - A)K}{2c_2^{1/2}} \tag{84}$$

Equations 79 yield

$$\frac{1}{2}\frac{\partial^2 Q_B}{\partial \dot{A}^2} = \frac{\pi K}{2M} \int_0^\infty \frac{R\left[c_1^{3/2} \exp(-c_1 R^2) - c_2^{3/2} \exp(-c_2 R^2)\right]^2}{Ac_1^{5/2} \exp(-c_1 R^2) + (1 - A)c_2^{5/2} \exp(-c_2 R^2)} dR \tag{85}$$

$$\frac{1}{2}\frac{\partial^2 Q_B}{\partial \dot{A}\partial \dot{c}_1} =$$

$$\frac{\pi A K c_1^{1/2}}{4M} \int_0^\infty \frac{R\left[c_1^{3/2}\exp(-c_1 R^2) - c_2^{3/2}\exp(-c_2 R^2)\right](3 - 2c_1 R^2)\exp(-c_1 R^2)}{A c_1^{5/2}\exp(-c_1 R^2) + (1 - A)c_2^{5/2}\exp(-c_2 R^2)}\,dR \qquad (86)$$

$$\frac{1}{2}\frac{\partial^2 Q_B}{\partial \dot{A}\partial \dot{c}_2} =$$

$$\frac{\pi(1 - A)K c_2^{1/2}}{4M} \int_0^\infty \frac{R\left[c_1^{3/2}\exp(-c_1 R^2) - c_2^{3/2}\exp(-c_2 R^2)\right](3 - 2c_2 R^2)\exp(-c_2 R^2)}{A c_1^{5/2}\exp(-c_1 R^2) + (1 - A)c_2^{5/2}\exp(-c_2 R^2)}\,dR \qquad (87)$$

$$\frac{1}{2}\frac{\partial^2 Q_B}{\partial \dot{c}_1^2} = \frac{\pi A^2 K c_1}{8M} \int_0^\infty \frac{R\left((3 - 2c_1 R^2)\exp(-c_1 R^2)\right)^2}{A c_1^{5/2}\exp(-c_1 R^2) + (1 - A)c_2^{5/2}\exp(-c_2 R^2)}\,dR \qquad (88)$$

$$\frac{1}{2}\frac{\partial^2 Q_B}{\partial \dot{c}_1 \partial \dot{c}_2} = \frac{\pi A(1 - A)K c_1^{1/2}c_2^{1/2}}{8M} \int_0^\infty \frac{R(3 - 2c_1 R^2)(3 - 2c_2 R^2)\exp(-(c_1 + c_2)R^2)}{A c_1^{5/2}\exp(-c_1 R^2) + (1 - A)c_2^{5/2}\exp(-c_2 R^2)}\,dR \qquad (89)$$

$$\frac{1}{2}\frac{\partial^2 Q_B}{\partial \dot{c}_2^2} = \frac{\pi(1 - A)^2 K c_2}{8M} \int_0^\infty \frac{R\left((3 - 2c_2 R^2)\exp(-c_2 R^2)\right)^2}{A c_1^{5/2}\exp(-c_1 R^2) + (1 - A)c_2^{5/2}\exp(-c_2 R^2)}\,dR \qquad (90)$$

The integrals in equations 85–90 must be calculated numerically. The kinetics of the system is then given by equations 77.

An initial bimodal grain radii distribution function is defined by choosing $A(0) = 0.003$, $c_1(0) = 25$ and $c_2(0) = 1$. The evolution of the parameters is depicted in Fig. 1. One can observe that the contribution corresponding to c_1 decreases gradually and disappears at time $t = 0.027741$ ($A \rightarrow 0$ at that time – see Fig. 1a). The value of the parameter c_1 increases drastically at the same time (see Fig. 1b), which correlates with a significant refinement of grains corresponding to the c_1 contribution. The c_2 contribution exhibits gradual growth (decreasing of the value of c_2 – see Fig. 1c) and survives as the only one.

The time evolution of the distribution function corresponding to the evolution of the parameters A, c_1 and c_2 (calculated along the distribution concept) is depicted by the solid lines for times $t = 0$, 0.005, 0.01, 0.015, 0.02, 0.025 and 0.027741 in Fig. 2. This evolution is compared with that based on the multi-object (here multigrain) concept (dotted lines) for the same starting bimodal distribution and the same time instants.

3.6. Concluding remarks

Based on the TEP, the multi-object and distribution concepts are developed. The equations for the evolution of individual objects, derived with the multi-object concept,

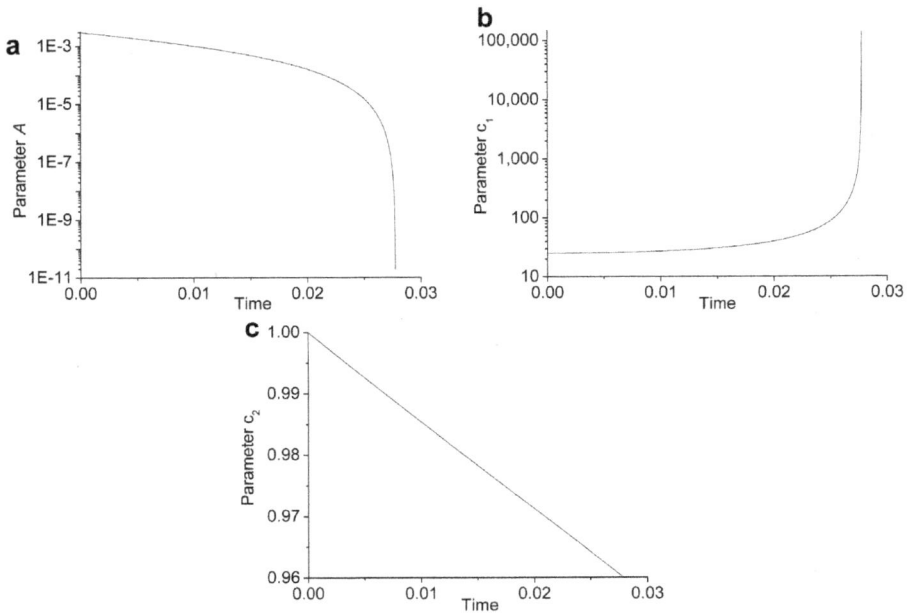

Figure 1. Evolution of parameters of the bimodal distribution function calculated up to the time 0.027741. (a) Parameter A, (b) Parameter c_1, (c) Parameter c_2. Reprinted from *Acta Materialia*, **55**, Svoboda, J. and Fischer, F.D., A new approach to modelling of non-steady grain growth, 4467–4474 (2007), with permission from Elsevier.

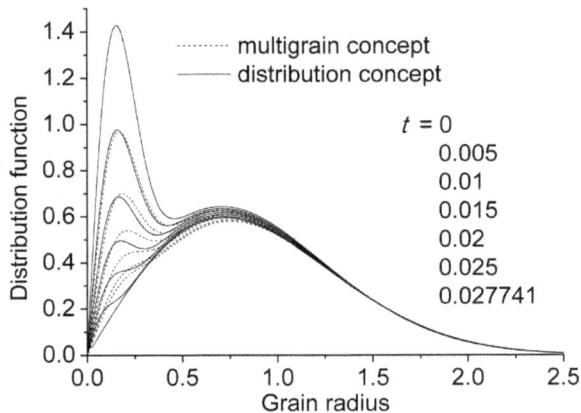

Figure 2. Comparison of the evolution of bimodal grain radii distribution function by application of multiobject (here multigrain) concept and distribution concept. Individual curves correspond to time instants denoted in the figure. Reprinted from *Acta Materialia*, **55**, Svoboda, J. and Fischer, F.D., A new approach to modelling of non-steady grain growth, 4467–4474 (2007), with permission from Elsevier.

coincide with Hillert's and Lifshitz-Slyozov-Wagner's classical equations. The distribution concept allows the treatment of the evolution of the object radii distribution without the need to know the evolution equations for individual objects.

The multi-object and distribution concepts are applied to the initially bimodal Rayleigh-type distribution function for grain growth. The results of simulations are compared, and a reasonable agreement is achieved.

Finally, note that the current concept is also applicable if, in addition to the surface energy, other energy terms such as the elastic strain energy and the energy stored in dislocations are included in the total Gibbs energy G. This opens the door to study recrystallization and grain refinement. Here we refer to a recent paper by Hackl and Renner (2013) in which a complementary thermodynamic extremal principle, equivalent to the TEP (Hackl and Fischer, 2008), was applied to recrystallization in olivine aggregates.

4. Application of the TEP to solid-state precipitation

4.1. Introduction

Precipitation is a phenomenon that every living being on our globe experiences more or less frequently. Precipitation occurs, for instance, as a consequence of a supersaturated state of water vapour in air, which we commonly refer to as "hot and humid weather". The precipitation process manifests itself as 'particles' or droplets of water that start falling out of clouds of supersaturated vapour. We call this phenomenon 'rain'. The typical sequence of conditions leading to precipitation are:

   (i)   Molecules of H_2O evaporate from liquid water in the form of molecules due to input of heat from the sun. This process is governed by a local equilibrium between the liquid and gaseous phases of water molecules directly at the liquid/gas interface. Local equilibrium is justified by the assumption that no thermodynamic force remains at the interface during the formation of a drop.

   (ii)   With the warm air, the H_2O molecules start rising in altitude until they eventually come into supersaturated condition. When temperature and pressure decrease, the solubility of H_2O molecules in the air which contains the water decreases and the molecules begin to form small aggregates. This is experienced commonly as clouds or fog.

   (iii)   Once the supersaturation reaches a critical level, the droplets of H_2O become supercritical and they grow spontaneously by attachment of more and more H_2O molecules. The increasingly large droplets start to 'fall out' of the clouds in the form of rain.

Interestingly, precipitation is not restricted to the formation of (liquid) water droplets from (gaseous) clouds of vapour. Exactly the same process occurs also in the solid state where supersaturated solid solutions are deliberately created by appropriate heat treatment of two- or more-component mixtures of atoms. The particles that fall out of the supersaturated solid solution are formed *via* the processes of nucleation and growth. Differences are certainly observed in the kinetics of movement of atoms in solid materials compared to molecules in vapour; the basic mechanisms of precipitate formation are, however, more or less, the same.

Therefore, we introduce a model that was originally developed for precipitation in metallic alloys. The methodology introduced there was generic however, and it can be applied to any kind of precipitation reaction independent of the specific nature of the material.

4.2. Problem formulation

Consider a representative volume element (RVE) of material with an arbitrary mixture of atoms. Furthermore, assume that spherical precipitates with various crystal structures and chemical compositions can form within this reference volume. A sketch of this situation is shown in Fig. 3.

For the sake of simplicity, we assume that each precipitate is characterized by its radius and mean chemical composition. The precipitates are considered to be randomly distributed. In the same sense, the matrix containing the precipitates is assumed to be constituted in an homogeneous manner and characterized by its mean composition only. Local variations in composition, chemical composition gradients as well as heterogeneities in precipitate distribution are ignored. In this case, the Gibbs energy G of the system can be expressed as

$$G = \sum_{i=1}^{n} N_{0i}\mu_{0i} + \sum_{k=1}^{m} \frac{4\pi\rho_k^3}{3}\left(\lambda_k + \sum_{i=1}^{n} c_{ki}\mu_{ki}\right) + \sum_{k=1}^{m} 4\pi\rho_k^2\gamma_k \qquad (91)$$

The first subscript 0 points to the matrix and k, $k = 1,...,m$, to individual precipitates. The second subscript i, $i = 1,...,n$ points to components. N_{0i} is the number of moles of component i in the matrix, μ_{0i} and μ_{ki} are the chemical potentials of component i in the matrix and in precipitate k, respectively, ρ_k is the radius of the precipitate k, c_{ki} is concentration of

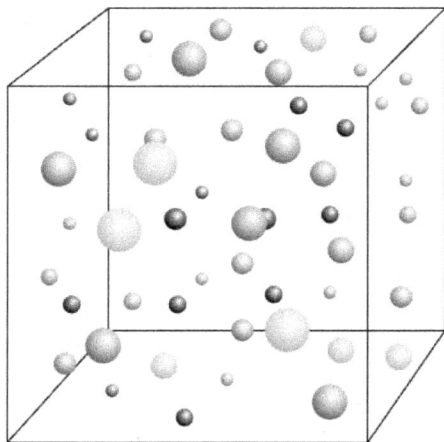

Figure 3. Sketch of the representative volume element (RVE) for precipitation kinetics modelling. The spheres depict precipitates of different crystal structure and chemical composition.

component i in precipitate k and γ_k is the specific interfacial energy. λ_k is a parameter accounting for the stress caused by volumetric misfit between precipitate k and matrix. In this formulation, the system state is fully determined on the basis of the set of characteristic parameters such as radii and composition of precipitates. The composition of the matrix is then determined by the mass-conservation relation

$$N_{01} = N_i - \sum_{k=1}^{m} \frac{4\pi\rho_k^3 c_{ki}}{3}, \quad i = 1, ..., n \tag{92}$$

where N_i is the total number of moles of component i in the system. Further details of the model are described in the corresponding key publications by Svoboda *et al.* (2004) and Kozeschnik *et al.* (2004a,b).

Once the Gibbs energy of the system is defined as a function of the characteristic parameters, the evolution of the system can then be determined after definition of the dissipation processes that apply to the specific system. In the present case, three dissipative processes are considered.

The first dissipative process is due to the migration of the precipitate/matrix interface during precipitate growth/shrinkage. The related dissipation function Q_1 reads

$$Q_1 = \sum_{k=1}^{m} \frac{4\pi\rho_k^2}{M_k^{IF}} \dot{\rho}_k^2 \tag{93}$$

where the parameter M_k^{IF} denotes the mobility of the precipitate/matrix interface; compare also equation 57. Note that $1/M_k^{IF}$ can be interpreted as a friction coefficient, thus the energy dissipated by interface migration corresponds to a friction coefficient multiplied by velocity squared (based on the standard assumption of a linear relation between the velocity and the thermodynamic force).

The second dissipative process considered in the model is related to the change in chemical composition of the precipitates. Practically, this term accounts for the fact that diffusion inside the precipitate is necessary for accommodation of the chemical composition of the precipitate as a reaction to changes in the matrix composition and/or temperature. The corresponding dissipation function reads as

$$Q_2 = \sum_{k=1}^{m} \sum_{i=1}^{r} \int_{0}^{\rho k} \frac{R_g T}{c_{ki} D_{ki}} 4\pi r^2 j_{ki}^2 \, dr \tag{94}$$

where D_{ki} is the tracer diffusion coefficient of element i in the precipitate k and j_{ki} is the corresponding diffusional flux.

The third dissipative process comes from diffusion of elements in the matrix as a consequence of atomic transport towards or away from the precipitate during growth or shrinkage. The corresponding dissipation function Q_3 with J_{ki} as diffusional flux in the matrix reads as

$$Q_3 = \sum_{k=1}^{m} \sum_{i=1}^{n} \int_{\rho_k}^{Z} \frac{RT}{c_{0i}D_{0i}} 4\pi r^2 J_{ki}^2 \mathrm{d} \approx \sum_{k=1}^{m} \sum_{i=1}^{n} \frac{4\pi RT \rho_k^3 (\dot{\rho}_k(c_{ki} - c_{0i}) + \rho_k \dot{c}_{ki}/3)^2}{c_{0i}D_{0i}} \quad (95)$$

Compare also equation 59.

The total dissipation function is then given by the sum

$$Q = Q_1 + Q_2 + Q_3 \quad (96)$$

The Gibbs energy G together with dissipation function Q allows us to apply TEP as described, particularly for discrete characteristic parameters in section 3.2. The details can be found in Svoboda *et al.* (2004a) for systems with interstitial and substitutional elements and in Kozeschnik *et al.* (2004b) for the thermodynamic sublattice model that is commonly utilized in CALPHAD-type databases (*e.g.* Lukas *et al.*, 2006).

In the mean-field formulation of the precipitation problem as introduced here, application of the TEP delivers evolution equations for the radius and chemical composition of the precipitates as a function of the actual state of the thermodynamic system. The progress of the precipitation process can then be obtained by numerical integration. A corresponding procedure was suggested by Kozeschnik *et al.* (2004a) and is depicted in Fig. 4.

In the actual integration procedure, which is implemented in the software package MatCalc (http://matcalc.at), the entire heat-treatment process is divided into appropriate isothermal time slices within an automatic time step routine.

In the practical simulation, the manifold of precipitates in an RVE is grouped into size classes for the benefit of computational efficiency. Accordingly, each size class represents N precipitates of identical radius and chemical composition. Typically, the number of size classes is chosen between 15 and 100 to maintain a reasonable balance between calculation speed and numerical accuracy. The list of size classes for each precipitate and their relation to the embedding matrix are depicted in Fig. 5.

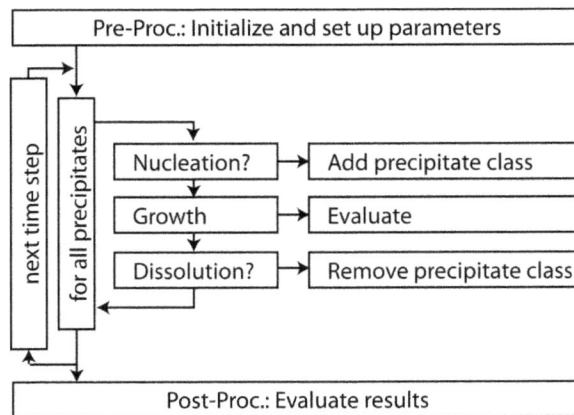

Figure 4. Schematic of the numerical integration procedure for solution of the precipitation kinetics evolution equations.

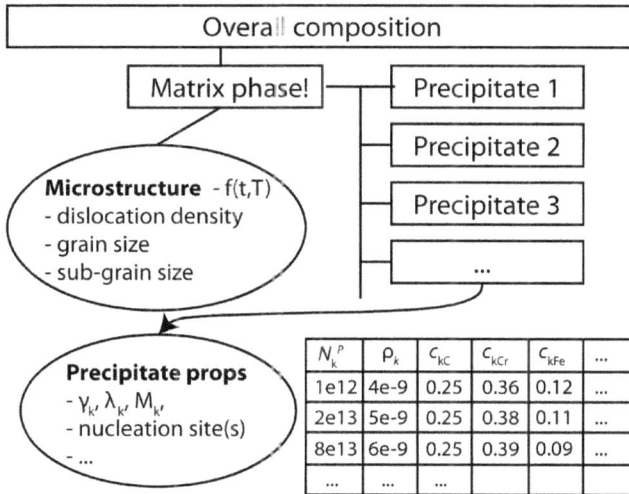

Figure 5. Relation between precipitates and matrix as realized in *MatCalc*.

4.3. Precipitate nucleation and interface energy modelling

In the numerical integration scheme for precipitate evolution, in each time step, the nucleation rate J, corresponding to a given precipitate type, is evaluated with the relations

$$J_S = N_{nuc} Z \beta^* \exp\left(-\frac{G^*}{k_B T}\right)$$ (97)

and

$$J = J_S \exp\left(-\frac{\tau}{t}\right)$$ (98)

The number of potential nucleation sites is denoted as N_{nuc}, Z is the Zeldovich factor, β^* describes the kinetics of atomic attachment to the critical nucleus, k_B is the Boltzmann constant and G^* is the nucleation barrier. The subscript 'S' points to steady state conditions. The transient nature of nucleation is accounted for in equation 98 with the incubation time τ.

The most important quantity describing the nucleation rate is the nucleation barrier G^*. In the framework of classical nucleation theory and assuming spherical symmetry, it is given as

$$G^* = \frac{16\pi}{3} \frac{\gamma^3}{(\Delta G_{vol})^2}$$ (99)

The term ΔG_{vol} is the volume free energy change due to nucleus formation (identical to the driving force for precipitation), and γ (earlier γ_k) is the specific energy of the interface between the precipitate and matrix. Because the interface energy occurs with an exponent value of 3 in the nucleation barrier G^*, which itself occurs in the exponent of the nucleation rate expression, see equation 97, it denotes the most critical and sensitive quantity describing nucleation. In many modelling attempts related to precipitation kinetics, this quantity is used as a fitting parameter between experiment and simulation. This is argued on the basis that even small variations in γ may lead to huge variations in the nucleation rate. Alternatively, the interface energy can also be calculated from appropriate models, and the precipitation kinetics simulation is then carried out on a predictive basis. A corresponding model has been developed recently by Sonderegger and Kozeschnik (2009a), where the interface energy has been related to the enthalpy of solution ΔH_{sol} of the precipitate with

$$\gamma = \frac{n_S z_{S,eff}}{N_A z_{L,eff}} \cdot \Delta H_{sol} \tag{100}$$

where n_S is the number of atoms on a unit area of interface, $z_{S,eff}$ is the effective number of broken bonds across the interface, N_A is the Avogardo number and $z_{L,eff}$ is the effective coordination number of the crystal structure. In their treatment of the generalized broken bond model, Sonderegger and Kozeschnik (2009a) evaluated the effective quantities with $z_{S,eff}/z_{L,eff} = 0.328$ for the bcc and $z_{S,eff}/z_{L,eff} = 0.329$ for the fcc structure.

The interface energy expression given in equation 100 is firstly calculated for the case of a planar and sharp interface. For application to the case of solid-state precipitation, two additional effects must be accounted for. First, in the early stages of precipitation, particularly in the nucleation stage, precipitates can be small and the effect of interface curvature must be considered properly. This has been carried out by Sonderegger and Kozeschnik (2009b). Second, the interface can become diffuse at higher temperatures, thus introducing entropic effects and decreasing the interfacial energy. A corresponding treatment was reported by Sonderegger and Kozeschnik (2010). With the generalized broken bond model and accounting for curvature and diffuse interfaces, rather accurate estimates for mean precipitate/matrix interfacial energies can be made and precipitation kinetics can be simulated on a predictive basis.

Details of the specific formulation of all quantities discussed in this section were reported by Kozeschnik (2013). In the following section, an application of the solid-state precipitation model to continuous casting of low-alloy steel is presented.

4.4. Solid-state precipitation during continuous casting of low-alloy steel

Today, most structural steel production is world-wide based on continuum casting technology. In this process, the molten steel is poured into a tundish and distributed further into an ingot mould. The strand solidifies from the outside due to the severe cooling applied. In the course of cooling, several types of precipitates are formed in the solid strand with the purpose of limiting grain growth of the metallic material at high temperature as well as increasing the material strength by precipitation hardening. For optimization of the production process, it is of immediate importance for the steel producer to know which types of precipitates form at which temperatures and in which size and number density. A corresponding computational and experimental study was performed by Zamberger et al. (2012), which is summarized next for illustration of the

Table 1. Chemical composition (wt.%) of the investigated steel grade.

C	Si	Mn	S	Al	Ti	Nb	N	Fe
0.22	0.24	1.24	0.025	0.025	0.0018	0.0356	0.0061	rest

applicability of the precipitation kinetics model. The steel considered in the study has a chemical composition as shown in Table 1.

In the simulations, the precipitation process has been investigated in full consideration of the dendritic solidified primary microstructure. Accordingly, the microstructure contains regions with enrichment of alloying elements due to segregation, as well as regions where the alloying elements have become depleted. Parts of these regions are then considered to be either defect free or considered as containing grain boundaries. The latter represent preferential heterogeneous nucleation sites for various precipitate types. A sketch of the assumed primary solidification microstructure is shown in Fig. 6, also indicating the two limiting conditions for precipitation, which are (i) precipitate formation in a region with preferential nucleation and growth of precipitates due to solute element

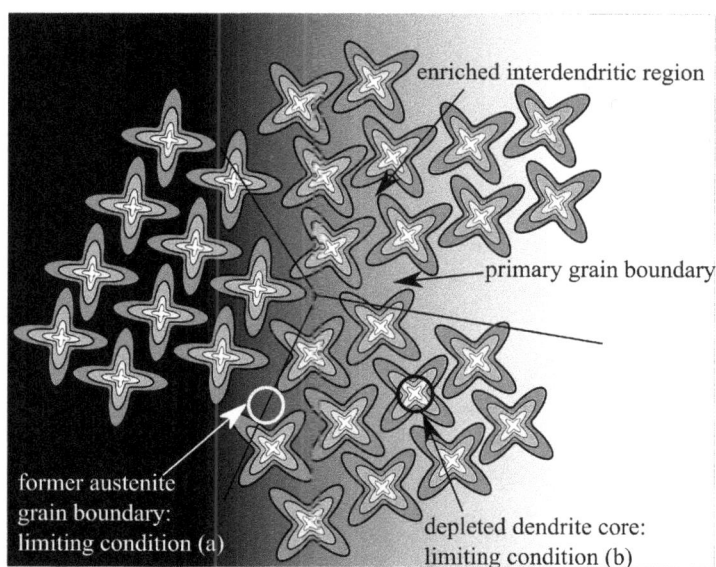

Figure 6. Primary solidified microstructure as considered in the precipitation simulation. Reproduced from Zamberger *et al.* (2012). Reprinted from *International Journal of Material Research*, **103**, Zamberger, S. *et al.*, Numerical simulation of the evolution of primary and secondary Nb(CN), Ti(CN) and AlN in Nb-microalloyed steel during continuous casting, 680–687, 2012, with permission from © Hanser Verlag, München.

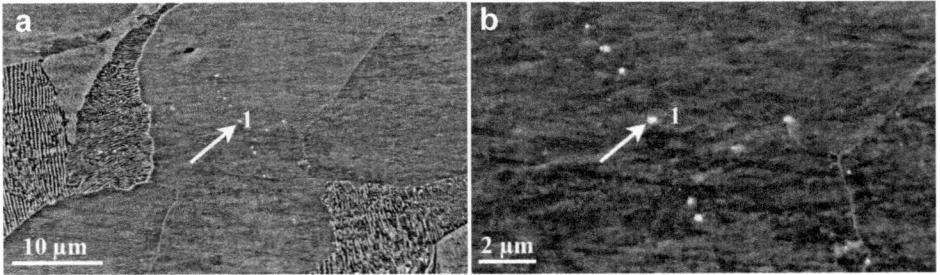

Figure 7. Typical precipitates of Nb(CN) type in micro-alloyed low-carbon steel as observed by scanning electron microscopy. Reproduced from Zamberger *et al.* (2012). Reprinted from *International Journal of Material Research*, **103**, Zamberger, S. *et al.*, Numerical simulation of the evolution of primary and secondary Nb(CN), Ti(CN) and AlN in Nb-microalloyed steel during continuous casting, 680–687, 2012, with permission from © Hanser Verlag, München.

enrichment and the availability of heterogeneous nucleation sites, as well as (ii) precipitate formation in a region with unfavourable nucleation conditions due to element depletion and the unavailability of heterogeneous sites.

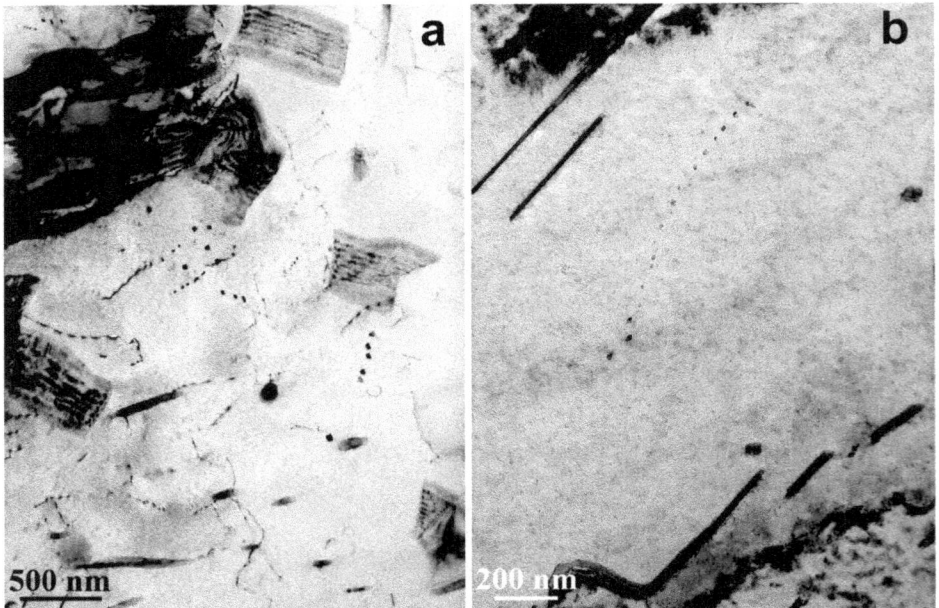

Figure 8. Typical precipitates of Nb(CN) type in micro-alloyed low-carbon steel as observed by transmission electron microscopy. Reproduced from Zamberger *et al.* (2012). Reprinted from *International Journal of Material Research*, **103**, Zamberger, S. *et al.*, Numerical simulation of the evolution of primary and secondary Nb(CN), Ti(CN) and AlN in Nb-microalloyed steel during continuous casting, 680–687, 2012, with permission from © Hanser Verlag, München.

Table 2. Results of the numerical simulations for the enriched interdendritic zone compared to the experimentally observed particles. γ denotes an austenitic matrix phase, α a ferritic matrix phase, NS a nucleation site (gb for grain boundaries and d for dislocations), r_{mean} mean radii and $r_{exp.}$ experimentally observed radii.

Matrix	Phase	NS	r_{mean} (nm)	Number density (m^{-3})	Predicted	Observed	$r_{exp.}$ (nm)
	AlN	gb	88	2×10^{11}	yes	no	–
	AlN	d	3	4×10^{16}			
	TiN	gb	73	3×10^{11}	yes	no	–
γ	TiN	d	21	1×10^{18}			
	NbC	gb	230	1×10^{15}	yes	yes	50–250
	NbC	d	143	4×10^{17}			
	AlN	gb	2	4×10^{20}	yes	no	–
	AlN	d	1	8×10^{18}			
α	TiN	gb	<1	2×10^{20}	yes	yes	2–15
	TiN	d	<1	1×10^{17}			
	NbC	gb	1	2×10^{19}	yes	yes	2–15
	NbC	d	<1	3×10^{14}			

Table 3. Results of the numerical simulations for the depleted dendrite core zone compared to the experimentally observed particles. Notations as in Table 2.

Matrix	Phase	NS	r_{mean} (nm)	Number density (m^{-3})	Predicted	Observed	$r_{exp.}$ (nm)
γ	TiN	d	2	1×10^{18}	yes	no	–
	NbC	d	6	1×10^{14}	yes	no	–
	AlN	gb	11	6×10^{19}	yes	yes	~15
	AlN	d	< 1	1×10^{14}			
α	TiN	gb	2	1×10^{19}	yes	yes	~5
	TiN	d	1	6×10^{17}			
	NbC	gb	4	2×10^{19}	yes	yes	2–15
	NbC	d	4	8×10^{17}			

In the experimental analysis, several samples of a solidified continuous cast bloom have been prepared for transmission electron microscopy (TEM) as well as scanning electron microscopy (SEM). The probed areas in the thin foils are randomly taken from either region of the heterogeneous microstructure and they show, for this reason, different precipitation states. Figure 7 shows the typical appearance on Nb(CN) precipitates as observed in SEM and Fig. 8 in TEM.

Finally, the calculated and observed precipitates are compared in Tables 2 and 3 in terms of their size, occurrence and microstructural location. Keeping in mind that the microstructure of the as-solidified steel is rather heterogeneous and that the simulations are performed without fitting parameters, the agreement is good and confirms the applicability of the concept for solid-state precipitation kinetics simulation introduced in this chapter.

4.5. Concluding remarks

The TEP is utilized for development of a computationally efficient and flexible model for precipitation-kinetics simulation in multi-component multi-phase multi-particle environments. Application of the TEP in the proposed mean-field framework delivers evolution equations for the radius and chemical composition of the precipitates in the form of a decoupled set of linear equations.

Combining the growth model for the precipitates with a suitable model for nucleation and implementing this framework into a computer code for performing the numerical integration of the evolution equations, allow simulation of precipitation reactions in solid materials on a predictive basis.

Application of the approach is finally demonstrated in the example of precipitation of carbides and nitrides during continuous casting of low-alloy steel. Recalling that the simulations are performed without fitting parameters, good agreement with the experiments is observed.

References

Atkinson, H.V. (1988) Overview no. 65: Theories of normal grain growth in pure single phase systems. *Acta Metallurgica*, **36**, 469–491.

Callen, H.B. (1966) *Thermodynamics*. John Wiley & Sons, New York.

Colemen, B.D. and Gurtin, M.E. (1967) Thermodynamics with internal state variables. *Journal of Chemical Physics*, **47**, 597–613.

de Groot, S.R. and Mazur P. (2013) *Non-Equilibrium Thermodynamics*. (Dover Books on Physics). Dover Publications Inc., Mineola, New York.

Fischer, F.D. and Svoboda, J. (2010) Substitutional diffusion in multicomponent solids with non-ideal sources and sinks for vacancies. *Acta Materialia*, **58**, 2698–2707.

Fischer, F.D. and Svoboda, J. (2011) Chemically and mechanically driven creep due to generation and annihilation of vacancies with non-ideal sources and sinks. *International Journal of Plasticity*, **27**, 1384–1390.

Fischer, F.D. and Svoboda, J. (2014) Diffusion of elements and vacancies in multi-component systems. *Progess in Materials Science*, **60**, 338–367.

Fischer, F.D. and Svoboda, J. (2015) Stress deformation and diffusion interactions in solids – a simulation study. *Journal of the Mechanics and Physics of Solids*, **78**, 427–442.

Fischer, F.D., Svoboda, J. and Fratzl, P. (2003) A thermodynamical approach to grain growth and coarsening. *Philosophical Magazaine*, **83**, 1075–1093.

Fischer, F.D., Svoboda, J. and Petryk, H. (2014) Thermodynamic extremal principles for irreversible processes in materials science. *Acta Materialia*, **67**, 1−20.

Fischer, F.D., Hackl, H. and Svoboda J. (2016) Improved thermodynamic treatment of vancancy-mediated diffusion and creep. *Acta Materialia*, **108**, 347−354.

Greenwood, G.W. (1956) The growth of dispersed precipitates in solutions. *Acta Metallurgica*, **4**, 243−248.

Hackl, K. and Fischer, F.D. (2008) On the relation between the principle of maximum dissipation and inelastic evolution given by dissipation potentials. *Proceedings of the Royal Society A*, **464**, 117−132.

Hackl, K. and Renner, J. (2013) High-temperature deformation and recrystallization: A variational analysis and its application to olivine aggregates. *Journal of Geophysical Research: Solid Earth*, **118**, 943−967.

Hackl, K., Fischer, F.D. and Svoboda, J. (2011a) A study on the principle of maximum dissipation for coupled and non-coupled non-isothermal processes in materials. *Proceedings of the Royal Society A*, **467**, 1186−1196.

Hackl, K., Fischer, F.D. and Svoboda, J. (2011b) Addendum to "A study on the principle of maximum dissipation for coupled and non-coupled non-isothermal processes in materials". *Proceedings of the Royal Society A*, **467**, 2422−2426.

Hillert, M. (1965) On the theory of normal and abnormal grain growth. *Acta Metallurgica*, **13**, 227−238.

Hillert, M. (1996) Analytical treatments of normal grain growth. *Materials Science Forum*, **204−206**, 3−18.

Kozeschnik, E. (2013) *Modeling Solid-state Precipitation*. Momentum Press, New York.

Kozeschnik, E., Svoboda, J., Fratzl, P. and Fischer, F.D. (2004a) Modelling of kinetics in multi-component multi-phase multi-particle systems with spherical precipitates II. − Numerical solution and application. *Materials Science and Engineering A*, **385**, 157−165.

Kozeschnik, E., Svoboda, J. and Fischer, F.D. (2004b) Modified evolution equations for the precipitation kinetics of complex phases in multi-component systems. *CALPHAD*, **28**, 379−382.

Lifshitz, L.M. and Slyozov, V.V. (1958) On kinetics of diffusion decay of oversaturated solid solutions. *Journal of Experimental and Theoretical Physics*, **35**, 479−491.

Lifshitz, L.M. and Slyozov, V.V. (1961) The kinetics of precipitation from supersaturated solid solutions. *Journal of Physics and Chemistry of Solids*, **19**, 35−50.

Lukas H.L., Fries S.G. and Sundman, B. (2007) *Computational Thermodynamics: The Calphad Method*. Cambridge University Press, Cambridge, UK.

Manning, J.R. (1970) Cross terms in the thermodynamic diffusion equations for multicomponent alloys. *Metallurgical and Materials Transactions B*, **1**, 499−505.

Marthinsen K., Hunderi O. and Ryum N. (1996) The influence of spatial grain size correlation and topology on normal grain growth in two dimensions. *Acta Materialia*, **44**, 1681−1689.

Onsager, L. (1931) Reciprocal relations in irreversible processes. I. *Physical Review*, **37**, 405−426.

Rios, P.R. (2004) Irreversible thermodynamics, parabolic law and self-similar state in grain growth. *Acta Materialia*, **52**, 249−256.

Rios, P.R., Gottstein, G. and Shvindlerman, L.S. (2001) Application of the thermodynamic theory of irreversible processes to normal grain growth. *Scripta Materialia*, **44**, 893−897.

Rios, P.R., Gottstein, G. and Shvindlerman, L.S (2002) An irreversible thermodynamic approach to normal grain growth with a pinning force. *Materials Science and Engineering A*, **A332**, 231−235.

Saetre, T.O. (2004a) A mean field theory of grain growth: I. *Modelling and Simulation in Materials Science Engineering*, **12**, 1267−1277.

Saetre, T.O. (2004b) A mean field theory of grain growth: II. *Modelling and Simulation in Materials Science Engineering*, **12**, 1279−1292.

Sonderegger, B. and Kozeschnik, E. (2009a) Generalized nearest-neighbor broken-bond analysis of randomly oriented coherent interfaces in multicomponent fcc and fcc structures, *Metallurgical and Materials Transactions A*, **40**, 499−510.

Sonderegger, B. and Kozeschnik, E. (2009b) Size dependence of the interfacial energy in the generalized nearest-neighbor broken-bond approach. *Scripta Materialia*, **60**, 635−638.

Sonderegger, B. and Kozeschnik, E. (2010) Interfacial energy of diffuse phase boundaries in the generalized broken-bond approach. *Metallurgical and Materials Transactions A*, **41**, 3262−3269.

Suwa, Y. and Saito, Y. (2003) Computer simulation of grain growth by the phase field model. Effect of interfacial energy on kinetics of grain growth. *Materials Transactions*, **44**, 2245−2251.

Suwa, Y., Saito, Y. and Onodera, H. (2007) Phase-field simulation of abnormal grain growth due to inverse pinning. *Acta Materialia*, **55**, 6881−6894.

Svoboda, J. and Fischer, F.D. (2007) A new approach to modelling of non-steady grain growth. *Acta Materialia*, **55**, 4467−4474.

Svoboda, J. and Fischer, F.D. (2009) Vacancy-driven stress relaxation in layers. *Acta Materialia*, **57**, 4649−4657.

Svoboda, J. and Fischer, F.D. (2011) Modelling of the influence of sources and sinks for vacancies and stress state on diffusion in crystalline solids. *Acta Materialia*, **59**, 1212−1219.

Svoboda, J. and Lukáš, P. (2000) Creep deformation modelling of superalloy single crystals. *Acta Materialia*, **48**, 2519−2528.

Svoboda, J. and Riedel, H. (1995) New solutions describing the formation of interparticle necks in solid-state sintering. *Acta Metallurgica et Materialia*, **43**, 1−10.

Svoboda, J. and Turek, I. (1991) On diffusion-controlled evolution of closed solid-state thermodynamic systems at constant temperature and pressure. *Philosophical Magazine Part B*, **64**, 749−759.

Svoboda, J., Fischer, F.D., Fratzl, P. and Kroupa, A. (2002) Diffusion in multi-component systems with no or dense sources and sinks for vacancies. *Acta Materialia*, **50**, 1369−1381.

Svoboda, J., Gamsjäger, E., Fischer, F.D. and Fratzl, P. (2004a) Application of the thermodynamic extremal principle to the diffusional phase transformations. *Acta Materialia*, **52**, 959−967.

Svoboda, J., Fischer, F.D., Fratzl, P. and Kozeschnik, E. (2004b) Modelling of kinetics in multi-component multi-phase systems with spherical precipitates I. − Theory. *Materials Science and Engineering A*, **385**, 166−174.

Svoboda, J., Turek, I. and Fischer, F.D. (2005) Application of the thermodynamic extremal principle to modeling of thermodynamic processes in material sciences. *Philosophical Magazine*, **85**, 3699−3707.

Svoboda, J., Fischer, F.D. and Fratzl, P. (2006) Diffusion and creep in multi-component alloys with non-ideal sources and sinks for vacancies. *Acta Materialia*, **54**, 3043−3053.

Svoboda, J., Gamsjäger, E., Fischer, F.D., Liu, Y. and Kozeschnik, E. (2011) Diffusion processes in a migrating interface − The thick interface model. *Acta Materialia*, **59**, 4775−4786.

Vorhees, P.W. (1985) The theory of Ostwald ripening. *Journal of Statistical Physics*, **38**, 231−252.

Vorhees, P.W. (1992) Ostwald ripening of two-phase mixtures. *Annual Review of Materials Science*, **22**, 197−215.

Wagner, C. (1961) Theorie der Alterung von Niederschlägen durch Umlösen (Ostwald-Reifung). *Zeitschrift für Elektrochemie*, **65**, 581−591.

Zamberger, S., Pudar, M., Spiradek-Hahn, K., Reischl, M. and Kozeschnik, E. (2012) Numerical simulation of the evolution of primary and secondary Nb(CN), Ti(CN) and AlN in Nb-microalloyed steel during continuous casting. *International Journal of Materials Research*, **103**, 680−687.

Ziegler, H. (1963) Some extremum principles in irreversible thermodynamics with applications to continuum mechanics. Chapter 2 in: *Progress in Solid Mechanics* Vol. IV (I.N. Sneddon and R. Hill, editors). North-Holland, Amsterdam.

Ziegler, H. (1977) *An Introduction to Thermomechanics*. North-Holland, Amsterdam.

Zöllner, D. and Streitenberger, P. (2006) Three-dimensional normal grain growth: Monte Carlo Potts model simulation and analytical mean field theory. *Scripta Materialia*, **54**, 1697−1702.

Atomic-scale modelling of crystal defects, self-diffusion and deformation processes

SANDRO JAHN[1,2] and XIAO-YU SUN[3]

[1]GFZ German Research Centre for Geosciences, Telegrafenberg, 14473 Potsdam, Germany
[2]Present address: University of Cologne, Institute of Geology and Mineralogy, Greinstrasse 4-6, 50939 Cologne, Germany, e-mail: s.jahn@uni-koeln.de
[3]Université Lille 1, UMET, UMR CNRS 8207, 59655 Villeneuve d'Ascq Cedex, France, e-mail: Xiaoyu.Sun@univ-lille1.fr

This chapter introduces basic concepts of numerical modelling of materials on the atomic scale. An atomic interaction potential is at the core of each simulation. Static energy calculations and molecular dynamics simulations are common approaches to study crystal defects, such as point defects, dislocations and grain boundaries, phase transitions and diffusion processes in solids and liquids. The application of these powerful methods to geological materials is demonstrated with a number of examples. An important challenge for the future appears to be the crossover between accurate and predictive atomic-scale models and the continuum scale at which material properties are conventionally described. Some recent developments in this field are discussed.

1. Introduction

Numerical simulations have become a powerful tool for studying material behaviour on very different scales. They are used in the geosciences, *e.g.* in geodynamic models of mantle convection. Many geological processes take place over long time and length scales, *i.e.* up to billions of years and thousands of kilometres. On the other hand, material properties are determined by the chemical bonds between atoms, which are only a few Ångstroms (10^{-10} m) apart and which vibrate with frequencies in the terahertz region (10^{12} s^{-1}). The large-scale behaviour of real materials is often controlled by crystal defects on the atomic scale, which open pathways for diffusion and are necessary for plastic deformation. Thus, the key for a deep understanding of macroscopic material behaviour lies on the atomic scale. This chapter introduces concepts and applications of atomic-scale simulation approaches that are used to model the structure and dynamics of Earth materials based on explicit treatment of atom and electron interactions. The focus will be on modelling crystal defect structures and energies, on diffusive transport and on deformation processes.

Before starting an atomic-scale simulation one should consider the following questions:

DOI: 10.1180/EMU-notes.16.8

- Which properties should be studied and what is the required accuracy? Is the atomic structure of the material already known and if so, to what extent? For transport properties or kinetic processes, what are possible transport or reaction mechanisms?
- What is the minimum number of atoms that must be contained in the simulation cell? Is a single unit cell or a supercell of a few unit cells large enough to study the problem of interest?
- What are the available computational resources and are they sufficient for the type of problem to be addressed?
- Which codes are already available and how well suited are they to the problem to be solved? How much code development will potentially be needed?

In the following, we will give a basic introduction into the computational methodology, review some of the recent activities in the field of defect structures and transport properties, and provide some practical advice on how to perform atomic-scale simulations. Finally, we will discuss approaches to how discrete atomic structures may be used to feed larger-scale continuum models.

2. Molecular modelling approaches

The key ingredient for any atomic-scale simulation is a model of the particle interactions, which is usually represented by an interaction potential. In the most general case, the potential may be based on a quantum-mechanical treatment of the material in terms of its electronic structure. Such an approach is referred to as first-principles or *ab initio* method as it does not require any empirical data but is built exclusively from basic physical principles. First-principles calculations provide an accurate description of chemical bonding for many different classes of materials and in a wide range of pressures and temperatures. However, they are computationally demanding, which restricts their applicability to systems of usually not more than a few hundred atoms in the simulation box (with the exception of linear scaling codes, see below). The spatial extent of defect structures often requires system sizes of thousands or even millions of atoms. This can be achieved by using classical force field interaction potentials which do not consider electrons explicitly but rather describe atomic interactions by more or less empirical functions.

Once an interaction potential is chosen, one can explore the structure, thermodynamic and physical properties of a system of atoms. As we are usually interested in condensed phases (crystals, melts, fluids), simulations are performed with periodic boundary conditions, *i.e.* the simulation cell is periodically repeated to form an infinite system in the three spatial dimensions. In the case of non-crystalline phases (melts, fluids, glasses), the artificial periodicity introduced by the boundary conditions may cause some problems and has to be kept in mind, especially for small simulation boxes where atoms may interact with their periodic images. To study free surface properties, periodic boundary conditions are sometimes reduced to two dimensions. Periodic boundary conditions can be switched off in some codes, *e.g.* when small molecules or nanoparticles are investigated.

The total energy of a static system of atoms is obtained by adding up all particle interactions. By moving individual atoms in or by changing the dimensions of the simulation cell, one can search for the lowest energy state, which represents the structure of the phase at a temperature of zero Kelvin. Often, the zero point motion of atoms present even at zero Kelvin is neglected, but this effect can be accounted for if required. Temperature may be added in molecular dynamics (MD) simulations, where the system of atoms evolves according to Newton's equations of motion. MD simulations can follow the system in 'real time' up to some nano- or even microseconds. However, many dynamical processes such as atom jumps between vacancies in crystals at low temperatures occur much less frequently, and therefore accelerated dynamics methods are required. Eventually, the choice of interaction potential and of the specific modelling strategy has to be a compromise between accuracy, general applicability and computational efficiency. In the remainder of this section, some of the most important simulation methods are presented. For a more detailed and formal introduction to molecular simulation methods, the reader is referred, *e.g.* to textbooks by Allen and Tildesley (1987), Frenkel and Smit (2002) or Marx and Hutter (2012). Molecular modelling in the context of Earth materials has been in the focus of dedicated volumes or chapters in Reviews in Mineralogy and Geochemistry (Cygan and Kubicki, 2001; Wentzcovitch and Stixrude, 2010; Jahn and Kowalski, 2014) and EMU Notes in Mineralogy (Gramaccioli, 2002).

2.1. Density-functional theory

As already mentioned above, the most general and most accurate approach to describe chemical bonding between atoms is based on quantum mechanics or more explicitly on electronic structure theory. Within this theory, the Born-Oppenheimer approximation states that the nuclear motion can be separated from the electronic motion due to the large mass difference. This means that for any atomic structure the electronic structure is assumed to remain always in its lowest energy state, *i.e.* in its ground state. In the electronic structure calculations, the atomic nuclei are usually considered as classical particles with a mass and a charge, whereas the electrons interact as quantum particles, which exhibit properties of particles and of waves. While nuclei interact essentially *via* electrostatic forces described by Coulomb's Law, the interaction between electrons is much more complicated. Electrons are fermions and behave according to the Pauli exclusion principle, which states that two electrons cannot occupy the same quantum state simultaneously. This leads to an additional electronic interaction energy, the so-called exchange energy. All other electron–electron interactions that are not accounted for by the classical Coulomb and by the Pauli exchange interactions are described by the correlation energy.

The goal of electronic-structure calculations is to find the electronic ground state for a set of nuclei and electrons. The dimensionality of the problem is given by the number of electrons, N. As an example, a unit cell of Mg_2SiO_4 forsterite contains 4 formula units ($Z = 4$), 28 nuclei and 280 electrons. An exact determination of the electronic ground state of an atomic system would require a solution of the $3N$ dimensional

Schrödinger wave equation. Even on state-of-the-art supercomputers, the Schrödinger equation cannot be solved exactly for systems with many electrons. Density-functional theory (DFT) was proposed in the 1960s by Hohenberg, Kohn and Sham (Hohenberg and Kohn, 1964; Kohn and Sham, 1965) as an alternative formulation, in which the ground state energy of a system is expressed uniquely as a functional of the three-dimensional ground state electron density. For practical calculations, a number of approximations is needed, which, in the case of the DFT approach, are more efficient than for wavefunction-based methods that seek approximate solutions of the Schrödinger equation. The first and most crucial approximation of DFT is the choice of the exchange-correlation functional, which essentially describes the quantum-mechanical part of the interaction of an electron with the charge density due to all other electrons. The most commonly used exchange-correlation functionals are either based on the local-density approximation (LDA) or on the generalized gradient approximation (GGA). Whereas in LDA, the exchange-correlation functional is expressed in terms of a local uniform electron gas, the GGA also considers local gradients in the charge density. Both approximations lead to systematic errors which, in the case of crystal structure optimizations, usually result in predictions of lattice constants that are too small for LDA and too large for GGA. The development of advanced and more accurate exchange-correlation functionals has been a field of active research in electronic structure theory for many years (Martin, 2004; Perdew and Ruzsinszky, 2010).

To account for the wave-like character, the state of each electron is described by a complex function which, in the case of DFT, is called the Kohn-Sham orbital. The sum of the squares of all Kohn-Sham orbitals gives the electron charge density, which can be computed at each position inside the simulation cell. Mathematically, each Kohn-Sham orbital is expressed as a superposition of so-called basis functions. Localized basis functions such as Gaussian-type orbitals are often used for molecules or atomic clusters but they are also employed in the CRYSTAL code (www.crystal.unito.it; Dovesi et al., 2014). Due to their periodicity, planewave basis sets are especially suited to condensed phases and they are used e.g. in the ABINIT (www.abinit.org; Gonze et al., 2009), CPMD (www.cpmd.org; Marx and Hutter, 2000) and QUANTUM ESPRESSO (www.quantum-espresso.org; Giannozzi et al., 2009) codes. Special localized basis sets are being developed to improve scaling of DFT calculations from about N^3 to N to achieve linear scaling with the number of electrons. This approach has the potential to perform DFT calculations with thousands up to millions of atoms. Promising case studies have been published using linear scaling codes such as SIESTA (www.icmab.es/siesta/; Artacho et al., 2008), ONETEP (www.onetep.org; Skylaris et al., 2005) or CONQUEST (www.order-n.org; Bowler and Miyazaki, 2012). For efficiency reasons, core electrons are often not treated explicitly but they are incorporated together with the nucleus into effective pseudopotentials (Payne et al., 1992). This reduces the calculations to those electrons that contribute to the chemical bonding.

Standard DFT does not always perform well for molecular systems and for compounds containing elements with partially filled d- and f-shells, which includes transition metals, lanthanide and actinide elements. Common exchange-correlation functionals do not correctly describe very weak forces, such as van der Waals or dispersion interactions, which are important, *e.g.* to describe the interaction between layered materials, such as graphite or clay minerals. In the case of atoms or ions with partially filled d- or f-shells, interactions between d- or f-electrons are stronger than described by standard DFT and often DFT wrongly predicts metallic behaviour for transition metal oxides and silicates. In practice, correction schemes have been very useful to at least partly account for these deficiencies. For instance, van der Waals corrections have proven very useful for molecular materials (see *e.g.* Klimeš and Michaelides, 2012). Modelling of transition metal oxides or silicates has often been improved by adding a Hubbard-*U* correction, which adds an additional repulsion between d-electrons (see *e.g.* Cococcioni, 2010).

As a general piece of advice, one should always test how well the approximations and possible corrections apply to the specific problem of interest before starting heavy calculations. In addition, one has to make sure that all other parameters of a DFT calculation lead to sufficiently well converged energies, forces or other properties of interest. This concerns the number of basis functions to represent the charge density, the convergence criteria or the sampling of the reciprocal space in case of small simulation cells (the so-called *k*-point sampling). Some DFT codes (*e.g.* the ABINIT code) provide comprehensive tutorials, in which practical issues of a DFT calculation are explained in detail.

2.2. Classical interaction potentials

Due to the high computational demand, system sizes of DFT calculations are usually restricted to a few hundred atoms and simulation times to a few tens of picoseconds. For larger systems, needed *e.g.* for the simulation of melts, mineral–melt interfaces, grain-boundary or dislocation structures, classical interatomic potentials may be employed. The smallest particles of such models are atoms or ions, which are described as classical particles. Particle interactions are represented by more or less empirical force fields. For ionic systems, the simplest potentials $V(r_{ij})$ of Born-Mayer type are pairwise additive (r_{ij} being the distance between ions i and j) and contain Coulomb attraction or repulsion due to the ionic charges q_i and q_j, a repulsive term, which accounts for charge density overlap of ions at small separation, and a van der Waals interaction term (A_{ij}, b_{ij} and C_{ij} are constants)

$$v(r)_{ij} = \frac{q_i q_j}{r_{ij}} + A_{ij} \exp(-b_{ij} r_{ij}) - \frac{C_{ij}}{r_{ij}^6} \qquad (1)$$

A typical pair potential between anions and cations is shown in Figure 1.

Such potentials work quite well for certain simple systems (*e.g.* Ogarov *et al.*, 2000; Guillot and Sator, 2007) but even in these cases they can be used in only a restricted range of properties, temperatures and pressures. Often, they are optimized for a specific problem, *e.g.* to reproduce densities or elastic properties of a small number of crystal structures. However, their predictive power at conditions not included in the model

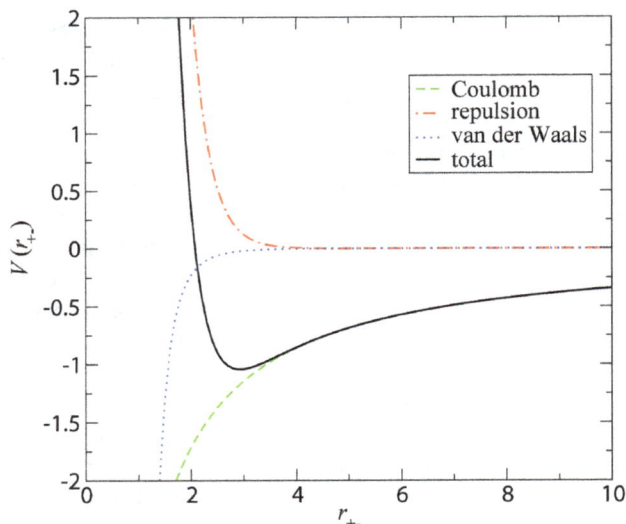

Figure 1. Typical Born-Mayer potential for anion–cation interaction, $V(r_{+-})$, as a function of interionic distance r_{+-} between anion and cation. Also shown are the relative contributions of attractive and repulsive terms. The equilibrium distance between nearest neighbours corresponds to the minimum of the total $V(r_{+-})$.

parameterization is usually rather limited. A more careful analysis of the interatomic forces reveals that important contributions to the interaction potentials are missing from the Born-Mayer pair potentials. The reason is that the electronic structure of an ion is strongly influenced by the confining potential of the surrounding ions. Local electric fields or field gradients lead to polarization effects, *i.e.* the formation of dipoles and higher-order multipoles. For instance, Wilson *et al.* (1996) demonstrated that an ionic potential including dipole polarizability was needed to obtain the correct number of infrared absorption peaks for amorphous SiO_2. Also, the change of the size and the shape of an ion in different coordination environments may have to be considered to obtain a reliable interaction model.

The first attempt to go beyond the rigid ion Born-Mayer-type potential and to include polarization effects was the shell model by Dick and Overhauser (1958). Other types of polarizable ion models were parameterized, *e.g.* by Madden and Wilson (2000) and Tangney and Scandolo (2002). Breathing shell models that include polarization and a flexible anion radius were explored by Matsui *et al.* (2000) and Zhang *et al.* (2005). Even more advanced ionic interaction potentials were parameterized by reference to DFT calculations (*e.g.* Jahn and Madden, 2007; Salanne *et al.*, 2012). For systems with molecular units such as CO_3^{2-}, H_2O or OH^-, covalent bonds have been accounted for by spring-like interaction models (*e.g.* Rohl *et al.*, 2003). References to other classical potentials were given by Jahn and Kowalski (2014). Ongoing efforts are being made to increase the transferability of classical force fields to allow their application to a wide range of materials, to different coordination environments, temperatures and pressures as well as to solids and liquids. Born-Mayer potentials are implemented in most of the available codes that use classical potentials. For the more advanced potentials, popular codes include GULP (gulp.curtin.edu.au; Gale and Rohl, 2003) and CP2K (www.cp2k.org).

2.3. Static energy calculations

Having chosen a method to represent the atomic interactions (DFT or classical potential) one can readily compute the total energy of an atomic structure. In the general case, after such a calculation there will be finite forces acting on individual atoms and the computed stress tensor will be more or less non-hydrostatic. To find the lowest energy of an ideal structure at a temperature of zero Kelvin, all atoms should be placed in the minimum of their local potential and the stress tensor should usually be hydrostatic. Many codes have implemented algorithms for geometry optimization, which can include atomic positions, simulation cell parameters (*i.e.* cell lengths and angles) or both. To obtain the structure and energy of a system under uniaxial compression the target stress tensor is set accordingly. A full geometry optimization (atoms and cell) yields the total energy and volume of the simulation cell for the specific interaction potential at a given stress or pressure and a temperature of 0 Kelvin.

2.4. Molecular dynamics simulations

Molecular dynamics simulations are used to model materials at finite temperature and pressure (Allen and Tildesley, 1987; Frenkel and Smit, 2002; Marx and Hutter, 2012). Starting with an initial atomic structure, each atom is assigned a random velocity, v. In statistical mechanics, the temperature T is related to the kinetic energy E_{kin} of a particle by

$$E_{kin} = \frac{m}{2} v^2 = \frac{3}{2} k_B T \tag{2}$$

with m being the atom mass and k_B the Boltzmann constant. In the following, the movement of atoms through time and space is determined by Newton's equations of motion

$$\mathbf{F} = m \frac{\partial \mathbf{v}}{\partial t} = -\frac{\partial V}{\partial \mathbf{r}} \tag{3}$$

where the force $\mathbf{F}$ acting on a particle leads to a change in its velocity. The force is computed from the gradient of the interaction potential V, which is computed using either DFT or classical potentials. The time integration is done numerically by replacing ∂t by a small Δt, usually of the order of 1 fs (10^{-15} s). At each time step of an MD simulation, atom positions and velocities are updated according to their interaction with surrounding particles. Depending on the type of interaction model used and the problem of interest, typical MD simulations sample the atomic dynamics at the picosecond (thousands of time steps) to microsecond (billions of time steps) time scale.

During an MD simulation, the system equilibrates itself within the thermodynamic ensemble used. If no exchange of heat and work with the surrounding is allowed, the simulation samples an isolated system (NVE) at constant number of particles N, constant volume V and constant total energy E. To keep the temperature constant, the simulation cell is coupled to a heat bath (thermostat) in the NVT ensemble. Simulations at constant pressure and temperature (NPT) involve coupling to a thermostat and to a barostat. In the latter case, the shape of the simulation cell (cell lengths and angles) is allowed to change.

When running *ab initio* MD simulations, one has to keep in mind that only the electronic interactions are computed quantum mechanically. The molecular dynamics of the atoms is modelled using the classical equations of motion. If quantum dynamics of light atoms needs to be considered, *e.g.* for certain H-bearing systems, path integral MD simulations could be employed (Marx and Parrinello, 1996). *Ab initio* MD simulations may be performed in two different modes: Born-Oppenheimer (BO) or Car-Parrinello (CP) MD (Marx and Hutter, 2012). Whereas in BOMD simulations the electronic ground state energy is computed in each simulation time step, in the CPMD method (Car and Parrinello, 1985) each electron is assigned a fictitious mass and a fictitious temperature. Besides the nuclei (including inner-shell electrons when pseudopotentials are used), the explicitly considered electrons are moved as quasi-particles with their own MD, which is much faster than the MD of the nuclei due to the big mass difference. CPMD is usually somewhat faster than BOMD and yields very similar atomic structures, although a much smaller time step needs to be employed. On the other hand, the fictitious electron dynamics may have significant effects, *e.g.* on diffusion coefficients. Therefore, the choice of method may depend on the specific problem.

2.5. Other atomic-scale sampling techniques

Molecular dynamics simulation approaches appear very attractive as they sample system properties in 'real' time with the system evolving according to physical laws. However, in many cases, the system gets stuck in metastable states due to processes that are slow compared to the pico- or nanosecond time scale of MD. For instance, transport processes with diffusion coefficients much smaller than about 10^{-10} m^2/s or deformations with shear rates much smaller than 10^8 s^{-1} are difficult to obtain in conventional MD. The same kinetic reasons are responsible for not achieving a true equilibrium state, *e.g.* for solid solutions, viscous melts or for systems close to phase boundaries. To overcome these kinetic problems, a number of methods are available, including metadynamics, sampling methods of transition state theory, kinetic Monte-Carlo and others (Frenkel and Smit, 2002; Marx and Hutter, 2012).

Metadynamics is an accelerated dynamics method often used to explore the multi-dimensional free-energy surface of a system of atoms (Marx and Hutter, 2012). This technique may be employed in cases where conventional MD is too slow to sample sufficiently many configurations of interest, *e.g.* different coordination environments of certain atoms, different crystal or related defect structures. The major kinetic barrier that hinders the transition of interest between different states is the required activation energy. For instance, a temperature of well above 2000 K is needed to observe directly the breakage of strong Si–O bonds in MD simulations of tens to hundreds of picoseconds (Spiekermann *et al.*, 2016). The idea of metadynamics is to force the system out of its actual state by adding an 'energy penalty' to all configurations that the system already visited (Fig. 2). To reduce the dimensionality of the problem a set of dynamic variables is defined. If a single bond-breaking process is studied this set of dynamic variables may be reduced to the anion–cation bond length. For investigations of the molecular structure of ions in melts or aqueous solutions, the coordination number may be a good dynamic variable. In the course of the metadynamics simulation, extra energies of Gaussian shape are added, *e.g.* to configurations of specific bond lengths or coordination numbers. This raises the total energy of the system in the specific region of configurational space and hence lowers the activation energy for a transition to a different state. In section 5, some examples of metadynamics simulations in the context of shear deformation and structural phase transitions will be presented. In these cases, the dynamic variables are the six parameters of the simulation cell, *i.e.* the three cell lengths and the three cell angles (Martoňák *et al.*, 2003, 2005b). During these simulations, the cell box is increasingly strained in a more or less random way until non-elastic deformations of the atomic structure occur.

The various methods of transition state theory are used to make predictions about the rates of a specific process (*e.g.* diffusion processes or structural phase transitions). A transition rate depends not only on the height of the lowest transition barrier but also on the detailed shape of the many-dimensional potential energy surface. For instance, transition path sampling MD (Dellago *et al.*, 1998) starts from a potential transition that

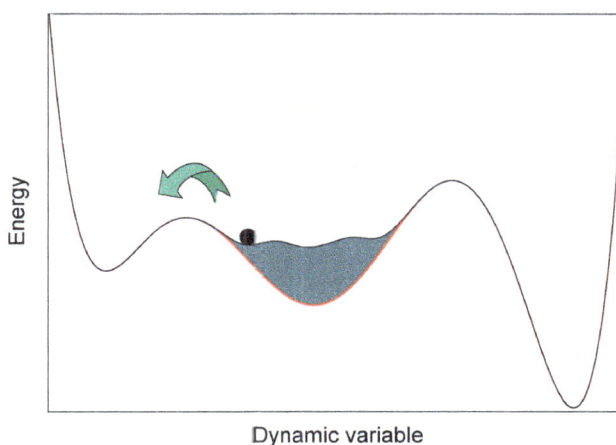

Figure 2. Principle of metadynamics illustrated for a single dynamic variable. The system (black circle) is initially in a state represented by the central energy minimum. After raising its energy it will first transform to the state with the local energy minimum on the left (arrow). Eventually, it will also cross the barrier on the right to reach the state with the global energy minimum.

may be derived from a metadynamics simulation. After slightly modifying the particle velocities and running the MD backward and forward in time, different particle trajectories are generated. Statistical analysis of whether or not the individual trajectories lead to a successful transition is used to derive the time dependence of the transition and furthermore provides insight into the transition mechanism without bias from a pre-defined reaction coordinate. In harmonic transition state theory (Vineyard, 1957) the rate factor, e.g. of an atom hopping to a vacant lattice site, depends on the curvature of the potential energy surface, which is related to the vibrational frequencies of the atoms in the ground state and in the transition state (for examples see section 4.3 and Ammann et al., 2010a). Once transition rates for different processes are determined it may be interesting to quantify the correlation between different mechanisms and to derive, e.g. an effective diffusion coefficient. Kinetic Monte-Carlo simulations are employed to study the time evolution of such a complex system by using statistical methods. Generally, in Monte-Carlo methods configurations or events are sampled by using random numbers (Frenkel and Smit, 2002). Acceptance of a random state or event depends on a physical law. In the case of kinetic Monte-Carlo methods of interest here, probabilities of randomly generated events are determined by the known transition rates. A few applications of this method are cited below.

2.6. Cross-over from the atomic to the continuum scale

Most existing models of crystal-defect mobility (including diffusion processes) and plastic deformation have been derived from a continuum description of materials. The atomic-scale simulation approach has the potential to be a very powerful tool to refine such models and to provide a detailed understanding of underlying processes on the sub-nanometre scale (de Leeuw et al., 2000; Adjaoud et al., 2012; Ghosh and Karki, 2013). This requires a working interface between the more or less discrete atomistic models and a continuum description. Continuous fields can be derived from atomic-scale calculations provided that the current (C) configuration is compared to a reference (R). If the vector $\mathbf{X}$ denotes the position of a material point in a body with respect to a reference coordinate system and the vector $\mathbf{x}$ provides the new position in the current configuration, the transformation gradient writes:

$$F_{ij} = \frac{\partial \mathbf{x}_i}{\partial \mathbf{X}_j} \tag{4}$$

As sketched in Fig. 3, an atom m located at a position $\mathbf{X}^m$ in the reference configuration moves to a new position $\mathbf{x}^m$ in the current configuration. The relative position of a neighbouring atom n with respect to atom m is given by the finite difference:

$$\Delta \mathbf{X}^{mn} = \mathbf{X}^n - \mathbf{X}^m \tag{5}$$

in the reference configuration and by

$$\Delta \mathbf{x}^{mn} = \mathbf{x}^n - \mathbf{x}^m \tag{6}$$

in the deformed configuration. The component of the transformation gradient at atom m induced by the variation of the relative position of atom n is then approximated by:

Reference state   Deformed state

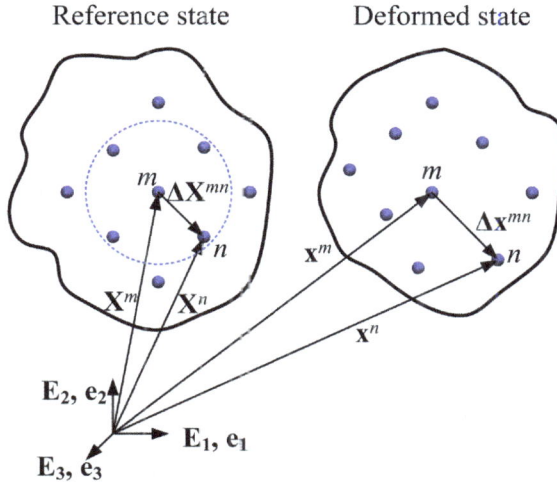

Figure 3. Illustration of the transformation gradient associated with the motion of material particles from the reference state to the current state.

$$F_{ij}^{mn} \approx \frac{\Delta x_i^{mn}}{\Delta X_j^{mn}} \qquad (7)$$

Because there is more than one neighbouring atom near atom m, the component of transformation gradient at the atom location is averaged over all neighbouring atoms n:

$$F_{ij}^{m} = \frac{1}{N} \sum_{r_{mn} \leq r_{cutoff}} F_{ij}^{mn} \qquad (8)$$

where r_{mn} is the distance between atoms m and n in the reference configuration, r_{cutoff} is a cut-off distance introduced to consider whether or not an atom n is in the neighbourhood of atom m, and N is the total number of neighbouring atoms identified within the cut-off distance. This cut-off distance is illustrated in Fig. 3 by the blue circle.

In continuous media, the total displacement vector field $\mathbf{u} = \mathbf{x} - \mathbf{X}$ describes changes in the position of matter. This field is assumed to be single-valued and continuous, possibly between atoms and below inter-atomic distances, so matter is assumed to be able to transmit stresses and couple stresses at this scale. The displacement gradient, also referred to as the distortion tensor, writes:

$$\mathbf{U} = \text{grad}(\mathbf{u}) = \text{grad}(\mathbf{x}) - \mathbf{I} = \mathbf{F} - \mathbf{I} \qquad (9)$$

$$U_{ij} = u_{i,j} = x_{i,j} - I_{ij} = F_{ij} - I_{ij} \qquad (10)$$

where $\mathbf{I}$ is the second-order identity tensor. The distortion is generally decomposed into strain and rotation. In a small perturbation approximation, the strain tensor ε is the symmetric part of the distortion, and the rotation tensor ω is its skew-symmetric part:

$$\varepsilon_{ij} = \frac{1}{2}(U_{ij} + U_{ji}) \tag{11}$$

$$\omega_{ij} = \frac{1}{2}(U_{ij} - U_{ji}) \tag{12}$$

A three-dimensional finite-difference method is used to calculate the components F_{ij}^{mn} of the transformation gradient, as well as all other derivatives. Once local values at atomic positions are obtained, all the atoms are projected onto plane ($\mathbf{e}_2$, $\mathbf{e}_3$), then two-dimensional linear interpolation is used to generate spatial field distributions in between atoms.

Application of this technique will be illustrated below for the case of a grain boundary in forsterite.

3. Crystal defect modelling

3.1. Point defects

The simplest (zero-dimensional) defects in crystals are point defects. There is a whole range of possible point defects, especially in more complex minerals. Two of the basic defects are shown in Fig. 4. A *Frenkel defect* is created by removing one atom from its equilibrium lattice site and placing it in the interstitial lattice (Fig. 4a). In ionic materials, another typical point defect is a so-called *Schottky defect*. In this case, an anion vacancy is balanced by a cation vacancy to maintain local charge neutrality (Fig. 4b).

In addition, minerals may contain a range of chemical impurities, *i.e.* atomic substitutions of other elements onto sites of the perfect structure. The degree to which substitution is possible depends on the specific structure and the thermodynamic state. In ionic crystals, anion vacancies such as chlorine vacancies in NaCl lead to the formation of colour centres (F-centres). At the anion vacancy position, an electron is trapped. This electron can get excited into different energy levels with some of the transitions being in the visible range of the electromagnetic spectrum.

Point defects are present in any crystal at $T > 0$ K, which is explained by

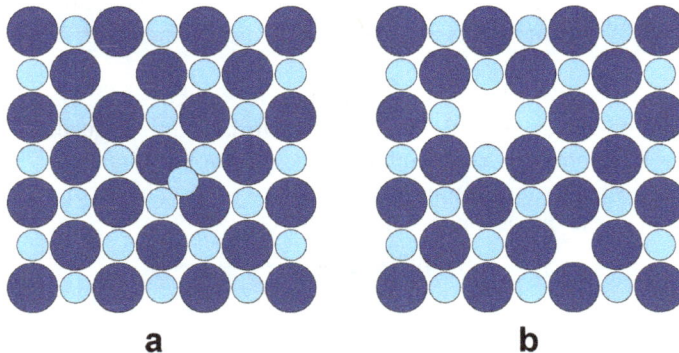

a   **b**

Figure 4. (a) A Frenkel defect consists of an atom in the interstitial lattice and a vacancy on a regular lattice site. (b) A Schottky defect can occur in ionic crystals as two charge-neutralizing vacancies (anion-cation vacancy pair).

thermodynamics. The equilibrium of a system at finite temperature is found by minimizing the Gibbs energy G

$$G = H - TS \tag{13}$$

Because the formation of defects requires energy, the enthalpy H rises with increasing defect concentration. However, the entropic term TS (entropy S) that is a measure of the disorder in the crystal, is always positive and hence decreases G. The resulting minimum of G is found at non-zero defect concentration (Putnis, 1992). With increasing temperature the number of defects N_d in crystals increases exponentially

$$N_d = N\exp(\Delta S_d/k)\exp(-\Delta H_d/k_BT) = N\exp(-\Delta G_d/k_BT) \tag{14}$$

where ΔS_d, ΔH_d and ΔG_d are the entropy, enthalpy and Gibbs energy of defect formation, respectively. N is the total number of sites. In molecular simulations, ΔG_d is usually approximated by computing defect-formation energies or enthalpies in the static limit (*i.e.* at $T = 0$ K).

The most direct way to compute defect energies is by using a supercell approach. The defect is contained in a relatively large simulation cell with periodic boundary conditions applied. While the latter are useful to sample the bulk of condensed phases and to avoid surface perturbations, their application to defect modelling has an intrinsic problem, namely the interaction of the defect with its periodic images. Charged defects have a strong interaction, which decays only slowly with the inverse of the particle distance $1/r$ according to Coulomb's law. For charge-neutral defects, the most significant perturbations to consider are electric dipoles and quadrupoles induced by the defect as well as elastic distortions of the lattice. One way to reduce the effect of defect interaction is to increase the size of the simulation cell, which is equivalent to increasing the number of atoms in the cell. This, in turn, leads to a considerable increase in computing time, which for DFT calculations in particular is feasible only to a certain extent. A correction scheme for charged defects is to compensate first the defect charge by an homogeneous background charge, which leads to an overall charge-neutral simulation cell. A second correction term is introduced to account for the defect–defect interaction of periodic cells (Leslie and Gillan, 1985).

This approach was used, for example, by Brodholt (1997) who performed DFT calculations to study the defect energies of charged point defects (either Mg^{2+}, O^{2-} or Si^{4+}) in forsterite Mg_2SiO_4. The simulation cell contained 28 atoms for the perfect crystal and 27 atoms for the defect structures. Simulations with single-charged defects were corrected for overall charge of the cell and for defect–defect interactions. To understand the practical aspects of the simulation, the major steps of such a calculation are summarized in the following. The first step of the calculation is to check convergence of the simulation parameters. For example, if the electron density is represented by a plane-wave expansion, the number of plane-waves accepted in the basis set has to be large enough. After that, the cell parameters (lattice constants, atomic positions) are optimized. Usually, the deviation of the DFT results from experimental parameters is within 1 to 2%. The defects are introduced by removing one of the atoms from the perfect crystal. In forsterite, six different atomic positions exist (see Fig. 5).

After the removal of an atom, the atoms in the vicinity of the defect are disturbed. Therefore the atomic positions have to be relaxed again. In a relaxed structure the individual atoms sit in a potential minimum and do not experience a force.

The main result of the Brodholt (1997) study is that there is a large difference in defect formation energy between atoms of the same atomic species sitting on different lattice sites. An O vacancy on the O3 site is more than 1 eV lower in energy than on the O2 site and 2.75 eV lower than on the O1 site. Mg vacancies on the Mg1 sites are more favourable than on the Mg2 site. The energy differences are so large, that almost all vacancies will be expected on the O3 and the Mg1 sites. This has implications for the diffusion behaviour in forsterite. Although the heights of the actual energy barriers for jump diffusion (see section on diffusion below) were not calculated explicitly here, the vacancy occupation numbers already suggest possible diffusion paths along the vacancy sites O3 and Mg1. If, for example, Mg2 sites are not available for diffusing Mg ions and the dominant diffusion mechanism is by jumps between vacancies, the jump distances increase in [100] and [001] directions, which has direct consequences for the diffusion anisotropy. As M1 octahedra are connected along [010], [010] is the direction for the fastest Mg diffusion.

More recent DFT investigations of point defects in minerals used a similar approach with an increased number of atoms in the simulation cell. For instance, Karki and co-workers studied point defects in MgO at high pressures using a supercell of 216 atoms (Karki and Khanduja, 2006), in $MgSiO_3$ bridgmanite and post-bridgmanite (Karki and Khanduja, 2007) and in the Mg_2SiO_4 polymorphs forsterite, wadsleyite and ringwoodite (Verma and Karki, 2009). The latter study also included modelling of

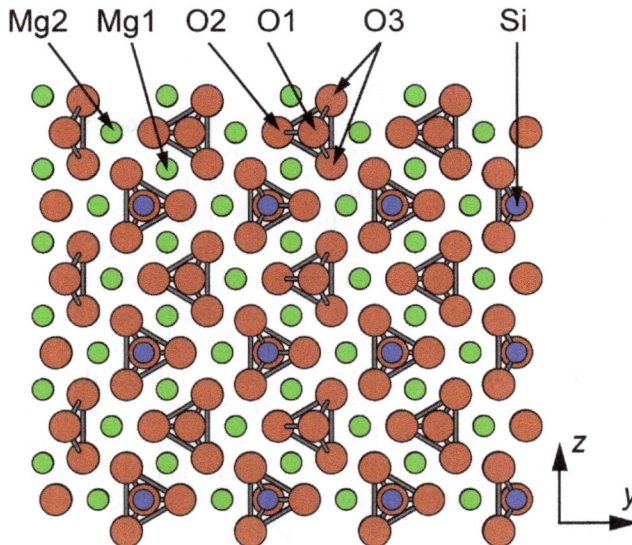

Figure 5. Structure of forsterite (space group *Pbnm*) viewed along [100] and distinct atomic sites.

charge-neutral Frenkel and Schottky defects as well as protonated defects. A relatively large amount of computational work was dedicated to the investigation of OH-defects in minerals, *e.g.* in Mg_2SiO_4 polymorphs (Verma and Karki, 2009; Blanchard *et al.*, 2009; Li *et al.*, 2009; Tsuchiya and Tsuchiya, 2009; Jahn *et al.*, 2013) and in $MgSiO_3$ enstatite (Balan *et al.*, 2013).

Alternatives to the supercell approach are embedded cluster models that avoid defect interactions between periodic supercell images. Generally, these models consist of an atomic cluster that contains an isolated defect. The cluster should be large enough to minimize interactions between the defect and the atoms at the edge of the cluster, *i.e.* increasing further the size of the cluster should not change the defect formation energy anymore. To account for long-range electrostatic interactions the region outside the cluster is modelled by a (simplified) classical potential with fixed atomic positions. The method used in many studies of this kind was first suggested by Mott and Littleton (1938). The defect formation energy is computed from the difference between the energies of the defective and the perfect crystal. Typically, the cluster contains at least a few hundred atoms and therefore this approach has been used mainly in conjunction with classical potentials. As in the supercell simulations all atoms of the inner regions are allowed to relax to their minimum potential energy. The other region of fixed point charges or polarizable ions typically contains up to a few thousand atoms. The Mott-Littleton method for computing defect formation energies is implemented in the classical force field lattice dynamics code GULP (Gale and Rohl, 2003). It was used *e.g.* to model point defects in Al_2O_3 and TiO_2 (Catlow *et al.*, 1982), carbonates (Fisler *et al.*, 2000), olivine (Walker *et al.*, 2003) or Mg_2SiO_4 and Mg_2GeO_4 spinels (Blanchard *et al.*, 2005).

The QM/MM method is an embedding technique where a cluster of atoms that interact quantum-mechanically is embedded into a region of classical particles. QM/MM stands for quantum mechanics/molecular mechanics. Combining ideas of QM/MM with the Mott-Littleton method allows us to compute defect energies with DFT precision. Usually, there are three regions in such calculations: the inner region that is modelled *e.g.* by DFT, the surrounding region for which classical potentials are used and the outer region of fixed point charges. Braithwaite *et al.* (2002, 2003) and Walker *et al.* (2009) used this embedded cluster approach to model various OH and Mg defects in forsterite and compared the results to those from the Mott-Littleton method and to experimental data. Alfe and Gillan (2005) used diffusion Monte-Carlo simulations to obtain defect formation energies in MgO going beyond the accuracy of DFT. Further information and references to defect modelling in crystals are provided in the review papers by Parker *et al.* (2004) and Freysoldt *et al.* (2014).

3.2. Dislocations

Dislocations are one-dimensional or line defects. They are key agents for plastic deformation processes and therefore the understanding of their structure and energy is of great importance. The elementary amount of shear carried by a dislocation is expressed by the so-called Burgers vector **b**. It can be determined from the closure vector of a loop of vectors connecting different lattice sites around the dislocation core compared to the corresponding loop in the perfect crystal. If the Burgers vector is perpendicular to the dislocation line, the dislocation is called an edge dislocation (Fig. 6). If the Burgers vector and dislocation line are parallel to each other, the dislocation is called screw dislocation. The notation for slip systems is $\mathbf{b} = [uvw](hkl)$, where [uvw] is the slip direction (or the Burgers vector) and (hkl) is the slip plane.

Far from the dislocation line, the presence of the defect is well described by its elastic fields. However, these fields are found to diverge in the immediate vicinity of the dislocation line. In this region, referred to as the dislocation core, a description at the atomic scale is mandatory. This region is important because the dislocation core has significant implications for the mobility of dislocation. When modelling a dislocation at the atomic scale the boundary condition problem arises and it is similar to that discussed above in the case of point defects. Generally, two different approaches have been employed: cluster models and dipole models (Walker *et al.*, 2010; Goryaeva *et al.*, 2015). The general idea of cluster models is similar to that of the Mott-Littleton method. As an ideal dislocation line is infinite in one dimension, periodic boundary conditions are applied in the direction of the dislocation line. Perpendicular to the line, only a finite cluster of atoms is considered in the vicinity of the dislocation core. Regions far away from the core are represented by a simplified elastic solid. The interface between the atomic and the elastic model has to be set up with care (see *e.g.* Walker *et al.*, 2010 for details). This approach has been used to study the dislocation cores of different materials including forsterite, zeolite A (Walker *et al.*, 2004), paracetamol-II, MgO (Walker *et al.*, 2005a,b; Zhang *et al.*, 2010) and wadsleyite (Walker, 2010). Dipole models use periodic boundary conditions. Due to the geometry

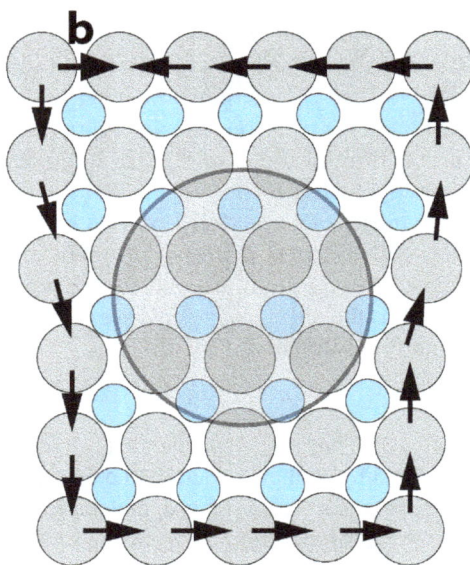

Figure 6. Schematic drawing of an edge dislocation viewed along the dislocation line. **b** represents the Burgers vector of the dislocation and the shaded circle highlights the dislocation core, *i.e.* the region in the immediate vicinity of the dislocation line where the theory of elasticity does not apply.

of a dislocation, the simulation cell has to contain at least two dislocations, which produces a 'dipole'. As in the case of point defects, the use of periodic boundary conditions allows us to perform standard DFT or MD simulations. However, this type of modelling results in a dense infinite sequence of dislocations and the interaction between dislocation cores needs to be accounted for (Walker *et al.*, 2010). An example for the dipole model is the investigation of bcc Fe screw dislocations (Clouet *et al.*, 2009, 2011). Recently, this approach has been applied to dislocations in $MgSiO_3$ bridgmanite at 30 GPa (Kraych *et al.*, 2016), and in $MgSiO_3$ post-perovskite at 120 GPa (Goryaeva *et al.*, 2015).

3.2.1. Gamma surfaces
A gamma surface is a theoretical concept, which is very important for understanding the plasticity of crystalline solids. A Generalized Stacking Fault (GSF) is produced when a rigid body shear (characterized by a vector **u**) is applied within a plane (Fig. 7, left). All possible vectors **u** within this plane are considered. As most configurations are unstable, the shear displacement **u** is imposed on the atoms, which can usually relax their positions in the direction perpendicular to the plane only. Each configuration is characterized by an excess energy, γ, which can be mapped as a function of the shear vector **u**; this is the so-called gamma surface. The example of Mg_2SiO_4 ringwoodite presented in Fig. 7, right, shows that a gamma surface allows us to find the easiest shear paths in a given plane. Here, for Mg_2SiO_4 ringwoodite in the (110) plane, it is found that shear is much easier along the $½[1\bar{1}0]$ direction than along any other direction within this plane (see in particular the shear path along [001]). The gamma surface is also an important ingredient in dislocation core modelling using the Peierls-Nabarro model as illustrated below.

3.2.2. Continuous dislocation models derived from atomic-scale simulations

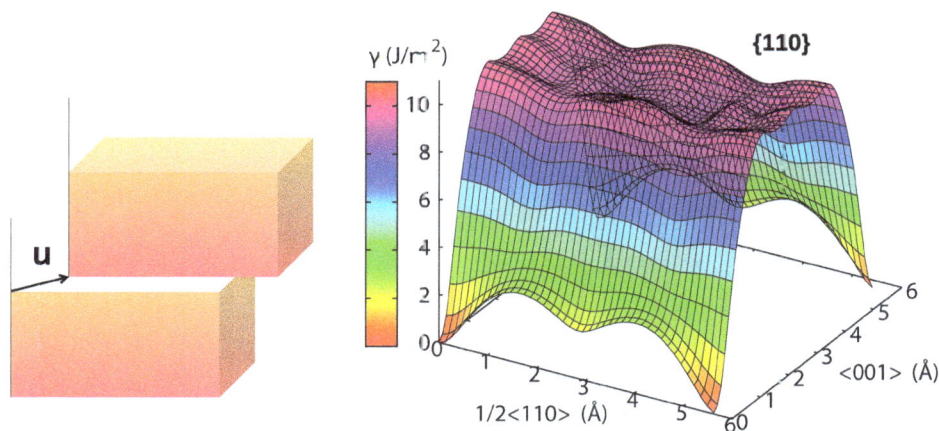

Figure. 7. Gamma surface. (*Left*) Geometry of the atomic system used for gamma surface calculations. The upper part of the crystal is sheared with respect to the lower part by a constant vector **u**. (*Right*) Example of a gamma surface calculated for the (110) plane in Mg_2SiO_4 ringwoodite at 20 GPa (modified after Ritterbex *et al.*, 2015).

During the construction of Volterra dislocations, material is cut into two parts, and then these two parts of the crystal separated by the cut relative to each other are moved by a translation vector. After that, the surfaces of the cut are welded together. Elastic deformation of the lattice in the region around the defect are allowed. Far from the dislocation line, the properties of a dislocation are well described by the theory of elasticity. However, the dislocation line is a singularity and all fields (stress, strain) are found to diverge when the dislocation line is approached. The aim of the Peierls-Nabarro model (Peierls, 1940; Nabarro, 1947) is, within a continuum approach, to describe the extension of the core of a dislocation opposed to the singularity that is associated with the Volterra dislocation (Carrez et al., 2007). The Peierls-Nabarro model assumes that the misfit region of inelastic displacement is restricted to a single plane where the dislocation will glide. Linear elasticity applies far from it. The dislocation corresponds to a continuous distribution of shear $S(x)$ along the glide plane (x is the coordinate along the displacement direction of the dislocation in the glide plane). $S(x)$ represents the disregistry across the glide plane.

The stress generated by such a displacement can be represented by a continuous distribution of infinitesimal dislocations with density $\rho(x)$ for which the total summation is equal to the Burgers vector $\mathbf{b}$. The restoring force, F, resulting from inelastic forces in the planar core is balanced by the stress resulting from the distribution of infinitesimal dislocations, which leads to the well known Peierls-Nabarro equation:

$$\frac{K}{2\pi} \int_{-\infty}^{+\infty} \frac{1}{x - x'} \left(\frac{dS(x')}{dx'} \right) dx' = \frac{K}{2\pi} \int_{-\infty}^{+\infty} \frac{\rho(x')}{x - x'} dx' = F(S(x)) \qquad (15)$$

where K, the energy coefficient, is a function of the dislocation character and the elastic stiffness coefficients. This coefficient can be calculated within the framework

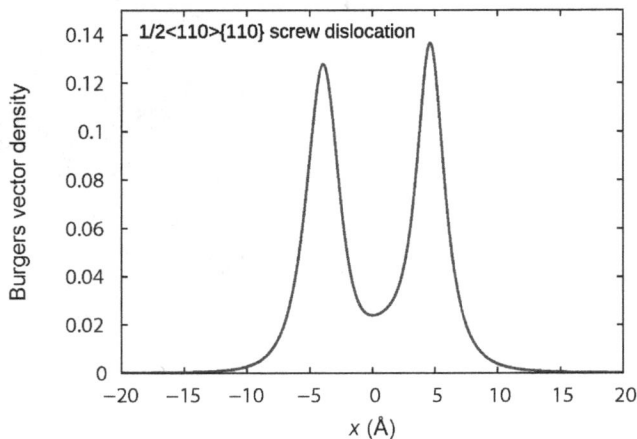

Figure. 8. Density of Burgers vectors corresponding to the Peierls-Nabarro model of the ½ <110>{110} screw dislocation in Mg$_2$SiO$_4$ ringwoodite (modified after Ritterbex et al., 2015).

of anisotropic elasticity. In the original model, an analytical solution of the Peierls-Nabarro equation is obtained by introducing a sinusoidal restoring force $F(S(x))$. A much more realistic description can be obtained if the restoring force introduced in the Peierls-Nabarro model is calculated from the gradient of the gamma surface calculated at the atomic scale (Vitek, 1968):

$$F(S) = -\text{grad}\gamma(S) \tag{16}$$

Figure 8 shows the result of the Peierls-Nabarro model for the ½<110>{110} screw dislocation in Mg_2SiO_4 ringwoodite. The dislocation core is found to spread into the {110} plane. The spreading is such that it leads to the formation of two partial dislocations separated by a stacking fault. This core structure was already suggested by the camel-hump shape of the gamma line along the <110> direction (Fig. 7, right) suggesting the presence of a stable stacking fault at 1/4<110> shear.

The Peierls-Nabarro equation (equation 15) is one-dimensional and as such can just describe colinear core spreading or dissociation. Further versions of the model are, however, able to describe non-colinear dissociations and even multi-planar core spreadings (Metsue *et al.*, 2010).

3.3. Grain and phase boundaries

Grain and phase boundaries are two-dimensional defects at the interfacial plane of two adjacent grains of the same or different types of crystals (see Petrishcheva and Abart, 2017b, Chapter 10, this volume). Usually, the atomic structure at the interface is more or less distorted, which leads to an enhanced grain boundary or interfacial energy compared to perfect crystals. As in the case of dislocations, atomic-scale modelling of grain and phase boundaries is challenging due to the constraints of periodic boundary conditions and the finite number of atoms in the simulation cell. For a general grain boundary, there are five macroscopic degrees of freedom: three describing the relative orientation of the two grains and two for the orientation of the grain boundary plane. At the atomic scale, there are additional degrees of freedom related to the termination layer of each grain at the interface, to the relative displacement of the two grains in the grain-boundary plane and to possible non-stoichiometry. Thus, the search for the lowest-energy structure of a specific grain boundary in complex oxides or silicates can be extremely tedious and is often restricted to a few special orientations.

Previous investigations of grain-boundary structures and energies in oxides include classical potential studies of MgO and other MO oxides at low and high P (Harris *et al.*, 1996, 1999; Harding *et al.*, 1999), Al_2O_3 (Buban *et al.*, 2006), (Ni,Fe)Cr_2O_4 spinel (Chartier *et al.*, 2013) and $Y_3Al_5O_{12}$ (YAG) (Aschauer *et al.*, 2008; Jiang *et al.*, 2012a,b) grain boundaries. DFT studies of grain boundaries have so far been restricted to relatively simple structures, such as MgO (McKenna and Shluger, 2009), ZnO (Körner *et al.*, 2011), Al_2O_3 (Fabris and Elsässer, 2001, 2003; Elsässer and Elsässer, 2005; Lei *et al.*, 2013), Cr_2O_3 (Van Der Geest *et al.*, 2013) and Fe_3O_4 (McKenna *et al.*, 2014). Grain boundaries in silicates are much less studied with most of the focus on forsterite, Mg_2SiO_4 (de Leeuw *et al.*, 2000; Gurmani *et al.*, 2011; Adjaoud *et al.*, 2012;

Ghosh and Karki, 2013). As an example, the study of Adjaoud *et al.* (2012) is presented in more detail.

Mg_2SiO_4 forsterite is the magnesium end-member of olivine, which is considered the major component of the Earth's upper mantle. The crystal-defect structures of forsterite have therefore been of special interest. Heinemann *et al.* (2005) studied synthetic bicrystals of forsterite using high-resolution transmission electron microscopy (TEM). The grains were misoriented by rotation about the crystallographic [100] direction which resulted in (0*kl*)/[100] symmetric tilt grain boundaries with tilt angles ranging between 9.9° and 21.5°. (0*kl*) describes the orientation of the crystals in the grain-boundary plane and the indices *k* and *l* depend on the tilt angle. Low-angle grain boundaries can be described as a periodic stacking of dislocations. With increasing tilt angle, the distance between individual dislocations decreases and eventually the dislocation cores overlap. This transition is described in the literature as the transition from low- to high-angle grain boundaries. Heinemann *et al.* (2005) investigated the tilt angle up to which individual dislocations could still be identified by high-resolution TEM. They found arrays of *c* dislocations (with Burgers vectors aligned with the crystallographic *c* direction) up to the largest tilt angle of 21.5° and concluded that, in forsterite, the transition from low- to high-angle grain boundaries happens at rather large tilt angles.

Using the classical potentials of Jahn and Madden (2007), Adjaoud *et al.* (2012) investigated a series of [100] symmetric tilt grain boundaries within a range of tilt angles from 9.6 to 60 degrees. The setup of the simulation cell included rotation of a supercell of forsterite, cutting the crystal in the desired orientation and sticking two bicrystals together. For the latter, the relative translation of the two bicrystals was chosen to obtain the lowest grain-boundary energy after full geometry relaxation. For small misorientations up to 22°, the resulting grain-boundary structures consisted of arrays of partial edge dislocations with Burgers vectors of 1/2 [001] associated with stacking faults. Their respective grain-boundary energies were described by the Read-Shockley dislocation model. For misorientation angles larger than 30°, the dislocation cores overlapped and formed repeated structural units (Fig. 9). Analysis of the atomic displacements relative to the bulk crystal suggested a spatial extent of the forsterite dislocation cores of ~2 nm. TEM image simulations using the MD structures (Adjaoud *et al.*, 2012) compared well with the TEM images taken by Heinemann *et al.* (2005).

3.3.1. Continuous modelling of grain boundaries

The crossover from atomic to continuum scale presented above can be used to analyse the structure of grain boundaries. Here we present, as an illustration, an application of this technique to the (011)/[100] tilt grain boundary with misorientation of 60.80° presented in Fig. 9. In Fig. 10 we present two components of the strain field, which show, in a narrow layer along the boundary, alternating zones of tension-compression. In this figure, the oxygen sublattice, which forms the skeleton of the forsterite structure, is used to build a continuous description.

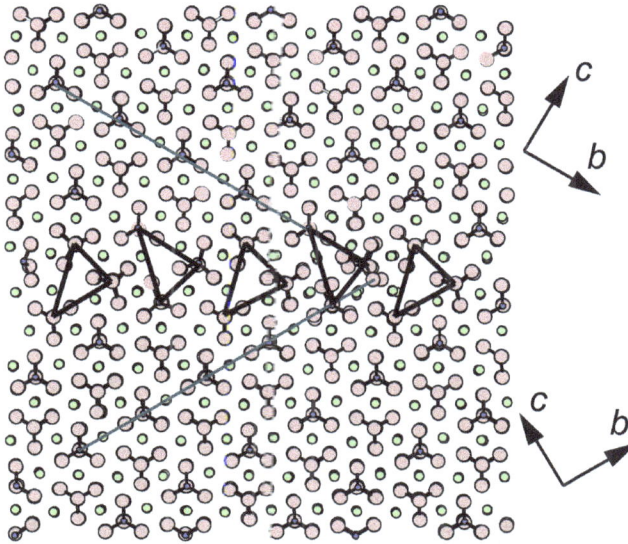

Figure 9. Structure of the large angle (011)/[100] symmetric tilt grain boundary of Mg_2SiO_4 forsterite from the study of (Adjaoud *et al.*, 2012). SiO_4 tetrahedra are drawn as balls and sticks, the Mg cations as small green balls. Structural units characterizing this kind of grain boundary are indicated by the black triangles. The tilt angle of ~60° is shown by the grey lines. Also shown are the crystallographic orientations of the two grains. The view is along the tilt axis [100].

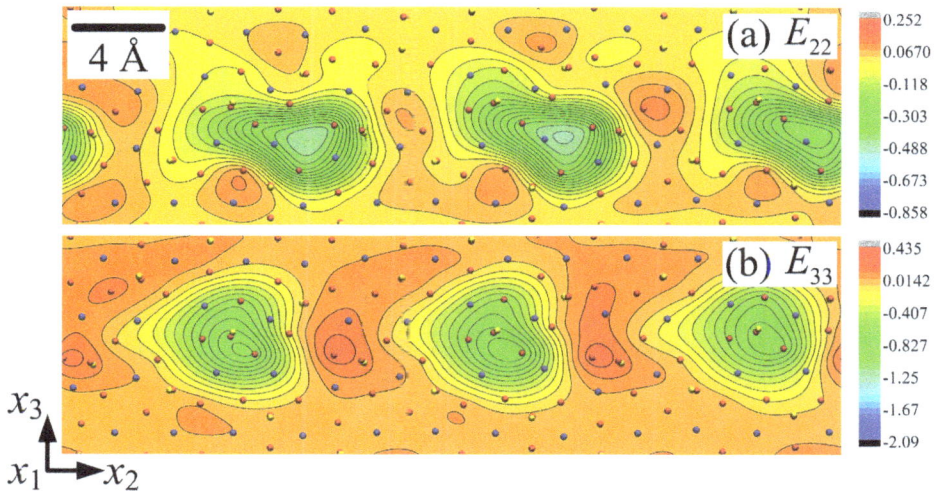

Figure 10. Strain fields E_{22} and E_{33} associated with the oxygen sublattice of the (011)/[100] tilt grain boundary in forsterite represented in Fig. 9 The atomic structure is represented by small spheres in the background.

Figure 11 shows the rotation vector field **ω** associated with this grain boundary. From its gradient, one can derive the so-called curvature tensor κ:

$$\kappa = \mathrm{grad}\,\omega \quad \text{or} \quad \kappa_{ij} = \mathrm{grad}\,\omega_{i,j} \tag{17}$$

In compatible elasticity, *i.e.* in the absence of defects, the curvature tensor would be curl-free. However, one notices in Fig. 11 that the rotation field exhibits very strong local variations at the atomic scale, which can be regarded as discontinuities (Fressengeas *et al.*, 2011, 2014). Plastic deformation does not result only from the dislocation motion. Dislocations are the crystal defects induced by translational lattice

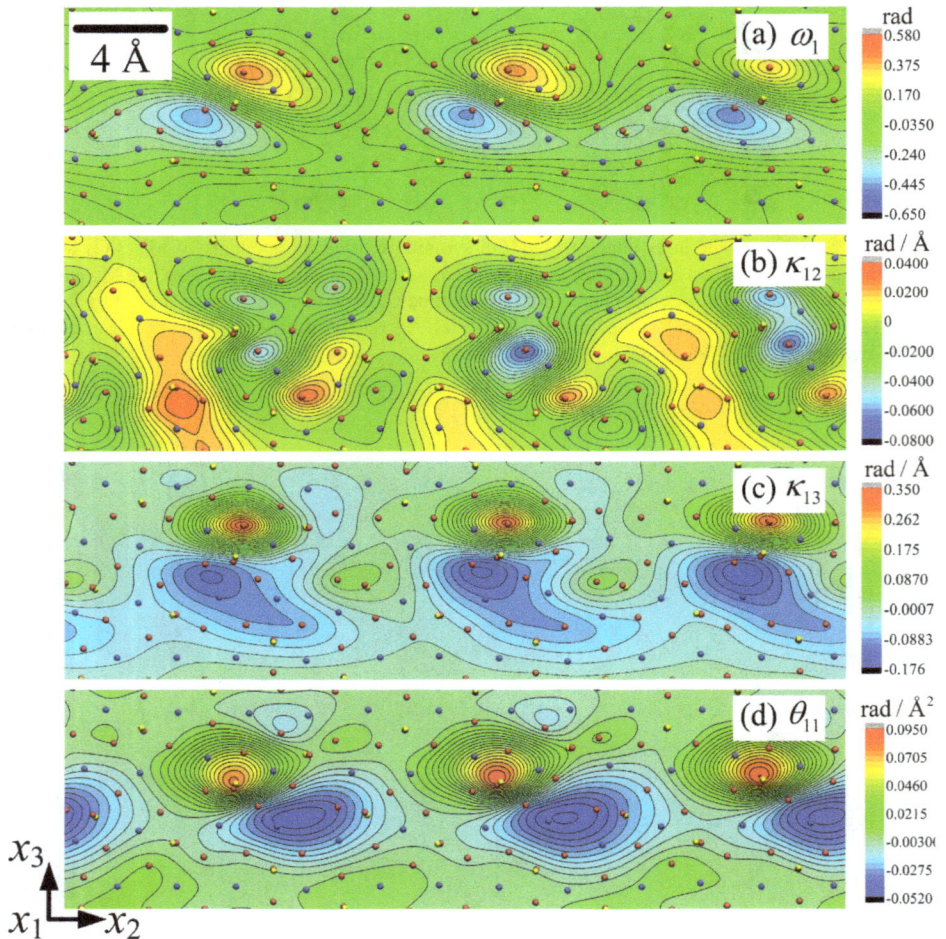

Figure 11. (011)/[100] tilt grain boundary in forsterite. All fields represented here have been calculated on the oxygen sublattice. (a) Rotation field ω_1. Curvature fields κ_{12} (b) and κ_{13} (c). (d) Disclination density field θ_{11}.

incompatibility, which can be measured by using Nye's dislocation density or the Burgers vector. For a finite volume of material, Nye's dislocation density tensor is defined by integrating along all dislocation lines

$$a_{ij} = \frac{1}{V} \int b_i t_j ds \qquad (18)$$

where V is the volume of the element, b is the Burgers vector, and t is the unit vector tangent to the dislocation line. Similarly, the rotational incompatibility can be related to defects, which are referred to as disclinations. Figure 12 illustrates the Volterra defect models. In Fig. 12a, a straight cut is made along a plane containing the axis of a cylinder in an elastic medium to create a defect surface. Then the surface of the cut is shifted. After the surfaces are welded, different deformation states are produced. There is only a limited and small number of possible independent cuts and shifts. All the other cuts plus some deformations can always be expressed as a linear superposition of the elementary cuts, which are shown in Figs 12b–g. If the shift is a translation, the defect is referred to as a dislocation (Figs.12b–d), the strength of which is the Burgers vector. If the shift is a pure rotation (Figs. 12e–g), the defect is referred to as a disclination, and the strength of the disclination is defined as the Frank vector. The curvature tensor is not curl-free and must contain an incompatible part, which is conveniently described by the continuous disclination density tensor:

$$\theta = \text{curl}\kappa \quad \text{or} \quad \theta_{ij} = e_{jkl}\kappa_{jl,k} \qquad (19)$$

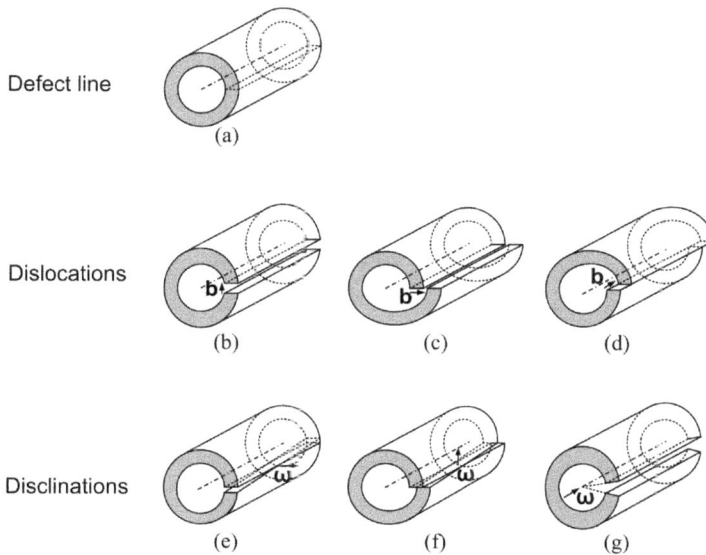

Figure 12. Volterra's distortions. (a) Defect line and cut surface in a reference cylinder. (b, c, d) Dislocations indicate the discontinuities of displacement. (e, f, g) Wedge disclinations show the discontinuities of rotation.

Figure 11d shows that the forsterite (011)/[100] tilt grain boundary can be described by a periodic distribution of disclination dipoles confined in a thin 8 Å layer along the boundary.

4. Modelling of self-diffusion

4.1. Atomic-scale mechanisms of self-diffusion

Basic concepts and definitions of diffusion are introduced in Chapter 9 of this volume, Petrishcheva and Abart (2017a). Molecular modelling approaches are usually concerned with the mobility of individual atoms, *i.e.* with self-diffusion. In crystals, self-diffusion results from a series of random atomic jumps between structural sites in the lattice. Several atomic mechanisms for diffusion have been proposed in the literature, *e.g.* Mehrer (2007; Fig. 13A):

- *vacancy mechanism:* the most common mechanism in solids, diffusion by atomic jumps into neighbouring vacant lattice sites
- *indirect interstitial mechanism:* an interstitial atom replaces a substitutional atom on a regular lattice site; in turn the latter takes the place of another regular atom, pushing it into an interstitial site
- *direct interstitial mechanism:* foreign atoms diffuse by jumping between interstitial sites in an otherwise perfect host crystal
- exchange mechanism without involving vacancies: *e.g.* octahedral rotation

As minerals often have complex crystal structures, the identification of the relevant diffusion mechanisms for the specific case as well as theoretical calculations are still

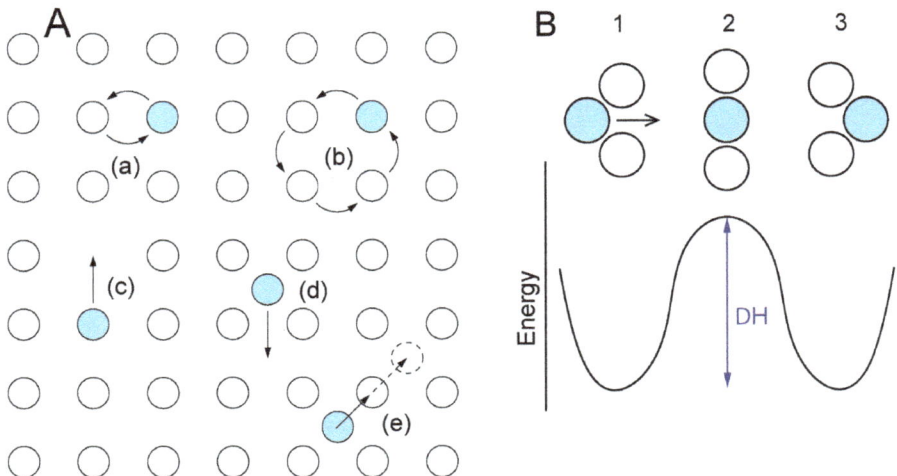

Figure 13. (A) Illustration of possible diffusion mechanisms: (a–b) particle exchange, (c) vacancy mechanism, (d) direct and (e) indirect interstitial mechanism. (B) Three stages of migration of the coloured atom from one site to another: (1) starting position, (2) intermediate position at the maximum of the energy barrier where the lattice has to be distorted to allow passing of the diffusive particle, (3) final position.

quite challenging. In general, there may be more than one diffusion mechanism active at a time. Whatever the mechanism of diffusion is, to initiate the diffusive jump from one site to another, an energy barrier, ΔH, has to be overcome (Fig. 13A,B). ΔH is referred to as the enthalpy of migration. The required activation energy to overcome the barrier comes from the thermal motion of the atoms. Together with the enthalpy of defect formation ΔH_d, which was introduced in section 3.1, the migration enthalpy is related to the self-diffusion coefficient D_S by (Poirier, 1985)

$$D_S = D_0\exp(-(\Delta H + \Delta H_d)/k_BT) \qquad (20)$$

with D_0 being a pre-exponential factor that will be discussed below in more detail.

4.2. Self-diffusion from molecular dynamics simulations

Diffusion processes with self-diffusion coefficients larger than $\sim 10^{-11}-10^{-10}$ m^2/s can be modelled directly in MD simulations. This applies in particular to melts and fluids where atoms are usually much more mobile than in crystals or glasses. There are also examples for direct MD simulations of ion self-diffusion in crystals at high temperatures, *e.g.* for fast ion conductors (*e.g.* Wilson *et al.*, 2004; Foy and Madden, 2006; Adams and Rao, 2012). Some example studies are discussed below.

In practice, the self-diffusion coefficient, D_S, of an atom is derived from the slope of its mean square displacement, MSD, plotted over time using the Einstein relation (Allen and Tildesley, 1987; Frenkel and Smit, 2002)

$$D_S = \frac{1}{6t}\langle r^2(t)\rangle \qquad (21)$$

The MSD is the averaged square of the distance r travelled by a particle in a given time t. In case of bulk systems (melts or single crystals), D_S is often computed as an average over all atoms or ions of the same element. However, the simulations also allow us to follow individual atoms in the course of the simulation, *e.g.* along grain boundaries. This approach is similar to measurements of tracer diffusion. A linear relation between t and MSD is only reached in the diffusive regime at long enough times, where the exact time requirement depends on the actual mobility of the diffusing particle. The factor six in equation 21 is valid for three-dimensional diffusion. It would have to be replaced by a factor two for linear and four for two-dimensional diffusion (Frenkel and Smit, 2002).

An instructive example of solid-state diffusion relates to Na diffusion in cryolite, Na_3AlF_6. This mineral crystallizes in a double perovskite structure with alternating Na and Al in octahedral positions. The remaining Na ions are eight-fold coordinated. At $T = 567°C$, the crystal structure changes from monoclinic ($P2_1/n$) to cubic ($Fm\bar{3}m$). This phase transition is related to a significant increase in ionic conductivity (Landon and Ubbelohde, 1957) and Na self-diffusion measured by quasi-elastic neutron scattering (Jahn *et al.*, 2008; Demmel *et al.*, 2009).

Using a polarizable ion potential in conjunction with classical MD, Foy and Madden (2006) studied the ion mobility in cryolite as a function of temperature. Here, we repeated some of these simulations using the same potential to illustrate two processes. The first MD simulation was performed with a stoichiometric crystal at high T (727°C). As shown in Fig. 14a, the MSD of the fluoride ion rises up to ~24 a.u. (*i.e.* ~6.5 Å^2, a.u. refers to atomic units) after a 10 to 15 picosecond MD run. The MSDs for Na and Al remain below 4 a.u. (1 Å^2) even after tens of picoseconds. Remarkably, all three MSDs reach a plateau at simulation times above 15 ps. As the slope of these curves in the long time limit seem to be close to zero, there is no significant self-diffusion. The small MDSs of the Na and Al ions are simply explained by the vibrational motion of these ions. Sodium ions are thereby displaced more from their original position than Al ions due to the weaker Na^+-F^- interaction compared to $Al^{3+}-F^-$. Fluorine anions are displaced much further, *i.e.* on average by ~2.5 Å. It turns out that the large MSD is due to reorientations and rotations of AlF_6 octahedra (Foy and Madden, 2006), but Al−F bonds remain essentially intact in the course of this simulation of only 20 ps.

The picture changes significantly if a NaF Schottky defect is introduced (Fig. 14b). In this case, the MSD of Na quickly reaches a linear slope, which indicates a translational motion of the Na ions. Al ions remain immobile while the F anions seem to diffuse slightly. Looking at the MD trajectories (*i.e.* the evolution of atomic positions during the MD simulation), a jump diffusion mechanism of Na ions becomes apparent. Na ions jump between octahedral and eight-fold coordinated positions into the vacancy sites. The anions also jump from one octahedron to the other but less frequently due to the stronger bond to the Al ions. Hence, in the case of cryolite, the large self-diffusion and the related large ionic conductivity (Landon and Ubbelohde, 1957; Jahn et al., 2008) can be explained by the presence of vacancies in the structure.

From studies of the atomistic diffusion mechanisms in melts it is known that the simple model of binary collisions of atoms followed by periods of free motion fails to

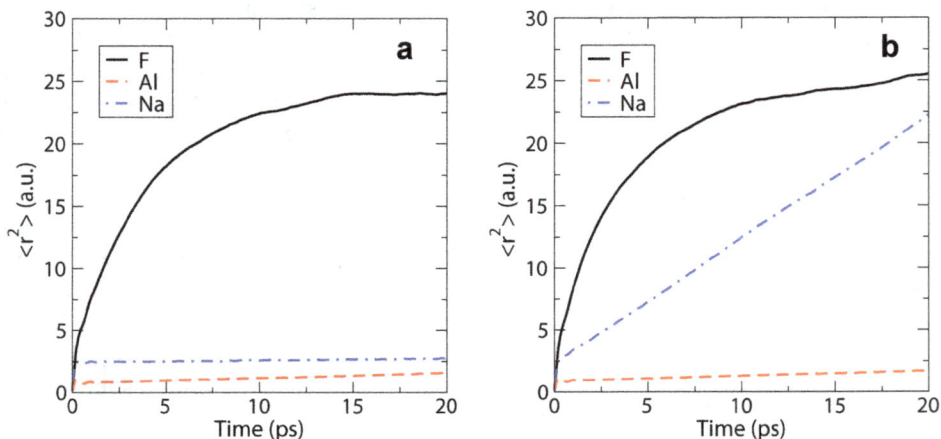

Figure 14. Mean square displacements of Na^+, Al^{3+} and O^{2-} in stoichiometric cryolite, Na_3AlF_6 (a) and in cryolite with a NaF Schottky defect, $Na_{2.94}AlF_{5.94}$ (b). The MSDs are given in atomic units ($Bohr^2$).

reproduce measured diffusion constants. Correlation effects due to the atomic interactions have to be taken into account. Atoms may be trapped in a 'cage' formed by their surrounding atoms. Due to the significant atomic mobility, these cages are formed and destroyed continuously. In glasses, the liquid structure is basically frozen in. Diffusion is again dominated by atomic jumps. However, the free volume in a glass is usually much larger than in a crystal. Because there are no lattice sites, the jump distances and directions are much more isotropic. As an illustration for self-diffusion in melts we show the dependence of D_S on composition for the binary system $CaO-Al_2O_3$ (Fig. 15). Again, a polarizable ion model (Jahn and Madden, 2007) was used for the classical MD simulations. The structure models predicted for these melts seem to be realistic, which is concluded from the very good agreement between measured and computed total neutron and X-ray structure factors (Drewitt *et al.*, 2011).

Self-diffusion coefficients along melt-wetted grain boundaries of Mg_2SiO_4 forsterite were studied by Gurmani *et al.* (2011). Melt layers of $MgSiO_3$ composition and a variable thickness between 0.8 and 7.0 nm were confined between slabs of forsterite with different surface terminations. This system was set up as a simplified model system of a partially molten peridotite. Experimental evidence for ultra-thin melt layers between olivine grains was given, *e.g.* by Wirth (1996) and de Kloe *et al.* (2000). Classical MD simulations using the ionic potential of Jahn and Madden (2007) were performed at ambient pressure and a constant temperature of 2000 K (see snapshot in Fig. 16). The goal of this study was to investigate how the ionic mobility depends on the thickness of the melt layer and on the crystal orientation. Self-diffusion coefficients were obtained using the mean square displacements of the atoms in the centre of the melt layer only. Results are shown in Fig. 16. The strongest dependence of the oxygen self-diffusion coefficient on melt thickness is observed for melt layers of <3 nm. For the crystal surface termination (010), which has the lowest surface energy, the reduction of the self-diffusion coefficient is least pronounced. Modelling of diffusion in dry grain boundaries using the direct MD approach is much more challenging due to the significantly reduced atomic mobility. Increasing the

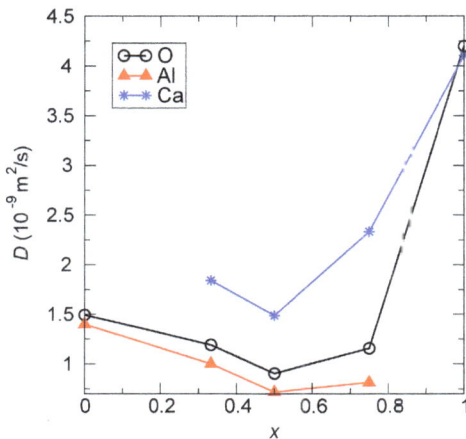

Figure 15. Self-diffusion coefficients of $(CaO)_x-(Al_2O_3)_{1-x}$ obtained from classical MD simulations at $T = 2500$ K and ambient pressure (modified from Drewitt *et al.*, 2011).

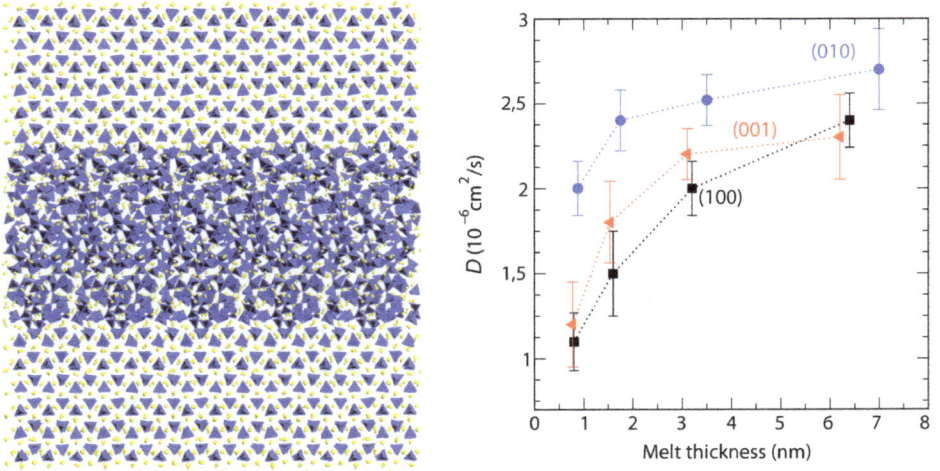

Figure 16. (*Left*) Snapshot from a simulation of a melt-wetted grain boundary of forsterite, Mg_2SiO_4. The composition of the melt layer is close to $MgSiO_3$. SiO_4 tetrahedra are shown in blue, whereas Mg ions are drawn as yellow balls. The melt wets the (010) surface of forsterite and the view is along [100]. (*Right*) Dependence of oxygen self-diffusion coefficient in the melt as a function of crystal orientation and of melt thickness.

temperature may be one option to accelerate the dynamics but diffusion mechanisms may change with temperature. As a consequence, the simulations may not sample the transport properties of interest correctly. An alternative could be to couple the MD simulation with the metadynamics technique, see *e.g.* the study of Ag and Ce diffusion in SiC grain boundaries (Rabone *et al.*, 2014).

4.3. Self-diffusion coefficients of crystalline phases

With the exception of solid-state ionic conductors (see examples in the previous section), diffusion coefficients in solids are far too small to allow direct investigation of the diffusion process by MD simulations. Instead, one has to evaluate the individual contributions to the diffusion process separately. In crystals, diffusion is usually described by a sequence of individual jumps of atoms or vacancies between sites in the crystal structure. The self-diffusion coefficient is then defined as (Poirier, 1985)

$$D_S = n_d \frac{Z}{6} l^2 \Gamma \qquad (22)$$

where $n_d = N_d/N$ is the fraction of point defects (see equation 14) that contribute to the diffusion process, Z is a geometrical factor incorporating the number of possible jumps, l is the jump distance and Γ is the jump frequency. Both Γ and l depend in the crystallographic direction, which often leads to an anisotropy of the self-diffusion coefficient. The jump frequency Γ is given by

$$\Gamma = \nu \exp\left(-\frac{\Delta G}{k_B T}\right) = \nu \exp\left(\frac{\Delta S}{k_B}\right) \exp\left(-\frac{\Delta H}{k_B T}\right) \tag{23}$$

with ν being the attempt frequency that describes how often the diffusing atom tries to cross the diffusion barrier. ΔG, ΔH and ΔS are the Gibbs energy, enthalpy and entropy of migration, respectively. The enthalpy of migration ΔH is obtained from the difference of the saddle point energy at the top of the barrier and the ground state energy before the jump (see Fig. 13B). Finding the exact saddle point is not trivial for complex 3-dimensional structures such as silicates (*e.g.* Ammann *et al.*, 2009, 2010a).

The climbing image nudged elastic band approach (Henkelman *et al.*, 2000) has been shown to be more reliable than previous attempts by trial and error or by a grid searching method. The method works as follows. First, a virtual line is drawn between the initial and final ground state positions of the diffusing atom. Then, the atom is placed on a number of possible positions on this line and corresponding energies and forces are computed. Subsequently, the position of the line is optimized in order to minimize the forces acting on the diffusing atom at each point of the line. Eventually, the minimum energy pathway between the initial and the final ground state position is sampled out (Ammann *et al.*, 2010a).

The attempt frequency ν is evaluated together with the migration entropy using harmonic transition-state theory (Vineyard, 1957). The method requires us to compute the $3N$ vibrational frequencies f_i (here, N is the number of atoms in the simulation cell) at the equilibrium point and the $3N-1$ non-negative vibrational frequencies f_i' with the diffusing atom placed at the saddle point. The product of the first two terms of equation 23 is then given by (Ammann *et al.*, 2010a)

$$\nu^* = \nu \exp\left(\frac{\Delta S}{k_B}\right) = \frac{\prod_{j=1}^{3N} f_j}{\prod_{j=1}^{3N-1} f_j'} \tag{24}$$

If the saddle point is determined well enough there should be only one negative frequency due to the atom sitting at the saddle point. Vibrational spectra are derived from the second derivative of the total energy with respect to atomic displacements. In the case of pairwise interaction potentials this calculation may be done analytically or by computing energies after finite displacement of each atom in the three cartesian directions. Some DFT codes rely also on the finite displacement method and others use density functional perturbation theory (Baroni *et al.*, 2001).

The transition-state theory approach in conjunction with the nudged elastic band method and DFT calculations has been applied to predict diffusion coefficients of the Mg end members of the lower mantle minerals $MgSiO_3$ bridgmanite and post-brigmanite as well as MgO periclase (Vocadlo *et al.*, 1995; Ammann *et al.*, 2010b), of ferrous iron in (Mg,Fe)O ferro-periclase (Ammann *et al.*, 2011) and of Al diffusion in MgO periclase (Ammann *et al.*, 2012). This method was also used to study lattice and grain boundary diffusion in α-Al_2O_3 (Lei *et al.*, 2013). With a somewhat different approach, classical potential and combined DFT/classical calculations were used to investigate the oxygen and magnesium diffusion in forsterite (Walker *et al.*, 2003, 2009). A kinetic Monte-Carlo method was used to study grain-boundary diffusion in NiO and Al_2O_3 (Harding and Harris, 2001).

5. Mechanisms of structural phase transitions and shear deformation

5.1. Displacive structural phase transitions

Some pressure- or temperature-induced phase transformations may be investigated directly by MD simulations. This approach is especially suited for displacive phase transitions where the initial and the final structures are related by a rather small re-arrangement of atoms and where no strong bonds need to be broken. So far, most studies of this kind have been performed using classical MD. Examples include transitions in carbonates (Liu *et al.*, 2001; Kawano *et al.*, 2009) and pyroxenes (Miyake *et al.*, 2002; Shimobayashi *et al.*, 2001; Jahn and Madden, 2007). Technically, the MD simulations follow a protocol that is similar to the experimental one. They are usually run in the *NPT* ensemble. Depending on the type of phase transition, temperature or pressure (or both) are re-set at regular intervals using, *e.g.* a temperature and/or pressure ramp. Of course, the rates of $T{-}P$ change in the simulations have to be much higher than in experiments due to the accessible simulation time scale. Similar to experiments, the phase transitions are often subject to hysteresis effects, especially if they happen at low temperatures.

As an example, Fig. 17 shows the compression and decompression curves of a classical MD simulation of clinoenstatite $MgSiO_3$ at room temperature (Jahn and Madden, 2007). On the compression path, low clinoenstatite transforms into high-pressure clinoenstatite at ~10 GPa. The transition is accompanied by a drop in volume of ~2%. The reverse transition is observed during decompression at ~1 GPa. This significant hysteresis of $\Delta P = 9$ GPa is reduced to ~1 GPa when the temperature of the simulation is increased to 1000 K. The mean transition pressure observed in the simulations is between 5 and 6 GPa. Experimentally, the transition was bracketed between 5.3 and 8.0 GPa at room temperature (Angel *et al.*, 1992).

5.2. Shear-induced deformation processes und phase transformations

Unlike the displacive transitions described above, most other structural phase transitions as well as non-elastic deformation processes require a more or less significant re-organization of the atomic structure. In some cases, however, only part of the structure has to be re-built, *e.g.* by changing the stacking sequence of close-packed atomic layers. Unique insight into the mechanisms of such martensitic-type phase transitions and of some shear deformation processes may be obtained by applying external stresses to the simulation cell. A successful approach in this respect is a combination of conventional MD with metadynamics, which was introduced briefly in the methods section above. In this case, the dynamic variables of metadynamics are the six parameters of the simulation cell, *i.e.* the three cell lengths and the three cell angles. At each metadynamics step, short MD simulations are performed in the *NVT* ensemble

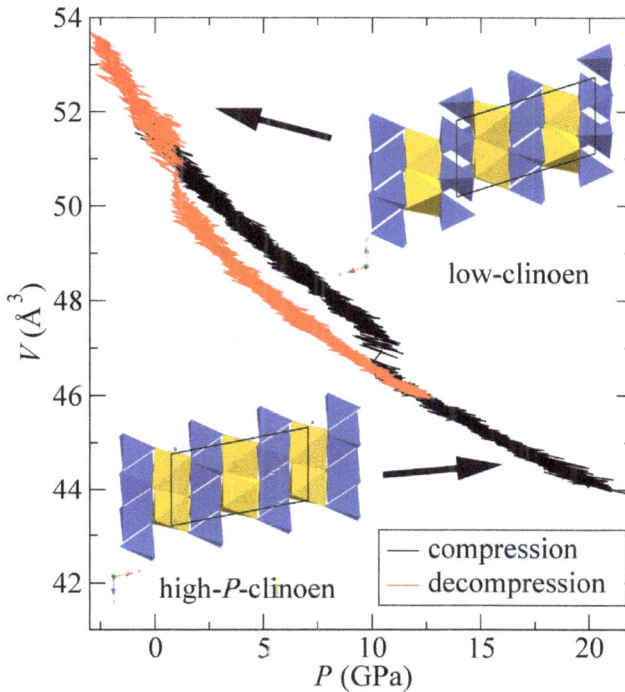

Figure 17. Evolution of cell volume during compression and decompression of $MgSiO_3$ clinoenstatite from classical MD simulations at 300 K (Jahn and Madden, 2007). The jump in volume is related to the low- to high-P clinoenstatite phase transition. Due to kinetic effects, a hysteresis of the transition pressure is observed.

to relax the atomic positions and to obtain an average stress tensor of this specific configuration, which is required to derive the simulation cell parameters of the following metadynamics step. While filling up a local energy valley (see Fig. 2), the structure is deformed elastically, *i.e.* it returns to the original structure at the energy minimum once the stress is set back to hydrostatic conditions. When the barrier to another energy minimum is crossed, the atomic structure rearranges in a non-elastic way, *i.e.* it relaxes to a different structure when the stress is released.

Metadynamics simulations in conjunction with DFT calculations showed possible transition paths between polymorphs of important oxides and silicates, including the $MgSiO_3$ bridgmanite to post-bridgmanite transition (Oganov *et al.*, 2005) as well as transitions between SiO_2 (Martoňák *et al.*, 2006, 2007; Donadio *et al.*, 2008) and H_2O ice (Martoňák *et al.*, 2005a) polymorphs. In metadynamics simulations combined with classical MD, the phase transitions between $MgSiO_3$ ortho-, proto- and clinoenstatites were investigated (Jahn and Martoňák, 2008, 2009; Jahn, 2010). Recently, the same method was used in a combined experimental and computational study that found two new high-pressure phases of Mg_2SiO_4 (Finkelstein *et al.*, 2014). Due to the auxiliary

Gaussian energies added to the system in the course of the metadynamics simulation, some information about the height and width of the transition barrier is obtained. Also, intermediate and metastable stacking faults that may be relevant to the transition mechanism, are frequently observed. Zahn (2013) used transition path sampling MD simulations to explore the rate of the $MgSiO_3$ bridgmanite to post-bridgmanite transition and concluded that once the transition starts it proceeds rapidly with a velocity similar to the speed of sound.

If the metadynamics simulations are performed at pressures and temperatures where no phase transition to a different phase takes place (for thermodynamic or kinetic reasons), the crystal in the simulation box may be deformed by homogeneous shear, *i.e.* the final structure after deformation is identical to the initial structure but the cell parameters have changed according to the shear vectors. As an example, Fig. 18 illustrates the two-stage shear deformation mechanism of Mg_2SiO_4 forsterite observed in metadynamics simulations. The simulation cell contained 672 atoms and the ionic interactions were described by an advanced polarizable potential (Jahn and Madden, 2007). Pressure and temperature were set to 15 GPa and 2000 K. During the first metadynamics steps the simulation cell was deformed elastically. At some point, a shear in [001] direction in the (100) plane was observed. The respective displacement vector was about half of the c lattice parameter. As a result, a stacking fault in the (100) plane was created (Fig. 18, centre). A second shear happened a few metadynamics steps later. The stacking fault was removed and the perfect crystal structure of forsterite was obtained again. The total displacement of the two half-crystals in the shear plane corresponded to one unit cell in the [001] direction. During the shear, SiO_4 tetrahedra remained intact.

While shear of an infinite crystal is not realistic by itself, the activation of a certain shear deformation seems to correspond well to activation of the respective slip systems observed in experiments, which for olivine at high pressures are dominated by slip in [001]($hk0$) (*e.g.* Mainprice *et al.*, 2005). Starting from the shear mechanism found in

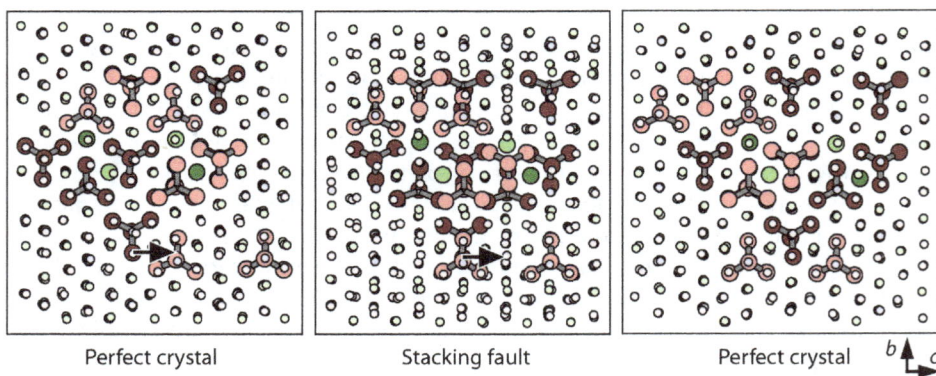

Perfect crystal   Stacking fault   Perfect crystal

Figure 18. Mechanism of two-step shear deformation of Mg_2SiO_4 forsterite during homogeneous shear in [001](100). The view is along [100] and the shear is in the horizontal direction. Displacement vectors in the shear plane are indicated by arrows. For better visualization, a few SiO_4 tetrahedra above the shear plane are drawn as light, large circles and tetrahedra below the shear plane are shown as dark large circles.

the metadynamics simulation (which is not necessarily realistic), transition path sampling could provide further insight into the detailed path and rate of the deformation mechanism. Further, it will probably be necessary to include a detailed account of dislocation cores and to model the mechanism of dislocation glide with more confidence. Explicit modelling of dislocation climb in molecular simulations has been performed using a kinetic Monte-Carlo approach for body-centred cubic iron (Kabir *et al.*, 2010) and found to to consistent with continuum theory (Clouet, 2011). Similar advanced methods have also been used to study grain-boundary motion in metals (*e.g.* Mishin *et al.*, 2007; Deng and Schuh, 2011). Eventually, molecular simulations provide unique information to parameterize advanced continuum models, which was demonstrated recently for MgO (Cordier *et al.*, 2012). The application of such multi-scale approaches to silicates or other complex Earth materials will be a formidable task for the future and may transform *e.g.* our understanding of mantle rheology and other deformation processes in the Earth.

Acknowledgements

The authors thank Patrick Cordier for useful discussions and critical comments. Andrew Walker and Rainer Abart provided thoughtful reviews. Financial support from DFG in the framework of the Research Unit FOR741 'Nanoscale processes and geomaterials properties' is gratefully acknowledged.

References

Adams, S. and Rao, R.P. (2012) Ion transport and phase transition in $Li_{7-x}La_3(Zr_{2-x}M_x)O_{12}$ (M = Ta^{5+}, Nb^{5+}, *x*=0, 0.25). *Journal of Materials Chemistry*, **22**, 1426−1434.

Adjaoud, O., Marquardt, K. and Jahn, S. (2012) Atomic structures and energies of grain boundaries in Mg_2SiO_4 forsterite from atomistic modeling. *Physics and Chemistry of Minerals*, **39**, 749−760.

Alfe, D. and Gillan, M.J. (2005) Schottky defect formation energy in MgO calculated by diffusion Monte Carlo. *Physical Review B*, **71**, 220101(R).

Allen, M.P. and Tildesley, D.J. (1987) *Computer Simulations of Liquids*. Oxford University Press, Oxford, UK.

Ammann, M.W., Brodholt, J.P. and Dobson, D.P. (2009) DFT study of migration enthalpies in $MgSiO_3$ perovskite. *Physics and Chemistry of Minerals*, **36**, 151−158.

Ammann, M.W., Brodholt, J.P. and Dobson, D.P. (2010a) Simulating diffusion. Pp. 201−224 in: *Theoretical and Computational Methods in Mineral Physics: Geophysical Applications* (R. Wentzcovitch and L. Stixrude, editors). Reviews in Mineralogy and Geochemistry, **71**. Mineralogical Society of America and the Geochemical Society. Chantilly, Virginia, USA.

Ammann, M.W., Brodholt, J.P., Wookey, J. and Dobson, D.P. (2010b) First-principles constraints on diffusion in lower-mantle minerals and a weak D'' layer. *Nature*, **465**, 462−465.

Ammann, M.W., Brodholt, J.P. and Dobson, D.P. (2011) Ferrous iron diffusion in ferro-periclase across the spin transition. *Earth and Planetary Science Letters*, **302**, 393−402.

Ammann, M.W., Brodholt, J.P. and Dobson, D.P. (2012) Diffusion of aluminium in MgO from first principles. *Physics and Chemistry of Minerals*, **39**, 503−514.

Angel, R.J., Chopelas, A. and Ross, N.L. (1992) Stability of high-density clinoenstatite at upper-mantle pressures. *Nature*, **358**, 322−324.

Artacho, E., Anglada, E., Dieguez, O., Gale, J.D., Garcia, A., Junquera, J., Martin, R.M., Ordejon, P., Pruneda, J.M., Sanchez-Portal, D. and Soler, J.M. (2008) The SIESTA method; developments and applicability. *Journal of Physics: Condensed Matter*, **20**, 064208.

Aschauer, U., Bowen, P. and Parker, S.C. (2008) Surface and mirror twin grain boundary segregation in

Nd:YAG: An atomistic simulation study. *Journal of the American Ceramic Society*, **91**, 2698–2705.

Balan, E., Blanchard, M., Yi, H. and Ingrin, J. (2013) Theoretical study of OH-defects in pure enstatite. *Physics and Chemistry of Minerals*, **40**, 41–50.

Baroni, S., de Gironcoli, S., Dal Corso, A. and Giannozzi, P. (2001) Phonons and related crystal properties from density-functional perturbation theory. *Reviews of Modern Physics*, **73**, 515–562.

Blanchard, M., Wright, K. and Gale, J.D. (2005) A computer simulation study of OH defects in Mg_2SiO_4 and Mg_2GeO_4 spinels. *Physics and Chemistry of Minerals*, **32**, 585–593.

Blanchard, M., Balan, E. and Wright, K. (2009) Incorporation of water in iron-free ringwoodite: A first-principles study. *American Mineralogist*, **94**, 83–89.

Bowler, D.R. and Miyazaki, T. (2012) O(n) methods in electronic structure calculations. *Reports on Progress in Physics*, **75**, 036503.

Braithwaite, J.S., Sushko, P.V., Wright, K. and Catlow, C.R.A. (2002) Hydrogen defects in forsterite: A test case for the embedded cluster method. *Journal of Chemical Physics*, **116**, 2628–2635.

Braithwaite, J.S., Wright, K. and Catlow, C.R.A. (2003) A theoretical study of the energetics and IR frequencies of hydroxyl defects in forsterite. *Journal of Geophysical Research*, **108**, doi:10.1029/2002JB002126.

Brodholt, J. (1997) Ab initio calculations on point defects in forsterite (Mg_2SiO_4) and implications for diffusion and creep. *American Mineralogist*, **82**, 1049–1053.

Buban, J.P., Matsunaga, K., Chen, J., Shibata, N., Ching, W.Y., Yamamoto, T. and Ikuhara, Y. (2006) Grain boundary strengthening in alumina by rare earth impurities. *Science*, **311**, 212–215.

Car, R. and Parrinello, M. (1985) Unified approach for molecular dynamics and density functional theory. *Physical Review Letters*, **55**, 2471–2474.

Carrez, P., Ferré, D. and Cordier, P. (2007) Peierls stresses for dislocations in $MgSiO_3$ post-perovskite calculated at 120 GPa from first principles. *Philosophical Magazine*, **87**, 1–19.

Catlow, C.R.A., James, R., Mackrodt, W.C. and Steward, R.F. (1982) Defect energetics in α-Al_2O_3 and rutile TiO_2. *Physical Review B*, **25**, 1006–1026.

Chartier, A., Golovchuk, B., Gossé, S. and Van Brutzel, L. (2013) Disordering and grain boundaries of $(Ni,Fe)Cr_2O_4$ spinels from atomistic calculations. *Journal of Chemical Physics*, **139**, 134702.

Clouet, E. (2011) Predicting dislocation climb: Classical modeling versus atomistic simulations. *Physical Review B*, **84**, 092106.

Clouet, E., Ventelon, L. and Willaime, F. (2009) Dislocation core energies and core fields from first principles. *Physical Review Letters*, **102**, 055502.

Clouet, E., Ventelon, L. and Willaime, F. (2011) Dislocation core field. II. Screw dislocation in iron. *Physical Review B*, **84**, 224107.

Cococcioni, M. (2010) Accurate and efficient calculations on strongly correlated minerals with the LDA+U method. Pp. 147–167 in: *Theoretical and Computational Methods in Mineral Physics: Geophysical Applications* (R. Wentzcovitch and L. Stixrude, editors). Reviews in Mineralogy and Geochemistry, **71**. Mineralogical Society of America and the Geochemical Society, Chantilly, Virginia, USA.

Cordier, P., Amodeo, J. and Carrez, P. (2012) Modelling the rheology of MgO under Earth's mantle pressure, temperature and strain rates. *Nature*, **481**, 177–181.

Cygan, R.T. and Kubicki, J.D. (editors) (2001) *Molecular Modeling Theory: Applications in the Geosciences.* Reviews in Mineralogy and Geochemistry, **42**. Mineralogical Society of America and the Geochemical Society, Chantilly, Virginia, USA.

de Kloe, R., Drury, M.R. and van Roermund, H.L.M. (2000) Evidence for stable grain boundary melt films in experimentally deformed olivine-orthopyroxene rocks. *Physics and Chemistry of Minerals*, **27**, 480–494.

de Leeuw, N.H., Parker, S.C., Catlow, C.R.A. and Price, G.D. (2000) Proton-containing defects at forsterite 010 tilt grain boundaries and stepped surfaces. *American Mineralogist*, **85**, 1143–1154.

Dellago, C., Bolhuis, P.G., Csajka, F.S. and Chandler, D. (1998) Transition path sampling and the calculation of rate constants. *Journal of Chemical Physics*, **108**, 1964–1977.

Demmel, F., Seydel, T. and Jahn, S. (2009) Sodium diffusion in cryolite at elevated temperatures studied by quasielastic neutron scattering. *Solid State Ionics*, **180**, 1257–1260.

Deng, C. and Schuh, C.A. (2011) Diffusive-to-ballistic transition in grain boundary motion studied by atomistic simulations. *Physical Review B*, **84**, 214102.

Dick, B.G. and Overhauser, A.W. (1958) Theory of the dielectric constants of alkali halide crystals. *Physical Review*, **112**, 90–103.

Donadio, D., Martoňák, R., Raiteri, P. and Parrinello, M. (2008) Influence of temperature and anisotropic pressure on the phase transitions in α-cristobalite. *Physical Review Letters*, **100**, 165502.

Dovesi, R., Orlando, R., Erba, A., Zicovich-Wilson, C.M., Civalleri, B., Casassa, S., Maschio, L., Ferrabone, M., De La Pierre, M., D'Arco, P., Noel, Y., Causa, M., Rerat, M. and Kirtman, B. (2014) CRYSTAL14: A program for the ab initio investigation of crystalline solids. *International Journal of Quantum Chemistry*, **114**, 1287–1317.

Drewitt, J.W.E., Jahn, S., Cristiglio, V., Bytchkov, A., Leydier, M., Brassamin, S., Fischer, H.E. and Hennet, L. (2011) The structure of liquid calcium aluminates as investigated by neutron and high-energy X-ray diffraction in combination with molecular dynamics simulation methods. *Journal of Physics: Condensed Matter*, **23**, 155101.

Elsässer, C. and Elsässer, T. (2005) Codoping and grain-boundary cosegregation of substitutional cations in α-Al_2O_3: A density-functional-theory study. *Journal of the American Ceramic Society*, **88**, 1–14.

Fabris, S. and Elsässer, C. (2001) Σ13 ($10\bar{1}4$) twin in α-Al_2O_3: A model for a general grain boundary. *Physical Review B*, **64**, 245117.

Fabris, S. and Elsässer, C. (2003) First-principles analysis of cation segregation at grain boundaries in α-Al_2O_3. *Acta Materialia*, **51**, 71–86.

Finkelstein, G.J., Dera, P.K., Jahn, S., Oganov, A.R., Holl, C.M., Meng, Y. and Duffy, T.S. (2014) Phase transitions and equation of state of forsterite to 90 GPa from single-crystal x-ray diffraction and molecular modeling. *American Mineralogist*, **99**, 35–43.

Fisler, D.A., Gale, J.D. and Cygan, R.T. (2000) A shell model for the simulation of rhombohedral carbonate minerals and their point defects. *American Mineralogist*, **85**, 217–224.

Foy, L. and Madden, P.A. (2006) Ionic motion in crystalline cryolite. *Journal of Physical Chemistry B*, **110**, 15302–15311.

Frenkel, D. and Smit, B.J. (2002) *Understanding Molecular Simulation: From Algorithms to Applications.* Academic Press, San Diego, California, USA.

Fressengeas, C., Taupin, V. and Capolungo, L. (2011) An elasto-plastic theory of dislocation and disclination fields. *International Journal of Solids and Structures*, **48**, 3499–3509.

Fressengeas, C., Taupin, V. and Capolungo, L. (2014) Continuous modeling of the structure of symmetric tilt boundaries. *International Journal of Solids and Structures*, **51**, 1434–1441.

Freysoldt, C., Grabowski, B., Hickel, T., Neugebauer, J., Kresse, G., Janotti, A. and Van de Walle, C.G. (2014) First-principles calculations for point defects in solids. *Reviews of Modern Physics*, **86**, 253–305.

Gale, J.D. and Rohl, A.L. (2003) The general utility lattice program (GULP). *Molecular Simulation*, **29**, 291–341.

Ghosh, D.B. and Karki, B.B. (2013) First principles simulations of the stability and structure of grain boundaries in Mg_2SiO_4 forsterite. *Physics and Chemistry of Minerals*, **41**, 163–171.

Giannozzi, P., Baroni, S., Bonini, N., Calandra, M., Car, R., Cavazzoni, C., Ceresoli, D., Chiarotti, G.L., Cococcioni, M., Dabo, I., Dal Corso, A., de Gironcoli, S., Fabris, S., Fratesi, G., Gebauer, R., Gerstmann, U., Gougoussis, C., Kokalj, A., Lazzeri, M., Martin-Samos, L., Marzari, N., Mauri, F., Mazzarello, R., Paolini, S., Pasquarello, A., Paulatto, L., Sbraccia, C., Scandolo, S., Sclauzero, G., Seitsonen, A.P., Smogunov, A., Umari, P. and Wentzcovitch, R.M. (2009) QUANTUM ESPRESSO: a modular and open-source software project for quantum simulations of materials. *Journal of Physics: Condensed Matter*, **21**, 395502.

Gonze, X., Amadon, B., Anglade, P.-M., Beuken, J.-M., Bottin, F., Boulanger, P., Bruneval, F., Caliste, D., Caracas, R., Cote, M., Deutsch, T., Genovese, L., Ghosez, P., Giantomassi, M., Goedecker, S., Hamann, D.R., Hermet, P., Jollet, F., Jomard, G., Leroux, S., Mancini, M., Mazevet, S., Oliveira, M.J.T., Onida, G., Pouillon, Y., Rangel, T., Rignanese, G.-M., Sangalli, D., Shaltaf, R., Torrent, M., Verstraete, M.J., Zerah, G. and Zwanziger, J.W. (2009) ABINIT: First-principles approach of materials and nanosystem properties. *Computer Physics Communications*, **180**, 2582–2615.

Goryaeva, A.M., Carrez, P. and Cordier, P. (2015) Modeling defects and plasticity in $MgSiO_3$ post-perovskite: Part 2 - screw and edge [100] dislocations. *Physics and Chemistry of Minerals*, **42**, 793–803.

Gramaccioli, C.M. (editor) (2002) *Energy Modelling in Minerals*. Vol. **4**, EMU Notes in Mineralogy. Eötvös University Press, Budapest.

Guillot, B. and Sator, N. (2007) A computer simulation study of natural silicate melts. part I: Low pressure properties. *Geochimica et Cosmochimica Acta*, **71**, 1249–1265.

Gurmani, S.F., Jahn, S., Brasse, H. and Schilling, F.R. (2011) Atomic scale view on partially molten rocks: Molecular dynamics simulations of melt-wetted olivine grain boundaries. *Journal of Geophysical Research*, **116**, B12209.

Harding, J.H. and Harris, D.J. (2001) Simulation of grain-boundary diffusion in ceramics by kinetic Monte Carlo. *Physical Review B*, **63**, 094102.

Harding, J.H., Harris, D.J. and Parker, S.C. (1999) Computer simulation of general grain boundaries in rocksalt oxides. *Physical Review B*, **60**, 2740–2746.

Harris, D.J., Watson, G.W. and Parker, S.C. (1996) Atomistic simulation of the effect of temperature and pressure on the [001] symmetric tilt grain boundaries of MgO. *Philosophical Magazine A*, **74**, 407–418.

Harris, D.J., Watson, G.W. and Parker, S.C. (1999) Computer simulation of pressure-induced structural transitions in MgO [001] tilt grain boundaries. *American Mineralogist*, **84**, 138–143.

Heinemann, S., Wirth, R., Gottschalk, M. and Dresen, G. (2005) Synthetic [100] tilt grain boundaries in forsterite: 9.9 to 21.5°. *Physics and Chemistry of Minerals*, **32**, 229–240.

Henkelman, G., Uberuaga, B.P. and Jonsson, H. (2000) A climbing image nudged elastic band method for finding saddle points and minimum energy paths. *Journal of Chemical Physics*, **113**, 9901–9904.

Hohenberg, P. and Kohn, W. (1964) Inhomogeneous electron gas. *Physical Review*, **136**, B864–B871.

Jahn, S. (2010) Integral modeling approach to study the phase behavior of complex solids: Application to phase transitions in MgSiO$_3$ pyroxenes. *Acta Crystallographica*, **A66**, 535–541.

Jahn, S. and Kowalski, P.M. (2014) Theoretical approaches to structure and spectroscopy of Earth materials. Pp. 691–743 in: *Spectroscopic Methods in Mineralogy and Material Sciences* (G.S. Henderson, D.R. Neuville and R.T. Downs, editors). Reviews in Mineralogy and Geochemistry, **78**. Mineralogical Society of America and the Geochemical Society, Chantilly, Virginia, USA.

Jahn, S. and Madden, P.A. (2007) Modeling Earth materials from crustal to lower mantle conditions: A transferable set of interaction potentials for the CMAS system. *Physics of the Earth and Planetary Interiors*, **162**, 129–139.

Jahn, S. and Martoňák, R. (2008) Plastic deformation of orthoenstatite and the ortho- to high-pressure clinoenstatite transition: A metadynamics simulation study. *Physics and Chemistry of Minerals*, **35**, 17–23.

Jahn, S. and Martoňák, R. (2009) Phase behavior of protoenstatite at high pressure studied by atomistic simulations. *American Mineralogist*, **94**, 950–956.

Jahn, S., Ollivier, J. and Demmel, F. (2008) Fast ionic mobility in cryolite studied by quasielastic neutron scattering. *Solid State Ionics*, **179**, 1957–1961.

Jahn, S., Rahner, R., Dachs, E., Mrosko, M. and Koch-Müller, M. (2013) Thermodynamic properties of anhydrous and hydrous wadsleyite, β–Mg$_2$SiO$_4$. *High Pressure Research*, **33**, 584–594.

Jiang, S., Chen, J., Long, Y. and Lu, T. (2012a) Atomic structure, electronic structure, and optical properties of YAG (110) twin grain boundary. *Journal of the American Ceramic Society*, **95**, 3894–3900.

Jiang, S., Chen, J., Lu, T., Long, Y. and Sun, K. (2012b) Correlation of the atomic and electronic structures and the optical properties of the Σ5(210)/[001] symmetric tilt grain boundary in yttrium aluminum garnet. *Acta Materialia*, **60**, 7041–7050.

Kabir, M., Lau, T.T., Rodney, D., Yip, S. and Van Vliet, K.J. (2010) Predicting dislocation climb and creep from explicit atomistic details. *Physical Review Letters*, **105**, 095501.

Karki, B.B. and Khanduja, G. (2006) Vacancy defects in MgO at high pressure. *American Mineralogist*, **91**, 511–516.

Karki, B.B. and Khanduja, G. (2007) A computational study of ionic vacancies and diffusion in MgSiO$_3$ perovskite and post-perovskite. *Earth and Planetary Science Letters*, **260**, 201–211.

Kawano, J., Miyake, A., Shimobayashi, N. and Kitamura, M. (2009) Molecular dynamics simulation of the rotational order-disorder phase transition in calcite. *Journal of Physics: Condensed Matter*, **21**, 095406.

Klimeš, J. and Michaelides, A. (2012) Perspective: Advances and challenges in treating van der Waals dispersion forces in density functional theory. *Journal of Chemical Physics*, **137**, 120901.

Kohn, W. and Sham, L.J. (1965) Self-consistent equations including exchange and correlation effects. *Physical Review*, **140**, A1133−A1138.

Körner, W., Bristowe, P.D. and Elsässer, C. (2011) Density functional theory study of stoichiometric and nonstoichiometric ZnO grain boundaries. *Physical Review B*, **84**, 045305.

Kraych, A., Carrez, P., Hirel, P., Clouet, E. and Cordier, P. (2016) Peierls potential and kink-pair mechanism in high-pressure MgSiO$_3$ perovskite: An atomic scale study. *Physical Review B*, **93**, 014103.

Landon, G.J. and Ubbelohde, A.R. (1957) Melting and crystal structure of cryolite (3NaF,AlF$_3$). *Proceedings of the Royal Society of London*, **240**, 160−172.

Lei, Y., Gong, Y., Duan, Z. and Wang, G. (2013) Density functional calculation of activation energies for lattice and grain boundary diffusion in alumina. *Physical Review B*, **87**, 214105.

Leslie, M. and Gillan, M.J. (1985) The energy and elastic dipole tensor of defects in ionic crystals calculated by the supercell method. *Journal of Physics C: Solid State Physics*, **18**, 973−982.

Li, L., Brodholt, J. and Alfe, D. (2009) Structure and elasticity of hydrous ringwoodite: A first principle investigation. *Physics of the Earth and Planetary Interiors*, **177**, 103−115.

Liu, J., Duan, C., Ossowski, M.M., Mei, W.N., Smith, R.W. and Hardy, J.R. (2001) Simulation of structural phase transition in NaNO$_3$ and CaCO$_3$. *Physics and Chemistry of Minerals*, **28**, 586−590.

Madden, P.A. and Wilson, M. (2000) 'Covalent' effects in 'ionic' liquids. *Journal of Physics: Condensed Matter*, **12**, A95−A108.

Mainprice, D., Tommasi, A., Couvy, H., Cordier, P. and Frost, D.J. (2005) Pressure sensitivity of olivine slip systems and seismic anisotropy of Earth's upper mantle. *Nature*, **433**, 731−733.

Martin, R.M. (2004) *Electronic Structure: Basic Theory and Practical Methods*. Cambridge University Press, Cambridge, UK.

Martoňák, R., Laio, A. and Parrinello, M. (2003) Predicting crystal structures: The Parrinello-Rahman method revisited. *Physical Review Letters*, **90**, 075503.

Martoňák, R., Donadio, D. and Parrinello, M. (2005a) Evolution of the structure of amorphous ice: From low-density amorphous through high-density amorphous to very high-density amorphous ice. *Journal of Chemical Physics*, **122**, 134501.

Martoňák, R., Laio, A., Bernasconi, M., Ceriani, C., Raiteri, P., Zipoli, F. and Parrinello, M. (2005b) Simulation of structural phase transitions by metadynamics. *Zeitschrift für Kristallographie*, **220**, 489−498.

Martoňák, R., Donadio, D., Oganov, A.R. and Parrinello, M. (2006) Crystal structure transformations in SiO$_2$ from classical and ab initio metadynamics. *Nature Materials*, **5**, 623−626.

Martoňák, R., Donadio, D., Oganov, A.R. and Parrinello, M. (2007) From four- to six-coordinated silica: transformation pathways from metadynamics. *Physical Review B*, **76**, 014120.

Marx, D. and Hutter, J. (2000) Ab initio molecular dynamics: Theory and implementation. Pp. 301−449 in: *Modern Methods and Algorithms of Quantum Chemistry* (J. Grotendorst, editor). Forschungszentrum Jülich, NIC Series. Vol. **1**.

Marx, D. and Hutter, J. (2012) *Ab initio molecular dynamics: Basic Theory and Advanced Methods*. Cambridge University Press, Cambridge, UK.

Marx, D. and Parrinello, M. (1996) Ab initio path integral molecular dynamics: Basic ideas. *Journal of Chemical Physics*, **104**, 4077−4082.

Matsui, M., Parker, S.C. and Leslie, M. (2000) The MD simulation of the equation of state of MgO: Application as a pressure calibration standard at high temperature and high pressure. *American Mineralogist*, **85**, 312−316.

McKenna, K.P. and Shluger, A.L. (2009) First-principles calculations of defects near a grain boundary in MgO. *Physical Review B*, **79**, 224116.

McKenna, K.P., Hofer, F., Gilks, D., Lazarov, V.K., Chen, C., Wang, Z. and Ikuhara, Y. (2014) Atomic-scale structure and properties of highly stable antiphase boundary defects in Fe$_3$O$_4$. *Nature Communications*, **5**, 5740.

Mehrer, H. (2007) *Diffusion in Solids*. Springer, Berlin Heidelberg.

Metsue, A., Carrez, P., Denoual, C., Mainprice, D. and Cordier, P. (2010) Plastic deformation of wadsleyite: IV Dislocation core modelling based on the Peierls-Nabarro-Galerkin model. *Acta Materialia*, **58**, 1467−1478.

Mishin, Y., Suzuki, A., Uberuaga, B.P. and Voter, A.F. (2007) Stick-slip behavior of grain boundaries studied by accelerated molecular dynamics. *Physical Review B*, **75**, 224101.

Miyake, A., Shimobayashi, N., Miura, E. and Kitamura, M. (2002) Molecular dynamics simulations of phase transition between high-temperature and high-pressure clinoenstatite. *Physics of the Earth and Planetary Interiors*, **129**, 1−11.

Mott, N.F. and Littleton, M.J. (1938) Conduction in polar crystals. I. Electrolytic conduction in solid salts. *Transactions of the Faraday Society*, **34**, 485−499.

Nabarro, F.R.N. (1947) Dislocations in a simple cubic lattice. *Proceedings of the Physical Society of London*, **59**, 256−272.

Oganov, A.R., Brodholt, J.P. and Price, G.D. (2000) Comparative study of quasiharmonic lattice dynamics, molecular dynamics and Debye model applied to $MgSiO_3$ perovskite. *Physics of the Earth and Planetary Interiors*, **122**, 277−288.

Oganov, A.R., Martoňák, R., Laio, A., Raiteri, P. and Parrinello, M. (2005) Anisotropy of Earth's D″ layer and stacking faults in the $MgSiO_3$ post-perovskite phase. *Nature*, **438**, 1142−1144.

Parker, S.C., Cooke, D.J., Kerisit, S., Marmier, A.S., Taylor, S.L. and Taylor, S.N. (2004) From HADES to PARADISE – Atomistic simulation of defects in minerals. *Journal of Physics: Condensed Matter*, **16**, S2735−S2749.

Payne, M.C., Teter, M.P., Allan, D.C., Arias, T.A. and Joannopoulos, J.D. (1992) Iterative minimization techniques for ab initio total-energy calculations: molecular dynamics and conjugate gradients. *Reviews of Modern Physics*, **64**, 1045−1097.

Peierls, R. (1940) The size of a dislocation. *Proceedings of the Physical Society of London*, **52**, 34−37.

Perdew, J.P. and Ruzsinszky, A. (2010) Density functional theory of electronic structure: A short course for mineralogists and geophysicists. Pp. 1−18 in: *Theoretical and Computational Methods in Mineral Physics: Geophysical Applications* (R. Wentzcovitch and L. Stixrude, editors). Reviews in Mineralogy and Geochemistry, **71**. Mineralogical Society of America and the Geochemical Society, Chantilly, Virginia, USA.

Petrishcheva, E. and Abart, R. (2017a) Diffusion: some mathematical foundations and applications in mineralogy. Pp. 255−294 in: *Mineral Reaction Kinetics: Microstructures, Textures, Chemical and Isotopic Signatures* (R. Abart and W. Heinrich, editors). EMU Notes in Mineralogy, **16**. European Mineralogical Union and Mineralogical Society of Great Britain & Ireland, London.

Petrishcheva, E. and Abart, R. (2017b) Interfaces. Pp. 295−345 in: *Mineral Reaction Kinetics: Microstructures, Textures, Chemical and Isotopic Signatures* (R. Abart and W. Heinrich, editors). EMU Notes in Mineralogy, **16**. European Mineralogical Union and Mineralogical Society of Great Britain & Ireland, London.

Poirier, J.P. (1985) *Creep of Crystals*. Cambridge University Press, Cambridge, UK.

Putnis, A. (1992) *Introduction to Mineral Sciences*. Cambridge University Press, Cambridge, UK.

Rabone, J., López-Honorato, E. and Van Uffelen, P. (2014) Silver and cesium diffusion dynamics at the β-SiC Σ5 grain boundary investigated with density functional theory molecular dynamics and metadynamics. *Journal of Physical Chemistry A*, **118**, 915−926.

Ritterbex, S., Carrez, P., Gouriet, K. and Cordier, P. (2015) Modeling dislocation glide in Mg_2SiO_4 ringwoodite: Towards rheology under transition zone conditions. *Physics of the Earth and Planetary Interiors*, **248**, 20−29.

Rohl, A.L., Wright, K. and Gale, J.D. (2003) Evidence from surface phonons for the (2x1) reconstruction of the $(10\bar{1}4)$ surface of calcite from computer simulation. *American Mineralogist*, **88**, 921−925.

Salanne, M., Rotenberg, B., Jahn, S., Vuilleumier, R. and Madden, P.A. (2012) Including many-body effects in models for ionic liquids. *Theoretical Chemistry Accounts*, **131**, 1143.

Shimobayashi, N., Miyake, A., Kitamura, M. and Miura, E. (2001) Molecular dynamics simulations of the phase transition between low-temperature and high-temperature clinoenstatites. *Physics and Chemistry of Minerals*, **28**, 591−599.

Skylaris, C.-K., Haynes, P.D., Mostofi, A.A. and Payne, M.C. (2005) Introducing ONETEP: Linear-scaling density functional simulations on parallel computers. *Journal of Chemical Physics*, **122**, 084119.

Spiekermann, G., Wilke, M. and Jahn, S. (2016) Structural and dynamical properties of supercritical H_2O-SiO_2 fluids studied by ab initio molecular dynamics. *Chemical Geology* , **426**, 85−94.

Tangney, P. and Scandolo, S. (2002) An ab initio parametrized interatomic force field for silica. *Journal of*

Chemical Physics, **117**, 8898−8904.

Tsuchiya, J. and Tsuchiya, T. (2009) First principles investigation of the structural and elastic properties of hydrous wadsleyite under pressure. *Journal of Geophysical Research*, **114**, B02206.

Van Der Geest, A.G., Islam, M.M., Couvant, T. and Diawara, B. (2013) Energy ordering of grain boundaries in Cr_2O_3: insights from theory. *Journal of Physics: Condensed Matter*, **25**, 485005.

Verma, A.K. and Karki, B.B. (2009) Ab initio investigations of native and protonic point defects in Mg_2SiO_4 polymorphs under high pressure. *Earth and Planetary Science Letters*, **285**, 140−149.

Vineyard, G.H. (1957) Frequency factors and isotope effects in solid state rate processes. *Journal of the Physics and Chemistry of Solids*, **3**, 121−127.

Vitek, V. (1968) Intrinsic stacking faults in body-centered cubic crystals. *Philosophical Magazine A*, **18**, 773−786.

Vocadlo, L., Wall, A., Parker, S.C. and Price, G.D. (1995) Absolute ionic diffusion in MgO-computer calculations via lattice dynamics. *Physics of the Earth and Planetary Interiors*, **88**, 193−210.

Walker, A.M. (2010) Simulation of screw dislocations in wadsleyite. *Physics and Chemistry of Minerals*, **37**, 301−310.

Walker, A.M., Wright, K. and Slater, B. (2003) A computational study of oxygen diffusion in olivine. *Physics and Chemistry of Minerals*, **30**, 536−545.

Walker, A.M., Slater, B., Gale, J.D. and Wright, K. (2004) Predicting the structure of screw dislocations in nanoporous materials. *Nature Materials*, **3**, 715−720.

Walker, A.M., Gale, J.D., Slater, B. and Wright, K. (2005a) Atomic scale modelling of the cores of dislocations in complex materials part 1: methodology. *Physical Chemistry Chemical Physics*, **7**, 3227−3234.

Walker, A.M., Gale, J.D., Slater, B. and Wright, K. (2005b) Atomic scale modelling of the cores of dislocations in complex materials part 2: applications. *Physical Chemistry Chemical Physics*, **7**, 3235−3242.

Walker, A.M., Woodley, S.M., Slater, B. and Wright, K. (2009) A computational study of magnesium point defects and diffusion in forsterite. *Physics of the Earth and Planetary Interiors*, **172**, 20−27.

Walker, A.M., Carrez, P. and Cordier, P. (2010) Atomic-scale models of dislocation cores in minerals: progress and prospects. *Mineralogical Magazine*, **74**, 331−413.

Wentzcovitch, R. and Stixrude, L. (editors) (2010) *Theoretical and Computational Methods in Mineral Physics: Geophysical Applications*. Reviews in Mineralogy and Geochemistry, **71**. Mineralogical Society of America and the Geochemical Society, Chantilly, Virginia, USA.

Wilson, M., Jahn, S. and Madden, P.A. (2004) The construction and application of a fully flexible computer simulation model for lithium oxide. *Journal of Physics: Condensed Matter*, **16**, S2795−S2810.

Wilson, M., Madden, P.A., Hemmati, M. and Angell, C.A. (1996) Polarization effects, network dynamics, and the infrared spectrum of amorphous SiO_2. *Physical Review Letters*, **77**, 4023−4026.

Wirth, R. (1996) Thin amorphous films (1−2 nm) at olivine grain boundaries in mantle xenoliths from San Carlos, Arizona. *Contributions to Mineralogy and Petrology*, **124**, 44−54.

Zahn, D. (2013) Nucleation mechanism and kinetics of the perovskite to post-perovskite transition of $MgSiO_3$ under extreme conditions. *Chemical Physics Letters*, **573**, 5−7.

Zhang, F., Walker, A.M., Wright, K. and Gale, J.D. (2010) Defects and dislocations in MgO: atomic scale models of impurity segregation and fast pipe diffusion. *Journal of Materials Chemistry*, **20**, 10445−10451.

Zhang, Y., Zhao, D. and Matsui, M. (2005) Anisotropy of akimotoite: A molecular dynamics study. *Physics of the Earth and Planetary Interiors*, **151**, 309−319.

EMU Notes in Mineralogy, Vol. 16 (2017), Chapter 9, 255–294

Diffusion: Some mathematical foundations and applications in mineralogy

ELENA PETRISHCHEVA* and RAINER ABART

*[1]Department of Lithospheric Research, University of Vienna,
UZA II - Universitätszentrum Althanstrasse, Althanstraße 14, 1090 Vienna, Austria,
e-mail: elena.petrishcheva@univie.ac.at
* corresponding author*

Diffusion equations studied in this chapter apply to many physical situations: mass transport, particle transport such as self-diffusion or binary diffusion, heat transfer, potential and current diffusion, and even expansion of biological populations. To be specific, in what follows we consider transport of matter that arises from the random movement of atoms, ions or molecules. Diffusion occurs in all states of aggregation, and it is the only mode of mass transfer in solids. Diffusion is one of the most fundamental kinetic processes underlying phase change. There are several excellent textbooks giving detailed accounts of diffusion including the mathematics of diffusion (Crank, 1975) and diffusion in solids (Glicksman, 2000; Mehrer, 2007). There are also a number of comprehensive reviews summarizing the current state of knowledge about diffusion in geological materials (Zhang and Cherniak, 2010).

In this chapter we present the mathematical framework that we consider indispensible for dealing with the complex diffusion problems in mineral and rock systems. This chapter is by no means a comprehensive presentation of the mathematics of diffusion, rather it is a selection of topics that we deem particularly relevant in the context of geo-materials research. We begin with a brief account of the historical evolution of diffusion theory and derive the basic relations from both the macroscopic and the microscopic point of view. We then make the important distinction between linear diffusion, where the diffusion coefficient is independent of composition, and non-linear diffusion, where the opposite is true. In recent work on mineral systems composition-dependent diffusivity has turned out to be the rule rather than the exception; ample room thus is given to this phenomenon. In a first step, non-linearity caused by relatively weak composition dependence of diffusivity that may arise from defects introduced with the diffusing species or from strongly contrasting self-diffusion coefficients in multi-component diffusion are addressed. In a second step, the links between diffusion and thermodynamics are explored. Non-ideal thermodynamic behaviour of solution phases may cause effective diffusion coefficients to become negative leading to uphill diffusion and eventually producing phase separation. The basic mathematical concepts for describing diffusion phenomena in mineral and rock systems and for solving inverse problems such as extracting diffusivities from experiment or time scales from diffusion profiles in minerals are derived. Due to space restrictions we refrain from including phenomena such as the Kirkendall effect and reactive diffusion. These phenomena are addressed in the chapter on solid-state reactions in mineral systems by Gaidies *et al.* (2017, this volume).

©Copyright 2017 the European Mineralogical Union and the Mineralogical Society of Great Britain & Ireland
DOI: 10.1180/EMU-notes.16.9

1. Landmarks of diffusion

Studies of diffusion have a long history that dates back to ancient Greece and Aristotle. The description of diffusion in mathematical terms, one that results in quantitative predictions which can be either accepted or rejected on the basis of experiments, is much younger. We start this Chapter with a brief description of the most important milestones of the quantitative theory.

- In 1833 Thomas Graham summarized his investigations of gases escaping through a small hole and formulated the following law (Graham, 1833): "The diffusion or spontaneous intermixture of two gases in contact, is effected by an interchange in position of indefinitely minute volumes of the gases, which volumes are not necessarily of equal magnitude, being, in the case of each gas, inversely proportional to the square root of the density of that gas." He thus provided an insight into the microscopic origin of the diffusion process. Later, Graham's experiments were explained in the framework of the kinetic theory of gases. Thomas Graham also pioneered the experimental investigation of diffusion in liquids (Graham, 1864).

- In 1855 the diffusion equation first appeared in a paper by Adolf Fick (Fick, 1855). Fick pointed out that the same mathematical equation appears in Fourier's theory of the conduction of heat and Ohm's theory describing diffusion of electrical charge in a conductor. According to Fick: "the transfer of salt and water occurring in a unit of time, between two elements of space filled with differently concentrated solutions of the same salt, must be, caeteris paribus, directly proportional to the difference of concentration, and inversely proportional to the distance of the elements from one another". In mathematical terms Fick's (first) law reads

$$j(x,t) = -D\frac{\Delta n}{\Delta x} = -D\frac{\partial n(x,t)}{\partial x} \tag{1}$$

where $j(x,t)$ denotes the diffusion flux, the total number of atoms that cross a unit area in the Oyz plane at point x at time t per second, and the parameter $n(x,t)$ denotes concentration, *e.g.* number of atoms per unit volume. The presupposed distribution of atoms is inhomogeneous in x direction but is homogeneous in the y and z directions. The diffusivity parameter $D > 0$, *i.e.* the inhomogeneity-induced diffusion flux is directed downhill with respect to spatial concentration variations and tends to remove the inhomogeneity. The latter may exist due to the initial conditions, boundary conditions, or source and sink terms. In what follows, we will take advantage from the following remarks going beyond Fick's work:

1. If the distribution of the diffusing atoms is inhomogeneous in more than one direction, the scalar flux is replaced by a flux vector

$$\mathbf{j}(\mathbf{r},t) = -D\nabla n(\mathbf{r},t) \tag{2}$$

where $\mathbf{r} = (x,y,z)$. The components of ∇ in the chosen coordinate system are

$$\nabla n = \left(\frac{\partial n}{\partial x}, \frac{\partial n}{\partial y}, \frac{\partial n}{\partial z}\right) \tag{3}$$

Whenever appropriate, equation 2 will be written in components

$$j_i = -D\frac{\partial n}{\partial x_i}, \quad 1 \le i \le 3$$

where we do not distinguish between

$$\mathbf{r} = (x,y,z) \text{ and } \mathbf{x} = (x_1,x_2,x_3)$$

2. It is important to realize that the diffusivity D connects two vectors $-\nabla n$ and $\mathbf{j}$, which are parallel to each other in accordance with equation 2. On the other hand, in some situations (*e.g.* in crystals and in systems with non-uniform pressure or temperature distribution) $-\nabla n$ may act in one direction and the resulting flux $\mathbf{j}$ may point in a different direction. In such case, a single constant diffusivity D is replaced by a second-rank tensor (Nye, 1957; Fleisch, 2011) which is represented by its nine components D_{ij}. The generalized Fick's first law reads

$$j_i = -\sum_j D_{ij}\frac{\partial n}{\partial x_j}, \quad 1 \le i,j \le 3 \tag{4}$$

where the involved components of the diffusivity tensor D_{ij} and both vectors $-\nabla n$ and $\mathbf{j}$ refer to the same coordinate frame. In what follows, we will examine such a situation in more detail, but for now the diffusivity parameter in Fick's law is just a scalar quantity.

3. Note that equations 1–4 are linear: an increase in $-\nabla n$ leads to a proportional increase in $\mathbf{j}$. As it appears, diffusivity is more than just a constant parameter in many important systems. For instance, D may depend on concentration and on gradient of concentration, it may even become negative resulting in a highly peculiar uphill diffusion. These issues lead to the so-called *non-linear diffusion* and will be discussed below, but for now $D = \text{const} > 0$.

• In 1905 Karl Pearson wrote a short letter, "The problem of the random walk", to the journal *Nature* (Pearson, 1905), where he formulated the following question: "A man starts from the point O and walks l yards in a straight line; he then turns through any angle whatever and walks another l yards in a second straight line. He repeats this process n times. Inquire the probability that after n stretches he is at a distance between r and $r+ \delta r$ from his starting point O". Alternatively, we may consider an ensemble of moving particles and ask for the number of particles that are located near a given point in space at a given time, *e.g.* in the context of diffusing atoms. Taking this point of view, A. Einstein (1905) introduced a general integral equation of the following type

$$n(\mathbf{r}, t + \Delta t) = \int p(\mathbf{r}, \mathbf{r}')n(\mathbf{r}', t)d^3\mathbf{r}' \tag{5}$$

where the time interval parameter Δt corresponds to a single "jump", $n(\mathbf{r}', t)d^3\mathbf{r}'$ is the number of atoms in an elementary volume at position $\mathbf{r}'$, and $p(\mathbf{r}, \mathbf{r}')$ describes the probability for an atom at position $\mathbf{r}'$ to jump to the position $\mathbf{r}$. Note, that $p(\mathbf{r}, \mathbf{r}')$ does not depend on the number of atoms that are currently present at $\mathbf{r}$ and $\mathbf{r}'$. Therefore equation 5 is linear and can be used to derive the linear diffusion equation. Equation 5 provides a better insight into the microscopic origins of diffusion than the phenomenological equation 2.

- Pearson's problem is closely related to the chaotic behaviour of small particles suspended in a liquid or gas, which was first observed by Robert Brown (Brown, 1828). Instead of considering all diffusing particles together with an unknown jump-probability distribution, as in equation 5, one can start with a single particle. Consider, for simplicity, one spatial dimension. The brownian particle with mass m and velocity $v(t)$ is described by the Langevin equation (Langevin, 1908)

$$m\frac{dv}{dt} = -\frac{v}{b} + F \tag{6}$$

where b is the mobility parameter and $-v/b$ is the macroscopic damping force that results from the viscosity of the liquid or gas. $F = F(t)$ is a chaotic microscopic force attributed to the molecular motion of the surrounding atoms.

Equation 6 is a stochastic differential equation. It cannot be "solved", because there is no explicit expression for $F(t)$. The only available information is hidden in the statistical properties of the microscopic force, which are encoded by the mean values

$$\langle F(t) \rangle, \quad \langle F(t_1)F(t_2) \rangle, \quad \langle F(t_1)F(t_2)F(t_3) \rangle$$

and so on. The notation $\langle \rangle$ applies to the so-called ensemble average: an average over a large number of different realizations of equation 6, each with its own $F(t)$. It is natural to assume that the averaged microscopic force vanishes. Moreover, both quick and unpredictable motions of atoms surrounding and hitting the brownian particle suggest that $F(t_1)$ and $F(t_2)$ are statistically independent as long as $t_1 \neq t_2$. In mathematical terms

$$\langle F(t) \rangle = 0, \quad \langle F(t_1)F(t_2) \rangle = \frac{2D}{b^2}\delta(t_1 - t_2) \tag{7}$$

where $\delta()$ denotes Dirac's delta-function. The origin of the pre-factor is not clear at this point; note, however, that parameter D has physical dimension of the diffusion coefficient. For instance, for $m = 0$ we just have $v = bF$ and

$$\langle v(t_1)v(t_2) \rangle = 2D\delta(t_1 - t_2)$$

such that $[D] = (\text{length})^2/(\text{time})$. As we will see in what follows, the above choice of the pre-factor makes equation 7 compatible with the well established relations (8) and (9). The parameter D turns out to be the diffusion coefficient for the brownian particles.

Now, following Langevin's ideas, we calculate the ensemble averaged value $\langle v(t_1)v(t_2)\rangle$ for arbitrary $m > 0$. We start with the formal solution of equation 6

$$v(t) = v(0)e^{-t/(mb)} + \frac{e^{-t/mb}}{m} \int_0^t F(t')e^{t'/(mb)}dt'$$

and assume that $t \gg mb$, such that influence of the initial condition can be ignored. Therefore we obtain

$$v(t_1)v(t_2) = \frac{e^{-(t_1+t_2)/mb}}{m^2} \int_0^{t_1} \int_0^{t_2} F(t')F(t'')e^{(t'+t'')/(mb)}dt'dt''$$

By averaging over many realizations of the Langevin equation one derives that

$$\langle v(t_1)v(t_2)\rangle = \frac{e^{-(t_1-t_2)/mb}}{m^2} \int_0^{t_1} \int_0^{t_2} \langle F(t')F(t'')\rangle e^{(t'+t'')/(mb)}dt'dt''$$

Replacing $\langle F(t')F(t'')\rangle$ in accord with equation 7, one can perform integration and derive that

$$\langle v(t_1)v(t_2)\rangle = \frac{D}{mb} e^{|t_1+t_2|/(mb)}$$

Recall that both t_1 and t_2 are considerably larger than mb such that the system has enough time to "forget" about the initial condition, whereas $t_1 - t_2$ is arbitrary. An immediate consequence is that

$$\langle v^2(t)\rangle = \frac{D}{mb}$$

and after equalizing $m\langle v^2\rangle/2$ and $k_BT/2$, as appropriate for one spatial dimension, we see that Langevin's approach is compatible with the famous Einstein relation between diffusivity and mobility

$$b = \frac{D}{k_BT} \tag{8}$$

where k_BT is temperature in energetic units. Furthermore, the standard relation between position and velocity

$$x(t) = \int_0^t v(t')dt'$$

cannot be used directly, because $v(t)$ is not available. One can, however, consider the product

$$x(t_1)x(t_2) = \int_0^{t_1} \int_0^{t_2} v(t')v(t'')dt'dt''$$

and use the above expression for $\langle v(t_1)v(t_2)\rangle$ to calculate $\langle x(t_1)x(t_2)\rangle$. In particular, it appears that

$$\langle x^2(t) \rangle = 2Dt \tag{9}$$

which is a well known link between random walk and diffusivity. Both equations 8 and 9 indicate that the parameter D, which seems to be an intrinsic property of the random force (7), is the standard diffusivity.

- An important contribution to our understanding of diffusion and, more generally, of a way by which a perturbed macroscopic system returns to its equilibrium state was made by Lars Onsager (1931). To describe it we return to the anisotropic Fick's law (4) and reformulate as follows. A concentration gradient in x_i direction, $\partial n/\partial x_i$, induces a diffusion flux in x_i direction, $-D_{ii}\partial n/\partial x_i$, and also in x_j direction, $-D_{ji}\partial n/\partial x_i$. In the same way: a concentration gradient in x_j direction, $\partial n/\partial x_j$, induces a diffusion flux in x_j direction, $-D_{jj}\partial n/\partial x_j$, and also in x_i direction, $-D_{ij}\partial n/\partial x_j$. As it appears, the involved cross-diffusivities must be equal

$$D_{ij} = D_{ji}, \ 1 \leqslant i,j \leqslant 3 \tag{10}$$

such that any diffusivity tensor contains six independent parameters at most. The latter equation is an exemplary application of a very general principle first formulated by Onsager (1931). The principle will play an important role for describing the so-called uphill diffusion in what follows.

2. Linear diffusion equation

Probably the most intriguing property of diffusion is its universality. The same model appears in seemingly different physical contexts, as was first mentioned by Fick (1855). In this section the universal diffusion equation is first derived from the macroscopic Fick's (first) law and then from the microscopic treatment.

2.1. Phenomenological derivation

The simplest way to derive the diffusion equation is to combine the phenomenological Fick's law (2) with the continuity equation (see *e.g.* Landau and Lifshitz, 1987)

$$\frac{\partial n}{\partial t} + \nabla \mathbf{j} = 0 \tag{11}$$

where $\nabla \mathbf{j}$ denotes the divergence

$$\nabla \mathbf{j} = \frac{\partial j_x}{\partial x} + \frac{\partial j_y}{\partial y} + \frac{\partial j_z}{\partial z} \tag{12}$$

Recall, that $n = n(\mathbf{r},t)$ and $\mathbf{j} = \mathbf{j}(\mathbf{r},t)$ denote particle density and particle flux density, respectively. The general continuity equation 11 indicates that moving particles are neither born nor destroyed in the system. They are just rearranged in space by fluxes. To be specific, let us consider some finite domain Ω in space with the smooth boundary $\partial\Omega$, and let $N(t)$ be the total number of atoms inside Ω given by

$$N(t) = \int_\Omega n(\mathbf{r}, t) d^3\mathbf{r} \Rightarrow \frac{dN}{dt} = \int_\Omega \frac{\partial n}{\partial t} d^3\mathbf{r}$$

Now, let dS be an elementary area on $\partial\Omega$, and let $\mathbf{m}$ be a unit vector $\mathbf{m} \perp dS$ directed outward. The quantity $(\mathbf{j} \cdot \mathbf{m}) dS$ is the number of atoms that leave Ω through dS per second. Here, both $\mathbf{j}$ and $\mathbf{m}$ are taken at the same point on dS. The total rate at which $N(t)$ changes is

$$\frac{dN}{dt} = -\int_{\partial\Omega} (\mathbf{j} \cdot \mathbf{m}) dS = -\int_\Omega (\nabla \mathbf{j}) d^3\mathbf{r}$$

where Gauss's law was used in the last step. Combining the last two equations for dN/dt we get

$$\int_\Omega \left(\frac{\partial n}{\partial t} + \nabla \mathbf{j} \right) d^3\mathbf{r} = 0$$

the latter equation is valid for any Ω, which implies equation 11.

Using equation 11 together with equation 2 one immediately derives the fundamental diffusion equation

$$\frac{\partial n}{\partial t} = D\nabla^2 n \tag{13}$$

where the Laplace operator $\nabla^2 n$ is defined as $\nabla (\nabla n)$ in accord with equations 3 and 12

$$\nabla^2 n = \frac{\partial}{\partial x} (\nabla n)_x + \frac{\partial}{\partial y} (\nabla n)_y + \frac{\partial}{\partial z} (\nabla n)_z = \frac{\partial^2 n}{\partial x^2} + \frac{\partial^2 n}{\partial y^2} + \frac{\partial^2 n}{\partial z^2}$$

It is worth mentioning that the seemingly simple form of the Laplace operator is due to the Cartesian coordinates. The coordinates are, however, dictated by the geometry of the problem. For example, one may face spherical or ellipsoidal coordinates. If the Cartesian coordinates (x,y,z) are replaced by a general orthogonal coordinate system (ξ, η, ζ), the elementary length

$$ds^2 = dx^2 + dy^2 + dz^2$$

needs to be recalculated to derive the so-called Lamé coefficients h_ξ, h_η, h_ζ (see, *e.g.* Arnold, 1989), where the new expression for the elementary length reads

$$ds^2 = (h_\xi d\xi)^2 + (h_\eta d\eta)^2 + (h_\zeta d\zeta)^2.$$

The Laplace operator is then given by

$$\nabla^2 n = \frac{1}{h_\xi h_\eta h_\zeta} \frac{\partial}{\partial \xi} \left(\frac{h_\eta h_\zeta}{h_\xi} \frac{\partial n}{\partial \xi} \right) + \frac{1}{h_\xi h_\eta h_\zeta} \frac{\partial}{\partial \eta} \left(\frac{h_\zeta h_\xi}{h_\eta} \frac{\partial n}{\partial \eta} \right) + \frac{1}{h_\xi h_\eta h_\zeta} \frac{\partial}{\partial \zeta} \left(\frac{h_\xi h_\eta}{h_\zeta} \frac{\partial n}{\partial \zeta} \right)$$

For instance, in spherical coordinates

$$x = r\cos\varphi\sin\theta$$
$$y = r\sin\varphi\sin\theta \Rightarrow ds^2 = dr^2 + r^2\sin^2\theta d\varphi^2 + r2d\theta^2$$
$$z = r\cos\theta$$

such that

$$h_r = 1, \; h_\varphi = r\sin\theta, \; h_\theta = r$$

and

$$\nabla^2 n = \frac{1}{r^2}\frac{\partial}{\partial r}\left(r^2\frac{\partial n}{\partial r}\right) + \frac{1}{r^2\sin^2\theta}\frac{\partial^2 n}{\partial\varphi^2} + \frac{1}{r^2\sin\theta}\frac{\partial}{\partial\theta}\left(\sin\theta\frac{\partial n}{\partial\theta}\right)$$

For example, radially symmetric diffusion in three dimensions is described by the equation

$$\frac{\partial n}{\partial t} = D\left(\frac{\partial^2 n}{\partial r^2} + \frac{2}{r}\frac{\partial n}{\partial r}\right)$$

Another important remark is that equation 13 appears naturally in different contexts and has different names including Fick's second law (diffusion), Fourier's law (conduction of heat), Darcy's law (hydraulic flow) and Ohm's law (charge transport). In the case of one-dimensional geometry, which is presupposed in equation 1, the general equation 13 takes a particularly simple form

$$\frac{\partial n}{\partial t} = D\frac{\partial^2 n}{\partial^2 x} \qquad (14)$$

For an anisotropic system we use equation 11 with equation 4 and obtain

$$\frac{\partial n}{\partial t} = \sum_{i,j=1}^{3} D_{ij}\frac{\partial^2 n}{\partial x_i\partial x_j} \qquad (15)$$

If the diffusivity varies with space coordinates due to some inhomogeneity, equation 13 and equation 15 are replaced by

$$\frac{\partial n}{\partial t} = \nabla[D(\mathbf{r})\nabla n] \quad \text{and} \quad \frac{\partial n}{\partial t} = \sum_{i,j=1}^{3} \frac{\partial}{\partial x_i}\left[D_{ij}(\mathbf{r})\frac{\partial n}{\partial x_j}\right]$$

We are in a good position to stress that, although the presupposed linear dependence of the diffusion flux on ∇n greatly simplifies mathematics, it may be insufficient in some cases. As an alternative, compare the latter equation to the following one

$$\frac{\partial n}{\partial t} = \sum_{i,j=1}^{3} \frac{\partial}{\partial x_i}\left[D_{ij}(n,\mathbf{r})\frac{\partial n}{\partial x_j}\right] \qquad (16)$$

Now we included a dependence of the diffusivity on concentration that makes equation 16 non-linear and extremely difficult for analytical treatment. Some non-linear diffusion equations will be considered later, but for now all considerations are for the linear equations.

2.2. Microscopic derivation

Equation 13 can also be derived directly from the microscopic equation 5 if some natural assumptions are made (Einstein, 1905). Namely, consider a homogeneous system in which

$$p(\mathbf{r},\mathbf{r}') = p(\mathbf{r} - \mathbf{a},\mathbf{r}' - \mathbf{a})$$

where $\mathbf{a}$ is an arbitrary displacement. Taking $\mathbf{a} = \mathbf{r}$ and introducing

$$\mathbf{R} = \mathbf{r}' - \mathbf{r} \quad \text{and} \quad P(\mathbf{R}) = p(0,\mathbf{r}' - \mathbf{r})$$

we reduce equation 5 to the form

$$n(\mathbf{r},t + \Delta t) = \int P(\mathbf{R})n(\mathbf{r} + \mathbf{R},t)d^3\mathbf{R} \tag{17}$$

where we changed the integration variable from $\mathbf{r}'$ to $\mathbf{R}$. Note, that the total number of the diffusing atoms remains constant

$$\int n(\mathbf{r},t + \Delta t)d^3\mathbf{r} = \int n(\mathbf{r},t)d^3\mathbf{r}.$$

An important property of $P(\mathbf{R})$ is derived by multiplying equation 17 with $d^3\mathbf{r}$ and integrating it over the whole space. Thus, we obtain

$$\int P(\mathbf{R})d^3\mathbf{R} = 1$$

simply because the total number of the diffusing particles should remain constant between t and $t + \Delta t$.

Another natural assumption is that $p(\mathbf{r},\mathbf{r}') = p(\mathbf{r}',\mathbf{r})$, *i.e.* $P(\mathbf{R}) = P(-\mathbf{R})$. In particular, the so-called first-order moments of $P(\mathbf{R})$ must vanish

$$\int XP(\mathbf{R})d^3\mathbf{R} = \int YP(\mathbf{R})d^3\mathbf{R} = \int ZP(\mathbf{R})d^3\mathbf{R} = 0$$

where $\mathbf{R} = (X,Y,Z)$. In contrast, the second-order moments

$$M_{ij} = \int X_iX_jP(\mathbf{R})d^3\mathbf{R}, \, 1 \leqslant i,j \leqslant 3$$

generally do not vanish, where as usual we do not distinguish between (X,Y,Z) and (X_1,X_2,X_3). To proceed, we naturally assume that changes in $n(\mathbf{r},t)$ are slow on the time-scale Δt so that, to a good approximation

$$n(\mathbf{r},t + \Delta t) = n(\mathbf{r},t) + \frac{\partial n(\mathbf{r},t)}{\partial t}\Delta t \tag{18}$$

Moreover, $P(\mathbf{R})$ decreases quickly with an increase in $\mathbf{R}$. Therefore only a small neighbourhood of $\mathbf{r}$ needs to be accounted for when considering $n(\mathbf{r} + \mathbf{R},t)$ in equation 17. Adopting Taylor expansion up to the second order we set

$$n(\mathbf{r} + \mathbf{R},t) = n(\mathbf{r},t) + \sum_{i=1}^{3} X_i\frac{\partial n(\mathbf{r},t)}{\partial x_i} + \frac{1}{2}\sum_{i,j=1}^{3} X_iX_j\frac{\partial^2 n(\mathbf{r},t)}{\partial x_i\partial x_j} \tag{19}$$

Inserting the Taylor expansions (18) and (19) into equation 17 and using the just-derived properties of the moments of $P(\mathbf{R})$ we obtain

$$\frac{\partial n(\mathbf{r},t)}{\partial t} = \frac{1}{2\Delta t}\sum_{i,j=1}^{3} M_{ij}\frac{\partial^2 n(\mathbf{r},t)}{\partial x_i\partial x_j}$$

That is, we have just derived the general anisotropic equation 15 together with the microscopic representation of all diffusivities

$$D_{ij} = \frac{1}{2\Delta t} \int X_i X_j P(\mathbf{R}) d^3 R, \quad 1 \le i, j \le 3$$

For instance, Onsager's property (10) is evident because of the last equation. The isotropic diffusion equation appears if the only non-vanishing second-order moments are

$$M_{11} = M_{22} = M_{33}$$

This applies when $P(\mathbf{R})$ is isotropic implying that the jump probability actually depends on $|\mathbf{R}|$ but is independent of direction.

The present derivation of the diffusion equation (Einstein, 1905) is somewhat longer than that based on Fick's law, but it provides a clear picture of the necessary assumptions. Both derivations can be generalized for non-linear systems; in the microscopic derivation $p(\mathbf{r}, \mathbf{r}')$ is replaced by

$$p(\mathbf{r}, \mathbf{r}') \rightarrow p(\mathbf{r}, \mathbf{r}', n(\mathbf{r}), n(\mathbf{r}'))$$

2.3. Fundamental properties

In this section we discuss some important properties of the linear diffusion equation 13. To begin with, we note that the physical dimension of the diffusion coefficient is $[D] = m^2/s$. In particular, if the problem in question has a characteristic space scale L, the expected time for diffusion to develop is L^2/D. If there is a characteristic time scale T, the expected dimension of the diffusion pattern is $\sqrt{DT}$. If both L and T are available, one can use L^2/T to estimate D. Finally, if D is already known and L^2/T differs from it considerably, the problem at hand is more complicated than equation 13, and additional processes must be accounted for. They may be given by a new transport process or by a non-linear dependence of diffusivity on concentration, stresses, temperature, etc.

An interesting property of the diffusion equation is that an arbitrary time dependence of the diffusivity can be removed by a proper time-dependent rescaling. Indeed, consider the following generalization of equation 13:

$$\frac{\partial n}{\partial t} = D(t)\nabla^2 n, \quad D(t) = D_0 f(t)$$

where D_0 is the initial value of the diffusivity such that $f(t)$ starts from $f(0) = 1$ and, e.g. decreases corresponding to progressive suppression of diffusive transport. The question arises of whether it is possible that diffusion ceases completely? To address this problem we introduce a new rescaled time τ

$$\tau = \int_0^t f(t') dt'$$

and replace $n = n(\mathbf{r}, t)$ by $n = n(\mathbf{r}, \tau)$ such that

$$\frac{\partial n}{\partial t} = \frac{\partial n}{\partial \tau}\frac{\partial \tau}{\partial t} = \frac{\partial n}{\partial \tau}f(t)$$

Note, that $f(t)$ is dimensionless and therefore indeed $[\tau] = s$. The time-inhomogeneity in the diffusion equation seems to disappear because now

$$\frac{\partial n}{\partial \tau} = D_0\nabla^2 n$$

The next key question regards the behaviour of τ, when $t \rightarrow \infty$. If $\tau \rightarrow \infty$ together with $t \rightarrow \infty$, the time-dependent diffusivity does not introduce new features except for new time-scales. On the other hand, if

$$\int\limits_0^\infty f(t)dt < \infty$$

our new time variable is bounded

$$\tau \leq \int\limits_0^\infty f(t)dt$$

In this case a new feature appears: diffusion completely stops at the spatial scale

$$\sqrt{D_0 \int\limits_0^\infty f(t)dt}$$

This phenomenon is referred to as *diffusive closure*. For instance, if

$$D(t) = \frac{D_0}{1 + (t/t_0)^\alpha}, \qquad t_0 = \text{const.}$$

such that $D(t)$ decreases as $1/t^\alpha$ for $t \rightarrow \infty$, diffusion occurs as usual, if $0 < \alpha < 1$ and stops, if $\alpha > 1$. For the critical $\alpha = 1$ case the diffusion, being extremely slow, formally never stops.

In a typical application the diffusivity decreases with decreasing temperature and stops completely at some blocking temperature (Rollinson, 1993; Chakraborty, 2008). This determines the *closure* of a geochronological or petrological system where diffusive re-equilibration is essentially "frozen" during cooling (Dodson, 1973). Given that the diffusivity D follows the Arrhenius relation

$$D(T) = D_\infty \exp\left(-\frac{E}{RT}\right)$$

where D_∞ is the diffusion coefficient for $RT \gg E$, and the temperature varies with time, $T = T(t)$, then

$$D(T) = D_\infty \exp\left[-\frac{E}{RT(t)}\right]$$

A typical cooling history could be

$$T(t) = \frac{T_0}{1 + st/T_0}$$

where T_0 is the starting temperature. This corresponds to cooling at a constant cooling rate s in degrees per second for $st \ll T_0$ during the initial stages of cooling so that $T \approx T_0 - st$. During later stages of cooling successively lower cooling rates prevail. For this specific cooling history the time-dependent diffusivity reads

$$D(t) = D_\infty \exp\left[-\frac{E}{RT_0} - \frac{Est}{RT_0^2}\right] = D_0 \exp\left[-\frac{Est}{RT_0^2}\right]$$

where $D_0 = D_\infty \exp(-E/RT_0)$ has been used. Let $\gamma = Es/RT_0^2$ then this becomes

$$D(t) = D_0 e^{-\gamma t}$$

Given that $D(t) = D_0 f(t)$ we have $f(t) = e^{-\gamma t}$. Recalling that rescaled time is given by $\tau = \int_0^t f(t')dt'$ we obtain

$$\tau = \frac{1}{\gamma}(1 - e^{-\gamma t})$$

with

$$\lim_{t \to \infty}(\tau) = \tau_{\max} = \frac{1}{\gamma} = \frac{RT_0^2}{Es}$$

so that τ is bounded.

For estimating the critical temperature T_c when *diffusive closure* occurs in a cooling system, a characteristic length a in the system, for example the radius of a spherical mineral grain, or the half width of a tabular grain, *etc.* is compared to the characteristic diffusion length. For diffusion during cooling, the characteristic diffusion length is determined from the standard Einstein relation but in terms of D_0 and rescaled time τ

$$l = \sqrt{2D_0\tau}$$

Employing the upper bound of τ the maximum diffusion length reads

$$l_{\max} = \sqrt{2D_0\tau_{\max}} = \sqrt{\frac{2D_\infty RT_0^2}{Es} \exp\left(-\frac{E}{RT_0}\right)}$$

Setting $l_{\max} = a$ and $T_0 = T_c$ and re-arranging we obtain

$$T_c = \frac{E/R}{\ln\left[2D_\infty RT_c^2/(a^2 Es)\right]}$$

from which the *closing temperature* T_c is calculated in recursive procedure (Dodson, 1973).

Another principal feature of the diffusion equation comes into play, when we address, *e.g.* equation 13, in order to calculate $n(\mathbf{r},t)$ from $n(\mathbf{r},0)$ in a bounded domain $\Omega \in \mathfrak{R}^3$. The solution cannot be found unless some additional information on $n(\mathbf{r},t)$ at the boundary $\partial\Omega$ is provided (Crank, 1975). Indeed, a direct computation of the term $\nabla^2 n(\mathbf{r},t)$ for $\mathbf{r} \in \partial\Omega$ requires knowledge of $n(\mathbf{r},t)$ outside Ω, which is not available. A boundary condition in the form of either prescribed concentration or prescribed flux for all $\mathbf{r} \in \partial\Omega$ is required. The most useful boundary conditions are that of Dirichlet

$$n(\mathbf{r},t)|_{\mathbf{r}\in\partial\Omega} = \text{const}$$

and that of Neumann

$$\mathbf{m}\cdot \nabla n(\mathbf{r},t)|_{\mathbf{r}\in\partial\Omega} = 0$$

where we recall that $\mathbf{m}$ denotes a unit vector that is orthogonal to the boundary $\partial\Omega$ and is directed outward. Even if one considers equation 13 in the whole $\mathfrak{R}^3$ space without any boundaries, one typically implies restrictions like

$$n(\mathbf{r},t) \to 0 \text{ for } |\mathbf{r}| \to \infty \tag{20}$$

and these restrictions replace the boundary conditions.

Let us now recall that diffusion leads to an irreversible homogenization of the system in question. The irreversibility can be recognized from the analysis of some observables, which are derived from $n(\mathbf{r},t)$ and which permanently increase or decrease as the system evolves. The observables never return to their original values. For instance, consider diffusion in $\mathfrak{R}^3$ and the so-called free-energy functional

$$F[n] = \int (\nabla n)^2 d^3\mathbf{r}$$

with the additional condition (20). Using partial integration and equation 13 one derives that

$$\frac{dF}{dt} = -2 \int \frac{\partial n}{\partial t}(\nabla^2 n)d^3\mathbf{r} = -2D \int (\nabla^2 n)^2 d^3\mathbf{r} \leq 0$$

such that $F[n]$ indeed decreases. For the above specified problem posing, the permanent decrease of the free energy goes along with the conservation of the particle number

$$N = \int n(\mathbf{r}, t)d^3\mathbf{r} = \text{const.}$$

As a second example let us consider behaviour of the entropy-like quantity

$$S[n] = -\int n\ln(n)d^3\mathbf{r}$$

which is defined following Boltzmann. It is easy to calculate that

$$\frac{dS}{dt} = \frac{dN}{dt} - D \int (\nabla^2 n)\ln(n)d^3\mathbf{r} = D \int \frac{(\nabla n)^2}{n}d^3\mathbf{r} \geq 0$$

such that $S[n]$ increases.

After the diffusive evolution is accomplished, the system enters into a stationary state in which

$$\frac{\partial^2 n}{\partial x^2} + \frac{\partial^2 n}{\partial y^2} + \frac{\partial^2 n}{\partial z^2} = 0$$

solutions of the latter equation are referred to as harmonic functions. For instance, particle concentration may take a constant value, as in the case of full thermodynamic equilibration. A less trivial harmonic solution is given by a linear function for a stationary flux of particles between two reservoirs. Non-trivial solutions are also possible in a bounded domain, *e.g.* the simplest cylindrically symmetric harmonic function is given by the expression $x^2 + y^2 - 2z^2$.

3. Self-similar solutions

There are hundreds of solutions to the linear diffusion equation in different contexts, a simple list of such solutions would fill an entire book (see Jost, 1965; Crank, 1975, for example). Instead of presenting an exhaustive list we focus on the special class of so-called *self-similar solutions*. They are of particular importance for solving inverse problems such as extraction of diffusivities from experimental data or reconstruction of time scales from frozen diffusion profiles in minerals and rocks. Moreover, self-similar solutions can also be obtained for the non-linear diffusion equation.

To illustrate the idea, we consider equation 13 in which we rescale variables by introducing

$$t' = \frac{t}{T} \qquad \text{and} \qquad \mathbf{r}' = \frac{\mathbf{r}}{L}$$

Equation 13 is transformed to the form

$$\frac{\partial n}{\partial t'} = \frac{TD}{L^2} \left(\frac{\partial^2 n}{\partial x'^2} + \frac{\partial^2 n}{\partial y'^2} + \frac{\partial^2 n}{\partial z'^2} \right)$$

where the space scale L may be chosen so that the numerical value of the rescaled diffusion coefficient is equal to 1. This is a convenient way to do numerics. The self-similar solutions are somewhat similar to the above rescaling, however the variables are now rescaled through each other. For instance, one can rescale x using y or rescale both x and y using $\sqrt{Dt}$. Below we discuss briefly the mathematics and then provide several practical applications of the technique.

3.1. Birkhoff's idea

Reduction of a complex mathematical structure to a more simple one without a noticeable change of its properties is a welcome step in natural sciences. The most important example is, of course, thermodynamics, where motion of myriads of atoms is reduced to a few simple laws. In this section we deal with another fundamental reduction that is related to the symmetries of partial differential equations and results in the so-called *self-similar solutions*. The latter were systematically considered by Birkhoff (1950).

The simplest and most important case is the so-called *scaling symmetry*. For instance, numerical values of the standard coordinates x, y, z depend on units, whereas their ratios are invariant. One may assume that some dimensionless observable, e.g. $\Theta = \Theta(x,y,z)$ that formally depends on three variables, actually depends only on two invariant combinations, $\Theta = \Theta(x/z, y/z)$. Therefore one geometric dimension is eliminated.

In a more complex situation, the reduced function has some physical dimension and may require a pre-factor. For instance, the internal energy U of a one-component thermodynamic system depends on three variables: entropy, S, volume, V and number of particles, N. The concept of extensive and intensive variables suggests the reduction

$$U = U(S, V, N) = Nu\left(\frac{S}{N}, \frac{V}{N}\right)$$

where the energy per particle u depends on only two variables, the entropy per particle and the volume per particle. Again, one degree of freedom is eliminated.

Let us return to the diffusion problem. Although the general theory behind symmetries and reductions is sophisticated (Dresner, 1983), the final recipe is remarkably simple. For the sake of brevity we consider only one space dimension. Assume that solution of the diffusion equation should yield the yet unknown distribution $n = n(x,t)$. One can then try the following substitution

$$n = t^{\alpha}f\left(\frac{x}{t^{\beta}}\right) \tag{21}$$

After proper choice of the two free parameters α and β one can hope to reduce a partial differential equation for $n(x,t)$ to a simpler ordinary differential equation for $f(\xi)$, where $\xi = x/t^{\beta}$ is the so-called *self-similar variable*. In the following we show several typical examples.

3.2. Boltzmann's variable

As a model example of the scaling technique let us consider the non-linear diffusion equation in one spatial dimension

$$\partial_t n = \partial_x[D(n)\partial_x n] \tag{22}$$

where the diffusion coefficient depends on concentration. A general solution of equation 22, $n = n(x,t)$, involves both space and time variables. As first suggested by Boltzmann, we consider a special class of self-similar solutions and set

$$n = f(\xi) \quad \text{and} \quad \xi = \frac{x}{\sqrt{t}} \tag{23}$$

One can check directly that the partial differential equation 22 is reduced to a more tractable ordinary differential equation

$$-\frac{1}{2}\xi\frac{df}{d\xi} = \frac{d}{d\xi}\left[D(f)\frac{df}{d\xi}\right] \tag{24}$$

It is important to realize, that there are many possible self-similar variables, *e.g.* one could use $\xi = x/(2\sqrt{t})$, or $\xi = \sqrt{t}/x$, or even $\xi = x^2/t$. The corresponding semi-scale equations may look different; still they describe the same physics and we will not distinguish between them. If a typical reference value D_r of the diffusion coefficient is available, we will prefer the following dimensionless definition

$$\xi = \frac{x}{2\sqrt{D_r t}} \tag{25}$$

where the related ordinary differential equation for $f(\xi)$ reads

$$-2\xi \frac{df}{d\xi} = \frac{d}{d\xi}\left[\frac{D(f)}{D_r}\frac{df}{d\xi}\right] \tag{26}$$

In particular, equation 25 will be used if $D(n)$ is just a constant, then

$$-2\xi \frac{df}{d\xi} = \frac{d^2 f}{d\xi^2}$$

the latter equation is the simplest representative of the general family of semi-scale solutions to the linear diffusion equation [see equation 28].

Another important remark is that the self-similar solutions are just special solutions of equation 22, the general solution need not necessarily be self-similar. Still, self-similar solutions naturally appear, if they are compatible with the initial and boundary conditions. For instance, equation 22 is expected to "start" from a known initial distribution $n(x,t=0)$ and to yield $n(x,t>0)$. With respect to Boltzmann's variable (23) we have:

$$\text{for a given } x \neq 0 \text{ and } t \to 0 \quad n = f(\xi) \to \begin{cases} f(+\infty) & \text{if } x > 0 \\ f(-\infty) & \text{if } x < 0 \end{cases}$$

Therefore, $f(+\infty)$ is the only possible initial value for all $n(x>0,t=0)$ and $f(-\infty)$ is the only possible initial value for all $n(x<0,t=0)$. The compatible initial conditions for the diffusion equation read

$$n(x < 0,t = 0) = n_- = \text{const}, \qquad n(x > 0,t = 0) = n_+ = \text{const} \tag{27}$$

these conditions are mapped onto the following condition for $f(\xi)$

$$f(-\infty) = n_-, \qquad f(+\infty) = n_+$$

Equation 27 naturally applies when two reservoirs with constant but different concentrations of the diffusing atoms are put into contact with each other. Without loss of generality we describe the contact plane by $x = 0$ and apply the self-scaling substitution. Other situations in which the self-similar substitution "respects" all boundary conditions will be considered in what follows.

It is worth mentioning that the ordinary differential equation 24 for $f(\xi)$ is of second order and requires two initial conditions. In a standard formulation one starts with, say, $f(0)$ and $f'(0)$, and directly calculates $f(\xi)$. In our case the conditions $f(\pm\infty) = n_\pm$ are given at different points and the straightforward numerical solution is not possible.

Some kind of shooting method should be used instead. We now turn to specific examples of the self-similar solutions.

3.3. Linear diffusion

In this section we consider self-similar solutions of the linear diffusion equation 14 in one spatial dimension. In accord with both, the general substitution (21) and the definition of Boltzmann's variable (25), we set

$$n(x,t) = (Dt)^{\varepsilon/2} f(\xi), \quad \xi = x/2\sqrt{Dt} \tag{28}$$

where ε is an arbitrary parameter. Actually, we have parametrized a family of self-similar solutions by ε. All members of the family are compatible with the linear diffusion equation 14. The related ordinary differential equation reads

$$\frac{d^2 f}{d\xi^2} + 2\xi \frac{df}{d\xi} - 2\varepsilon f = 0 \tag{29}$$

The specific choice of ε is dictated by the problem at hand. Three typical situations are as follows.

1. Integral restriction: Assume that no boundaries are present, and therefore the diffusing atoms just redistribute in the entire space while maintaining their total number

$$N = \int_{-\infty}^{\infty} n(x, t) dx = \text{const.} \tag{30}$$

   Equation 28 indicates that $\varepsilon = -1$ is the only value compatible with equation 30, because in this case

   $$\int_{-\infty}^{\infty} n(x, t) dx = \frac{1}{\sqrt{Dt}} \int_{-\infty}^{\infty} f\left(\frac{x}{2\sqrt{Dt}}\right) dx = 2 \int_{-\infty}^{\infty} f(\xi) d\xi$$

   and we have an integral restriction

   $$\int_{-\infty}^{\infty} f(\xi) d\xi = \frac{N}{2}$$

   for a suitable solution of equation 29.

2. Dirichlet condition: Assume that we have a prescribed concentration value n_b at the boundary $x = 0$. Equation 28 implies that

   $$\varepsilon = 0 \text{ and } f(0) = n_b \Rightarrow n(0,t) = n_b \tag{31}$$

   Therefore equation 29 must be solved with the Dirichlet boundary condition at $\xi = 0$. The second condition is that $f(\xi)$ vanishes for $\xi \to \infty$.

3. Neumann condition: Assume that we have a prescribed flux j_b at the boundary $x = 0$. Equation 28 implies that

$$\varepsilon = 1 \quad \text{and} \quad f'(0) = -\frac{2j_b}{D} \Rightarrow -D\left(\frac{\partial n}{\partial x}\right)_{x=0} = j_b \tag{32}$$

Therefore equation 29 must be solved with the Neumann boundary condition at $\xi = 0$. The second condition is that $f(\xi)$ vanishes for $\xi \to \infty$.

We now can derive the three most important self-similar solutions of equation 14.

3.3.1. The point-source solution

Assume that N atoms are inserted at a small region near $x = 0$ at $t = 0$. The corresponding initial condition is given by Dirac's delta-function

$$n(x,t = 0) = N\delta(x) \tag{33}$$

which indicates simply that $n(x,t = 0)$ vanishes everywhere but at $x = 0$. The total number of atoms is conserved, therefore we impose equation 30, the latter is valid for $t = 0$

$$\int_{-\infty}^{\infty} n(x, t = 0)dx = N \int_{-\infty}^{\infty} \delta(x)dx = N$$

by definition of the delta-function. The initial condition (33) is compatible with equation 27 and requires $f(\pm\infty) = 0$. In accord with the first example from the previous section we set $\varepsilon = -1$ in equation 28

$$n(x, t) = \frac{1}{\sqrt{Dt}}f(\xi), \qquad \xi = \frac{x}{2\sqrt{Dt}}$$

and in equation 29. Altogether, the self-similar substitution reduces equation 14 to the following set

$$\frac{d}{d\xi}\left(\frac{df}{d\xi} + 2\xi f\right) = 0, \quad f(\pm\infty) = 0, \quad \int_{-\infty}^{\infty} f(\xi)d\xi = \frac{N}{2}$$

An immediate consequence of the above equations is that

$$f(\xi) = \frac{N}{2\sqrt{\pi}}e^{-\xi^2} \tag{34}$$

and that

$$n(x, t) = \frac{N}{2\sqrt{\pi Dt}}\exp\left(\frac{x^2}{4Dt}\right) \tag{35}$$

which is the famous point-source solution.

It is worth mentioning that solutions (35) can be combined, as long as we deal with the linear equation 14. For instance, if N_a atoms are inserted at $x = a$ and N_b at $x = b$, the suitable solution of equation 14 reads

$$n(x,t) = \frac{N_a}{2\sqrt{\pi Dt}} \exp\left[-\frac{(x-a)^2}{4Dt}\right] + \frac{N_b}{2\sqrt{\pi Dt}} \exp\left[-\frac{(x-b)^2}{4Dt}\right]$$

Moreover, if $n(x,t=0) = n_0(x)$, one can perform a virtual discretization of the Ox axis with the step da and assume that $n_0(a)da$ atoms are inserted at $x = a$ for each discrete a, therefore

$$n(x,t) = \sum_a \frac{n_0(a)da}{2\sqrt{\pi Dt}} \exp\left[-\frac{(x-a)^2}{4Dt}\right]$$

Replacing the discrete sum of point-source solutions by integration, we find that the time evolution of the initial distribution $n_0(x)$ is given by

$$n(x,t) = \int_{-\infty}^{\infty} \frac{n_0(a)}{2\sqrt{\pi Dt}} \exp\left[-\frac{(x-a)^2}{4Dt}\right] da$$

Thus, the general solution of the initial-value problem for equation 14, which is not a self-similar solution as such, is still a simple combination of self-similar solutions. The latter, of course, are of fundamental importance.

3.3.2. Linear Dirichlet problem

Assume that all diffusing atoms are initially located in some "reservoir", in the half-space $x < 0$, where their concentration is kept constant. For $t > 0$ they start to penetrate into the $x > 0$ domain. The profile $n(x,t)$ is expected to be a decreasing function that takes some prescribed boundary value n_b at $x = 0$ and vanishes for $x \to \infty$. The corresponding set of boundary and initial conditions reads

$$n(0,t) = n_b, \qquad n(x \to \infty,t) = 0 \tag{36}$$

$$n(t = 0, x > 0) = 0 \tag{37}$$

and applies to equation 14 which should be solved in the half-space $x \geqslant 0$. The solution appears to be self-similar.

Indeed, the initial condition (37) is compatible with equation 27, if we consider only positive ξ and set $f(\infty) = 0$. In accord with equation 31 we set $\varepsilon = 0$ both in equation 28

$$n(x,t) = f(\xi), \qquad \xi = \frac{x}{2\sqrt{Dt}}$$

and in equation 29. Altogether, the self-similar substitution reduces equation 14 with the additional conditions (36)–(37) to the following set

$$\frac{d^2 f}{d\xi^2} + 2\xi \frac{df}{d\xi} = 0, \qquad f(0) = n_b, \qquad f(\infty) = 0$$

A simple integration yields

$$f(\xi) = \frac{2n_b}{\sqrt{\pi}} \int\limits_{\xi}^{\infty} \exp(-s^2)ds = n_b \mathrm{erfc}(\xi) \tag{38}$$

where we use the standard notations for the error function

$$\mathrm{erf}(\xi) = \frac{2}{\sqrt{\pi}} \int\limits_{0}^{\xi} \exp(-s^2)ds, \qquad \mathrm{erfc}(\xi) = 1 - \mathrm{erf}(\xi)$$

The final result reads

$$n(x,t) = n_b \mathrm{erfc}\left(\frac{x}{2\sqrt{Dt}}\right) \tag{39}$$

and represents an important self-similar solution of the linear diffusion equation, the so-called erfc solution.

The total number of particles that leave the reservoir by time t is given by

$$N(t) = \int\limits_{0}^{\infty} n(x,t)dx = 2n_b\sqrt{Dt} \int\limits_{0}^{\infty} \mathrm{erfc}(\xi)d\xi = 2n_b\sqrt{\frac{Dt}{\pi}}$$

and increases as opposed to the point-source problem. Measurements of $N(t)$ provide a simple way to estimate D. If the values of D calculated for different times differ too much from each other, the diffusion process may be non-linear.

In full analogy with the point-source solution (35), expressions of the type (39) can be combined and yield explicit solutions for many important diffusion problems.

3.3.3. Linear Neumann problem

As in the previous section, we consider atoms that penetrate into the halfspace $x > 0$ for $t > 0$. However, now we assume that the particle flux is kept constant, $j = j_b$, at the boundary $x = 0$. The corresponding set of boundary and initial conditions reads

$$-D\left(\frac{\partial n}{\partial x}\right)_{x=0} = j_b, \qquad n(x \to \infty, t) = 0 \tag{40}$$

$$n(t = 0, x > 0) = 0 \tag{41}$$

and applies to equation 14 which should be solved in the half-space $x \geqslant 0$. The solution turns out to be self-similar.

Indeed, the initial condition (41) is compatible with equation 27, if we consider only positive ξ and set $f(\infty) = 0$. In accord with equation 32 we set $\varepsilon = 1$ in equation 28

$$n(x,t) = \sqrt{Dt}\, f\left(\frac{x}{2\sqrt{Dt}}\right), \qquad \xi = \frac{x}{2\sqrt{Dt}}$$

and in equation 29. Altogether, the self-similar substitution reduces equation 14 with the additional conditions (40)–(41) to the following set

$$\frac{d^2f}{d\xi^2} + 2\xi\frac{df}{d\xi} - 2f = 0, \quad f'(0) = -\frac{2j_b}{D}, \quad f(\infty) = 0$$

Direct integration yields

$$f(\xi) = \frac{2j_b}{D}\left[\frac{1}{\sqrt{\pi}}e^{-\xi^2} - \xi\,\mathrm{erfc}(\xi)\right] \tag{42}$$

we write the result as

$$n(x,t) = \frac{2j_b}{D}\left[\sqrt{\frac{Dt}{\pi}}\exp\left(-\frac{x^2}{4Dt}\right) - \frac{x}{2}\mathrm{erfc}\left(\frac{x}{2\sqrt{Dt}}\right)\right]$$

In particular, the boundary concentration reads

$$n(0,t) = 2j_b\sqrt{\frac{t}{\pi D}}$$

The total number of particles that has left the reservoir by time t is given by $j_b t$

$$N(t) = \int_0^\infty n(x,t)dx = 2Dt\int_0^\infty f(\xi)d\xi = j_b t$$

3.3.4. Summary
The most important properties of the self-similar solutions derived above for the linear diffusion equation 14 are given in Table 1 and shown in Fig. 1.

4. Non-linear diffusion
In this section we return to the general non-linear diffusion equation 22. We consider typical models for $D(n)$ and discuss how $D(n)$ can be reconstructed from experimental data.

Table 1. In all three types of self-similar solutions to equation 14 the self-similar variable is $\xi = x/(2\sqrt{Dt})$. The solution domain is $-\infty < x < \infty$ for the first row and $0 \leqslant x < \infty$ for the other rows. The parameters n_b and j_b refer to the Dirichlet and Neumann boundary conditions at $x = 0$.

Process	$f(\xi)$	$n(x,t)$	$n(0,t)$	$\int n(x,t)dx$
Spreading	$\frac{N}{2\sqrt{\pi}}e^{-\xi^2}$	$\frac{1}{\sqrt{Dt}}f\left(\frac{x}{2\sqrt{Dt}}\right)$	$\frac{N}{2\sqrt{\pi Dt}}$	N
Dirichlet	$n_b\mathrm{erfc}(\xi)$	$f\left(\frac{x}{2\sqrt{Dt}}\right)$	n_b	$2n_b\sqrt{\frac{Dt}{\pi}}$
Neumann	$\frac{2j_b}{D}\left[\frac{e^{-\xi^2}}{\sqrt{\pi}} - \xi\,\mathrm{erfc}(\xi)\right]$	$\sqrt{Dt}f\left(\frac{x}{2\sqrt{Dt}}\right)$	$2j_b\sqrt{\frac{t}{\pi D}}$	$j_b t$

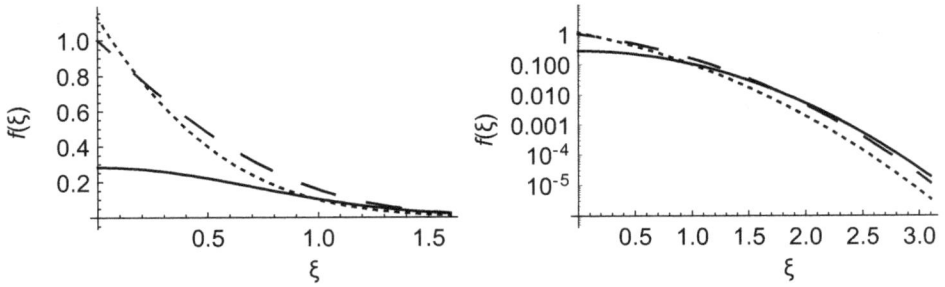

Figure 1. Left: plot of the characteristic functions $\frac{1}{2\sqrt{\pi}}e^{-\xi^2}$ (solid), erfc(ξ) (dashed), and $\frac{1}{\sqrt{\pi}}e^{-\xi^2} - \xi\text{erfc}(\xi)$ (dotted) for the typical solutions from Table 1. Right: logarithmic plot of the same functions; here all solutions look rather similar, thus revealing the diffusion character from experimental data may be demanding.

4.1. Boltzmann equation

Let us consider the following simple model for a concentration-dependent diffusion coefficient

$$D(n) = kn, \quad k = \text{const} > 0 \tag{43}$$

The constitutive law for diffusion flux is then

$$j = -kn\partial_x n$$

In this scenario the diffusion flux decreases quickly with decreasing concentration resulting in a closure-type effect: all diffusing particles may be trapped in some space domain, and the particle density vanishes outside this domain. However, the spatial scale of the domain increases permanently; the increase, while slow, never stops (see equation 46). The nonlinear diffusion equation corresponding to the model (43) reads

$$\partial_t n = k\partial_x(n\partial_x n) \tag{44}$$

which was first considered by Boltzmann (see Dresner (1983) and references therein).

We assume that at $t = 0$ the diffusing atoms start to penetrate into the half-space $x > 0$ from the boundary at $x = 0$, where their concentration is kept constant, $n = n_b$. Moreover, the half-space $x > 0$ is populated initially by atoms of the same kind with the concentration $n = n_+$. Therefore,

$$n(x = 0,t) = n_b, \quad n(x > 0, t = 0) = n_+$$

where it is natural to assume that

$$n(x \rightarrow \infty, t) = n_+$$

cf. equation 27.

As in the previous section, one may try the self-similar solution (28) having in mind the Dirichlet boundary condition, *i.e.* with $\varepsilon = 0$

$$n(x, t) = n_b f(\xi), \quad \xi = \frac{x}{2\sqrt{D_r t}}$$

where the boundary value n_b is used for both to normalize $f(\xi)$ and to define the reference value of the diffusivity

$$D_r = kn_b$$

Now we insert the above self-similar substitution into equation 44 and obtain the following ordinary differential equation for $f(\xi)$

$$f\partial_\xi^2 f + (\partial_\xi f)^2 + 2\xi\partial_\xi f = 0 \qquad (45)$$

where

$$f(0) = 1, \quad f(\infty) = \frac{n_+}{n_b}$$

Equation 45 can be solved numerically starting from the boundary at $\xi = 0$ with $f(0) = 1$ and taking some test initial slope $f'(0)$. The numerical solution yields the end-value $f(\xi \to \infty)$, see Fig. 2. Finally the initial slope is adjusted to match the end value and n_+/n_b.

In particular, Fig. 2 indicates that, if particles penetrate into the empty domain, $n_+ = 0$, we should use the most negative initial slope that happens to be $f'(0) \approx -0.887$. Here all diffusing particles are completely trapped in the domain $|\xi| < 0.808$ such that

$$n(x,t) \neq 0 \quad \text{for} \quad |x| < 1.616\sqrt{kn_b t} \qquad (46)$$

Further decrease of $f'(0)$ leads to unphysical solutions with negative values of concentration.

Concentration-distance data resembling the curves in Fig. 2 are known from diffusion in ionic crystals involving heterovalent substitution. A typical situation is the in-diffusion of cations with higher valence state than the majority cation on some sub-lattice of a crystal. When the extra charge is compensated by the formation of cation vacancies, this enhances diffusion, where diffusion enhancement increases with the extent of the heterovalent substitution. This phenomenon has been described from diffusion of divalent cations in alkali halides (Lidiard, 1955), and from the diffusion of trivalent cations in periclase (Van Orman et al., 2009).

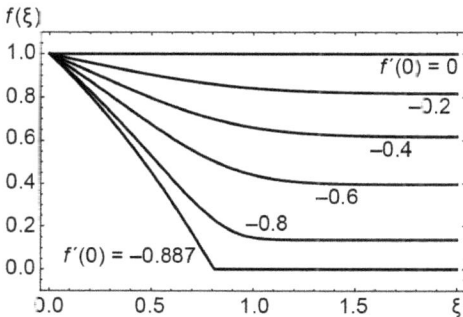

Figure 2. Numerical solutions of equation 45 for $f(0) = 1$ and different slopes at $\xi = 0$. Each solution yields its own limiting value $f(+\infty)$. The most negative slope corresponds to trapped particles: $f(\xi)$ strictly vanishes outside some domain in space.

4.2. Boltzmann-Matano method

A standard mathematical application of the diffusion equation is to start from some initial distribution $n(\mathbf{r},t = 0)$ and to calculate $n(\mathbf{r},t > 0)$; the result can be compared to observations. Such calculation can be performed, if all equation parameters, e.g. diffusivity and boundary values of concentration, are already known. In this section we will deal with the inverse problem, where one starts from measurements and uses a kind of "inverse engineering" to reconstruct unknown equation parameters. Specifically we will start from $n(\mathbf{r},t)$ and quantify diffusivity.

For instance, let us reconstruct the parameter D in the simplest diffusion equation 14 assuming that all diffusing particles penetrate into the halfspace $x > 0$ with the prescribed boundary value $n(x = 0,t) = n_b$. We know from Table 1 that the total number of the diffusing particles by time t (per unit area in the YZ) plane is given by

$$N_{tot}(t) = \int_0^\infty n(x,t)dx = 2n_b\sqrt{\frac{Dt}{\pi}}$$

This can be used to estimate D. If a time-series for $N_{tot}(t)$ is available, one can plot $N_{tot}(t)$ against $\sqrt{t}$ and use least squares fitting to quantify the slope $2n_b\sqrt{D/\pi}$. Moreover, if the full spatial distribution $n(x,t)$ of the diffusing atoms at time t is known, one can apply the explicit solution from Table 1 that we put into the form

$$\mathrm{erfc}^{-1}\left[\frac{n(x,t)}{n_b}\right] = \frac{x}{2\sqrt{Dt}}$$

where erfc^{-1} denotes the inverse function. Now, one can plot $\mathrm{erfc}^{-1}(n/n_b)$ vs. $x/\sqrt{t}$ to derive the slope and to quantify D. Similarly, the inverse approach can be used for extracting time scales of a diffusion event or cooling rate from concentration-distance data observed in minerals, if the relevant diffusivity and its temperature dependence, in the case of diffusion during cooling, are known. Then $\mathrm{erfc}^{-1}(n/n_b)$ is plotted vs. x/D, or vs. x/D_0 in the case of cooling, to quantify t or τ, respectively. This provides the basis for applications in geo-speedometry (Chakraborty, 2008).

The inverse engineering becomes more complicated for non-linear diffusion with $D = D(n)$. Consider, for instance, the same Dirichlet problem as above, but for equation 22. Using Boltzmann's ξ-variable (25) and the self-similar substitution $n(x,t) = f(\xi)$ we arrive at the ordinary differential equation 26

$$\frac{d}{d\xi}\left[D(f(\xi))\frac{df(\xi)}{d\xi}\right] = -2D_r\xi\frac{df(\xi)}{d\xi}$$

We are now interested in deriving $D(f)$ from $f(\xi)$. We recall that D_r is a reference diffusivity that yields dimensionless $\xi = x/(2\sqrt{D_r t})$.

The latter ordinary differential equation is integrated over the self-similar variable starting from some arbitrary ξ-value and till infinity

$$D(f(\xi))\frac{df(\xi)}{d\xi} = 2D_r \int_{\xi}^{\infty} \xi' \frac{df(\xi')}{d\xi'} d\xi' \tag{47}$$

where we naturally assume that both $f(\xi)$ and its first derivative vanish for $\xi \to \infty$. Now the key idea, first introduced by Matano (1933), is to change from $f(\xi)$ to the inverse function $\xi(f)$. Parameter f becomes the new independent variable; according to the standard rules of analysis $d\xi$ and $df/d\xi$ are replaced by

$$d\xi \mapsto \frac{d\xi(f)}{df} df \qquad \frac{df(\xi)}{d\xi} \mapsto 1 / \frac{d\xi(f)}{df}$$

The integration limits $\xi < \xi' < \infty$ are transformed to $f > f' > 0$. After transformation to $\xi(f)$ equation 47 takes the form

$$D(f) = -2D_r \frac{d\xi(f)}{df} \int_{0}^{f} \xi(f') df' \tag{48}$$

and solves the inverse problem.

Being formally exact, equation 48 has its limitations. Firstly, assume that $n = n(x,t)$ is at our disposal at some time t. Taking a suitable D_r one can define $\xi = x/(2\sqrt{D_r t})$, plot concentration on the (n,ξ) plane, and name it a semi-scale solution $n = f(\xi)$. The inverse function and equation 48 yield $D = D(f)$. The result can be tested by direct numerical solution of equation 22 with the just-derived $D(f)$ or $D(n)$. As a rule, the newly calculated $n = n(x,t)$ is in perfect agreement with the old one. A problem arises, however, when several representatives of the time series $n = n(x,t)$ are available, and reconstructions of diffusivity for different times provide different results. If this is the case, the diffusive particle transport must be more complex than what is described by the self-similar solution of equation 22. For instance, either the initial or the boundary condition may be incompatible with the self-similar solution. Moreover, the experiment may involve additional processes such as trapping or liberation of particles, which are not accounted for by the model equation 22.

Another difficulty is that calculation of $d\xi/df$ is questionable for scattered experimental data such as in the exemplary Fig. 3. Here, one obtains different results for different "nice" smoothing functions $f(\xi)$ and $\xi(f)$. The reason is that for small values of concentration the integral in equation 48 tends to zero, whereas the derivative $d\xi/df$ tends to infinity. The product $0 \cdot \infty$ yields arbitrary results for diffusivity.

4.3. Anisotropic diffusivities
In this section we deal with the diffusion in a spatially uniform but anisotropic system, *e.g.* in a crystal. The basic equation is then the general equation 16, in which we ignore dependence of diffusivities on space variables but keep its dependence on direction and concentration

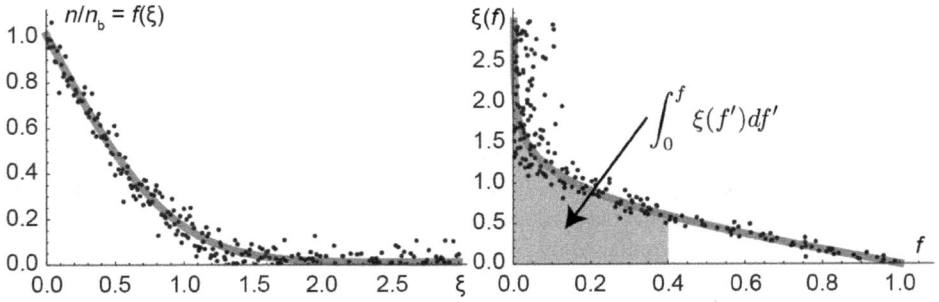

Figure 3. Left: typical set of measured concentrations versus the semi-scale variable (points) and a smooth average $f(\xi)$ (solid grey line). Right: corresponding inverse function and integration domain (light grey) in equation 48. Evidently for $f \to 0$ both the integral value and the slope $d\xi/df$ are not well defined and depend strongly on the procedure used to derive the smooth averaged $f(\xi)$ function (solid grey line).

$$\frac{\partial n}{\partial t} = \sum_{i,j=1}^{3} \frac{\partial}{\partial x_i}\left[D_{ij}(n)\frac{\partial n}{\partial x_j}\right] \tag{49}$$

Our goal is to reconstruct $D_{ij}(n)$ following the ideas from the previous section.

In order to investigate anisotropic diffusion in a crystal, one may prepare geometrically simple samples, *e.g.* polished plates with different orientations relative to the crystal lattice. The surfaces of the plate are put into contact with a large reservoir of the diffusing particles at sufficiently high temperature, so that some atoms (ions) from the reservoir exchange with the corresponding atoms (ions) in the crystal and penetrate into the crystal plate. After some time, the samples are quenched, and the concentration of the penetrated atoms is measured (Petrishcheva *et al.*, 2014; Schäffer *et al.*, 2014). For the chosen geometry, the diffusive transport in each sample is essentially one-dimensional. On the other hand, the diffusion rate is different for samples with different crystallographic orientation (see Fig. 4). Here mathematics comes into play and yields both the values of D_{ij} and their possible dependence on concentration.

To address the problem mathematically we note that the numerical values of the components of any tensor, *e.g.* D_{ij}, depend on the chosen frame of reference (Fleisch, 2011; Nye, 1957), such that we have to fix one. After the frame is fixed, each plate sample can be characterized by three components of the orientation vector **m**, the latter is perpendicular to the polished surface of the plate and indicates the direction of diffusion flux in each experiment (see Fig. 4). Thereafter one can introduce a new variable

$$\mathfrak{X} = xm_x + ym_y + zm_z$$

and utilize the fact that the diffusion is one-dimensional along **m** and therefore $n(\mathbf{r},t)$ can be replaced by $n(\mathfrak{X},t)$. Equation 49 is then replaced by

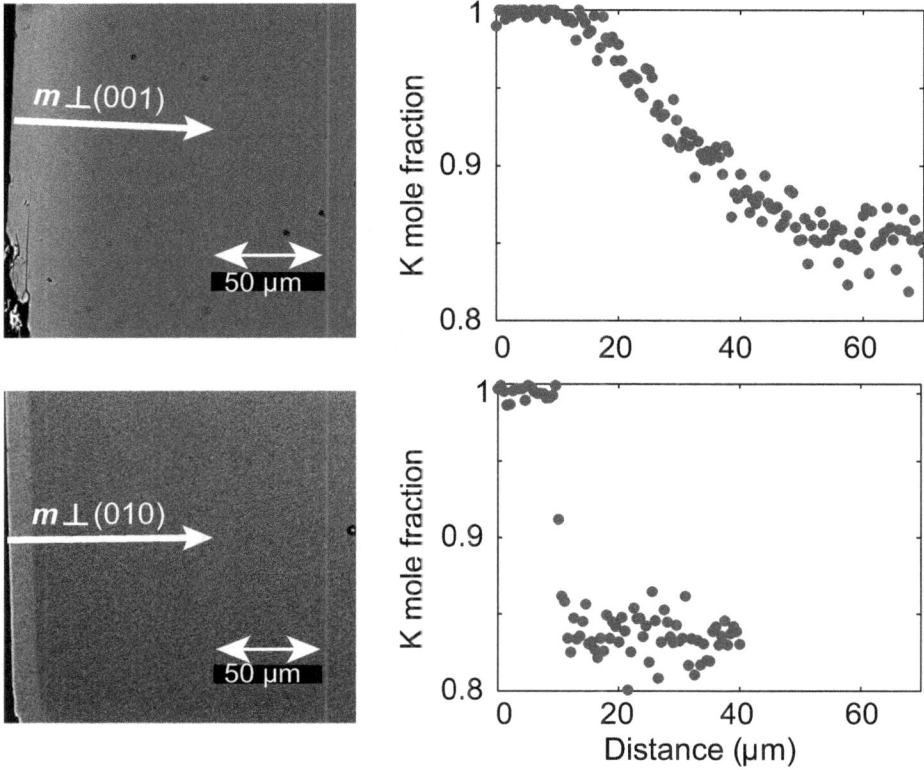

Figure 4. Backscattered electron (BSE) images (left) and composition profiles (right) produced from Na-K exchange between potassium-rich alkali feldspar ($X_K = 0.85$) and a KCl salt melt at 850°C and 1 bar for 8 days; top panel: diffusion perpendicular to (001), lower panel: diffusion perpendicular to (010); diffusion penetration depth and profile shapes are different for the two directions indicating substantial direction-dependence of Na-K interdiffusion. Reprinted with the permission of *American Journal of Science* from Petrishcheva *et al.* (2014).

$$\frac{\partial n}{\partial t} = \frac{\partial}{\partial \mathfrak{X}}\left[\mathfrak{D}(n)\frac{\partial n}{\partial \mathfrak{X}}\right]; \quad \mathfrak{D}(n) = \sum_{i,j=1}^{3} D_{ij}(n)m_i m_j \tag{50}$$

where $\mathfrak{D}(n)$ is the effective diffusion coefficient that results from the general tensor D_{ij} for the direction **m** (see Nye, 1957).

The function $\mathfrak{D}(n)$ can be reconstructed from the measured concentration profiles $n(\mathfrak{X},t)$ using the standard Boltzmann-Matano method as described in the previous section. Assume that S such reconstructions for S different directions $\mathbf{m}^{(s)}$ with $1 \leqslant s \leqslant S$ are performed and S scalar diffusivities $\mathfrak{D}^{(s)}(n)$ are available. Returning to the full tensor $D_{ij}(n)$ and using the second equation 50 we obtain the following system of equations

$$\sum_{i,j=1}^{3} D_{ij}(n) m_i^{(s)} m_j^{(s)} = \mathfrak{D}^{(s)}(n), \qquad s = 1, 2, ... S \tag{51}$$

Here $D_{ij}(n)$ are unknowns that should be determined from the known quantities $\mathbf{m}^{(s)}$ and $\mathfrak{D}^{(s)}(n)$. In other words, equation 51 is a system of linear algebraic equations that can be addressed using the standard methods of linear algebra. For the most generic triclinic system, the tensor D_{ij} contains six independent components and at least six sets of concentration-distance data for six different directions are required for their determination. For example, for a monoclinic system one can take the $(\mathbf{a}, \mathbf{b}, \mathbf{c}^*)$ axes as the reference coordinate frame, in which $\mathbf{b}$ is set parallel to the crystallographic diad axis. The diffusivity tensor then takes the form

$$D_{ij} = \begin{pmatrix} D_{11} & 0 & D_{13} \\ 0 & D_{22} & 0 \\ D_{31} & 0 & D_{33} \end{pmatrix}$$

where $D_{13} = D_{31}$ such that at least four measurements are required. Of course, a larger number of measurements is beneficial; the components D_{ij} can then be obtained from the same (now over-determined) linear system (51) by using the least-squares method.

In the experiments mentioned above (Petrishcheva et al., 2014; Schäffer et al., 2014) composition-distance data were obtained for Na-K interdiffusion along six different crystallographic directions of a monoclinic alkali feldspar. From this the components of the diffusivity tensor were obtained by applying the procedure described. The resulting components for the diffusivity tensor and their composition dependence are shown in Fig. 5. The representation quadric for the K-mole fraction $X_K = 0.86$ is shown in Fig. 6. For a monoclinic crystal the representation quadric is a triaxial ellipsoid with one major axis parallel to the crystallographic diad axis $\mathbf{b}$ and the other two major axes lying in the $\mathbf{a} - \mathbf{c}^*$ plane. The length of a radius vector is proportional to $1/\sqrt{D}$. It is seen that the major anisotropy is within the $\mathbf{a} - \mathbf{c}^*$ plane, which contains the direction of fastest diffusion parallel to [101] and slowest diffusion perpendicular to $(10\bar{1})$. Diffusion parallel to [010] (intermediate major axis of the quadric) is nearly as slow as perpendicular to $(10\bar{1})$. It must be noted that the inverse approach described here is the

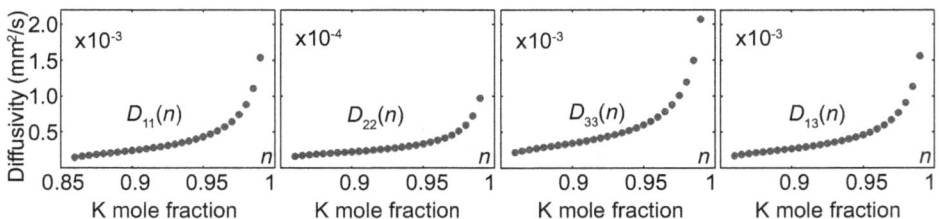

Figure 5. Exemplary application of equation 51 for reconstructing the diffusivity tensor for Na-K interdiffusion in monoclinic alkali feldspar, including its composition dependence. Reprinted with the permission of American Journal of Science from Petrishcheva et al. (2014).

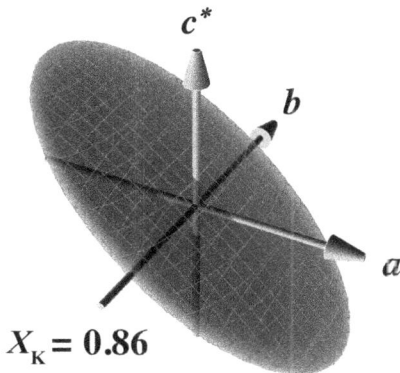

$X_K = 0.86$

Figure 6. The representation quadric of the diffusivity tensor is a three-axed ellipsoid, the lengths of the radius vectors are proportional to $1/\sqrt{D}$. Reprinted with the permission of *American Journal of Science* from Petrishcheva *et al.* (2014).

method of choice for quantifying Na-K interdiffusion including its direction- and composition-dependence. It does not, however, give any direct insight into the diffusion mechanisms.

5. Uphill diffusion

In this section we deal with a specific variant of non-linear diffusion, which is based on the intimate link between thermodynamics and diffusion. Actually, the resulting equation is so general that it is capable of describing such complex phenomena as phase separation. To explain the underlying concepts let us return to Fick's law (equation 1)

$$j(x, t) = -D(n) \frac{\partial n(x, t)}{\partial x}, \qquad D(n) > 0 \tag{52}$$

where we assume that diffusivity depends on concentration and consider one spatial dimension for brevity. Equation 52 indicates that the diffusive flux is directed towards lower values of concentration; the flux tends to make the distribution $n = n(x,t)$ spatially uniform. The imposed non-linearity $D = D(n) > 0$ does not change this behaviour qualitatively, only the rates of homogenization are different for different values of concentration. In short: we still deal with the standard downhill diffusion.

On the other hand, many natural systems show an opposite behaviour: particle flux is directed from domains of low concentration to domains with higher ones. Consider, for instance, a one-component two-phase system: small inclusions of the second phase with $n = n_2$ are embedded in the first phase — a bulk substance with the equilibrium concentration $n = n_1$, let $n_2 > n_1$. Assume that system temperature decreases and passes through a critical value such that the second — more dense — phase becomes energetically more favoured. Above the transition threshold the inclusions should dissolve and particle transport would occur in the downhill direction; one may hope to describe the transport in terms of a possibly non-linear diffusion equation. Below the transition threshold the second phase should grow. The particles should then move against the existing concentration gradient. This raises the question whether it is possible to describe this so-called "uphill diffusion" in terms of the standard equation 52?

At first glance, we should just allow diffusivity to be negative, $D(n) < 0$, for some interval of concentrations that are larger than n_1 but smaller than n_2. The values $D(n_1)$ and $D(n_2)$ should remain positive to keep both phases stable. The problem is that, e.g. equation 14 becomes mathematically invalid for $D < 0$. This is evident without any knowledge of mathematics just because the fundamental relations (8) and (9) do not make any sense for $D < 0$ (see also discussion of numerical solutions to the non-linear diffusion equation with negative diffusivity in Nauman and He, 2001). To resolve this difficulty we need to return to the fundamental principles and generalize equation 52. We will do this by adopting Onsager's approach.

5.1. Onsager's approach

We state that from the thermodynamic point of view there is no special reason that favours system evolution towards a spatially uniform state. The only requirement is that rearrangement of atoms pushes the system towards the state with the lowest possible free energy. The latter must be chosen in accord with the imposed conditions. If, for example, externally imposed pressure and temperature are kept constant, the Gibbs energy must decrease during any irreversible process. The final equilibrium state requires homogenous chemical potentials, such that further decrease of the Gibbs energy due to the rearrangement of atoms is impossible.

As long as the system in question is not in an equilibrium, special forces with thermodynamic origin act on atoms. These forces are directed opposite to the spatial gradients of the chemical potentials. Note, that if $n_1 = n_2$ then $\mu(n_1) = \mu(n_2)$ as well, but the reverse is not true as illustrated in Fig. 7. Atomic fluxes driven by gradients of chemical potentials in a multicomponent system are given by the Onsager-type relation

$$\mathbf{j}_\alpha = -\sum_\beta L_{\alpha\beta} \nabla \mu_\beta \qquad (53)$$

where Greek letters $\alpha, \beta = A, B, C...$ enumerate components, and $\mu_\beta = \mu_\beta(\mathbf{r}, t)$ is the chemical potential of component β such that $-\nabla \mu_\beta$ is the thermodynamic force per B-atom. The

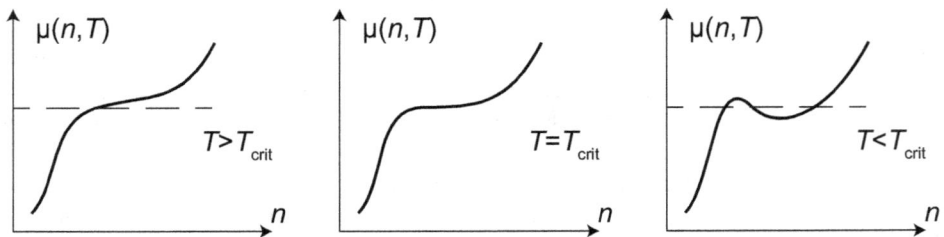

Figure 7. Schematic profiles of $\mu(n,T)$ for three different temperatures. Left: if $\mu(n_1) = \mu(n_2)$ then $n_1 = n_2$. Right: even if $\mu(n_1) = \mu(n_2)$ it is still possible that $n_1 \neq n_2$. Middle: for the critical temperature the derivative $\partial\mu/\partial n$ vanishes at some concentration.

parameters $L_{\alpha\beta}$ are yet unknown kinetic coefficients. In full analogy to equation 10 we require the reciprocal relation

$$L_{\alpha\beta} = L_{\beta\alpha}$$

One naturally expects that $L_{\alpha\alpha} > 0$. The atomic diffusivities are not present in equation 53 directly; these derived quantities are related to the kinetic coefficients in what follows. Having in mind further applications to solid-state diffusion we presuppose that atoms do not leave the system, occupy identical volumes and only rearrange within some fixed sub-lattice of a crystal. Therefore, $L_{\alpha\beta} < 0$ holds for $\alpha \neq \beta$, which implies that the driving force $-\nabla\mu_\beta$ in equation 53 yields a flux of component β in the same direction as the force ($L_{\beta\beta} > 0$) and fluxes of all the other components $\alpha \neq \beta$ in the opposite direction.

Here we consider only substitutional interdiffusion, which is a special case of interdiffusion for which the total volume concentration of atoms does not change

$$n_\Sigma = \sum_\alpha n_\alpha = \text{const.}$$

To address substitutional interdiffusion quantitatively, we impose an additional zero-sum condition for the kinetic coefficients:

> The sum of all kinetic coefficients in each column (and in each row) of the matrix $L_{\alpha\beta}$ vanishes.

A priori, the matrix of the kinetic coefficients for an N-component system contains N^2 elements, the reciprocal relation reduces the number of independent coefficients to $1/2N(N+1)$. The zero-sum condition yields N additional restrictions such that

[number of independent kinetic coefficients] $= (N(N-1))/2$

Note that all equations in what follows apply to a special frame of reference. Namely, we presuppose that atoms do not leave the system, they only rearrange (because of presence of vacancies) within some fixed crystalline sub-lattice. The latter defines the "laboratory" frame of reference in which the total flux vanishes. Let us now consider typical examples.

5.2. Flux models
In a two-component system with $\alpha = A,B$ only one kinetic coefficient is relevant

$$L = \begin{pmatrix} -L_{AB} & L_{AB} \\ L_{AB} & -L_{AB} \end{pmatrix}, \quad L_{AB} < 0$$

such that the diagonal elements of L are positive and

$$\mathbf{j}_A = L_{AB}\nabla(\mu_A - \mu_B), \qquad \mathbf{j}_B = L_{AB}\nabla(\mu_B - \mu_A) \tag{54}$$

As expected, the total flux vanishes, $\mathbf{j}_A + \mathbf{j}_B = 0$, *i.e.* the flux model is compatible with the interdiffusion condition $n_A + n_B = \text{const.}$ Combining the just derived expression for $\mathbf{j}_A$ with the continuity equation 11 we obtain

$$\frac{\partial n_A}{\partial t} + \nabla[L_{AB}\nabla(\mu_A - \mu_B)] = 0 \tag{55}$$

where the second equation for n_B is redundant because $n_B = n_\Sigma - n_A$. In what follows we will specify the involved chemical potentials and transform the latter equation to that for both the downhill and the uphill diffusion.

In a three-component system with $\alpha = A,B,C$ only three kinetic coefficients L_{AB}, L_{AC} and L_{BC} are relevant and we have

$$L = \begin{pmatrix} -L_{AB} - L_{AC} & L_{AB} & L_{AC} \\ L_{AB} & -L_{AB} - L_{BC} & L_{BC} \\ L_{AC} & L_{BC} & -L_{AC} - L_{BC} \end{pmatrix}$$

where we recall that all off-diagonal kinetic coefficients are negative; the positive diagonal elements are derived from the zero-sum condition. These kinetic coefficients yield fluxes

$$\mathbf{j}_A = L_{AB}\nabla(\mu_A - \mu_B) + L_{AC}\nabla(\mu_A - \mu_C) \tag{56}$$

$$\mathbf{j}_B = L_{AB}\nabla(\mu_B - \mu_A) + L_{BC}\nabla(\mu_B - \mu_C) \tag{57}$$

$$\mathbf{j}_C = L_{AC}\nabla(\mu_C - \mu_A) + L_{BC}\nabla(\mu_C - \mu_B) \tag{58}$$

As expected, the total flux vanishes, $\mathbf{j}_A + \mathbf{j}_B + \mathbf{j}_C = 0$. Such a three component system is governed by two continuity equations

$$\frac{\partial n_A}{\partial t} + \nabla[L_{AB}\nabla(\mu_A - \mu_B) + L_{AC}\nabla(\mu_A - \mu_C)] = 0 \tag{59}$$

$$\frac{\partial n_B}{\partial t} + \nabla[L_{AB}\nabla(\mu_B - \mu_A) + L_{BC}\nabla(\mu_B - \mu_C)] = 0 \tag{60}$$

where the missing third equation is redundant because the remaining concentration $n_C = n_\Sigma - n_A - n_B$.

Equation 55 and equations 59–60 can be used only after certain expressions for the chemical potentials are specified. Before doing that we will replace the kinetic coefficients by diffusivities. Note, that, e.g. in a one-component system, the expression $\mathbf{j} = -L\nabla\mu$ (which yields particle flux induced by the thermodynamic force $-\nabla\mu$ per atom) should be equivalent to the expression

$$\mathbf{j} = n\mathbf{v} = nb \cdot (-\nabla\mu) = -\frac{D}{k_B T}n\nabla\mu \Rightarrow L = \frac{D}{k_B T}n$$

where n and b denote particle density and mobility, respectively, and b is related to the diffusivity D in accord with Einstein's relation (8). Now we return to the general equation 53. In accord with the reciprocal relation, the kinetic coefficient $L_{\alpha\beta}$ describes: (i) the contribution of $-\nabla\mu_\beta$ to $\mathbf{j}_\alpha$; and (ii) the contribution of $-\nabla\mu_\alpha$ to $\mathbf{j}_\beta$. That is, for $\alpha \neq \beta$ the kinetic coefficient $L_{\alpha\beta}$ should be proportional to both n_α and n_β. Therefore we introduce a set of diffusivities $D_{\alpha\beta}$, where, by construction

$$L_{\alpha\beta} = -\frac{D_{\alpha\beta}}{k_B T}\frac{n_\alpha n_\beta}{n_\Sigma}, \quad \alpha \neq \beta$$

Here, the factor n_Σ is just a constant, it ensures a correct physical dimension of the kinetic coefficient. The minus sign ensures $D_{\alpha\beta} > 0$, because $L_{\alpha\beta} < 0$ for $\alpha \neq \beta$.

Note that there is no need for a special equation for $L_{\alpha\alpha}$ due to the zero-sum condition. Moreover, equation 53 can be transformed to the form

$$j_\alpha = -\sum_\beta L_{\alpha\beta} \nabla\mu_\beta = \underbrace{\sum_\beta L_{\alpha\beta} \nabla\mu_\alpha - \sum_\beta L_{\alpha\beta} \nabla\mu_\beta}_{=0} = -\sum_{\beta \neq \alpha} L_{\alpha\beta} \nabla(\mu_\alpha - \mu_\beta) = -\sum_{\beta \neq \alpha} \frac{D_{\alpha\beta}}{k_B T} \frac{n_\alpha n_\beta}{n_\Sigma} \nabla(\mu_\alpha - \mu_\beta)$$

where, without loss of generality, the summation over all β is replaced by the summation over all $\beta \neq \alpha$ (cf. equation 54 and equations 56–58).

Altogether, binary diffusion in an N-component system is described by the following equations

$$\frac{\partial n_\alpha}{\partial t} = \nabla \left[\sum_{\beta \neq \alpha} \frac{D_{\alpha\beta}}{k_B T} \frac{n_\alpha n_\beta}{n_\Sigma} \nabla(\mu_\alpha - \mu_\beta) \right] \tag{61}$$

in which only $N - 1$ of N possible equations are relevant. To apply this generalized transport model we have to specify all chemical potentials.

5.3. Chemical potentials

For simplicity we consider a binary system in which $\alpha = A, B$. In a standard thermodynamic formulation, the chemical potentials are given by the relations

$$\mu_A = \frac{\partial \mathfrak{G}}{\partial N_A}, \qquad \mu_B = \frac{\partial \mathfrak{G}}{\partial N_B} \Rightarrow \delta\mathfrak{G} = \mu_A \delta N_A + \mu_B \delta N_B$$

where derivatives of the free energy $\mathfrak{G} = \mathfrak{G}(N_A, N_B)$ are taken with respect to the number of particles of each component (for simplicity we keep both temperature and pressure constant and only allow for changes in N_A and N_B). If the system in question is non-uniform, it is natural to replace N_A and N_B by the number of atoms per unit volume $n_A(\mathbf{r})$ and $n_B(\mathbf{r})$ and write

$$\mathfrak{G} = \int G(n_A(\mathbf{r}), n_B(\mathbf{r})) d^3\mathbf{r} \quad \Rightarrow \quad \delta\mathfrak{G} = \int (\mu_A \delta n_A + \mu_B \delta n_B) d^3\mathbf{r}$$

where $G(n_A, n_B)$ is the free energy density and integration is performed over the whole volume occupied by the system. Now, the usual derivatives are replaced by the variational ones

$$\mu_A(\mathbf{r}) = \frac{\delta\mathfrak{G}}{\delta n_A(\mathbf{r})}, \qquad \mu_B(\mathbf{r}) = \frac{\delta\mathfrak{G}}{\delta n_B(\mathbf{r})}$$

in accord with the definition of variational derivatives and physical meaning of chemical potentials (see Feynman *et al.* (1964) for an informal introduction in variational calculus). For instance, we can employ the scaling properties of the free energy and assume that

$$G(n_A, n_B) = (n_A + n_B) g(a), \quad \text{where} \quad a(\mathbf{r}) = \frac{n_A(\mathbf{r})}{n_A(\mathbf{r}) + n_B(\mathbf{r})}$$

Here $g(a)$ is just the free energy per particle in a spatially uniform system. For the case at hand there is no difference between, *e.g.* $\delta\mathfrak{G}/\delta n_A$ and $\partial G/\partial n_A$, such that

$$\mu_A = g(a) + (1-a)\frac{\partial g}{\partial a}, \quad \mu_B = g(a) - a\frac{\partial g}{\partial a}$$

When deriving the latter equations, it is important to stress that n_A and n_B are just two independent variables that enter into $G(n_A,n_B)$, we do not impose any preconditions. The interdiffusion condition ($n_A + n_B = $ const or $\mathbf{j}_A + \mathbf{j}_B = 0$) has nothing to do with the general definition of chemical potentials; the interdiffusion is a direct consequence of the specific flux model that yields expressions like (54) or (56−58). Other flux models are also possible.

A seminal step made by Cahn and Hilliard (1958) was to consider a generalized expression for the free energy density in which G depends on spatial gradients of concentrations and this way accounts for non-uniformity of $n_A(\mathbf{r})$ and $n_B(\mathbf{r})$. Among other things, this makes the standard and the variational derivatives of the free energy different. The simplest representative of the class of generalized free energies is given by the expression

$$G(n_A,n_B) = (n_A + n_B)g(a) + \kappa/2(\nabla a)^2, \quad \kappa = \text{const}$$

Following the technique described by Feynman *et al.* (1964), one can take variational derivatives of $\mathfrak{G}$ and check that

$$\mu_A = g(a) + (1-a)\left(\frac{\partial g}{\partial a} - \kappa\nabla^2 a\right), \quad \mu_B = g(a) - a\left(\frac{\partial g}{\partial a} - \kappa\nabla^2 a\right)$$

Inserting

$$\mu_A - \mu_B = \frac{\partial g}{\partial a} - \kappa\nabla^2 a$$

into equation 61 we obtain

$$\frac{\partial a}{\partial t} = \frac{D_{AB}}{k_B T}\nabla\left[a(1-a)\nabla\left(\frac{\partial g}{\partial a} - \kappa\nabla^2 a\right)\right] \tag{62}$$

where we take advantage of the fact that n_Σ can be taken constant for any valid solution of equation 61. Equation 62 is a general representative of the Cahn-Hilliard-type equations (Nauman and He, 2001; Steinbach, 2013) and is discussed below.

5.4. Cahn-Hilliard-type equation
In this section we consider equation 62 in more detail. First, equation 62 reduces to the standard non-linear diffusion equation (22) for vanishing κ

$$\kappa \to 0 \Rightarrow \frac{\partial a}{\partial t} = \nabla[D_{\text{eff}}(a)\nabla a]$$

with

$$D_{\text{eff}}(a) = D_{AB}\frac{a(1-a)}{k_B T}\frac{\partial^2 g}{\partial a^2}$$

Consider, for instance, the free energy per atom of an ideal solution first ignoring interactions between the components

$$g(a) = \mu_A^0 a + \mu_B^0 b + k_BT(a\ln a + b\ln b), \quad b = 1 - a \tag{63}$$

Here the parameters $\mu_A^0 = g(0)$ and $\mu_B^0 = g(1)$ denote chemical potentials of the pure phase components. Calculating $\partial^2 g/\partial a^2$ we immediately find that

$$D_{\text{eff}}(a) = D_{AB} = \text{const}$$

so that an ordinary linear diffusion takes place. Furthermore, consider the following generalization of equation 63

$$g(a) = \mu_A^0 a + \mu_B^0 b - k_BT(a\ln a + b\ln b + wab) \tag{64}$$

where $w = \varepsilon/(k_BT) > 0$ is the normalized interaction parameter ε in the regular solution model. Now we derive that

$$\frac{D_{\text{eff}}(a)}{D_{AB}} = 1 - 2wa(1 - a)$$

The lowest, possibly negative value of $D_{\text{eff}}(a)$ is obtained for $a = 1/2$

$$\min_a D_{\text{eff}}(a) = D_{AB}\left(1 - \frac{w}{2}\right)$$

so that uphill diffusion is expected when $w > 2$ (see convex regions with $g''(a) < 0$ in Fig. 8). In this case, the κ-term in equation 62 cannot be ignored anymore and it becomes necessary to deal with the full Cahn-Hilliard equation

$$\frac{\partial a}{\partial t} = D_{AB}\nabla\left[(1 - 2wa(1 - a))\nabla a - L^2 a(1 - a)\nabla(\nabla^2 a)\right] \tag{65}$$

where we introduced a characteristic length L such that $L^2 = \kappa/(k_BT)$.

Due to the complexity of equation 65, the general Cahn-Hilliard equation can only be solved numerically. It is convenient to use normalized variables: space and time are measured using L and L^2/D_{AB} units. The behaviour of $a(\mathbf{r},t)$, where initially a takes a constant value such that $g''(a) < 0$ (plus a small random perturbation), is shown in

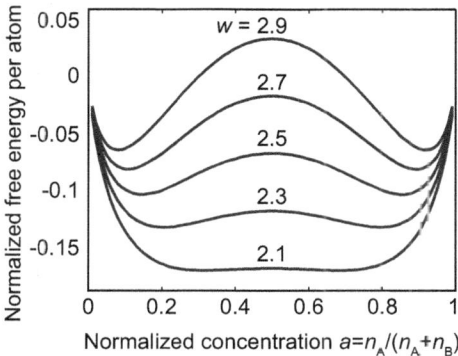

Figure 8. Behaviour of $g(a) - \mu_A^0 a - \mu_B^0 b$ normalized by k_BT for different values of w (equation 64). Regions with downward convex $g(a)$ appear when $w > 2$ indicating uphill diffusion and phase separation.

Fig. 9. Equation 65 describes uphill diffusion correctly, which produces phase separation. The model that originates from the seminal work by Cahn and Hilliard (1958) can be used to solve specific problems in mineral and rock systems (Petrishcheva and Abart, 2009, 2012).

The Cahn-Hilliard approach was applied for modelling the unmixing of alkali feldspar during slow cooling (Abart *et al*., 2009; Petrishcheva and Abart, 2012). To this end, an asymmetric regular solution model was adopted from the literature and the gradient energy parameter κ was formulated as a second rank tensor to account for direction-dependent interfacial energy. Modelling results are shown in Fig. 10. An alkali feldspar with K mole fraction $X_K = 0.6$ is cooled at different cooling rates ranging from 1°C per day down to 5°C per million years. Thereby it was assumed that unmixing starts when the coherent spinodal is hit during cooling and that the interfaces between the Na- and the K-rich domains produced by unmixing quickly loose coherency and compositions follow the strain-free solvi until diffusion becomes so inefficient that chemical re-equilibration essentially ceases. As expected, chemical separation extends down to successively lower temperatures with decreasing cooling rates. It is interesting to note that the exsolution lamellae start small and coarsen during cooling, where most of the coarsening occurs immediately after primary exsolution. Rapid cooling leads to

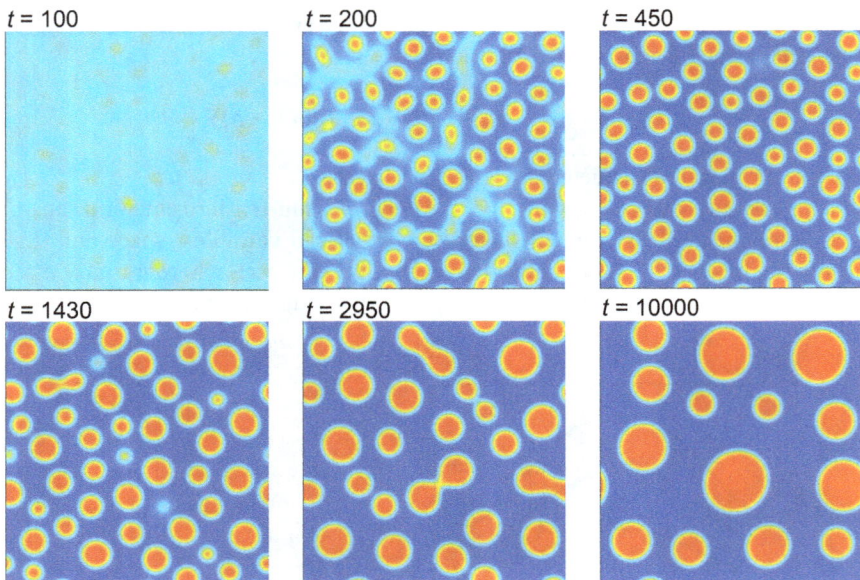

Figure 9. Exemplary numerical solution of the Cahn-Hilliard equation 65 in two dimensions. Density plots of $a(x,y,t)$ for different times are calculated at $w = 2.5$ and $a|_{t=0} = 0.35 +$ (small noise). Plots are given in normalized units (see Petrishcheva and Abart (2009) for details). Uphill diffusion resulting in phase separation is clearly observed. Reprinted with the permission of the *American Journal of Science* from Petrishcheva and Abart (2009).

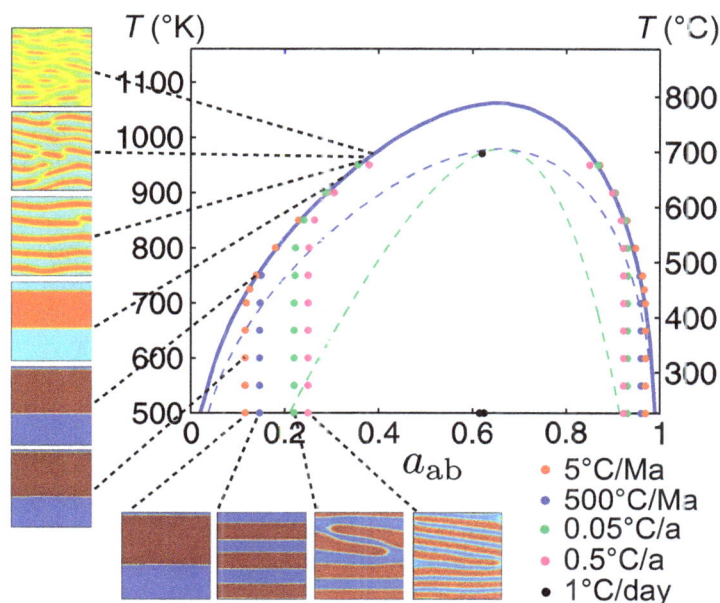

Figure 10. Composition-temperature diagram for the albite-orthoclase binary solid-solution calculated for a pressure of 800 MPa: solid blue line: strain free solvus, dashed blue and green lines: coherent solvus and coherent spinodal curves; colour-coded dots indicate different cooling paths; vertical column of colour maps: exsolution microstructures and phase compositions for successive stages along the slowest cooling path, bottom row of colour maps: microstructures and phase compositions preserved after cooling (see Abart *et al.* (2009) for details).

the small characteristic size of the lamellae and incomplete chemical separation. Slow cooling leads to coarse lamellae with more complete chemical separation. The compositions that are ultimately preserved in the albite- and orthoclase-rich lamellae of a perthite reflect the cooling rate rather than the peak temperature and should thus not be used for geothermometric purposes.

The approach has been extended to ternary solution phases and applied to the composition evolution during exsolution of ternary feldspar (Petrishcheva and Abart, 2012). Application of the Cahn-Hilliard theory to exsolution and phase separation in mineral systems has provided new insights into the associated dynamics. It has proven an elegant link between thermodynamics and diffusion with great potential for further applications. From generalization of Cahn-Hillliard theory, in particular from accounting for non-conservative *phase field variables* an important method has emerged in the form of *phase-field modelling* (Steinbach, 2009), which has become the method of choice for modelling microstructure evolution in material sciences.

6. Summary

In this chapter we have presented the mathematical framework of diffusion. Both a phenomenological and a microscopic derivation of the diffusion equation as well as some of its fundamental properties were discussed. We introduced the concept of self-similar solutions and from this derived some solutions for specific cases of linear diffusion. We extended the formalism to account for non-linear diffusion involving composition-dependent diffusion coefficients and to the case of direction-dependent diffusion. Both composition- and direction-dependent diffusion are the rule rather than the exception for diffusion in minerals. The inverse approach that is followed when extracting diffusion coefficients from experimentally produced composition-distance data or time scales from "frozen" diffusion profiles in minerals was outlined for the most general case of direction- and composition-dependent diffusion. The concept was then generalized and linked to thermodynamics. It was shown that for solution phases with ideal mixing properties ordinary linear diffusion or diffusion with composition- and direction-dependent diffusivity occurs. Finally, the most general case, where phases with non-ideal mixing properties are involved, was addressed, and the phenomenon of spontaneous unmixing in the spinodal region of the phase diagram of a solution phase was addressed based on Cahn-Hilliard theory. Despite taking several steps towards increasing complexity, the formalism presented in this chapter is restricted in that it does not explicitly account for the flux of vacancies (vacancy wind), which may yield additional phenomena such as the Kirkendall effect or plastic deformation. Generalizations of diffusion theory accounting for such additional effects were presented in the literature (see for example the review by Fischer and Svoboda (2014) and have been treated in the framework of irreversible thermodynamics (see Svoboda *et al.*, 2017, this volume).

Without doubt, the phenomenological analysis has proven very versatile in describing a multitude of diffusion phenomena in minerals and rocks. It does, however, not provide direct information on the underlying mechanisms. Diffusion occurs by movement and rearrangement of atoms, molecules or ions on the atomic scale, and a mechanistic analysis of diffusion must explicitly account for its microscopic nature. The atomistic approach towards modelling diffusion processes has been described by Jahn and Sun (2017, this volume). Understanding diffusion in minerals requires that both the phenomenological and the microscopic approach are combined. Linking the two at a satisfactory level of scrutiny is one of the major challenges in diffusion research. Finally, diffusion is an important process in diffusive phase transformations. Some related phenomena in mineral systems and their model description have been addressed by Gaidies *et al.* (2017, this volume).

Acknowledgements

The authors acknowledge gratefully the constructive reviews by Nico Stolwijk and Ingo Steinbach, which helped to improve this chapter substantially.

References

Abart, R., Petrishcheva, E., Wirth, R. and Rhede, D. (2009) Exsolution by spinodal decomposition II: Perthite formation during slow cooling of anatexites from Ngoronghoro, Tanzania. *American Journal of Science*, **309**, 540–475.

Arnold, V.I. (1989) *Mathematical Methods of Classical Mechanics*, 2nd edition. Springer, Berlin.

Birkhoff, G. (1950) *Hydrodynamics*. Princeton University Press, New Jersey, USA.

Brown, R. (1828) A brief account of microscopical observations made in the months of June, July and August, 1827, on the particles contained in the pollen of plants; and on the general existence of active molecules in organic and inorganic bodies. *Philosophical Magazine*, **4**, 161–173.

Cahn, J. and Hilliard, J. (1958) Free energy of a nonuniform system. I. Interfacial free energy. *Journal of Chemical Physics*, **28**, 258–267.

Chakraborty, S. (2008) Diffusion in solid silicates: A tool to track timescales of processes comes of age. *Annual Review of Earth and Planetary Sciences*, **36**, 153–190.

Crank, J. (1975) *The Mathematics of Diffusion*. 2nd edition. Oxford University Press, Oxford, UK

Dodson, M. (1973) Closure temperature in cooling geochronological and petrological systems. *Contributions to Mineralogy and Petrology*, **40**, 259–274.

Dresner, L. (1983) *Similarity Solutions of Nonlinear Partial Differential Equations*. Pitman, Boston, Massachusetts, USA.

Einstein, A. (1905) Über die von der molekularkinetischen Theorie der Wärme geforderte Bewegung von in ruhenden Flussigkeiten suspendierten Teilchen. *Annalen der Physik*, **322**, 549–560.

Feynman, R.P., Leighton, R.B. and Sands, M. (editors) (1964) *The Feynman Lectures on Physics*, volume 2, chapter 19. Addison-Wesley, Reading, UK.

Fick, A. (1855) Über Diffusion. *Poggendorff's Annalen der Physik und Chemie*, **94**, 59–86.

Fischer, F. and Svoboda, J. (2014) Diffusion of elements and vacancies in multicomponent systems. *Progress in Materials Science*, **60**, 338 – 367.

Fleisch, D.A. (2011) *A Student's Guide to Vectors and Tensors*. Cambridge University Press, Cambridge, UK.

Gaidies, F., Milke, R., Heinrich, W. and Abart, R. (2017) Metamorphic mineral reactions: Porphyroblast, corona and symplectite growth. Pp. 469–540 in: *Mineral Reaction Kinetics: Microstructures, Textures and Chemical Compositions* (R. Abart and W. Heinrich, editors). EMU Notes in Mineralogy, **16**. European Mineralogical Union and Mineralogical Society of Great Britain & Ireland, London.

Glicksman, M. (2000) *Diffusion in Solids*. John Wiley and Sons.

Graham, T. (1833) On the law of the diffusion of gases. *Philosophical Magazine*, **2**, 175–190, 269–276, 351–358.

Graham, T. (1864) The Bakerian lecture: on the diffusion of liquids. *Philosophical Transactions of the Royal Society of London*, **140**, 1–46.

Jahn, S. and Sun, X.-Y. (2017) Atomic-scale modelling of crystal defects, self-diffusion and deformation processes. Pp. 215–253 in: *Mineral Reaction Kinetics: Microstructures, Textures and Chemical Compositions* (R. Abart and W. Heinrich, editors). EMU Notes in Mineralogy, **16**. European Mineralogical Union and Mineralogical Society of Great Britain & Ireland, London.

Jost, W. (1965) *Diffusion in Solids, Liquids, Gases*, 4th edition. Academic Press, New York.

Landau, L.D. and Lifshitz, E.M. (1987) *Fluid Mechanics* 2nd edition. Butterworth-Heinemann, Oxford, UK

Langevin, P. (1908) Sur la théorie du mouvement brownien. *Comptes Rendus de l'Académie des Sciences. (Paris)*, **146**, 530–533.

Lidiard, A. (1955) Impurity diffusion in crystals (mainly ionic crystals with the sodium chloride structure). *Philosophical Magazine*, **46**, 1218–1237.

Matano, C. (1933) On the relation between the diffusion-coefficients and concentrations of solid metals (the nickel-copper system). *Japanese Journal of Physics*, **8**, 109–113.

Mehrer, H. (2007) *Diffusion in Solids – Fundamentals, Methods, Materials, Diffusion-Controlled Processes*. Springer Series in Solid State Science, volume **155**, Berlin.

Nauman, E.B. and He, D.Q. (2001) Nonlinear diffusion and phase separation. *Chemical Engineering Science*, **56**, 1999–2001.

Nye, J. (1957) *Physical Properties of Crystals*. Oxford at the Clarendon Press, UK.

Onsager, L. (1931) Reciprocal relations in irreversible processes. I. *Physical Review*, **37**, 405–426.

Pearson, K. (1905) The problem of the random walk. *Nature*, **72(1865)**, 294.

Petrishcheva, E. and Abart, R. (2009) Exsolution by spinodal decomposition I: evolution equation for binary mineral solutions with anisotropic interfacial energy. *American Journal of Science*, **309**, 431–449.

Petrishcheva, E. and Abart, R. (2012) Exsolution by spinodal decomposition in multicomponent mineral solutions. *Acta Materialia*, **60**, 5481–5493.

Petrishcheva, E., Abart, R., Schäffer, A.-K., Habler, G. and Rhede, D. (2014) Sodium-potassium interdiffusion in potassium-rich alkali feldspar I: Full diffusivity tensor at 850°C. *American Journal of Science*, **314**, 1284–1299.

Rollinson, H. (1993) *Using Geochemical Data: Evaluation, Presentation, Interpretation*. Longman, Essex, UK.

Schäffer, A.-K., Petrishcheva, E., Habler, G., Abart, R., Rhede, D. and Giester, G. (2014) Sodium-potassium interdiffusion in potassium-rich alkali feldspar II: Composition- and temperature-dependence obtained from cation exchange experiments. *American Journal of Science*, **314**, 1300–1318.

Steinbach. I. (2009) Phase-field models in materials science. *Modelling and Simulation in Materials Science and Engineering*, **17**, 1–31.

Steinbach, I. (2013) Phase-field model for microstructure evolution at the mesoscopic scale. *Annual Review of Materials Research*, **43**, 89–107.

Svoboda, J., Fischer, F.D. and Kozeschnik, E. (2017) Thermodynamic modelling of irreversible processes. Pp. 181–214 in: *Mineral Reaction Kinetics: Microstructures, Textures and Chemical Compositions* (R. Abart and W. Heinrich, editors). EMU Notes in Mineralogy, **16**. European Mineralogical Union and Mineralogical Society of Great Britain & Ireland, London.

Van Orman, J., Li, C. and Crispin, K. (2009) Aluminum diffusion and Al-vacancy association in periclase. *Physics of the Earth and Planetary Interiors*, **172**, 34–42.

Zhang, Y. and Cherniak, D. (editors) (2010) *Diffusion in Minerals and Melts*. Reviews in Mineralogy and Geochemistry, **72**. Mineralogical Society of America and the Geochemical Society, Washington, D.C.

EMU Notes in Mineralogy, Vol. 16 (2017), Chapter 10, 295–345

Interfaces

ELENA PETRISHCHEVA* and RAINER ABART

Department of Lithospheric Research, University of Vienna,
UZA II - Universitätszentrum Althanstrasse, Althanstraße 14, 1090 Vienna, Austria
e-mail: elena.petrishcheva@univie.ac.at
* corresponding author

All condensed phases (solids and liquids) have interfaces where they are in contact with other phases. Materials such as ceramics, metals and rocks comprise a multitude of grains of a single phase or of several different phases that are connected through a three-dimensional network of interfaces. Although the interfaces usually occupy only a negligible fraction of the overall volume, they control many of the material's physical properties such as mechanical strength, toughness, electrical conductivity, magnetic susceptibility, effective diffusivities, corrosion resistance and many more.

We refer to an aggregate of crystals of a single phase as a 'polycrystal'. A composite material that is composed of grains of different phases is referred to as a 'polyphase aggregate'. An interface, where two crystals of the same phase that only differ in their lattice orientations are in contact, is referred to as a 'grain boundary'. An interface, where two different phases meet, constitutes a 'phase boundary'. Solid–liquid, solid–vapour and liquid–vapour interfaces as well as interfaces between two different solids are phase boundaries. The topology and geometry of the grain- and phase-boundary network is controlled by a compromise between micro-structural equilibrium and space filling. Furthermore, the network of interfaces may evolve, *e.g.* under elevated temperatures. In the absence of mineral reactions and deformation, this evolution is driven by the reduction of the free energy associated with interfaces. Both grain and phase boundaries are essentially two-dimensional objects. For example, the generally employed formulation of the diffusion equation that approximates diffusion along an interface involves two spatial coordinates. The third dimension, which is perpendicular to the interface, is represented by a small characteristic width, which may be only a few inter-atomic distances. The effective diffusion coefficient that enters into the surface diffusion equation depends on the width of the surface.

The microstructures and textures observed in rocks exhibit an overwhelming diversity (Vernon, 2004). They bear important information on rock formation and have significant control over bulk-rock properties and behaviour. Interfaces are the locations where both reactions and chemical mass transfer take place, and they are thus central in mediating mineral reactions, rock alteration and deformation. Furthermore, epitaxial and topotaxial orientation relations are established at interfaces (see Habler and Griffths, 2017, this volume).

In this chapter we introduce the phenomenological treatment of interfaces. The concepts of interfacial energy and surface tension are discussed and conditions for micro-structural equilibrium are derived. Thereafter, capillary force driven interface motions including coarsening and grain-growth are addressed. We investigate thermodynamic equilibrium at curved interfaces and its implication for the evolution of heterogeneous systems. Finally, we treat some specific properties and the motion of interfaces in crystalline materials. Although most of the considerations made in this chapter are of a general nature, the applications we have in mind are related to minerals and rocks.

1. Interfacial energy

A mathematical definition of the interfacial energy and physical reasons why interfaces possess an excess energy are discussed in this section using the fluid/gas interface as an example. Consider a liquid phase in contact with air. It is well known that the liquid atoms at the interface are subject to a net force directed away from the interface into the liquid phase. The force results from interactions between the atoms at the interface and the neighbouring atoms in the liquid (Fig. 1), neglecting their weak interactions with the atoms in the air. A certain amount of work has to be done against this net force to raise an atom from the liquid phase to the interface. The amount of work that is required to increase the interface by unit area is referred to as the 'surface tension' or 'surface tension coefficient'. It is presupposed that the increase takes place in a quasi-static manner such that both the temperature and volume of the liquid phase are kept constant. As usual, the isothermal work is associated with a change of the free energy, which we denote by $\mathcal{F}$. All interfaces such as crystal surfaces, grain boundaries and phase boundaries possess an excess free energy, proportional to the interface area A, *i.e.* the surface tension can be defined as the excess free energy per unit area of the interface. In mathematical terms

$$d\mathcal{F} = \gamma dA \qquad (1)$$

where $d\mathcal{F}$ and dA refer to small changes in the excess free energy and surface area, respectively. The parameter γ is the surface tension coefficient, *i.e.* the interfacial energy per unit area, $[\gamma] = \mathrm{J/m^2}$.

All systems with interfaces have the tendency to minimize the free energy either by minimizing the free energy per unit area of the interface or by minimizing the interfacial area as such. For instance, the spherical shape of liquid droplets results from minimizing the liquid–gas interfacial area for a given volume of liquid. The preferential removal of material from the areas near grain and phase boundaries during thermal or chemical etching is an experimental manifestation of the excess energy associated with an interface.

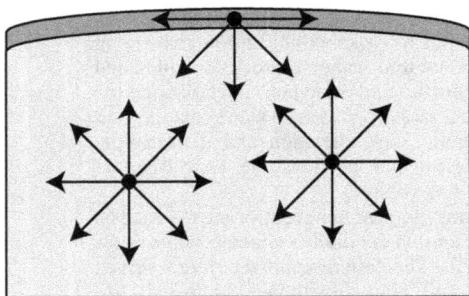

Figure 1. Schematic sketch of atomic interactions in a condensed phase; atoms at the surface experience a net force towards the interior of the condensed phase.

2. Surface tension and capillary force

As a simple application of the surface energy concept we consider a soap film that is attached to a U-shaped wire frame as shown in Fig. 2. The thin film has only a small amount of liquid between its two surfaces so that the contribution of the bulk liquid free energy to the total free energy of the system can be ignored. Let one edge of the film be attached to a sliding piece of wire A—B with length l. The surface tension tends to decrease the surface area and the resulting force $2f$ acts on the line A—B (the factor 2 accounts for the two surfaces on either side of the film). Static equilibrium requires a compensating external force $F = 2f$.

Now, consider a small shift, say dx, of the sliding wire. The shift may result from a small fluctuation. The total surface area increases by $2ldx$ and the related increase of the surface free energy reads $\gamma \cdot 2ldx$. The increase is equal to the mechanical work performed by the external force, that is Fdx or $2fdx$. Equating the above expressions we derive that

$$f = \gamma l \tag{2}$$

i.e. surface tension coefficient as an alternative to (1) can be defined as a tension force acting on a unit length of the interface boundary with the unit N/m ($=$J/m^2).

3. Applications

In this section the basic concepts of surface energy and surface forces as defined in the section above are illustrated by several typical examples.

3.1. Force balance at a triple junction

A line along which three crystals meet in a polycrystal is referred to as a 'triple junction'. Cross-sections on planes that are perpendicular to the triple junctions are shown in Fig. 3. From a triple junction three grain boundaries emanate, each of which has an associated interfacial energy and tension force. At a triple junction the tension forces must balance, requiring the so-called 'Young's condition'

$$\frac{F_{12}}{\sin\theta_3} = \frac{F_{23}}{\sin\theta_1} = \frac{F_{31}}{\sin\theta_2} \tag{3}$$

to hold in a plane perpendicular to the line defining the triple junction. Here, *e.g.* F_{12} is the value of the surface tension force for the interface between the first and the

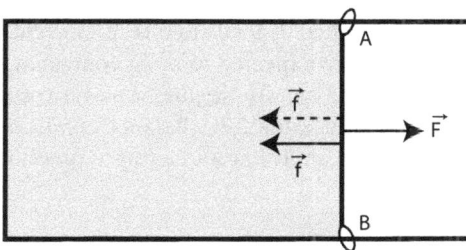

Figure 2. Schematic sketch of a soap film (grey) attached to a U-shaped wire frame with a sliding wire piece (A—B) on the right.

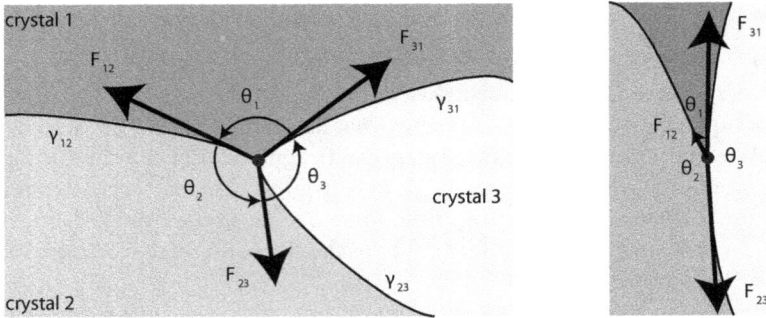

Figure 3. Left: cross-section of a polycrystal at a triple junction (a line along which three crystals meet locally normal to the plane of drawing). Here, three surface tension forces must balance, as expressed by Young's condition (3) or (4). *Right:* if $\gamma_{12} \ll \gamma_{23} \approx \gamma_{31}$, the corresponding angle θ_3 tends to π.

second crystal, θ_3 denotes the angle between the directions of the vectors $\mathbf{F}_{23}$ and $\mathbf{F}_{31}$ in the third crystal (Fig. 3, left). Static equilibrium requires

$$\mathbf{F}_{12} + \mathbf{F}_{23} + \mathbf{F}_{31} = 0$$

One can vectorially multiply the latter relation by, *e.g.* $\mathbf{F}_{23}$ and present the result as

$$\mathbf{F}_{12} \times \mathbf{F}_{23} = \mathbf{F}_{23} \times \mathbf{F}_{31}$$

such that

$$F_{12} F_{23} \sin \theta_2 = F_{23} F_{31} \sin \theta_3 \quad \text{and} \quad \frac{F_{12}}{\sin \theta_3} = \frac{F_{31}}{\sin \theta_2}$$

which proves Young's condition (3). Note, that in 3D a triple junction corresponds to a grain edge that is shared by three grains. Let the length of the common grain edge be l, then it holds

$$F_{12} = \gamma_{12} l, \qquad F_{23} = \gamma_{23} l \qquad F_{31} = \gamma_{31} l$$

implying that Young's condition can be formulated in the tension coefficients involved as

$$\frac{\gamma_{12}}{\sin \theta_3} = \frac{\gamma_{23}}{\sin \theta_1} = \frac{\gamma_{31}}{\sin \theta_2} \tag{4}$$

In the case of identical tension coefficients all angles equal $2\pi/3$ or $120°$. If one surface tension coefficient is larger than the sum of the other two, equilibrium is not possible. On the other hand, if, *e.g.* $\gamma_{12} \ll \gamma_{23} \approx \gamma_{31}$, then θ_3 tends to π (Fig. 3, right).

To conclude this section we note that four phases are expected to be in contact at a point, this is referred to as a 'quadruple junction'. At a quadruple junction four triple lines meet. If all surface tension coefficients are identical, the force balance condition requires that all edge angles of the involved trihedral angles at a quadruple junction equal $109.5°$.

3.2. Force balance at a pore tip

We consider a pore located at a boundary between two crystals in a polycrystal (see Fig. 4 for a cross-section). The pore may be filled with a fluid or gas, the equilibrium shape of the pore results from the force-balance condition. Applying Young's condition to the pore tip we obtain

$$\frac{\gamma_{12}}{\sin(\Phi_1 + \Phi_2)} = \frac{\gamma_{31}}{\sin(\pi - \Phi_2)} = \frac{\gamma_{23}}{\sin(\pi - \Phi_1)}$$

such that

$$\gamma_{23} = \frac{\sin \Phi_1}{\sin \Phi} \gamma_{12} \quad \text{and} \quad \gamma_{31} = \frac{\sin \Phi_2}{\sin \Phi} \gamma_{12}$$

where $\Phi = \Phi_1 + \Phi_2$ is referred to as the *dihedral angle* (the angle between the traces of two surfaces on a plane that orthogonally cuts the line of intersection). For pores filled with a wetting fluid, Φ is small and

$$\gamma_{23} \approx \frac{\Phi_1}{\Phi} \gamma_{12} \quad \text{and} \quad \gamma_{31} \approx \frac{\Phi_2}{\Phi} \gamma_{12}$$

If $\gamma_{12} > \gamma_{13} + \gamma_{23}$, a pore cannot exist. For non-wetting fluids, Φ is large. Measurement of dihedral angles has been used to estimate wetting properties of geological fluids (*e.g.* Watson and Brenan, 1987; Holness and Graham, 1991).

3.3. Excess pressure under a curved interface

Surface tension is responsible for an 'excess pressure' under a curved interface. As an illustration of this effect, consider a soap bubble with radius r. The bubble consists of two thin soap films with a small amount of liquid between them. A spherical segment, *e.g.* at the top of the bubble (Fig. 5), has a tendency to reduce its area, yet it stays in an equilibrium state when an external force is applied to its boundary line from the rest of the bubble. This force produces an excess pressure inside the bubble.

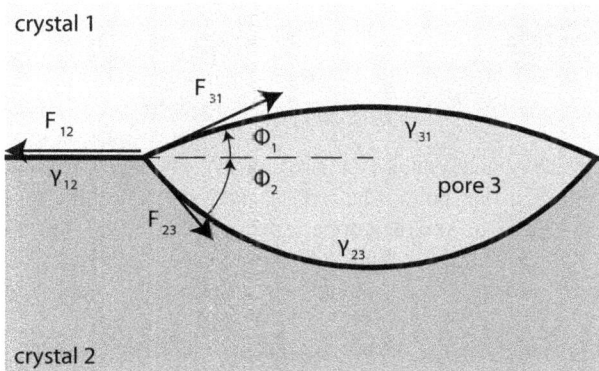

Figure 4. Force equilibrium at a pore tip.

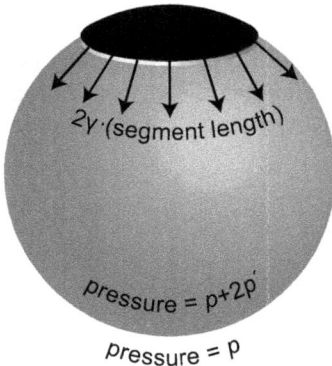

Figure 5. Explanation of the origin of the excess pressure p' in a soap bubble.

To quantify the excess pressure we assume that the volume part of the free energy is small compared to the contribution from the surfaces. Let p and $p + 2p'$ be the pressure outside and inside the bubble, respectively. We are interested in the excess pressure p' due to a single surface. Now, assume that r experiences a small increase dr due to some fluctuation. The related increases in bubble area and volume read

$$dA = 4\pi(r + dr)^2 - 4\pi r^2 = 8\pi r\, dr$$

and

$$dV = \frac{4\pi}{3}(r + dr)^3 - \frac{4\pi}{3}r^3 = 4\pi r^2\, dr$$

respectively. The bubble's free energy increases by $2\gamma dA$, where the factor 2 accounts for the two sides of the soap film. The increase must be equal to the work done by the pressure

$$2\gamma dA = -p\, dV + (p + 2p')dV \Rightarrow p' = \gamma \frac{dA}{dV}$$

Using the above expressions for dA and dV we obtain the excess pressure

$$p' = 2\gamma/r \tag{5}$$

The latter equation is a special case of a more general expression

$$p' = \gamma \left(\frac{1}{r_1} + \frac{1}{r_2} \right) = \gamma(\kappa_1 + \kappa_2) \tag{6}$$

where $r_{1,2}$ refer to two main radii (for a definition of main curvature/radius see next section) of the surface at the point where p' is measured. The inverse quantities $\kappa_{1,2}$ are referred to as main curvatures. For a spherical bubble $\kappa_1 = \kappa_2 = r^{-1}$ and we return to equation 5. The full derivation of the equation for excess pressure p' is very similar to the simplified one shown here, but requires appreciation of the general geometric relation

$$\frac{dA}{dV} = \kappa_1 + \kappa_2$$

The excess pressure is also referred to as 'Laplace pressure'. It is positive for convex interfaces (like a spherical droplet or a bubble) and negative for concave interfaces. It may even vanish if κ_1 and κ_2 have opposite signs and compensate each other, as is true for so-called 'minimal surfaces'. Soap films are familiar examples of minimal surfaces.

To conclude this section we summarize that the surface tension coefficient can be defined using several points of view. One possible definition concerns excess free energy related to a surface, see equation 1. Another possibility is to use surface tension force (equation 2). Alternatively one can define γ from the excess pressure (equation 6). Of course, all possibilities are compatible with each other. For instance, consider the spherical segment shown in Fig. 5; the effect of the excess pressure applied to its surface is equivalent to the effect of the surface tension force applied to its boundary.

3.4. Capillary action

An interesting consequence of the surface tension effect and of wetting behaviour is the 'capillary action' (lat. *capillaris* – pertaining to the hair). Consider a thin tube of radius r that is dipped into a fluid at one end (normal to the interface, see Fig. 6). We consider a wetting fluid that is inclined to maintain contact with the tube material such that a concave fluid interface is created, e.g. a glass tube that is dipped into water. If r is sufficiently small, this interface can be approximated by a segment of a spherical surface, say with radius R. The excess pressure $2\gamma/R$ is directed upward and contributes to the excess elevation h of the water such that

$$\rho g h = \frac{2\gamma}{R} \quad \text{and} \quad h = \frac{2\gamma}{\rho g R} \tag{7}$$

where ρ is the density of the fluid, and $\rho g h$ accounts for the hydrostatic pressure at depth h. If the wetting angle θ (see Fig. 6) is known, then $R\cos\theta = r$ and therefore

$$h = \frac{2\gamma \cos\theta}{\rho g r}$$

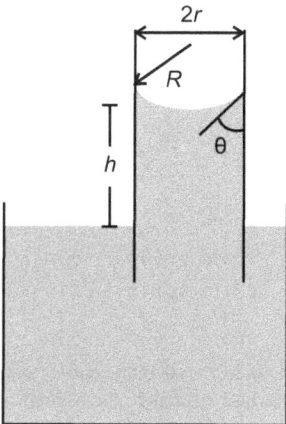

Figure 6. Schematic drawing of capillary action.

Consider, *e.g.* a glass tube with $r = 0.2$ mm in water at atmospheric pressure and 20°C. Then the surface tension is $\gamma = 0.0728$ N/m, density $\rho = 1000$ kg/m^3, and $\theta \approx 0$. The height is h $\approx$ 70 mm.

4. Interface movement

Foams (soap films) have frequently been studied as analogue systems for polycrystals (*e.g.* MacPherson and Srolovitz , 2007). As in a polycrystal, the interfaces in a foam arrange to ensure force balance and mechanical equilibrium. The corresponding dihedral angles are 120° at triple lines and 109.5° for edge angles of the trihedral angles involved at quadruple junctions. A single polyhedral shape with flat surfaces cannot meet these criteria simultaneously and ensure space filling. As a consequence, the bubbles of a foam have curved interfaces. Also the interfaces in a polycrystal or polyphase aggregate are generally curved that results in interface motion. In the following we first give the definition of curvature and then discuss both the forces and the dynamics associated with curved interfaces.

4.1. Curvature of a curve on a plane

Consider an imaginary point particle that moves with unit velocity along a smooth curve on a plane. The curve can be specified by a functional dependence $y = f(x)$ in a suitable coordinate frame. The particle's velocity is then a unit vector (say $\boldsymbol{\tau}$) that is tangent to the curve. Let **n** be the unit normal vector to the curve. The particle's acceleration is pointed to **n**, but it is not necessarily of unit extent. The ($\boldsymbol{\tau}$, **n**) pair is a natural basis adopted for the description of the curvilinear particle motion, as introduced by Frenet. Note, that the basis vectors follow the particle. In contrast to the standard basis vectors $\mathbf{e}_x$, $\mathbf{e}_y$, the Frenet basis is not always the same: the basis rotates, as the particle moves (Fig. 7). The curvature parameter quantifies how fast this rotation is. By construction, the velocity of the imaginary particle following the curve is always unity such that "fast" or "slow" rotation of Frenet's basis is determined solely by the curve.

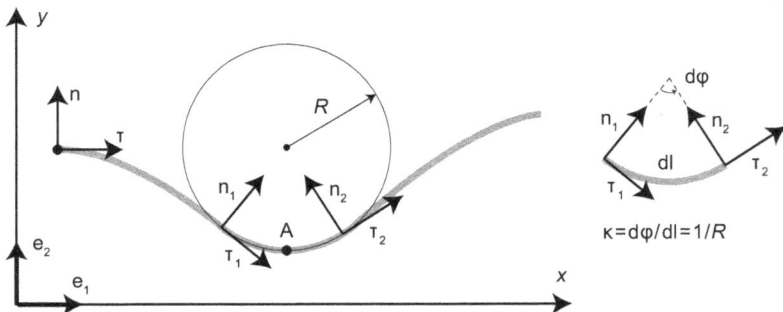

Figure. 7. Frenet's basis vectors $\boldsymbol{\tau}$ and **n** are naturally defined at each point of a smooth plane curve (thick grey line). The curvature of the curve at some point A is $1/R$, where the radius R refers to the osculating circle shown by the thin line.

More specifically, let l be the path length along the curve and let $\phi(l)$ be the basis rotation angle that is accumulated after motion of length l. The curvature is defined as the derivative

$$\kappa = \frac{d\phi}{dl}$$

a relation that can be simplified using the osculating circle, as shown for point A in Fig. 7. Consider a small part of the curve between $(\tau_1, \mathbf{n}_1)$ and $(\tau_2, \mathbf{n}_2)$ around A, its length $dl = Rd\phi$, where $d\phi$ is the corresponding change of the rotation angle, therefore

$$\kappa = \frac{d\phi}{Rd\phi} = \frac{1}{R}$$

Alternatively, the curvature can be expressed in Cartesian coordinates in terms of the functional dependence $y = f(x)$, namely (Dubrovin *et al.*, 1984)

$$\kappa = \frac{\partial_x^2 f}{[1 + (\partial_x f)^2]^{3/2}}$$

where $\partial_x f$ is used as a short-hand notation for $\partial f/\partial x$. The latter expression for κ is especially useful when the curve is basically parallel to a suitably chosen OX axis and experiences only small changes; then

$$|\partial_x f| \ll 1 \qquad \text{and} \qquad \kappa \approx \partial_x^2 f$$

a situation (the so-called small slope approximation) that is particularly relevant when an initially straight interface is slightly perturbed due to some fluctuation or instability for example.

4.2. Curvature of a surface

The curvature parameter κ defined above is the only essential parameter for a smooth curve embedded in a plane (Dubrovin *et al.*, 1984). Now we turn to a smooth surface embedded in the physical three-dimensional space. Two curvature parameters, the so-called main or principal curvatures $\kappa_{1,2}$, are necessary in this case. These parameters enter into, *e.g.* the general expression (6) for the Laplace pressure.

In general, a surface in 3D can be described in different ways. First, the surface can be defined by some implicit equation $F(x,y,z) = \text{const}$, *e.g.*

$$x^2 - y^2 + z^2 = 1$$

for a unit sphere. Another possibility is to specify how Cartesian coordinates of the surface points depend on two suitable parameters

$$x = x(u,v), \qquad y = y(u,v) \qquad \text{and} \qquad z = z(u,v)$$

For instance, for the unit sphere one can use the two angles from spherical coordinates, *i.e.*

$$x = \cos u\ \sin v, \qquad y = \sin u\ \sin v \qquad \text{and} \qquad z = \cos v$$

Another possible way to describe a surface is to choose a suitable coordinate frame and associate the surface with a function

$$z = f(x,y)$$

a useful point of view that results in the simplest expression for the sum of the main curvatures (see equation 8 in what follows).

To explain the geometric meaning of the main curvatures we consider some surface, take a point P on it, and construct a normal vector $\mathbf{n}$ that is attached to P (Fig. 8). Now we introduce a family of planes all running through P and containing $\mathbf{n}$. The intersection between the surface and a plane from the family produces a plane curve. Let κ denote its curvature at P. The main or principal curvatures $\kappa_{1,2}$ are just the minimal and the maximal values of κ that would be obtained from the entire family of planes. For instance, for a sphere with the radius R all cross sections have the same curvature, therefore

$$\kappa_1 = \kappa_2 = 1/R$$

In the same way for a cylinder with radius R

$$\kappa_1 = 0 \quad \text{and} \quad \kappa_2 = 1\backslash R$$

Although the general expressions for $\kappa_{1,2}$ might be complicated, their sum (the only quantity of interest in the framework of Laplace pressure) is given by the following simple expression (Dubrovin *et al.*, 1984)

$$\kappa_1 + \kappa_2 = \nabla\left(\frac{\nabla f}{\sqrt{1 + (\nabla f)^2}} \right) \tag{8}$$

where operator $\nabla = (\partial_x, \partial_y)$ refers to two dimensions. The sum $\kappa_1 + \kappa_2$ is referred to as the mean curvature. For instance, the partial differential equation that describes minimal surfaces reads

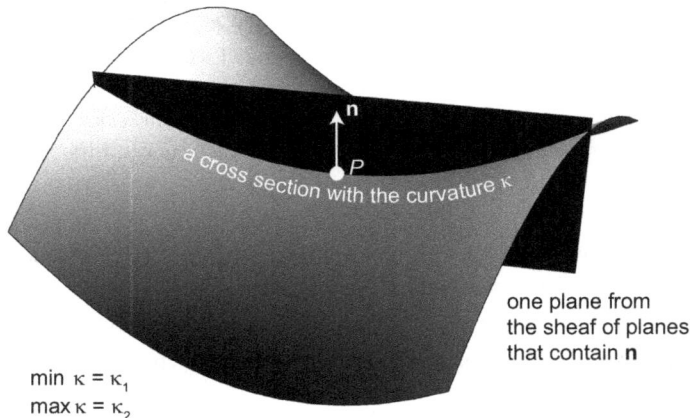

min $\kappa = \kappa_1$
max $\kappa = \kappa_2$

Figure 8. Taking cross-sections by different planes running through $\mathbf{n}$, we obtain curves with different curvatures κ at P. The largest and the smallest κ are the main curvatures at P.

$$\nabla\left(\frac{\nabla f}{\sqrt{1+(\nabla f)^2}}\right) = 0$$

To conclude this section we return to the small slope approximation introduced above for curves. The approximation applies to a surface, when it is approximately parallel to the properly chosen Oxy plane, such that both derivatives $|\partial_x f|$ and $|\partial_y f|$ are small and $(\nabla f)^2$ in equation 8 can be neglected. In this case

$$\kappa_1 + \kappa_2 \approx \nabla^2 f = \partial_x^2 f + \partial_y^2 f$$

i.e. the mean curvature is defined by a two-dimensional Laplacian. In this approximation the minimal surfaces, which require $\kappa_1 + \kappa_2 = 0$, correspond to solutions of the equation $\nabla^2 f = 0$, *i.e.* they are given by harmonic functions.

4.3. Interface mobility

Thus far we have discussed forces and shapes related to a stationary interface in a polycrystal. Now we ask how the shape and position of an interface change due to surface tension and other driving forces.

In general, evolution of a polycrystal towards its equilibrium state is driven by reduction of a properly chosen free energy. The contribution of the interfaces to the free energy is proportional to the total interface area. If capillary effects dominate, the polycrystal tends to reduce its total interface area and the interfaces tend to flatten with time. To quantify these considerations we utilize the fact (Balluffi *et al.*, 2005) that, if a particle, *e.g.* a nucleus of a second phase or a fluid-filled pore that is embedded in some matrix, migrates due to a drag force K_{drag}, then the particle velocity is proportional to the drag

$$v = M_P K_{drag} \tag{9}$$

where the index P stands for a particle-like inclusion and M_P refers to mobility. The latter equation cannot be applied directly to interfaces, rarely moving as a whole with a constant velocity. One can, however, split the interface into a large set of small segments, each with its own velocity, and look for a relation, in the spirit of equation 9, between the local interface velocity and local drag per unit area of the interface. Assume that a small segment of an interface migrates with a constant velocity v directed normally to the segment due to some drag force per unit area K_{drag}. Then the so-called interface or boundary mobility M_I may be introduced such that

$$v = M_I K_{drag} \tag{10}$$

No matter how complicated the geometry of the interface, equation 10 can be applied to any small segment of that interface. If the driving force results from the surface tension effect, then

$$v = M_I \gamma (\kappa_1 + \kappa_2) \tag{11}$$

where γ refers to the surface tension coefficient. Both the driving force and the resulting velocity are orthogonal to the respective interface segment. Each segment may have its own velocity, all the elementary motions contribute to the complicated motion of the interface as such.

As a simple example, consider 'collapse'of a spherical pore with the decreasing radius $R(t)$, embedded in a homogeneous matrix (Fig. 9). The velocity of each surface segment has the following value

$$v = \frac{2M_I\gamma}{R}$$

and is directed to the centre of the pore. We therefore derive

$$\frac{dR}{dt} = -v \quad \Rightarrow \quad \frac{dR}{dt} = -\frac{2M_I\gamma}{R}$$

which is integrated to

$$R(t) = \sqrt{R(0)^2 - 4M_I\gamma t} \tag{12}$$

The spherical pore disappears at $t = R(0)^2/(4M_I\gamma)$. Another simple example is the collapse of a cylindrical inclusion for which

$$R(t) = \sqrt{R(0)^2 - 2M_I\gamma t} \tag{13}$$

Of course, such simple analytical solutions are possible only for simple pore geometries. In the general case, motion of an interface $z = f(x,y,t)$ is determined by the following nonlinear partial differential equation

$$\partial_t f = M_I\gamma\sqrt{1 + |\nabla f|^2} \left[\frac{\partial}{\partial x} \frac{\partial_x f}{\sqrt{1 + |\nabla f|^2}} + \frac{\partial}{\partial y} \frac{\partial_y f}{\sqrt{1 + |\nabla f|^2}} \right]$$

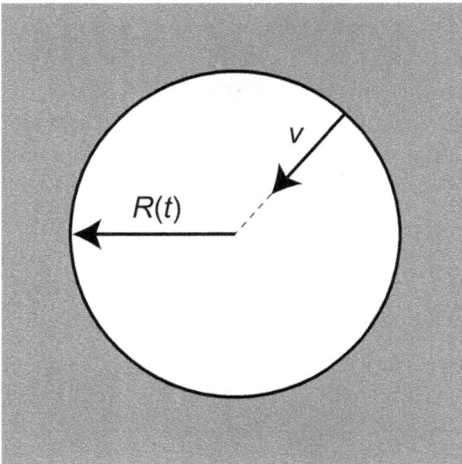

Figure 9. Cross section of a shrinking spherical (cylindrical) pore.

where $|\nabla f|^2 = (\partial_x f)^2 + (\partial_y f)^2$. In the small slope approximation we ignore $|\nabla f|$ and obtain

$$\partial_t f = M_I \gamma (\partial_x^2 f + \partial_y^2 f)$$

such that interface motion is mathematically equivalent to diffusion and $M_I \gamma$ has physical dimension of diffusivity. For further details see Petrishcheva and Renner (2005, 2010).

To conclude this section we recall that interface mobility, M_I, is a local quantity that refers to a small-area element of the interface. In general, M_I may depend on position and orientation of the element. For the sake of simplicity we consider a uniform and isotropic matrix, such that M_I is a constant. In addition to surface tension force, several other driving forces are conceivable for interface motion, *e.g.* gradients in temperature, jumps in chemical potential across the interface, and jumps in the defect density; for further reading see (Sutton and Balluffi, 1995).

4.4. Zener drag

In this section we consider a migrating boundary and a particle with the mobility, M_P, *e.g.* a second phase inclusion, which is attached to the boundary, as shown in Fig. 10. It is observed that grain-coarsening in carbonate rocks during metamorphism may be suppressed when finely dispersed graphite or sulfide phases are present (*e.g.* Olgaard and Evans, 1986; Abart, 1995). It is commonly observed that grey, graphite or sulfide bearing marble is more fine-grained than white marble that is devoid of second-phase particles (Herwegh *et al.*, 2011). On the other hand, small particles may be pushed aside by a propagating grain or phase boundary. In the following we consider drag and drop of a spherical non-deformable second-phase particle at a moving grain boundary.

We consider a grain boundary between two matrix grains that sweeps over a second-phase particle. The particle may be 'dragged' along with the moving boundary. When the velocity of grain-boundary motion exceeds a critical value, the particle is 'dropped' and stays behind the moving grain boundary forming an isolated second-phase particle within a matrix grain. Following Zener (Smith, 1948; Nes *et al.*, 1985), we calculate the critical boundary velocity at which a spherical particle is dropped from a moving boundary.

Let the surface tension coefficient of the interface between the second-phase particle and the matrix grains be much larger than, say γ, of the interface between the two matrix

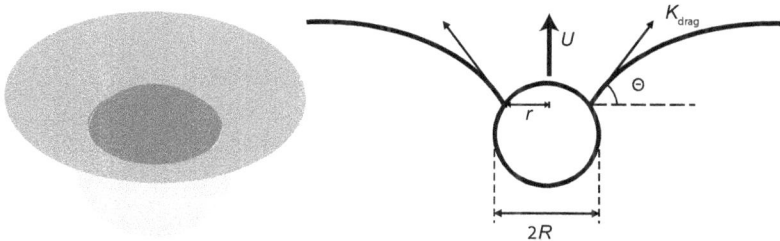

Figure 10. Spherical particle with radius R attached to a grain boundary between two matrix grains along a contact circle with radius r; arrows indicate drag force K_{drag} and particle velocity U, Θ is the drag angle.

grains. In terms of Young's condition we are dealing with the situation shown in Fig. 3 (right) such that the corresponding dihedral angle is close to π and the second-phase particle can be approximated by a spherical inclusion with radius R (see Fig. 10).

The total drag force, as shown in Fig. 10 is directed upwards and reads

$$K_{\text{drag total}} = 2\pi r \gamma \sin\Theta = 2\pi R \gamma \sin\Theta \cos\Theta,$$

where Θ is the 'drag angle'. According to equation 9, the above drag yields the velocity

$$U = M_P K_{\text{drag total}} = \pi M_P R \gamma \sin 2\Theta$$

with a maximum

$$U_{\max} = \pi M_P R \gamma$$

when $\Theta = \pi/4$. Particle drag at this condition is referred to as 'Zener drag'. If the velocity of the boundary exceeds this limit, the particle is separated from the boundary. It stays behind the moving boundary and remains as an isolated inclusion in the interior of the grain.

The phenomenon of drag and drop of deformable pores by a moving interface was investigated by Petrishcheva and Renner (2005, 2010) who also solved for the pore shape, which adjusts itself to the local interface velocity.

5. Coarsening

Capillary forces play a crucial role in the microstructural evolution of polyphase materials, for example, in those formed by precipitation. In this context, different steps are discerned:

1. Nuclei of a second phase, which are generated due to sufficiently large fluctuations, appear in the supersaturated homogeneous medium (nucleation stage).
2. The precipitates grow. After an initial stage of "free" growth, the precipitates start to encounter each other and to compete. Some of the precipitates continue to grow at the expense of others that must shrink and eventually disappear (coarsening stage).
3. The entire volume is filled with the growing precipitate. Geometric considerations come into play, *e.g.* stresses, deformations and rearrangement of grain boundaries becomes important (grain-growth stage).

Nucleation theory is dealt with in detail by Gaidies *et al.* (2017). In this section, we consider only the coarsening- and grain-growth stages. To this end, we first discuss the equilibrium conditions at a curved interface.

5.1. Equilibrium vapour pressure at a curved surface

5.1.1. One-component system

We consider fluid–vapour equilibrium in a one-component system, *e.g.* in water. More precisely, we assume that an otherwise empty closed container is partially filled with water. The system temperature is kept constant. The water starts to evaporate so that after a short while the previously empty part of the container will be filled with water vapour. Eventually precipitation comes into play. At some vapour concentration n_{sv}

evaporation and precipitation compensate each other. Water and saturated vapour reach equilibrium in which n_{sv} is determined by temperature.

The dependence $n_{sv}(T)$ was measured and is tabulated. The values provided in textbooks refer to a flat interface; the equilibrium concentration for a curved interface is different and will be denoted by $n_{sv}(\kappa,T)$, where $\kappa = \kappa_1 + \kappa_2$ denotes the mean curvature, see equation 8. The reason for the variability of n_{sv} with κ is demonstrated in Fig. 11, which shows schematically a vapour molecule near a curved vapour–fluid interface. We see that a concave interface has better chances of absorbing the vapour molecule. Due to enhanced precipitation the equilibrium vapour density should be smaller at a concave interface. In the same way, a convex surface hampers precipitation and yields a larger equilibrium vapour density. In both cases, the evaporation process is less affected because on the atomic scale water molecules cannot "feel" any difference between flat, concave, or convex interfaces shown in Fig. 11. It is natural to assume that for small curvatures, $n_{sv}(\kappa,T)$ can be approximated by its Taylor expansion

$$n_{sv}(\kappa,T) = n_{sv}(0,T)(1 - \Lambda\kappa) \tag{14}$$

where $\Lambda > 0$ is an unknown factor. Our goal is to quantify Λ.

Assume that a thin vertical tube is dipped into water. From equation 9 the elevation of the water inside the tube due to capillary action reads $h = \gamma\kappa/(\rho g)$, where $\kappa = 2/R$ describes the curvature of the elevated water surface. The vapour concentration n_B at point B in Fig. 12 is smaller than $n_{sv}(0,T)$; according to the well known Boltzmann relation it reads

$$n_B = n_{sv}(0, T) \exp\left(-\frac{mgh}{kT}\right)$$

where m denotes the mass of the molecule constituting the vapour, and $n_{sv}(0,T)$ is the saturated vapour concentration at a flat water surface outside the tube. The concentration in point B should be equal to the concentration in point A $n_{sv}(\kappa,T)$ because otherwise either continuous precipitation or continuous condensation at the curved water surface would result in a water flux along the tube. The flux could then be used for a direct transformation of heat into work, which would violate the second law of thermodynamics. Therefore

$$n_{sv}(\kappa, T) = n_{sv}(0, T) \exp\left(-\frac{\Omega\gamma\kappa}{kT}\right)$$

where $\Omega = m/\rho$ is the volume per molecule in water. For small curvatures

Figure 11. Interaction of a vapour molecule with neighbouring molecules near flat, convex and concave interfaces. The interaction area is shown schematically by a dashed circle.

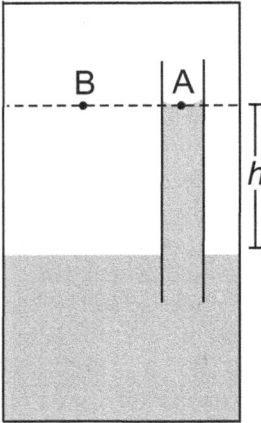

Figure 12. The derivation of equation 15.

$$n_{\mathrm{sv}}(\kappa, T) = n_{\mathrm{sv}}(0, T)\left(1 - \frac{\Omega\gamma\kappa}{kT}\right)$$

in accord with equation 14. For instance, the equilibrium vapour concentration at the surface of a water bubble with radius R reads

$$n_{\mathrm{sv}}(0, T)\exp\left(\frac{2\Omega\gamma}{RkT}\right) \approx n_{\mathrm{sv}}(0, T)\left(1 + \frac{2\Omega\gamma}{RkT}\right) \tag{15}$$

and is larger than the concentration at the plane interface.

5.1.2. Two-component system

In this section we reconsider equilibrium vapour pressure at a curved interface and extend results of the previous section to a two-phase (α and β) two-component (A and B) system (see Fig. 13). We make two key assumptions:

1. The α phase is composed of both components. It can be considered as a dilute solution of the B component such that the corresponding chemical potential can be calculated as

$$\mu_{\mathrm{B}}^{\alpha} = (\text{pure component part}) + kT\ln(hX_{\mathrm{B}}^{\alpha})$$

where h denotes the activity coefficient and X_{B}^{α} is the mole fraction of B in α. We assume that h is independent of X_{B}^{α} (Henry's law for dilute solutions).

2. The β phase is rich in the B component. We assume that the partial molar volume per B component in the β phase, denoted by Ω, is independent of changes in pressure that are possibly induced by curved interfaces.

All notations are summarized in Fig. 14. The first assumption is used to quantify the difference between $\mu_{\mathrm{B}}^{\alpha}$ near a curved and near a flat α–β interface, namely

$$\mu_{\mathrm{B}}^{\alpha}\big|_{\mathrm{curved}} - \mu_{\mathrm{B}}^{\alpha}\big|_{\mathrm{plane}} = kT\ln\frac{X_{\mathrm{B}}^{\alpha}\big|_{\mathrm{curved}}}{X_{\mathrm{B}}^{\alpha}\big|_{\mathrm{plane}}} \tag{16}$$

a phase

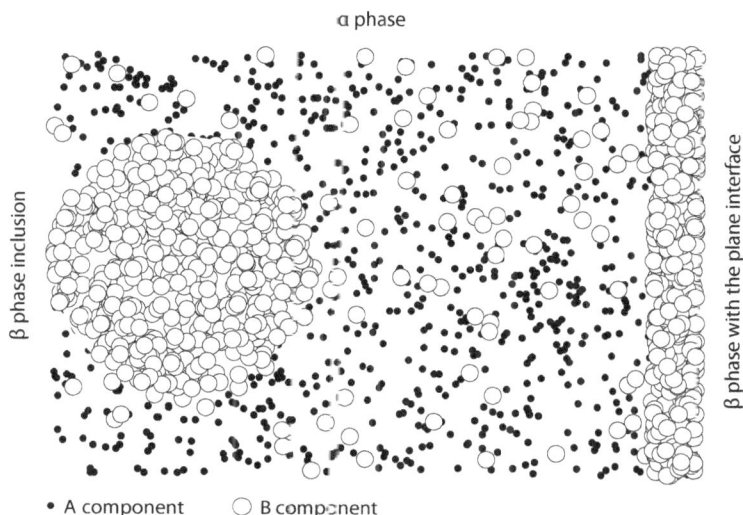

β phase inclusion

β phase with the plane interface

• A component   ○ B component

Figure 13. Schematic representation of a two-phase (α and β) two-component system (A and B) containing a spherical inclusion of β embedded in a matrix of α and a planar $\alpha-\beta$ interface.

The second assumption is used to quantify $\mu_B^\beta|_{\text{curved}} - \mu_B^\beta|_{\text{plane}}$. We assume that several B atoms have been absorbed by one of the inclusions of the β phase in an isothermal manner (Fig. 13). We apply the Gibbs-Duhem relation

$$SdT - Vdp^\beta + N_B d\mu_B^\beta = 0$$

where N_B refers to the total number of the B atoms inside the inclusion and p^β is the pressure in the inclusion. With $dT = 0$ it follows

$$d\mu_B^\beta = \Omega dp^\beta$$

The second assumption, $\Omega = V/N_B = \text{const}$, makes it possible to integrate this differential relation to

$$\mu_B^\beta = \Omega p^\beta + (\text{pressure independent part})$$

which we use for $\mu_B^\beta|_{\text{curved}}$ and $\mu_B^\beta|_{\text{plane}}$. Taking their difference and applying equation 6 for the excess pressure yields

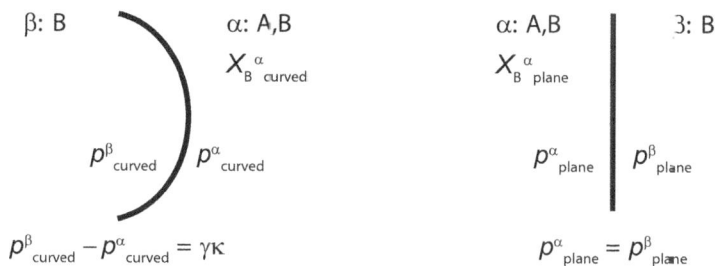

β: B   α: A,B   α: A,B   β: B

$X^\alpha_{B\ \text{curved}}$ $X^\alpha_{B\ \text{plane}}$

p^β_{curved} p^α_{curved} p^α_{plane} p^β_{plane}

$p^\beta_{\text{curved}} - p^\alpha_{\text{curved}} = \gamma\kappa$ $p^\alpha_{\text{plane}} = p^\beta_{\text{plane}}$

Figure 14. Summary of notations used throughout this section (*cf.* Fig. 13).

$$\mu_B^\beta|_{curved} - \mu_B^\beta|_{plane} = \Omega(p^\beta|_{curved} - p^\beta|_{plane}) = \Omega\gamma\kappa \qquad (17)$$

Note that, strictly speaking, the system shown in Fig. 13 cannot be in a state of thermodynamic equilibrium because a small difference between $X_B^\alpha|_{curved}$ and $X_B^\alpha|_{plane}$ will cause a small diffusive flux across the α phase. The flux will be investigated in detail in the next section. Right now it is safe to assume that the system is in a local thermodynamic equilibrium such that

$$\mu_B^\alpha|_{curved} = \mu_B^\beta|_{curved} \text{ and } \mu_B^\alpha|_{plane} = \mu_B^\beta|_{plane}$$

and therefore

$$\mu_B^\alpha|_{curved} - \mu_B^\alpha|_{plane} = \mu_B^\beta|_{curved} - \mu_B^\beta|_{plane}$$

By substituting (16) and (17), we obtain

$$kT \ln \frac{X_B^\alpha|_{curved}}{X_B^\alpha|_{plane}} = \Omega\gamma\kappa$$

such that finally

$$\frac{X_B^\alpha|_{curved}}{X_B^\alpha|_{plane}} = \exp\left(\frac{\Omega\gamma\kappa}{kT}\right)$$

For a spherical β bubble with radius R we obtain

$$\frac{X_B^\alpha|_{curved}}{X_B^\alpha|_{plane}} = \exp\left(\frac{2\Omega\gamma}{RkT}\right) \approx 1 + \frac{2\Omega\gamma}{RkT}$$

by analogy with equation 15 from the previous section.

The relation between the equilibrium concentration and curvature is illustrated in Fig. 15, where the free energy per particle for two phases α and β is plotted *vs.* the mole fraction of the component B. The phase β, which is rich in B component, is represented by points. The chemical potential of component B in the phase β increases with curvature, $\mu_B^\beta|_{plane} < \mu_B^\beta|_{curved}$, resulting in a larger equilibrium mole fraction of component B in α at a curved interface when compared to that at a plane interface, $X_B^\alpha|_{plane} < X_B^\alpha|_{curved}$. The difference in concentrations yields a small diffusive flux from the spherical inclusion to the plane interface, the flux finally exhausts the spherical inclusion in favour of the plane one.

5.2. Single droplet

In this section we investigate non-equilibrium dynamics of an isolated droplet of one phase surrounded by and interacting with the second phase. To be specific, we consider a one-component system and a spherical fluid droplet surrounded by vapour. The latter may, but need not necessarily, be saturated. The results of this section can be generalized also for a two-component system such as that shown in Fig. 14.

According to equation 15, the vapour concentration that ensures balanced evaporation and precipitation at the surface of a droplet with radius R can be written as

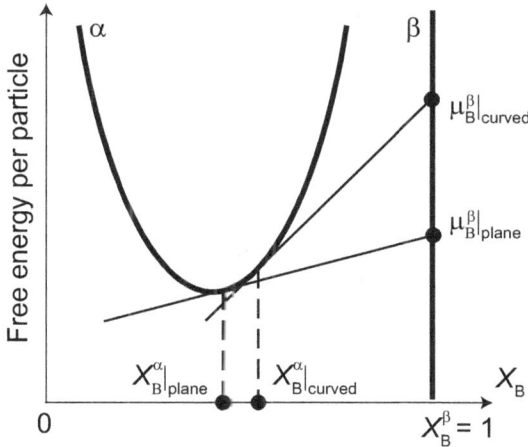

Figure 15. The difference between chemical potentials at plain and curved interfaces, $\mu_B^\beta|_{plane}$ and $\mu_B^\beta|_{curved}$, yields the difference between equilibrium concentrations of components $X_B^\alpha|_{plane}$ and $X_B^\alpha|_{curved}$.

$$n = n_0 \left(1 + \frac{a}{R}\right), \qquad a = \frac{2\Omega\gamma}{kT} \qquad (18)$$

Here n_0 abbreviates $n_{sv}(\kappa = 0, T)$ and corresponds to the saturated vapour concentration for a plane liquid–vapour interface. The parameter a provides a typical space scale at which surface curvature is important. The vapour concentration well away from the droplet is denoted by n_{out}; it differs from n_0 if the vapour is not saturated for some reason.

When $n_0(1 + a/R)$ differs from n_{out}, the difference induces a diffusive flux either towards the droplet or from the droplet outwards. The flux supplies or removes atoms from the droplet and results in slow changes of its radius, $R = R(t)$. Specifically, the droplet dissolves if initially

$$n_0 \left(1 + \frac{a}{R(0)}\right) > n_{out} \qquad (19)$$

Note that a decrease in $R(t)$ leads to successively faster dissolution. The droplet inevitably dissolves if $n_0 > n_{out}$, irrespective of $R(0)$. When $n_0 < n_{out}$ the droplet still dissolves if $R(0) < R_{crit}$, where

$$n_0 \left(1 + \frac{a}{R_{crit}}\right) = n_{out} \qquad (20)$$

yielding an (unstable) equilibrium with the vapour. A larger than critical droplet such that

$$n_0 \left(1 + \frac{a}{R(0)}\right) < n_{out} \qquad (21)$$

just grows in size. Note that an increase in radius favours the inequality 21, *i.e.* a large enough droplet keeps growing but with the permanently decreasing rate.

Now let us refine the above considerations and derive a differential equation for $R(t)$. We assume that the spatial distribution of vapour atoms is controlled by volume diffusion such that (see Petrishcheva and Abart, 2017, this volume)

$$\partial_t n = D\nabla^2 n \qquad \mathbf{j} = -D\nabla n$$

where D is diffusivity and $\mathbf{j}$ denotes diffusive flux. As long as the other droplets are well separated from that in question, the diffusive flux is radially symmetric and $n = n(r,t)$ is constrained by the radial version of the diffusion equation

$$\partial_t n = D\left(\partial_r^2 n + \frac{2}{r}\partial_r n\right)$$

The radial flux

$$j_r = -D\partial_r n$$

permanently supplies (removes) particles from the droplet environment and results in continuous growth (shrinkage) of the droplet, such that $R = R(t)$. We assume that j_r self-organizes on a shorter time scale than the time scale at which $R(t)$ changes. Under this (adiabatic) approximation, the flux is governed by a stationary, radially symmetric solution $n = n(r)$ of the diffusion equation

$$n(r) = C_1 + C_2/r \text{ with } C_{1,2} = \text{const for } r \geqslant R(t)$$

with the boundary conditions

$$n|_{r=R} = n_0\left(1 + \frac{a}{R}\right) \qquad \text{and} \qquad n|_{r=\infty} = n_{\text{out}}$$

The boundary conditions constrain $C_{1,2}$ such that

$$n(r) = n_{\text{out}} + \frac{R}{r}\left[n_0\left(1 + \frac{a}{R}\right) - n_{\text{out}}\right]$$

and the radial flux

$$j_r = \frac{DR}{r^2}\left[n_0\left(1 + \frac{a}{R}\right) - n_{\text{out}}\right]$$

which is negative if atoms are supplied to the droplet. The number of atoms dN that is supplied over the time span dt seconds reads

$$dN = (-j_r) \cdot 4\pi r^2 \cdot dt = 4\pi DR\left[n_{\text{out}} - n_0\left(1 + \frac{a}{R}\right)\right]dt$$

The resulting increase of the droplet volume is ΩdN, where Ω denotes the volume per atom in the fluid phase as before. Therefore

$$\Omega dN = dV \text{ with } V = (4\pi R^3)/3 \Rightarrow \Omega dN = 4\pi R^2 dR$$

and inserting the above expression for dN we obtain the following differential relation

$$\Omega \cdot 4\pi D R \left[n_{\text{out}} - n_0 \left(1 + \frac{a}{R} \right) \right] dt = 4\pi R^2 dR$$

It follows that $R(t)$ is given by an ordinary differential equation as first derived by Greenwood (1956)

$$\frac{dR}{dt} = D\Omega n_0 \left(\frac{\Delta}{R} - \frac{a}{R^2} \right), \qquad \Delta = \frac{n_{\text{out}} - n_0}{n_0} \tag{22}$$

In the rest of this section we describe solutions of equation 22, applications to the systems with many droplets (Lifshitz and Slyozov, 1961) are described in the next section. Note that particle density in the fluid phase $n_{\text{fl}} = 1/\Omega$ and therefore $\Omega n_0 = n_0/n_{\text{fl}} \ll 1$, which also applies for the two-component case as shown in Fig. 13. Therefore the 'effective' diffusion coefficient $D\Omega n_0$ in equation 22 is much smaller than D in accord with the adiabatic approximation. Using a from equation 18 as a typical space-scale for a droplet, we find the following characteristic time

$$t_c = a^2/(D\Omega n_0) \tag{23}$$

where the index c stands for 'coarsening'.

According to equation 22, the evolution of $R(t)$ depends on the sign of Δ. Negative Δ means that $n_0 > n_{\text{out}}$ such that the droplet dissolves. Decrease of $R(t)$ makes the right-hand-side of equation 22 more and more negative and finally $R(t)$ vanishes at some finite time. When, for instance, $R(0)$ is so small that the Δ/R term in equation 22 can be ignored from the very beginning [*cf.* equations 12 and 13] then

$$t_c \frac{dR}{dt} = -\frac{a^2}{R^2} \Rightarrow R(t) = \sqrt[3]{R^3(0) - 3a^3 \frac{t}{t_c}}$$

Taking the Δ/R term in equation 22 into account makes the mathematics more complicated but does not change the conclusion: the droplet dissolves in a finite time for $\Delta < 0$.

For $\Delta > 0$ we define the critical radius

$$R_{\text{crit}} = a/\Delta \tag{24}$$

the definition is just a reformulation of equation 20. The droplet still dissolves if $R(0) < R_{\text{crit}}$ and it grows if $R(0) > R_{\text{crit}}$. If $R(0)$ is so large that the a/R^2 term in equation 22 can be ignored from the very beginning, then

$$t_c \frac{dR}{dt} = \Delta \frac{a^2}{R} \Rightarrow R(t) = \sqrt{R^2(0) - 2a^2\Delta \frac{t}{t_c}}$$

Again, taking the a/R^2 term in equation 22 into account makes the solution more tedious but does not change the conclusion: the droplet radius increases and it is asymptotically proportional to $\sqrt{t}$.

To conclude this section we note that an isolated droplet grows at the expense of vapour molecules and when the accompanying decrease of n_{out} cannot be ignored, it stops the growth. A different situation arises for many different droplets: the larger droplets can still grow at the expense of the smaller ones. This is the subject of the coarsening theory presented in the next section.

5.3. Theory of coarsening

In this section we will deal only with a one-component liquid-vapour system. All results can, however, be generalized to a two-component system as for that shown in Fig. 13. We follow the classical paper (Lifshitz and Slyozov, 1961) and describe coarsening, the capillary-force driven growth of some liquid droplets at the expense of others. Let us first summarize the main assumptions:

1. We consider a part, say a unit volume, of a larger system such that an influence of system boundaries can be ignored (Fig. 16). The temperature is kept constant. The system consists of numerous liquid droplets of different sizes and some amount of vapour between them. The total number of droplets changes with time, because some of them ultimately dissolve.

2. The total number of droplets warrants introduction of a distribution function $f(R,t)$, where $f(R,t)dR$ is the number of droplets with radius between R and $R + dR$. We consider R as an abstract variable that will be used in the partial differential equation for $f(R,t)$. Dealing with a specific droplet we will use the notation R_i or $R_i(t)$, where index i enumerates droplets.

3. The parameter n_{out} denotes the vapour concentration outside of the droplets. As opposed to the previous section, it is time dependent, $n_{out} = n_{out}(t)$, resulting from matter conservation. As a consequence, the parameter

$$\Delta(t) = \frac{n_{out}(t) - n_0}{n_0}$$

is also time dependent and will be quantified in the following.

4. The evolution of a droplet is described by the adiabatic equation 22. Throughout this section we will use two special units: a from equation 18 to measure distances and t_c from equation 23 to measure time. In particular, the normalized equation 22 reads

$$\frac{dR_i}{dt} = \frac{\Delta(t)}{R_i} - \frac{1}{R_i^2} \tag{25}$$

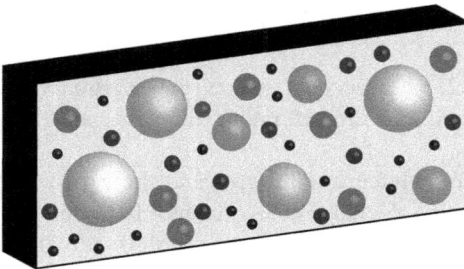

Figure 16. Schematic representation of a system experiencing coarsening: liquid droplets are surrounded by vapour, the larger droplets mainly precipitate vapour and grow, whereas the smaller ones mainly evaporate, dissolve, and indirectly supply material to the larger ones.

where t and $R_i(t)$ are now dimensionless, equation 25 uses the above defined time-dependent $\Delta(t)$.

5. The system evolves in such a way that 'small' droplets dissolve and supply atoms to the vapour phase, whereas 'large' droplets grow by assimilating atoms from the vapour. One can distinguish between the small and large droplets using a time-dependent critical radius

$$R_{\mathrm{crit}}(t) = 1/\Delta(t) \qquad (26)$$

the latter equation is just the normalized version of equation 24. A droplet that grows initially may start to dissolve later.

6. The vapour part of the system serves as a transmitting medium; the great majority of atoms are contained in the condensed phase. Thus, the total volume occupied by the liquid phase remains constant

$$\frac{4\pi}{3} \sum_i R_i^3(t) = \text{const.} \qquad (27)$$

where summation over all droplets is applied. To a good approximation, the above constant is just the total number of atoms multiplied by volume per atom in the condensed phase.

Firstly, let us demonstrate that $R_{\mathrm{crit}}(t)$ corresponds to the average droplet radius $\bar{R}(t)$. Taking the time derivative of equation 27 and using equation 25 yields

$$\sum_i R_i^2 \left[\frac{\Delta(t)}{R_i} - \frac{1}{R_i^2} \right] = 0$$

that can be transformed to

$$\frac{1}{\Delta(t)} = \frac{\sum_i R_i(t)}{\sum_i 1} = \bar{R}(t)$$

such that $R_{\mathrm{crit}}(t) = \bar{R}(t)$ in accordance with equation 26.

Now let us exploit the distribution function $f(R,t)$, that is governed by the following partial differential equation

$$\frac{\partial f}{\partial t} + \frac{\partial}{\partial R} \left[\left(\frac{\Delta(t)}{R} - \frac{1}{R^2} \right) f \right] = 0 \qquad (28)$$

which represents a kind of continuity equation directly following from the definition of $f(R,t)$ and the ordinary differential equation 25. In terms of the distribution function the mean radius is given by

$$\bar{R}(t) = \frac{\int_0^\infty f(R,t) R\, dR}{\int_0^\infty f(R,t)\, dR} \qquad (29)$$

and conservation of volume yields

$$\frac{4\pi}{3}\int_0^\infty f(R,t)R^3dR = \text{const.} \tag{30}$$

as a replacement of equation 27.

Now, following Lifshitz and Slyozov (1961) we obtain a solution of equation 28 in three steps. Firstly, we expect that an acceptable distribution function is regular for $R \to 0$. The $1/R^2$ factor in equation 28 indicates that $f(R,t)$ should be proportional to R^2 for small R. Secondly, we look for self-similar solutions that depend on a self-similar variable $R/\bar{R}(t)$. To this end we introduce

$$\xi = \Delta(t)R$$

where $\Delta(t)$ is still unknown. Thirdly, we use a generalized semi-scale substitution (see Petrishcheva and Abart (2017, this volume) for further examples of such substitutions)

$$f(R,t) = \Delta^n(t)\xi^2 g(\xi)$$

Here $g(\xi)$ will be found in what follows together with $\Delta(t)$; the prefactor ξ^2 ensures the required behaviour of $f(R,t)$ for small R, and the index n results from the volume-conservation condition (37). It is easy to see that

$$\int_0^\infty f(R,t)R^3dR = \Delta^{n-4}(t)\int_0^\infty g(\xi)\xi^5 d\xi$$

is constant if $n = 4$.

After these preliminary considerations we derive $\Delta(t)$ and $g(\xi)$ by inserting $\Delta^4(t)\xi^2 g(\xi)$ instead of $f(R,t)$ in equation 28

$$\frac{\partial f}{\partial t} = 4\Delta^3\frac{d\Delta}{dt}\xi^2 g(\xi) + \Delta^3\frac{d\Delta}{dt}\xi\frac{d}{d\xi}\left[\xi^2 g(\xi)\right] \quad \text{and}$$

$$\frac{\partial}{\partial R}\left[\left(\frac{\Delta}{R} - \frac{1}{R^2}\right)f\right] = \Delta^7\frac{d}{d\xi}[(\xi - 1)g(\xi)]$$

Then equation 28 takes the form

$$\frac{1}{\Delta^4}\frac{d\Delta}{dt} + \frac{(\xi - 1)g'(\xi) + g(\xi)}{\xi^3 g'(\xi) + 6\xi^2 g(\xi)} = 0 \tag{31}$$

where $g'(\xi)$ denotes $dg/d\xi$. Equation 31 holds if and only if

$$\frac{(\xi - 1)g'(\xi) + g(\xi)}{\xi^3 g'(\xi) + 6\xi^2 g(\xi)} = B \quad \text{and} \quad \frac{1}{\Delta^4}\frac{d\Delta}{dt} = -B$$

where B is a constant. An immediate consequence is the following explicit expression for $\Delta(t)$

$$\frac{1}{\Delta^3(t)} = \frac{1}{\Delta^3(0)} + 3Bt$$

such that for long time periods the mean droplet radius increases proportionally to $t^{1/3}$. This prediction is in good agreement with observations and provides strong support for the semi-scale approach. Another immediate observation is that $g(\xi)$ is obtained from the following ordinary differential equation

$$(1 - \xi + B\xi^3)\bar{g}'(\xi) = (1 - 6B\xi^2)g(\xi)$$

that has the explicit solution

$$g(\xi) = g(0)\exp\left(\int_0^\xi \frac{1 - 6B\xi'^2}{1 - \xi' + B\xi'^3}d\xi'\right) \tag{32}$$

We require that equation 32 provides a physically reasonable $g(\xi)$ without singularities for $\xi > 0$. We furthermore require that $g(\xi)$ vanishes for $\xi \to \infty$ in such a way that the total volume of droplets per unit space volume, which in accord with equation 37 equals $(4\pi/3)\int_0^\infty g(\xi)\xi^5 d\xi$, is finite.

To find the only appropriate value of B we investigate roots of the polynomial $1 - \xi + B\xi^3$ in equation 32, as each such root can potentially result in an unacceptable singularity of $g(\xi)$. It is sufficient to consider positive roots and positive values of B. Three possibilities exist

1. If $0 < B < 4/27$ the polynomial $1 - \xi + B\xi^3$ has two positive roots and any $g(\xi)$ yielded by equation 32 is singular for these values of ξ.
2. If $B > 4/27$ the polynomial $1 - \xi + B\xi^3$ has no positive root and $g(\xi)$ yielded by equation 32 is a regular function. However, it does not vanish rapidly enough for $\xi \to \infty$ and $\int_0^\infty g(\xi)\xi^5 d\xi$ diverges logarithmically.
3. If $B = 4/27$ the polynomial $1 - \xi + B\xi^3$ has a double root at $\xi = 3/2$. One can construct an appropriate smooth solution of equation 32 for $0 \leqslant \xi < 3/2$

$$g(\xi) = \frac{g(0)\exp\left(-\frac{2\xi}{3-2\xi}\right)}{\left[1 + \frac{\xi}{3}\right]^{\frac{7}{3}}\left(1 - \frac{2\xi}{3}\right)^{\frac{11}{3}}}$$

Moreover, one can check that $g(\xi)$ and all its derivatives tend to zero for $\xi \to 3/2$. Therefore we set $g(\xi) = 0$ for $\xi > 3/2$.

We conclude that the evolution of the mean (or critical) radius given in terms of normalized quantities reads

$$\bar{R}(t) = \frac{1}{\Delta(t)} = \sqrt[3]{\frac{1}{\Delta^3(0)} + \frac{4t}{9}} \tag{33}$$

and that the distribution function is given by

$$f(R,t) = C\Delta^4(t) \left[\frac{\xi^2 \exp\left(-\frac{2\xi}{3-2\xi}\right)}{\left(1+\frac{\xi}{3}\right)^{\frac{7}{3}} \left(1-\frac{2\xi}{3}\right)^{\frac{11}{3}}} \right]_{\xi=\Delta(t)R}$$

where $C = g(0) = $ const. The total number of droplets reads

$$\int_0^\infty f(R,t)dR = C\Delta^3(t) \int_0^{\frac{3}{2}} \frac{\xi^2 \exp\left(-\frac{2\xi}{3-2\xi}\right)}{\left(1+\frac{\xi}{3}\right)^{\frac{7}{3}} \left(1-\frac{2\xi}{3}\right)^{\frac{11}{3}}} d\xi = \frac{9C}{4}\Delta^3(t)$$

These equations provide straightforward usage. Assume that a snapshot of all droplets at some time is available. One can calculate the mean (same as critical) radius $\bar{R} = \Delta^{-1}$ and change to the self-similar variable $\xi = \Delta R$. Let $p(\xi)d\xi$ be the probability for a droplet to belong to the interval between ξ and $\xi + d\xi$. The physical radius should then be between ξ/Δ and $(\xi + d\xi)/\Delta$. According to the definition of $f(R,t)$, the number of such droplets will be $f(\xi/\Delta,t)d\xi/\Delta$. The latter quantity divided by the total number of droplets gives the required probability, therefore

$$p(\xi) = \frac{4\xi^2 \exp\left(-\frac{2\xi}{3-2\xi}\right)}{9\left(1+\frac{\xi}{3}\right)^{\frac{7}{3}} \left(1-\frac{2\xi}{3}\right)^{\frac{11}{3}}} \tag{34}$$

a universal expression that does not depend on any system-specific details. This distribution (Lifshitz and Slyozov, 1961) is the main result of the coarsening theory. A plot of $p(\xi)$ is shown in Fig. 17. Note that $p(\xi) = 0$ for $\xi \geq 3/2$. For the asymptotic solution, a droplet radius cannot be larger than $1.5 \times$ the current mean radius. In a real system, the number of such 'over-sized' droplets decreases, as the system evolves towards the asymptotic solution and $R_{crit}(t)$ increases.

6. Grain growth

In this section we consider briefly so-called grain growth, which we understand as the successive increase of the average grain size in a situation where different grains

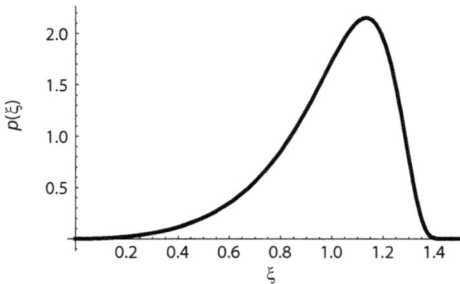

Figure 17. Probability distribution of the self-similar variable for the droplet size during coarsening from equation 34.

contact directly and interact with each other as shown in Fig. 18. Of course, some grains have to disappear to support those that grow. One should distinguish between (1) 'normal grain growth', which occurs in a uniform manner such that after a proper rescaling all snapshots in Fig. 18 look similar, and (2) 'abnormal grain growth'. In the latter case one cannot expect a universal distribution function like that shown in Fig. 17. Still, initially abnormal growth may transform to normal later on.

Even the simplest normal grain growth is more complicated than the coarsening process from the previous section. However, as suggested by experimental observations (Burke and Turnbull, 1952; Atkinson, 1988; Evans *et al.*, 2001; Renner *et al.*, 2002), increase of the mean grain size $\bar{R}(t)$ is still described by the power law

$$\bar{R}^s(t) - \bar{R}^s(0) \sim t \text{ for } t \to \infty$$

where the exponent may differ from $s = 3$ obtained for coarsening. For instance, so-called parabolic growth corresponds to $s = 2$ driven by reduction in surface energy.

Fortunately, in spite of the high complexity of possible grain shapes, one can sacrifice some of the rigour and describe normal grain growth similar to the evaporation-precipitation process of coarsening. The key step is to find an approximate replacement of the Greenwood equation 22 which applies only to an ideal spherical droplet. Consider, *e.g.* grain growth in a single-phase material and assume that the growth process is supported by the decrease of the interface energy. In accord with equation 11, the local velocity of all grain boundaries is determined by the general expression

$$v(t) = M_I \gamma \left(\frac{1}{r_1} + \frac{1}{r_2} \right) \tag{35}$$

where M_I is the common mobility and γ denotes the common surface tension coefficient. The parameters $r_{1,2}$ refer to two main radiuses of the boundary at the point where $v(t)$ is measured, the radiuses may have different signs. Now, following Hillert (1965), we postulate for the evolution of the radius of the i-th grain

$$\frac{dR_i}{dt} = \alpha M_I \gamma \left(\frac{1}{R_{\text{crit}}} - \frac{1}{R_i} \right) \tag{36}$$

see Fischer *et al.* (2003) for a rigorous derivation. Equation 36 is structurally similar to equation 22, but exhibits important differences. The parameter $R_i(t)$ should be

Figure 18. Different stages of grain growth. As opposed to coarsening, the growing grains interact directly with each other. Image courtesy of J. ter Heege.

understood as an effective radius that characterizes a possibly non-spherical grain in question. We assume that $4\pi R_i^3/3$ still approximates the volume of the non-spherical grain such that relation (27) holds. It is convenient to use a generalized equation 27

$$\sum_i R_i^{\mathcal{D}}(t) = \text{const.}, \quad \mathcal{D} = 2,3 \tag{37}$$

where $\mathcal{D}$ refers to the dimension of the problem.

The factor on the right-hand-side of equation 36 is similar to that in equation 35, but contains an additional dimensionless scaling multiplier α that will be estimated later. The as yet unknown time-dependent critical radius $R_{\text{crit}}(t)$ dynamically separates growing 'large' grains and shrinking 'small' ones. Taking the time derivative of equation 37 and using equation 36 we derive that

$$R_{\text{crit}}(t) = \frac{\sum_i R_i^{\mathcal{D}-1}(t)}{\sum_i R_i^{\mathcal{D}-2}(t)} \tag{38}$$

For two spatial dimensions, we retrieve that the critical radius equals the mean radius. For three spatial dimensions, equation 38 might be considered as a new definition of the mean radius $\bar{R}(t)$.

Now let us estimate the yet unknown parameter α first in two spatial dimensions. One may use the famous result of Neumann (1952) and Mullins (1956) that area $S_i(t)$ of a 2D grain is defined by the exact equation

$$\frac{dS_i}{dt} = 2\pi M_I \gamma \left(\frac{n_i}{6} - 1\right)$$

where n_i refers to the number of neighbouring grains. A critical grain that neither grows nor shrinks has six neighbours, all three-, four- and five-corner grains shrink and eventually disappear; all other grains grow. When small grains with an insufficient number of neighbours disappear, the number of neighbours for the growing grains generally decreases. Thus, an initially growing grain may stop its growth and start to shrink. Grain area is related to effective radius, $S_i = \pi R_i^2$, such that

$$\frac{dR_i}{dt} = \frac{M_I \gamma}{R_i}\left(\frac{n_i}{6} - 1\right)$$

which we write in the following equivalent form

$$\frac{dR_i}{dt} = \alpha M_I \gamma \left(\frac{\frac{n_i}{6\alpha} - \frac{1}{\alpha} + 1}{R_i} - \frac{1}{R_i}\right)$$

To get equation 36 one needs to introduce some averaging procedure and replace

$$\frac{\frac{n_i}{6\alpha} - \frac{1}{\alpha} + 1}{R_i} \quad \text{by} \quad \frac{1}{\bar{R}} \tag{39}$$

because in two spatial dimensions $R_{\text{crit}} = \bar{R}$. Note that a small vanishing grain typically has three neighbours. Singular behaviour in (39) for $n_i \rightarrow 3$ and $R_i \rightarrow 0$ is avoided, when

$$\frac{n_i}{6\alpha} - \frac{1}{\alpha} + 1 \rightarrow 0 \quad \text{for} \quad n_i \rightarrow 3$$

that is, $\alpha = 1/2$. In the opposite case of a large grain with many neighbours $n_i \gg 1$ we should have

$$\frac{n_i}{6\alpha} \approx \frac{R_i}{\bar{R}}$$

The perimeter of the large grain in question is estimated as $2\pi R_i$, the part of the perimeter occupied by a typical neighbour grain is estimated as $2\bar{R}$; therefore we expect that

$$n_i \approx \frac{\pi R_i}{\bar{R}}$$

Combining the last two equations we derive $\alpha = \pi/6 = 0.52... \approx 1/2$. Altogether one can just set $\alpha = 1/2$ in equation 36 and consider

$$\frac{dR_i}{dt} = \frac{M_I \gamma}{2} \left(\frac{1}{\bar{R}} - \frac{1}{R_i} \right) \tag{40}$$

as the grain-growth equivalent of the coarsening equation (22). Calculation of α in the three-dimensional system is more complicated and results in a numerical value of order 1 (Hillert, 1965).

Now, applying equation 36 one can repeat all basic steps from the previous section and transform the above results for the coarsening process to the new results for grain growth, as was first done by Hillert (1965). In particular, grain growth appears to be parabolic and governed by a universal distribution function qualitatively similar to that shown in Fig. 17.

7. Interfaces in crystalline material

The considerations made so far did not account for anisotropy of surface tension and interface mobility. Anisotropic interface properties are, however, inherent in crystalline materials. In the following, some specific features associated with interfaces in crystalline material are discussed. Firstly, the anisotropy of interface energy and its implication for the equilibrium shape of crystals is addressed. Then models for predicting the microscopic structure of grain- and phase boundaries are reviewed. Finally interface motion is addressed. The rich literature on interfaces in crystalline materials includes excellent textbooks (Sutton and Balluffi, 1995; Howe, 1997; Balluffi et al., 2005). This section represents a small selection of topics that the authors deem salient in the context of geological materials.

7.1. Anisotropy of surface energy and equilibrium shape

When a crystal is cut and new surfaces are generated, bonds must be broken. To a first approximation the number and strength of broken bonds per surface atom determine the excess free energy of an atom at the new surface. Both the number and strength of broken bonds depend on the orientation of the surface, and the surface excess energy is different for differently oriented crystal surfaces. We refer to the planar segments of a crystal surface generated during crystallization as 'a facet'. A convenient way of representing the anisotropic surface energy of a crystal is the so-called 'γ plot'. Hereby a vector $\mathbf{n}_{hkl}$ is taken as the normal to the crystal facet with Miller indices (hkl). The length of $\mathbf{n}_{hkl}$ is proportional to the surface energy γ per unit area of this facet.

The surface free energy of a faceted crystal is given by the weighted sum of the surface free energies of all its facets

$$\mathcal{F} = \Sigma_j \gamma_j S_j$$

where γ_j is the constant surface free energy per area of facet j, and S_j is the area of facet j. It was proposed by Gibbs that the equilibrium shape of a crystal minimizes $\mathcal{F}$ for a given volume of the crystal. Later Wulff (1901) suggested that the length of the plane normal to a crystal facet drawn through the centre of an isometric crystal is proportional to the surface free energy of this facet, *i.e.* $n_j \propto \gamma_j$. Wulff's conjecture can be proven by minimizing $\mathcal{F}$ at constant volume:

$$\text{minimize} \left(\sum_j \gamma_j S_j - \lambda \frac{1}{3} \left(\sum_j n_j S_j \right) \right)$$

where $1/3(\Sigma_j n_j S_j)$ is the volume of a pyramid with basal plane S_j and height n_j, and λ is a Lagrange multiplier. The volume of the crystal may be thought of as the sum of all such pyramids with the crystal facets as their basal planes. The constrained minimization reads

$$\text{d}\left(\sum_j \gamma_j S_j \right) - \frac{\lambda}{3} \text{d}\left(\sum_j n_j S_j \right) = 0$$

leading to

$$\sum_j \gamma_j \text{d}S_j - \frac{\lambda}{3} \underbrace{\sum_j n_j \text{d}S_j}_{\text{shape change}} - \frac{\lambda}{3} \underbrace{\sum_j n_j S_j \text{d}n_j}_{\text{volume change}} = 0$$

Due to the constant volume constraint the term $\Sigma_j S_j \text{d}n_j$ must vanish, and thus

$$\sum_j \gamma_j \text{d}S_j = \frac{\lambda}{3} \sum_j n_j \text{d}S_j$$

As the minimization must be valid for an arbitrary shape change, we conclude for each j

$$\gamma_j = \frac{\lambda}{3} n_j$$

which confirms the original conjecture of Wulff.

The equilibrium shape of a crystal can be found from the γ plot using the so called 'Wulff construction'. The vectors **n**$_j$ are drawn from the origin so that their endpoints correspond to the polar plot of γ(**n**). The planes normal to the vectors **n**$_j$ drawn through the endpoints are referred to as the 'Wulff planes'. The inner hull of the Wulff planes represents the equilibrium shape of the crystal (see Fig. 19). This construction always yields convex bodies. Cusps of low γ usually correspond to low-indexed planes with high atomic density, which are prone to forming facets. These planes are in a local energy minimum and are therefore referred to as 'singular surfaces'. They exist at low temperatures relative to the material's melting point. Surfaces with orientations close to the singular orientations are called 'vicinal surfaces' and comprise discrete steps and terraces, which exhibit the structure of the nearby singular surfaces. Surfaces with orientations far from singular orientations have highly stepped and irregular surface structures of relatively high energy and are referred to as 'general surfaces'. In the following, some basic concepts for describing grain- and phase-boundaries in crystalline materials are presented.

7.2. Orientation relations at a grain boundary

The properties of a grain boundary depend on its microscopic structure, which is, in turn, determined by the orientation relations among the two neighbouring grains and the grain-boundary plane.

Any two orientations of two crystals of the same kind are related by a single rotation operation. A convenient way of describing this rotation is by specifying the rotation axis and the rotation angle. The rotation axis is referred to as the 'misorientation axis' and the rotation angle is referred to as the 'misorientation angle' or the 'misorientation' for short. The orientation relation between two grains may be described by a set of three parameters, namely two direction cosines for specifying the direction of the

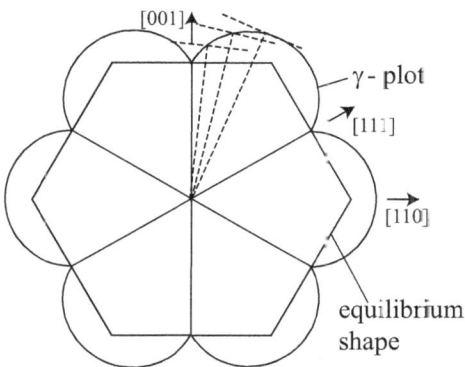

Figure 19. Schematic drawing of a 2D γ plot and Wulff construction of the equilibrium crystal shape for an f.c.c. crystal viewed along the [1$\bar{1}$0] direction. Curved line represents γ(**n**), and radial solid lines indicate plane normals of stable crystal facets as indicated by the hexagon. Dashed lines indicate plane normals and facets that do not pertain to the minimum energy configuration.

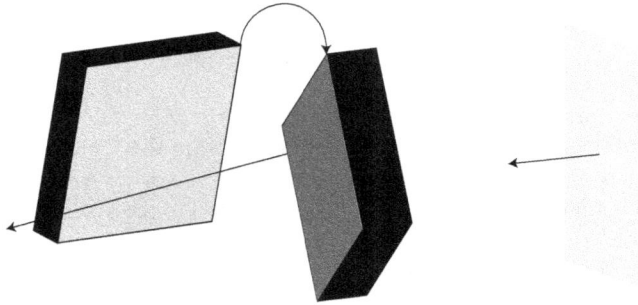

Figure 20. Left: two crystals of the same kind with different orientations. The two orientations are related through a single rotation around the rotation axis. Two direction cosines define the direction of the rotation axis. One rotation angle defines the extent of rotation between the two grains around the axis. *Right*: the orientation of the grain-boundary plane is given by the direction of the plane normal as defined by two direction cosines.

misorientation axis, and the misorientation angle. The orientation of the grain-boundary plane is described by two additional direction cosines that specify the orientation of the grain-boundary normal. The orientation relations between two crystals of the same phase and their common grain boundary are thus uniquely defined by specifying five parameters, which are referred to as the 'five macroscopic degrees of freedom' (see Fig. 20). On the microscopic level, three more degrees of freedom become relevant, which are related to the displacement of a certain lattice position in one crystal with respect to the equivalent position in the other crystal.

7.3. Tilt and twist grain boundaries

Two special cases are distinguished with respect to the orientation relations between grain-boundary plane and misorientation axis. When the misorientation axis lies in the grain-boundary plane, the grain boundary is referred to as a 'tilt grain boundary'. When the misorientation axis is oriented perpendicular to the grain-boundary plane, the grain boundary is referred to as a 'twist grain boundary' (see Fig. 21). Tilt and twist grain boundaries are special cases. Twin boundaries are typical examples. General grain boundaries do not pertain to either one of these special cases. In a general grain boundary,

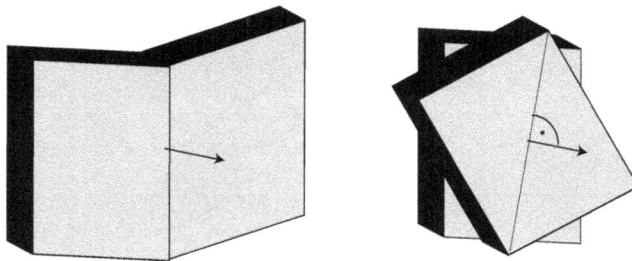

Figure 21. Left: for a tilt-grain boundary, the rotation axis lies in the grain-boundary plane. *Right*: for a twist-grain boundary, the rotation axis is perpendicular to the grain-boundary plane.

the grain boundary plane neither contains the rotation axis nor is it perpendicular to it. Nevertheless, it is instructive to study the special grain boundary types.

The microscopic structure of a symmetric tilt grain boundary is the simplest to visualize (see Fig. 22). In a symmetric tilt grain boundary the grain boundary plane corresponds to identical crystallographic planes in both crystals sharing the common boundary. Microscopically, a cut through a crystal oblique to a low-index plane consists of an array of ledges. When two crystals of the same type with symmetric oblique cuts are combined into a bicrystal, the ledges become edge dislocations in the boundary plane. A symmetrical tilt grain boundary may be viewed as an array of edge dislocations. As an example we show a Fourier-filtered high-resolution transmission electron microscope image (HRTEM) of a synthetic symmetric tilt grain boundary in a forsterite bicrystal (see Fig. 23). In this grain boundary the two crystals share a common [100] direction around which they are rotated by ~10°. The grain-boundary plane contains the [100] direction and is symmetric with respect to the two grains. It comprises an array of edge dislocations extending in the [100] direction.

The spacing d between the edge dislocations in the boundary depends on the misorientation angle θ and the Burgers vector b of the edge dislocation

$$d = b/(2\sin(\theta/2))$$

In general, the spacing d between the edge dislocations decreases with increasing misorientation angle θ. As long as the edge dislocations are separated by dislocation-free boundary segments, the boundary is referred to as a 'low-angle grain boundary' (Warrington and Boon, 1975). When the spacing between the edge dislocations is so small that the elastic strain fields around individual dislocations overlap, and dislocation-free boundary segments are absent, this is called a 'high-angle grain boundary'. A simple criterion to decide whether two edge dislocations are separated or overlap is lacking, and thus the transition from low-angle to high-angle grain boundary may occur over a range of misorientation angles from about 10° to 20°.

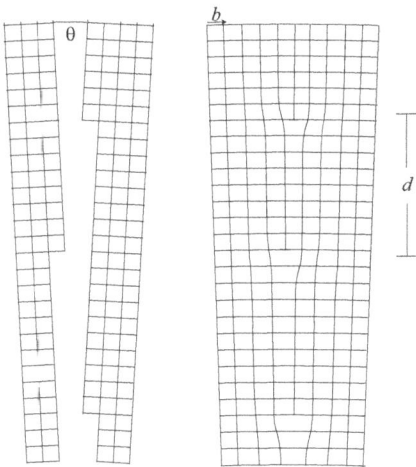

Figure 22. Schematic drawing of a symmetrical tilt grain boundary, redrawn after Howe (1997).

Figure 23. Energy-filtered HRTEM image of a grain boundary in a synthetic forsterite bicrystal (image courtesy of R. Wirth). The view direction is parallel to [100]. Edge dislocations extending in view direction are indicated by ⊥ signs. Greyscale contrasts portray columns of SiO_4 tetrahedra aligned parallel to [100].

7.4. Interface energy of a low-angle grain boundary

For low-angle grain boundaries the grain-boundary energy can be calculated from the excess energy associated with the individual edge dislocations. For a symmetric tilt grain boundary composed of dislocations with Burgers vector b and spacing d between dislocations, the grain-boundary energy reads (Read and Shockley, 1950)

$$\gamma_S = \gamma_0 \theta (A - \ln\theta)$$

where for small misorientation angles $\theta \approx b/d$, and $\gamma_0 = Gb/4\pi \, (1 - v)$, with G denoting the shear modulus and v the Poisson's ratio; γ_0 accounts for the elastic strain energy associated with a single line dislocation; $A = 1 + \ln(b\pi r_0/2)$, where r_0 is the radius of the dislocation core. The Read-Shockley equation predicts a strictly monotonic increase in grain-boundary energy with increasing misorientation angle and applies to misorientations up to the transition to high-angle grain-boundaries.

The defect structure at a grain boundary with general orientation relations is much more complex than that of a symmetrical tilt grain boundary. The defect structure of a general grain boundary can usually not be constructed from crystal structure and misorientation relations using simple geometrical models. The detailed defect structure of an interface in a crystalline material can be obtained from atomistic modelling, though (see Jahn and Sun, 2017, this volume). Simple geometrical models can, however, provide important constraints on the defect structure. In the following, three such concepts, namely the coincidence site lattice model, Frank's theory, and the O-lattice concept are introduced. In doing so, we follow the presentation of Howe (1997). For a detailed account of the subject, the reader is referred to this textbook.

7.5. Coincident site lattice model

When, in a thought experiment, two identical crystal lattices, which coincide perfectly, are rotated with respect to each other around a common lattice point, a fraction of the lattice points of both lattices will coincide for specific rotation angles. The coinciding lattice points form a superstructure, which has translational symmetry and constitutes a new lattice referred to as the 'coincident site lattice' (CSL). Consider two identical cubic lattices projected down their common [100] direction and rotated with respect to one another by $\phi = 36.87°$ from initial coincidence around an axis, which is parallel to the [100]-direction and runs through a specific lattice point that is common to both lattices (see Fig. 24). For this rotation $\cot(\phi/2) = 3$, and the lattice vector [013] of the white lattice is identical to the [0$\bar{1}$3] lattice vector of the black lattice (dashed grey arrow) defining a coincident site of the two lattices. Application of the [013] and the [03$\bar{1}$] lattice vectors of the white lattice defines the coincident site lattice (dashed grey square), where two out of ten lattice points of the white lattice coincide with lattice points of the black lattice. Alternatively, the primitive unit cell of the CSL can be defined in terms of the [0$\bar{1}$2] and the [021] lattice vectors of the white lattice. In the primitive unit cell of the CSL, one out of five lattice points of the white lattice coincides with lattice points of the black lattice. The reciprocal of the fraction of lattice points of one lattice that coincides with lattice points in the other lattice is denoted as Σ; $\Sigma = 5$ in the case at hand. Alternatively, Σ may be defined as the volume of the smallest possible CSL unit cell in units of the primitive unit cell of the generating lattices, *i.e.*

$$\Sigma = \frac{\text{volume of smallest unit cell of the CSL}}{\text{volume of primitive unit cell of crystal lattice}}$$

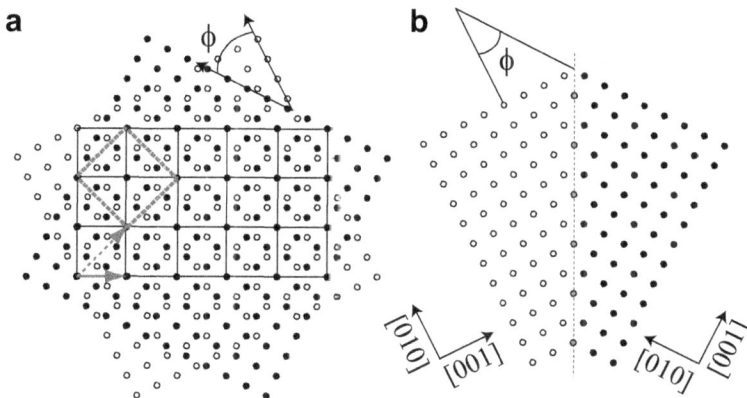

Figure 24. (a) Schematic drawing of two cubic lattices projected down their common [100] direction; the two lattices are rotated with respect to each other by 36.87° around an axis parallel to [100] running through a lattice point of both lattices; dashed grey arrow: [013] lattice vector of the white and [0$\bar{1}$3] lattice vector of the back lattice, grey dashed square: unit cell of corresponding coincident site lattice; grey arrow: [0$\bar{1}$2] lattice vector of the white lattice and [0$\bar{2}$1] lattice vector of the back lattice; the primitive unit cells of the corresponding coincident site lattice are indicated as black squares; (b) Σ5 CSL tilt grain boundary obtained from orientation relations shown in (a) with boundary plane (0$\bar{1}$2) of the white lattice and (02$\bar{1}$) of the black lattice.

For a rotation of 36.87° around [100] in a cubic lattice, we have

$$\sum = \frac{a(a\sqrt{5})^2}{a^3} = 5$$

It can be shown that Σ is always an odd integer for cubic crystals (Bollmann, 1970). In general, any lattice vector can be used to define a CSL, and a large number of low-Σ CSL exist. Brandon *et al.* (1964) listed all 61 CSL with $\Sigma \leqslant 19$ for cubic lattices.

Grain boundaries tend to follow densely packed planes of the CSL. Grain boundaries with a high density of coincident sites on the grain-boundary plane are referred to as 'CSL grain boundaries'. Grain boundaries with grain-boundary planes slightly deviating from the orientation of densely packed CSL-planes still tend to follow these planes, and the orientation deviation is accommodated by steps, where the grain-boundary jumps between two neighbouring densely packed CSL-planes (Brandon *et al.*, 1964). Low-Σ CSL grain boundaries have specific properties and are referred to as 'special grain-boundaries'. For example, it is generally observed that fewer impurities segregate at special than at general grain boundaries (Doherty *et al.*, 1997). Furthermore, low-Σ CSL grain boundaries in Pb-Sn alloys were found to migrate much faster than general grain boundaries (Aust and Rutter, 1959). A simple relation between Σ value and grain-boundary energy does not exist. In several cases, low-Σ grain boundaries have been found to have low interface energy (Sutton and Balluffi, 1995; Rohrer, 2011). In their seminal review on models of interface energy, Sutton and Balluffi (1987) concluded, however, that the CSL concept does not, in general, provide useful criteria for estimating grain-boundary energy.

The CSL concept is purely geometrical and coincidence of lattice points is immediately lost when the misorientation between two lattices slightly deviates from the special misorientation that ensures coincidence of lattice sites. In fact, it is generally observed that grain boundaries with misorientations slightly deviating from special angles may show similar properties as low-Σ CSL grain boundaries. Geometrically, general high-angle grain boundaries may be treated as deviations from the nearest CSL orientation relation. Similar to low-angle grain boundaries, where deviations from perfect match ($\Sigma = 1$ CSL) are accommodated by misfit dislocations separated by boundary segments with perfect coincidence of the two crystal lattices, deviations from perfect CSL are accommodated by dislocations separated by boundary segments with perfect CSL relations between the two crystal lattices. The misfit dislocations associated with low-angle grain boundaries are referred to as 'primary dislocations', the misfit dislocations associated with near CSL boundaries are referred to as 'secondary dislocations'. The maximum angular deviation $\Delta\phi$ from a low-Σ CSL orientation relation, for which a grain boundary exhibits special properties, may be expressed as (Brandon, 1966)

$$\Delta\phi = \frac{\phi_0}{\sqrt{\Sigma}}$$

where $\phi_0 \approx 15°$ is the upper limit of misorientation for a low-angle boundary.

7.6. Dislocations at a general grain boundary

The dislocation content of a general grain boundary can be determined based on the
theory of Frank (1951). In Fig. 25a, the lattice of a cubic crystal viewed down its [100]
direction is cut by a plane parallel to [100] and running along the line $A-A'$. The left
side of the lattice is labelled "$-$", the right side is labelled "$+$", and the plane normal is
defined by the unit vector **n** pointing from lattice "$-$" to lattice "$+$". The lattice on the
left is rotated over an angle $\phi/2$ to the left and the lattice on the right is rotated over an
angle $\phi/2$ to the right, where the rotation axis is defined by the unit vector **l** $\parallel$ [100]
running through point O and pointing into the page (see Fig. 25b). The two misoriented
lattices are then extended so that they meet at the position of the original cut (Fig. 25c).
A vector **x** is drawn within the boundary plane from point O extending over several unit
cells, where **x** may have any direction within the boundary plane. The net Burgers
vector of the dislocations that are intersected by **x** is obtained by constructing a Burgers
circuit around an axis (**x** $\times$ **n**) and containing **x** (see Fig. 25c). The Burgers circuit starts
at point S, which is the endpoint of vector **x**, extends within crystal "$+$" to point O and
then extends within crystal "$-$" to point F, which coincides with point S. The
equivalent circuit in the reference lattice is shown in Fig. 25d. The closure failure of the
Burgers circuit in Fig. 25d corresponds to the net Burgers vector **B** of the dislocations
intersected by **x**. In general, **x** encloses an angle ϕ with the rotation axis **l**. No

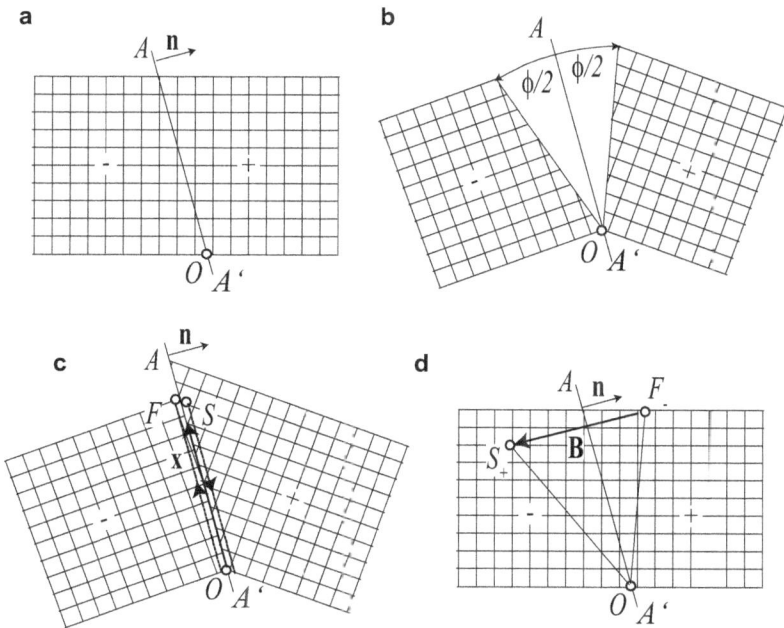

Figure 25. Schematic illustration of the theory of Frank for determining dislocation density in a general
grain boundary, redrawn from Howe (1997).

dislocations are generated with Burgers vectors having components $\parallel$ **l**. Only the projection of **x** on a plane normal to **l**, that is $x\sin\phi$, is relevant. The length of vector **B** is thus

$$B = x\sin\phi\, 2\sin(\phi/2)$$

and its direction is given by the cross product $\mathbf{x} \times \mathbf{l}$. Noting that the length of this cross product is $x/\sin\phi$, and **l** is a unit vector we find

$$\mathbf{B} = (\mathbf{x} \times \mathbf{l})2\sin(\phi/2)$$

the net Burgers vector of the dislocations that are intersected by **x**. The net Burgers vector may be regarded as the sum of the Burgers vectors of the individual dislocations that are intersected by **x**

$$\mathbf{B} = \sum_i n_i \mathbf{b}_i$$

where n_i is the number of dislocations with Burgers vector $\mathbf{b}_i$.

In general, at least three arrays of non coplanar Burgers vectors $\mathbf{b}_i$ are required to build the net Burgers vector **B**. Let $\mathbf{b}_1$, $\mathbf{b}_2$, $\mathbf{b}_3$ be three non-coplanar Burgers vectors of dislocations with their line directions defined by the unit vectors ξ_1, ξ_2, ξ_3 with spacing D_1, D_2, D_3. Then we have

$$n_1\mathbf{b}_1 + n_2\mathbf{b}_2 + n_3\mathbf{b}_3 = (\mathbf{x} \times \mathbf{l})2\sin(\phi/2)$$

where n_i is given by $n_i = \mathbf{N}_i \cdot \mathbf{x}$ with

$$\mathbf{N}_i = (\mathbf{n} \times \xi_i)/D_i$$

being a vector in the boundary plane that is normal to the line direction of the i'th dislocation array and with a magnitude of $1/D_i$.

Note that Frank's theory merely provides the net Burgers vector. It does not, however, define the dislocation structure at a given grain boundary. The net Burgers vector may be built from different dislocation arrays. Given that at least three dislocation arrays with three non-coplanar Burgers vectors are necessary, Frank's theory provides an important constraint on possible combinations of dislocation arrays. For example, when a set of dislocations is identified by TEM at a general grain boundary, Frank's theory allows us to decide whether this set of dislocations is sufficient or whether additional dislocations are necessary to build the net Burgers vector that corresponds to the macroscopic orientation relations at the grain boundary.

7.7. Dislocations at a general interface

With reference to Fig. 25d, the net Burgers vector **B** may be written as

$$\mathbf{B} = \overrightarrow{OS_+} - \overrightarrow{OF_-}$$

Let us define $\mathbf{R}_+$ and $\mathbf{R}_-$ as the rotation tensors transforming the reference lattice into lattices "+" and "−", respectively so that

$$\overrightarrow{OS_+} = \mathbf{R}_+^{-1}\mathbf{x} \quad \text{and} \quad \overrightarrow{OF_-} = \mathbf{R}_-^{-1}\mathbf{x}$$

Then we have

$$\mathbf{B} = (\mathbf{R}_+^{-1} - \mathbf{R}_-^{-1})\mathbf{x}$$

where both rotations are expressed with respect to the reference lattice (Fig. 25a). If instead lattice "+" is taken as the reference lattice, we obtain

$$\mathbf{B} = (\mathbf{I} - \mathbf{R}^{-1})\mathbf{x}$$

where $\mathbf{R}$ is the rotation tensor transforming lattice "+" into lattice "−", and $\mathbf{I}$ is the identity matrix. This formulation can be extended readily to interfaces between different phases by replacing the rotation tensor $\mathbf{R}$ by a tensor $\mathbf{A}$, which may include both a general strain and a rotation. The net Burgers vector of the dislocations intersected by vector $\mathbf{x}$ in a general interface reads

$$\mathbf{B} = (\mathbf{I} - \mathbf{A}^{-1})\mathbf{x}$$

Given that the net Burgers vector is composed of at least three dislocation arrays with non-coplanar Burgers vectors, this may be written as

$$n_1\mathbf{b}_1 + n_2\mathbf{b}_2 + n_3\mathbf{b}_3 = (\mathbf{I} - \mathbf{A}^{-1})\mathbf{x}$$

This extension of Frank's theory thus allows us to decide whether a set of dislocations observed at a general interface is complete, or whether additional dislocations are required to accommodate the misfit between the lattices of two different phases for a given set of macroscopic orientation relations.

7.8. The O lattice

When two crystal lattices are allowed to interpenetrate, a regular array of points is generated which have the same internal coordinates with respect to a subset of the overlapping unit cells of both lattices. This phenomenon was referred to as the O *lattice* by Bollmann (1970). The basic idea of the O lattice can be formulated as follows: Consider a lattice α and a lattice β which is generated from lattice α by a general deformation that may include rotation, volume strain (expansion/shrinkage) and/or shape change (shearing). The lattices are now thought to interpenetrate, and we look for coincidence points between the two lattices. The search for the coincidence points is not restricted to the lattice points of the α and β lattices. Every point that has the same coordinates in a unit cell of both crystal lattices constitutes a coincidence. Such a point may, but need not necessarily, be a lattice point of the α and β lattices.

A vector $\mathbf{x}^\beta$ in lattice β is generated from the corresponding vector $\mathbf{x}^\alpha$ in lattice α through (see Fig. 26)

$$\mathbf{x}^\beta = \mathbf{A}\mathbf{x}^\alpha$$

A point in the O lattice is defined as a vector $\mathbf{x}^{(O)}$, where $\mathbf{x}^\beta$ differs from $\mathbf{x}^\alpha$ by a translation vector $\mathbf{b}^{(L)}$ of the α lattice such that

$$\mathbf{x}^{(O)} = \mathbf{x}^\beta = \mathbf{x}^\alpha + \mathbf{b}^{(L)} = \mathbf{A}\mathbf{x}^\alpha$$

From this set of equations, we obtain

$$\mathbf{b}^{(L)} = (\mathbf{I} - \mathbf{A}^{-1})\mathbf{x}^{(O)} \tag{41}$$

which is the basic equation of O lattice theory. Rearrangement of equation 41 yields an explicit expression for $\mathbf{x}^{(O)}$, which defines the O lattice with respect to lattice α

$$\mathbf{x}^{(O)} = (\mathbf{I} - \mathbf{A}^{-1})^{-1}\mathbf{b}^{(L)} \tag{42}$$

The strategy for finding the basal vectors of the O lattice is now to insert the basal vectors of the lattice α for $\mathbf{b}^{(L)}$

As an example consider two simple cubic lattices that are rotated with respect to each other around their common [001] direction over an angle ϕ. The rotation tensor reads

$$\mathbf{A} = \begin{pmatrix} \cos\phi & -\sin\phi & 0 \\ \sin\phi & \cos\phi & 0 \\ 0 & 0 & 1 \end{pmatrix}, \text{ its inverse is } \mathbf{A}^{-1} = \begin{pmatrix} \cos\phi & \sin\phi & 0 \\ -\sin\phi & \cos\phi & 0 \\ 0 & 0 & 1 \end{pmatrix}$$

and

$$(\mathbf{I} - \mathbf{A}^{-1}) = \begin{pmatrix} 1-\cos\phi & -\sin\phi & 0 \\ \sin\phi & 1-\cos\phi & 0 \\ 0 & 0 & 0 \end{pmatrix}$$

and finally

$$(\mathbf{I} - \mathbf{A}^{-1})^{-1} = \begin{pmatrix} \frac{1}{2} & -\frac{1}{2}\cot\frac{\phi}{2} & 0 \\ \frac{1}{2}\cot\frac{\phi}{2} & \frac{1}{2} & 0 \\ 0 & 0 & 0 \end{pmatrix}$$

Inserting into equation 42 yields

$$\begin{pmatrix} x_1 \\ x_2 \\ x_3 \end{pmatrix} = \begin{pmatrix} \frac{1}{2} & -\frac{1}{2}\cot\frac{\phi}{2} & 0 \\ \frac{1}{2}\cot\frac{\phi}{2} & \frac{1}{2} & 0 \\ 0 & 0 & 0 \end{pmatrix} \begin{pmatrix} b_1 \\ b_2 \\ b_3 \end{pmatrix}$$

where x_1, x_2, x_3 are the components of the O-lattice vector $\mathbf{x}^{(O)}$, and b_1, b_2, b_3 are the components of the vector $\mathbf{b}^{(L)}$. It follows that $x_3 = 0$ and b_3 may have any value. If we consider a pure twist grain boundary, possible base vectors of the α lattice are

$$\mathbf{b}_1 = \begin{pmatrix} 1 \\ 0 \\ 0 \end{pmatrix} \text{ and } \mathbf{b}_2 = \begin{pmatrix} 0 \\ 1 \\ 0 \end{pmatrix}$$

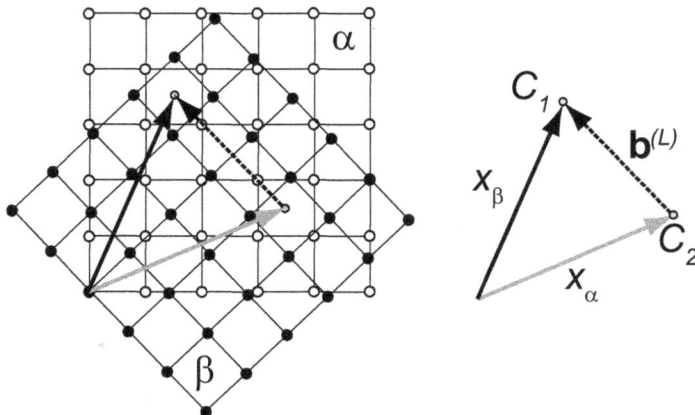

Figure 26. [001] view of a dichromatic pattern of two simple cubic lattices rotated with respect to each other over 43.6° around their common [001] axis; small grey point C_2 is obtained from small grey point C_1 and has identical internal coordinates (0.5, 0.5) in both lattices α and β, and C_2 is displaced from C_1 by $\mathbf{b}^{(L)}$, which is a translation vector in lattice α.

The corresponding O-lattice vectors are

$$\mathbf{x}_1^{(O)} = \begin{pmatrix} \frac{1}{2} \\ \frac{1}{2}\cot\frac{\phi}{2} \\ 0 \end{pmatrix}, \quad \text{and} \quad \mathbf{x}_2^{(O)} = \begin{pmatrix} -\frac{1}{2}\cot\frac{\phi}{2} \\ \frac{1}{2} \\ 0 \end{pmatrix}$$

Vectors $\mathbf{x}_1^{(O)}$ and $\mathbf{x}_2^{(O)}$ are continuous functions of ϕ. Special cases arise when $\cot(\phi/2)$ is rational. For example, when $\cot(\phi/2)$ is an even integer, the O-lattice points correspond to lattice points of the α and β lattices. When $\cot(\phi/2)$ is an odd integer, the O-lattice points are at the centre of edges of unit cells or in the centre of the unit cells.

The O lattice generated by superposition of two simple cubic lattices that are rotated with respect to each other by 8.17° around their common [001] direction is shown in Fig. 27. The O-lattice vectors for base vectors (100) and (010) in the reference lattice are $\mathbf{x}_1^{(O)} = (1/2,7,0)$ and $\mathbf{x}_1^{(O)} = (-7,1/2,0)$ with respect to the reference lattice (black lattice in Fig. 27). The solid lines indicate the projection of the O-lattice. For a rotation

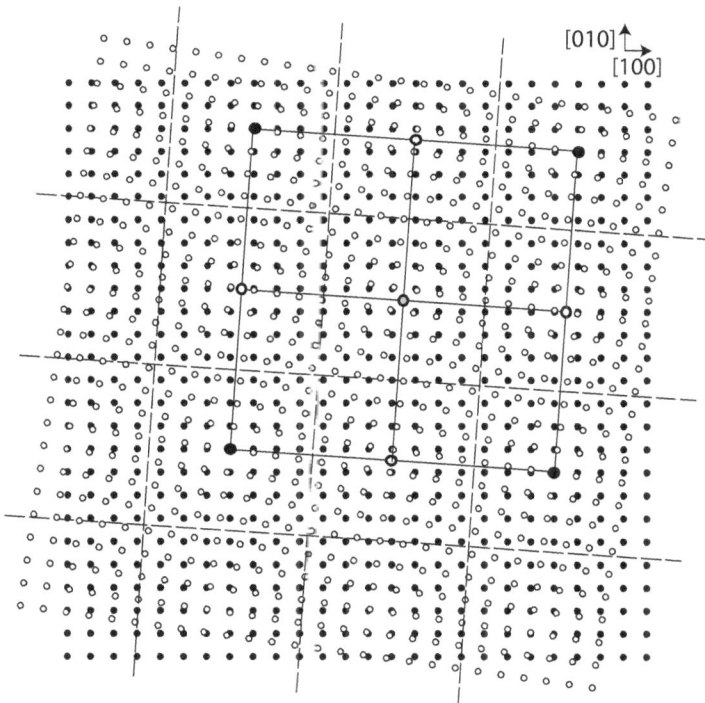

Figure 27. [001] view of the dichromatic pattern of two simple cubic lattices which are rotated with respect to one another by 8.17° around their common [001] direction; the solid lines show the O lattice, the heavy black circles indicate O-lattice points coinciding with lattice points of both cubic lattices, the heavy white circles are O-lattice points located at the centre of an edge of a unit cell of both cubic lattices; the grey heavy circle is an O-lattice point located at the centre of a unit cell of both cubic lattices.

angle of 8.17°, the cot($\phi/2$) is an integer and the O-lattice points either coincide with lattice points in both cubic lattices (black circles), fall on the edge of a unit cell of both cubic lattices (white circles), or fall into the centre of a unit cell of both cubic lattices (grey circle). At the O-lattice points there is perfect match between the two lattices. The mismatch between the two lattices increases away from the O-lattice points and is at a maximum half way between two O-lattice points. The dashed lines trace the locations of maximum mismatch between the two lattices. These are the locations where mismatch can be accommodated most efficiently by introducing misfit dislocations.

In Fig. 28, a variant of the dichromatic pattern of the two lattices of Fig. 27 is shown, where the orientation mismatch has been concentrated into narrow zones. Domains with perfect fit between the two lattices are centred on the O-lattice points and are separated from one another by a network of dislocation arrays half way between the O-lattice points, that is along the planes of maximum mismatch. The dislocation array on a grain boundary between the two lattices is obtained by intersecting the grain-boundary plane with these dislocation arrays. For a pure twist grain boundary the misfit dislocations have pure screw character.

Another simple relation between two crystal lattices is illustrated in Fig. 29. In this case two cubic lattices with similar orientation but with different lattice parameters are superimposed. The lattice indicated in black has a lattice parameter that is larger than

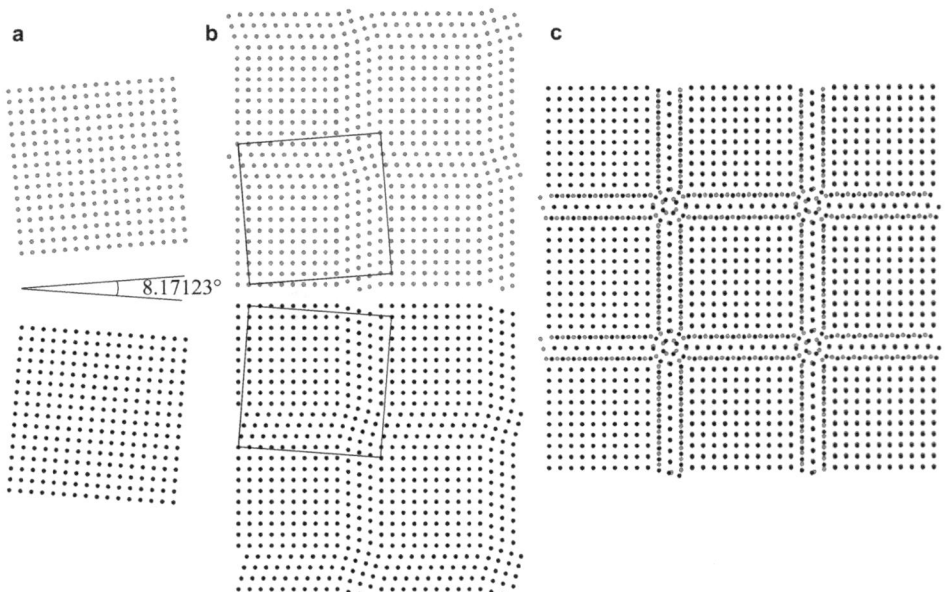

Figure 28. Same situation as in Fig. 27; (a) two simple cubic lattices viewed down their common [001] direction rotated through 8.17° around [001]; (b) the two lattices with localized lattice strain; (c) dichromatic pattern of superimposed lattices with regions of perfect match centred on the O-lattices points and a network of screw dislocations half way between the O-lattices points.

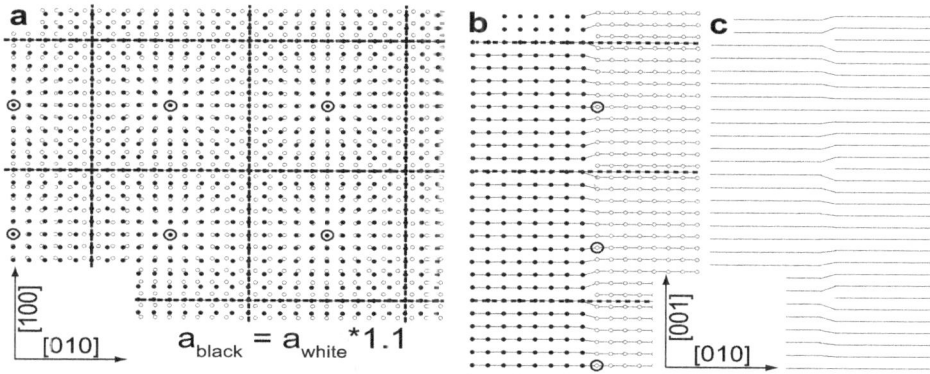

Figure 29. (a) Dichromatic pattern generated by superposition of two cubic lattices with a difference in lattice parameters of 10%, $a_{black} = 1.1 a_{white}$ and similar orientation viewed down the [001] direction, large open circles indicate O-lattice sites, heavy dashed lines indicate planes of maximum mismatch between the two lattices; (b) [001] view of the same configuration as in (a), where the white and the black lattice have been removed from the left and right half of the image; (c) traces of (100) planes showing edge dislocations with every 12th lattice planes of the white lattice ending at the interface–misfit dislocation.

the lattice parameter of the white lattice by a factor of 1.1, $a_{black} = 1.1 a_{white}$. The transformation matrix A that generates the black lattice from the white one reads

$$A = \begin{pmatrix} 1.1 & 0 & 0 \\ 0 & 1.1 & 0 \\ 0 & 0 & 1.1 \end{pmatrix}$$

Application of equation 42 and using

$$b_1 = \begin{pmatrix} 1 \\ 0 \\ 0 \end{pmatrix} \quad \text{and} \quad b_2 = \begin{pmatrix} 0 \\ 1 \\ 0 \end{pmatrix} \quad \text{and} \quad b_3 = \begin{pmatrix} 0 \\ 0 \\ 1 \end{pmatrix}$$

as the base vectors of the reference lattice yields the base vectors of the O lattice:

$$x_1^{(O)} = \begin{pmatrix} 11 \\ 0 \\ 0 \end{pmatrix} \quad \text{and} \quad x_2^{(C)} = \begin{pmatrix} 0 \\ 11 \\ 0 \end{pmatrix} \quad \text{and} \quad x_3^{(O)} = \begin{pmatrix} 0 \\ 0 \\ 11 \end{pmatrix}$$

In Fig. 29b,c, a view along the [001] direction of a (010)-oriented interface between the black and the white lattice, which was generated by removing the white lattice from the left and the black lattice from the right half of the image, is shown. In between two perfectly aligned (100) planes there are ten (100) lattice planes in the white and only nine (100) lattice planes in the black lattice. The lattice mismatch is accommodated by introducing an edge dislocation at every 11th (100) lattice plane of the white lattice half way between the O-lattice points. The corresponding edge dislocations have a dislocation line parallel to [001] and a Burgers vector in [100] direction.

A similar situation was documented for a (001)-oriented interface between spinel (MgAl$_2$O$_4$) and periclase (MgO) with simple cube-to-cube topotactic orientation relation (Li *et al.*, 2016). It is seen in Fig. 30 that about every 23rd (010) lattice plane of the spinel (lattice on the right) ends with an edge dislocation at the interface. All the other (010) lattice planes run across the interface without any defects. The difference in lattice parameters of strain-free periclase and spinel is such that $2a_{per} = 1.043a_{sp}$. Using equation 42 the base vectors of the O-lattice in the (001)-oriented interface are calculated as $x_1^{(O)} = x_2^{(O)} = 23.8$, which matches the observation perfectly.

To conclude this section we remark that, when two lattices α and β with different orientations and/or different lattice parameters, are allowed to interpenetrate, a periodic array of points arises, where the internal coordinates in the unit cell of lattice α are the same as the internal coordinates in the unit cell of lattice β. These points constitute the O lattice (Bollmann, 1970). The points of the O lattice may but, in general, are not lattice points of the α and β lattices. The O-lattice points represent locations of perfect match between the two lattices. The planes half way between the O-lattice points represent a 3D interconnected network of domains with poor match

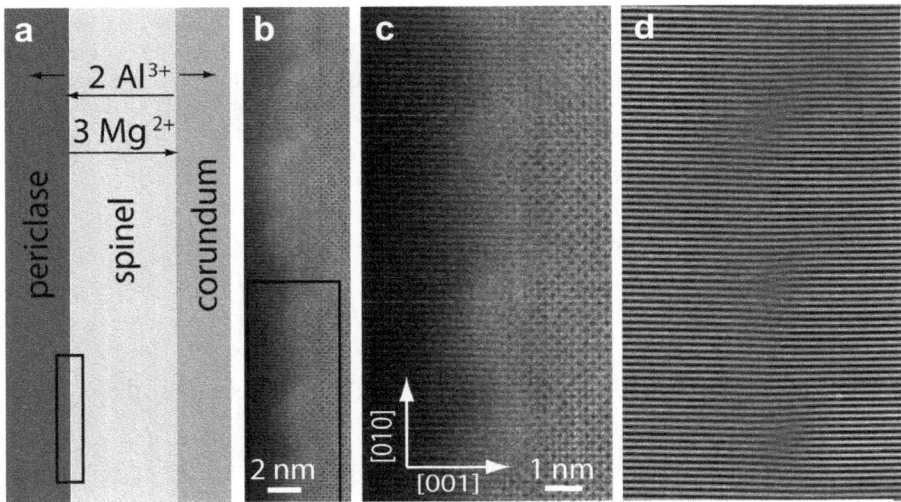

Figure 30. (a) Schematic drawing of a diffusion-reaction setting, where spinel (MgAl$_2$O$_4$) grows at a periclase (MgO)–corundum (Al$_2$O$_3$) contact due to the inter diffusion of Mg^{2+} and Al^{3+} (Jerabek *et al.*, 2014), both, the periclase–spinel as well as the corundum–spinel interface move into the reactant phases; (b) scanning transmission electron microscopy (STEM) image of the domain indicated by the black rectangle in (a) with the periclase (left) – spinel (right) contact viewed down the [100] direction, note the scalloped shape of the interface with domes protruding into the periclase and cusps pointing towards spinel; (c) zoom into the area indicated by the black rectangle in (b); (d) Fourier-filtered image of the domain shown in (c) highlighting the traces of (010) lattice planes, note that about every 23rd (010) lattice plane of spinel ends at the interface (indicated by thin dashed white lines), the corresponding edge dislocations are close to the cusps of the scalloped interface; STEM images courtesy Chen Li.

between the lattices and constitute the most likely locations, where dislocations may be introduced. The dislocations at a general interface are obtained by intersecting the interface plane with the 3D network of maximum-mismatch domains. The O lattice is a geometrical concept, which provides a continuum theory for the dislocations at a general interface.

7.9. Grain-boundary character distribution

Monomineralic rocks, such as quartzites or marbles, may be regarded as polycrystals comprising a multitude of crystals that are in contact along a three-dimensional grain-boundary network. The grain-boundary network determines many of the physical properties and reactivity of a polycrystal. In this context the question arises up to the relative abundances of the different grain-boundary types in a polycrystal.

In standard petrographic analysis usually only 2D sections are available, and out of the five macroscopic parameters needed to fully specify the orientation relations at a grain boundary, only four can be determined. For example, with conventional crystal orientation microscopy by electron backscatter diffraction (EBSD) the misorientation between neighbouring grains (three degrees of freedom) and the direction of the trace of the grain-boundary plane on the sample surface (one degree of freedom) can be determined (see Zaefferer and Habler 2017, this volume). The information on the inclination of the grain-boundary plane with respect to the sample surface remains undetermined. This information is, however, indispensable for a complete determination of the five macroscopic parameters characterizing the orientation relations at a grain boundary. The missing information can be accessed if EBSD is performed on serial cuts, produced either by conventional grinding or by using the focused ion beam technique. Several workers have used 3D EBSD to obtain both the crystal orientation relations as well as the orientations of grain-boundary planes. Considering the five degrees of freedom, the number of possible orientation relations at a grain boundary is large. If the five-dimensional parameter space is discretized in $10°$ intervals, there are roughly 6×10^3 different grain-boundary orientation relations for a cubic material and even more for materials with lower symmetry. Accordingly a large number of grain boundaries must be sampled in order to obtain a statistically meaningful distribution of grain-boundary orientation relations. The distribution of area fractions of grain boundaries with specific lattice misorientation and grain-boundary plane orientation is referred to as the 'grain boundary-character distribution' (GBCD). Due to progress in automated 3D EBSD (see Zaefferer and Habler 2017, this volume) obtaining statistically meaningful GBCDs has become feasible. GBCDs have been determined primarily for metals but also for oxides such as MgO (Saylor *et al.*, 2003a,b), TiO_2 (Pang and Wynblatt, 2006), Al_2O_3 (Kitayama and Glaeser, 2002), and for olivine (Marquardt *et al.*, 2015), and fundamental insights have been obtained. For a detailed review see Rohrer (2011). Here, only a few salient points are summarized.

Grain-boundary energy is strongly influenced by the orientation of the grain boundary plane. For a given misorientation between two grains, some orientations of the grain-boundary plane exist that have significantly lower energies than others. For example Saylor *et al.* (2003a) applied 3D EBSD to determine all five macroscopic orientation parameters for about 4.1×10^6 boundary plane segments and derived the full GBCD for hot-pressed magnesia (MgO). Furthermore, from the geometric and crystallographic characteristics of approximately 10^4 triple junctions, the relative grain-boundary energy of the different grain-boundary types was determined (Saylor *et al.*, 2003b). It was found that the relative energy of a grain boundary is inversely proportional to its frequency of occurrence. For all given misorientations between two grains those boundaries that are oriented normal to the [100] direction of one or both grains occur at least twice as frequently as the average boundary. For all misorientations, these grain boundaries have relatively low energies. Rohrer (2011) concluded that for low-angle grain boundaries, dislocation models provide good estimates for grain-boundary energy. At high misorientation angles, the energy is correlated to the sum of the energies of the free surfaces comprising the boundary. In other words, lattice planes that form facets on a free-standing crystal also tend to form the grain boundaries in a polycrystal.

We conclude this section by summarizing that grain boundaries with boundary planes parallel to a low-index plane of at least one crystal have lower energies and appear more frequently in the microstructure than other boundaries. The most frequent grain-boundary plane orientations correspond to low index, low energy surface planes. The most favoured grain-boundary types are thus related to configurations in which at least one side of the grain boundary is terminated by a low index plane. The energy anisotropy resulting from variations in the grain-boundary plane orientation are greater than the anisotropy resulting from variations in the lattice misorientation. This corroborates the view that the energy cost for making a grain boundary may be thought of as the energy needed to create the two crystal surfaces reduced by the binding energy that is recovered by bringing the two surfaces together.

7.10. Motion of an interface in crystalline material

The motion of an interface in a crystalline material is intimately linked to the motion of its associated dislocations. Dislocations can move by either one or a combination of two end-member processes namely 'dislocation glide' and 'dislocation climb'. Dislocation glide is the motion of a dislocation along its 'glide plane' (also called 'slip plane'), which is the plane containing the dislocation line and the Burgers vector. Dislocation climb is the motion of a dislocation in the direction perpendicular to the glide plane. Dislocation glide is a conservative mode of motion in the sense that it occurs by local rearrangements of atoms or bonds, and no material needs to be delivered to or removed from the dislocation core. In contrast, dislocation climb requires that the material which is involved in extending or shortening the extra plane associated with an edge dislocation is delivered to or removed from the dislocation core. As the extra plane is composed of the major components of the bulk phase, the necessary diffusion is likely to occur by a vacancy mechanism implying that an atomic flux to or from the dislocation core is compensated by a corresponding flux of vacancies in the opposite direction. Depending on the direction of motion a climbing dislocation may thus act as a source or as a sink for atoms or vacancies.

Very much like the motion of dislocations, the motion of interfaces in crystalline materials can be conservative or non-conservative. Conservative interface motion is possible only when the atomic proportions on both sides of the interface are the same, *i.e.* when the crystals on either side of the interface have the same composition. The conservative motion of a sharp interface occurs by simple shuffling of atoms across the boundary without long-range diffusion of species to or from the interface. The motion of grain boundaries during grain growth or the glissile motion of the edge dislocations associated with a low-angle symmetrical tilt grain boundary are examples of conservative interface motion. A low-angle symmetrical tilt grain boundary is composed of an array of well separated edge dislocations (see Fig. 22). When the edge dislocations move in a coordinated manner, for example in response to a mechanical force acting on the interface, this leads to a macroscopic shape change of the bicrystal (see Fig. 31a). The highly correlated shuffling of atoms is referred to as 'military motion'. The glissile motion of such an interface is possible only when all dislocations have the same glide plane. When several families of dislocations are present, glissile motion of the interface is still possible, when the respective glide planes lie in the same zone with the zone axis pointing out of the boundary plane. If this is not the case, interface motion can only occur by a combination of dislocation glide and climb. Thereby the necessary diffusive fluxes can be envisaged to simultaneously aid extension and shortening of lattice planes at edge dislocations (see Fig. 31a). During the non-conservative motion of an interface, atoms or vacancies must be added to or removed from one or both crystals adjacent to the interface. The non-conservative motion of an interface implies the creation or destruction of lattice sites in one or both crystals sharing the interface. Thus, the interface acts as a source or sink for atoms or vacancies. An example of non-conservative interface motion is the growth or

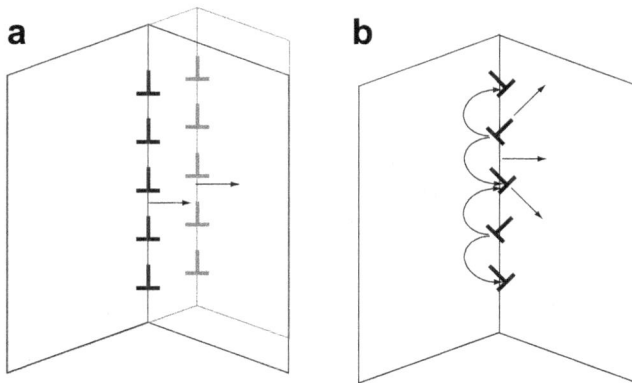

Figure 31. (a) Schematic drawing of a bicrystal with symmetrical tilt-grain boundary with only a single family of edge dislocations; the arrows indicate the direction of interface movement; in this case, interface motion implies a change in the overall shape; (b) bicrystal with symmetrical tilt-grain boundary with two different families of edge dislocations; the horizontal arrow indicates the direction of interface motion; straight arrows pointing up and down indicate the directions of movement of the edge dislocations; the curved arrows indicate material transfer aiding dislocation climb.

shrinkage of a phase in a polyphase material, which requires long-range diffusion of components such as during porphyroblast or interlayer growth (see Gaidies *et al.*, 2017, this volume).

An example of interface motion by pure climb was documented by Li *et al.* (2016). For the spinel-periclase interface shown in Fig. 30 to move into the reactant periclase, the edge dislocations at the termination of about every 23rd spinel (010) and (100) lattice plane must climb. Dislocation climb implies that new unit cells of $MgAl_2O_4$ spinel are added to the terminations of the (100) and (010) lattice planes. At the segments between the edge dislocations, the (010) and (100) lattice planes of the periclase can be traced into the corresponding (010) and (100) lattice planes of the spinel without any intervening dislocation at the interface. Propagation of these interface segments requires that out of four formula units of periclase, $3Mg^{2+}$ cations are liberated at the interface and transferred to the spinel–corundum interface by long-range diffusion and are replaced by $2\,Al^{3+}$ cations, which are supplied from the spinel-corundum interface by long-range diffusion. At the dislocation-free segments of the interface the oxygen atoms remain in place. The corresponding local mass balance equation may be written as

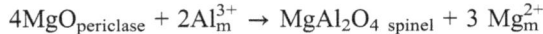

$$4MgO_{periclase} + 2Al_m^{3+} \rightarrow MgAl_2O_{4\ spinel} + 3\ Mg_m^{2+}$$

where Al_m^{3+} and Mg_m^{2+} denote mobile components, which are transferred between the periclase-spinel and the spinel-corundum interfaces by long-range diffusion. In contrast, the local mass balance at a climbing edge dislocation reads

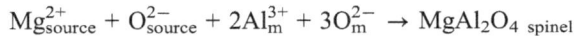

$$Mg_{source}^{2+} + O_{source}^{2-} + 2Al_m^{3+} + 3O_m^{2-} \rightarrow MgAl_2O_{4\ spinel}$$

where it is implied that Al_m^{3+} and O_m^{2-} are supplied from the spinel-corundum interface. The correlated long-range diffusion of Al_m^{3+} and O_m^{2-} is required for maintaining charge balance at the spinel–corundum interface, because for the fraction of Al_m^{3+} supplied to the edge-dislocations no corresponding flux of Mg_m^{2+} exists. The Mg_{source}^{2+} and O_{source}^{2-} species must be supplied by an independent source. The only conceivable candidate is the reactant periclase from which these species can be liberated, if paired anion-cation vacancies (Schottky defects) are generated.

In summary, motion of the dislocation-free segments of the interface occurs by $3Mg^{2+}$ *vs.* $2Al^{3+}$ cation exchange and involves long-range diffusion of Mg^{2+} *vs.* Al^{3+} through the spinel layer. In contrast, climb of the edge dislocations at the termination of every 23rd (010) and (100) lattice plane of the spinel requires correlated long-range diffusion of Al^{3+} and O^{2-} as well as the generation of Schottky defects in the reactant periclase including the transport of the emerging Mg_{source}^{2+} and O_{source}^{2-} species to the dislocation core. The processes involved in dislocation climb appear to be considerably more energetically expensive than simple cation exchange and interdiffusion at the dislocation-free interface segments. This view is corroborated by the fact that the edge dislocations lag behind the dislocation-free segments of the interface (see Fig. 30). The edge dislocations act as obstacles to interface motion and have a pinning effect very much like the effect of second phase particles at moving interfaces (Zener drag). In addition, the dislocation-free segments of the interface cannot annihilate the vacancies formed in the reactant periclase, which eventually may become supersaturated with respect to the Schottky defects and pores are nucleated. Such pores are a typical feature of the spinel–periclase interface (Li *et al.*, 2016; Jerabek *et al.*, 2014) further corroborating the mechanism of interface motion described.

Acknowledgement

The authors are grateful to Jörg Renner for his detailed and constructive comments which helped to clarify and improve this chapter.

References

Abart, R. (1995) Phase-equilibrium and stable-isotope constraints on the formation of metasomatic garnet-vesuvianite veins (SW Adamello, N Italy). *Contributions to Mineralogy and Petrology*, **122**, 116–133.

Atkinson, H.V. (1988) Theories of normal grain growth in pure single phase systems. *Acta Metallurgica*, **36**, 469–491.

Aust, K. and Rutter, J. (1959) Grain boundary migration in high-purity lead and dilute lead-tin alloys. *Transactions of the American Institute of Mining and Metallurgical Engineers*, **215**, 119–127.

Balluffi, R.W., Allen, S.M. and Carter, C.W. (2005) *Kinetics of Materials*. John Wiley & Sons, Inc. Chichester, UK.

Bollmann, W. (1970) *Crystal Defects and Crystalline Interfaces*. Springer, Heidelberg-Berlin.

Brandon, D. (1966) The structure of high-angle grain boundaries. *Acta Metallurgica*, **14**, 1479–1484.

Brandon, D., Ralph, B., Ranganathan, S. and Wald, M. (1964) A field ion microscope study of atomic configuration at grain boundaries. *Acta Metallurgica*, **12**, 81–821.

Burke, J. E. and Turnbull, D. (1952) Recrystallization and grain growth. *Progress in Metal Physics*, **3**, 220–292.

Doherty, R., Hughes, D., Humphreys, F., Jonas, J., Jensen, D., Kassner, M., King, W., McNelley, T., McQueen, H. and Rollett, A. (1997) Current issues in re-crystallization: a review. *Materials Science and Engineering A-Structural Materials Properties Microstructure and Processing*, **238**, 219–274.

Dubrovin, B.A., Fomenko, A. and Novikov, S. (1984) *Modern Geometry Methods and Applications: Part I: The Geometry of Surfaces, Transformation Groups, and Fields*. Springer, Berlin-Heidelberg.

Evans, B., Renner, J. and Hirth, G. (2001) A few remarks on the kinetics of static grain growth in rocks. *International Journal of Earth Sciences*, **90**, 88–103.

Fischer, F.D., Svoboda, J. and Fratzl, P. (2003) A thermodynamic approach to grain growth and coarsening. *Philosophical Magazine*, **83**, 1075–1093.

Frank, F. (1951) Crystal dislocations – elementary concepts and definition. *Philosophical Magazine*, **42**, 809–819.

Gaidies, F., Milke, R., Heinrich, W. and Abart, R. (2017) Metamorphic mineral reactions: Porphyroblast, corona and symplectite growth. Pp. 469–540 in: *Mineral Reaction Kinetics: Microstructures, Textures and Chemical Compositions* (R. Abart and W. Heinrich, editors). EMU Notes in Mineralogy, **16**. European Mineralogical Union and the Mineralogical Society of Great Britain & Ireland, London.

Greenwood, G. (1956) the growth of dispersed precipitates in solutions. *Acta Metallurgica*, **4**, 243–248.

Habler, G. and Griffiths, T. (2017) Crystallographic orientation relationships Pp. 541–585 in: *Mineral Reaction Kinetics: Microstructures, Textures and Chemical Compositions* (R. Abart and W. Heinrich, editors). EMU Notes in Mineralogy, **16**. European Mineralogical Union and the Mineralogical Society of Great Britain & Ireland, London.

Herwegh, M., Linckens, J., Ebert, A., Berger, A. and Brodhag, S.H. (2011) The role of second phases for controlling microstructural evolution in polymineralic rocks: A review. *Journal of Structural Geology*, **33**, 1728–1750.

Hillert, M. (1965) On theory of normal and abnormal grain growth. *Acta Metallurgica*, **13**, 227–238.

Holness, M. and Graham, C. (1991) Equilibrium dihedral angles in the system H_2O-CO_2-NaCl-calcite, and implications for fluid-flow during metamorphism. *Contributions to Mineralogy and Petrology*, **108**, 368–383.

Howe, J. (1997) *Interfaces in Materials*. John Wiley & Sons Inc. Chichester, UK.

Jahn, S. and Sun, X.-Y. (2017) Atomic-scale modelling of crystal defects, self-diffusion and deformation processes. Pp. 215–253 in: *Mineral Reaction Kinetics: Microstructures, Textures and Chemical Compositions* (R. Abart and W. Heinrich, editors). EMU Notes in Mineralogy, **16**. European Mineralogical Union and the Mineralogical Society of Great Britain & Ireland, London.

Jerabek, P., Abart, R., Rybacki, E. and Habler, G. (2014) Microstructure and texture evolution during growth of magnesio-aluminate spinel at corundum-periclase interfaces under uniaxial load: the effect of loading on reaction progress. *American Journal of Science*, **314**, 1–26.

Kitayama, M. and Glaeser A. (2002) The Wulff shape of alumina: III, undoped alumina. *Journal of the American Ceramic Society*, **85**, 611–622.

Li, C., Griffiths, T., Pennycook, T., Mangler, C., Jerabek, P., Meyer, J., Habler, G. and Abart, R. (2016) The structure of a propagating $MgAl_2O_4/MgO$ interface: Linked atomic and μm-scale mechanisms of interface motion. *Philosophical Magazine*, **96**, 2488–2503.

Lifshitz, I.M. and Slyozov, V.V. (1961) The kinetics of precipitation from supersaturated solid solutions. *Journal of the Physics and Chemistry of Solids*, **19**, 35–50.

MacPherson, R.D. and Srolovitz, D.J. (2007) The von Neumann relation generalized to coarsening of three-dimensional microstructures. *Nature*, **446**, 1053–1055.

Marquardt, K., Rohrer, G.S., Morales, L., Rybacki, E., Marquardt, H. and Lin, B. (2015) The most frequent interfaces in olivine aggregates: the GBCD and its importance for grain boundary related processes. *Contributions to Mineralogy and Petrology*, **170**, Article No. 40, DOI: 10.1007/s00410-015-1193-9.

Mullins, W.W. (1956) Two dimensional motion of idealized grain boundaries. *Jounral of Applied Physics*, **27**, 900–904.

Nes, E., Ryum, N. and Hunderi, O. (1985) On the Zener drag. *Acta Metallurgica*, **33**, 11–22.

Neumann, J.V. (1952) *Metal Interfaces*. American Society for Testing of Materials, Cleveland, Ohio, USA.

Olgaard, D.L. and Evans, B. (1986) Effect of second-phase particles on grain growth in calcite. *Journal of the American Ceramic Society*, **69**, 272–277.

Pang, Y. and Wynblatt, P. (2006) Effects of Nb doping and segregation on the grain boundary plane distribution in TiO_2. *Journal of the American Ceramic Society*, **89**, 666–671.

Petrishcheva, E. and Abart, R. (2017) Diffusion: some mathematical foundations and applications in mineralogy Pp. 255–294 in: *Mineral Reaction Kinetics: Microstructures, Textures and Chemical Compositions* (R. Abart and W. Heinrich, editors). EMU Notes in Mineralogy, **16**. European Mineralogical Union and the Mineralogical Society of Great Britain & Ireland, London.

Petrishcheva, E. and Renner, J. (2005) Two-dimensional analysis of pore drag and drop. *Acta Materialia*, **53**, 2793–2803.

Petrishcheva, E. and Renner, J. (2010) Characteristics of pore migration controlled by diffusion through the pore-filling fluid. *Physics and Chemistry of Minerals*, **37**, 601–611.

Read, W. and Shockley, W. (1950) Dislocation models of crystal grain boundaries. *Physical Review*, **78**, 275–289.

Renner, J., Evans, B. and Hirth, G. (2002) Grain growth and inclusion formation in partially molten carbonate rocks. *Contributions to Mineralogy and Petrology*, **142**, 501–514.

Rohrer, G.S. (2011) Grain boundary energy anisotropy: a review. *Journal of Materials Science*, **46**, 5881–5895.

Saylor, D., Morawiec, A. and Rohrer, G. (2003a) Distribution of grain boundaries in magnesia as a function of five macroscopic parameters. *Acta Materialia*, **51**, 3663–3674.

Saylor, D., Morawiec, A. and Rohrer, G. (2003b) The relative free energies of grain boundaries in magnesia as a function of five macroscopic parameters. *Acta Materialia*, **51**, 3675–3686.

Smith, C.S. (1948) Grains, phases, and interfaces – an interpretation of microstructure. *Transactions of the Metallurgical Society A.I.M.E.*, **175**, 15–51.

Sutton, A. and Balluffi, R. (1987) Overview no. 61. On geometric criteria for low interfacial energy. *Acta Metallurgica*, **35**, 2177–2201.

Sutton, A. and Balluffi, R. (1995) *Interfaces in Crystalline Materials*. Oxford Scientific Publications, Oxford, UK.

Vernon, R. (2004) *A Practical Guide to Rock Microstructure*. Cambridge University Press, Cambridge, UK.

Warrington, D. and Boon, M. (1975) Ordered structures in random grain-boundaries - some geometrical probabilities. *Acta Metallurgica*, **23**, 599−607.

Watson, B.E. and Brenan, J.M. (1987) Fluids in the lithosphere, 1. Experimentally-determined wetting characteristics of CO_2-H_2O fluids and their implications for fluid transport, host-rock physical properties, and fluid inclusion formation. *Earth and Planetary Science Letters*, **85**, 497−515.

Wulff, G. (1901) On the question of speed of growth and dissolution of crystal surfaces. *Zeitschrift für Krystallographie und Mineralogie*, **34**, 449−530.

Zaefferer, S. and Habler, G. (2017) Scanning electron microscopy and electron backscatter diffraction Pp. 37−95 in: *Mineral Reaction Kinetics: Microstructures, Textures and Chemical Compositions* (R. Abart and W. Heinrich, editors). EMU Notes in Mineralogy, **16**. European Mineralogical Union and the Mineralogical Society of Great Britain & Ireland, London.

Nucleation in geological materials

FRED GAIDIES

Department of Earth Sciences, Carleton University, 1125 Colonel By Drive, Ottawa, Ontario, K1S 5B6, Canada, e-mail: Fred.Gaidies@carleton.ca

Nucleation is the initial process of most phase transformations and is of fundamental importance for the kinetics of mineral reactions. A departure from equilibrium is required to overcome the energy barrier associated with nucleation, which is a function of the structural and compositional differences between the nucleus and the metastable reactant, and the level of elastic deformation experienced by the nucleus as it forms in the host lattice. Nucleation in geological materials almost always takes place at grain boundaries, crystal defects or impurities, which catalyse the nucleation process and influence the chemical composition, size, shape, lattice orientation and spatial distribution of nuclei with important implications for the texture and microstructure evolution of rocks. Nuclei are microscopic in most systems and thus are too small to be observed in experiment. This is why nucleation has been intensively studied theoretically and through numerical simulations. In those treatments, nucleation integrates more elementary processes such as chemical diffusion and interface motion.

This chapter provides the essential physics of the thermodynamics and kinetics of nucleation. It reviews the fundamentals of the classical nucleation theory including the chemical driving force for nucleation in partitioning systems and the interfacial area of clusters, discusses possible nucleus/substrate interactions and their influence on the free energy of the nucleus and the energy barrier to nucleation, and presents expressions for the classical nucleation rate. In a second part, extensions to CNT are outlined that couple long-range diffusion with the kinetics of interface processes in order to address the formation of nucleation exclusion or depleted zones around supercritical clusters and the enrichment of the precipitated components in the vicinity of subcritical clusters. Finally, the reader is introduced to non-classical gradient-energy continuum approaches to nucleation in inhomogeneous systems, and to the phase field method for the simulation of microstructure evolution.

1. Classical nucleation theory

Classical nucleation theory (CNT) is a model developed to quantify and predict the initial process during phase transformations associated with an energy barrier caused by structural differences between a reacting and product system. This type of phase transformation is referred to as discontinuous phase transformation. In its initial form, CNT is based on the work of Gibbs (1928) who introduced the concept of interfaces of zero thickness between homogenous phases in his attempt to develop a formulation for the thermodynamics of heterogeneous systems. Kaischew and Stranski (1934) extended CNT in their application to the formation of crystals from supersaturated vapour, based on work done previously by Volmer and Weber (1926) on vapour

DOI: 10.1180/EMU-notes.16.11

condensation. Contributions by Becker and Döring (1935), Volmer (1939), Frenkel (1946), and Turnbull and Fisher (1949) expanded CNT for application to condensed systems. CNT may now be the best known quantitative approach to nucleation as it successfully predicts the main features of nucleation in appropriate systems irrespective of its inherent approximations.

At the heart of CNT is the description of the energy barrier associated with the formation of the interface between reactants and products of a discontinuous phase transformation, and a formalism that links this energy barrier to the kinetics of molecular attachment and detachment processes at the interface. Because long-range diffusion through the reacting system is not accounted for, CNT is an inherently interface-limited model.

1.1. Energy barrier to nucleation

The nucleation barrier is assumed to be a function of the energy required to form the structural transition across the product-reactant interface per unit interfacial area, σ, and the chemical force that drives nucleation, ΔG_V. For a given pressure (P), temperature (T) and bulk chemical composition (X), ΔG_V is the difference in bulk Gibbs energy between reacting system and nucleating phase, per product molecule. For an interfacial area that is composed of i individual regions, A_i, the Gibbs energy of formation of a product cluster with n molecules, ΔG_n, is written

$$\Delta G_n = n\Delta G_V + \sum_i A_i \sigma_i \tag{1}$$

Because ΔG_V is negative in a thermodynamically favoured phase transformation and $\Sigma \sigma_i$ is always positive, and because the contribution of the interfacial energy term to ΔG_n is larger than that of the bulk term for relatively small n given the surface to volume relationship, ΔG_n develops a maximum before it decreases and becomes negative with an increase of cluster size.

If σ is direction-independent, a spherical cluster geometry can be assumed so that equation 1 simplifies to

$$\Delta G_n = n\Delta G_V + (36\pi)^{1/3}(\bar{v}n)^{2/3}\sigma \tag{2}$$

where $\bar{v}$ is the volume of a molecule in the cluster. Figure 1 illustrates the relationship between ΔG_n and n assuming a spherical cluster geometry.

In general, clusters with a size that maximizes ΔG_n are referred to as nuclei (or critical clusters), and their formation is called nucleation. The number of molecules contained in a spherical critical cluster, n^*, can be found by differentiation of (2) and is given by

$$n^* = \frac{-32\pi\bar{v}^2\sigma^3}{3(\Delta G_V)^3} \tag{3}$$

so that its radius, r^*, is equal to

$$r^* = \frac{-2\bar{v}^2\sigma}{\Delta G_V} \tag{4}$$

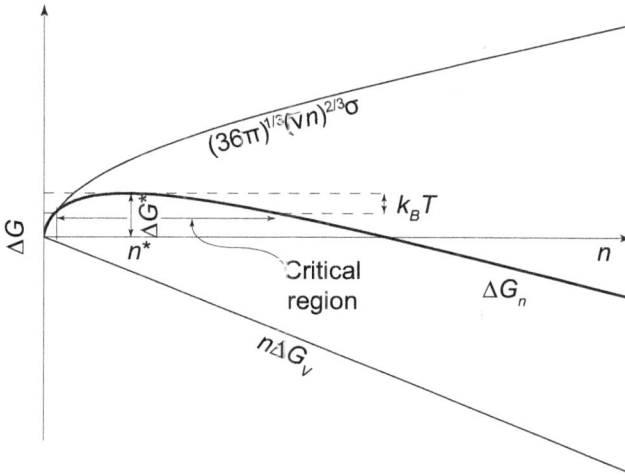

Figure 1. The Gibbs energy of cluster formation, ΔG_n, as a function of the number of molecules per product cluster, n, assuming an isotropic interfacial energy, σ, and a negative chemical driving force for nucleation, ΔG_V. ΔG^* is the critical energy barrier to nucleation, and n^* refers to the number of molecules in the critical cluster.

The maximum of ΔG_n (Fig. 1) can be obtained by combining equations 2 and 3, and is given by

$$\Delta G^* = \frac{16\pi\bar{v}\sigma^3}{3(\Delta G_V)^2} \tag{5}$$

ΔG^* represents the energy barrier which has to be overcome to form critical clusters of the product of the discontinuous phase transformation. These clusters are in unstable equilibrium with the reacting system because the decomposition of subcritical clusters ($n < n^*$) and the attachment of additional molecules to the surfaces of supercritical clusters ($n > n^*$) are energetically favoured as that decreases ΔG_n (Fig. 1). Cluster sizes for which

$$\Delta G^* \geqslant \Delta G_n \geqslant \Delta G^* - k_B T \tag{6}$$

where k_B is Boltzmann's constant, mark the critical region of nucleation (Fig. 1). According to equations 3 and 5, the critical cluster region as well as n^* and ΔG^* increase with a decrease in ΔG_V so that discontinuous phase transformations close to equilibrium require relatively fast transfer processes at the interface for nucleation to take place. Note that CNT does not define n^* and ΔG^* for $\Delta G_V = 0$. An increase in ΔG_V results in a decreasing energy barrier to nucleation and smaller critical clusters as fewer molecules are required for their formation. In the case of exceedingly small critical clusters, the separation of bulk and interfacial energy terms, expressed in equation 1 and referred to as capillarity approximation of CNT, cannot be applied and alternative methods such as the Cahn-Hilliard approach (section 3) may be used.

1.1.1. Chemical driving force and nucleus composition in partitioning systems
The chemical driving force for nucleation, ΔG_V, is the Gibbs energy change associated with the formation of a small amount of the product phase out of the metastable reacting system, per molecule of the new phase. It may be seen as the Gibbs energy per product molecule that drives the processes underlying nucleation such as molecular attachment and detachment processes at the interface and migration of the interface into the reactant. In a partitionless discontinuous phase transformation, the nucleus grows with the same composition as the reactant. However, nucleation in natural systems is commonly a partitioning process so that the composition of the nucleus differs from that of the reacting system. In such a case, the Gibbs energy change is maximized if the differences in chemical potentials between metastable reacting system and nucleus of the components involved in the phase transformation are identical.

For a binary system with components A and B the composition of the nucleus, x^P_{nuc}, can be determined graphically by the point of tangency on the $G–x$ function of the product, P, of a line that is parallel to the tangent that corresponds to the energy state of the reactant, R (stippled lines in Fig. 2c,d). As shown in Fig. 2c, x^P_{nuc} in partitioning systems differs from the equilibrium composition of the product, x^P_{equ}. The difference

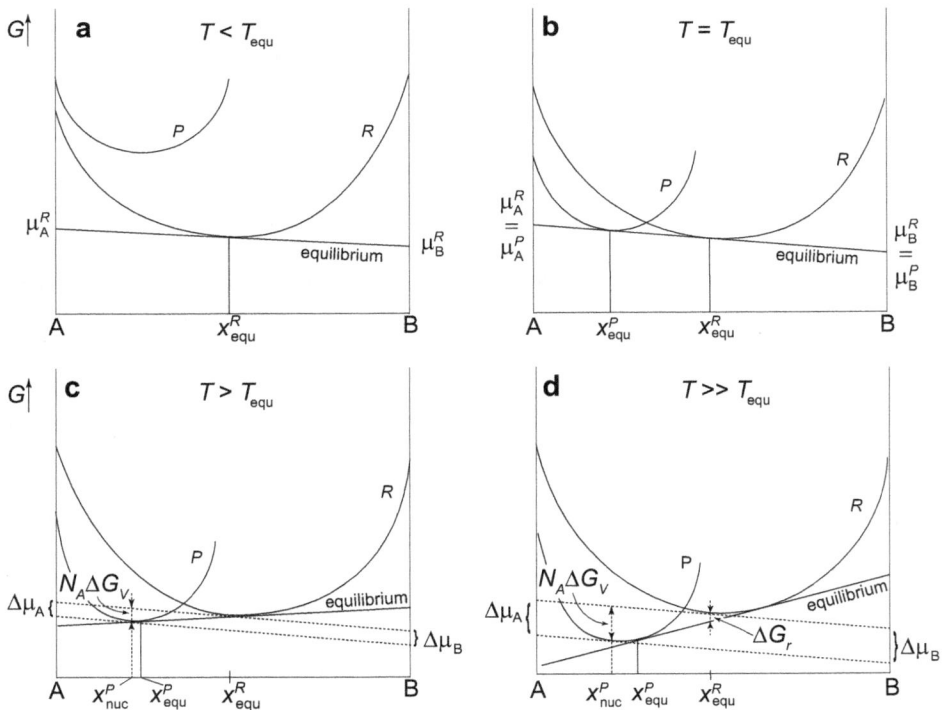

Figure 2. Schematic molar Gibbs energy diagrams for the binary system $A–B$ illustrating the relationship between ΔG_V, ΔG_r, X^P_{nuc}, X^P_{equ}, X^R_{equ} and the departure from equilibrium. N_A is Avogadro's constant.

between x_{nuc}^P and x_{equ}^P increases with departure from equilibrium required to overcome the energy barrier to nucleation, and depends on the $G-x$ relationships of the phases involved (Fig. 2d), provided that the bulk composition of the system, x_{equ}^R, does not change during the nucleation. Once nucleated, the product will change its composition towards equilibrium with the surrounding reactant if diffusion within the product and reacting system and across their interface is efficient. The molar Gibbs energy difference between the initial and final equilibrium states, ΔG_r (Fig. 2d), is the maximum available force that drives all the processes of the overall phase transformation. Diffusion transfers material between R and P resulting in the growth of P and the dissipation of Gibbs energy until equilibrium is established This may be visualized by rotating the two stippled parallel lines in Fig. 2 tangential to R and P until they coincide with the single equilibrium tangent to R and P (equilibrium line in Fig. 2c, d). ΔG_r may be obtained through integration over the Gibbs energy dissipated during the diffusion. The total amount of ΔG_r that can be dissipated depends on the bulk composition of the system, x_{equ}^R, and reaches a maximum if x_{equ}^R approaches x_{equ}^P.

The tangent-method may also be applied to nucleation in multicomponent systems and was utilized by Gaidies *et al.* (2011) to estimate ΔG_V associated with the contact metamorphic nucleation of garnet in a metapelite of the aureole of the Nelson Batholith (British Columbia, Canada). The reacting rock matrix was modelled as a MnO–Na$_2$O–CaO–K$_2$O–FeO–MgO–Al$_2$O$_3$–SiO$_2$–H$_2$O–TiO$_2$ (MnNCKFMASHT) system using the thermodynamic data of Holland and Powell (1998). The matrix was assumed to be equilibrated at any point in P-T space, but the nucleation of garnet was assumed to be kinetically hindered. Such a scenario may be envisioned for the nucleation of phases with structural properties significantly different from those of the phases in the reacting system, and where diffusion in the reacting system is faster than across the interface with the nucleus. ΔG_V is calculated based on the tangent-method outlined above and is illustrated in Fig. 3a as a function of the departure from equilibrium.

Note that the coloured P-T regions correspond to the conditions where ΔG_V is negative so that the formation of garnet is thermodynamically favoured. In general, ΔG_V increases with thermal overstep of the low-T conditions of the stability field of garnet-bearing assemblages, and decreases towards the high-T limit. Similar to the exemplarily binary system, ΔG_V associated with the crystallization of garnet in multicomponent systems is a function of the $G-x$ relationships of the matrix phases and garnet, and how these thermodynamic properties vary with temperature and pressure. The lower T limits at which staurolite and andalusite enter the matrix assemblages are characterized by a "channel" in ΔG_V-P-T space and mark the conditions with the highest garnet nucleation probability (stippled line in Fig. 3a).

Figure 3b shows ΔG_r associated with the formation of garnet from the respective phase assemblages in the modelled bulk chemical system. It can be seen that ΔG_r differs from ΔG_V not only quantitatively but also with respect to its dependence on P-T. Whereas ΔG_V reflects the force on a product molecule involved in the formation of a critical cluster, ΔG_r corresponds to the integrative molar Gibbs energy that dissipates through the entire transformation process. Hence, ΔG_V can be understood as part of ΔG_r, but because relatively few molecules are involved in nucleation compared to the

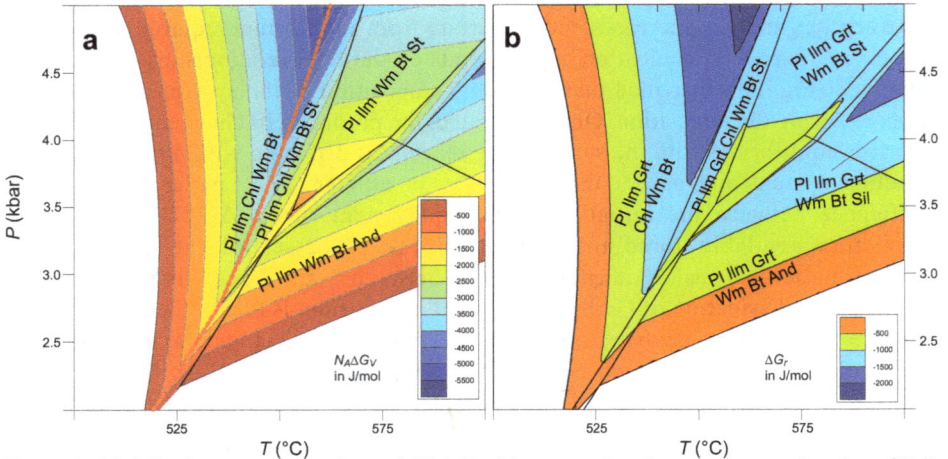

Figure 3. (a) ΔG_V, for garnet nucleation and (b) ΔG_r of the garnet-forming reaction as a function of *P-T*. Modified after Gaidies *et al.* (2011). And – andalusite, Bt – biotite, Chl – chlorite, Grt – garnet, Ilm – ilmenite, Pl – plagioclase, Sil – sillimanite, St – staurolite, Wm – white mica.

growth of the particles to macroscopic crystal sizes, the majority of ΔG_r is used for the diffusion associated with crystal growth.

Similar to ΔG_V, the differences between x^P_{nuc} and x^P_{equ} of garnet increase with departure from equilibrium. Whereas the differences between x^P_{nuc} and x^P_{equ} are small with respect to the pyrope and grossular contents of garnet, the spessartine and almandine contents of a nucleus differ significantly from the equilibrium composition (Fig. 4). Even though chemical fractionation during garnet crystallization is common, reflecting relatively low rates of diffusion in garnet, and irrespective of its influence on the thermodynamically effective bulk composition, the impact of chemical fractionation on ΔG_r, ΔG_V and garnet chemistry cannot be shown in a *P-T* diagram because it is dependent on the *P-T* path of crystallization.

1.1.2. Elastic strain energy

If there are differences in the volume or shape between the nucleating phase and the reacting host then the energetics of nucleation may involve elastic strain energy, ΔG_E. This instance may be particularly relevant for discontinuous phase transformations at the solid state. In the case of lattice misfits, ΔG_E will be positive and will scale with the volume of the nucleus so that equation 1 changes to

$$\Delta G_n = n(\Delta G_V + \Delta_E) + \sum_i A_i \sigma_i \qquad (7)$$

Assuming a spherical nucleus geometry and an isotropic interfacial energy, ΔG_n can be written as

$$\Delta G_n = \frac{4\pi r^3}{3\bar{v}}(\Delta G_V + \Delta G_E) + 4\pi r^2 \sigma \qquad (8)$$

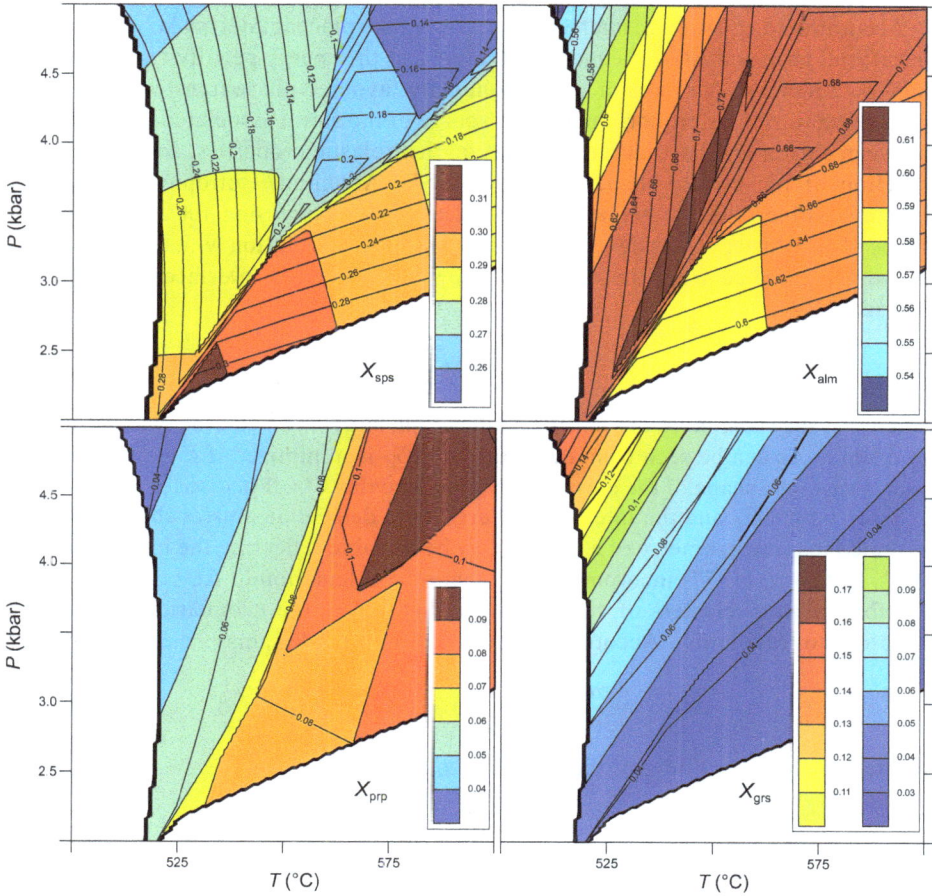

Figure 4. Equilibrium composition of garnet (isopleths) as a function of *P-T*, and chemical composition of a garnet nucleus (coloured) for different degrees of overstepping. After Gaidies *et al.* (2011). X_{sps} = Mn/(Fe^{2+}+Ca+Mg+Mn), X_{alm} = Fe^{2+}/(Fe^{2+}+Ca+Mg+Mn), X_{prp} = Mg/(Fe^{2+}+Ca+Mg+Mn), and X_{grs} = Ca/(Fe^{2+}+Ca+Mg+Mn).

where the volume of a cluster corresponds to $n\bar{v}$. Hence, r^* can be written as

$$r^* = \frac{-2\bar{v}\sigma}{\Delta G_V + \Delta G_E} \tag{9}$$

so that

$$\Delta G^* = \frac{16\pi\bar{v}^2\sigma^3}{3(\Delta G_V + \Delta G_E)^2} \tag{10}$$

According to equation 8, nucleation can only take place if $\Delta G_E < |\Delta G_V|$. Hence, the interfacial term and ΔG_E of equation 10 represent the energy penalty to nucleation as they both increase ΔG^* in a thermodynamically favoured phase transformation.

In general, ΔG_E is proportional to the square of the strain associated with the misfit between cluster and host lattice, depends on their elastic properties, and is a complex function of the cluster shape (Eshelby, 1957). Assuming isotropic elasticity and incompressible clusters, Nabarro (1940) derived an expression for ΔG_E where the shape of the cluster can be described by an ellipsoid of revolution with the semi-axes a, a and c, separated from the host lattice by an incoherent interface. According to Nabarro (1940)

$$\Delta G_E = 6S\bar{v}e^2 F\left(\frac{c}{a}\right) \tag{11}$$

where S is the shear modulus, e is the misfit strain, and F is a shape-dependent function of a and c. As can be seen from equation 11, ΔG_E is minimized if F approaches 0. In such a case, the shape of the cluster corresponds to that of a thin disc for which $c \ll a$.

Since both ΔG_E and the interfacial energy term depend on cluster shape, it may be assumed that the geometry of a cluster may be variable reflecting the minimization of ΔG^*. Lee et al. (1977) incorporated ΔG_E into the calculation of ΔG^* associated with the formation of clusters of varying shape that follow the geometric constraints of ellipsoids of revolution. According to Lee et al. (1977), ΔG_n may be expressed as

$$\Delta G_n = \frac{4\pi a^3}{3\bar{v}}\beta(\Delta G_V + \Delta G_E) + [2 + g(\beta)]\pi a^2 \sigma \tag{12}$$

where β is the aspect ratio of the cluster, c/a, and $g(\beta)$ is given by

$$g(\beta) = \begin{cases} \frac{2\beta^2}{\sqrt{1-\beta^2}}\tanh^{-1}(\sqrt{1-\beta^2} & \text{for } \beta<1 \\ 2 & \text{for } \beta=1 \\ \frac{2\beta}{\sqrt{1-\beta^{-2}}}\sin^{-1}(\sqrt{1-\beta^2} & \text{for } \beta>1 \end{cases} \tag{13}$$

Differentiation of equation 12 with respect to a results in an expression for the critical radius that accounts for the influence of cluster shape on ΔG^*, and can be written as

$$a^* = \frac{-\bar{v}\sigma[2 + g(\beta)]}{2\beta(\Delta G_V + \Delta G_E)} \tag{14}$$

Substituting equation 14 into equation 12, one obtains:

$$\Delta G^*(\beta) = \frac{\pi\bar{v}^2\sigma^3[2 + g(\beta)]^3}{12\beta^2(\Delta G_V + \Delta G_E)^2} \tag{15}$$

Normalization of $\Delta G^*(\beta)$ with respect to the critical energy barrier for a spherical cluster shape in the absence of strain energy (ΔG^* in equation 5) results in

$$\frac{\Delta G^*(\beta)}{\Delta G^*} = \frac{[2 + g(\beta)]^3}{[8\beta(1 + \Delta G_E/\Delta G_V)]^2} \qquad (16)$$

Figure 5 shows the variation of the normalized energy barrier, $\Delta G^*(\beta)/\Delta G^*$, with β for different ratios of $\Delta G_E/\Delta G_V$ for coherent nucleation (Lee *et al.*, 1977). In general, the normalized energy barrier rises with an increase of ΔG_E and a decrease of β. If ΔG_E contributes less than ~85% to the total bulk energy of cluster formation, a spherical cluster geometry ($\beta = 1$) results in a reduced energy barrier compared to cluster geometries with $\beta < 1$. It is only for higher relative contributions of ΔG_E that the energy barrier is minimized if the cluster geometry approaches the shape of an oblate spheroid. In those cases, the barrier is minimized if the cluster shape is that of a thin disk with β ranging between 0.2 and 0.3. In other words, if $|\Delta G_V|$ is large compared to ΔG_E, a spherical cluster shape may be a reasonable approximation of nucleus geometry provided that σ is isotropic.

1.1.3. Interfacial area
Critical cluster formation in the uniform region of the bulk matrix is referred to as homogenous nucleation. Heterogeneous nucleation refers to critical cluster formation in contact with a surface where interactions between cluster and surface reduce the energy barrier for nucleation. Heterogeneous nucleation commonly occurs at special sites such as foreign particles, crystal defects or grain boundaries. At these sites nucleation is catalysed because the interfacial energy and, hence, ΔG^*, are reduced.

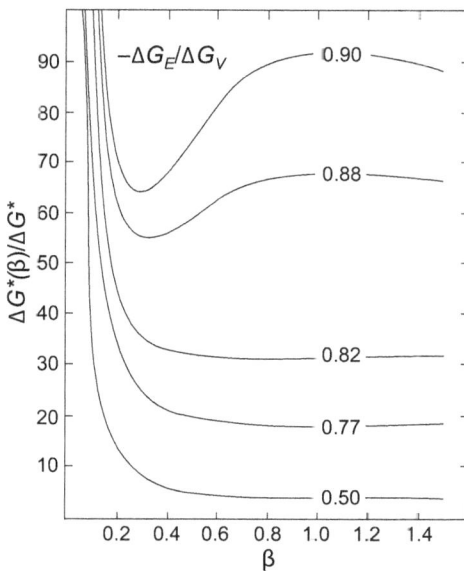

Figure 5. The normalized energy barrier, $\Delta G^*(\beta)/\Delta G^*$, as a function of β for different ratios of $\Delta G_E/\Delta G_V$ for isotropic coherent nucleation (after Lee *et al.*, 1977).

In the case of cluster formation on a flat surface in the absence of ΔG_E and considering isotropic interfacial energies, the cluster has the shape of a spherical cap the size of which is determined by the contact angle ϕ between the cluster and the reacting system (Fig. 6a). In this case, the critical energy barrier is reduced as a function of ϕ and is given by (Christian, 1975)

$$\Delta G^*_{flat} = f(\phi)\Delta G^* \tag{17}$$

where

$$f(\phi) = \tfrac{1}{4}(2 - 3\cos\phi + \cos^3\phi) \tag{18}$$

and

$$\cos\phi = (\sigma^{S\alpha} - \sigma^{S\beta})/\sigma^{\alpha\beta} \tag{19}$$

with

$$0 \leqslant \phi \leqslant 180° \tag{20}$$

so that

$$\Delta G^*_{flat} = \frac{4\pi\bar{v}^2(\sigma^{\alpha\beta})^3(2 - 3\cos\phi + \cos^3\phi)}{3(\Delta G_V)^2} \tag{21}$$

If $\phi = 90°$, the energy barrier to nucleation on the flat surface, ΔG^*_{flat}, is half of that required for homogenous nucleation, ΔG^*. For larger ϕ, ΔG^*_{flat} increases until it is equal to ΔG^* at $\phi = 180°$. As ϕ approaches 0, ΔG^* decreases until it disappears at $\phi = 0$. This phenomenon is referred to as wetting of the surface or grain boundary.

Equation 21 may be a reasonable approximation for nucleation of a liquid from a gas at a flat surface, or precipitation on a mineral surface in contact with an aqueous solution. However, nucleation in most geological materials probably occurs on heterogeneities such as grain boundaries, grain edges or corners. For nucleation at a grain boundary the interfacial area of the cluster will be twice that of the $\alpha-\beta$ interface of the spherical cap (Fig. 6b) so that

$$\Delta G^*_{gb} = 2f(\phi)\Delta G^* = \frac{8\pi\bar{v}^2(\sigma^{\alpha\beta})^3(2 - 3\cos\phi + \cos^3\phi)}{3(\Delta G_V)^2} \tag{22}$$

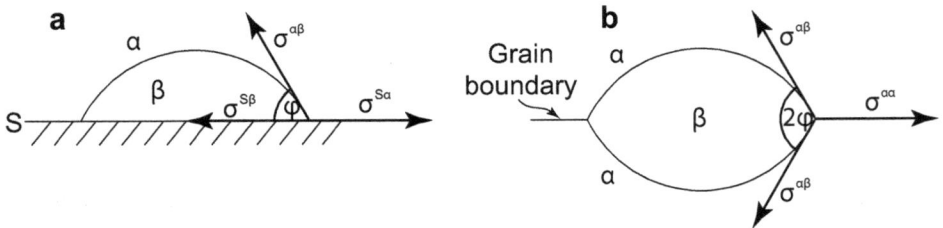

Figure 6. Geometry of (a) a cluster of phase β shaped as a spherical cup on a flat surface, surrounded by phase α, and (b) a lens-shaped cluster of phase β on a grain boundary in phase α.

Expressions similar to equations 21 and 22 can be derived for nucleation at grain edges or corners using their geometric relations for volume and interfacial area. According to Cahn (1956), the energy barrier decreases with a decrease in the dimensionality of the nucleation site. In other words, the barrier for nucleation along grain edges (1D) is less than that for nucleation along grain boundaries or cracks (2D) but larger than the energy barrier associated with nucleation at grain corners (0D). However, both the energy barrier to nucleation and the availability of the respective nucleation site in the reaction system dictate the rate of nucleation and how it changes in the course of the discontinuous phase transformation.

1.1.4. Nucleation at dislocations

In addition to grain boundaries, grain edges or corners, dislocations in host crystals are favourable sites for nucleation. However, their catalytic effect is fundamentally different from that exerted by the surfaces of grain boundaries or edges because the contact area between a dislocation and a cluster is too small to effectively reduce the interfacial area of a cluster. Similar to ΔG_V, the elastic strain energy of a dislocation released during nucleation is negative and may counteract the elastic misfit strain energy, ΔG_E, reducing the energy barrier to nucleation. The common migration of dislocations into sub-grain or small-angle grain boundaries indicates that nucleation at dislocations may not only take place in isolated areas in the interior of reacting crystals but that it may be a rather common type of nucleation, particularly during synkinematic metamorphic discontinuous phase transformations.

Cahn (1957) developed a formalism to treat nucleation at dislocations for the case that the interphase structure between cluster and reacting system is incoherent. According to Cahn (1957), the Gibbs energy per unit length of a cylindrical cluster forming on a dislocation considering an isotropic interfacial energy can be written as

$$\Delta G_D(r) = \frac{\pi r^2}{\bar{v}} \Delta G_V + 2\pi r\sigma - A \ln r \tag{23}$$

where the first term is the (negative) bulk contribution, r is the radial distance of the cylinder surface from the dislocation line, and $A\ln r$ reflects the elastic strain energy associated with the dislocation. For edge dislocations, A is given by

$$A_{edge} = \frac{Sb^2}{4\pi(1-v)}$$

and for screw dislocations

$$A_{screw} = \frac{Sb^2}{4\pi}$$

where b is the length of the Burgers vector and v is Poisson's ratio. Because v is ~0.3 for many solids, the difference in $\Delta G_D(r)$ between clusters at edge and screw dislocations may be ignored.

From equation 23 it follows that $\Delta G_D(r)$ develops a minimum at

$$r_0 = \frac{\bar{v}\sigma}{2\Delta G_V}(1 - \sqrt{1 - \alpha}) \tag{24}$$

and a maximum at

$$r^* = \frac{\bar{v}\sigma}{2\Delta G_V}(1 + \sqrt{1 - \alpha}) \tag{25}$$

where $\alpha = 2A\Delta G_V/\pi\bar{v}\sigma^2 < 1$ (Fig. 7a). r_0 corresponds to the radius of a metastable cylindrical cluster and is smaller than r^*, the radius of the critical cylindrical cluster that forms along the dislocation line. Beyond r^*, addition of molecules to the cluster surface will reduce the Gibbs energy of the system. Note that expressions (24) and (25) are based on the assumption that the length of the cluster corresponds to that of the dislocation line. A more realistic relationship is derived by Cahn (1957) using variational calculus methods in order to identify the cluster shape that reduces $\Delta G_D(r)^*$ the most. The resulting values for $\Delta G_D(r)^*$ are normalized with respect to the energy barrier to homogeneous nucleation, $\Delta G_H(r)^*$, and plotted against α in Fig. 7b. As can be seen, $\Delta G_D(r)^*/\Delta G_H(r)^*$ decreases drastically as α approaches unity. For $\alpha > 1$, the bulk energy term and the energy associated with the strain field around the dislocation balance the surface energy for all cluster sizes so that $\Delta G_D(r)$ does not contain extrema (Fig. 7a). In this case nucleation is effectively barrier-free and the discontinuous phase transformation depends entirely on the energetics and kinetics of crystal growth.

1.2. Rate of nucleation

In CNT, expressions for the rate of nucleation link the thermodynamics associated with the energy barrier to the kinetics of molecular attachment and detachment processes at the cluster/matrix interface. CNT predicts that once ΔG_V is negative, a finite number of

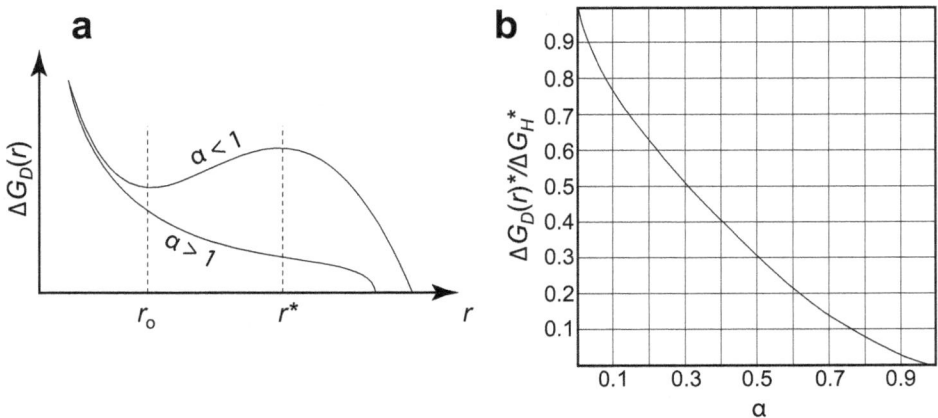

Figure. 7. (a) $\Delta G_D(r)$ vs. r of a cylindrical cluster surrounding a dislocation line. (b) $\Delta G_D(r)^*/\Delta G_H(r)^*$ vs. α for nucleation on a dislocation. After Cahn (1957).

clusters of various sizes is present, and that smaller ones are more abundant than larger ones (*e.g.* Christian, 1975; Kelton, 2006). Clusters of n molecules, E_n, are assumed to shrink or grow by loss or addition of a single molecule, E_1, through a series of bimolecular reactions

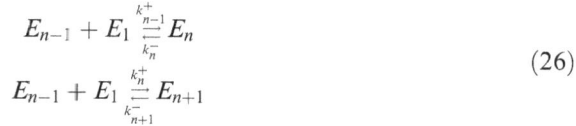

$$E_{n-1} + E_1 \underset{k_n^-}{\overset{k_{n-1}^+}{\rightleftharpoons}} E_n$$

$$E_{n-1} + E_1 \underset{k_{n+1}^-}{\overset{k_n^+}{\rightleftharpoons}} E_{n+1}$$

(26)

where k_n^+ is the rate with which a molecule is attached to the surface of a cluster of size n and k_n^- is the rate of detachment of molecules from E_n. Impingement of clusters are assumed to be rare compared to the bimolecular reactions and are, therefore, not considered.

Based on reaction rate theory (Turnbull and Fisher, 1949) molecular attachment and detachment rates may be approximated by

$$k_n^+ = O_n \gamma \exp\left(\frac{-\delta g_n}{2k_B T}\right)$$

$$k_{n+1}^- = O_n \gamma \exp\left(\frac{+\delta g_n}{2k_B T}\right)$$

(27)

where O_n is a geometrical factor reflecting the number of molecules that can be attached to the surface of a cluster composed of n molecules, $\delta g_n = \Delta G_{n+1} - \Delta G_n$ and γ is the molecular jump rate into and out of the cluster surface

$$\gamma = \frac{6D}{\lambda^2}$$

with the jump distance, $\lambda = \bar{v}^{1/3}$, and D, the molecular mobility at the interface. For spherical clusters, $O_n \approx 4n^{2/3}$ (Kelton *et al.*, 1983).

The cluster size frequency distribution $N_n(t)$ for homogenous nucleation can be derived by solving a modified system of Becker-Döring equations (Becker and Döring, 1935; Kelton, 1991)

$$\frac{dN_1}{dt} = -\sum_{n=1}^{v-1} k_n^+ N_n + \sum_{n=2}^{v} (k_n^+ + k_n^-) N_n - \sum_{n=3}^{v+1} k_n^- N_n$$

$$\frac{dN_n}{dt} = k_{n-1}^+ N_{n-1} - (k_n^- + k_n^+) N_n + k_{n+1}^- N_{n+1} \quad \text{for} \quad 1 < n < v$$

(28)

where $v + 1$ is the upper limit of the range of cluster sizes that is considered. The numerical value of v depends on ΔG^* and is chosen to represent a cluster size that is bigger than the upper root of the $\Delta G(n)$ function (see Fig. 1). Following Kelton *et al.* (1983), $N_{n \geq v}(t)$ may be assumed to be zero to ensure that there is no backward flux of molecules from stable clusters. $N_1(t)$ is assumed to be the number of particles with single molecules which may decrease with time due to the formation of stable clusters, in particular in condensed geological materials. The modified system of Becker-Döring equations 28 accounts for the decrease in $N_1(t)$. It represents a numerically stiff set of ordinary differential equations with the initial conditions (Gaidies *et al.*, 2011)

$$N_n(0) = \begin{bmatrix} N_1(0) \\ N_2(0) \\ \vdots \\ N_v(0) \end{bmatrix} = \begin{bmatrix} N^{\mathrm{eq}} \exp\left[\frac{-\Delta G_V}{k_B T}\right] \\ 0 \\ \vdots \\ 0 \end{bmatrix} \tag{29}$$

where N^{eq} is the number of molecules predicted to form at given *P-T-X* conditions during thermodynamic equilibrium. The nucleation rate can then be calculated as the flux of clusters past the critical cluster size and is given by

$$I^* = k_{n*}^+ N_{n*} - k_{n*+1}^- N_{n*+1} \tag{30}$$

where N_{n*} is the number of critical clusters.

The form of the heterogeneous nucleation rate equation is identical to that for homogenous nucleation (equations 28 to 30). However, for heterogeneous nucleation, and assuming that nucleation barriers are similar irrespective of the type of nucleation site, $N_1(t)$ does not correspond to the number of available monomers but reflects the number density of nucleation sites instead. Because the number of nucleation sites during heterogeneous nucleation is significantly smaller than the number of unclustered molecules, heterogeneous nucleation rates may be relatively low irrespective of the reduced energy barriers. In other words, the rate of heterogeneous nucleation depends on the number of available nucleation sites and their respective energy barriers and, hence, reflects the changes to the microstructure of the reacting system, such as grain-size reductions or changes to the dislocation density, that may evolve during crystallization. Considering that energy barriers in geological materials reflect a range of nucleation sites, and, hence, vary across the reacting system, different nucleation mechanisms may operate simultaneously at different locations in the system with the fastest mechanism dominating the overall nucleation rate.

It is important to note that, in addition to changes in ΔG^* and D, variations in the nucleation site density during crystallization result in changes to the overall nucleation rate. Therefore, a formalism similar to equation 28 may be best suited to numerically model time-dependent cluster-formation rates during heterogeneous nucleation in geological materials. This formalism also allows us to predict whether a steady-state nucleation rate may be established during the evolution of the cluster-size distribution.

Given that N_1 during homogenous nucleation in dilute systems is significantly larger than the number of critical clusters that form during the phase transformation, a steady-state nucleation rate may be approached in such a system. In this case, the steady-state nucleation rate, I^S, can be approximated by

$$I^S = O_{n*} \gamma \exp\left(\frac{-\Delta G^*}{k_B T}\right) N_1 Z$$

where Z is the dimensionless Zeldovich factor (Kelton, 2006)

$$Z = \sqrt{\frac{|\Delta G^*|}{3\pi k_B T n^*}}$$

which ranges commonly between 0.01 and 0.1. Z may be interpreted to correct for the backward flux of molecules associated with the decay of stable clusters.

Steady-state nucleation in condensed systems, such as during solid-state phase transformations associated with metamorphism, may be rather unlikely given the relatively small number of available nucleation sites. In fact, the number of energetically preferable nucleation sites may decrease significantly and approach the amount of critical clusters as metamorphic crystallization proceeds resulting in a reduction of the nucleation rate.

2. Coupled-flux analysis

Because CNT is an inherently interface-limited model, it cannot be applied to study nucleation in partitioning systems in which long-range diffusion across the reacting system is comparable to or slower than the molecular mobility at the cluster/matrix interface. If the capillarity approximation of CNT holds and the bulk term of the Gibbs energy of cluster formation can be separated from the interfacial term, the coupling of the flux across the interface with long-range diffusion may be simulated. Originally introduced by Russell (1968) and expanded by Kelton (2000), the coupled-flux approach considers a transitional domain that separates the cluster from the matrix (Fig. 8). At the domain/cluster interface a formalism similar to CNT is used where the energy barrier to homogenous nucleation is linked to the rates of molecular attachment and detachment processes. However, the bimolecular reaction rates applied in the coupled-flux analysis also consider the number of molecules available in the transitional domain, ρ, and the relative rates of the exchange of molecules between domain and cluster and between domain and matrix.

The exchange rates between transitional domain and matrix may be written as

$$\begin{aligned}
\alpha(n, \rho - 1) &= \xi\rho \frac{D_m}{\lambda^2} \left(\frac{\rho_r^m - \rho + 1}{\rho}\right)^{1/2} \left(\frac{N_0}{N_S - N_0}\right)^{1/2} \\
\beta(n, \rho) &= \xi\rho \frac{D_m}{\lambda^2} \left(\frac{\rho_r^m - \rho + 1}{\rho}\right)^{-1/2} \left(\frac{N_0}{N_S - N_0}\right)^{-1/2}
\end{aligned} \qquad (31)$$

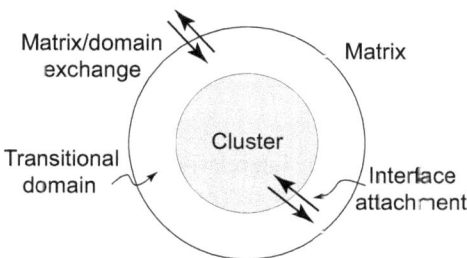

Figure 8. Schematic illustration of the coupled-flux model. After Russell (1968) and Kelton (2000).

where $\alpha(n,\rho - 1)$ is the rate with which a molecule diffuses into the domain around a cluster of size n, and $\beta(n,\rho)$ is the rate with which it diffuses out of that domain into the matrix. D_m is the diffusion coefficient in the matrix, ρ_n^m is the maximum number of molecule sites available in the transitional domain surrounding a cluster of size n, and N_0 is the number of molecules distributed among N_S sites in the matrix per unit volume. ξ is a normalization constant. For a spherical cluster geometry, ρ_n^m increases approximately with cluster size as $4n^{2/3}$.

Similar to equation 27 but also accounting for the availability of molecules in the domain surrounding the cluster, the attachment and detachment rates at the cluster/domain interface can be approximated by

$$k_{n,\rho}^+ = \rho\gamma \exp\left(\tfrac{-\delta g_n}{2k_B T}\right) G(n,\rho)$$
$$k_{n+1,\rho-1}^- = \rho\gamma \exp\left(\tfrac{+\delta g_n}{2k_B T}\right) \tfrac{1}{G(n,\rho)} \tag{32}$$

where the correction factor $G(n,\rho)$ considers the changes in entropy in the transitional domain and matrix caused by the attachment of a molecule to the cluster surface

$$G(n,\rho) = \cfrac{1}{\sqrt{\cfrac{\alpha_n \rho_n^m!(\rho_{n+1}^m - \rho + 1)!}{\alpha_{n+1}\rho_{n+1}^m!\rho(\rho_{n+1}^m - \rho)!}}\left(\cfrac{N_0}{N_S - N_0}\right)}$$

For a fast exchange between matrix and domain relative to γ, $G(n,\rho) = 1$ and $\rho = \rho_n^m$. In this case, equation 32 is identical to equation 27 of CNT, and the nucleation rate is controlled by the rate of the interface processes.

Note that ΔG_V is considered independent of the rates of interface processes and matrix diffusion. For the calculation of ΔG_V it is assumed that there are no chemical potential gradients across the matrix so that the cluster composition for all sizes is fixed and defined by the maximum Gibbs energy dissipation after some departure from equilibrium. The coupled differential equations required to quantify the cluster size frequency distribution $N_{n,\rho}(t)$ can then be written as

$$\frac{dN_{n,\rho}}{dt} = I_{n-1} - I_n + J_{\rho-1} - J_\rho \tag{33}$$

where

$$I_{n-1} = N_{n-1,\rho+1} - N_{n,\rho}k_{n,\rho}^-$$
$$I_n = N_{n,\rho}k_{n,\rho}^+ - N_{n+1,\rho-1}k_{n+1,\rho-1}^-$$
$$J_{\rho-1} = N_{n,\rho-1}\alpha_{n,\rho-1} - N_{n,\rho}\beta_{n,\rho}$$
$$J_\rho = N_{n,\rho}\alpha_{n,\rho} - N_{n,\rho+1}\beta_{n,\rho+1}$$

Solution of equation 33 for the case of slow matrix diffusion relative to the chemical mobility at the cluster surface indicates that the transitional domains around subcritical clusters become enriched during the discontinuous phase transformation while the domains surrounding supercritical clusters become deplete with respect to the number of molecules available for cluster formation compared to the average matrix composition. In addition, nucleation rates become smaller than calculated from CNT, and the time required to establish a steady state during homogenous nucleation increases as long-range diffusion becomes more important.

The spatial extent of the depletion zone around a supercritical cluster that develops during diffusion-controlled nucleation cannot be predicted with the coupled-flux model. However, the extent of depletion must reflect the matrix diffusivity and the increase in volume of the growing cluster. The addition of molecules to the cluster may reduce the chemical potential of the diffusing component in the vicinity of the cluster depending on the $G-x$ relationship of the matrix. If there are local sinks in the chemical potential around the supercritical cluster, the chemical driving force for nucleation available in the depletion zone would be lower than in the matrix, and the energy barrier to nucleation in this zone may not be overcome. Nucleation in the depletion zone may only be possible if there are reductions to the energy barrier to nucleation across this zone. Possible scenarios that may explain nucleation within the depletion zone include the drastic increase in ΔG_V as a response to changes in P, T, X during crystallization, a decrease in ΔG_E, or a decrease of the interfacial area and an increase in the number of nucleation sites caused by the reduction of grain size in the reacting system.

3. Nucleation in inhomogeneous systems

The separation of bulk and interface terms in the energetics of cluster formation is referred to as capillarity approximation and limits the applicability of CNT to discontinuous phase transformations close to equilibrium. In such a case, the reacting system and nucleating phase may be considered homogenous separated by a sharp, thin interface. However, nucleation at significant departure from equilibrium may result in critical sizes too small for the capillarity approximation to be valid. In this case, the interface may be diffuse and reactant and nucleus may be considered inhomogeneous instead. Non-classical gradient-energy continuum approaches to nucleation allow prediction of the properties of a critical cluster also at significant departure from equilibrium.

3.1. Gradient-energy continuum approach

If an initially homogeneous thermodynamically stable system is quenched into a two-phase field through a rapid variation of pressure or temperature, it may become unstable or metastable depending on its composition (Fig. 9a). In case the system is quenched into an unstable state, it tends toward equilibrium through the process of spinodal decomposition. The $G-x$ relationship of the unstable system is characterized by a curvature that is convex towards higher values of G so that $\partial^2 G/\partial x^2 < 0$ and infinitesimal fluctuations in x reduce G initiating the phase transformation (Fig. 9a). This type of phase transformation is referred to as *continuous* and is not associated with any energy barrier other than those required for chemical diffusion of components between the coexisting phases.

If the $G-x$ relationship within the two-phase region is characterized by a curvature that is concave towards higher values of G so that $\partial^2 G/\partial x^2 > 0$ (Fig. 9a), the system

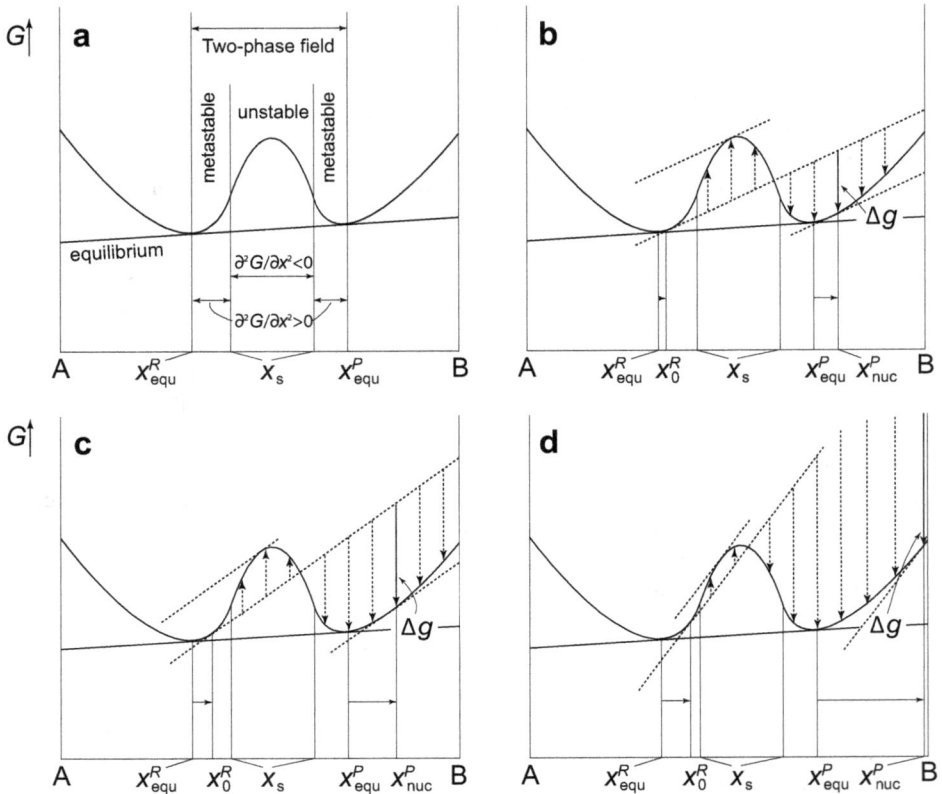

Figure 9. Schematic molar *G-x* diagrams showing the chemical driving force for nucleation, Δg, for different degrees of supersaturation. Nucleation takes place in the metastable region of the two-phase field.

experiences a metastable state and small fluctuations in x increase the Gibbs energy of the system. The extent of metastability is limited by the chemical spinodal ($\partial^2 G/\partial x^2 = 0$) and binodal ($x_s$ and x_{equ}^R, respectively, in Fig. 9). In order to re-establish equilibrium, large fluctuations in x are required to decrease the Gibbs energy. These fluctuations may be compared to critical clusters, and the energy barrier that has to be overcome for their formation may be referred to as the energy barrier to nucleation characteristic for *discontinuous* phase transformations. The energy barrier to nucleation is largest close to equilibrium ($x_0^R \rightarrow x_{equ}^R$ in Fig. 9b) and decreases towards the spinodal ($x_0^R \rightarrow x_s$ in Fig. 9d). The energy barrier to nucleation vanishes at the chemical spinodal.

    The chemical driving force for nucleation in a system with double-well potential can be obtained through the parallel tangent method (Δg in Fig. 9b–d). It is the maximum difference in Gibbs energy between the critical cluster and the reacting system and increases with departure from equilibrium ($x_0^R \rightarrow x_s$). Whereas the properties of reacting system and nucleus for phase transformations close to equilibrium do not vary significantly with x, so that the phases may be approximated as homogeneous, relatively small variations in x influence significantly G of the reacting system and nucleus as the departure from equilibrium increases (Fig. 9).

The influence of chemical heterogeneity on the energetics of nucleation can be estimated using the gradient-energy approach of Cahn and Hilliard (1958)

$$G = \int_V \left[g(x) + \kappa_x (\nabla x)^2 \right] dV \tag{34}$$

where $g(x)$ is the Gibbs energy of a homogeneous system at composition x, and κ_x is a positive materials constant referred to as the compositional gradient energy coefficient. κ_x corrects for the spatial chemical heterogeneity and is derived through a Taylor series expansion of g in powers of ∇x where the series is truncated after the quadratic term and the linear terms neglected due to symmetry requirements. Because x describes a field in an inhomogeneous system, G is a functional of the composition field integrated over the volume of the system.

According to Cahn and Hilliard (1959), the energy barrier to nucleation in an inhomogeneous system can be expressed as

$$\Delta G^* = \int_V \left[\Delta g(x) + \kappa_x (\nabla x)^2 \right] dV \tag{35}$$

and, assuming a spherical nucleus geometry,

$$\Delta G^* = 4\pi \int_0^\infty \left[\Delta g(x) + \kappa_x \left(\frac{dx}{dr} \right)^2 \right] r^2 dr \tag{36}$$

subject to the boundary conditions

$$\frac{dx}{dr} \to 0 \text{ as } r \to 0 \text{ and } r \to \infty$$

and

$$x \to x_0^R \text{ as } r \to \infty$$

where x_0^R is the composition of the reactant (Fig. 9). ΔG^* is a saddle point of the G functional and approaches the classical energy barrier to nucleation only for phase transformations close to equilibrium. In this case, interface curvature and thickness are minimized and the size of a critical cluster approaches infinity. However, with increasing departure from equilibrium, the critical energy barrier determined through the Cahn-Hilliard model decreases more than that of CNT and becomes zero at the chemical spinodal. In contrast, CNT predicts that the energy barrier vanishes only for an infinitely large degree of departure from equilibrium (*e.g.* equation 5). Another fundamental difference between the predictions of CNT and the Cahn-Hilliard gradient energy approach concerns the size of the critical cluster. Whereas CNT predicts an exceedingly small critical size far from equilibrium (*e.g.* equation 4), the non-classical gradient energy approach predicts that the size of the critical fluctuation diverges at the spinodal after it experiences a minimum at intermediate degrees of supersaturation (Fig. 10).

In the Cahn-Hilliard gradient energy model, the energy barrier to nucleation is a function of the properties of the "hump" in the $G-x$ function of the system (Fig. 9), and the gradient energy term. The energy barrier increases and the interface thins and gets sharper as the size of the "hump" increases. The gradient energy term increases the thickness of the interface and smoothens the compositional gradient across it. In contrast to CNT where interfaces are characterized by sharp structural differences between the reacting system and nucleus, interfaces in the Cahn-Hilliard gradient-energy model are defined as locations with significant compositional gradients. The excess energy associated with these compositional gradients constitutes the interfacial energy, and there are no structural differences between reacting system and nucleus.

If a phase transformation is isochemical and associated only with a transition in structure then the Cahn-Hilliard approach cannot be used to quantify the energy barrier to nucleation and the size of a nucleus. Instead, the Allen-Cahn approach (Allen and Cahn, 1979) may be applied which allows us to model continuous order-disorder phase transformations. The Gibbs energy may then be written

$$G = \int_V \left[g(\eta) + \kappa_\eta (\nabla \eta)^2 \right] dV \qquad (37)$$

where $g(n)$ is the Gibbs energy of the homogeneous system, η is a long-range order parameter and κ_η is the gradient energy coefficient for that order parameter.

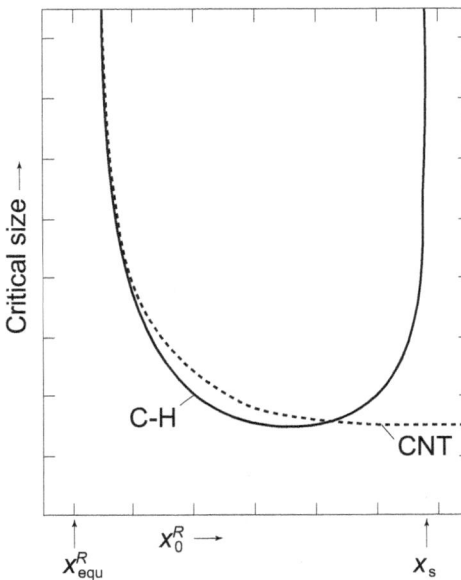

Figure 10. Schematic relationship between the size of a critical cluster and the departure from equilibrium according to the classical nucleation theory (CNT) and the Cahn-Hilliard gradient energy approach (C-H). Modified after Cahn and Hilliard (1959).

Poduri and Chen (1996) developed a model that describes phase transformations involving both compositional and structural changes combining the Cahn-Hilliard and Allen-Cahn approaches. According to Poduri and Chen (1996), the increase in Gibbs energy arising from the formation of a spherical nucleus is given by

$$\Delta G^* = 4\pi \int \left[\Delta g(\eta, x) + \kappa_\eta \left(\frac{d\eta}{dr} \right)^2 + \kappa_x \left(\frac{dx}{dr} \right)^2 \right] r^2 dr \qquad (38)$$

Similar to the Cahn-Hilliard approach, the non-classical critical cluster size determined by Poduri and Chen (1996) diverges at the spinodal. However, according to Binder (1991), the divergence of the critical size is an artefact of the gradient energy approach if thermal fluctuations are ignored. It has been suggested (Poduri and Chen, 1996) that the incorporation of a thermal noise term in the calculation of the nucleation energetics, as initially proposed by Cook (1970), does result in finite critical sizes also for nucleation close to the spinodal.

The Cahn-Hilliard gradient energy approach is similar to a model initially developed by van der Waals (1893). More advanced models of nucleation derived by Oxtoby and Evans (1988), Bagdassarian and Oxtoby (1994) and Granasy *et al.* (2002) are based on the pioneering work of van der Waals and Cahn-Hilliard and are similar in that the gradient energy terms act as a penalty for sharp compositional or structural gradients allowing for the modelling of interfacial tension between reacting system and nucleus. The treatment of Bagdassarian and Oxtoby (1994), however, predicts a decrease in both, ΔG^* and r^* with departure from equilibrium, with ΔG^* vanishing and r^* taking on a finite size at the spinodal. In the work by Bagdassarian and Oxtoby (1994), a thermal noise term, as suggested by Binder (1991), was not required to predict more realistic nucleus sizes at the spinodal. Instead, the homogeneous Gibbs energy functional was modelled with two parabolas, each centred with their minimum on the reacting and nucleating phase, respectively, and intersecting at a Gibbs energy related to the driving force for nucleation (Bagdassarian and Oxtoby, 1994). According to this work, r^* increases away from the spinodal and is identical to the classical critical radius close to equilibrium if the curvatures of the two parabolas are the same. In case the Gibbs energy curvature of the parabola about the nucleus is greater than that about the reacting system, the predicted critical size close to equilibrium is greater than the classical one. The Bagdassarian and Oxtoby (1994) model of intersecting Gibbs energy parabolas may be a valuable approach to study the energetics of nucleation in geological materials if the curvatures of the parabolas are fit to the realistic Gibbs energy profiles of the geological phases. The phase field method may then be used to predict the dynamics of nucleation and microstructure evolution.

3.2. Nucleation simulations with the phase field method

The gradient energy approach allows us to model the microstructural evolution of a system during a discontinuous phase transformation if linked to rate expressions for nucleation and growth. Such evolution models constitute the phase field method (*e.g.*

Chen, 2002) which requires the description of the Gibbs energy density as a function of field variables, commonly referred to as order parameters, such as system composition and structure. The two most common approaches to the dynamics of nucleation in a phase field model are the introduction of a random noise term to simulate thermal fluctuations (*e.g.* Wang *et al.*, 1995), and the explicit nucleation method (*e.g.* Simmons *et al.*, 2000). In any case, the phase field method is a computational tool applied to the study of the dynamics of nucleation where the microstructure evolution of the model system is driven by an overall reduction of Gibbs energy. Because it considers the influence of spatial variations in order parameters on the Gibbs energy of the system, the phase field method allows us to study potential interactions between neighbouring nuclei such as the competition for nutrients during diffusion-controlled cluster formation or the coalescence of nuclei. It is important to note that the capillarity approximation and the interface-controlled kinetics inherent in CNT do not allow for a similar treatment limiting it to a 0-dimensional theory.

The random noise phase field method (*e.g.* Wang *et al.*, 1995) adds Langevin random fluctuation terms (Landau and Lifshitz, 1969) both to the stochastic Cahn-Hilliard time evolution equation (Cahn and Hilliard, 1958) and to the Allen-Cahn time evolution equation (Allen and Cahn, 1979) so that

$$\frac{\partial x}{\partial t} = M \left[\frac{\partial^2 g(\eta, x)}{\partial x^2} \nabla^2 - 2\kappa_x \nabla^4 x \right] + \xi$$

and

$$\frac{\partial \eta}{\partial t} = -L \left[\frac{\partial g(\eta, x)}{\partial \eta} - 2\kappa_\eta \nabla^2 \eta \right] + \xi$$

where t is time, M and L are the positive kinetic coefficients related to the diffusional mobility and the microscopic rearrangement kinetics, respectively, and ξ is the Langevin noise term satisfying the fluctuation-dissipation theorem (Landau and Lifshitz, 1969). $g(\eta, x)$ is usually approximated by a Landau expansion polynomial

$$g(\eta, x) = \frac{A_1}{2}(x - x_1)^2 + \frac{A_2}{2}(x - x_2)(\eta_1^2 + \eta_2^2) - \frac{A_3}{4}(\eta_1^4 + \eta_2^4) + \frac{A_4}{6}(\eta_1^2 + \eta_2^2)^3$$

where x_1, x_2, A_i ($i = 1,...4$) are positive constants and η_1 and η_2 are the long-range order parameters. x_1 and x_2 are close to the equilibrium compositions of the reactant and nucleus, and the A_i values define the shape of the Gibbs energy surface.

The random noise phase field method has been shown to allow appropriately for the simulation of homogeneous (Wang *et al.*, 1995) and heterogeneous (Castro, 2003) nucleation if the energy barrier to nucleation is relatively small. However, unphysically large random noise is required to simulate discontinuous phase transformations in systems with significant energy barriers to nucleation which results in unrealistic nucleation-rate predictions. In addition, the high spatial and temporal resolution of the simulations required by the Langevin noise terms is computationally prohibitive unless they are applied to very small model systems.

The explicit nucleation model (*e.g.* Simmons *et al.*, 2000; Jokisaari *et al.*, 2016) may be considered a valuable alternative to the random noise phase field method as it is computationally less expensive. It allows us to link the time evolution equations of the phase field method with the nucleation energetics and nucleus sizes derived either through CNT (*e.g.* Simmons *et al.*, 2000) or the gradient-energy approach (*e.g.* Heo *et al.*, 2010). The Langevin noise terms are not included in the explicit nucleation method. Instead, the nuclei are introduced explicitly to the system assuming that the time required for their formation is significantly shorter than the temporal resolution of the nucleation simulation, Δt. At random locations, r, nucleation probabilities, P, are calculated through

$$P(r,t) = 1 - \exp(-I^*\Delta t)$$

and compared to a uniform random number, R, which ranges between 0 and 1. In case $P > R$, a nucleus is placed at the respective location, and the local composition field is depleted accordingly to account for the conservation of the overall composition if the phase transformation is partitioning.

The time evolution equations of the phase field method are coupled partial differential equations that can be solved numerically in various ways, including finite differences and spectral methods, and the finite element method. These numerical approaches allow the simulation of nucleation even in complex geometries, such as those associated with dendrite formation (*e.g.* Granasy *et al.*, 2004a,b), and have been successfully used to study nucleation in multiphase systems (*e.g.* Steinbach *et al.*, 1996; Nishida *et al.*, 2014). The additional consideration of growth kinetics (see Gaidies *et al.* 2017, this volume) allows the analysis of the complete microstructure evolution of the system associated with the discontinuous phase transformation.

4. Summary and potential future research

Nucleation is a fundamental process during most phase transformations in geological materials. CNT may be used to characterize the energetics of the critical barrier and the rate of nucleation for reactions close to equilibrium provided that long-range diffusion is fast relative to attachment and detachment processes at the interface between nucleus and metastable reactant. Non-classical gradient energy approaches to nucleation, such as the Bagdassarian and Oxtoby (1994) model, may be valuable alternatives to CNT as they allow the consideration of the influence of chemical heterogeneities on nucleation kinetics. Such heterogeneities may form in the metastable matrix in the vicinity of a nucleus during diffusion-controlled nucleation, or across the nucleus/matrix interface during nucleation far from equilibrium where the nucleus is assumed to be so small that interface properties are rather diffuse. The most promising approach to study the spatial distribution and evolution of nuclei in multiphase systems, such as most geological materials, may be the phase field method. Application of this method to appropriate systems may result in a better understanding of nucleation in geological materials and its influence on micro-structure and rock-texture formation.

Acknowledgements

The author thanks R. Abart for numerous discussions and his editorial advice, and F. George for valuable comments on an earlier draft of this chapter. Detailed comments and suggestions by E. Petrishcheva significantly helped to improve the presentation of this chapter.

References

Allen, S.M. and Cahn, J.W. (1979) A microscopic theory for antiphase boundary motion and its application to antiphase domain coarsening. *Acta Metallurgica*, **27**, 1085–1095.

Bagdassarian, C.K. and Oxtoby, D.W. (1994) Crystal nucleation and growth from the undercooled liquid: A nonclassical piecewise parabolic free-energy model. *Journal of Chemical Physics*, **100**, 2139–2148.

Becker, R. and Döring, W. (1935) Kinetische Behandlung der Keimbildung in übersättigten Dämpfen. *Annalen der Physik*, **24**, 719–752.

Binder, K. (1991) Spinodal decomposition in materials science and technology. Pp. 405–471 in: *Phase Transformations in Materials* (P. Haasen, editor), **5**. Wiley-VCH Verlag GmbH, Weinheim, Germany.

Cahn, J.W. (1956) The kinetics of grain boundary nucleated reactions. *Acta Metallurgica*, **4**, 449–459.

Cahn, J.W. (1957) Nucleation on dislocations. *Acta Metallurgica*, **5**, 169–172.

Cahn, J.W. and Hilliard, J.E. (1958) Free energy of a nonuniform system. I. Interfacial free energy. *The Journal of Chemical Physics*, **28**, 258–267.

Cahn, J.W. and Hilliard, J.E. (1959) Free energy of a nonuniform system. III. Nucleation in a two-component incompressible fluid. *The Journal of Chemical Physics*, **31**, 688–699.

Castro, M. (2003) Phase-field approach to heterogeneous nucleation. *Physical Review B*, **67**, 035412.

Chen, L.Q. (2002) Phase-field models for microstructure evolution. *Annual Review of Materials Research*, **32**, 113–140.

Christian, J.W. (1975) *The Theory of Transformations in Metals and Alloys: Part 1 – Equilibrium and General Kinetic Theory*. Pergamon Press, Oxford, UK.

Cook, H.E. (1970) Brownian motion in spinodal decomposition. *Acta Metallurgica*, **18**, 297–306.

Eshelby, J.D. (1957) On the determination of the elastic field of an ellipsoidal inclusion, and related problems. *Proceedings of the Royal Society*, **A 241**, 376–396.

Frenkel, J. (1946) *Kinetic Theory of Liquids*. Oxford University Press, Oxford, UK.

Gaidies, F., Pattison, D.R.M. and de Capitani, C. (2011) Toward a quantitative model of metamorphic nucleation and growth. *Contributions to Mineralogy and Petrology*, **162**, 975–993.

Gaidies, F., Milke, R., Heinrich, W. and Abart, R. (2017) Metamorphic mineral reactions: Porphyroblast, corona and symplectite growth. Pp. 469–540 in: *Mineral Reaction Kinetics: Microstructures, Textures and Chemical Compositions* (R. Abart and W. Heinrich, editors). EMU Notes in Mineralogy, **16**. European Mineralogical Union and Mineralogical Society of Great Britain & Ireland, London.

Gibbs, J.W. (1928) *The Collected Works. Vol. 1 Thermodynamics*. Longmans, Green & Co, New York.

Granasy, L., Pusztai, T. and James, P.F. (2002) Interfacial properties deduced from nucleation experiments: A Cahn-Hilliard analysis. *Journal of Chemical Physics*, **117**, 6157–6168.

Granasy, L., Pusztai, T., Börzsonyi, T., Warren, J.A. and Douglas, J.F. (2004a) A general mechanism of polycrystalline growth. *Nature Materials*, **3**, 645–650.

Granasy, L., Pusztai, T. and Warren, J.A. (2004b) Modelling polycrystalline solidification using phase field theory. *Journal of Physics: Condensed Matter*, **16**, R1205–R1235.

Heo, T.W., Zhang, L., Du, Q. and Chen, L.Q. (2010) Incorporating diffuse-interface nuclei in phase-field simulations. *Scripta Materialia*, **63**, 8–11.

Holland, T.J.B. and Powell, R. (1998) An internally consistent thermodynamic data set for phases of petrological interest. *Journal of Metamorphic Geology*, **16**, 309–343.

Jokisaari, A.M., Permann, C. and Thornton, K. (2016) A nucleation algorithm for the coupled conserved–nonconserved phase field model. *Computational Materials Science*, **112**, 128–138.

Kaischew, R. and Stranski, I.N. (1934) Concerning the mechanism of the equilibrium of small crystals. *Zeitschrift für Physikalische Chemie-Abteilung B- Chemie der Elementarprozesse Aufbau der Materie*, **26**, 312–316.

Kelton, K.F. (1991) Crystal nucleation in liquids and glasses. Pp. 75–178 In: *Solid State Physics* (H. Ehrenreich, and D. Turnbull, editors), **45**. Academic Press, London.

Kelton, K.F. (2000) Time-dependent nucleation in partitioning transformations. *Acta Materialia*, **48**, 1967–1980.

Kelton, K.F. (2006) Nucleation. Pp. 6388–6393 in: *Encyclopedia of Materials: Science and Technology* (K.H. Buschow, R.W. Cahn, M.C. Flemings, B. Ilschner, E.J. Kramer, S. Mahajan and P. Veyssiere, editors), **7**. Elsevier, Amsterdam.

Kelton, K.F., Greer, A.L. and Thompson, C.V. (1983) Transient nucleation in condensed systems. *Journal of Chemical Physics*, **79**, 6261–6276.

Landau, L.D. and Lifshitz, E.M. (1969) *Statistical Physics*, 2nd edition. Pergamon Press, New York.

Lee, J.K., Barnett, D.M. and Aaronson, H.I. (1977) The elastic strain energy of coherent ellipsoidal precipitates in anisotropic crystalline solids. *Metallurgical Transactions A*, **8A**, 963–970.

Nabarro, F.R.N. (1940) The influence of elastic strain on the shape of particles segregating in an alloy. *Proceedings of the Physical Society*, **52**, 90–104.

Nishida, Y., Aiga, F. and Itoh, S. (2014) Microstructural path analysis of polycrystalline solidification by using multi-phase-field method incorporating a nucleation model. *Journal of Crystal Growth*, **405**, 110–121.

Oxtoby, D.W. and Evans, R. (1988) Nonclassical nucleation theory for the gas-liquid transition. *Journal of Chemical Physics*, **89**, 7521–7530.

Poduri, R. and Chen, L.Q. (1996) Non-classical nucleation theory of ordered intermetallic precipitates – application to the Al-Li alloy. *Acta Materialia*, **44**, 4235–4259.

Russell, K.C. (1968) Linked flux analysis of nucleation in condensed phases. *Acta Metallurgica*, **16**, 761–769.

Simmons, J.P., Shen, C. and Wang, Y. (2000) Phase field modeling of simultaneous nucleation and growth by explicitly incorporating nucleation events. *Scripta Materialia*, **43**, 935–942.

Steinbach, I., Pezzolla, F., Nestler, B., Sesselberg, M., Prieler, R., Schmitz, G.J., Rezende, J. L.L. (1996) A phase field concept for multiphase systems. *Physica D*, **94**, 135–147.

Turnbull, D. and Fisher, J.C. (1949) Rates of nucleation in condensed systems. *Journal of Chemical Physics*, **17**, 71–73.

van der Waals, J.D. (1893) Thermodynamische Theorie der Kapillarität unter Voraussetzung stetiger Dichteänderung. *Zeitschrift für Physikalische Chemie*, **13**, 675–725.

Volmer, M. (1939) *Kinetik der Phasenbildung*. Steinkopf, Dresden, Germany.

Volmer, M. and Weber, A. (1926) Keimbildung in übersättigten Gebilden. *Zeitschrift für Physikalische Chemie*, **119**, 277–301.

Wang, Y., Wang, H.-Y., Chen, L.Q. and Khachaturyan, A.G. (1995) Microstructural development of coherent tetragonal precipitates in Magnesium-partially-stabilized zirconia: a computer simulation. *Journal of the American Ceramic Society*, **78**, 657–661.

EMU Notes in Mineralogy, Vol. 16 (2017), Chapter 12, 373–418

Dynamic crystallization in magmas

SILVIO MOLLO[1,2] and JULIA EVE HAMMER[3]

[1] *Department of Earth Sciences, Sapienza - University of Rome, P. le Aldo Moro 5, 00185 Roma, Italy*
[2] *Instituto Nazionale di Geofisica e Vulcanologia, Via di Vigna Murata 605, 00143 Roma, Italy, e-mail: silvio.mollo@ingv.it*
[3] *Department of Geology and Geophysics, SOEST, University of Hawaii 96822, Honolulu, Hawaii, USA, e-mail: jhammer@hawaii.edu*

Undercooling and crystallization kinetics are recognized increasingly as important processes controlling the final textures and compositions of minerals as well as the physicochemical state of magmas during ascent and emplacement. Within a single volcanic unit, phenocrysts, microphenocrysts and microlites can span a wide range of compositions, develop complex zoning patterns, and show intricate textures testifying to crystallization far from equilibrium. These petrographic complexities are not associated necessarily with magma chamber processes such as mixing or mingling of distinctly different bulk compositions but, rather, may be caused by variable degrees of initial magma-undercooling and the evolution of undercooling through time. Heat-dissipation and decompression are the most effective driving forces of cooling and volatile loss that, in turn, exert a primary control on the solidification path of magma. Understanding these kinetic aspects over the temporal and spatial scales at which volcanic processes occur is therefore essential to interpret correctly the time-varying environmental conditions recorded in igneous minerals.

This contribution aims to summarize and integrate experimental studies pertaining to the crystallization of magmas along kinetic or time-dependent pathways, where solidification is driven by changes in temperature, pressure and volatile concentration. Fundamental concepts examined in the last decades include the effect of undercooling on crystal nucleation and growth as well as on the transition between interface- and diffusion-controlled crystal growth and mass transfer occurring after crystals stop growing. We summarize recent static and dynamic decompression and cooling experiments that explore the role of undercooling in syn-eruptive crystallization occurring as magmas ascend in volcanic conduits and are emplaced at the surface. The ultimate aim of such studies is to decode the textural and compositional information within crystalline phases to place quantitative constraints on the crustal transport, ascent and emplacement histories of erupted and intrusive magmas.

Magma crystallization under dynamic conditions will be assessed also through a comparative description of the disequilibrium features in minerals found in experimental and natural materials. A variety of departures from polyhedral growth, including morphologies indicating crystal surface instability, dendritic structures, sector zoning and growth twins are linked to the rate at which crystals grow. These have implications for the entrapment of melt inclusions and plausibility for interpreting the growth chronology of individual crystals. A simple "tree-ring" model, in which the oldest part of the crystal lies at the centre and the youngest at the rim, is not an appropriate description when growth is non-concentric. Further, deviation from chemical

DOI: 10.1180/EMU-notes.16.12

equilibrium develops in response to kinetically controlled cation redistributions related to the partitioning of major and trace elements between rapidly growing crystal and melt. The incorporation into the crystal lattice of chemical components in non-stoichiometric or non-equilibrium proportions has important implications for the successful interpretation of the conditions under which magmas crystallize and for the development of new equilibrium models based on mineral compositional changes.

Finally, it is important to stress that the main purpose of this contribution is to ignite research exploring the causes and consequences of cooling and decompression-driven crystal growth kinetics in order to appreciate in full the evolutionary paths of volcanic rocks and interpret the textural and compositional characteristics of their mineral constituents.

1. Crystal nucleation, growth and undercooling in laboratory experiments

This section builds on previous summaries of crystal nucleation and growth in laboratory experiments (Hammer, 2008; Kirkpatrick, 1981) in order to examine insights from recent publications, beginning with the laboratory techniques that impose superheating and undercooling, and their influences on solidification. In order to facilitate comparison across the breadth of recent work, we summarize trends and correlations relating experimental variables (temperature, pressure, cooling and decompression rate) to the most commonly reported compositional (major and minor element) and textural characteristics (*e.g.* crystal number density and volume fraction). The reader is referred to the original publications for additional data.

1.1. The significance of undercooling

Nucleation conditions dictate overall rock texture; growth conditions control crystal morphology, whether melt inclusions are likely to be enclosed and isolated, and the spatial heterogeneity of major and minor elements within crystals. Whether growth conditions permit incorporation of elements from a melt according to equilibrium partitioning or reflect kinetics of melt transport and crystal–melt interface reaction is a distinction of fundamental importance to the interpretation of magmatic history from rocks, and one that is amenable to laboratory study and modern microanalysis. The need to understand the time-evolution of magma crystallinity and evolving mineral compositions in the context of magma movement through the Earth's crust motivates experiments in which temperature, pressure and fluid composition are varied. The concept of undercooling is central to making sense of the time evolution of crystal nucleation and growth within these experimental and natural samples. Undercooling is defined as the difference between the system temperature and the temperature at which a phase saturates in the given liquid (Kirkpatrick, 1981). The degree of undercooling decreases during crystallization because the phase-saturation temperature decreases as the liquid composition changes. Undercooling is used inclusively here to refer to the thermodynamic driving force associated with either a decrease in temperature or compositionally induced increase in a melt's liquidus temperature driven by decompression under H_2O-saturated conditions. In studies of decompression-driven crystallization (*e.g.* Hammer and Rutherford, 2002; Couch *et al.*, 2003; Waters *et al.*, 2015), the term 'effective undercooling' is used to distinguish the driving force induced by compositional change from that of temperature change. However, in both cases, it is quantified by the difference between the phase-in and system temperatures. The difficulty in ascertaining the actual undercooling during an experiment is addressed in some studies by the 'nominal undercooling' (Faure et *al.*, 2003) which ignores compositional changes and refers to the difference between the phase-in temperature of the starting liquid and the quench temperature. The distinction is important; during a cooling experiment the nominal undercooling is fixed, while the actual undercooling declines.

This driving force is balanced against component mobility, typically taken to scale with liquid viscosity (Zhang, 2010; Vetere et *al.*, 2015), to control the rates at which crystals nucleate and grow. Small growth rates associated with low undercooling require exceptional methodologies to resolve, exemplified by the *in situ* observations of olivine growing inside plagioclase-hosted melt inclusions of Jambon et *al.* (1992), and are relatively scarce in recent literature. Determining the crystal-growth rate, nucleation rate and overall magma solidification rate at moderate to high undercooling is a task amenable to quench-based laboratory techniques and relevant to a wide variety of volcano-magmatic situations.

Although numerical values for the rates of nucleation and growth vary by mineral, system composition, temperature with respect to the liquidus (*e.g.* undercooling as well as 'superheating' above the phase-in temperature), dwell time and possibly other variables, several features of plagioclase nucleation and growth kinetics (Hammer and Rutherford, 2002), illustrated schematically in Fig. 1 for decompressing rhyolite melt, appear to be generally applicable. The positions of the nucleation and growth-curve maxima are phase and system specific. The overall variation in nucleation rate is several orders of magnitude, compared with a factor of only <3 for growth rate (Swanson, 1977; Kirkpatrick, 1981; Hammer and Rutherford, 2002). Thus, nucleation rate exerts the fundamental control on the spatial distribution of solid in a crystallizing system: a high initial nucleation rate produces a large number-density of small crystals, and low nucleation rate produces fewer, larger crystals. Secondly, the separation between the maxima of the nucleation and growth curves, with the nucleation maximum occurring at higher undercooling, increases the likelihood that initial crystallization proceeds in either a nucleation-dominated or growth-dominated regime, associated with higher or lower undercooling, respectively. The tendency is manifest in rapidly decompressed synthetic and natural materials as a characteristic ratio of crystal-number density, N_A, to volume fraction of crystals, ϕ. Large ratios are associated with rapid ascent rates and moderate-to-high eruption intensities, because undercooling increases progressively with time. Low ratios are associated with slower decompression and vulcanian-to-dome-forming eruption styles (Brugger and Hammer, 2010a). In their study of crystallization in synthetic CaO–Na_2O–Al_2O_3–SiO_2 haplotonalite melt (CNAS rhyolite) following variable degrees of rapid decompression, Mollard et *al.* (2012) observe a strong correlation between undercooling and the number density of feldspar.

Undercooling may decrease, increase or remain constant, depending on whether the degree of solidification at earlier points along the pressure (or temperature) path is

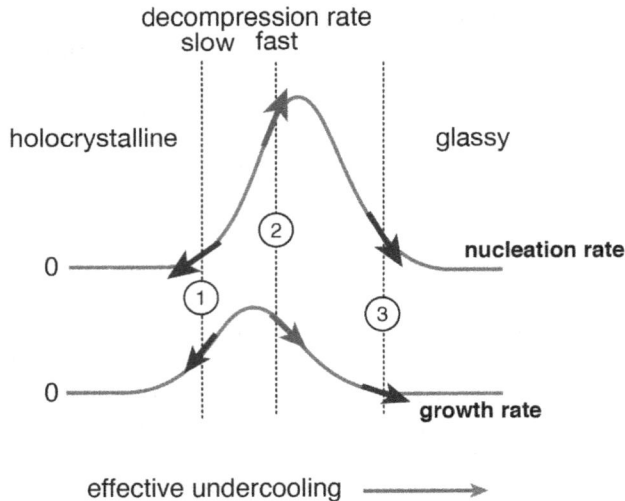

Figure 1. Schematic illustration of the functional dependence of crystal nucleation and growth rate on thermodynamic undercooling, based on classical theory (Tamman and Mehl, 1925; Dowty, 1980) and relevant experiments (*e.g.* Swanson, 1977; Hammer and Rutherford, 2002; Couch *et al.*, 2003). Three trajectories (among many possible) are highlighted. From a position of relatively low undercooling associated with slow magma decompression (or cooling), declining undercooling (arrows at 1) with time leads to coarsening of sparse crystals and holocrystalline texture. From a position of intermediate undercooling associated with low-intensity volcanic eruption, increasing undercooling (2) leads to a second nucleation event and phenocrystic texture. From a position of high initial undercooling, increasing undercooling (3) produces glassy texture.

sufficient to approach equilibrium, as summarized in Fig. 2. Intense volcanic eruptions that produce glassy pyroclasts are systems in which undercooling increases over the short time available for crystallization. Declining undercooling with time ultimately produces the coarse textures associated with intrusive rocks.

1.2. Assessing crystal nucleation in laboratory experiments

Actual nuclei of rock-forming minerals are too small for resolution by backscattered electron microscopy, and the refractory containers utilized in experimental techniques are typically opaque to *in situ* observation (with notable exceptions discussed below). Thus, the process of nucleation is typically approached indirectly, by assessing the number of crystals that grow to observable size (~1 μm) in a given time following a perturbation from an initial crystal-poor or crystal-free state. Knowledge of the rate at which crystals form in response to an applied solidification-inducing thermodynamic driving force, which is typically cooling or decompression and referenced with respect to the liquidus temperature and H_2O-pressure (P_{H_2O}), thus constitutes a functional understanding of nucleation. Even by this modest definition of success, laboratory approaches to the study of nucleation in silicate melts involve challenges. Difficulties to be overcome include: (1) synthetic materials composed of oxide and carbonate

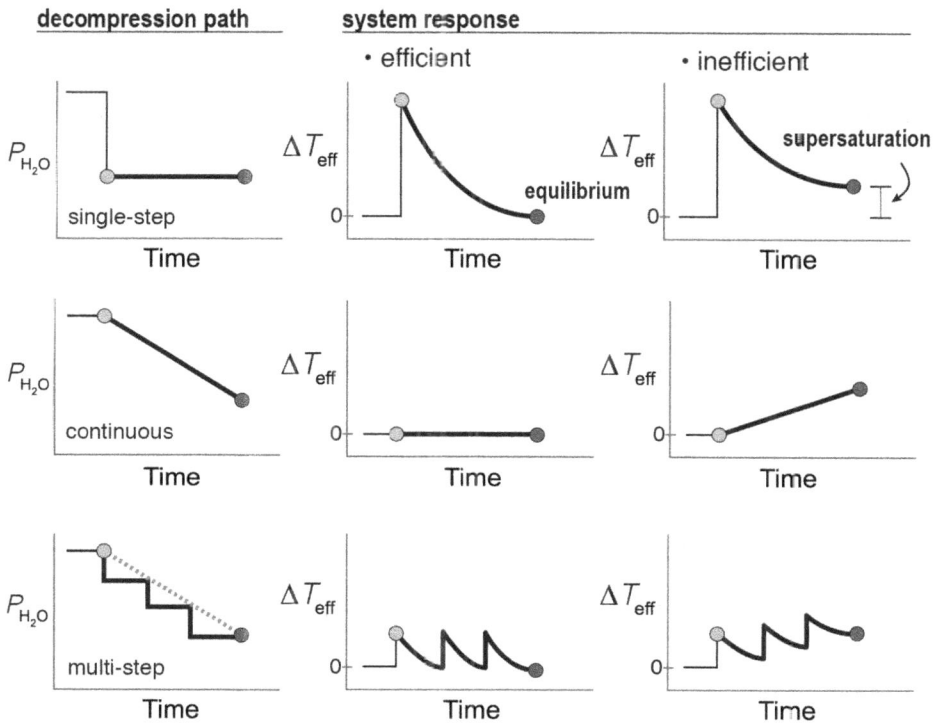

Figure 2. Relationships between experimental decompression paths and system response for instantaneous (single-step), continuous, and stepwise (multistep) changes in pressure. If change is imposed infinitesimally slowly, or at a rate at which solidification mechanisms (crystal nucleation and growth) are relatively rapid, the system maintains equilibrium and arrives at final pressure (P_f) at a state of zero undercooling. If the rate of solidification is unable to keep pace with pressure decrease, undercooling increases with time. Stepwise decompression produces spikes in undercooling and potentially different textures and phase compositions at P_f, despite similar time-integrated rates of change, because the rates of transformation differ along the path. Most of the decompression experiments described in this review represent the third scenario. This conceptual framework is also relevant to experiments in which temperature decrease drives solidification.

reagents must first react in order to become silicate melts; failure to homogenize the liquid could provide sites for heterogeneous nucleation that are difficult to characterize (Iezzi *et al.*, 2008; Vetere *et al.*, 2013, 2015); (2) natural starting materials must be crushed in order to load them into capsules, and the pores and grain interfaces that result from grinding natural glass and crystal-rich starting materials might provide energetically advantageous surfaces that are not present in nature (Vetere *et al.*, 2015); (3) the entire thermal history of the material, including run-up to initial conditions (Ni *et al.*, 2014) and superliquidus heat treatments (Sato, 1995; Pupier *et al.*, 2008; Waters *et al.*, 2015), may influence the distribution of subcritical clusters poised to become nuclei; and (4) containers and exterior bead surfaces provide energetically favourable sites for heterogeneous nucleation of spinel-structured oxide phases (Berkebile and Dowty, 1982; Mollo *et al.*, 2012a), which can influence the formation of silicates through the formation of boundary layers or heterogeneous nucleation assisted by epitaxy (*e.g.* clinopyroxene on spinel-structured oxides; *cf.* Hammer *et al.*, 2010; Vetere *et al.*, 2015).

Noting that the time scale of structural relaxation, τ, in silicate melt is $<10^{-7}$ s, as estimated using reasonable values for η (viscosity of the melt) and G (shear modulus) by means of the Maxwell (Dingwell and Webb, 1990) relation,

$$\tau = \eta/G \qquad (1)$$

Vetere *et al.* (2013) suggested that melt structure at the liquidus can be considered path-independent. They strive to eliminate extrinsic defects such as chemical and physical heterogeneities (*e.g.* bubbles or relict crystals) by preparing six starting materials for one-atmosphere cooling experiments as combinations of natural basalt (T_l = 1230°C) and rhyolite (T_l = 1045°C), repeatedly powdering and fusing the mixtures to 1600°C ($+\Delta T$ = 370−555°C) and annealing the homogenized glasses for 2 h at 1400°C ($+\Delta T$ = 170−355°C) before performing the cooling experiments, which begin after a 2 h dwell period at 1300°C ($+\Delta T$ = 70−255°C). The authors evaluate the success of this sequence for producing structurally relaxed liquids at 1300°C by demonstrating reproducibility of final crystal texture in duplicate runs and comparing the run products against similarly prepared and cooled samples that had undergone an extra 40 h dwell period at 1400°C ($+\Delta T$ = 170−355°C) prior to the dwell at 1300°C. The consistency of mineralogy, crystal shape and abundance in the two series demonstrates that the additional dwell at 1400°C had no effect, and this is interpreted as a consequence of having eliminated extrinsic defects in the previous steps. It is unknown what effect, if any, the cumulative superheating (16 h by $T > 370$°C imposed on all samples) had on nucleation, and whether a similar material would have nucleated crystals at a higher rate had it been held nearer the liquidus before cooling. That is, the experiments performed are not capable of evaluating the relevance of the Maxwell structural relaxation time in describing a system's readiness for nucleation. Are intrinsic artefacts introduced by superheating, even as extrinsic artefacts are eliminated?

If diffusive homogenization over the length scale of experimental charges provides the relevant time scale for the formation of the equilibrium, structurally-relaxed distribution of subcritical clusters (Dingwell, 1995; Vetere *et al.*, 2013), it should be possible to isolate the extrinsic and intrinsic effects and study homogeneous nucleation with a matrix of experiments with high- and low-temperature dwell periods of varying duration and superheating degree, that eliminate possible foreign substrates (assuming they are soluble in the melt), and subsequently erase intrinsic impediments to nucleation.

1.3. Crystal nucleation as a means of examining volcanic processes

The studies described in the previous section target the kinetics of crystal nucleation as an explicit goal and necessarily consider the degree and duration of superheating as factors influencing subsequent crystallization. Other studies manipulate nucleation kinetics in order to achieve a particular texture in the run products. In their study of crystallization in H_2O-obsidians, rhyolite and dacite, Waters *et al.*, (2015) modulated the starting pressure in order to control nucleation of plagioclase during subsequent decompression. Decompression experiments initiated from 44 MPa above the plagioclase-in stability curve at the experimental temperature produce fewer, larger crystals in aphyric obsidian than those held 6 MPa below the plagioclase-in curve, even though the subliquidus dwell period (48 h) produced no discernable feldspar and the crystal volume fraction for a given final pressure was similar. This degree of superheating was a requirement for matching the texture of natural samples, perhaps because it reduced the number of sites available for immediate growth. A similar technique was employed by Lofgren *et al.* (2006), on the basis of earlier work (*e.g.* Lofgren and Lanier, 1992), in a dynamic cooling study of rare earth element partitioning in sector-zoned pyroxenes The experiments were initiated <5°C above the liquidus temperature, with the goal of reducing, but not eliminating, the number of sites for immediate growth.

Given the importance of solidification kinetics (*via* control on nucleation and growth rates) on undercooling and the importance of crystallization in prior time increments in controlling undercooling (Fig. 2), the manner in which samples are decompressed would logically exert a fundamental control on final texture. Experiments in which undercooling is applied instantaneously are uniquely capable of resolving nucleation and growth rates as a function of driving force; incremental cooling or decompression allows investigation of crystallization in a manner more likely to occur in nature. Decompression experiments prior to 2009 were executed by bleeding in a series of small steps the gas or water exerting pressure on the capsules in the hydraulic line (Hammer and Rutherford, 2002; Couch *et al.*, 2003; Martel and Schmidt, 2003). Brugger and Hammer (2010a) used a specially designed automated screw-jack mechanism to impose decompression quasi-continuously on crystal-poor rhyodacite, and included two experiments in which the same integrated decompression rate (1 MPa h^{-1}) was achieved continuously and in steps of 10 MPa each. The multi-step decompression produced qualitatively different crystal texture and composition from the continuous decompression. In the multi-step runs, crystals are more skeletal, more inhomogenously distributed, and lower in number density; all indications that nucleation was inhibited and growth occurred at relatively high undercooling. Evidently the higher driving force did not translate into more rapid crystallization, and the system strayed further from equilibrium during decompression (*e.g.* third scenario in Fig. 2). Nowak *et al.* (2011) introduced a valve capable of metering incremental gas in the IHPV setup. This device was used to study degassing kinetics and was employed in crystallization experiments of Fiege *et al.* (2015), although the results were not compared with multi-step style decompression. Preliminary experiments suggest that stepwise decompression produces similar plagioclase contents regardless of the number of steps (*n*, for *n* from 1 to 25), appreciably different plagioclase and pyroxene compositions if *n* < 5, and plagioclase morphology that is sensitive to *n* ⩽ 25 (Hammer *et al.*, unpublished data).

1.4. Cooling and decompression, imposed separately and together

Shea and Hammer (2013) address the question of whether decompression-driven crystallization produces a fingerprint distinct from cooling-driven crystallization by using pairs of single-step decompression (SSD) and single-step cooling (SSC) experiments (Fig. 3a). Dwell pressures in the SSD were selected to match effective undercooling values of 52, 82, 112, 137 and 155°C imposed by the corresponding dwell temperatures in the SSC series. Natural lava from the Mascota volcanic field, Mexico (previously characterized in phase equilibrium experiments by Moore and Carmichael, 1998) was crushed lightly prior to hydration and equilibration at initial conditions below the stability curves of plagioclase, clinopyroxene, olivine and Cr-spinel. The two series produced similar plagioclase abundances, similar nucleation and growth rates during dwell periods of 12, 24 and 48 h, and similar compositions up to relatively high undercooling ($\Delta T = 112$°C, final pressure $P_f = 45$ MPa in the SSD series).

The insensitivity of outcome to solidification driving force at conduit temperatures and pressures could be useful in numerical modelling of magma crystallization paths. However, consistent differences emerge at even higher undercooling. Feldspar number density and abundance are higher in the decompression series, peaking at the largest undercooling examined (155°C). Because of more dissolved H_2O content throughout the cooling runs, amphibole and olivine are the dominant ferromagnesian phases; Ca-clinopyroxene dominates in the decompression runs. This mineralogical difference has an unexpected implication regarding feldspar texture. Conspicuous clustering of plagioclase crystals around clinopyroxene and correlated number densities of these phases suggest that clinopyroxene provides a substrate for heterogeneous nucleation of plagioclase. Vetere et al. (2015) report a similar result for cooling of nominally anhydrous basaltic liquids. Heterogeneous nucleation of feldspar on clinopyroxene is not recognized in silicic magma compositions, and could be an important factor controlling the formation and texture of feldspar in mafic-intermediate magmas.

Arzilli and Carroll (2013) studied crystallization of trachyte melt in response to a variety of supersaturation mechanisms, including isobaric cooling, isothermal decompression, and a combination of cooling and decompression using a natural obsidian from the Breccia Museo Member of the Campanian Ignimbrite, Campi Flegrei containing 10 vol.% crystals (alkali feldspar, clinopyroxene, magnetite and biotite). Initial conditions include a variety of temperature-pressure combinations above the feldspar stability curve, and consequently, a range in effective superheating from 15 to 131°C. The temperature drops and decompressions were performed rapidly and the charges quenched after $2-14$ h dwell periods at static (isothermal, isobaric) conditions. Because superheating and undercooling are not isolated as experimental variables (Fig. 3b), it is uncertain whether differences in feldspar abundance in run products at the same final conditions are due to path through P-T space, H_2O content at initial conditions, magnitude of initial superheat, or possibly failure to achieve equilibrium at the initial conditions during the 2 h hydration period. However, feldspar textures in the run products indicate consistently that nucleation dominates crystallization at low pressure and temperature, whereas growth dominates at conditions closer to the

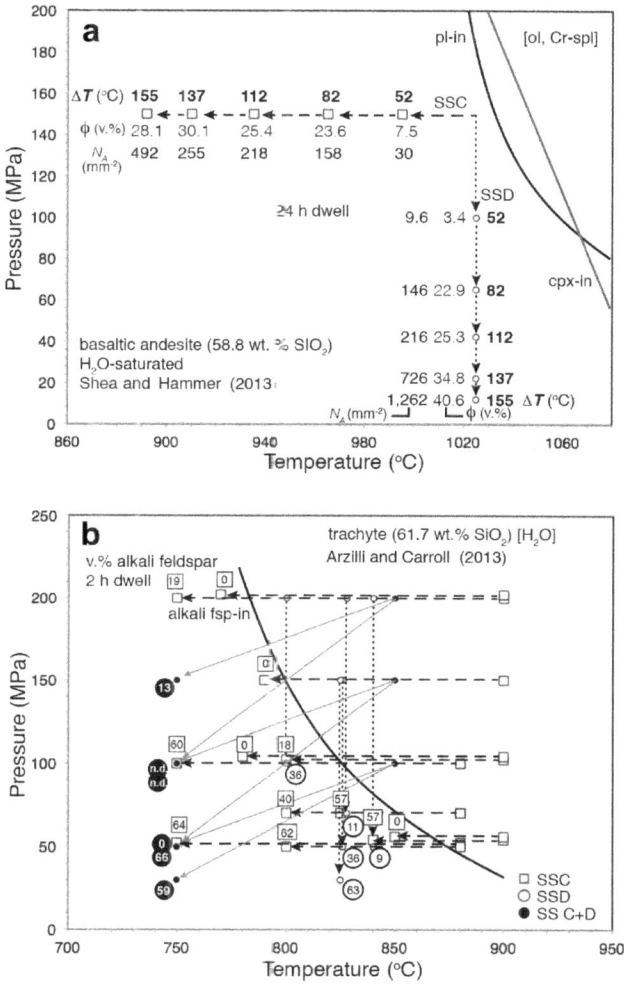

Figure 3. Pressure-temperature diagrams summarizing experiments on basaltic andesite and trachyte by Shea and Hammer (2013) and Arzilli and Carroll (2013), respectively. In (a), values of plagioclase feldspar abundance and number density are shown beside each set of final conditions. In (b), the mineral-in curves for biotite, titanomagnetite and clinopyroxene are omitted for clarity. All experimental conditions occur within the stability field of titanomagnetite, but only experiments above 100 MPa are within the biotite stability field, and experiments at high initial superheating begin outside of the clinopyroxene stability field and end within it. Experimental paths performed at the same pressure and temperature are separated slightly, for legibility. The values of alkali feldspar crystallized abundance in decompression-only experiments are indicated with open circles positioned to the lower right of the T_f, P_f coordinates; cooling experiment values are shown with squares above the T_f, P_f points, and decompression+cooling values are in shaded circles, positioned to the lower left of the T_f, P_f points. The two SSC experiments labelled 57 and 62 vol.% feldspar correspond to the path type and conditions producing a second nucleation burst at long dwell times (described in text).

feldspar stability curve. Moreover, the time series provide constraints on the changes in the rates of crystal growth, nucleation and crystallization rate (for fixed initial conditions and superheat) following identical perturbations in environmental conditions. Interestingly, although the nucleation rate decreases progressively during the dwell periods of the decompression runs, several of the longer-duration cooling-only runs undergo a second feldspar nucleation event. A difference in the time-evolution of effective undercooling (*e.g.* different versions of inefficient system response; Fig. 3a,b) may be correlated with supersaturation mechanism.

1.5. Novel apparatus

The vast majority of experimental studies involve sealing reactants inside (opaque) noble metal capsules capable of withstanding high temperatures and pressures, bringing the capsules to magmatic conditions inside a furnace and pressure vessel, rapidly cooling them to ambient conditions in an attempt to quench-in the magmatic state, and dissecting the run product to analyse the crystals. Schiavi *et al.* (2009) departed from this approach and achieved continuous *in situ* observation of magma crystallization by containing sample material in an externally heated moissanite cell at ambient pressure. Although temperature control is an issue with this methodology (Hammer, 2009), its strength is that continuous observation alleviates concern about quench effects and allows observation of crystal populations and individual crystals through time. Schiavi *et al.* (2009) observed growth of prismatic feldspar crystals in nominally dry synthetic high-K andesitic basalt (in which FeO was replaced by MnO for optical transmissivity) during 15–50 min dwell periods at 900°C, ~330°C below its liquidus, noting that coalescence is the dominant process of grain-size increase. When neighbouring crystals impinge, the result is continued growth of the coalesced units, whether or not the neighbours were aligned initially. Using an improved version of the apparatus, Ni *et al.* (2014) observe olivine and clinopyroxene crystals growing from initially superheated synthetic basalt along a variety of thermal paths involving continuous cooling as well as isothermal dwelling.

Because the systems examined with the *in situ* technique (Schiavi *et al.*, 2009; Ni *et al.*, 2014) crystallize far from equilibrium, and the degree of phase undercooling is not constrained, the geologic relevance of the plagioclase, olivine and clinopyroxene growth-rate values obtained ($2.3-4.8 \times 10^{-9}$ m s^{-1}, $2-7 \times 10^{-9}$ m s^{-1} and $6-17 \times 10^{-9}$ m s^{-1}, respectively) is difficult to assess. Insights about growth processes, rather than absolute growth rate values, are the chief contributions of the *in situ* studies. Intriguing results from Ni *et al.* (2014) included: (1) melt-inclusion entrapment and closure in the absence of measurable thermal cycling; (2) size-proportionate growth (described earlier by Kile and Eberl, 2003), in which larger crystals grow faster than smaller crystals; (3) unchanging rate of crystal growth during cooling by 100°C, despite evolution from tabular to hopper crystal morphology; and (4) coarsening of the population by size-proportionate growth and dissolution of small crystals. The last observation is consistent with results from hydrous magma decompression experiments (*e.g.* Brugger and Hammer, 2010a; Mollard *et al.*, 2012; Arzilli and Carroll, 2013), and has important implications for interpreting crystal-size distributions (CSDs). Because coarsening is typically associated with near-equilibrium conditions (Cabane *et al.*, 2005; Higgins and Roberge, 2003), a decrease in the frequency of crystals in the smallest size classes encountered by Brugger and Hammer (2010b) was interpreted as inadequate stereological correction of cut effects. The *in situ* observations suggest that reduction of crystal–melt interfacial energy among silicate phases is sufficient to drive redistribution of mass from small to larger crystals, even at conditions far from equilibrium.

2. Interface- *vs.* diffusion-controlled reactions

In igneous settings, the interpretation of rock textures and compositions is commonly based on the assumption of equilibrium between minerals and the host magmas. This general expectation justifies, most of the time, the use of thermodynamically derived formalisms to reconstruct the intensive variables of the plumbing systems, as well as to model the geochemical evolution of erupted magmas (*e.g.* Putirka, 2008 and references therein). During recent decades, however, a gamut of studies has reported clear evidence of disequilibrium features in both intrusive and effusive volcanic rocks on scales ranging from the hand specimen to the outcrop (*e.g.* DePaolo and Getty, 1996; Roselle *et al.*, 1997; Müller *et al.*, 2004; Nabelek, 2007; Baker, 2008). It has also documented that both the textural evolution and compositional change of minerals are not necessarily related to the attainment of equilibrium, but rather the development of reaction kinetics along most of the crystallization path of magma leading to final disequilibrium textures and compositions (Baker and Grove; 1985; Hammer, 2006, 2008; Iezzi *et al.*, 2008, 2011; Del Gaudio *et al.*, 2010; Mollo *et al.*, 2010, 2011a,b).

With respect to the superliquidus state of magma, a certain amount of undercooling is always necessary for the onset of crystal nucleation and growth. The degree of undercooling (ΔT) is the difference between the liquidus temperature (T_l) and the final resting temperature (T_r) of the system (Fig. 4a). Low-to-large ΔT can be achieved prior to crystallization in response to variable degassing and cooling paths experienced by magmas. Once crystal nucleation occurs, crystal growth may proceed with variable rates as a function of the magnitude of ΔT. For the simplest case represented by a diopsidic liquid ($CaO \cdot MgO \cdot 2SiO_2$), the growth rate of clinopyroxene has been determined experimentally over the effect of increasing undercooling (Nascimento *et al.*, 2004 and references therein). The data span a range of about seven orders of magnitude with a variety of undercoolings between the glass transition temperature (T_g) and liquidus temperature of the diopsidic liquid (T_l) (Fig. 4b). As T_r approaches T_l, the crystal growth rate (G) increases to a maximum value (G_{max}) corresponding to the shift of the crystal-growth regime from diffusion-controlled to interface-controlled (*e.g.* Dowty, 1980; Lofgren and Smith, 1980; Kirkpatrick, 1981; Hammer, 2008). Small undercoolings and high T_r (approaching T_l) promote interface-controlled crystal growth mechanisms where the diffusion of chemical elements in the melt prevails over the crystal growth rate.

Minerals develop euhedral textures (*e.g.* prismatic, tabular, well faced, *etc.*), core-to-rim crystal zoning is generally lacking and the composition of the melt next to the

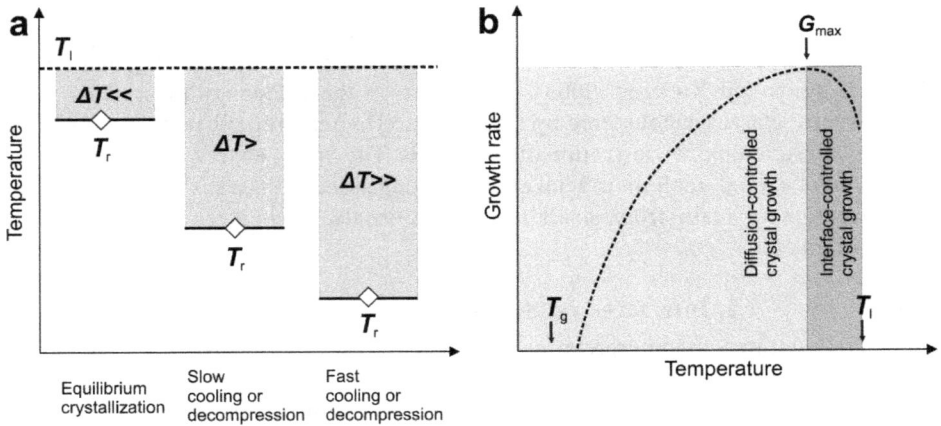

Figure 4. Equilibrium and dynamic crystallization of magma as a function of undercooling (a). The degree of undercooling (ΔT) is the difference between the liquidus temperature (T_l) and the final resting temperature (T_r) of the system. Small-to-large ΔT can be achieved prior to crystallization in response to variable degassing and cooling paths experienced by magmas. Schematic sketch showing the relationship between growth rate and undercooling for a diopsidic liquid derived by Nascimento *et al.* (2004) (b). The undercooling is confined between the glass transition temperature (T_g) and liquidus temperature of the melt ($T_{l,}$). As T_r approaches T_l, the crystal growth rate (G) increases up to a maximum value (G_{max}) corresponding to the shift of the crystal-growth regime from diffusion-controlled to interface-controlled.

crystal surface is homogeneous (Fig. 5a). In contrast, large undercoolings and low T_r (approaching T_g) lead to diffusion-controlled crystal-growth processes where the crystal growth rate exceeds the ability of chemical elements to diffuse through the melt. The crystal texture is anhedral showing a wide spectrum of disequilibrium features (*e.g.* dendritic, skeletal, swallowtail, hopper, sieve-textured, *etc.*), minerals are chemically zoned and compositional gradients develop in the melt next to the crystal surface (Fig. 5b). Thus the attachment and detachment of atoms at the crystal–melt interface exert a primary control on the diffusion-limited crystal growth which occurs especially if component mobility in the melt is low (*i.e.* low cation diffusivity) owing to high melt viscosity. For the stoichiometric (polymorphic) crystallization of the diopsidic liquid, the effective diffusion mechanism controlling diopside crystal growth is related to the slow diffusivity of Si and O with respect to Ca and Mg (D_{Ca}, $D_{Mg} \gg D_{Si}$, D_O). Silicon and O diffuse at almost the same rate through the melt next to the crystal boundary and their slow mobility is rate-limiting for Ca and Mg (*cf.* Nascimento *et al.*, 2004). Relative to chemical species with weaker cation–oxygen bonds (*i.e.* network-modifier cations), the addition of tetrahedral groups (*i.e.* network-former cations) to the surface of a crystal nucleus is the rate-controlling step of the crystallization reaction. The crystal growth rate is limited by the ability of melt components to diffuse towards or away from the crystallizing front. Chemical elements incompatible in the crystal lattice are rejected by the advancing crystal surface and concentrated in the diffusive boundary layer surrounding the mineral. Unlike long-lived intrusive bodies and metamorphic

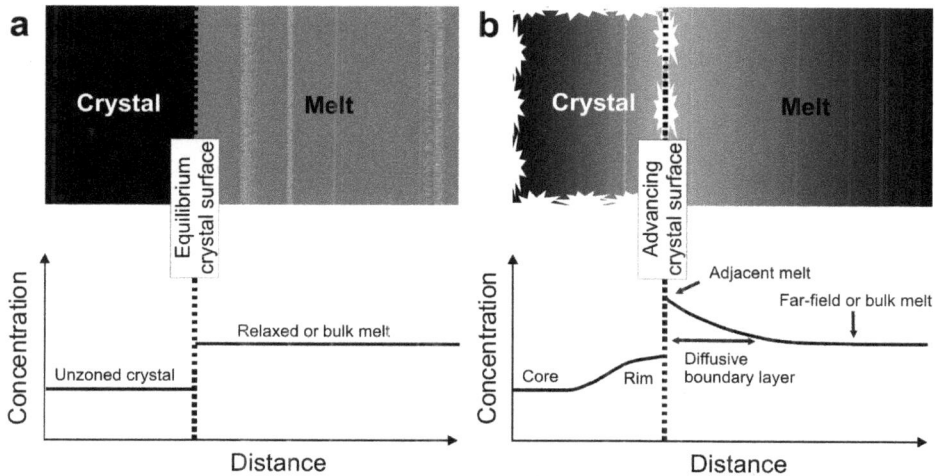

Figure 5. Equilibrium crystallization due to interface-controlled crystal growth (a). The diffusion of chemical elements in the melt prevails over the crystal growth rate. Euhedral and unzoned crystals are surrounded by a compositionally homogeneous melt. Disequilibrium crystallization due to diffusion-controlled crystal growth (b). Rapidly growing crystals are anhedral and chemically zoned, as well as compositional gradients develop in the melt next to the crystal surface.

reactions, fast crystal growth during rapid magma cooling and decompression precludes lattice diffusion as a cause of enrichments of cations in the near-surface or interfacial regions of crystals (*e.g.* Watson and Liang, 1995).

2.1. Crystal–melt boundary layer *vs.* olivine chemistry

Several authors investigating basaltic melt compositions (Kushiro, 1974; Kring and McKay, 1984; Kring, 1985; Thornber and Huebner, 1985) observe that diffusion profiles adjacent to rapidly growing olivines are depleted in MgO and FeO, and enriched in CaO, TiO_2, SiO_2 and Al_2O_3 (Fig. 6a). The width of the enrichment and depletion zones decreases in the order MgO > CaO > FeO > TiO_2 > SiO_2 ≈ Al_2O_3, reflecting different cation diffusivities in the melt (*e.g.* Zhang, 2010). As a first approximation, MgO and FeO in the boundary zone decrease by ~30% and ~10%, respectively, whereas CaO and Al_2O_3 increase by ~10% and ~20%, respectively (Kring, 1985). The diffusive boundary layer is prevalently depleted in MgO rather than FeO, suggesting that the melt more easily supplies fresh Fe cations to the advancing crystal surface (*i.e.* $D_{Fe} > D_{Mg}$). Similarly, the smaller increase of CaO relative to Al_2O_3 and SiO_2 at the crystal–melt interface indicates that the mobility of Ca in the melt is faster than that of network-former cations (*i.e.* $D_{Ca} > D_{Al}$). This diffusion-controlled process is corroborated by dynamic experiments conducted on a trachybasalt cooled from the liquidus down to 1150, 1125 and 1000°C at 1 and 900°C/h (Conte *et al.*, 2006). As the quench temperature decreases, the concentration of MgO in olivine decreases, counterbalanced by increasing FeO and CaO contents (Fig. 6b). In contrast, the

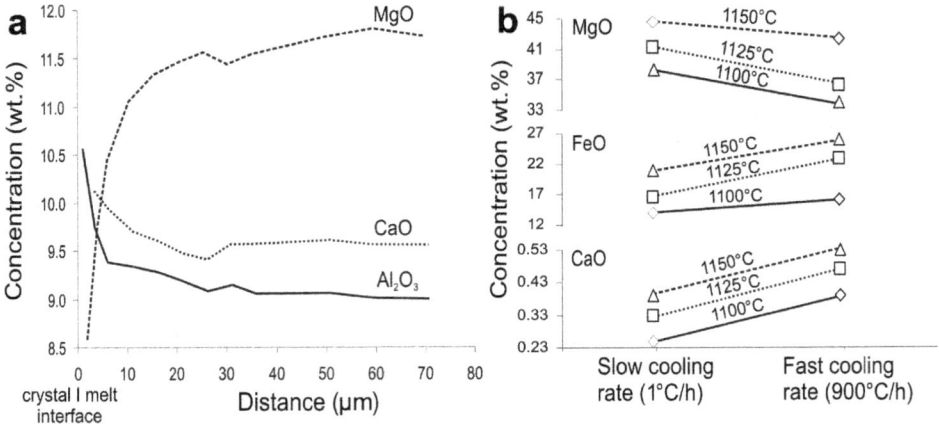

Figure 6. The rapid growth of olivine is accompanied by element diffusion in the melt adjacent to the crystal surface showing depletions in MgO and FeO, and enrichments in CaO (a). After Kring and McKay (1984) in which a synthetic low-Ti basalt was heated above its liquidus at atmospheric pressure, cooled to below olivine saturation (1200–1250°C), held at constant temperature for up to 1000 h, and air-quenched at >600°C/min. With increasing cooling rate, MgO decreases in olivine, whereas FeO and CaO invariably increase whether the experimental charge is quenched at 1100, 1125 or 1150°C (b). Data from Conte *et al.* (2006) in which a trachybasalt was melted for 3 h above the liquidus temperature, then cooled to 1150, 1125 and 1100°C at linear cooling rates of 900°C/h (fast cooling rate) or 1°C/h (slow cooling rate), then held at these temperatures for 12 h, and finally quenched at atmospheric pressure. If not observable, error bars are within symbols.

development of a chemically depleted boundary zone at the crystal–melt interface means that MgO is less favourably incorporated in olivine with increasing cooling rate. To compensate this cation deficiency, FeO and CaO increase remarkably (Fig. 6b) through a cation redistribution in which the amount of forsterite (Fo: Mg_2SiO_4) component decreases in favour of larger amounts of fayalite (Fa: Fe_2SiO_4), Ca-bearing monticellite (Mtc; $CaMgSiO_4$), and kirschsteinite (Kir; $CaFeSiO_4$) schematically illustrated as:

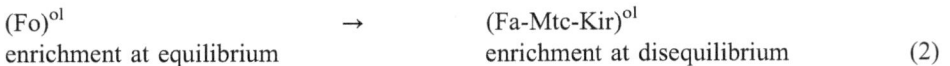

$(Fo)^{ol}$ $\rightarrow$ $(Fa\text{-}Mtc\text{-}Kir)^{ol}$

enrichment at equilibrium   enrichment at disequilibrium   (2)

Equation 2 implies that, in natural crystallization conditions, growth at large undercoolings may produce disequilibrium olivines that are more evolved than required by equilibrium with the host magma (Maaløe and Hansen, 1982; Welsch *et al.*, 2013). In this regard, the composition of diffusion-controlled olivine growth trends in the opposite sense from diffusion-controlled growth of other magmatic minerals. Rapidly grown clinopyroxene, plagioclase and titanomagnetite tend to form compositions that are more primitive than the corresponding equilibrium compositions (Mollo *et al.*, 2012b).

2.2. Crystal–melt boundary layer *vs.* clinopyroxene chemistry

In the late 1970s and early 1980s, many laboratory studies were conducted to better understand the influence of undercooling on clinopyroxene textural and compositional variations (Lofgren *et al.*, 1974; Walker *et al.*, 1976; Grove and Bence, 1977; Grove and Raudsepp, 1978; Gamble and Taylor, 1980; Shimizu, 1981; Kouchi *et al.*, 1983; Baker and Grove, 1985; Tsuchiyama, 1985). The overall finding was that departure from equilibrium arises mainly because of a difference in the rate of crystal–melt interface advance and the rate of chemical diffusion in the melt. In cases where clinopyroxene was surrounded by homogeneous melt, suggesting interface-control over crystal growth, the sluggishness of short-range diffusion of Al was nevertheless capable of causing disequilibrium in the solid (Grove and Raudsepp, 1978; Gamble and Taylor, 1980; Mollo *et al.*, 2010, 2012b). This result is explained by the presence of a thin diffusive boundary layer enriched in chemical species less compatible in clinopyroxene (*i.e.* Al_2O_3, TiO_2 and Na_2O), and relatively depleted in highly compatible chemical elements (*i.e.* FeO, CaO and MgO) (Fig. 7a). Concentration-dependent partitioning produces clinopyroxene growth layers that respond to the chemical gradients in the melt, producing compositions that are progressively enriched in ^{IV}Al, Ti and Na and depleted in Fe^{2+}, Ca and Mg with increasing cooling rate (Fig. 7b). Therefore, under dynamic crystallization conditions, diopside (Di: $MgCaSi_2O_6$) and hedenbergite (Hd: $FeCaSi_2O_6$) components both decrease, whereas the opposite occurs for ferrosilite (Fs: $FeSiO_3$), enstatite (En: $MgSiO_3$), jadeite (Jd: $NaAlSi_2O_6$), and calcium-, ferri- and titanium-Tschermak components (Ts: $CaAl_2SiO_6$, $CaFe_2SiO_6$ and $CaTiAl_2O_6$, respectively), summarized in the scheme:

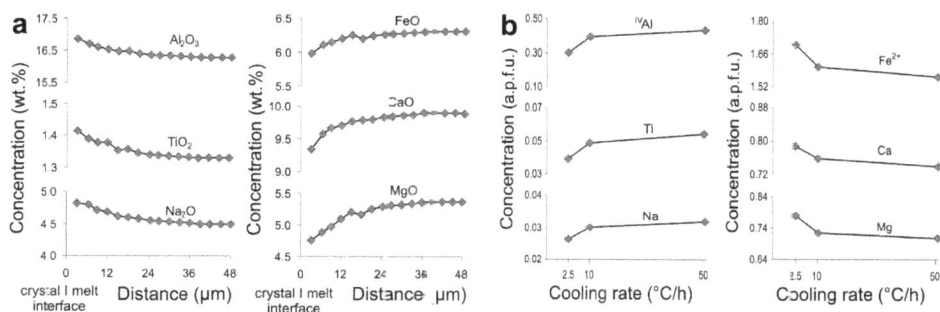

Figure. 7. During rapid growth of clinopyroxene, the diffusive boundary layer in the melt is enriched in chemical species incompatible with the crystal lattice (*i.e.* Al_2O_3, TiO_2 and Na_2O), whereas the concentrations of compatible chemical elements decrease (*i.e.* FeO, CaO and MgO) (a). Disequilibrium clinopyroxene crystals are progressively enriched in ^{IV}Al, Ti and Na counterbalanced by depletions in Fe^{2+}, Ca and Mg with increasing cooling rate (b). After Mollo *et al.* (2013c) in which a trachybasalt was cooled at atmospheric pressure using cooling rates of 2.5, 10 and 50°C/h from 1250°C down to 1100°C. A superliquidus temperature (1250°C) was reached starting from room temperature (with a heating ramp of 100°C/min), and it was maintained for 30 min before the start of cooling. If not observable, error bars are within symbols.

(Di-Hd)^{cpx} →   (Ts-Jd-En-Fs)^{cpx}
enrichment at equilibrium   enrichment at disequilibrium   (3)

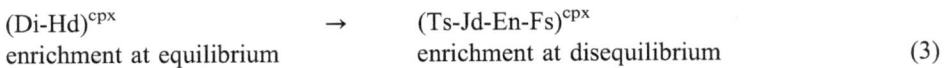

   According to equation 3, clinopyroxene chemistry is controlled by the substitution of ^{M2}(Mg, Fe^{2+}) with ^{M1}(Al, Fe^{3+}) coupled with the substitution of Si with Al in the tetrahedral site to form the Tschermak components. Highly charged cations, such as titanium and trivalent iron, are accommodated in the $M1$ site of clinopyroxene to balance the charge deficiency caused by the increasing concentration of aluminium (Mollo et al., 2013a). In naturally cooled clinopyroxenes, remarkable TiO$_2$ and Al$_2$O$_3$ enrichments have been recognized as a response to rapid crystal growth caused by heat dissipation at the chilled margins of dykes (Ujike, 1982; Mollo et al., 2011b; Scarlato et al., 2014) and the outermost parts of lava flows (Mevel and Velde, 1975; Coish and Taylor, 1979; Mollo et al., 2015), as well as during fast cooling of extraterrestrial materials like lunar basalts (Schnare et al., 2008) and basaltic meteorites (Mikouchi et al., 1999).

2.3. Crystal–melt boundary layer vs. plagioclase chemistry

More than perhaps any other magmatic mineral, plagioclase is characterized by marked sensitivity to pressure, temperature and changes in the concentration of H$_2$O dissolved in the melt. The kinetics of CaAl-NaSi intracrystalline diffusion is extremely slow (Tsuchiyama, 1985) and, consequently, diffusive chemical equilibration for plagioclase crystals 1 to 100 mm in size may require durations longer than the typical lifetimes of magma chambers. Since whole-crystal equilibration by subsolidus diffusion is virtually impossible for plagioclase, the ubiquitous core-to-rim compositional variations observed in natural crystals record the physicochemical perturbations of the magmatic systems. For this reason, numerous cooling and decompression studies explore the effect of undercooling and magma mixing on the crystallization kinetics of plagioclase (Kirkpatrick et al., 1976; Corrigan, 1982; Smith and Lofgren, 1983; Muncill and Lasaga, 1987, 1988; Sato, 1995; Hammer and Rutherford, 2002; Couch, 2003; Pupier et al., 2008; Iezzi et al., 2008, 2011, 2014; Brugger and Hammer, 2010a, 2015; Mollo et al., 2011a,b). If crystal growth-rate is large compared with the rate of component diffusion in the melt, a diffusive boundary layer depleted in compatible elements (Ca and Na), and enriched in relatively incompatible elements (Mg and Fe) results (Fig. 8a). Additionally, the strength of the Si−O bonds in the melt (443 kJ/mol) is greater than that of Al−O bonds (330−422 kJ/mol) due to the influence of charge-balancing metal cations (e.g. Ca, Na, etc.) on the bonding forces of aluminium in tetrahedral coordination and the lower charge of Al^{3+} compared to Si^{4+} (Kirkpatrick, 1983). This difference in bonding energy means that the transfer rates of Al cations from the melt to the surface of rapidly growing crystals is facilitated more than that of Si cations (Iezzi et al., 2011, 2014). Under diffusion-controlled crystal-growth conditions, Si-rich plagioclases are therefore delayed with respect to Si-poor ones, favouring systematic Al enrichments in the crystal lattice (Iezzi et al., 2014). Although the melt supplies Na cations to the advancing plagioclase surface at rates faster than Ca ($D_{Na} > D_{Ca}$), the transport of Al across the crystal–melt interface is rate-limiting for the growth of plagioclase. In order

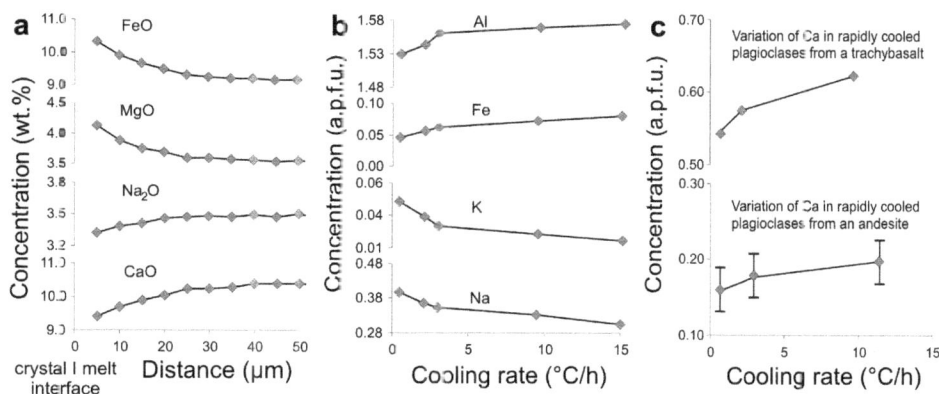

Figure 8. The disequilibrium growth of plagioclase is associated with the development of a diffusive boundary layer in the melt depleted in Ca and Na, and enriched in Mg and Fe (a). The plagioclase crystal lattice is progressively depleted in Si, Na and K, and enriched in Al, Ca, Fe and Mg (b). After Mollo *et al.* (2011a) in which a trachybasalt was held at 500 MPa and a superliquidus temperature of 1250°C for 30 min, and then cooled using cooling rates of 15, 9.4, 3, 2.1 and 0.5°C/min from 1250°C down to 1000°C. Variation of Ca in rapidly cooled plagioclases from trachybasalt and andesitic melts (c). If trachybasaltic and andesitic melts are cooled at comparable rates, plagioclases from the more viscous andesitic liquid become mantled by thick diffusive boundary layers that impinge upon one another as crystallization proceeds. As a consequence, each single crystal competes with its near neighbours for chemical nutrients. For a single cooling rate, calcium in andesitic plagioclases shows variations much greater than those observed for trachybasaltic crystals. Data from Mollo *et al.* (2011a) and Iezzi *et al.* (2014) in which an andesite was held at atmospheric pressure and 1300°C, and then cooled at 12.5, 3 and 0.5°C/min to the final quenching temperature of 800°C. If not observable, error bars are within symbols.

to compensate for the cation-charge imbalance due to Al-Si substitution, plagioclase becomes preferentially enriched in Ca rather than Na, also incorporating Fe and Mg at non-stoichiometric proportions (Mollo *et al.*, 2011a).

Undercooling exerts a powerful control on the composition of plagioclase crystals. Consequently, the ions available for uptake by growing plagioclase causes Si, Na and K to decrease progressively in plagioclase, whereas Al, Ca, Fe and Mg increase (Fig. 8b). However, with respect to primitive magmas, cooling and decompression experiments conducted on more evolved andesitic and rhyodacitic melts (quenched at relatively low temperatures) have indicated strong calcium variations in plagioclases formed from the same charge (Brugger and Hammer, 2010a; Iezzi *et al.*, 2014). It is apparent that the compositional variability of plagioclase depends on the magnitude of concentration gradients in the melt, which in turn scale with undercooling. On the other hand, the diffusivity of chemical elements scales with the melt viscosity that, in turn, changes as a function of melt differentiation and temperature. A thick diffusive boundary layer develops more favourably around crystals growing from high-viscosity (and/or low-temperature) melts and this thickness gradually increases as long as crystallization continues (Watson and Müller, 2009). The combination of an enlarged boundary layer and a fast growth rate may overwhelm the control of linear cooling rate on mineral chemistry leading to the formation of highly heterogeneous crystals. If trachybasaltic and andesitic melts are cooled at comparable rates, plagioclases from the more viscous andesitic liquid become mantled by thick diffusive boundary layers that impinge upon one another as crystallization proceeds. Diffusion effects extend completely across the melt regions among growing plagioclases so that each single crystal competes with its near neighbours for chemical nutrients. This competitive-diffusion mechanism is more intense where plagioclases cluster in close proximity to one another, but is less intense where crystals are relatively widely and evenly spaced. As a consequence, for a single cooling rate, calcium in andesitic plagioclases shows variations much greater than those observed for trachybasaltic crystals (Fig. 8c).

Experimentally derived plagioclase compositions also indicate that the rearrangement of the Al-Si tetrahedral framework and the occurrence of more calcic crystals both imply that the amount of anorthite component (An: $CaAl_2Si_2O_8$) increases at the expense of albite (Ab: $NaAlSi_3O_8$) and orthoclase (Or: $KAlSi_3O_8$) through the scheme (Mollo *et al.*, 2011b):

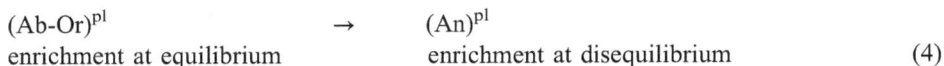

$$(Ab\text{-}Or)^{pl} \qquad\qquad \rightarrow \qquad (An)^{pl}$$
enrichment at equilibrium   enrichment at disequilibrium   (4)

It is interesting to note that the direction of equation 4 is the same, regardless of whether cation redistributions are driven by equilibrium or disequilibrium crystallization conditions. Indeed, increasing temperature and/or melt-water concentration reduce the activity of both Ab and Or relative to An in plagioclase. Therefore, in naturally cooled magmas, it may be difficult to distinguish effects of disequilibrium that are induced by rapid growth rate from effects of closed-system crystallization, which have an identical influence on plagioclase components (Singer *et al.*, 1995; Kent *et al.*, 2010). The difficulty of assessing and quantifying disequilibrium growth of plagioclase can be mitigated by considering the non-stoichiometric uptake of incompatible elements (*e.g.* Mg and Fe) and/or detailed crystal textural analyses. For example, core-to-rim compositional profiles in plagioclase phenocrysts are important for distinguishing thermally and chemically closed magmatic systems from open systems, constraining patterns of magmatic heat loss and convection, and quantifying the rates at which magma ascent occurs. In this framework, abrupt increases in Mg and Fe measured in plagioclase phenocrysts may reveal episodes of rapid disequilibrium due to undercooling at pre-eruptive magma chamber conditions and/or during magma ascent to the surface (Singer *et al.*, 1995).

2.4. Crystal–melt boundary layer *vs.* titanomagnetite chemistry

Only a few laboratory studies have investigated the effect of undercooling on the crystallization of accessory minerals typical of igneous rocks, such as titanomagnetite (Mollo *et al.*, 2012a,b). The rapid growth of titanomagnetite is closely related to the development of a diffusive boundary layer enriched in Al_2O_3 and MgO, and depleted in

TiO$_2$ and FeO (Fig. 9a). These latter components are both compatible in titanomagnetite, but chemical diffusion does not supply Ti and Fe to the advancing crystal surface at the same rate (*i.e.* $D_{Fe} > D_{Ti}$). Consequently, TiO$_2$ is observed to decrease drastically by ~50% in the diffusive boundary layer, whereas FeO increases by only ~10%. This disequilibrium fractionation between Ti and Fe leads to the growth of TiO$_2$-poor titanomagnetite crystals (Fig. 9b), showing Ti/Fe$_{tot}^{2+}$ ratios that gradually decrease as a function of undercooling (*e.g.* Zhou *et al.*, 2000). Since magmatic titanomagnetites exhibit a complete solid solution between magnetite (Mt: Fe$_3$O$_4$) and ulvöspinel (Usp: Fe$_2$TiO$_4$) end-members, this kinetically controlled cation redistribution can be expressed as:

$$(Usp)^{timt} \rightarrow (Mt)^{timt}$$

enrichment at equilibrium   enrichment at disequilibrium   (5)

It is worth stressing that the Usp-Mt replacement described by equation 5 parallels the preferential stability of Mt during equilibrium crystallization of titanomagnetite at higher temperature, except that it is more exaggerated (Mollo *et al.*, 2013b). By analogy with the disequilibrium growth of plagioclase discussed above, the incorporation of Al and Mg cations into the titanomagnetite crystal lattice is the best evidence of departure

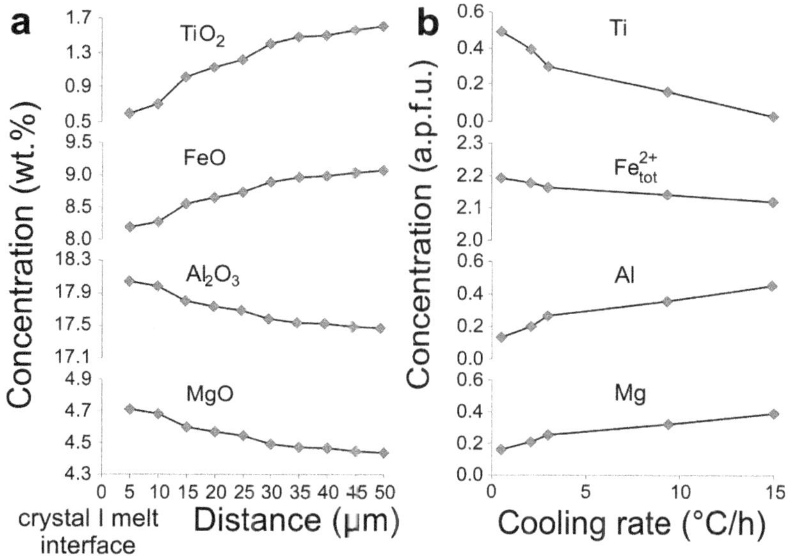

Figure 9. The rapid growth of titanomagnetite leads to the development of a diffusive boundary layer enriched in Al$_2$O$_3$ and MgO, and depleted in TiO$_2$ and FeO (a). After Mollo *et al.* (2013b) in which a trachybasalt was held at 500 MPa and a superliquidus temperature of 1250°C for 30 min, and then cooled using cooling rates of 15, 9.4, 3, 2.1 and 0.5°C/min from 1250°C down to 1000°C. With increasing cooling rate, Ti and Fe decrease in the titanomagnetite crystal lattice, whereas Al and Mg increase (b). Data from Mollo *et al.* (2013b). If not observable, error bars are within symbols.

from equilibrium under diffusion-controlled crystal growth conditions (Fig. 9b). It is not a coincidence that euhedral Ti-rich, Al-Mg-poor titanomagnetites are found generally in the innermost and less efficiently cooled parts of dykes and lavas, whereas anhedral Ti-poor, Al-Mg-rich crystals characterize the outermost and rapidly cooled portions (Zhou *et al.*, 2000; Mollo *et al.*, 2011b).

3. Effect of crystal growth rate on element partitioning

When crystal growth shifts from interface-controlled to diffusion-controlled the transport of chemical elements in the melt is not sufficiently rapid to supply nutrients in the same proportion as exists in the far-field melt. A diffusive boundary layer develops in the melt adjacent to the crystal surface showing composition different from that of the bulk melt far from the crystal surface where chemical gradients cease and the system returns to homogeneous concentrations. An important consequence of this process is that the partition coefficient measured between the advancing crystal surface and the diffusive boundary layer interface changes substantially with the effect of chemical gradients in the melt. The partition coefficient is expressed as:

$$K_i = C_i^{xls}/C_i^{melt} \tag{6}$$

where C_i^{xls} is the concentration of a chemical element i in the crystal and C_i^{melt} is the concentration of the same element in the melt (both crystal and melt concentrations expressed as wt.%). In order to understand better some magmatic reactions, it is also convenient to use the exchange partition coefficient that represents the partitioning of two elements, i and j, between crystal and melt:

$$K_d(i-j)^{xls-melt} = K_i/K_j = (C_i^{xls}/C_i^{melt})/(C_j^{xls}/C_j^{melt}) = (C_i/C_j)^{xls}/(C_i/C_j)^{melt} \tag{7}$$

When the compositions of both crystal and melt result from interface-controlled growth and are supposed to be in equilibrium with respect to element exchange at the crystal–melt interface, equations 6 and 7 provide the 'true' (or 'actual' or 'real') partition coefficient K_i^{true}. In contrast, the effect of diffusion-controlled growth is to change K_i^{true} in an 'apparent' (or 'effective') partition coefficient K_i^{app} when the chemistry of the advancing crystal surface reflects the chemical concentrations in the diffusive boundary layer rather than the composition of the original bulk melt. It is surprising to find that, in studies dealing with magmatic processes, K_i^{app} is calculated as the ratio between: (1) the concentrations of an element at a given analytical spot on the crystal and matrix glass; (2) the compositions of a crystal rim and residual glass unaffected by chemical gradients; and (3) the averaged chemistry of a zoned crystal and the composition of the host melt represented by the bulk-rock analysis. Apart from this variety of calculation methods, Albarede and Bottinga (1972) provided the most complete definition for K_i^{app}, as each possible partition coefficient that differs from the value measured "between a freshly deposited crystalline layer and the portion of the melt right adjacent to this layer". Thus, during development of diffusion-controlled reactions, the simple partition coefficient K_i calculated exactly at the crystal–melt interface will represent the condition in which the advancing crystal surface is in near local equilibrium with the adjacent melt. When a steady state condition is achieved at the crystal–melt interface, both breakage and formation of chemical bonds proceed at the same rate, irrespective of the ever-changing compositions of the zoned crystal and diffusive melt layer. Any other situation will invariably lead to the attainment of an infinite number of K_i^{app} that more or less approximate to K_i^{true}, depending on the relative magnitudes of the crystal growth rate and chemical diffusivity in the melt. It is also expected that fast-diffusing species will show the least difference between K_i^{app} and K_i^{true}, whilst slow-diffusing species will show the greatest difference. Taking all of the above into account, in the present usage the term K_i refers to the measurement of element partitioning between crystal rim and coexisting melt (*i.e.* element partitioning at the crystal–melt interface), besides the fact that a certain uncertainty may be introduced by the spatial resolution of the analytical spot size.

3.1. Olivine–melt partitioning

Using experimental data from Conte *et al.* (2006), the partitioning of Mg, Fe and Ca between rapidly growing olivine crystals and a trachybasaltic melt have been calculated. Figure 10a shows that K_{Mg}, K_{Fe} and K_{Ca} increase by ~3%, ~10% and ~30%, respectively, with increasing cooling rate. Calcium is highly incompatible with olivine and its concentration rises in the contacting melt. The composition of the advancing crystal surface parallels the gradual Ca enrichment of the melt leading to a substantial increase in K_{Ca} as a function of cooling rate. Although both Mg and Fe are compatible in olivine, the increase observed for K_{Fe} is greater than that measured for K_{Mg} due to the preferential supply of fresh Fe cations from the melt to the advancing crystal surface. It is interesting to observe that the value of $K_d(Fe-Mg)^{ol-melt}$ does not change substantially at fast cooling rates (Fig. 10a), remaining close to the equilibrium range of 0.30 ± 0.03 estimated for equilibrium crystallization conditions (Roeder and Emslie, 1970; Falloon *et al.*, 2007). Comparable equilibrium values ranging from 0.27 to 0.33 were also found unexpectedly in cooling-rate studies conducted on lunar basalts (Walker *et al.*, 1976; Powell *et al.*, 1980; Kring and McKay, 1984), and terrestrial trachybasalts and basaltic andesites (Baker and Grove, 1985; Conte *et al.*, 2006).

Petrologists and volcanologists often use the Fe-Mg exchange to ascertain whether the olivine–melt pairs chosen for modelling of magmatic processes were in equilibrium at the time of crystallization. However, the fact that the value of $K_d(Fe-Mg)^{ol-melt}$ remains almost constant upon kinetic effects raises many questions about the fidelity of Fe-Mg exchange as a test for true multicomponent equilibrium crystallization and interface-controlled growth. Several motivations have been proposed to explain the tendency of olivine to maintain Fe-Mg equilibrium with the surrounding melt. In natural magmas, olivine is generally the liquidus phase and saturates the melt at a relatively high temperature and low degree of crystallization. The melt viscosity is low and chemical species diffuse quickly. The kinetic growth of olivine occurs in a restricted thermal path that is fairly close to the near-liquidus region of the melt

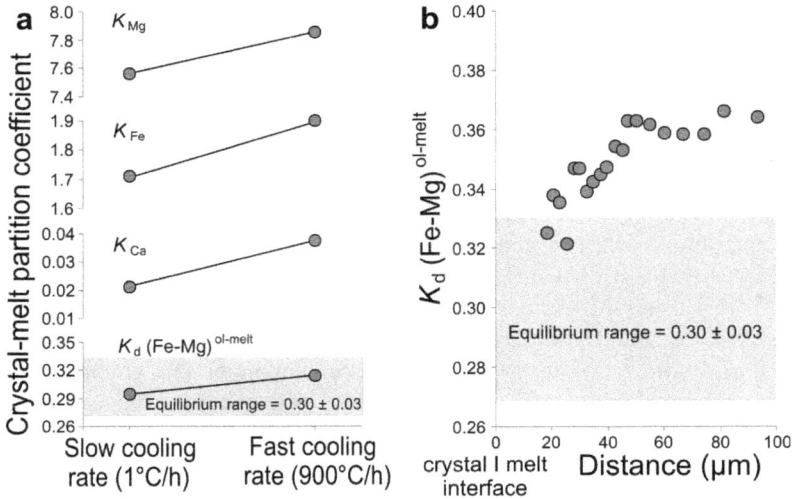

Figure 10. K_{Mg}, K_{Fe}, K_{Ca} and K_d(Fe−Mg)$^{ol−melt}$ measured at the interface between olivine crystals and a trachybasaltic melt are found to increase with increasing cooling rate (a). Notably, the value of K_d(Fe−Mg)$^{ol−melt}$ does not change substantially at fast cooling rates, falling into the equilibrium range of 0.30 ± 0.03. Data from Conte *et al.* (2006) in which a trachybasalt was melted for 3 h above the liquidus temperature, then cooled to 1150, 1125 and 1100°C at linear cooling rates of 900°C/h (fast cooling rate) or 1°C/h (slow cooling rate), then held at these temperatures for 12 h, and finally quenched at atmospheric pressure. K_d(Fe−Mg)$^{ol−melt}$ measured at the crystal−melt interface most closely approaches the equilibrium range but its value increases progressively towards disequilibrium levels as the glass farther and farther from the interface is used for the calculation (b). Data from Powell *et al.* (1980) in which a basalt was cooled from 1198 to 1050°C with a cooling rate of 64°C/h, and then quenched at atmospheric pressure. If not observable, error bars are within symbols.

corresponding to a few degrees of crystallization. These factors suggest that, at the early stage of olivine crystallization, the fresh molecular layer deposited on the surface of a growing crystal maintains Fe-Mg partitioning equilibrium with the contacting melt. Apparently, the development of a boundary layer favours local interface equilibrium (*i.e.* equilibrium of a reaction in a small volume) rather than kinetic effects due to diffusion-limited reactions. The crystal surface remains in equilibrium with the host melt through steady-state chemical gradients established at the onset of olivine nucleation, so that local equilibrium is maintained persistently in terms of Fe-Mg exchange between the advancing crystal surface and the diffusive boundary layer (Kring and McKay, 1984). This is confirmed by microprobe analytical traverses across a basaltic glass surrounding olivine crystals cooled rapidly at 64°C/h from 1198°C to 1050°C (Powell *et al.*, 1980). While the value of K_d(Fe−Mg)$^{ol−melt}$ measured at the crystal−melt interface approaches most closely the equilibrium range, the exchange partition coefficient is found to increase progressively towards disequilibrium levels as the glass further and further from the interface is used for the calculation (Fig. 10b). Interestingly, the skeletal and dendritic morphologies of rapidly growing olivine crystals (Faure *et al.*, 2003, 2007), and strong evidence for early dendritic morphology from phosphorus enrichments (Milman-Barris *et al.*, 2008; Welsch *et al.*, 2014) indicate that the advance of the solid–liquid interface is not within the interface-controlled regime, as explained by stability theory (Mullins and Sekerka, 1963). Small perturbations at the interface generate protrusions from the crystal that are exposed to slightly higher nutrient concentration, promoting their accelerated growth in a positive feedback. Further penetration of the boundary layer by the protrusion supplies fresh chemical components without the need for long-range chemical diffusion (Walker *et al.*, 1976). Evidently, morphologic manifestation of diffusion control is decoupled from Fe-Mg exchange equilibrium, perhaps because these elements are less unlikely to be fractionated by differences in melt diffusivity within the boundary layer.

3.2. Clinopyroxene–melt partitioning

The Fe-Mg exchange is also used as a test for equilibrium for the crystallization of clinopyroxene from magma. The equilibrium value of $K_d(Fe-Mg)^{cpx-melt}$ is assumed to be in the range of 0.27 ± 0.03, but is not perfectly invariant due to a slight temperature dependency (*cf.* Putirka, 2008). As for the case of olivine, the Fe-Mg exchange cannot be assumed to be a reliable indicator of clinopyroxene-melt equilibrium due to the fact that diffusion-controlled compositions yield a fairly close correspondence of $K_d(Fe-Mg)^{cpx-melt}$ to the equilibrium range (Baker and Grove, 1985; Conte *et al.*, 2006; Mollo *et al.*, 2010, 2012b). With increasing cooling rate, Mollo *et al.* (2013a) found that the Fe-Mg exchange coefficient measured at the crystal–melt interface depicts an asymptotic trend fitted by an exponential rise to a maximum model (Fig. 11a). This suggests that the increasing K_{Fe} and K_{Mg} values inevitably tend towards an almost constant K_{Fe}/K_{Mg} ratio when the crystal growth rate is much greater than the mobility of both Fe and Mg in the melt. The constancy of $K_d(Fe-Mg)^{cpx-melt}$ reflects the achievement of local interface equilibrium, especially at high temperature where clinopyroxene nucleation and growth occurs close to the near-liquidus region of magma.

The development of local equilibrium interface reactions also supports the hypothesis that incorporation of rare earth elements (*REE*) into rapidly growing clinopyroxene is controlled by the local charge-balanced configurations, bearing in mind the extent of disorder across the $M1$, $M2$ and T sites (Mollo *et al.*, 2013c). As clinopyroxene crystals grow more rapidly in response to increasing cooling rate, values of K_{REE} increase systematically (Fig. 11b) together with K_{Al} and K_{Na}, whereas K_{Si} and K_{Ca} decrease (Fig. 11c). The entry of *REE* into the $M2$ site is facilitated by a coupled substitution where either Na substitutes for Ca on the $M2$ site or ^{IV}Al substitutes for Si in the tetrahedral site (Blundy *et al.*, 1998; Wood and Blundy, 2001). This latter substitution reflects an increased ease of locally balancing the excess charge at $M2$ as the number of surrounding ^{IV}Al atoms increases (Hill *et al.*, 2000). Due to the lower concentration of Ca in rapidly cooled clinopyroxenes, divalent large ion lithophile elements (LILE) on $M2$ decrease (*i.e.* Ca, Sr and Ba) in concert with a concomitant increase of monovalent cations (*i.e.* Li, Na, Rb). This causes a decoupling effect

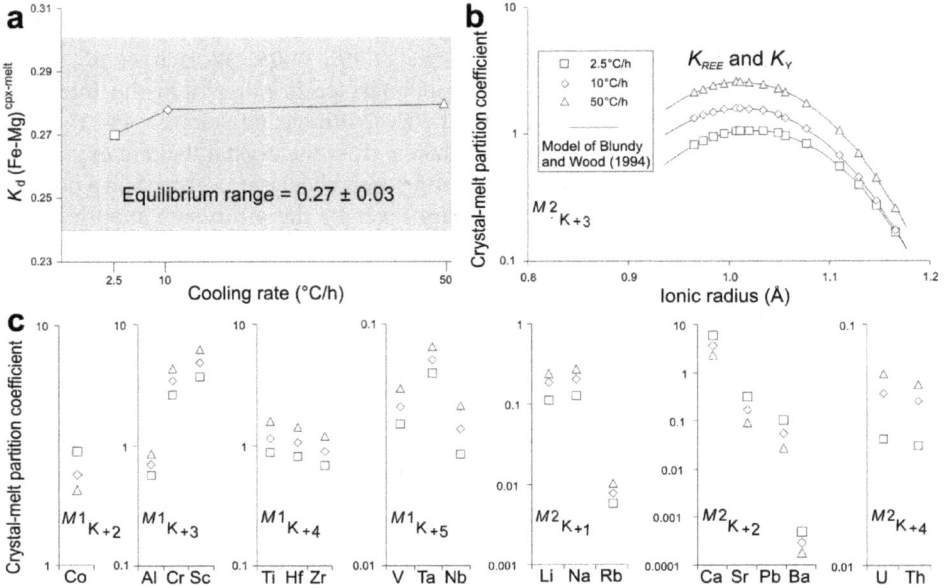

Figure 11. As the cooling rate is increased, the Fe–Mg exchange coefficient, *i.e.* $K_d(\text{Fe–Mg})^{\text{cpx-melt}}$, tends to increase to a maximum value that closely agrees with the equilibrium range of 0.27 ± 0.03 (a). After Mollo *et al.* (2013a) in which a trachybasalt was cooled at atmospheric pressure using cooling rates of 2.5, 10 and 50°C/h from 1250°C down to 1100°C. Plots of partition coefficients for *REE* and Y *vs.* ionic radii as a function of cooling rate (b). The lines show fits to the lattice strain model of Blundy and Wood (1994). After Mollo *et al.* (2013c) in which the starting composition and experimental conditions are the same as reported above. Partition coefficients for trace elements in the *M*1 and *M*2 sites of clinopyroxene as a function of cooling rate (c). Elements are grouped by ionic charge and arranged from smallest to largest ionic radius within each group. After Mollo *et al.* (2013c). If not observable, error bars are within symbols.

between partition coefficients for monovalent and divalent cations where K_{LILE} decrease and increase simultaneously with increasing cooling rate as a function of the cation charge (Fig. 11c). Conversely, high-field-strength elements enter the *M*1 (HFSE: *i.e.* Ti, Zr, Nb and Ta) and K_{HFSE} show a net increase as average charge on this site increases. Charge imbalances are caused by increasing Ca-Al-Tschermak ($CaAl_2SiO_6$) and Ca-ferri-Tschermak ($CaFe_2SiO_6$) substitutions driven by replacement of Mg^{2+} by Fe^{3+} and Al^{3+} into the octahedral site (Wood and Trigila, 2001). Therefore, to maintain charge balance, K_{HFSE} tends always to increase in rapidly cooled clinopyroxenes (Fig. 11c). Upon kinetic effects, the concentration of trivalent transition elements (TE) into the *M*1 site also increases (*i.e.* Cr and Sc), accompanied by greater $^{\text{VI}}Al$ content in the same structural site and the substitution of $^{M2}Fe^{2+}$ for $^{M1}Fe^{3+}$ (Mollo *et al.*, 2013c). Although both K_{Cr} and K_{Sc} increase with increasing cooling rate, K_{Co} shows the opposite behaviour. However, the *M*1 site requires an increased net charge (from 2+ to 3+) to balance the overall increase in aluminium (Hill *et al.*, 2000) and, consequently, trivalent-charged TE cations are incorporated preferentially over divalent ones. It is worth noting that, using the Onuma diagram (Onuma *et al.*, 1968) for *REE* and Y, partition coefficients of isovalent cations *vs.* ionic radius lie on parabola-like curves irrespective of equilibrium or dynamic crystallization conditions (Fig. 11b). In the Onuma diagram, the peak position of the parabola is due to the control of crystal structure on trace-element partitioning where the crystal structure (*i.e.* the size of the site occupied by the cation) is correlated with the crystal composition. As the cooling rate increases, ^{IV}Al in clinopyroxene increases and the parabola shifts upward, opens up more and becomes wider. The lattice strain in the crystal has been identified as the cause of this parabolic trend and Blundy and Wood (1994) provided a quantitative expression relating the variation in K_i with ionic radius to the physical properties of the host crystal. In the lattice strain model, the partitioning of a substituent cation (radius r_i in metres) onto a structural site is related to the site radius (r_0), its apparent Young modulus (E in GPa), and the strain-free partition coefficient K_0 for a fictive cation with radius r_0:

$$K_i = K_0 \exp\{-4\pi E N_A [r_0/2(r_i - r_0)^2 + 1/3(r_i - r_0)^3]/(RT)\} \qquad (8)$$

where N_A is Avogadro's number (6.022×10^{23} mol^{-1}), R is the universal gas constant (8.3145×10^3 kJ mol^{-1}), and T is the temperature (in Kelvin). There is good agreement between the partition coefficients predicted by equation 7 and those measured at both equilibrium and cooling-rate conditions (Fig. 11b). It is also found that the calculated value of r_0 exhibits a weak negative correlation with ^{IV}Al content, whereas K_0 and E increase significantly determining, respectively, the increasing height and curvature of the partitioning parabola (Mollo *et al.*, 2013c). It is important to note that Wood and Blundy (1997) developed also a method of predicting variations in partition coefficients using thermodynamic principles in combination with the lattice strain model. In practice, all *REE* and Y trends depicted in Fig. 11b can be described rigorously thermodynamically and are hence equally valid whether or not the growth of clinopyroxene is driven by equilibrium or disequilibrium exchanges. In light of the lattice strain model, kinetically controlled major cation substitution reactions affect K_i through charge-balance mechanisms comparable to those observed during equilibrium partitioning. Therefore, deviations from equilibrium partitioning are insufficient to change the ability of a trace element to be compatible (or incompatible) within the clinopyroxene crystal lattice and the measured partition coefficient is of the same order of magnitude as the equilibrium value (Fig. 11b). In other words, upon the effect of rapid crystal-growth conditions, local chemical equilibrium at the crystal–melt interface responds to the increasing probability of trace cations entering a locally charge-balanced site when the major cation concentrations change (Mollo *et al.*, 2013c). Rationally, as local equilibrium is achieved between rapidly growing clinopyroxene crystals and coexisting melts, cation substitution reactions can be treated in terms of the energetics of the various charge-imbalanced configurations (*cf.* Wood and Blundy, 2001). For the sake of completeness, it is worth stressing that trace-element partition coefficients many orders of magnitude greater than the equilibrium values have been measured during rapid growth of olivine and clinopyroxene. Pinilla *et al.* (2012) noted that lattice discontinuities or defects of all types (*e.g.* voids and channels) at crystal grain boundaries may serve as important storage sites for trace elements, although most trace elements remain incompatible in the bulk crystal lattice because of unfavourable incorporation energies. Additionally, Kennedy *et al.* (1993) documented that rapidly growing crystals may entrap small portions of the diffusive boundary layer that are found as minute melt inclusions randomly distributed in the mineral phase. However, under such circumstances, any form of trace-element uptake cannot be ascribed to the achievement of local chemical equilibrium. Charge-balanced configurations and the lattice-strain model are substantially inapplicable, and regular relationships between the ionic radius, the valence of the trace element, and the partition coefficient are never found.

3.3. Plagioclase–melt partitioning

In the case of plagioclase, dynamic crystallization experiments reproducing magma cooling and decompression have shown that the partition coefficients of compatible and incompatible elements may deviate from the equilibrium values in proportion to the rates of environmental change (Brugger and Hammer, 2010a; Mollo *et al.*, 2011a). For example, during crystallization of a trachybasaltic melt, K_{Ca}, K_{Na} and K_K are observed to decrease with increasing cooling rate, whereas K_{Fe}, K_{Mg} and K_{Al} increase significantly (Mollo *et al.*, 2011a). This is interpreted to be a consequence of the incompatible character of Fe and Mg in plagioclase causing build-up in a boundary layer and the slow progress of Al across the crystal–melt interface that is rate-limiting for the crystal growth (Fig. 12a). Using the lattice-strain model and thermodynamic principles, Dohmen and Blundy (2014) demonstrate that K_{Ca} and K_{Na} have a strong control on the partitioning behaviour of trace elements between plagioclase and melt. For a wide range of plagioclase and melt compositions, K_{Sr} increases with decreasing K_{Ca}, although Sr substitutes for Ca in plagioclase during magmatic processes. This apparent anomaly is explained in part by the greater elasticity of the albite lattice in which larger Sr cations are more easily accommodated with respect to anorthite-rich plagioclases (Blundy and Wood, 1991). It is important to note that Singer *et al.* (1995) observed that the inverse relationship between K_{Sr} and K_{Ca} is still valid for disequilibrium plagioclase phenocrysts formed from naturally cooled dacitic magmas. The authors modelled the progressive change of K_{Sr}^{app} due to the development of a hypothetical diffusive boundary layer in the melt adjacent to the rapidly growing crystal. K_{Sr}^{app} is expressed as the ratio of the concentration of Sr at a given analytical spot on the crystal to the concentration of Sr in the bulk melt (*i.e.* the bulk-rock analysis). Furthermore, the authors model the variation of K_{Sr}^{app} assuming that the advance of the crystal–melt interface into the crystal-poor dacitic melt does not violate the condition of crystal growth into a non-convecting infinite melt reservoir (Smith *et al.*, 1955):

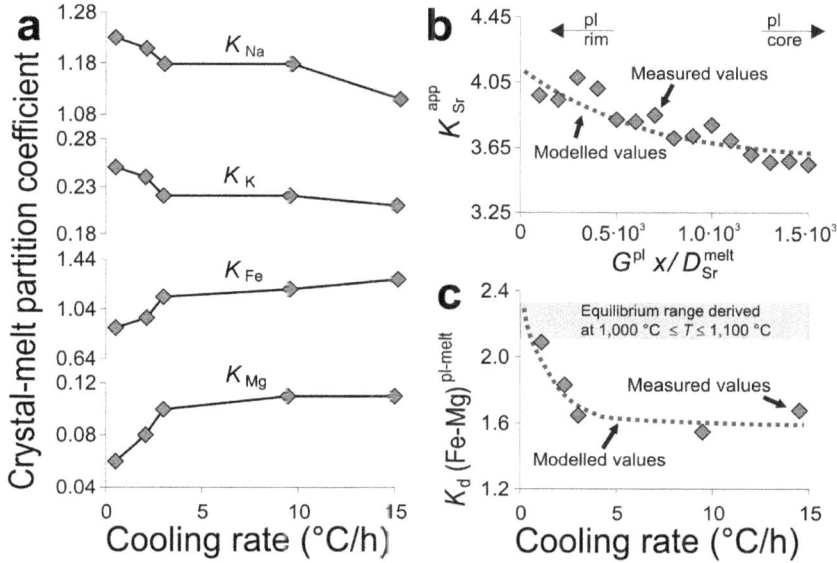

Figure 12. Element partitioning between plagioclase and melt showing that K_{Ca}, K_{Na} and K_K decrease with increasing cooling rate, whereas K_{Fe}, K_{Mg}, and K_{Al} increase (a). After Mollo *et al.* (2011a) in which a trachybasalt was held at 500 MPa and a superliquidus temperature of 1250°C for 30 min, and then cooled using cooling rates of 15, 9.4, 3, 2.1 and 0.5°C/min from 1250°C down to 1000°C. Progressive change of K_{Sr}^{app} due to the development of a hypothetical diffusive boundary layer in the melt adjacent to the rapidly growing plagioclase rim (b). Values 'modelled' using equation 9 and 'measured' for K_{Sr}^{app} are aligned along two almost identical trends, confirming the disequilibrium growth of plagioclase due to magma undercooling. Data from Singer *et al.* (1995). The value of $K_d(Fe-Mg)^{pl-melt}$ measured at the plagioclase−melt interface decreases progressively with increasing cooling rate according to a logarithmic function described by equation 10 (c). After Mollo *et al.* (2011a). If not observable, error bars are within symbols.

$$K_{Sr}^{app} = \frac{1}{2}\left\{1 + \mathrm{erf}\left[\sqrt{G^{pl}x_i/D_{Sr}^{melt}}/2\right]\right\}$$

$$+ (2K_{Sr}^{true} - 1)\exp\left\{\left[-K_{Sr}^{true}(1 - K_{Sr}^{true})(G^{pl}x/D_{Sr}^{melt})\right]\right.$$

$$\left.\mathrm{erfc}\left[(2K_{Sr}^{true} - 1)/2\sqrt{G^{pl}x/D_{Sr}^{melt}}\right]\right\} \qquad (9)$$

where K_{Sr}^{true} is the equilibrium partition coefficient expected during equilibrium crystallization, G^{pl} is the crystal growth velocity, D_{Sr}^{melt} is the diffusivity of element *i* in the melt, and *x* is the distance the crystal interface moves into the liquid phase and corresponds to the linear dimension of the mineral. Equation 9 is used by Singer *et al.* (1995) to model the effect of rapid crystal growth on the compositional zoning of natural plagioclase phenocrysts in terms of the dimensionless parameter $G^{pl}x/D^{melt}_{Sr}$. Figure 12b shows that values 'modelled' and 'measured' for K^{app}_{Sr} are aligned along two almost identical trends, confirming the disequilibrium growth of plagioclase phenocryst rims due to magma undercooling. There are no experimental data in which the effects of P, T, fo_2 and composition are effectively isolated during partitioning of Sr (and other trace elements) between plagioclase and melt at high degrees of undercooling. However, in analogy with the preferential incorporation of Sr in plagioclase, alkali feldspars obtained in the laboratory from a rapidly cooled silicic magma were found to be significantly depleted in Ba leading to lower K_{Ba} with increasing undercooling (Morgan and London, 2003). Besides partition coefficients of Ca and Sr, which are calculated for elements highly compatible with the main solid solution of plagioclase, of particular interest are those of Fe and Mg. These cations occur as minor components in the crystal lattice and, consequently, their partitioning behaviour may potentially reveal the disequilibrium uptake of incompatible elements. If the rate of plagioclase growth is slow enough to ensure that diffusion of elements through the melt maintains near-equilibrium concentrations at the crystal-melt interface, the value of $K_d(Fe-Mg)^{pl-melt}$ is almost constant for a range of isothermal temperatures (Fig. 12c). In contrast, under diffusion-controlled conditions, $K_d(Fe-Mg)^{pl-melt}$ decreases progressively with increasing cooling rate (CR in °C/min) according to the logarithmic function (Mollo et al., 2011a):

$$K_d(Fe-Mg)^{pl-melt} = 2.020 - 0.154(\ln CR + 0.185) \tag{10}$$

Although the fit of the data does not provide a very high correlation coefficient ($R^2 = 0.807$), equation 10 is a first attempt to describe quantitatively the effect of crystallization kinetics on element partitioning. The logarithmic decrease of $K_d(Fe-Mg)^{pl-melt}$ modelled for plagioclase is considerable in comparison with the nominal changes observed during the disequilibrium growth of olivine and clinopyroxene crystals. If chemical species are both highly compatible with the rapidly growing crystals, as in the case of Fe and Mg in olivine (Fig. 10a) and clinopyroxene (Fig. 11a), the achievement of local interface equilibrium between the advancing crystal surface and the diffusive boundary layer does not produce significant changes in the magnitude of the exchange partition coefficient. Conversely, elements incompatible with the crystal lattice can be incorporated into the rapidly growing crystals at non-stoichiometric and non-equilibrium proportions, causing strong deviations from the equilibrium partitioning. The contrasting behaviour observed for the Fe-Mg exchange between plagioclase and mafic minerals reflects the importance of incompatible elements as monitors of magma disequilibrium crystallization and their potential for calibrating new and more powerful tests for equilibrium.

4. Implications of disequilibrium partitioning on modelling magma processes

Igneous thermometers and barometers are important tools for estimating the saturation pressures and temperatures of minerals, as well as for elucidating magma storage conditions and transport (*e.g.* Ghiorso and Evans, 2008; Putirka, 2008). Generally, these equations are calibrated using experimental data obtained under equilibrium conditions, and are based on a number of crystal and melt parameters that may influence differently the ability prediction of the models. Other equations are derived using only the crystal components as predictors or, alternatively, the composition of the melt in equilibrium with the mineral of interest. Despite the different model calibrations, igneous thermometers and barometers nearly always assume that the investigated natural compositions approach equilibrium. Eruptive products are commonly associated with magma dynamics in which minerals attempt to re-equilibrate over a range of temperatures and pressures. Moreover, crystals may grow rapidly and incorporate elements from the melt in abundances that do not represent equilibrium partitioning. Given a crystal and melt (host lava or matrix glass) composition, it is not a trivial task to ascertain whether the two are appropriate for use in thermobarometry.

In Fig. 13a clinopyroxene–melt pairs from cooling rate experiments of Mollo *et al.* (2013a) are used as input data for two different barometric equations derived by Putirka (2008). The first barometer is based on both clinopyroxene and melt components. With increasing cooling rate, the error of estimate increases slightly but its value does not exceed the model error of ±200 MPa. In contrast, the second barometer is primarily dependent on Al_{tot} and calcium-Tschermak components in clinopyroxene. The

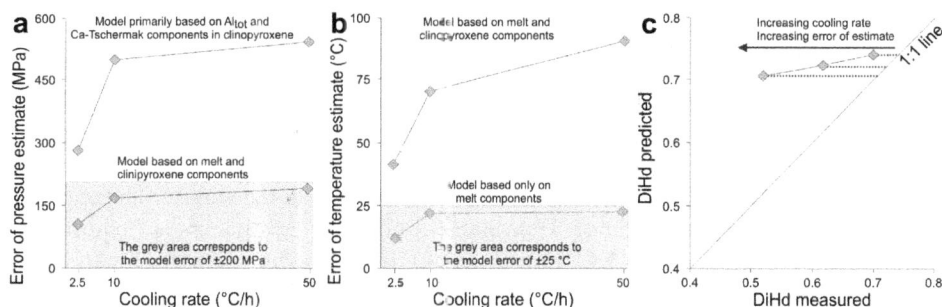

Figure 13. Clinopyroxene–melt pairs from cooling rate experiments of Mollo *et al.* (2013a) are used as input data for two different barometric equations derived by Putirka (2008) (a). The first barometer is based on both clinopyroxene and melt components, yielding a small error of estimate. The second barometer is primarily dependent on Al_{tot} and calcium-Tschermak components in clinopyroxene, yielding a large error of estimate. Cooling rate experiments of Mollo *et al.* (2013a) are also used as input data for two different thermometric equations proposed by Putirka (2008) (b). The thermometer calibrated using only Si, Mg and Ca cation fractions in the melt is affected by a small error of estimate. In contrast, the thermometer derived using both clinopyroxene and melt components, is affected by a large error of estimate. The equilibrium model proposed by Mollo *et al.* (2013a) is based on the difference between diopside+hedenbergite (ΔDiHd) components predicted for clinopyroxene *via* regression analysis of clinopyroxene–melt pairs in equilibrium conditions, and those measured in the analysed crystals (c). From a theoretical point of view, the equilibrium condition is achieved when the clinopyroxene compositions plot on the one-to-one line of the "DiHd measured" *vs.* "DiHd predicted" diagram. As the cooling rate increases, the experimental compositions of Mollo *et al.* (2013a) progressively depart from the one-to-one line as a consequence of disequilibrium crystallization conditions.

prediction ability of the model is evidently affected by a systematic error of estimate that substantially increases with increasing cooling rate. As discussed above, the disequilibrium growth of clinopyroxene is controlled by the substitution of $M2$(Mg, Fe^{2+}) with $M1$(Al, Fe^{3+}) coupled with the substitution of Si for Al in the tetrahedral site to form the Tschermak components. Consequently, the predictive power of the Tschermak-based barometer is remarkably poor, accounting for the kinetically controlled cation redistributions in clinopyroxene. As a further test, the cooling rate experiments of Mollo et al. (2013a) are also used as input data for two different thermometric equations proposed by Putirka (2008). Figure 13b shows that the thermometer calibrated using only Si, Mg and Ca cation fractions in the melt yields a relatively small error of estimate, and within the model error of ±25°C. This suggests that the error of estimate is minimized when the thermometer reflects the slow-to-intermediate diffusivity of some chemical species in the melt over the timescale of rapid crystal growth. On the other hand, if the thermometer is derived using both clinopyroxene and melt components, its predictive power is more sensitive to disequilibrium growth conditions, and the error of estimate increases systematically with increasing cooling rate. As expected, this is due to the combined effect of the development of a diffusive boundary layer in the melt and the preferential incorporation into the growing crystal of chemical components in non-equilibrium proportions.

Motivated to determine the depth of the magmatic plumbing system beneath a Hawaiian volcano in its post-shield stage of development, Hammer et al. (2016) apply clinopyroxene-liquid thermobarometry to an ankaramite from Haleakala, calculating liquids that satisfy Fe-Mg exchange equilibrium by mineral subtraction from the whole rock using mass balance. An earlier application of clinopyroxene thermobarometry to similar rocks from Haleakala (Chatterjee et al., 2005) concluded that polybaric crystallization began 45 km below the volcano's summit, in the mantle. The approach was to assume that growth of Haleakala clinopyroxenes occurs at equilibrium conditions and to transform Na X-ray concentration maps into pressure maps using control points obtained using thermometry (Putirka, 2008). The pressure range corresponding to intracrystalline compositional variation is 60–950 MPa. However, the presence of Na-rich domains associated with euhedral facets in contact with matrix is not consistent with concentric growth at near-equilibrium conditions of decreasing pressure, but rather co-crystallization of both domains under conditions of partial disequilibrium. Analysis of the spatial pattern of element covariation (Welsch et al., 2016) suggests that the crystals are sector zoned, and thus represent kinetic effects associated with rapid growth. Conservatively assuming that low-Na regions are less prone to kinetic partitioning, crystallization pressures for the Haleakala magmas correspond to crustal levels.

Petrologists often use K_d(Fe–Mg)$^{cpx-melt}$ to ascertain whether the clinopyroxene–melt pairs chosen for the estimate of magmatic pressures and temperatures were in equilibrium at the time of crystallization. However, it has long been demonstrated that Fe-Mg exchange has certain limitations and is not a reliable indicator of clinopyroxene–melt equilibrium for a wide range of compositions (*e.g.* Putirka, 2008). With this in mind, Mollo *et al.* (2013a) performed a new global regression analysis that improves upon one of the most suitable models that test for equilibrium, as presented in Putirka (1999). The equilibrium model is based on the difference between diopside + hedenbergite (ΔDiHd) components predicted for clinopyroxene *via* regression analysis of clinopyroxene-melt pairs in equilibrium conditions, and those measured in the analysed crystals. The value of ΔDiHd provides a more robust test for equilibrium than $K_d(Fe-Mg)^{cpx-melt}$ due to the fact that deviation from equilibrium occurs in response to the kinetically controlled exchange of equation 3, as diopside and hedenbergite are consumed to form ferrosilite, enstatite, jadeite and Tschermak components. From a theoretical point of view, the equilibrium condition is achieved when the clinopyroxene compositions plot on the one-to-one line of the 'DiHd measured' *vs.* 'DiHd predicted' diagram. At this condition, the value of ΔDiHd goes to zero minimizing the error of estimate caused by the accidental use of disequilibrium clinopyroxene–melt pairs as input data for the thermobarometric models (Mollo and Masotta, 2014). Figure 13c shows that, as the cooling rate increases, experimental data from Mollo *et al.* (2013a) progressively depart from the one-to-one line. As a consequence, the value of ΔDiHd increases from 0.05 to 0.19 revealing disequilibrium crystallization conditions that, in turn, cause substantial increase of the error of estimate for pressure (Fig. 13a) and temperature (Fig. 13b).

5. Textural outcomes

5.1. Decompression-driven crystallization in H_2O + CO_2 bearing systems

Decompression-driven feldspar crystallization is the focus of several recent experimental studies involving rhyolite and rhyodacite liquid compositions. The studies differ with respect to decompression path type, position of the initial dwell pressure with respect to the equilibrium plagioclase stability curve, the presence or absence of plagioclase crystals in the melt at the start of decompression, and fluid composition. Brugger and Hammer (2010a) examined the control of decompression rate on progressive feldspar crystallization in natural Aniakchak (Alaska) rhyodacite under conditions of H_2O-only fluid saturation (referred to hereafter as 'H_2O-saturated'), executing continuous decompression at rates of 10, 2, 1 and 0.5 MPa h^{-1}. Cichy *et al.* (2011) decompressed synthetic rhyodacite in contact with both H_2O-only and H_2O+CO_2 fluids starting from below the plagioclase-in curve and proceeding to the quench pressure of 50 MPa at a wide range in rates (72,000, 3,600, 360, 36, 3.6, 1.8 and 0.72 MPa h^{-1}). Using synthetic $CaO-Na_2O-Al_2O_3-SiO_2$ haplotonalite melt (CNAS rhyolite), Mollard *et al.* (2012) explored the effect of very rapid initial decompression (at rates of 1200, 100 and 30 MPa h^{-1}) on subsequent isobaric crystallization. Martel (2012) studied both isobaric and syn-decompression crystallization of synthetic multicomponent rhyolite relevant to the interstitial liquid in Mt. Pelée (MP) andesite in continuous and multistep experiments performed at 12 decompression rates ranging from 1500 to 0.18 MPa h^{-1}. The experiments included pairs of H_2O-only and H_2O+CO_2 fluid-bearing charges; at the initiation of decompression, the former were crystal-free while the latter contained 3 vol.% plagioclase. Waters *et al.* (2015) prepared initially crystal-free obsidian for relatively slow, continuous decompression (2.9, 1.0, 0.8 and 0.1 MPa h^{-1}) from an initial dwell pressure that was either above or below the plagioclase stability curve at the run temperature. Riker *et al.* (2015) decompressed synthetic rhyodacite relevant to Mount Saint Helens (MSH) interstitial liquid that was in contact with either H$_2$O-only or H$_2$O+CO$_2$ fluid, all containing initial crystals, along disparate paths but with similar total durations.

The decompression paths of these six studies are summarized in plots of log decompression time *vs.* pressure (Figs 14, 15 and 16), which resemble classical time-

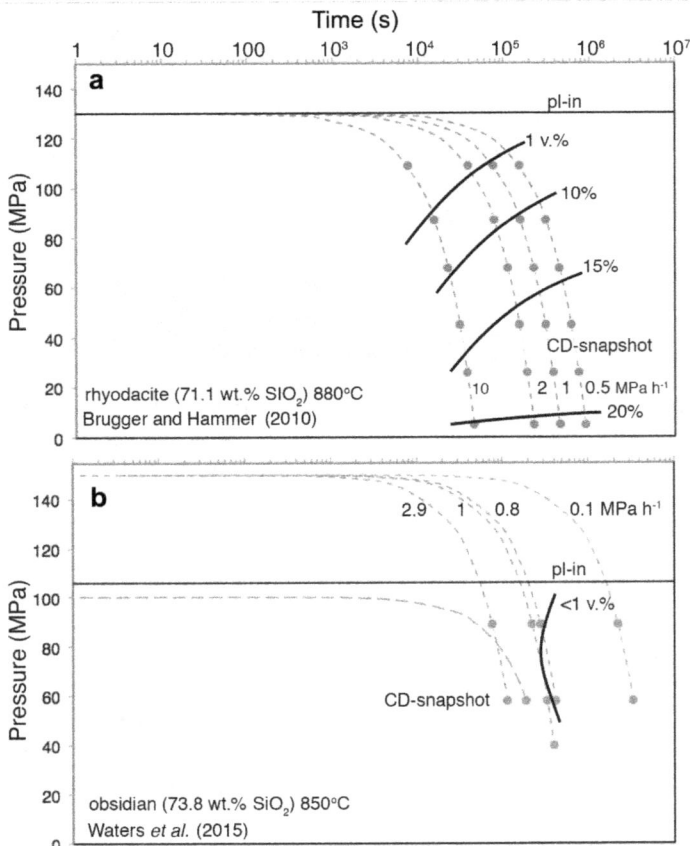

Figure 14. Experimental decompression paths employed by Brugger and Hammer (2010a) and Waters *et al.* (2015) in panels (a) and (b), respectively. In (b), experiment initiated at 100 MPa (6 MPa below plagioclase-in curve) was performed at the same decompression rate (0.8 MPa h^{-1}) as one of the superliquidus experiments. Although these charges spent the same amount of time below the liquidus these paths do not line up vertically on the plot because the charge that was decompressed from lower initial pressure reached the final pressure sooner.

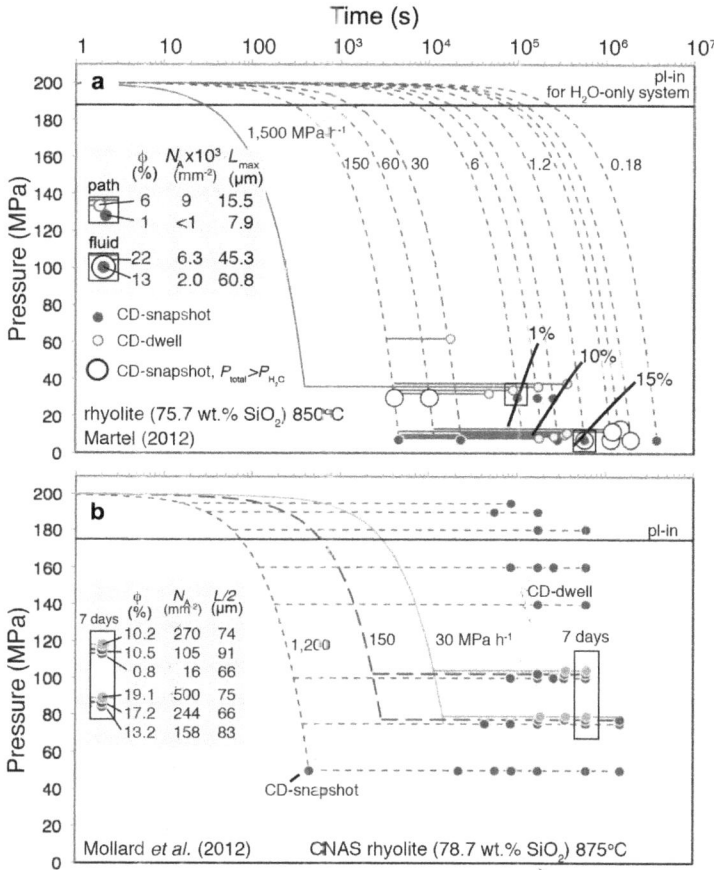

Figure 15. Experimental decompression paths employed by Martel (2012) and Mollard *et al.* (2012) in panels (a) and (b), respectively. Experiments in which cooling occurred in final steps are not shown in panel (a). In (b), the constant volume % plagioclase isograds are drawn using data from CD-snapshot experiments only. Note difference in units of N_A; feldspar number densities are $\geqslant 10\times$ smaller in the experiments of Martel (2012). Crystal lengths are reported by the authors as L_{max} and $L/2$ in the units shown.

temperature-transformation (TTT) diagrams. Constructed from experiments in which temperature is changed nearly instantaneously and then allowed to dwell isothermally, TTT plots allow phase-transformation kinetics to be compared among different materials (*e.g.* Putnis and Bish, 1983; Hammer and Rutherford, 2002; Bowles *et al.*, 2012; Vetere *et al.*, 2013). A similar approach, substituting P_{H_2O} for temperature, was taken in constructing an isobaric crystallization diagram for rhyolite interstitial liquid in natural dacite (Hammer and Rutherford, 2002). Because most of the experiments described here are not isobaric, the plots are not true transformation diagrams, but they do provide a convenient means for comparing experiments that span many orders of

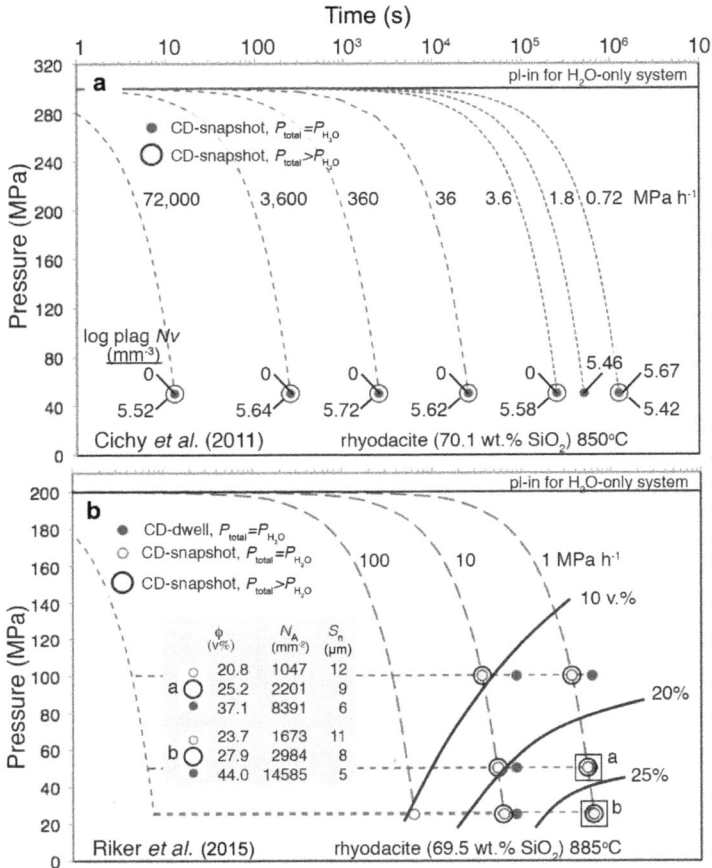

Figure 16. Experimental decompression paths employed by Cichy *et al.* (2011) and Riker *et al.* (2015) in panels (a) and (b), respectively. In (a), the long-dash indicates decompression was conducted continuously. The short dash line pattern indicates multistep decompression (step increment not reported).

magnitude in time scale. Two decompression path types dominate the recent studies. The first type is continuous or quasi-continuous decompression, with single-step decompression (SSD) being the end-member of rapid decompression, followed by a dwell period. This general type is termed CD-dwell, and was utilized by Martel (2012), Mollard *et al.* (2012) and Riker *et al.* (2015). The second type, termed CD-snapshot, is continuous decompression terminated without a final dwell period. The CD-snapshot style was applied by Brugger and Hammer (2010a), Cichy *et al.* (2011), Martel (2012) and Waters *et al.* (2015).

The results from the four studies employing CD-snapshot experiments at H_2O-saturated conditions are contoured for the amount of plagioclase present in run products (Figs 14, 15 and 16), allowing comparison of different materials subjected to similar decompression path times. The crystallization onset time, related to the glass-forming ability of a material (Vetere *et al.*, 2015), is resolvable from data sets that contain runs without any crystals. For example, crystal-free obsidian decompressed at 850°C from above the plagioclase stability curve (Waters *et al.*, 2015) is more sluggish to begin crystallizing than synthetic rhyolite (Martel, 2012) at the same temperature and decompression rate, as evident from the shorter times represented by the 1 vol.% plagioclase isograds at 30–40 MPa. Continuous decompression at 1 MPa h^{-1} to 75 MPa produces ~1 vol.% plagioclase in obsidian (Fig. 14b), ~12 vol.% plagioclase in Aniakchak rhyodacite (Fig. 14a), and 20 vol.% in the MSH rhyodacite (Fig. 16b). The more rapid solidification in the latter may be due to decompression from initially higher dissolved H_2O and thus lower melt viscosity. In addition, the obsidian results exhibit a 'knee' or minimum in the time at which crystallization begins (over the applied decompression rates) at 75 MPa. Both above and below this pressure, crystallization takes longer to start. In contrast, the natural Aniakchak rhyodacite and synthetic MSH rhyodacite exhibit progressive decrease in the crystallization onset time and monotonic increase in the volume % plagioclase as final pressure decreases. Higher temperature in the latter runs may play some role in this difference, but a more likely pair of related factors is the initial effective superheating and lack of pre-existing crystals in the obsidian experiments. The obsidian data set resembles the "C" shape of transformation curves in which pre-existing growth sites or nuclei are absent; the disappearance or softening of the lower half of the C may be characteristic of systems in which nucleation is not required (*e.g.* Hammer and Rutherford, 2002).

The CD-dwell experiments of Mollard *et al.* (2012) isolate the effect of decompression rate on ensuing crystallization, employing a variety of rates that are so rapid as to preclude syn-decompression crystallization, and then observing the system after dwell times of 2, 4, 7 and 17 days. Melts decompressed at 1200 MPa h^{-1} have consistently smaller feldspar abundances and smaller crystal-number densities than melts decompressed at 150 or 30 MPa h^{-1}. The results suggest a failure of melt to relax structurally during the decompression (in turn calling into question the Maxwell relaxation time (equation 1) as adequately capturing the 'structure' relevant to nucleation), lower rate of crystal nucleation, and insufficient dwell duration for rapidly decompressed materials to respond. This is interesting in the context of paired CD-dwell and CD-snapshot experiments on MP rhyolite (Martel, 2012) and MSH (Riker *et al.*, 2015) rhyodacite in which decompression rate and dwell times are selected to produce similar total time below initial pressure and isolate decompression path as a variable. As was reported for Pinatubo rhyolite interstitial melt (Hammer and Rutherford, 2002), both investigations observe greater crystallinities in the CD-dwell runs than in the rapidly decompressed CD-snapshot runs, and attribute these differences to the higher initial undercooling. Why, if higher initial undercooling stimulates solidification, does the 1200 MPa h^{-1} decompression not produce greater crystallinity than the 150 or 30 MPa h^{-1} runs in CNAS rhyolite (Mollard *et al.*, 2012)? Although the decompression rate applied to the MSH rhyodacite (80,000 MPa h^{-1}) was much higher than was applied to CNAS rhyolite (1200 MPa h^{-1}), the MSH rhyodacite achieved equilibrium feldspar crystallinity after a one-day dwell time at 50 MPa in a nucleation-dominated regime. Evidently, either structural relaxation required for crystal nucleation was faster in the rhyodacite, because of its lower SiO_2 or higher temperature, or the initial superheating of the CNAS rhyolite proved a critical impediment to subsequent crystallization.

5.2. Beyond H_2O and CO_2: Adding other volatiles

Recent experiments broaden the range of fluids in contact with melt during decompression to include species in the H_2O-CO_2-Cl system. Martel (2012) incorporates CO_2 in the MP rhyolite experiments to reduce the amount of dissolved H_2O and drive crystallization during the pre-decompression dwell. By final pressure of 5–10 MPa, charges with pre-decompression crystals have greater crystal-number densities and smaller mean sizes than the initially crystal-free charges, particularly at low decompression rates (Fig. 15a, inset). These features point toward higher rates of nucleation, and thus greater undercooling, in the samples with pre-existing crystals. However, noting the tabular (as opposed to anhedral) morphologies of syn-decompression crystals in the charges containing initial crystals (Martel, 2012) we can infer that the role of the pre-existing crystals is to decrease the degree of undercooling by providing surfaces on which crystallization occurs.

Cichy *et al.* (2011) and Riker *et al.* (2015) modulate the mole fraction of H_2O in the fluid ($X_{H_2O}^{fl}$) from 0.6 to 1.0 and from 0.8 to 1.0 by adding CO_2 in CD-snapshot experiments on Unzen rhyodacite and MSH rhyodacite, respectively (Fig. 16). Both studies also report lower mean feldspar microlite sizes and greater crystal-number densities (with negligible nucleation at the five fastest decompressions in the Unzen experiments and by a factor of 2 in the MSH experiments) in the H_2O-CO_2 runs. The existence of crystals prior to decompression cannot be implicated in these cases, since both capsule preparations (CO_2-free and CO_2-bearing) include pre-existing crystals. Riker *et al.* (2015) attributed the difference to an increase in the temperature of the nucleation rate maximum in the CO_2-bearing material, and yet another factor more likely contributing to the observed textural differences is the effect of lowering dissolved H_2O content, through the addition of CO_2, on the phase relations of the material at the initial dwell conditions. Decreasing dissolved H_2O at constant pressure has the same effect on plagioclase stability as decreasing H_2O during H_2O-saturated decompression: the plagioclase-in temperature increases (Hammer *et al.*, 2002; Médard and Grove, 2008; Fiege *et al.*, 2015). Taking the MP rhyolite as an example, phase equilibrium data under H_2O-saturated conditions (Martel, 2012) indicate that at the plagioclase-in pressure (188 MPa at 850°C), the melt has 5.8 wt.% dissolved H_2O. A reduction to 4.7 wt.% by adding CO_2 has the effect of reducing P_{H_2O} to ~132 MPa. In order to equilibrate (achieve zero undercooling) during the initial dwell period, ~5 wt.% plagioclase needs to nucleate and crystallize. Assuming the system does equilibrate, subsequent decompression drives dissolution of this feldspar. By the time the system reaches 132 MPa, the equilibrium system is crystal-free. Thus, decompression from an initially H_2O-undersaturated state is thermally equivalent to rapidly supercooling, heating to the liquidus, and then cooling the material. Given the influence of fluctuations about the liquidus for the distribution of crystal nuclei (Mills *et al.*, 2011; Mills and Glazner, 2013), it is not surprising that the crystallinity and texture of these final products differ from H_2O-saturated decompression from the liquidus. In addition, the lower dissolved H_2O at all pressures along the decompression path and consequently higher viscosity and instantaneous undercooling are consistent with the higher rate of crystal nucleation in the CO_2-bearing experiments observed by Cichy *et al.* (2011), Martel (2012) and Riker *et al.* (2015).

The presence of crystals at the start of decompression remains an important consideration, related but potentially separable from the degree of melt undercooling. Given the slow rates at which crystals form in evolved liquids, it is possible to initiate an experiment from below the plagioclase-in stability curve and yet have no crystals in the starting material, as is the case in the obsidian decompression experiments (Waters *et al.*, 2015). Future work manipulating initial crystallinity with and without varying the initial dissolved H_2O content will help resolve the contribution of the fluid composition.

The difficulty containing and moderating oxidation in higher-temperature melts within cold-seal pressure vessels suppresses the experimental pursuit of crystallization kinetics studies relative to high silica compositions. Fiege *et al.* (2015) studied crystallization of natural fused trachybasalt from the 2001 eruption of Mt. Etna, applying isothermal continuous decompression rates of 3600, 360 and 36 MPa h^{-1} from 300 to 70 MPa using a low-flow pressure metering valve for the internally heated pressure vessel (Nowak *et al.*, 2011). In general, a higher rate of decompression results in lower crystallinity (Fig. 17). Replicate decompression histories were applied to fluid-saturated melts coexisting with either H_2O+Cl only, $H_2O+Cl+CO_2$, and $H_2O+Cl+CO_2+S$, with the unexpected outcome that the S-bearing samples contain considerably more clinopyroxene (up to 50 vol.%) than the S-free runs (20 vol.%), which match the equilibrium abundances calculated by MELTS thermodynamic simulations (~23 vol.%; Gualda *et al.*, 2012). The run products contain nearly identical dissolved H_2O contents, ruling out a role for H_2O in modulating phase relations. The suggestion by Fiege *et al.* (2015), that a reduction in the activity of H_2O within the fluid phase accounts for the greater clinopyroxene content of the S-bearing system, requires further investigation.

5.3. Cooling-driven crystallization in nominally anhydrous systems

Vetere *et al.* (2015) augment the experimental data set of Vetere *et al.* (2013), which examines the effects of major-element composition, cooling rate, and thermal history on solidification kinetics in oxidized, nominally anhydrous silicate melts (Fig. 18). Together, these studies encompass six cooling rates (1, 7, 620, 3020, 8020 and 9000°C h^{-1}) applied to six glasses homogenized along a mixing line between Icelandic basalt to Aeolian rhyolite from superliquidus conditions. A particular focus is characterizing the glass-forming ability of the materials by determining the minimum cooling rate, or critical cooling rate, Rc, at which <2 vol.% crystallizes. The Rc increases monotonically with increasing basalt component, exhibiting 3.5 orders of magnitude increase between pure

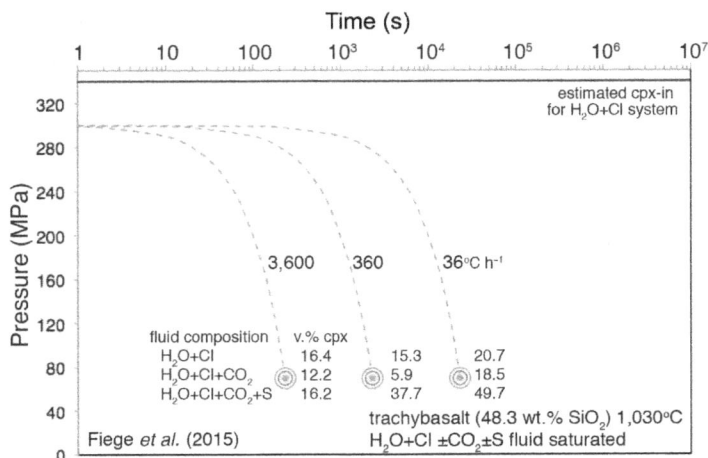

Figure 17. Experimental decompression paths employed by Fiege et al. (2015). Plagioclase-in and clinopyroxene-in pressures are not reported, but plotted as 340 MPa because (a) the H_2O+Cl experiment performed at 300 MPa and 1030°C contained 1.5 wt.% clinopyroxene; (b) The experiments with S and CO_2 in the fluid had more crystals (9 wt.% clinopyroxene), and also included spinel; and (c) MELTS indicates 30 wt.% crystals at 1030°C for the H_2O+Cl system. Either the experiments quenched from initial conditions were not equilibrated or MELTS is inaccurate. Either way, the liquidus is probably higher than 300 MPa and the experiments were initiated from below it. Fiege et al. (2015) report that the equilibrium clinopyroxene content in the H_2O+Cl system at final pressure of 70 MPa is ~23 wt.%.

rhyolite and the 60−40 wt.% hybrid ($Bas_{60}Rhy_{40}$) and an additional order of magnitude increase between $Bas_{60}Rhy_{40}$ and Bas_{100}. The results allow Rc (°C h^{-1}) to be expressed as an exponential function of the ratio of non-bridging to tetrahedrally coordinated cations, or NBO/T: (Mysen and Richet, 2005):

$$Rc = 9214/\{1 + \exp-[(NBO/T - 0.297)/0.04]\} \quad (11)$$

Vetere et al. (2015) suggested that parameterizations of Rc involving NBO/T may also describe the solidification kinetics of alkalic melts, although Iezzi et al. (2008) stressed that small differences in composition produce dramatic differences in Rc in their study of alkali melts. They prepared latite from the Fondo Riccio vent (Campi Flegrei) and trachyte from Agnano Monte Spina using similar procedures to the Vetere et al. (2013, 2015) studies. Differences in the SiO_2 (lower by 2.3 wt.% in the latite) and FeO_T (larger by 3.7 wt.%) are probably responsible for the formation of clinopyroxene in the latite at cooling rates of ⩽30°C h^{-1}, echoing an earlier result from cooling experiments on evolved silicic melts (Naney and Swanson, 1980). Interestingly the sub-alkaline andesite from Panarea (Aeolian Islands) studied by Iezzi et al., (2011) begins to crystallize at more rapid cooling rates than the latite or trachyte, but is more sluggish to complete solidification during slow cooling (Fig. 18b). Understanding the structural and compositional controls over the kinetics of crystallization in the last tens of weight% interstitial liquid may require a similarly intense empirical effort.

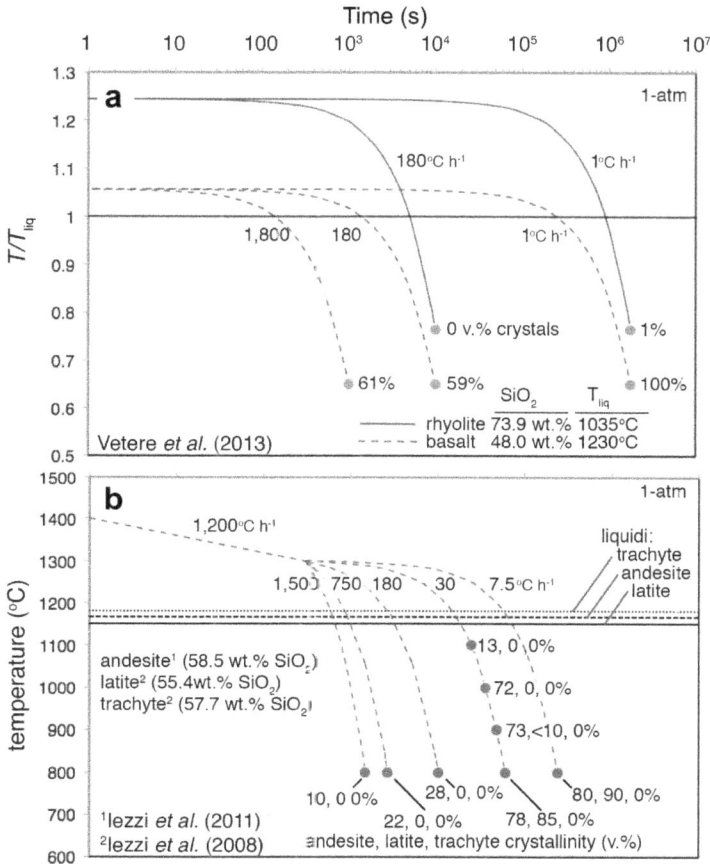

Figure 18. Experimental cooling paths employed by Vetere *et al.* (2013), in panel (a), and Iezzi *et al.* (2008, 2011) in panel (b). In (a), only the two end-member compositions of a suite of six compositions ranging from rhyolite to basalt are included. Temperatures are normalized by the liquidus temperatures in order to compare extents of superheating. The crystallinity for basalt at middle and fastest rates is the average of replicate experiments.

6. Selected applications to volcanological problems

Experimental studies may be designed to resolve a specific petrological or volcanological problem involving the environmental conditions or time scale of a magmatic event. For example, Waters *et al.* (2015) concluded that the melt which ascended to produce crystal-poor rhyolite obsidian must have separated from a multiply-saturated magma at slightly H_2O-undersaturated conditions, as this is the simplest way for the liquid to have been initially superheated in nature. In essence, a constraint from crystal nucleation kinetics is used to address the problem of obsidian petrogenesis. Martel (2012) resolved the magma ascent times associated with plinian and dome/block and ash flow eruptions of Mt. Pelée to <1 h and 2−5 days, respectively, having characterized the plagioclase textural and compositional characteristics in experiments spanning a wide range in decompression paths. In a similar approach, Andrews and Gardner (2010) study grey and white pumice from the 1,800 ^{14}C yr BP eruption of Ksudach Volcano, Kamchatka, using H_2O-saturated continuous decompression experiments and find that microlite textures observed in the white pumice require decompression rates of 0.01 MPa s^{-1}, whereas the higher crystallinities and more numerous microlites of grey pumice require decompression at ~0.0025 MPa s^{-1}. Andrews (2014) compares the microlite number densities in dacite from Santa Maria Volcano, Guatemala to decompression-texture relationships from Andrews and Gardner (2010) to constrain magma ascent rate of ~0.005 to 0.01 MPa s^{-1} during the 1902 Santa Maria eruption. Cichy *et al.* (2011) constrain the ascent rate of 1991–1995 Unzen magma by comparing natural and experimental plagioclase microlite size. The natural glass compositions and microlite lengths are reproduced in H_2O-saturated experiments at decompression rates ≤ 0.0005 MPa s^{-1} or ~50 m h^{-1} ascent; decreasing the H_2O activity reduces these rates. The studies described here provide encouraging basis for interpreting natural magmatic storage, ascent, and emplacement processes using experimental calibrations of crystal texture and composition. However, it is important to note the limitations of applicability associated with extrapolation beyond the bulk and phase compositions explicitly represented in an experimental data set, as well as the influence of starting material preparation and other elements of experiment design that influence the final result.

Acknowledgements

This chapter benefited greatly from thoughtful reviews and comments by Laura E. Waters and Harald Behrens. The authors are grateful for the essential editorial guidance of Wilhelm Heinrich and Rainer Abart. J. Hammer acknowledges NSF award EAR-1347887 and the Laboratoire Magmas et Volcans, University of Clermont-Ferrand, France, for support during preparation of this contribution. This is SOEST publication no. 9825.

References

Albarede, F. and Bottinga, Y. (1972) Kinetic disequilibrium in trace-element partitioning between phenocrysts and host lava. *Geochimica et Cosmochimica Acta*, **36**, 141−156.

Andrews, B.J. (2014) Magmatic storage conditions, decompression rate, and incipient caldera collapse of the 1902 eruption of Santa Maria Volcano, Guatemala. *Journal of Volcanology and Geothermal Research*, **282**, 103−114.

Andrews, B.J. and Gardner, J.E. (2010) Effects of caldera collapse on magma decompression rate: An example from the 1800 14C yr BP eruption of Ksudach Volcano, Kamchatka, Russia. *Journal of Volcanology and Geothermal Research*, **198**, 205−216.

Arzilli, F. and Carroll, M.R. (2013) Crystallization kinetics of alkali feldspars in cooling and decompression-induced crystallization experiments in trachytic melt. *Contributions to Mineralogy and Petrology*, **166**, 1011−1027.

Baker, D.R. (2008) The fidelity of melt inclusions as records of melt composition. *Contributions to Mineralogy and Petrology*, **156**, 377−395.

Baker, M.B. and Grove, T.L. (1985) Kinetic controls on pyroxene nucleation and metastable liquid lines of descent in a basaltic andesite. *American Mineralogist*, **70**, 279−287.

Berkebile, C.A. and Dowty, E. (1982) Nucleation in laboratory charges of basaltic composition. *American Mineralogist*, **67**, 886−899.

Blundy, J.D. and Wood, B.J. (1991) Crystal-chemical controls on the partitioning of Sr and Ba between plagioclase feldspar, silicate melts and hydrothermal solutions. *Geochimica et Cosmochimica Acta*, **55**, 193−209.

Blundy, J.D. and Wood, B.J. (1994) Prediction of crystal−melt partition coefficients from elastic moduli. *Nature*, **372**, 452−454.

Blundy, J.D., Robinson, J.A.C. and Wood B.J. (1998) Heavy REE are compatible in clinopyroxene on the spinel lherzolite solidus. *Earth and Planetary Science Letters*, **160**, 493−504.

Bowles, J.A., Tatsumi-Petrochilos, L., Hammer, J.E. and Brachfeld, S.A. (2012) Multicomponent cubic oxide exsolution in synthetic basalts: Temperature dependence and implications for magnetic properties. *Journal of Geophysical Research: Solid Earth*, **117**, B03202.

Brugger, C.R. and Hammer, J.E. (2010a) Crystallization kinetics in continuous decompression experiments: Implications for interpreting natural magma ascent processes. *Journal of Petrology*, **51**, 1941−1965.

Brugger, C.R. and Hammer, J.E. (2010b) Crystal size distribution analysis of plagioclase in experimentally decompressed hydrous rhyodacite magma. *Earth and Planetary Science Letters*, **300**, 246−254.

Brugger, C.R. and Hammer, J.E. (2015) Prevalence of growth twins among anhedral plagioclase microlites. *American Mineralogist*, **100**, 385−395.

Cabane, H., Laporte, D. and Provost, A. (2005) An experimental study of Ostwald ripening of olivine and plagioclase in silicate melts: Implications for the growth and size of crystals in magmas. *Contributions to Mineralogy and Petrology*, **150**, 37−53.

Chatterjee, N., Bhattacharji, S. and Fein, C. (2005) Depth of alkalic magma reservoirs below Kolekole cinder cone, Southwest rift zone, East Maui, Hawaii. *Journal of Volcanology and Geothermal Research*, **145**, 1−22.

Cichy, S.B., Botcharnikov, R.E., Holtz, F. and Behrens, H. (2011) Vesiculation and microlite crystallization induced by decompression: a case Study of the 1991−1995 Mt Unzen Eruption (Japan). *Journal of Petrology*, **52**, 1469−1492.

Coish, R.A. and Taylor, L.A. (1979) The effect of cooling rate on texture and pyroxene chemistry in DSDP Leg 34 basalt: a microprobe study. *Earth and Planetary Science Letters*, **42**, 389−398.

Conte, A.M., Perinelli, C. and Trigila, R. (2006) Cooling kinetics experiments on different Stromboli lavas: effects on crystal morphologies and phases compositions. *Journal of Volcanology and Geothermal Research*, **155**, 179−200.

Corrigan, G.M. (1982) Supercooling and crystallization of plagioclase, olivine and clinopyroxene from basaltic magmas. *Mineralogical Magazine*, **46**, 31−42.

Couch, S., Sparks, R.S.J. and Carroll, M.R. (2003) The kinetics of degassing-induced crystallization at Soufriere Hills volcano, Montserrat. *Journal of Petrology*, **44**, 1477−1502.

Del Gaudio, P., Mollo, S., Ventura, G., Iezzi, G., Taddeucci, J. and Cavallo, A. (2010) Cooling rate-induced differentiation in anhydrous and hydrous basalts at 500 MPa: implications for the storage and transport of magmas in dikes. *Chemical Geology*, **270**, 164−178.

DePaolo, D.J. and Getty, S.R. (1996) Models of isotopic exchange in reactive fluid−rock systems: implications for geochronology in metamorphic rock. *Geochimica et Cosmochimica Acta*, **60**, 3933−3947.

Dingwell, D.B. (1995) Relaxation in silicate melts: Some applications. Pp. 21−66 in: *Structure, Dynamics and Properties of Silicate Melts* (J.F. Stebbins P. McMillan and D.B. Dingwell, editors). Reviews in Mineralogy, **32**. Mineralogical Society of America, Washington, D.C.

Dingwell, D.B. and Webb, S.L. (1990) Relaxation in silicate melts. *European Journal of Mineralogy*, **2**, 427−449.

Dohmen, R. and Blundy, J. (2014) A predictive thermodynamic model for element partitioning between plagioclase and melt as a function of pressure temperature and composition. *American Journal of Science*, **314**, 1319−1372.

Dowty, E. (1980) Crystal growth and nucleation theory and the numerical simulation of igneous crystallization. Pp. 419−485 in: *The Physics of Magmatic Processes* (R.B. Hargraves, editor) Princeton University Press, Princeton, New Jersey, USA.

Falloon, T.J., Danyushevsky, L.V., Ariskin, A., Green, D.H. and Ford, C.E. (2007) The application of olivine geothermometry to infer crystallization temperatures of parental liquids: Implications for the temperatures of MORB magmas. *Chemical Geology*, **241**, 207-233.

Faure, F., Trolliard, G., Nicollet, C. and Montel, J.M. (2003) A development model of olivine morphology as a function of the cooling rate and the degree of undercooling. *Contributions to Mineralogy and Petrology*, **145**, 251–263.

Faure, F., Schiano, P., Trolliard, G., Nicollet, C. and Soulestin, B. (2007) Textural evolution of polyhedral olivine experiencing rapid cooling rates. *Contributions to Mineralogy and Petrology*, **153**, 405–416.

Fiege, A., Vetere, F., Iezzi, G., Simon, A. and Holtz, F. (2015) The roles of decompression rate and volatiles (H_2O + Cl ± CO_2 ± S) on crystallization in (trachy-) basaltic magma. *Chemical Geology*, **411**, 310–322.

Gamble, R.P. and Taylor, L.A. (1980) Crystal/liquid partitioning in augite: effects of cooling rate. *Earth and Planetary Science Letters*, **47**, 21–33.

Ghiorso, M.S. and Evans, B.W. (2008) Thermodynamics of rhombohedral oxide solid solutions and a revision of the Fe-Ti two-oxide geothermometer and oxygen-barometer. *American Journal of Science*, **308**, 957–1039.

Grove, T.L. and Bence, A.E. (1977) Experimental study of pyroxene–liquid interaction in quartz-normative basalt 15597. Pp. 1549–1579 in: *Proceedings of Lunar and Planetary Science Conference, VIII*. Lunar and Planetary Institute, Houston, Texas, USA.

Grove, T.L. and Raudsepp, M. (1978) Effects of kinetics on the crystallization of quartz-normative basalt 15597: an experimental study. Pp. 585–599 in: *Proceedings of Lunar and Planetary Science Conference, IX*. Lunar and Planetary Institute, Houston, Texas, USA.

Gualda, G.R., Ghiorso, M.S., Lemons, R.V. and Carley, T.L. (2012) Rhyolite–MELTS: a modified calibration of MELTS optimized for silica-rich, fluid-bearing magmatic systems. *Journal of Petrology*, **53**, 875–890.

Hammer, J.E. (2006) Influence of f_{O_2} and cooling rate on the kinetics and energetics of Fe-rich basalt crystallization. *Earth and Planetary Science Letters*, **248**, 618–637.

Hammer, J.E. (2008) Experimental studies of the kinetics and energetics of magma crystallization. Pp. 9–59 in: *Minerals, Inclusions and Volcanic Processes* (K.D. Putirka and F.J. Tepley editors). Reviews in Mineralogy and Geochemistry, **69**. Mineralogical Society of America and Geochemical Society, Washington, D.C.

Hammer, J.E. (2009) Capturing crystal growth. *Geology*, **37**, 1055–1056.

Hammer, J.E. and Rutherford, M.J. (2002) An experimental study of the kinetics of decompression-induced crystallization in silicic melt. *Journal of Geophysical Research* **107**.

Hammer, J.E., Rutherford, M.J. and Hildreth, W. (2002) Magma storage prior to the 1912 eruption at Novarupta, Alaska. *Contributions to Mineralogy and Petrology*, **144**, 144–162.

Hammer, J.E., Sharp, T.G. and Wessel, P. (2010) Heterogeneous nucleation and epitaxial crystal growth of magmatic minerals. *Geology* **38**, 367–370.

Hammer, J.E., Jacob, S., Welsch, B., Hellebrand, E. and Sinton, J.M. (2016) Clinopyroxene in postshield Haleakala ankaramite: 1. Efficacy of thermobarometry. *Contributions to Mineralogy and Petrology*, **171**, 7.

Higgins, M.D. and Roberge, J. (2003) Crystal size distribution of plagioclase and amphibole from Soufriere Hills Volcano, Montserrat: Evidence for dynamic crystallization – textural coarsening cycles. *Journal of Petrology* **44**, 1401–1411.

Hill, E., Wood, B.J. and Blundy, J.D. (2000) The effect of Ca-Tschermaks component on trace element partitioning between clinopyroxene and silicate melt. *Lithos*, **53**, 203–215.

Iezzi, G., Mollo, S., Ventura, G., Cavallo, A. and Romano, C. (2008) Experimental solidification of anhydrous latitic and trachytic melts at different cooling rates: The role of nucleation kinetics. *Chemical Geology*, **253**, 91–101.

Iezzi, G., Mollo, S., Torresi, G., Ventura, G., Cavallo, A. and Scarlato, P. (2011) Experimental solidification of an andesitic melt by cooling. *Chemical Geology*, **283**, 261–273.

Iezzi, G., Mollo, S., Shahini, E., Cavallo, A. and Scarlato, P. (2014) The cooling kinetics of plagioclase feldspars as revealed by electron-microprobe mapping. *American Mineralogist*, **99**, 898–907.

Jambon, A., Lussiez, P., Clocchiatti, R., Weisz, J. and Hernandez, J. (1992) Olivine growth rates in a tholeiitic basalt: An experimental study of melt inclusions in plagioclase. *Chemical Geology*, **96**, 277–287.

Kent, A.J.R., Darr, C., Koleszar, A.M., Salisbury, M.J. and Cooper, K.M. (2010) Preferential eruption of andesitic magmas through recharge filtering. *Nature Geoscience*, **3**, 631–636.

Kennedy, A.K., Lofgren, G.E. and Wasserburg, G.J. (1993) An experimental study of trace-element partitioning between olivine, orthopyroxene and melt in chondrules – equilibrium values and kinetic effects. *Earth and Planetary Science Letters*, **115**, 177–195.

Kile, D.E. and Eberl, D.D. (2003) On the origin of size-dependent and size-independent crystal growth: Influence of advection and diffusion. *American Mineralogist*, **88**, 1514–1522.

Kirkpatrick, R.J. (1981) Kinetics of crystallization of igneous rocks. Pp. 321–397 in: *Kinetics of Geological Processes* (A.C. Lasaga and R.J. Kirkpatrick, editors). Reviews in Mineralogy, **8**, Mineralogical Society of America, Washington, D.C.

Kirkpatrick, R.J. (1983) Theory of nucleation in silicate melts. *American Mineralogist*, **68**, 66–77.

Kirkpatrick, R.J., Robinson, G.R. and Hayes, J.F. (1976) Kinetics of crystal growth from silicate melts: anorthite and diopside. *Journal of Geophysical Research*, **81**, 5715–5720.

Kouchi, A., Sugawara, Y., Kashima, K. and Sunagawa, I. (1983) Laboratory growth of sector zoned clinopyroxenes in the system $CaMgSi_2O_2$–$CaTiAl_2O_6$. *Contributions to Mineralogy and Petrology*, **83**, 986–990.

Kring, D.A. (1985) Use of melt inclusions in determining liquid lines of descent. In *American Geophysical Union, Fall Meeting (San Francisco), Abstracts*, Eos, **66**, 404.

Kring, D.A. and McKay, G.A. (1984) Chemical gradients in glass adjacent to olivine in experimental charges and Apollo 15 green glass vitrophyres. Pp. 461–462 in: *Proceedings of Lunar and Planetary Science Conference, XV*. Lunar and Planetary Institute. Houston, Texas, USA.

Kushiro, I. (1974) Melting of hydrous upper mantle and possible generation of andesitic magma: an approach from synthetic systems. *Earth and Planetary Science Letters*, **22**, 294–299.

Lofgren, G.E. and Smith, D.P. (1980) The experimental determination of cooling rates of rocks: some complications. *Lunar and Planetary Science*, **XI**; 631–633, abstract.

Lofgren, G.E. and Lanier, A.B. (1992) Dynamic crystallization experiments on the Angra dos Reis achondritic meteorite. *Earth and Planetary Science Letters*, **111**, 455–466.

Lofgren, G.E., Donaldson, C.H., Williams, R.J., Mullins, O. Jr. and Usselman, T.M. (1974) Experimentally reproduced textures and mineral chemistry of Apollo 15 quartz-normative basalts. Pp. 549–567 in: *Proceedings of Lunar and Planetary Science Conference, V*. Lunar and Planetary Institute, Houston, Texas, USA.

Lofgren, G.E., Huss, G.R. and Wasserburg, G.J. (2006) An experimental study of trace–element partitioning between Ti–Al–clinopyroxene and melt: Equilibrium and kinetic effects including sector zoning. *American Mineralogist* **91**, 1596–1606.

Maaløe, S. and Hansen, B. (1982) Olivine phenocrysts of Hawaiian olivine tholeiite and oceanite. *Contributions to Mineralogy and Petrology*, **81**, 203–211.

Martel, C. (2012) Eruption dynamics inferred from microlite crystallization experiments: Application to Plinian and dome-forming eruptions of Mt. Pelee (Martinique, Lesser Antilles). *Journal of Petrology* **53**, 699–725.

Martel, C. and Schmidt, B.C. (2003) Decompression experiments as an insight into ascent rates of silicic magmas. *Contributions to Mineralogy and Petrology*, **144**, 397–415.

Médard, E. and Grove, T.L. (2008) The effect of H_2O on the olivine liquidus of basaltic melts: experiments and thermodynamic models. *Contributions to Mineralogy and Petrology*, **155**, 417–432.

Mevel, C. and Velde, D. (1976) Clinopyroxenes in Mesozoic pillow lavas from the French Alps: influence of cooling rate on compositional trends. *Earth and Planetary Science Letters*, **32**, 158–164.

Mills, R.D. and Glazner, A.F. (2013) Experimental study on the effects of temperature cycling on coarsening of plagioclase and olivine in an alkali basalt. *Contributions to Mineralogy and Petrology*, **166**, 97–111.

Mills, R.D., Ratner, J.J. and Glazner, A.F. (2011) Experimental evidence for crystal coarsening and fabric development during temperature cycling. *Geology* **39**, 1139–1142.

Mikouchi, T., Miyamoto, M. and McKay, G.A. (1999) The role of undercooling in producing igneous zoning trends in pyroxenes and maskelynites among basaltic martian meteorites. *Earth and Planetary Science Letters*, **173**, 235–256.

Milman-Barris, M.T, Beckett, J.R., Baker, M.B., Hofmann, A.E., Morgan, Z., Crowley, M.R., Vielzeuf, D. and Stolper, E. (2008) Zoning of phosphorus in igneous olivine. *Contributions to Mineralogy and Petrology*, **155**, 739–765.

Mollard, E., Martel, C. and Bourdier, J.L. (2012) Decompression-induced crystallization in hydrated silica-rich melts: Empirical models of experimental plagioclase nucleation and growth kinetics. *Journal of Petrology* **53**, 1743–1766.

Mollo, S. and Masotta, M. (2014) Optimizing pre-eruptive temperature estimates in thermally and chemically zoned magma chambers. *Chemical Geology*, **368**, 97–103.

Mollo, S., Del Gaudio, P., Ventura, G., Iezzi, G. and Scarlato, P. (2010) Dependence of clinopyroxene composition on cooling rate in basaltic magmas: Implications for thermobarometry. *Lithos*, **118**, 302–312.

Mollo, S., Putirka, K., Iezzi, G., Del Gaudio, P. and Scarlato, P. (2011a) Plagioclase–melt (dis)equilibrium due to cooling dynamics: implications for thermometry, barometry and hygrometry. *Lithos*, **125**, 221–235.

Mollo, S., Lanzafame, G., Masotta, M., Iezzi, G., Ferlito, C. and Scarlato, P. (2011b) Cooling history of a dike as revealed by mineral chemistry: A case study from Mt. Etna volcano. *Chemical Geology*, **283**, 261–273.

Mollo, S., Iezzi, G., Ventura, G., Cavallo, A. and Scarlato, P. (2012a) Heterogeneous nucleation mechanisms and formation of metastable phase assemblages induced by different crystalline seeds in a rapidly cooled andesitic melt. *Journal of Non-Crystalline Solids*, **358**, 1624–1628.

Mollo, S., Misiti, V., Scarlato, P. and Soligo, M. (2012b) The role of cooling rate in the origin of high temperature phases at the chilled margin of magmatic intrusions. *Chemical Geology*, **322–323**, 28–46.

Mollo, S., Putirka, K., Misiti, V., Soligo, M. and Scarlato, P. (2013a) A new test for equilibrium based on clinopyroxene–melt pairs: Clues on the solidification temperatures of Etnean alkaline melts at post-eruptive conditions. *Chemical Geology*, **352**, 92–100.

Mollo, S., Putirka, K., Iezzi, G. and Scarlato, P. (2013b) The control of cooling rate on titanomagnetite composition: Implications for a geospeedometry model applicable to alkaline rocks from Mt. Etna volcano. *Contributions to Mineralogy and Petrology*, **165**, 457–475.

Mollo, S., Blundy, J., Scarlato, P., Iezzi, G. and Langone, A. (2013c) The partitioning of trace elements between clinopyroxene and trachybasaltic melt during rapid cooling and crystal growth. *Contributions to Mineralogy and Petrology*, **166**, 1633–1654.

Mollo, S., Giacomoni, P.P., Andronico, D. and Scarlato, P. (2015) Clinopyroxene and titanomagnetite cation redistributions at Mt. Etna volcano (Sicily, Italy): Footprints of the final solidification history of lava fountains and lava flows. *Chemical Geology*, **406**, 45–54.

Moore, G. and Carmichael, I.S.E. (1998) The hydrous phase equilibria (to 3 kbar) of an andesite and basaltic andesite from western Mexico: Constraints on water content and conditions of phenocryst growth. *Contributions to Mineralogy and Petrology*, **130**, 304–319.

Morgan, G.B. and London, D. (2003) Trace-element partitioning at conditions far from equilibrium: Ba and Cs distributions between alkali feldspar and undercooled hydrous granitic liquid at 200 MPa. *Contributions to Mineralogy and Petrology*, **144**, 722–738.

Müller, T., Baumgartner, L.P., Foster, C.T. and Vennemann, T.W. (2004) Metastable prograde mineral reactions in contact aureoles. *Geology*, **32**, 821–824.

Mullins, W.W. and Sekerka R.F. (1963) Morphological stability of a particle growing by diffusion or heat flow. *Journal of Applied Physics*, **34**, 323–329.

Muncill, G.E. and Lasaga, A.C. (1987) Crystal-growth kinetics of plagioclase in igneous systems: one atmosphere experiments and approximation of a simplified growth model. *American Mineralogist*, **72**, 299–311.

Muncill, G.E. and Lasaga, A.C. (1988) Crystal-growth kinetics of plagioclase in igneous systems: isothermal H_2O-saturated experiments and extension of a growth model to complex silicate melts. *American Mineralogist*, **73**, 982–992.

Mysen, B.O. and Richet, P. (2005) *Silicate Glasses and Melts: Properties and Structure*. Elsevier, Amsterdam.

Nabelek, P.I. (2007) Fluid evolution and kinetics of metamorphic reactions in calc-silicate contact aureoles – from H_2O to CO_2 and back. *Geology*, **35**, 927–930.

Naney, M.T. and Swanson, S.E. (1980) The effect of Fe and Mg on crystallization in granitic systems. *American Mineralogist*, **65**, 639–653.

Nascimento, M.L.F., Ferreira, E.B. and Zanotto, E.D. (2004) Kinetics and mechanisms of crystal growth and diffusion in a glass-forming liquid. *Journal of American Physics*, **121**, 8924–8928.

Ni, H., Keppler, H., Walte, N., Schiavi, F., Chen, Y., Masotta, M. and Li, Z. (2014) In situ observation of crystal growth in a basalt melt and the development of crystal size distribution in igneous rocks. *Contributions to Mineralogy and Petrology*, **167**, 1003.

Nowak, M., Cichy, S.B., Botcharnikov, R.E., Walker, N. and Hurkuck, W. (2011) A new type of high-pressure low-flow metering valve for continuous decompression: First experimental results on degassing of rhyodacitic melts. *American Mineralogist*, **96**, 1373–1380.

Onuma, N., Higuchi, H., Wakita, H. and Nagasawa, H. (1968) Trace element partitioning between two pyroxenes and the host lava. *Earth and Planetary Science Letters*, **5**, 47–51.

Pinilla, C., Davis, S.A., Scott, T.B., Allen, N.L. and Blundy, J.D. (2012) Interfacial storage of noble gases and other trace elements in magmatic systems. *Earth and Planetary Science Letters*, **319–320**, 287–294.

Powell, M.A., Walker, D. and Hays, J.F. (1980) Experimental solidification of a eucrite basalt: microprobe studies. Pp. 896–898 in: *Proceedings of Lunar and Planetary Science Conference, XI*. Lunar and Planetary Institute, Houston, Texas, USA.

Pupier, E., Duchene, S. and Toplis, M.J. (2008) Experimental quantification of plagioclase crystal size distribution during cooling of a basaltic liquid. *Contributions to Mineralogy and Petrology*, **155**, 555–570.

Putirka, K.D. (1999) Clinopyroxene + liquid equilibria. *Contributions to Mineralogy and Petrology*, **135**, 151–163.

Putirka, K.D. (2008) Thermometers and barometers for volcanic systems. Pp. 61–120 in: *Minerals, Inclusions and Volcanic Processes* (K.D. Putirka and F.J. Tepley editors). Reviews in Mineralogy and Geochemistry, **69**. Mineralogical Society of America and Geochemical Society, Washington, D.C.

Putnis, A. and Bish, D. L. (1983). The mechanism and kinetics of Al, Si ordering in Mg-cordierite. *American Mineralogist* **68**, 60–65.

Riker, J.M., Cashman, K.V., Rust, A.C. and Blundy, J.D. (2015) Experimental constraints on plagioclase crystallization during H_2O- and H_2O-CO_2-saturated magma decompression. *Journal of Petrology*, **56**, 1967–1998.

Roeder, P.L. and Emslie, R.F. (1970) Olivine–liquid equilibrium. *Contributions to Mineralogy and Petrology*, **29**, 275–289.

Roselle, G.T., Baumgartner, L.P. and Chapman, J.A. (1997) Nucleation-dominated crystallization of forsterite in the Ubehebe peak contact aureole. California *Geology*, **25**, 823–826.

Sato, H. (1995) Textural difference of pahoehoe and aa lava of Izu–Oshima volcano, Japan – an experimental study on population density of plagioclase. *Journal of Volcanology and Geothermal Research*, **66**, 101–113.

Scarlato, P., Mollo, S., Blundy, J., Iezzi, G. and Tiepolo, M. (2014) The role of natural solidification paths on REE partitioning between clinopyroxene and melt. *Bulletin of Volcanology*, **76**, 810.

Schiavi, F., Walte, N. and Keppler, H. (2009) First in situ observation of crystallization processes in a basaltic–andesitic melt with the moissanite cell. *Geology*, **37**, 963–966.

Schnare, D.W., Day, J.M.D., Norman, M.D., Liu, Y. and Taylor L.A. (2008) A laser–ablation ICP–MS study of Apollo 15 low-titanium olivine-normative and quartz-normative mare basalts. *Geochimica et Cosmochimica Acta*, **72**, 2556–2572.

Shea, T. and Hammer, J.E. (2013) Kinetics of cooling- and decompression-induced crystallization in hydrous mafic–intermediate magmas. *Journal of Volcanology and Geothermal Research*, **260**, 127–145.

Shimizu, N. (1981) Trace element incorporation into growing augite phenocryst. *Nature*, **289(5798)**, 575–577.

Singer, B.S., Dungan, M.A. and Layne, G.D. (1995) Textures and Sr, Ba, Mg, Fe, K and Ti compositional profiles in volcanic plagioclase: clues to the dynamics of calcalkaline magma chambers. *American Mineralogist*, **80**, 776–798.

Smith, R.K. and Lofgren, G.E. (1983) An analytical and experimental study of zoning in plagioclase. *Lithos*, **16**, 153–168.

Smith, V.G., Tiller, W.A. and Rutter, J.W. (1955) A mathematical analysis of solute redistribution during solidification. *Canadian Journal of Physics*, **33**, 723–745.

Swanson, S.E. (1977) Relation of nucleation and crystal-growth rate to the development of granitic textures. *American Mineralogist*, **62**, 966–978.

Tammann, G. and Mehl, R.F. (1925) *The States of Aggregation*. Van Nostrand Reinhold, New York.

Thornber, C.R. and Huebner, J.S. (1985) Dissolution of olivine in basaltic liquids: experimental observations and applications. *American Mineralogist*, **70**. 934–945.

Tsuchiyama, A. (1985) Crystallization kinetics in the system $CaMgSi_2O_6$–$CaAl_2Si_2O_8$: development of zoning and kinetics effects on element partitioning. *American Mineralogist*, **70**, 474–486.

Ujike, O. (1982) Microprobe mineralogy of plagioclase, clinopyroxene and amphibole as records of cooling rate in the Shirotori–Hiketa dike swarm, northeastern Shikoku, Japan. *Lithos*, **15**, 281–293.

Vetere, F., Iezzi, G., Behrens, H., Cavallo, A., Misiti, V., Dietrich, M., Knipping, J., Ventura, G. and Mollo, S. (2013) Intrinsic solidification behaviour of basaltic to rhyolitic melts: A cooling rate experimental study. *Chemical Geology*, **354**, 233–242.

Vetere, F., Iezzi, G., Behrens, H., Holtz, F., Ventura, G., Misiti, V., Cavallo, A., Mollo, S. and Dietrich, M. (2015) Glass forming ability and crystallisation behaviour of sub–alkaline silicate melts. *Earth Science Reviews*, **150**, 25–44.

Walker, D., Kirkpatrick, R.J., Longhi, J. and Hays, J.F. (1976) Crystallization history of lunar picritic basalt samples 12002: phase equilibria and cooling rate studies. *Geological Society of America Bulletin*, **87**, 646–656.

Waters, L.E., Andrews, B.J. and Lange, R.A. (2015) Rapid crystallization of plagioclase phenocrysts in silicic melts during fluid-saturated ascent: Phase equilibrium and decompression experiments. *Journal of Petrology*, **56**, 981–1006.

Watson, E.B. and Liang, Y. (1995) A simple model for sector zoning in slowly grown crystals: implications for growth rate and lattice diffusion, with emphasis on accessory minerals in crustal rocks. *American Mineralogist*, **80**, 1179–1187.

Watson, E.B. and Müller, T. (2009) Non-equilibrium isotopic and elemental fractionation during diffusion-controlled crystal growth under static and dynamic conditions. *Chemical Geology*, **267**, 111–124.

Welsch, B., Faure, F., Famin, V., Baronnet, A. and Bachelery, P. (2013) Dendritic crystallization: A single process for all the textures of olivine in basalts? *Journal of Petrology*, **54**, 539–574.

Welsch, B., Hammer, J.E. and Hellebrand, E. (2014) Phosphorus zoning reveals dendritic architecture of olivine. *Geology*, **42**, 867–870.

Welsch, B., Hammer, J.E., Baronnet, A., Jacob, S., Hellebrand, E. and Sinton, J.M. (2016) Clinopyroxene in postshield Haleakala ankaramite 2. Texture, compositional zoning, and supersaturation in the magma. *Contributions to Mineralogy and Petrology*, **171**, 6.

Wood, B.J. and Blundy, J.D. (1997) A predictive model for rare earth element partitioning between clinopyroxene and anhydrous silicate melt. *Contributions to Mineralogy and Petrology*, **129**, 166–181.

Wood, B.J. and Blundy, J.D. (2001) The effect of cation charge on crystal–melt partitioning of trace elements. *Earth and Planetary Science Letters*, **188**, 59–72.

Wood, B.J. and Trigila, R. (2001) Experimental determination of aluminium clinopyroxene–melt partition coefficients for potassic liquids, with application to the evolution of the Roman province potassic magmas. *Chemical Geology*, **172**, 213–222.

Zhang, Y. (2010) Diffusion in minerals and melts: theoretical background. Pp. 5–60 in: *Diffusion in Minerals and Melts* (Y. Zhang and D. Cherniak, editors). Reviews in Mineralogy and Geochemistry, **72**. Mineralogical Society of America and Geochemical Society, Washington, D.C.

Zhou, W., Van der Voo, R., Peacor, D.R. and Zhang, Y. (2000) Variable Ti content and grain size of titanomagnetite as a function of cooling rate in very young MORB. *Earth and Planetary Science Letters*, **179**, 9–20.

Reactions between minerals and aqueous solutions

Encarnación RUIZ-AGUDO[1], Christine V. PUTNIS[2,3]
and Carlos RODRÍGUEZ-NAVARRO[1]

[1]*Departamento de Mineralogía y Petrología, Universidad de Granada,
18071 Granada, Spain, e-mail: encaruiz@ugr.es*
[2]*Institut für Mineralogie, Universität Münster, 48149 Münster, Germany,
e-mail: putnisc@uni-muenster.de*
[3]*Nanochemistry Research Institute, Curtin University, Perth 6102, Australia*

Kinetics of reactions between minerals and solutions govern a wide range of natural and technological processes including weathering and soil formation, nutrient availability, biomineralization, acid mine-drainage, the fate of contaminants, or nuclear waste disposal.

Theoretical and experimental studies performed in recent decades have changed our understanding of the mechanisms of mineral–solution reactions significantly. This chapter reviews recent results and advances in terms of non-classical mineral-growth processes (pre-nucleation clusters, liquid and amorphous precursor phases or the occurrence and participation of nanoclusters as building blocks in the growth process) as well as other mineral–solution equilibration processes occurring by interface-coupled dissolution-precipitation reactions, which lead to replacement of the original mineral assemblage.

1. Introduction

Reactions between minerals and solutions (growth, dissolution and replacement) are central to all geochemical natural and technological processes. Global rates of these processes govern a wide range of important issues such as weathering and soil formation, nutrient availability, biomineralization, acid mine drainage, the fate of contaminants or nuclear waste disposal (Putnis and Ruiz-Agudo, 2013). Theoretical and experimental studies performed in recent decades are dramatically changing our current understanding of the mechanisms of mineral–solution reactions. In the case of mineral growth, the classical view of crystal formation ("classical nucleation theory", Volmer and Weber, 1926; Becker and Doring, 1935) is currently challenged by an increasing body of computational and experimental studies that indicate the presence of alternative crystallization processes, including the existence of stable pre-nucleation clusters (Gebauer *et al.*, 2008), liquid and amorphous precursor phases (Wallace *et al.*, 2013) or the occurrence and participation of nanoclusters as building blocks in the growth process (Penn and Banfield, 1998; Gebauer *et al.*, 2008; Meldrum and Cölfen, 2008; Wallace *et al.*, 2013; Nielsen *et al.*, 2014; De Yoreo *et al.*, 2015). These processes are all considered as 'non-classical' crystallization routes.

The development of advanced analytical methods that allow direct, *in situ* observations of dissolving mineral surfaces at the nanoscale have stimulated new prospects for the elucidation of the mechanisms ruling mineral–solution interactions. These observations suggest a complex reactivity landscape of mineral surfaces that depends on both the crystallographic orientation of the surface and the specific site within a given mineral surface (Fischer *et al.*, 2012; Godinho *et al.*, 2012; Lüttge *et al.*, 2013; Smith *et al.* 2013). Moreover, *in situ* observations of reacting surfaces and high-resolution analysis of reaction interfaces are providing new insights into the mechanisms of apparent 'incongruent' dissolution and the formation of surface altered layers (SALs) during the dissolution of multicomponent minerals and glasses (*e.g.* Hellmann *et al.*, 2015; Ruiz-Agudo *et al.*, 2012). This new evidence suggests that 'incongruent' dissolution and SALs formation is essentially equal to other mineral–solution equilibration processes, including mineral-replacement processes occurring in the presence of a solvent, such as the replacement of KBr by KCl (Putnis and Mezger, 2004; Putnis *et al.*, 2005) or leucite by analcime (Putnis *et al.*, 2007b; Xia *et al.*, 2009a). The mechanism of these interactions can be interpreted in the framework of an interface-coupled dissolution-precipitation model (*e.g.* Putnis, 2002, 2009; Ruiz-Agudo *et al.*, 2014; Hellmann *et al.*, 2015). In this model, a mineral in contact with an aqueous solution dissolves stoichiometrically and leads to supersaturation of a fluid layer, in contact with the mineral surface, with respect to a secondary phase, which eventually precipitates on the surface of the pristine mineral. This chapter summarizes recent results and advances in research performed in these fields, which are increasing our ability to understand the mechanisms and predict the kinetics of mineral–solution reactions.

2. Quantitative *in situ*, nanoscale observations of reactions at the mineral–water interface by atomic force microscopy

Advances in the development of surface analySPtical techniques that enable *in situ* observations of dissolution and growth processes have represented a significant breakthrough in the study of mineral reactivity. In particular, the invention and widespread use of atomic force microscopy (AFM) has allowed *in situ*, real-time nanoscale observations of the topography evolution of mineral surfaces whilst they react to different fluids, obtaining detailed atomic information on these processes. In AFM a sharp tip attached to a cantilever scans the mineral surface, and the actual z position of the tip is recorded as a function of the lateral x-y position of the corresponding deflection of a laser beam (Fig. 1). The image is formed by monitoring the interatomic forces between the electrons in the outer shells of the atoms at the sample surface and the electrons of the atoms of the sharp tip that is scanned across the surface. A feedback loop controls the vertical position of the tip with respect to the sample surface, so that the reflected laser beam stays at the centre of a light detector. Scanners made of piezoelectric materials control the movement of the cantilever along the x, y and z directions and provide force/distance information enabling accurate height and x−y measurements.

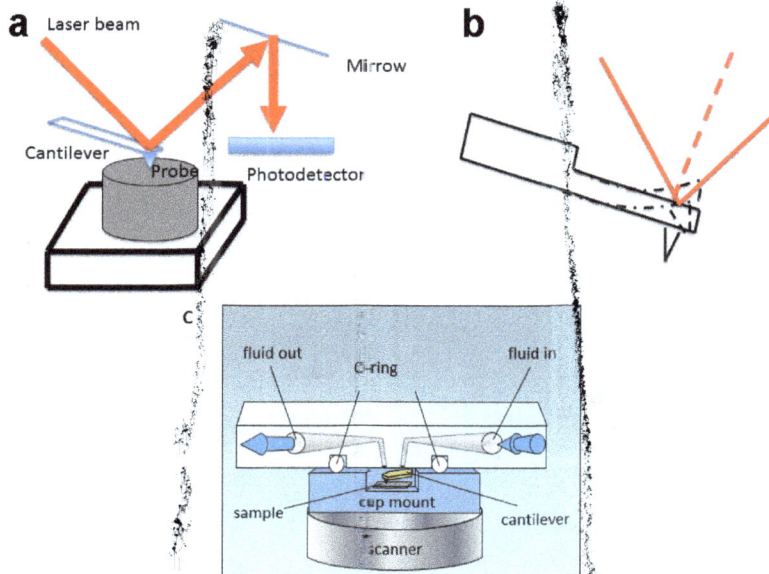

Figure 1. (a) Sketch of a possible implementation of an atomic force microscope. (b) Beam deflection scheme in an AFM. (c) O-ring sealed fluid cell to be used in AFM studies for *in situ* experiments on surface mineral reactivity (Image courtesy of J.M. Astilleros, Complutense University of Madrid, Spain).

Over the last two decades, AFM studies have contributed to a deeper understanding of the mechanisms of mineral reactivity from direct observations of mineral-fluid reactions *in situ* within a fluid cell. They enable direct evaluation and/or quantification of dislocation source activity, two- and three-dimension nucleation, step propagation rates, step roughness, terrace widths and impurity-step interactions (*e.g.* De Yoreo *et al.*, 2001). In these studies, dissolution and growth kinetics are inferred from the absolute rate of step propagation, calculated by measuring the distances between a step edge and a fixed reference point in sequential images scanned in the same direction (*e.g.* Teng, 2004; Xu and Higgins, 2011). Locating a fixed reference point on a dissolving or growing surface is not always feasible, and thus rates are normally reported as the length increase per unit time of opposite parallel steps (*i.e.* between parallel steps delimiting etch pits or growth islands) in sequential images (*e.g.* Liang *et al.*, 1996a,b). This can be done for example for steps parallel to the [001] or [101] directions in the case of etch pits formed during gypsum dissolution (*see* Fig. 2a) or for steps parallel to the <$\bar{4}$41> direction on growth islands formed during calcite growth (Fig. 2b). In the case of minerals such as carbonates, which frequently grow by the formation of growth hillocks nucleated at screw dislocations, measurements of geometric relationships of the hillocks in sequential images allow calculation of the absolute spreading velocities of opposite steps (Larsen *et al.*, 2010a,b), using the equations:

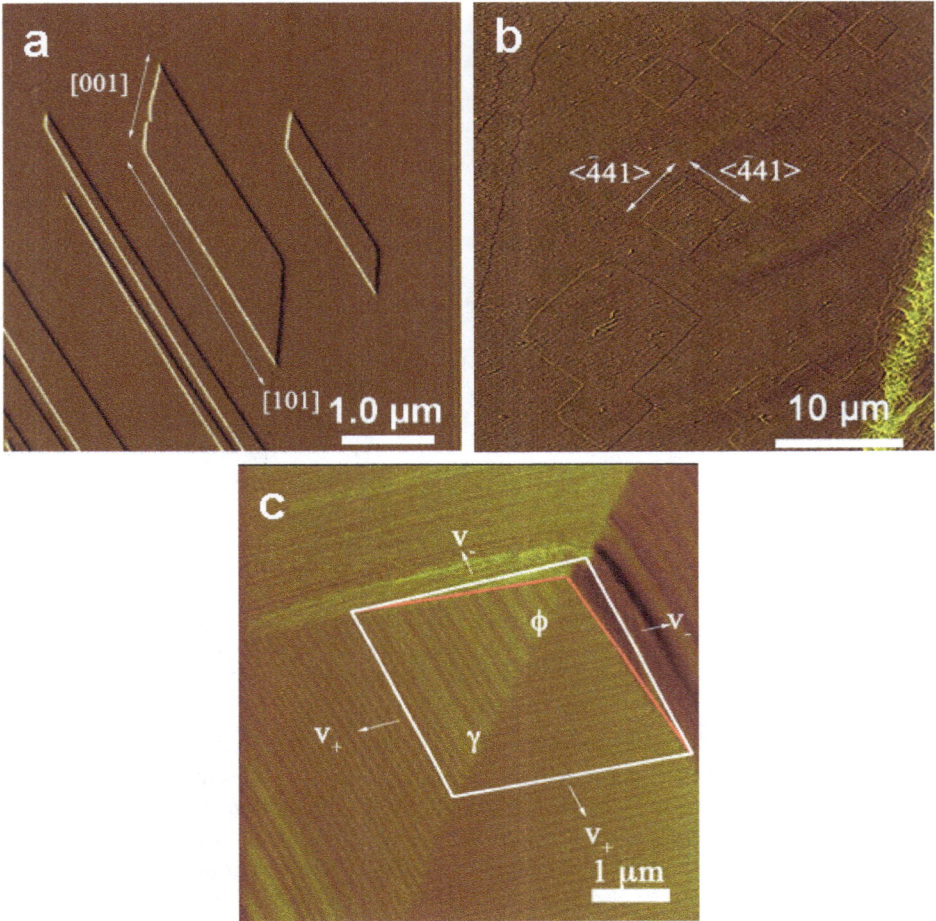

Figure 2. AFM deflection images showing: (a) etch pits on a gypsum {010} surface showing steps parallel to the [001] and [101] directions; (b) rhombohedral growth islands on a calcite {10$\bar{1}$4} surface delimited by steps parallel to the <$\bar{4}$41> direction and (c) spiral growth hillocks on a {10$\bar{1}$4} surface.

$$v_+ = \frac{\sin\left(\frac{\phi+\gamma}{2}\right)}{t} \cdot x \qquad (1)$$

$$v_- = \frac{\sin\left(\frac{\phi-\gamma}{2}\right)}{t} \cdot x \qquad (2)$$

where v_+ and v_- are the spreading velocities for the obtuse and acute <$\bar{4}$41> steps, t is the time lapse between AFM images and x and the angles ϕ and γ (defined in Fig. 2c).

Overall dissolution rates, R_{AFM} (mol cm^{-2} s^{-1}), can be also be calculated from AFM observations considering that during dissolution, etch pits may change both their depth and lateral dimensions. Thus, rates can be determined by measuring the volume increase of etch pits, ΔV, over a given area in a given time in sequential images:

$$R_{AFM} = \frac{\Delta V \cdot N_{pit}}{v_m \cdot (t_2 - t_1)} \qquad (3)$$

where V_m is the molar volume of the mineral, N_{pit} is the number of etch pits per unit area of surface, and $t_2 - t_1$ is the time difference between two sequences images. This calculation has limitations with respect to depth measurements in the case of the absence of a reference (unreacted) surface. Thus, AFM studies allow the investigation of the effects of parameters such as solution composition, supersaturation, pH, alkalinity and the presence of additives and background electrolytes on the kinetics and mechanisms of mineral growth and dissolution (*e.g.* Kowacz and Putnis, 2008; Larsen *et al.*, 2010a,b; Ruiz-Agudo *et al.*, 2010a,b; 2011a,b; Stack and Grantham, 2010). Additionally, classical crystal growth theories can be tested by direct observations of reacting surfaces using AFM (*e.g.* Pina *et al.*, 1998; Rashkovich *et al.*, 2006; Sleutel and Van Driessche, 2014). Furthermore, AFM studies allow the determination of the molecular mechanisms ruling the interactions between minerals and impurities during growth, through the quantification of parameters such as step velocity and terrace width and their comparison to standard impurity and thermodynamic models (*e.g.* Davis *et al.*, 2000; Fig. 3). Additionally, AFM observations have provided evidence of non-classical mineral-growth processes, such as the formation of precursor species or growth by the direct attachment of nanoparticles (*e.g.* Wang *et al.*, 2012a; Lupulescu and Rimer, 2014; Rodriguez-Navarro *et al.*, 2016). Finally, direct visualization of the initial stages of coupled dissolution-precipitation processes by *in situ* AFM gives quantitative information on the kinetics of the early stages of such coupled processes, as well as insights into the growth mechanism of the precipitating phase and potential epitaxial relationships between parent and product phases (Wang *et al.*, 2012b; Klasa *et al.*, 2013; Ruiz-Agudo *et al.*, 2013). In this review paper, a number of AFM investigations dealing with the different aspects of solution-induced mineral reactions (dissolution, growth and replacement) will be presented. Ruiz-Agudo and Putnis (2012) reviewed additional examples that show how AFM *in situ* studies provide information to enable quantification of mineral-solution reactions.

3. Mineral nucleation and growth: Classical and non-classical crystallization pathways

The formation of a mineral in solution involves two critical processes: (1) the nucleation of a solid phase and (2) its subsequent growth (Mullin, 2001). If the mineral is crystalline, that is, its structure displays three-dimensional (3D) long-range order, its crystallization takes place by the incorporation of ions with a high-energy solvated state into a crystal lattice, with lower bulk energy. Because the system experiences an

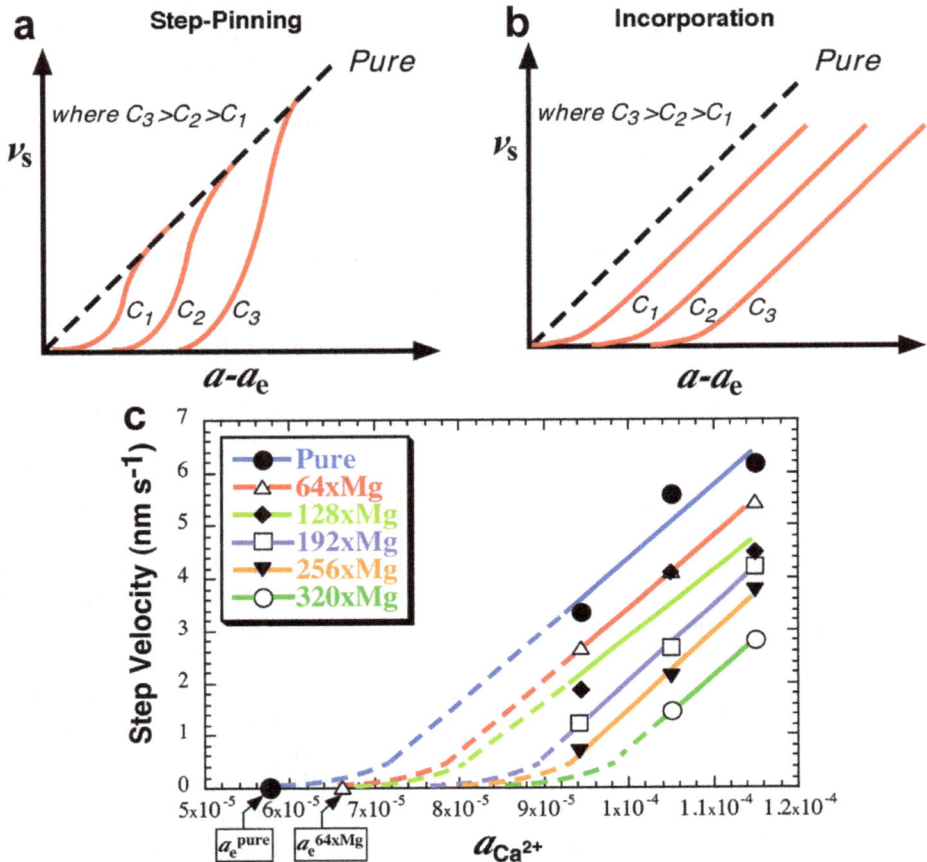

Figure 3. Plots showing the theoretical dependence of step velocity (v_s) on the supersaturation ($a-a_e$) for (a) step pinning and (b) impurity-incorporation models, for different values of impurity concentration (C_i) in solution. (c) Plot of the calcite <$\bar{4}41$> step velocity determined by *in situ* AFM experiments *vs.* calcium activity ($a_{Ca^{2+}}$) for different values of Mg^{2+} concentration in solution. Comparison of plots a, b and c indicates that magnesium inhibits calcite growth through an incorporation mechanism. From Davis *et al.* (2000) reprinted with permission from the American Association for the Advancement of Science (AAAS).

overall reduction in free energy, crystallization is a spontaneous irreversible process. However, as will be shown below, the situation is not always that simple (Dorvee and Veis, 2013). Experimentally, it is observed that for solutions that are supersaturated with respect to a crystalline phase, there is a metastable zone where no newly formed crystals are detected (Fig. 4; Mullin, 2001). Note that the width of the metastable zone depends entirely on the time of observation, making the designation of this area in the phase diagram arbitrary. The existence of the metastable zone is due to both thermodynamic and kinetics factors (García-Ruiz, 2003).

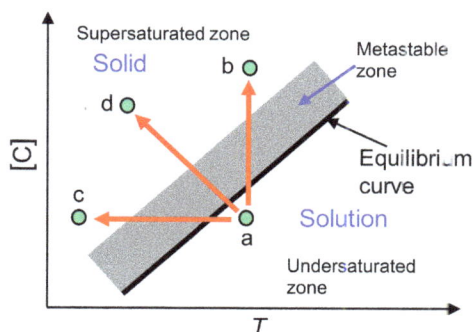

Figure 4. Solubility diagram of a mineral phase. The solubility curve and the metastable zone are indicated. Starting with a solute concentration [C] and temperature T at the point marked by **a**, cooling results in the crystallization of solid phase at point **c**, while through evaporation at a constant T, the solid phase is obtained at a concentration marked by **b**. Crystallization paths combining a change in T and evaporation are also possible (path **a**–**d**).

According to classical nucleation theory (CNT), which is rooted in the work of Gibbs (Gibbs, 1876, 1878) and was formally stated by, among others, Volmer and Weber (1926) and Becker and Doring (1935), the formation of a solid from a solution is a first-order phase transition occurring once an aggregate or cluster of monomers (atoms, ions or molecules) reaches a critical size that enables its subsequent spontaneous growth *via* monomer-by-monomer incorporation into the crystal lattice (Mutaftschiev, 2001). Below the critical size, clusters form (through random inelastic collisions) and disintegrate (*i.e.* they are metastable). This imposes the existence of a free energy barrier that has to be overcome for crystallization to take place. For clusters to reach and overcome the critical size, the whole system (*i.e.* liquid and solid phases) undergoing such a phase transition has to reduce its overall free energy, ΔG (De Yoreo and Vekilov, 2003), which is determined by

$$\Delta G = \frac{\frac{4}{3}\pi r^3}{v} kT \ln\left(\frac{a}{a_0}\right) + 4\pi r^2 \gamma \tag{4}$$

where r is the cluster radius, v is the molecular volume of the crystalline phase, a and a_0 are the ion activities of the ions in solution and in equilibrium (*i.e.* solubility product of the crystalline phase), respectively, and γ is the surface or interfacial energy of the solid in contact with the solution. The first term of the equation represents the free energy gain associated with the formation of a new phase (*i.e.* volume term, ΔG_v) and the second term represents the free energy penalty associated with the creation of a new surface (*i.e.* surface term, ΔG_s).

Figure 5 shows that the sum of surface and volume free energy terms has a maximum, ΔG^* when $d\Delta G/dr = 0$. Taking the derivative of equation 4 and setting it equal to zero, the value of the critical radius, r_c is obtained,

$$r_c = \frac{2v\gamma}{kT \ln\left(\frac{a}{a_0}\right)} \tag{5}$$

Equation 5 shows that the radius of the critical cluster in equilibrium with a supersaturated solution is proportional to its surface energy and inversely proportional to the supersaturation σ, ($\sigma = a/a_0$) and T. In other words, it would be

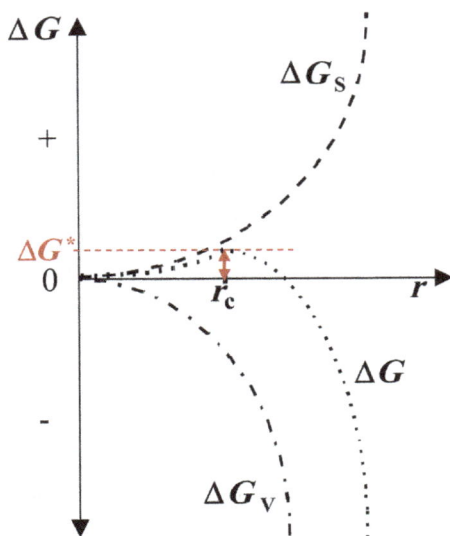

Figure 5. Variation of ΔG_v, ΔG_s and $\Delta G = \Delta G_v + \Delta G_s$ vs. cluster's radius, r. For a critical cluster radius, r_c, the value of ΔG reaches a maximum (ΔG^*) and for $r > r_c$ it decreases.

thermodynamically more favourable to nucleate a new phase (with smaller r_c) at high rather than at low σ. Note that there are several ways to express the supersaturation of a system, the most common, in addition to the one presented above, being:

$$\sigma = \ln\left(\frac{IAP}{k_s}\right) \tag{6}$$

or

$$SI = \log\left(\frac{IAP}{k_s}\right) \tag{7}$$

where SI is the so-called saturation index, and IAP and k_s are the ion activity product and solubility product of a relevant phase, respectively. Equations 6 and 7 are most appropriate for sparingly soluble minerals (Mullin, 2001).

From equations 4 and 5 the value of the free energy barrier for nucleation, ΔG^* can be calculated by

$$\Delta G^* = \frac{16\pi v^2 \gamma^3}{3\left[kT\ln\left(\frac{a}{a_0}\right)\right]^2} \tag{8}$$

The fact that the free energy barrier varies with γ^3 underlines the critical role this latter parameter plays during nucleation. The actual surface energy values of crystals at the relevant sizes for nucleation are, however, largely unknown, and are assumed to correspond to those of bulk crystals. This is the so-called 'capillary assumption' or 'capillary approximation' (Mutaftschiev, 2001; De Yoreo and Vekilov, 2003; Gebauer *et al.*, 2014). Indeed, the experimental determination of surface energies is difficult, as reflected by the spread of reported values for a given system, which strongly limits the applicability of CNT. For instance, the value of the critical size for homogeneous calcite nucleation can change by a factor of 7 if one considers the smallest and largest reported values of $\gamma_{(104)calcite-solution}$ (83 and 170 mJ/m^2; Travaille *et al.*, 2005).

The above discussion refers to homogeneous nucleation, that is, the formation of a crystal nucleus in the bulk solution. This is, however, not a common situation in nature or in the laboratory (García-Ruiz, 2003). The existence of a foreign surface (substrate) in the system favours a reduction of the free energy barrier for nucleation because the crystal-substrate interfacial energy is lower than that of the crystal-solution (Chernov, 1984; Mullin, 2001; De Yoreo and Vekilov, 2003). In this case nuclei will tend to form on the surface of such a substrate (which could be any kind of interface: solid-liquid or gas-liquid). This phenomenon is called heterogeneous nucleation and occurs at substantially lower σ and r_c values than homogeneous nucleation. Because it reduces the energetic penalty for the formation of a crystalline phase, heterogeneous nucleation is thought to play a critical role in many processes such as the formation of biominerals and their biomimetic counterparts (Meldrum and Cölfen, 2008). Indeed, a thoroughly explored biomimetic route to induce the heterogeneous crystallization of a particular crystalline phase involves the use of organic templates such as Langmuir-Blodgett films or self-assembled monolayers (Aizenberg *et al.*, 1999; Xu *et al.*, 2007; Sommerdijk and de With, 2008; Nudelman and Sommerdijk, 2012).

In addition to thermodynamics, kinetics also play a crucial role during crystal nucleation from a solution. Indeed, kinetics determine the nucleation rate, J^*, which is given by the following Ahrrenius-type equation (Gebauer *et al.*, 2014)

$$J^* = A \exp\left(-\frac{\Delta G^*}{kT}\right) \exp\left(-\frac{E_a}{kT}\right) \tag{9}$$

where A is a constant (the so-called pre-exponential factor, which depends on the properties of a particular system; De Yoreo and Vekilov, 2003). The first exponential term reflects the number of nuclei with the critical size formed per unit volume and unit time, while the second term reflects the existence of a kinetic barrier that depends on ion diffusion rates and other phenomena such as ion desolvation and structural rearrangements (Habraken *et al.*, 2013). J^* is thus dependent on T and supersaturation and can adopt different forms, depending on the specific system considered. The contribution of the second exponent in equation 9 is typically ignored due to the difficulty of determining the kinetic barrier (Baumgartner *et al.*, 2013; Gebauer *et al.*, 2014). Conversely, the first term, namely, the free energy barrier, can be determined readily, provided some key assumptions of CNT (*e.g.* the 'capillary assumption') hold true.

In many cases, however, nucleation rates calculated using equation 9 and experimental values differ by orders of magnitude (Vekilov, 2010). This is mostly due to the uncertainties associated with the determination of γ (*i.e.* 'capillary asumption', see above), and highlights the limited ability of CNT to predict the behaviour of real systems (Gebauer *et al.*, 2014). We will see below that in addition to the 'capillary assumption', such a handicap is in many cases also related to the fact that several systems (if not all) do not follow 'classical' crystallization routes.

Once a cluster with the critical size, now a crystal embryo, is formed, it can grow spontaneously. However, not all crystal embryos in a system will grow to a macroscopic size. If crystals with different sizes are present in a system at a given reaction time, and the amount of growth units is limited (*i.e.* as growth progresses, the overall reactant concentration of the system is reduced), the larger ones will grow at the expense of the smaller ones (provided the smallest ones have a size << 100 nm). This occurs because according to the Gibbs-Thomson equation, the solubility of the smaller crystals will be greater than that of the larger ones (Mullin, 2001). This coarsening process is called 'Ostwald ripening' (Ostwald, 1901), and its driving force is the overall reduction of surface free energy of the system (Adamson and Gast, 1997).

According to the classical picture, crystal growth proceeds *via* monomer-by-monomer incorporation of simple chemical species (atoms/ions/molecules) at energetically favourable sites (*e.g.* kinks present at steps) present on specific (*hkl*) surfaces of a crystal (Chernov, 1984; Mullin, 2001; De Yoreo and Vekilov, 2003). In contrast, non-classical crystal growth involves coarsening *via* a particle-mediated aggregation mechanism (see section 3.3.; De Yoreo *et al.*, 2015). As in the case of nucleation, the driving force for crystal growth is the overall reduction of the free energy of the system (*i.e.* that of the solvent molecules and ions/molecules in solution and in the newly formed crystal; De Yoreo and Vekilov, 2003). The presence of steps and kinks on the surface of a crystal (with size equal to or larger than the critical size) is a prerequisite for layer-by-layer growth according to the classical Terrace-Ledge-Kink (TLK) model (Burton *et al.*, 1951; De Yoreo *et al.*, 2009). Growth according to the TLK model involves the continuous creation of kinks either *via* movements of monomers on the step edge (thermal fluctuations) or attachment of new monomers from solution (1D nucleation; De Yoreo *et al.*, 2009; Fig. 6). Note that at the typical, relatively low *T* of crystallization from solutions, normal growth *via* continuous incorporation of ions or molecules into a specific (*hkl*) face is uncommon (unlike the case of crystallization from a melt; Chernov, 1984). To our knowledge, direct evidence (using *in situ* AFM) for such a growth mechanism in solution has only been suggested for the case of horse spleen apoferritin crystals grown at a very high supersaturation (Malkin *et al.*, 1995). For the standard layer-by-layer growth of faceted crystals in solution, ions/molecules can either be incorporated directly from solution into a kink site on a step or may have to diffuse towards the surface of a crystal where they are adsorbed, and then they can undergo 2D surface diffusion towards an energetically favourable site where they can be incorporated into the crystal lattice following desolvation (a key process which in many cases becomes the rate-determining step; Myerson, 2002). It could be argued that once sufficient ions or molecules have been incorporated at kink/step sites present on a particular (*hkl*) face (*e.g.* a 'flat' or F face; Mutaftschiev, 2001), so that a step has 'wiped' across the whole surface leaving a flat terrace, the growth in a perpendicular direction to such a face should stop as no further kinks or steps would be present for the ions/molecules to be incorporated into the crystal lattice at energetically favourable

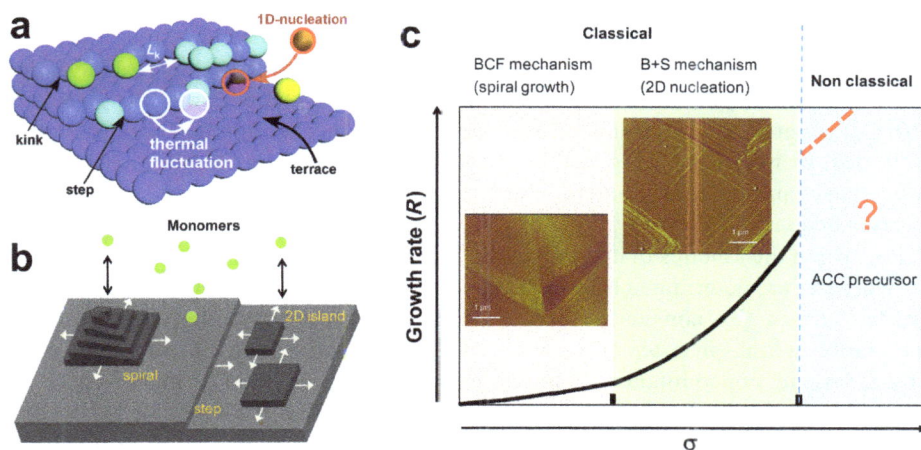

Figure 6. The mechanisms of crystal growth from solution. (a) Scheme of the TLK model for classical crystal growth *via* monomer-by-monomer incorporation into kinks (placed at a distance L_k), which are created by movement of ions along steps (thermal fluctuations) or *via* 1D nucleation; (b) growth *via* spreading of steps emerging from spirals (BCF mechanism) or 2D islands (2D surface nucleation, or birth-and-spread, B+S, growth model); (c) AFM images of calcite ($10\bar{1}4$) surface growing *via* spirals according to the BCF model (low supersaturation) and 2D surface nucleation according to the VKS model (moderate-high supersaturation). At a threshold supersaturation corresponding to the nucleation of amorphous calcium carbonate, a transition towards non-classical crystal growth can occur, and crystal can grow at a significant (yet, unknown) rate. Image (a) from De Yoreo *et al.* (2009), with permission of ACS, copyright 2009; images (b) and (c) modified from Rodriguez-Navarro *et al.* (2016), with permission of the American Chemical Society (ACS).

sites. According to classical crystallization theory, the stepwise growth of crystals can take place by two main mechanisms that allow the situation indicated above to be bypassed: (a) spiral growth, and (b) two-dimensional (2D) or 3D surface nucleation (Fig. 6), depending on the supersaturation of the system (Frank, 1949; Burton *et al.*, 1951; Chernov, 1984; Mullin, 2001; Mutaftschiev, 2001), as shown in Fig. 6c (Rodriguez-Navarro *et al.*, 2016).

In situ AFM observations have provided unambiguous evidence showing that crystals can grow *via* the continuous propagation of steps emanating from screw and edge dislocations outcropping at the crystal surface, as predicted by the model of Burton *et al.* (1951). For instance, the growth of calcite crystals at low supersaturation has been shown to proceed *via* the spreading of steps emanating from screw dislocations (*i.e.* the typical growth hillocks observed on $(10\bar{1}4)_{calcite}$; Teng *et al.*, 1998, 2000; Fig. 6c). Similarly, *in situ* AFM has shown that, at moderate/high supersaturation, crystal growth in solution can proceed *via* 2D (or 3D) surface nucleation and subsequent spreading of steps limiting these islands (Teng *et al.*, 2000). This second type of stepwise growth follows the surface nucleation model proposed by Volmer and Weber (1926), Volmer (1939), Kossel (1927) and Stranski (1928) (*i.e.* the so-called VKS model, also known as 'birth-and-spread' model). 2D (or 3D) surface nucleation requires overcoming a nucleation barrier (ΔG^*) which is dependent on σ (see equation 8) therefore, this growth mechanism is favoured at a relatively high supersaturation, while growth *via* screw dislocations, which does not require the overcoming of any additional free energy barrier, occurs at a low supersaturation. Figure 6c shows AFM images of growth hillocks and 2D islands on $(10\bar{1}4)_{calcite}$, and the relationship between growth rates/mechanisms and supersaturation. Note that, as reflected by Fig. 6c, for a sufficiently high supersaturation enabling the formation of an amorphous solid (or liquid) precursor, growth may become non-classical, as discussed later.

Regarding the kinetics of crystal growth in solution, experimental results show that growth rates are determined by one of the following elementary steps: transport to the crystal surface (by convection or diffusion), adsorption onto the crystal surface, integration at kinks in a step of a growth spiral, or integration at a kink present in a step of a 2D island, concomitant with ion desolvation (Nielsen, 1984). The existence of a particular rate-controlling step determines whether a linear, parabolic or exponential growth rate law with respect to supersaturation (typically expressed as $\exp(\Delta G/RT)$, where ΔG is the so-called 'chemical affinity', and R is the gas constant) is followed (Nielsen, 1984; Teng *et al.*, 2000). Transport- and adsorption-limited growth follows a linear rate law of the type (Teng et al., 2000)

$$R_m = k_+ \left[\exp\left(\frac{\Delta G}{RT}\right) - 1 \right] \tag{10}$$

where the net rate, R_m, (moles area^{-2} time^{-1}) depends on the rate constant of the forward reaction, k_+, and the free energy change of the overall reaction. In contrast, spiral growth follows a parabolic rate law, and growth *via* 2D surface nucleation follows an exponential rate law. For spiral growth and 2D surface nucleation, the non-linear rate law is of the type,

$$R_m = k_+ \left[\exp\left(\frac{\Delta G}{RT}\right) \right]^n \tag{11}$$

The term n is the order of the reaction, typically being 2 for growth *via* screw dislocations, and adopting higher orders for growth *via* 2D surface nucleation (Teng *et al.*, 2000).

Note that in most cases, crystals growing in solution typically display non-linear growth-rate laws with respect to supersaturation (*i.e.* parabolic or exponential rate laws; Nielsen, 1984; Teng *et al.*, 2000), which underlines the importance of spirals and 2D islands for crystal growth from solution. However, complex combinations of different growth processes can lead to rate laws, which may be difficult to interpret. This is the case for calcite for instance, whose growth rate dependence on supersaturation might be fitted to both parabolic and exponetial rate laws (with a dominant contribution of growth rate controlled by screw dislocation of relatively low supersaturation and rates controlled by 2D nucleation at higher supersaturation; Teng *et al.*, 2000).

Despite its enormous merit and the advances associated with the application of classical crystallization theory, experimental, computational and theoretical results have shown, however, that in several systems strong deviations from CNT and the classical picture above described for crystal growth occur (Nielsen *et al.*, 2014; De Yoreo *et al.*, 2015). The existence of stable pre-nucleation clusters (Gebauer *et al.*, 2008), as well as liquid and amorphous precursor phases (Wallace *et al.*, 2013), show that 'non-classical' nucleation pathways can take place in solution. Similarly, crystal growth *via* (nano)particle (amorphous or crystalline) aggregation or attachment has been demonstrated (Penn and Banfield, 1998; Meldrum and Cölfen, 2008; De Yoreo *et al.*, 2015), that does not fit with the 'classical' monomer-by-monomer crystal growth route indicated above. These 'non-classical' crystallization routes, which are described below and are represented schematically in Figs 7a and 7b, are dramatically changing our current understanding of crystallization from solution.

In the following sections, we will present an overview of different non-classical crystallization phenomena and pathways, that will be exemplified mostly considering the $CaCO_3$–H_2O system which has been the focus of extensive research due to its importance in a range of natural processes such as rock formation, the long-term C cycle, and biomineralization, as well as in technologically relevant processes such as drugs, paint and paper manufacturing (Rodriguez-Navarro and Ruiz-Agudo, 2013). We have chosen to focus on this system not only for its scientific and technological relevance, but also because much of the recent progress in our understanding of non-classical crystallization has taken place when studying such a system.

3.1. Prenucleation clusters

Gebauer *et al.* (2008) demonstrated that nucleation of calcium carbonate is preceded in undersaturated solutions by the formation of 'stable prenucleation clusters' (PNCs) with sizes below the critical radius. Their titration analyses show that the measured free calcium was systematically lower than the calcium dosed to a carbonate buffer solution (Fig. 8), an effect that could not be ascribed to ion pairing. The authors interpreted their

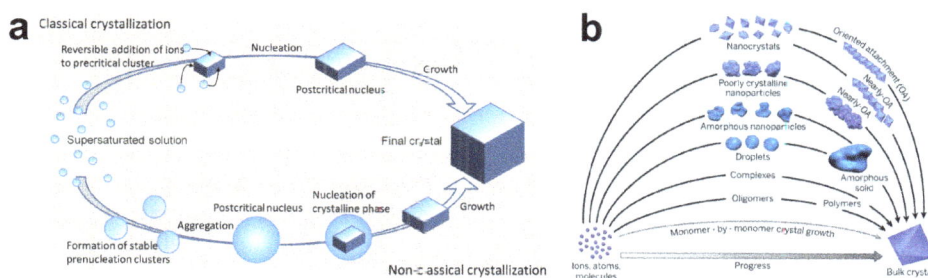

Figure. 7. Classical and non-classical crystallization. (a) According to classical nucleation theory, nucleation proceeds by addition of ions to a single cluster (upper). The non-classical mechanism involves nucleation of an amorphous solid following aggregation of stable prenucleation clusters (lower). The nucleated amorphous solid phase subsequently crystallizes to generate the final stable crystal product. Modified from Meldrum and Sears (2008) with permission from AAAS. (b) Classical monomer-by-monomer crystal growth *vs.* different routes for non-classical aggregation-bases crystal growth. From De Yoreo *et al.* (2015). Reprinted with permission from AAAS.

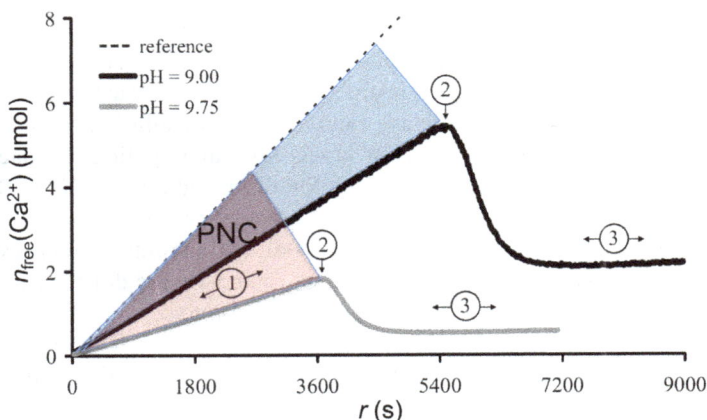

Figure 8. Evolution of free calcium concentration upon addition of $CaCl_2$ solution into 10 mM carbonate buffer (pH = 9.75 or pH = 9.00) at a rate of 10 μL min^{-1}. A smaller amount of free calcium than that dosed (marked by the dashed line) is detected systematically at the pre-nucleation stage (1), which is due to the formation of PNCs. The pH has a profound effect on the width of PNC stability field (colour-shaded areas) and on the onset of nucleation, which is marked by the inflection point (2) of the different curves, and the final ACC phase the solubility of which is marked by the horizontal line (3). Modified from Gebauer *et al.* (2009). With permission from Wiley.

titration results considering that (neutral) calcium and carbonate ion associates formed and were stable, because their formation resulted in a free energy minimum. Such stable ion associates (presumably highly hydrated) were called PNCs, and according to analytical ultracentrifugation analyses, PCNs were clusters ~2 nm in size. The authors also observed the subsequent formation of amorphous calcium carbonate (ACC), probably *via* PNCs aggregation. Using cryo-TEM, Pouget *et al.* (2009) observed clusters ~1.1 nm in size, which were believed to be calcium carbonate PNCs. PNCs are considered as solute species that precede the formation of solid phases during crystallization from solution (Gebauer *et al.*, 2008, 2014). As such, they do not display a surface, or an interfacial energy. This may cast some doubts about the actual nature of the particles observed by Pouget *et al.* (2009) using cryo-TEM, which were solids. Computer simulations have provided further insights into the nature of PNCs. Demichelis *et al.* (2011) showed that they are dynamically ordered liquid-like oxyanion polymers, referred to as DOLLOPS . Basically, they are polymers formed by bonding between Ca^{2+} and CO_3^{2-} ions, which are highly hydrated, and can display different structures (*e.g.* chains or rings). Such structures are not rigid, and can change over time. According to Wallace *et al.* (2013), PNCs (or DOLLOPS) can form dense liquid precursor phases that eventually can lead to the formation of solid ACC (see details in the following section).

Interestingly, PNCs can interact with additives (organic or inorganic) present in the solution, and which may dramatically alter the crystallization and growth of minerals (Song and Cölfen, 2011; Rodriguez-Navarro and Benning, 2013). The different types of interactions will be reflected on the actual path and kinetics of the precipitation process. Depending on the type of interaction, stabilization or destabilization of PNCs can occur. This in turn will lead to either promotion or inhibition of nucleation (Gebauer *et al.*, 2009). Such interactions define the effect additives have on the pre-nucleation stage (Verch *et al.*, 2011; Wolf *et al.*, 2015). In-depth knowledge of the effect different additives have on PNCs appears to be critical for a better understanding of biomineralization processes or for the optimization of biomimetic synthesis routes.

Overall, it is acknowledged that PNCs can play a key role in the formation of minerals (Gebauer and Cölfen, 2011). It is not known, however, what role they may play in their dissolution or in the transport of solutes in nature. These are critical issues that may have important implications not only for the weathering of rocks and the formation of sediments and ore deposits, but also for environmental problems, such as contaminant transport.

3.2. Liquid-like and solid amorphous precursors

According to classical crystallization theory, densification and long-range order (*i.e.* development of a crystalline structure) have to emerge simultaneously during precipitation (Sleutel and van Driessche, 2014). However, as stated by Ostwald (1897), in a system experiencing a transition from an unstable (or metastable) state to a stable one, and where multiple phases could form, the system does not go directly to the most stable conformation, instead the system may select intermediate stages where less stable phases form prior to the formation of the stable one. According to Ostwald, this first phase should be a (dense) liquid phase, while subsequent phases are typically solid (amorphous or crystalline). Note that the distinction between a dense liquid and an amorphous solid may not be straightforward. Such a distinction can, however, be made considering relevant physical-mechanical properties such as rheology/viscosity, contact angle and modulus. The above-described precipitation sequence is called the Ostwald step rule (also known as the Ostwald-Lussac or Ostwald-Weber rule of stages; De Yoreo and Vekilov, 2003; Gower, 2008; Meldrum and Cölfen, 2008; De Yoreo *et al.*, 2015). Its origin has been (and still is) a matter of controversy (Nyvlt, 1995), and several hypotheses, based either on thermodynamics or on kinetics, or both, have been proposed (Stranski and Totomanov, 1933; van Santen, 1984; Threlfall, 2003; Navrotsky, 2004).

One particular case of a non-classical multistage crystallization path defined by the Ostwald step rule involves the formation of dense liquid precursors. The formation of crystals *via* liquid-liquid phase separation (LLPS) of dense-liquid droplets has been suggested in the case of proteins (*e.g.* Velikov, 2004, 2010). In this so-called 'two-stage crystallization' (ten Wolde and Frenkel, 1997; Vekilov, 2004, 2010; Erdemir *et al.*, 2009), after the initial formation of the dense liquid precursor, nucleation of an ordered (crystalline) solid phase can take place (Chung *et al.*, 2010; Sleutel and van Driessche, 2014), involving or not a solid amorphous intermediate. Interestingly, this non-classical crystallization path is not restricted to the case of molecular crystals (*e.g.* proteins). Recent experimental and computational studies show that it is also followed in several inorganic systems, including the relevant $CaCO_3$-H_2O system (Wallace *et al.*, 2013). Wallace and co-workers (2013) used molecular dynamic computer simulations to demonstrate that the formation of ACC could be preceded by the formation of a dense liquid precursor phase after liquid-liquid phase binodal separation. Dehydration and coalescence of dense liquid droplets ultimately could lead to the formation of a solid amorphous phase. The same may occur at a very high supersaturation *via* spinodal decomposition leading to liquid-liquid phase separation preceding ACC formation (Fig. 9a; Faatz *et al.*, 2004). Indeed, it has been shown that the formation of ACC can involve an 'emulsion-like' or 'liquid-like' precursor phase (*e.g.* Rieger *et al.*, 2007; Wolf *et al.*, 2008; Bewernitz *et al.*, 2012). Unlike solid ACC, such a liquid precursor displayed features (necking among globules, difluence and/or spreading on surfaces) that reflected the typical wetting and/or flow characteristics of a

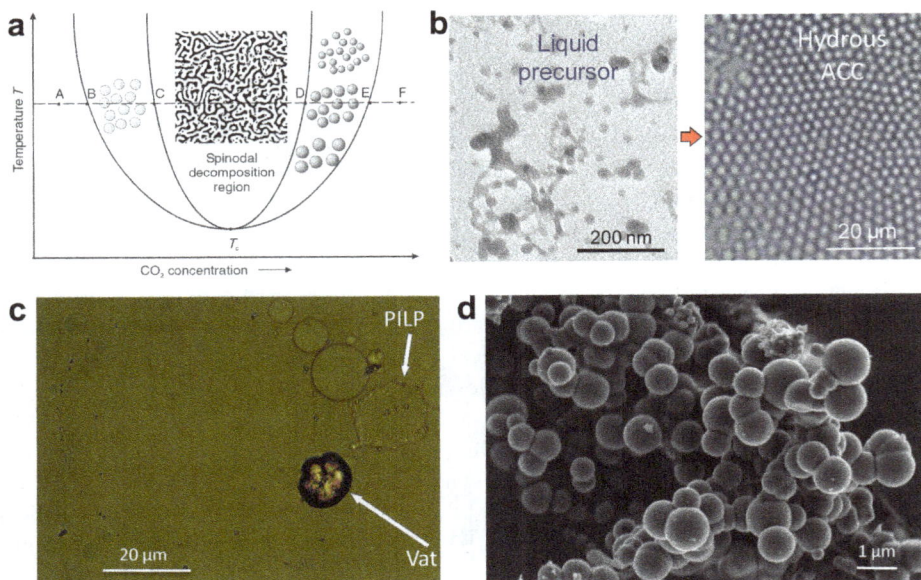

Figure 9. Formation of amorphous calcium carbonate *via* a liquid precursor. (a) Concentration *vs.* *T* in the system $CaCO_3$–H_2O. The phase diagram shows the binodal and spinodal regions, and the critical temperature, T_c. Morphologies of liquid-like precursor phases formed either by binodal separation (spherical droplets) or spinodal decomposition (bi-continuous phase) are depicted; (b) TEM image (*left*) of 'emulsion-like' (liquid-like) precursor formed *via* spinodal decomposition. This liquid evolves to an aggregate of ACC particles seen here under the optical microscope (*right*); (c) PILP phase formed in the $CaCO_3$–H_2O–Polyaspartic acid system. A vaterite precipitate formed after the PILP is indicated by an arrow; (d) representative SEM image of ACC nanoparticles. Image (a) from Faatz *et al.* (2004) reprinted with permission from Wiley; images (b) and (d) from Rodriguez-Navarro *et al.* (2015) reprinted with permission from the Royal Society of Chemistry; image in (c) from Gower and Odom (2000) reprinted with permission from Elsevier.

liquid (Xu *et al.*, 2005; Wolf *et al.*, 2008). Recently, Rodriguez-Navarro *et al.* (2015) showed that the formation of ACC nanoparticles in highly supersaturated additive-free solutions involved the formation of a dense liquid precursor phase after spinodal decomposition (Fig. 9b). Interestingly, the formation of liquid precursor phases seems to be promoted by organic additives, particularly poly(aspartic acid) and poly(acrylic acid; Gower, 2008). These are the so-called polymer-induced-liquid-precursors (PILPs; Gower and Odom, 2000). It has been claimed that a PILP-mediated mineralization could be general for the formation of biominerals (Oltzta *et al.*, 2003). Moreover, PILPs have been used for the biomimetic synthesis of several mineral structures (Fig. 9c; Gower, 2008). Ultimately, PILP-assisted synthesis routes may help understand the non-classical (from a crystallization point of view), complex interplay between organics and minerals forming in solution.

Typically, the solid phase formed after a liquid precursor is amorphous (Fig. 9d; Rodriguez-Navarro *et al.*, 2015). Again, the formation of an amorphous solid precursor phase does not fit within the classical crystallization picture where densification and long-range order have to emerge simultaneously, but it is consistent with the Ostwald step rule. Pioneering work by the Addadi and Weiner group showed that precursor amorphous phases precede the formation of crystalline phases in several biominerals (Addadi *et al.*, 2003). The existence of amorphous precursor phases has been reported for a range of biominerals, including calcium carbonates, iron oxihydroxides, and calcium phosphates (Towe and Lowenstam, 1967; Lowenstam and Weiner, 1985; Addadi *et al.*, 2003; Politi *et al.*, 2004; Mahamid *et al.*, 2008). Interestingly, the previous research group showed that ACC present in biominerals such as calcitic sea-urchin spines, displayed a short-range order which matched that of the final crystalline phase (Politi *et al.*, 2006). In other cases, ACC has been shown to display a vaterite or aragonite short-range order that matched that of the corresponding crystalline biomineral (Cartwright *et al.*, 2012). This has led to the concept of polyamorphism, that is, amorphous phases with similar composition but different protostructures (Cartwright *et al.*, 2012).

In the case of abiotic ACC, there is controversy regarding whether or not this amorphous precursor phase displays short- or medium-range order. While some researchers report that there is no close matching between the short-range order of synthetic ACC and that of any crystalline $CaCO_3$ phase (anhydrous and hydrated ones; Michel *et al.*, 2008, Nebel *et al.*, 2008), others claim that synthetic ACC may display short-range order corresponding to the structure of the crystalline phase formed after amorphous-to-crystalline transformation (Günther *et al.*, 2005; Gebauer *et al.*, 2010; Rodriguez-Navarro *et al.*, 2015), particularly when ACC precipitation occurs in the presence of additives (Lam *et al.*, 2007). Gebauer *et al.* (2010) concluded that the protostructure of ACC (with vaterite or calcite short-range order) is established at the prenucleation stage. These results suggest that PNCs can play a key role in the selection of a particular $CaCO_3$ polymorph.

In summary, liquid and amorphous solid precursor phases play critical roles in the non-classical crystallization of a range of minerals both in natural environments (biominerals) as well as in industrial or technologically relevant systems. There are, however, several aspects of their formation, transformation, structure and properties that are not currently fully understood, and are the focus of on-going research.

3.3. Oriented attachment and mesocrystal formation

Growing evidence shows that the growth (or coarsening) of crystals in solution does not necessarily take place *via* the classical layer-by-layer model proposed by Kossel (1927). Indeed, instead of the classical growth *via* monomer-by-monomer incorporation into a crystal lattice, a particle-mediated or aggregation-based growth commonly occurs (Zhang *et al.*, 2014; De Yoreo *et al.*, 2015). Such non-classical crystal growth can take place either *via* oriented attachment, or *via* mesocrystallization (Niederberger and Cölfen, 2006).

Penn and Banfield (1998) coined the term 'oriented attachment' (OA), equivalent to 'oriented aggregation' (Penn and Soltis, 2014). OA involves the particle-mediated growth of single crystals *via* self-assembly of nanocrystalline particles that 'attach' along equivalent crystallographic faces (Li *et al.*, 2012; Penn and Soltis, 2014; Zhang *et al.*, 2014). As a result, the whole system reduces its interfacial energy. OA is thus an energetically favourable growth mechanism. Cryo-TEM has shown that nanocrystals in solution can form aggregates limited by specific (*hkl*) faces, but, remarkably, each individual nanocrystal does not need to contact the surrounding nanocrystals (Yuwono *et al.*, 2010). These results suggest that the interfacial solution between nanocrystals can act as a 'bridge' transmitting the crystallographic information between nanocrystals undergoing OA. These results therefore provide strong evidence for the existence of ordered water molecules at the mineral-solution interface, and underline the fact that such ordered water might be critical for crystal coarsening *via* OA. Both experiments and molecular modelling show that water ordering/structuring can lead to its layering at the solid–solution interface (see Sathiyanarayanan *et al.*, 2011; and references therein). Such layering can in turn induce force oscillations both attractive and repulsive so that nanoparticles can reside in an attractive free energy minimum without aggregating prior to final fusion and formation of a single crystal (Sathiyanarayanan *et al.*, 2011). Direct *in situ* evidence demonstrating unambiguously that OA is a viable crystal-growth mechanism has been provided by fluid cell TEM observations for the case of iron oxyhydroxide nanoparticles (Li *et al.*, 2012). OA appears to be a very common growth mechanism, which under different names, over the last 100 years, has been reported to be operative in a range of systems (Ivanov *et al.*, 2014).

Cölfen and Antonietti (2005, 2008) coined the term 'mesocrystal' as an abbreviation for 'mesoscopically structured crystals'. They are colloidal crystals made up of individual crystalline nanoparticles arranged in crystallographic register *via* mesoscale aggregation and alignment (Niederberger and Cölfen, 2006; Meldrum and Cölfen, 2008). In this respect, mesocrystals have been considered as a required intermediate step of OA (Schwahn *et al.*, 2007). As a consequence, they diffract X-rays or electrons as a single crystal, but they can achieve forms that are not consistent with their crystal structure *(i.e.* they show 'forbidden' faces or curved surfaces; Fig. 10). Note however that mesocrystals may also form *via* colloidal assembly of amorphous nanoparticles, with the latter undergoing an amorphous-to-crystalline transformation (Song *et al.*, 2009; Song and Cölfen, 2010). Unlike crystals formed by OA, which ultimately will erase the boundaries between individual nanoparticles, mesocrystals typically display features that help to identify the individual nanoparticles that constitute the mesostructure. Indeed, the individual nanoparticle-building units can be identified because they are typically separated by amorphous organic or inorganic layers (Bergström *et al.*, 2015). Note that mesocrystals typically form in the presence of organic additives, although their presence is not a prerequisite for mesocrystallization (Song and Cölfen, 2010). In addition to entrapment of organics, the lack of perfect fusion among nanoparticles can also be due to the existence of mineral bridges or minor amorphous phases between nanocrystal subunits (Bergström *et al.*, 2015). Figure 10 shows an example of calcite mesocrystal formation (Song and Colfen, 2010).

Some biominerals, such as sea urchin spines, corals, eggshells and nacre, are thought to be mesocrystals (Seto *et al.*, 2012; Bergström *et al.*, 2015). Indeed, their outstanding physical-mechanical and optical properties have been related to their mesocrystalline structure (Bergström *et al.*, 2015). In parallel, biomimetic mesocrystals have been

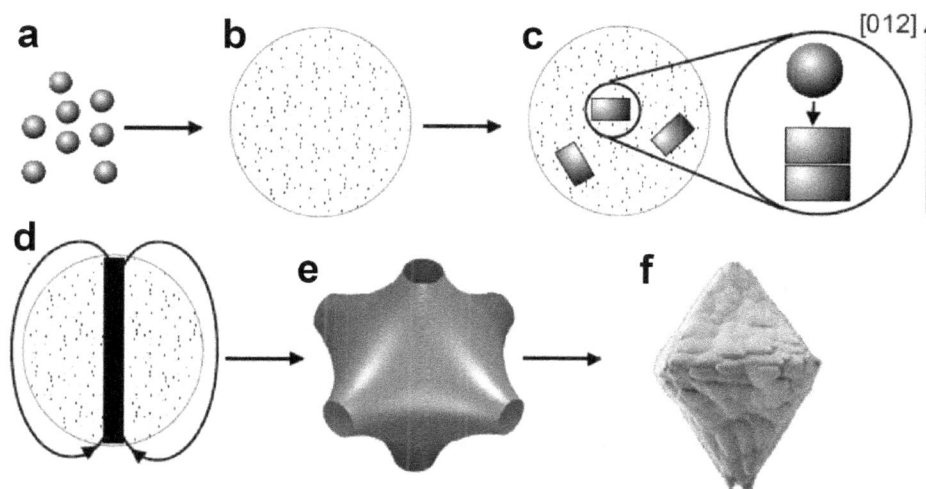

Figure 10. Pseudo-octahedral CaCO$_3$ mesocrystals formation in the presence of PSS-co-MA: (a) formation of ACC nanoparticles stabilized by the organic additive; (b) formation of liquid-like aggregates of ACC nanoparticles (dots); (c) crystallization of calcite nanoparticles with one charged high-energy face stabilized by the polyelectrolyte and further growth by oriented attachment of crystalline nanoparticles or ACC; (d) minimization of interaction between dipole fields of crystalline rods leads to three perpendicularly oriented rods (only one is depicted for the sake of clarity); (e) adhesion of liquid aggregate along the crystalline rod-like structure leads to P-surfaces where additional liquid-like ACC aggregates can attach; (f) further attachment of liquid-like phase leads to the scalloped surface of the mesocrystal. From Song *et al.* (2009) reproduced with permission from the American Chemical Society.

shown to display a range of exceptional properties that are finding applications in the rational design of, for instance, photocatalysts, electrodes, optoelectronics, and biomedical materials (Song and Cölfen, 2010; Zhou and O'Brien, 2012).

Another important but largely unknown aspect of the non-classical aggregation-based growth of crystals refers to the possibility of growth of an already formed crystal *via* the incorporation of nanoparticles, particularly amorphous ones. This situation is highly relevant in biomineralization, where mineral growth typically occurs sequentially as the organism develops (*e.g.* growth of the calcitic spicule of sea urchins or shell nacre). And, what is more important, most biominerals formed *via* amorphous precursors show a nanogranular texture (Gal *et al.*, 2015). This texture is characterized by aggregates of individual colloidal nanoparticles some tenths to a few hundred nm in size, and seems to be related to some of the unusual mechanical or optical properties of some biominerals (*e.g.* photonic crystals; Li *et al.*, 2015). In this respect, Gal *et al.* (2013, 2014) have shown that calcite rhombohedra growing in the presence of different organic additives typically display rough surfaces made up of nanogranular structures. Such nanogranular texture points to a growth process that does not proceed *via* a classical ion-mediated mechanism. The authors suggest that ACC nanoparticles stabilized by organic additives attached to calcite faces and after 2D diffusion, finally were incorporated onto macrosteps where they fused with the crystalline structure of the substrate. These observations led the authors to propose that such an ACC-mediated aggregation-based mechanism could explain the commonly observed nanogranular features of several biominerals (Gal *et al.*, 2015).

Recently, Rodriguez-Navarro *et al.* (2016) reported that calcite growth along specific crystallographic directions can occur *via* a non-classical particle-mediated colloidal growth mechanism that mirrors ion-mediated growth at the nanoscale, and involves the layer-by-layer attachment of amorphous calcium carbonate nanoparticles, followed by their restructuring and fusion with the substrate in perfect crystallographic registry (Fig. 11a). Such an ACC nanoparticle-mediated growth was observed directly using *in situ* AFM both in the absence and presence of organic additives (poly(acrylic acid); Fig. 11b). The authors also observed that faceted calcite crystals formed *via* ACC nanoparticle attachment in the absence of additives, initially displayed nanogranular rough surfaces. Upon aging involving dissolution-reprecipitation and ACC-to-calcite conversion, smooth surfaces resulted, similar to those observed following classical ion-mediated calcite growth. Conversely, crystals subjected to ACC nanoparticle-mediated growth in the presence of organic molecules retained nanogranular surface textures upon aging (and conversion of ACC into calcite). This particle-mediated growth mechanism may help to explain the nanogranular textural features of both biominerals and their biomimetics.

Finally, it should be noted that the existence of any of the above-mentioned non-classical crystal-growth mechanisms does not preclude the growth of minerals *via* a classical ion-mediated growth mechanism. Indeed, it has been shown that both non-classical and classical growth mechanism can take place in several mineral systems depending on their specific physical-chemical properties (*e.g.* pH, supersaturation, *T*

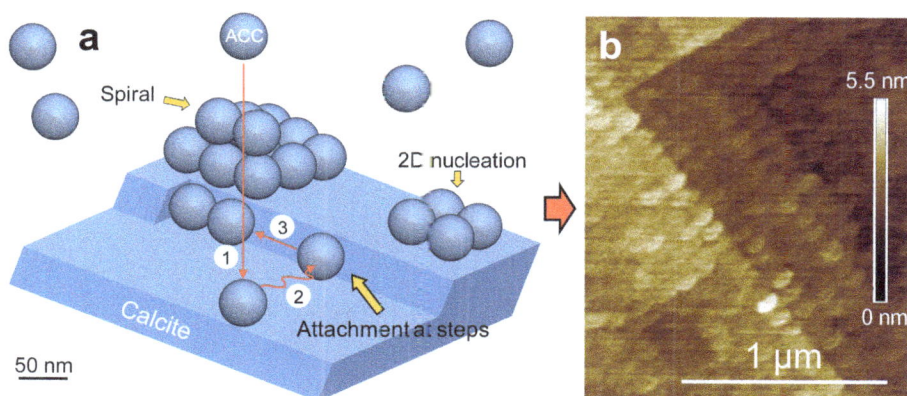

Figure 11. Calcite growth *via* ACC nanoparticle attachment. The diagram (a) shows macrospiral growth and 2D surface nucleation of ACC nanoparticles on (10$\bar{1}$4) calcite. Growth involves ACC nanoparticle bulk diffusion (1), surface diffusion (2) and final diffusion/attachment to a kink (3) present at a macrostep where restructuring and fusion with the calcite substrate occurs. The AFM height image (b) shows calcite growth *via* ACC attachment along steps according to this proposed non-classical growth mechanism. From Rodriguez-Navarro *et al.* (2016). Reproduced with permission from the American Chemical Society.

and absence/presence of additives; Rodriguez-Navarro *et al.*, 2016). Current on-going research is focused on shedding light on the actual conditions at which non-classical and classical mineral growth takes place, while trying to establish if these non-classical growth mechanisms are general in nature.

4. Dissolution

4.1. Mechanisms of mineral dissolution in aqueous solution: Congruent *vs.* incongruent dissolution

A mineral in contact with an aqueous solution with which it is out of equilibrium will continue to dissolve into the solution until it reaches saturation, *i.e.* the ion activity product (IAP) in the solution equals the solubility product of the mineral (K_{sp}). In general, the dissolution of a mineral A_cB_b can be described by

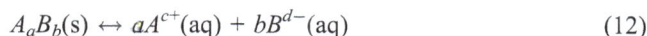

$$A_aB_b(s) \leftrightarrow aA^{c+}(aq) + bB^{d-}(aq) \tag{12}$$

The free energy difference (ΔG) between reactants and products in a system is defined

$$\Delta G = \Delta G_0 + RT\left(\frac{\{A^{c+}\}_{aq}^a \{B^{d-}\}_{aq}^b}{\{A_aB_b\}_s}\right) \tag{13}$$

where ΔG^o is the free energy difference of reaction at standard state conditions, R is the molar gas constant and T is temperature (Kelvin). $\{X\}_{aq}$ expresses the activity of the species X in the aqueous solution. For a pure solid, $\{X\} = 1$. $\{A^{c+}\}_{aq}^a\{B^{d-}\}_{aq}^b$ is IAP, the ion activity product. At the equilibrium, $\Delta G = 0$ and equation 14 results

$$\Delta G_0 = -RT\ln(\{A^{c+}\}^a\{B^{d-}\}^d)_{eq} \tag{14}$$

$(\{A^{c+}\}^a\{B^{d-}\}^d)_{eq}$ is the solubility product, K_{sp}, of the mineral A_aB_b under standard conditions. For a system that is not at equilibrium, the free energy can be written as

$$\Delta G = -RT\ln\left(\{A^{c+}\}^a\{B^{d-}\}^d\right)_{eq} + RT\ln\left(\{A^{c+}\}^a\{B^{d-}\}^d\right) = RT\ln\left(\frac{IAP}{K_{sp}}\right) \tag{15}$$

or

$$\frac{\Delta G}{RT} = \ln\left(\frac{IAP}{K_{sp}}\right) = \Omega \tag{16}$$

where Ω is the supersaturation of the solution with respect to the mineral A_aB_b and gives a measure of the deviation of the system from the equilibrium (*i.e.* $\Omega = 1$) and represents the driving force for dissolution. Since the 1970s, an increasing number of experimental studies aimed at formulating a universal rate law for dissolution (with the appropriate fitting parameters depending on the mineral considered). This rate law could thus be used to describe the dissolution behaviour of different minerals and provide the dependence of dissolution rates on solution parameters such as pH or concentration of specific solutes. Researchers have performed bulk measurements of mineral dissolution rates using mixed-flow reactors, determining the total fluxes of dissolved ions entering and leaving the reactor upon contact with the mineral(s) and normalizing the so-determined rates with respect to the initial total surface area, determined experimentally by gas adsorption techniques or, alternatively, calculated using geometric considerations. The derived rate equations, using assumptions from Transition State Theory (TST) and the Principle of Detailed Balancing (PDB), has the general form (Dove and Han, 2007):

$$r = k \cdot \left\{1 - \exp\left(\frac{\Delta G}{RT}\right)\right\}^n = k \cdot \left\{1 - \exp\left(\frac{IAP}{K_{sp}}\right)\right\}^n = k \cdot \{1 - \exp(\Omega)\}^n \tag{17}$$

where r is the rate of reaction expressed in moles of mineral reacted per unit time per unit area, k is the dissolution rate constant and the exponent n is a constant (not necessarily an integer) introduced to account for the control of surface defects on dissolution (Dove and Han, 2007).

These studies have undoubtedly provided key quantitative information regarding overall dissolution rates under specific experimental conditions (*e.g.* Oelkers, 2001; White and Brantley, 2003; Brantley, 2008; Brantley and Olsen, 2013). However, this experimental approach has problems and limitations. First, a critical issue in these studies is the accurate determination of the actual reactive surface area, as opposed to total surface area. It has been shown that grains of the same mineral with different crystallographic orientations exhibit different dissolution kinetics (Godinho *et al.*, 2012; Smith *et al.*, 2013; King *et al.*, 2014). Also, recent studies have shown that surface topography plays a major role in determining dissolution kinetics and mechanisms. Experimental studies performed using *in situ*, nanoscale imaging techniques that allow direct observations of reacting mineral surfaces such as atomic force microscopy (AFM) and interferometry have demonstrated that different sites on the mineral surface have different reactivity (Fischer *et al.*, 2012; Lüttge *et al.*, 2013; King *et al.*, 2014). The contribution of different sites to overall dissolution rates has been also studied by molecular modelling (*e.g.* Stack *et al.*, 2013).

One additional complication that may arise in dissolution studies is that many multicomponent minerals dissolve 'incongruently' or 'non-stoichiometrically' as deduced from the observation that the elemental molar ratios measured in solution during laboratory experiments differ from those in the bulk solid (see for example Murphy and Helgeson, 1989; Casey *et al.*, 1993; Pokrovsky and Schott, 2000; Brandt *et al.*, 2003; Hellmann *et al.*, 2003; Tisserand and Hellmann, 2008; Brantley, 2008; Daval *et al.*, 2011). There is a strong debate on the origin of such a phenomenon. Dolomite ($Ca_{0.5}Mg_{0.5}CO_3$) is an example of a mineral that shows this type of behaviour when dissolving in slightly acidified water, which has been commonly ascribed to the preferential release of calcium ions into the solution relative to magnesium and the subsequent development of a Mg-enriched surface layer (Busenberg and Plummer, 1982; Pokrovsky and Schott, 2001; Zhang *et al.*, 2007). This is claimed to be related to the lower stability of calcium at the dolomite surface due to its lower hydration energy compared to magnesium (Pokrovsky and Schott, 2001). However, recent *in situ*, nanoscale AFM observations of dolomite cleavage surfaces during dissolution in acidic solutions have challenged this traditionally accepted model (Ruiz-Agudo *et al.*, 2011c; Urosevic *et al.*, 2012; Putnis *et al.*, 2014; Fig. 12). These observations show that

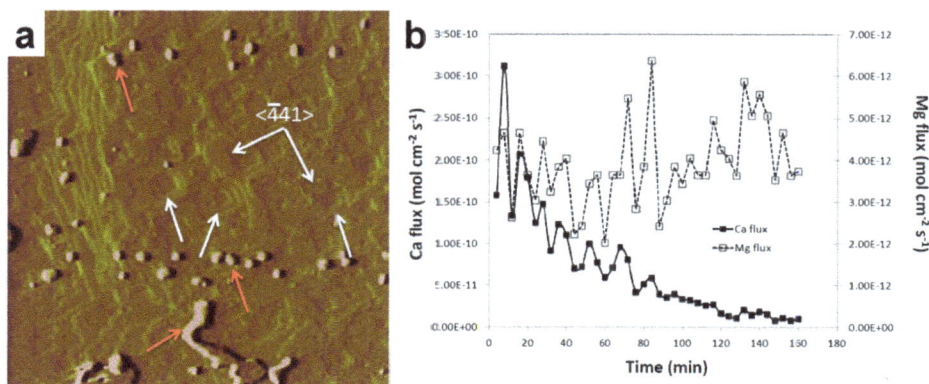

Figure 12. (a) AFM deflection image of dolomite cleavage surfaces dissolving in acidified water (HCl, pH = 3). Note the formation of regular etch pits with edges parallel to <$\bar{4}41$> directions (some of them are marked with a white arrow), simultaneously to the formation of a Mg-rich precipitate. (b) Plot of the Ca (left ordinate and solid line) and Mg (right ordinate and dotted line) fluxes in the outflow solution after reaction of dolomite with acidified water (pH = 3). Note the different flux scales for Ca and Mg. Both Ca and Mg show oscillating fluctuations with time reflected in the interface-coupled dissolution-precipitation processes occurring at the dolomite–water interface. Plot from Putnis *et al.* (2014), reprinted with the permission of the Mineralogical Society of Great Britain & Ireland.

dolomite dissolution occurs by the formation and spreading of regular rhombohedral etch pits the edges of which are parallel to the $<\bar{4}41>$ direction. This implies that dissolution is a stoichiometric process, because Ca and Mg ions are present in equal amounts along these steps edges. Furthermore, 'normal', stoichiometric dissolution of dolomite is followed by the precipitation of a Mg-rich phase (probably a hydrated Mg-carbonate phase) at the dissolving surface, an effect that is seen more clearly at acidic pH. Overall, these observations suggest that the 'non-stoichiometric' dissolution of dolomite is indeed related to a coupled dissolution−precipitation process occurring at the dolomite interface. This process may result in a cyclical variation of the solution chemistry measured in the solution leaving the reactor (Fig. 12; Putnis et al., 2014).

Similarly, most common silicate minerals and glasses seem to dissolve non-stoichiometrically or incongruently. Low-temperature chemical weathering of multicomponent silicate minerals is a key process for a wide range of geochemical processes (e.g. Weissbart and Rimstidt, 2000). Moreover, dissolution of silicate glasses also has implications in numerous processes such as the prediction of the durability of borosilicate glasses for long-term radionuclide retention in geological repositories or the weathering of volcanic glasses in contact with seawater (Hellmann et al., 2015). In this case, incongruent dissolution results in the formation of surface altered layers (SALs), which are zones at the solution−mineral interface enriched in one of the elements of the bulk mineral phase the thickness of which may reach up to several thousands of angstroms (Casey et al., 1993; Brantley, 2008; Hellmann et al., 2012) and that may significantly affect silicate dissolution rates. Due to their relevance for understanding a wide range of technological and geological processes, the mechanism of SALs formation has been the subject of much debate and controversy. As in the case of dolomite, a generally accepted model assumes that weakly bound elements are preferentially released to the solution (Weissbart and Rimstidt, 2000). According to this model these layers would form by the solid-state interdiffusion of protons into the solid and cations into the solution. Indeed, SALs are commonly referred to as 'leached' layers, a term that implies that they form by an element exchange through an otherwise inert silicate framework (Casey et al., 1989a,b, 1993; Petit et al., 1989, 1990; Schott et al., 2012).

However, multiple experimental observations reported during the last 15 years have questioned the validity of a leaching or diffusion mechanism for explaining the formation and main features of SALs found in experimentally and naturally altered silicate minerals and glasses, related to their 'incongruent' dissolution behaviour. Teng et al. (2001) suggested that the precipitation of a silica gel rather than a solid-state interdiffusion process could explain the apparent non-stoichiometric dissolution of orthoclase. Later on, Green and Lüttge (2006) performed VSI observations on wollastonite ($CaSiO_3$) dissolution at acidic pH and suggested that the observed 'swelling' of wollastonite upon contact with the acidic solution could be related to a precipitation event. Wollastonite has been selected commonly as a model system to study the formation of SALs due to its simple chemistry and the fact that it develops thick surface altered layers at room temperature at a fast rate. Few studies have followed this process *in situ*, though *in situ* AFM observations of the dissolution of wollastonite have provided evidence that SALs develop on silicate surfaces as a result of the congruent dissolution of wollastonite surfaces, as seen in the regular spreading of dissolution etch pits, followed by the precipitation of amorphous silica (Ruiz-Agudo *et al.*, 2012). This occurs in spite of the bulk solution being undersaturated with respect to amorphous silica; it has been experimentally shown that in fact it is just the fluid at the mineral interface that becomes supersaturated with respect to the secondary phase that then precipitates (see below, Ruiz-Agudo *et al.*, 2016; Fig. 13).

Together with these observations, the reported nm-sharp chemical and structural boundaries between the surface altered layer and the unaltered, pristine mineral do not

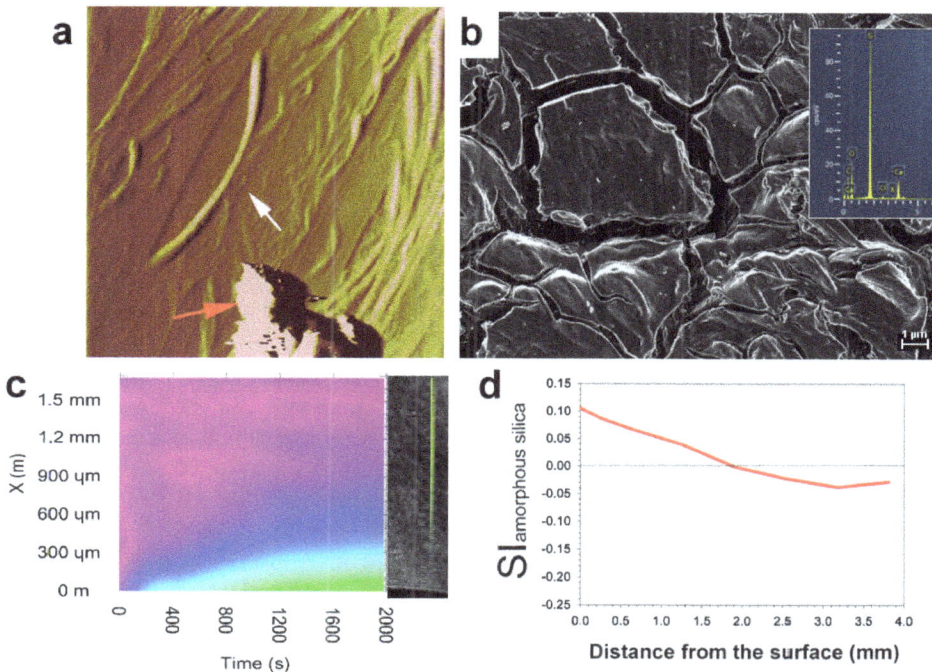

Figure 13. (a) AFM deflection image of a wollastonite, $CaSiO_3$ (001) surface. Note the formation of shallow dissolution pits (marked by a white arrow) as well as the simultaneous precipitation of a new phase (marked by a red arrow) on the dissolving surface. (b) FESEM image of a wollastonite surface reacted in an acidic solution (HCl, pH = 1.5), showing the formation of a surface layer with numerous cracks. EDX analysis (upper right corner) are in agreement with the altered layer being $SiO_{2(am)}$. (c) Time evolution of the refractive index of an acidic solution (HCl, pH = 1.5) in contact with a wollastonite surface along the green line marked in the image to the right. The refractive index is proportional to the concentration of the solution. Note the strong gradients established in the solution, with refractive index increasing from purple- to greenish colours. (d) Variation in the saturation index of an acidic solution (HCl, pH = 1.5) after 200 s of contact with a wollastonite surface as a function of the distance from the mineral surface, calculated from pH and calcium concentration measurements using microelectrodes and the stoichiometric amount of Si.

support a solid-state interdiffusion mechanism for the formation of SALs (Fig. 13). The significant progress in sample-preparation methods (especially focused ion beam, FIB) and analytical techniques (transmission electron microscopy (TEM) and atom probe tomography, APT) during recent decades allows the characterization of reaction interfaces with nm-scale spatial resolution. This in turn has provided additional experimental evidence challenging an interdiffusion mechanism (Hellmann et al., 2015). Most studies supporting such a solid-state mechanism (e.g. Casey et al., 1989a,b, 1993; Petit et al., 1989, 1990) use analytical techniques in which a wide beam hits the reaction interface, and this results commonly in false broad chemical profiles similar to diffusion profiles, but caused by the use of such wide beams. In addition, TEM results from the past 10 years are not in agreement with the expected broad sigmoidal profiles given by diffusion modelling (e.g. Hellmann et al., 2003, 2012). Moreover, dissolution experiments of silicate minerals (olivine, wollastonite) and glasses performed using ^{18}O-enriched acidic solution have shown that ^{18}O is incorporated in the silica-rich layer, as demonstrated by the peak shift in the Raman spectra (King et al., 2011; Geisler et al., 2015; Ruiz-Agudo et al., 2016). All in all, the more recent experimental evidence reported seems to indicate that the mechanism of 'incongruent dissolution' and SAL formation is equal to other mineral–solution equilibration processes. These observations can be interpreted in the framework of an interface-coupled dissolution-precipitation model, as a general mechanism for fluid–mineral interactions (e.g. Putnis, 2002, 2009; Ruiz-Agudo et al., 2014). Essentially, dissolution of multicomponent minerals such as dolomite or wollastonite in contact with aqueous solutions is a stoichiometric process that, depending on the fluid composition, may result in supersaturation of a thin layer of solution at the mineral surface with respect to a secondary phase (amorphous silica in the case of silicate minerals and glasses), while the bulk solution may remain undersaturated with respect to such a phase. This product phase then nucleates and grows on the surface of the pristine mineral. Further details and examples of systems in which this mechanism applies are given below.

Nevertheless, the debate about the mechanism of SAL formation is still on-going. Validation of the proposed dissolution-precipitation model would completely change current models explaining the mechanisms of chemical weathering, in the field and in the laboratory. Direct, in situ analysis of mineral surfaces and reacting interfacial fluids in a range of systems seems critical to obtain unequivocal experimental evidence that allows a better understanding of the mechanisms of incongruent mineral dissolution and leached layer formation, and thus unambiguous determination of the reaction mechanism. This is critical for an accurate evaluation of the dissolution kinetics of silicate glasses and major rock-forming minerals. This is particularly important for long-term prediction of the stability and durability of glasses intended for radioactive waste encapsulation and disposal (Grambow, 2006), as corrosion of the glass could occur if it comes into contact with ground water, thus enabling the mobilization of radioactive elements. Detailed knowledge of glass corrosion mechanisms is needed if the current proposals for nuclear waste storage are to provide safe and reliable long-term storage strategies.

Furthermore, accurate knowledge of mechanisms and rates of chemical weathering of silicate minerals on the Earth's surface is fundamental for precise modelling of the complex interplay between geochemical processes and Earth's climate evolution over geological time. C-cycle models used to estimate atmospheric CO_2 concentrations and global surface T over geological time scales depend on the calculations of CO_2 consumed during the chemical weathering of primary silicates (Walker *et al.*, 1981; Berner, 1990). A better understanding of the impact that SALs may have on silicate weathering and its dependence on T is required to extrapolate confidently dissolution rates determined in the laboratory to rates in natural systems, needed for accurate estimation of CO_2 fluxes and Earth's temperature over geological time. This is not straightforward, as recent results indicate that apparent incongruent dissolution and SAL formation depend on the relative rates of mass transport and surface reaction in the system (Ruiz-Agudo *et al.* 2016). Under conditions of fast fluid-flow rates used commonly in experimental studies, dissolution may be congruent and SALs might not form, which would result in the kinetics of the process being limited by the interfacial reaction; however, SALs could form under conditions closer to those of natural weathering systems, with low flow rates that would lead then to dissolution being rate-limited by transport. Moreover, variations in dissolution rate due to crystal anisotropy determine the formation or not of SALs under otherwise similar flushing conditions (*e.g.* Daval *et al.* 2013; Ruiz-Agudo *et al.* 2016). Calculations of palaeo-T and atmospheric CO_2 levels, performed using C-cycle models, result in higher CO_2 levels and temperatures when low activation energies (E_a) characteristic of a diffusion- or transport-controlled process (~5–10 kcal/mol; Brantley, 2008) are used, compared to those estimated assuming higher E_a determined in experiments in which SALs do not form (*i.e.* ~15–20 kcal/mol; Brantley, 2008; Brady, 1991). These calculations illustrate the effect that different weathering mechanisms and rate-controlling steps in experimental studies may have on calculations of palaeo-T and pCO_2 levels and thus for the prediction of future climate evolution.

4.2. The role of impurities in the kinetics of mineral dissolution

Minerals rarely grow or dissolve in pure environments. Water in contact with rock-forming minerals contains significant amounts of 'foreign' compounds, *i.e.* those different from the constituents of the solvent and crystal. Most aqueous fluids in the Earth contain Na^+, K^+ and Cl^- as well as many other ionic species. Mineral dissolution mechanisms and kinetics can vary greatly depending on the presence of ions in solution (*e.g.* Ruiz-Agudo *et al.*, 2010a). Components as different as small cations such as magnesium or large macromolecules such as polysaccharides and proteins can either modify the energetics of the mineral surface or the morphology of etch pits, cause enhancement or inhibition of rates of step retreat and etch-pit density, modify the anisotropy in step velocity propagation, provoke step bending, or become incorporated into the crystal (Lea *et al.*, 2001; Freij *et al.*, 2005; Vinson and Lüttge, 2005; Arvidson *et al.*, 2006; Teng *et al.*, 2006; Harstad and Stipp, 2007; Perez-Garrido *et al.*, 2007).

In recent decades, both macroscopic studies, and more importantly, studies performed using techniques such as AFM or interferometry have contributed to a deeper understanding of mineral-additive interactions during dissolution and growth. In particular, AFM studies have given direct insights into the terraces, edges or kinks that are affected by additives adsorbed or incorporated along specific crystallographic directions (*e.g.* Wang *et al.*, 2011). Thus, they provide critical kinetic and thermodynamic data which reflect additive-crystal interaction at the crystal surface during growth or dissolution. Dissolution processes can be strongly affected by the presence of small ions, even in ppb or ppm concentrations. The effects of ions on mineral dissolution have received much less attention than their influence on crystal growth, and have been explored mostly for the case of carbonates. In particular, much research has been devoted to study the effect of geologically relevant cations on calcite ($CaCO_3$) dissolution (*e.g.* Arvidson *et al.*, 2006; Ruiz-Agudo *et al.*, 2009; Xu and Higgins, 2011). Mg^{2+} inhibits calcite dissolution by reducing the etch-pit spreading rate, an effect that is enhanced under close-to-equilibrium conditions (Arvidson *et al.*, 2006; Xu and Higgins, 2011). Moreover, etch pit display a distorted morphology due to rounding of obtuse-obtuse corners in the presence of magnesium (Ruiz-Agudo *et al.*, 2009). Mn^{2+}, Sr^{2+} and Zn^{2+} also inhibit calcite dissolution and induce similar modifications in etch-pit morphologies (Lea *et al.*, 2001; Freij *et al.*, 2005; Arvidson *et al.*, 2006; Vinson *et al.*, 2007). Anions may also cause changes in the morphology of dissolution features, as has been reported for the dissolution of calcite in the presence of fluoride or sulfate ions, although these ions enhanced calcite dissolution kinetics (Ruiz-Agudo *et al.*, 2010b; Vavouraki *et al.*, 2010).

In addition, the interactions of heavy metals with carbonate surfaces have been studied in the framework of methods developed for the remediation of polluted waters. These methods rely on the fact that the adsorption or co-precipitation of toxic heavy metals upon contact with mineral surfaces strongly reduces their availability for the uptake by living organisms. Toxic elements such as nickel, cadmium, cobalt, mercury, selenium and arsenic show significant interactions with calcite surfaces during dissolution, which are reflected in changes in the morphology of the etch pits from their characteristic rhombohedral shape to elongated pseudo-elliptical morphology (Cd^{2+}), triangular (Hg^{2+}) or near-triangular (Co^{2+}) morphologies (Hoffmann and Stipp, 2001; Hay *et al.*, 2003; Godelitsas *et al.*, 2003; Freij *et al.*, 2004; Perez-Garrido *et al.*, 2009). Ultimately, at high metal concentrations, an additional mechanism, which consists of the precipitation of secondary phases including carbonate solid solutions, controls mineral–cation interaction (Godelitsas *et al.*, 2003; Lea *et al.*, 2003; Kamiya *et al.*, 2004; Chada *et al.*, 2005; Cubillas and Higgins, 2009; Perez-Garrido *et al.*, 2009). These secondary precipitates commonly inhibit dissolution, due to the formation of passivating layers on the pristine mineral surface.

Organic molecules also play a highly significant role in mineral-surface reactions. Many of the studies performed on this topic are related to the understanding of biomineralization, weathering processes in natural environments where organic molecules are ubiquitous and for the proper design of strategies to deal with scale formed in industrial processes. Knowledge of the influence of organics on mineral dissolution rates is critical for all these issues. Chelating agents such as EDTA increase significantly the dissolution kinetics of common scale minerals such as calcite or baryte

(Putnis *et al.*, 1995; Bosbach *et al.*, 1998; Kowacz *et al.*, 2009). On the contrary, other organic compounds such as phosphonate salts have been shown to inhibit dissolution of calcium sulfates (Pachon-Rodriguez and Colombani, 2013) and calcite (Ruiz-Agudo *et al.*, 2010a). Some of these studies conclude that changes in surface and solute hydration induced by organic molecules govern the effects of organic compounds on mineral growth and dissolution (Kowacz *et al.*, 2009; Ruiz-Agudo *et al.*, 2010a). Organic additives frequently exert a dual effect on mineral reactivity depending on their concentration. This has been shown for example in the case of brushite dissolution in the presence of citrate. Low concentrations of citrate (up to 100 µM) result in a decrease in step-retreat rates, while concentrations exceeding 0.1 mM result in acceleration in the rate of step retreat (Qin *et al.*, 2013).

Aside from these compounds, monovalent ions such as sodium, potassium or chloride are generally regarded as 'inert' towards mineral dissolution (*i.e.* they do not specifically interact with the dissolving mineral surface). Traditionally, the effect of these inert inorganic electrolytes on mineral-dissolution kinetics is thought to be related exclusively to a reduction in the activity of the building units of the crystal due to charge screening by the electric fields of the background ions ('ionic strength effect'; Buhmann and Dreybrodt, 1987). This results in decreased undersaturation (*i.e.* increased driving force for mineral dissolution) and, consequently, dissolution rate. However, recent studies have shown that the 'ionic strength effect' is ion-specific, and that ions with the same charge but different surface-charge densities produce different effects on mineral-dissolution kinetics, which cannot be exclusively understood considering the ions as point charges (Collins, 1997). These studies report differences in dissolution rates of carbonates (calcite and dolomite) and sulfates (baryte) measured in the presence of different 1:1 electrolytes at similar ionic strengths (Kowacz and Putnis, 2008; Ruiz-Agudo *et al.*, 2009, 2010b, 2011b). Crystal dissolution and growth kinetics are controlled by cation hydration-dehydration dynamics (Pokrovsky and Schott, 2002; Piana *et al.*, 2006), and thus dissolution and growth rates should be affected by any factor that can modify ion solvation, including the presence of other electrolytes. Cation hydration of ions building a crystal in electrolyte solutions is influenced by: (1) the stabilization of H_2O molecules in the hydration shells of calcite-building units by background ions present in solution; and (2) the interaction between background ions and water molecules (Kowacz and Putnis, 2008; Ruiz-Agudo *et al.*, 2009, 2010b, 2011b). Within this framework, trends observed in mineral-dissolution rates can be correlated to properties such as background ion separation in solution or the mobility of background ions (*i.e.* the volume of water affected by the presence of such ions; Kowacz and Putnis, 2008; Ruiz-Agudo *et al.*, 2009, 2010b, 2011b).

Additionally, background ions exert other effects such as enhancement in etch-pit nucleation density or changes in the morphology of dissolution features that can be interpreted considering the effect that ions in solution may have on surface hydration water. Enhancement in etch-pit nucleation density by strongly hydrated ions (*e.g.* F^- or Mg^{2+}) is related to a disruption of the surface hydration layer that facilitates the nucleation of etch pits (Ruiz-Agudo *et al.*, 2009, 2010b), while changes in etch-pit morphology can be interpreted considering that some background ions are able to increase hydration and thus stabilize non-stable polar faces that are normally not expressed in the morphology of the dissolution features formed when the mineral dissolves in pure water (Kowacz and Putnis, 2008; Ruiz-Agudo et al., 2010b).

5. Fluid-mediated replacement processes

5.1. Interface-coupled dissolution-reprecipitation as a mechanism of mineral-reequilibration in the presence of a fluid phase

Minerals, out of equilibrium with their environment typically within the rocks of the Earth, will constantly change in order to reach a new equilibrium. This may include phase changes by means of a solid-state diffusion mechanism (e.g. Abart et al., 2009; Petrishcheva and Abart, 2016). Given that atoms and ions are in constant motion, diffusion is a continuous process. This is a slow process although it increases significantly with increase in temperature. However, aqueous fluids are ubiquitous throughout the crust of the Earth. Whenever aqueous fluids are present, dissolution, precipitation and replacement reactions are orders of magnitude faster than solid-state diffusion and so exert a major control on the re-equilibration of minerals. Therefore solid-phase changes will not be considered in this chapter, except to note their on-going influence especially at higher temperatures.

Direct observations of mineral–fluid reactions have shown that the mechanism of most mineral re-equilibration processes in contact with aqueous fluids is common and can be described as a two-phase process: (1) on contact of the aqueous fluid with the mineral surface, the surface begins to dissolve, releasing ions to the fluid; (2) the release of these ions into the fluid will eventually cause it to approach saturation with respect to another phase. If supersaturation can be reached with respect to another phase, this new phase will precipitate. When the two processes of dissolution and precipitation are coupled at the interface, a pseudomorphic replacement process takes place (Putnis and Putnis, 2007; Putnis, 2009; Ruiz-Agudo et al., 2014). Examples of this type of observation are given in Fig. 14, which shows AFM images of the initial stages of calcite dissolution in the presence of a phosphate-rich solution (Wang et al., 2012a; Klasa et al., 2013) resulting in the precipitation of an apatite-type phase $(Ca_{10}(PO_4)_6(OH)_2)$. Similarly, during the interaction of phosphate-bearing solutions with cerussite $(PbCO_3)$, the dissolution of the latter is tightly coupled with the precipitation of a more stable phase (pyromorphite, $Pb_5(PO_4)_3Cl$). The kinetics of these coupled processes depends on parameters of the solution such as pH, ionic strength or the activity of phosphate species (Wang et al., 2012a). Moreover, quantitative measurements of brushite $(CaHPO_4 \cdot 2H_2O)$ growth kinetics onto gypsum $(CaSO_4 \cdot 2H_2O)$ dissolving surfaces clearly show the controlling effect of the solution composition on the coupling between growth and dissolution (e.g. Pinto et al., 2010).

The main features of the mechanism governing such processes can be visualized simply by using soluble salts whose thermodynamic properties have been well studied (Glynn et al., 1990; Lippmann, 1980; Dejewska, 1999) and reactions occur on a time scale enabling direct observations. For example, when a crystal of KBr is in contact

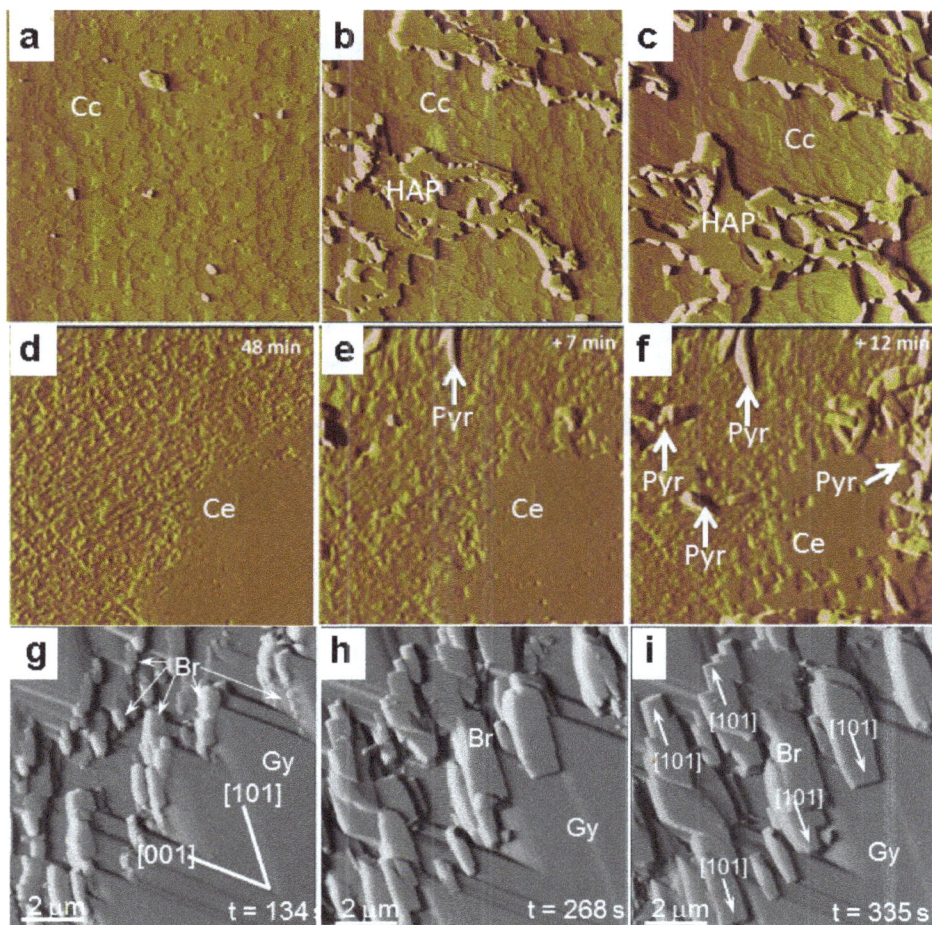

Figure 14. Sequences of AFM deflection images of the growth of (a–c) a Ca-phosphate phase (hydroxylapatite: HAP, $Ca_{10}(PO_4)_6(OH)_2$) on a dissolving calcite (Cc: $CaCO_3$) surface (modified from Klasa *et al.* (2013)), (d–f) pyromorphite (Pyr: $Pb_5(PO_4)_3Cl$) on a dissolving cerussite (Ce: $PbCO_3$) surface (modified from Wang *et al.* (2013)), and (g–i) brushite (Br: $CaHPO_4·2H_2O$) on a dissolving gypsum (Gy: $CaSO_4·2H_2O$) surface (modified from Pinto *et al.* (2013)).

with a saturated solution of KCl, the KBr crystal begins to dissolve but the interfacial fluid becomes supersaturated immediately with respect to a new KBr/KCl solid solution phase that precipitates simultaneously. This interfacial coupling of dissolution and precipitation of a new KBr/KCl phase occurs at the dissolving interface such that a crystal of KBr can be replaced pseudomorphically by a crystal of KCl. Figure 15 shows this reaction schematically (a–c) as well as SEM backscattered images (d–f) of the reaction arrested at various times to show the progression of the reaction (Putnis and Mezger, 2004; Putnis *et al.*, 2005). More KBr dissolves than KCl precipitates, *i.e.* this is

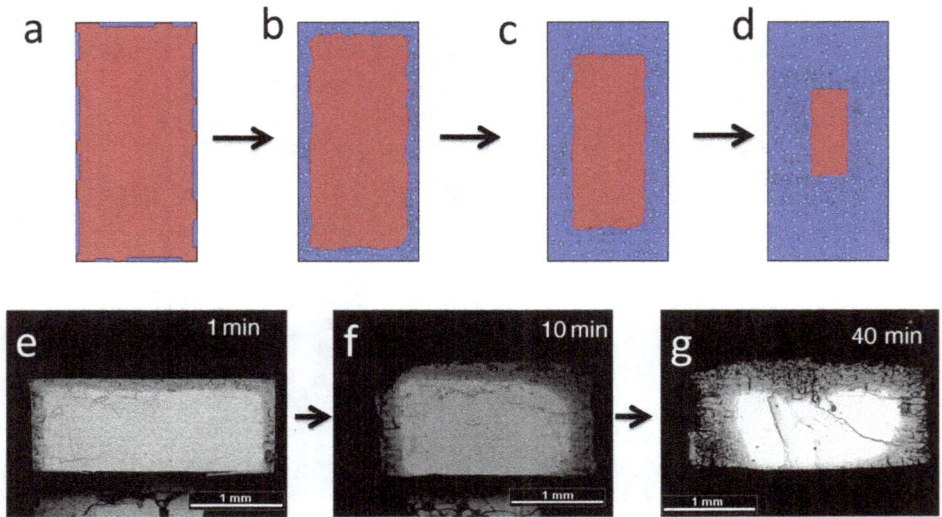

Figure 15. Pseudomorphic replacement by interface-coupled dissolution-precipitation. When a solid comes into contact with an aqueous fluid with which it is out of equilibrium, dissolution of even a few monolayers of this parent mineral may result in an interfacial fluid that is supersaturated with a product phase, which may nucleate on the surface. This is shown diagrammatically in (a) and (b−d) shows how continued dissolution and precipitation at the parent-product interface and the generation of interconnected porosity in the product phase allows the migration of the reaction interface from the surface through the parent phase, which is pseudomorphically replaced by the product. SEM electron backscattered images (e−g) of the pseudomorphic replacement of a KBr crystal by KCl, with the development of porosity seen in the product rim. Images taken from Ruiz-Agudo *et al.* (2014), reprinted by permission of Elsevier.

a volume-deficit reaction resulting in the formation of porosity in the new phase. This porosity generation is essential for the process of replacement to proceed as it allows access by the fluid to the reaction interface as it moves through the crystal being replaced. Inevitably more ions remain in solution after the reaction and these ions are transported in the fluid away from the reaction interface. The final result is a pseudomorphed crystal that is highly porous. This porosity is also a transient phase that may coarsen with time and even be eliminated (Putnis *et al.*, 2005). Textural re-equilibration within the reaction rim during calcite replacement by whewellite (King *et al.* 2014) has also been observed. Experiments have also shown that the porosity developed during a replacement reaction depends on two factors: (1) molar-volume differences between parent and product phases, and (2) solubility differences of the parent and product in the solution at the interface, this latter factor being important in determining if a coupled process can take place (Pollok *et al.*, 2011). If more solid volume is precipitated than dissolved, a porous product will not result and a dissolving mineral may be armoured by a non-porous layer of a new phase that would effectively passivate the surface from further reaction. As well as reaction-induced porosity, grain boundaries and any pre-existing fractures provide easy access of the fluid into the rock, and simultaneously coupled dissolution-precipitation replacement reactions occur normal to the grain boundary pathways (Fig. 16). This has been shown in the study of the replacement of both single crystals of calcite and mm-sized cubes of Carrara marble (almost pure calcite, $CaCO_3$) in hydrothermal experiments. In the latter case the marble is composed of compacted grains ~200 μm in diameter. In this example of a complete pseudomorphic replacement of a parent phase (calcite) by a product phase (apatite), the dissolution of the parent is coupled at the reaction interface with the precipitation of the product. Importantly the formation of an inter-connected porosity during the replacement enables fluid to continuously reach the reaction interface as it moves through the crystal (Jonas *et al.*, 2013, 2014; Etschmann *et al.*, 2014).

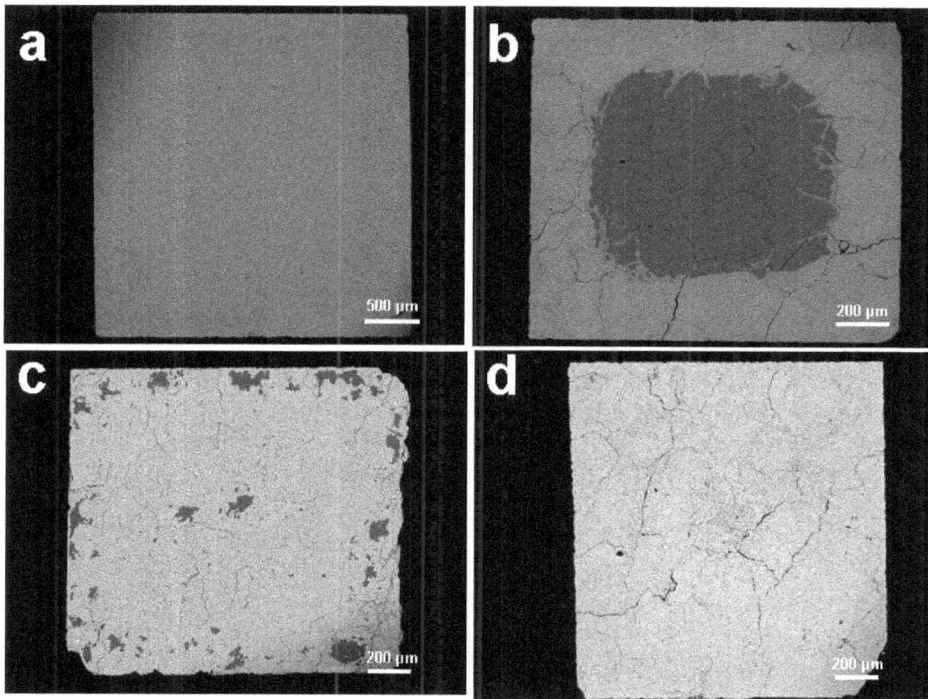

Figure 16. The pseudormorphic replacement of calcite by apatite. (a) 4 mm sized cube of Carrara marble (almost pure calcite $CaCO_3$) composed of compacted grains ~200 μm in diameter, (b) after 2 days in 0.5 M NH_4HPO_4 solution at 200°C, (c) reaction after 4 days, and (d) complete reaction after 6 days to form a pseudomorphed cube now consisting of apatite The grain boundaries, and any pre-existing fractures, clearly provide easy access by the fluid to the rock but simultaneously coupled dissolution-precipitation replacement reactions occur normal to the grain boundary pathways. Image courtesy of Elisabete Trindade Pedrosa.

5.2. The role of the boundary fluid–mineral interface in the control of replacement reactions.

The development of strong concentration gradients at the mineral–solution interface during replacement reactions has been demonstrated experimentally using real-time phase-shift interferometry during the replacement reaction of a KBr crystal in a KCl solution (Putnis et al., 2005). This apparently simple salt system has been shown to be useful as a model for more complex Earth systems involving mineral–fluid reactions during processes such as metasomatism, metamorphism and weathering. The main implication of the presence of such a gradient at the mineral–solution interface is that the dissolution of even a few monolayers of the parent solid may result in the fluid at the interface (boundary layer) becoming supersaturated with respect to the product phase/s, while the bulk solution may remain undersaturated. Indeed, this has been shown for the dissolution of a sparingly soluble mineral, such as wollastonite. During AFM flow-through dissolution experiments of wollastonite under acidic conditions, thermodynamic calculations indicate that the composition of effluent solutions is undersaturated with respect to amorphous silica (saturation indexes with respect to amorphous silica varied from -0.3 to -1.23), for both total Ca and the stoichiometric amount of Si, as well as the measured Si concentration. Yet, AFM observations and FESEM and MicroRaman spectroscopy analyses of the surface layers formed on wollastonite indicate that they are formed by an amorphous silica precipitate (Ruiz-Agudo et al., 2012b). Mach-Zhender Phase Shift Interferometry experiments and measurements performed using micro-ion selective electrodes show that the dissolution of even a few monolayers of the parent solid results in a steep compositional gradient at the mineral surface (Ruiz-Agudo et al., 2016). As a result, the fluid at the interface (boundary layer) becomes supersaturated with respect to amorphous silica that then precipitates, while the bulk solution remains undersaturated.

Apart from this direct evidence, the existence of a supersaturated fluid boundary layer has been inferred from experiments in a number of systems. For example, in situ flow-through AFM experiments of calcite dissolution in Se(IV) aqueous solutions show the formation of a surface precipitate while calcite is still dissolving. Geochemical modelling of these experiments demonstrate that even if equilibrium with calcite was reached, the bulk solution in the AFM fluid cell would be undersaturated with respect to any possible phase. Only assuming that calcite dissolves in a solution layer of thickness smaller than 10 µm, the solution becomes supersaturated with respect to $CaSeO_3 \cdot H_2O$, which could then form a surface precipitate (Putnis et al., 2013). The concept of an interfacial fluid that becomes supersaturated with respect to the product phase is essential in the description of interface-coupled dissolution-reprecipitation as a mechanism of mineral replacement, and relies heavily on the rate-limiting step in the reaction (Xia et al., 2009a,b; Ruiz-Agudo et al., 2016). This is ultimately given by properties of the interfacial fluid, including hydrodynamics of the solution (Ruiz-Agudo et al., 2016). The importance of understanding the properties of the fluid-mineral interface cannot thus be overestimated and has implications in every aspect of fluid-rock and fluid-mineral interaction as shown in some of the examples described below.

5.3. Coupled dissolution–precipitation reactions in the Earth's crust and applications in material synthesis and engineering processes

5.3.1. Coupled dissolution–precipitation reactions in the Earth's crust

Reactions similar to those described above occur in the Earth during such processes as metasomatism, metamorpism and weathering. One example is albitization (albite: $Na,AlSi_3O_8$), a common process when Ca- and K-rich feldspars come in contact with Na-rich aqueous fluids. Figure 17 shows two examples of albitization: (1) from the Curnamona Province in South Australia, and (2) the Bamble District in Southern Norway, respectively. Both of these areas are associated with significant mineralization. The former includes the major Broken Hill complex, one of the largest Pb-Zn-Ag deposits in the world, as well as many smaller Cu-Au mines alongside major uranium deposits. The latter area in Norway has many mineralized zones especially mined in the past for Fe-Cu-Ni sulfides as well as apatite. The proximity of albitized rocks to mineralized areas is significant. In both these cases Na-rich fluids have penetrated the rock through fractures and then reactions occur normal to these fractures. Fractures provide easy access for fluids into rocks. The albite phase appears pink in colour because of the presence of nano–micron-sized hematite crystals in pores within the

Figure 17. Images of two examples of albitization: (a) from the Curnamona Province in South Australia and (b) the Bamble District in southern Norway respectively. Fractures provide easy access pathways for fluid infiltration within the rocks and then replacement reactions have resulted in new albite precipitation and element mobilization. The pink colouration is caused by hematite nano-crystals in the albitized zones. Both these locations are associated with significant mineralization.

albite (Putnis *et al.*, 2007a) and the pink colour can be seen predominantly along fractures where fluids have first penetrated. This reaction has also been reproduced in the laboratory as seen in Fig. 18 from Hövelmann *et al.* (2010), when a Ca-rich plagioclase (oligoclase, $(Ca,Na),(Al,Si)_4O_8$) is reacted with an aqueous solution containing Na and Si under hydrothermal conditions, 2 kbar, 600°C, for 14 days. The original mineral is replaced by a Na-rich albite phase in a pseudomorphic reaction that proceeds from the grain surface inwards towards the centre of the grain at a very sharp (on a nm scale) interface. The new albite phase is porous, hence allowing for the reaction to continue within the mineral grain. As well as major-element mobilization during this process, minor elements have also been mobilized and hence are available for transport to other areas. The table of LA-ICP-MS (laser ablation–inductively coupled plasma-mass spectrometry) data presented by Hövelmann *et al.* (2010) shows the extent of mobilization of many minor elements during the interface-coupled dissolution-precipitation of albite. Within the rocks of the Earth, minerals may have

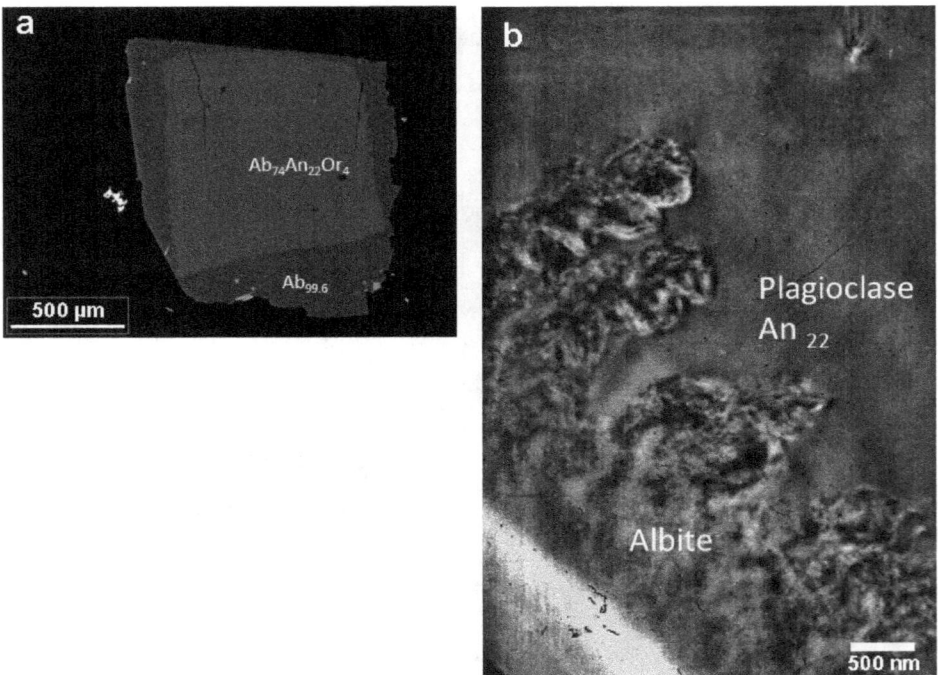

Figure 18. (a) Electron backscattered SEM image of a Ca-rich plagioclase (oligioclase: $(Ca,Na),(Al,Si)_4O_8$) reacted with an aqueous solution containing Na and Si under hydrothermal conditions, 2 kbar, 600°C for 14 days. The original mineral is replaced by a porous Na-rich albite phase in a pseudomorphic reaction that proceeds from the grain surface inwards towards the centre of the grain at a very sharp (on a nm scale) interface. (b) TEM image of the reaction interface. Turbidity in the albite rim results from the porous nature of the replacement. Images from Hövelmann *et al.* (2010) reprinted by permission of Springer.

been replaced, in which case the presence of porosity may be the only factor indicting that a fluid event and a replacement reaction has occurred. On the other hand interfaces are commonly observed when examining minerals at the macro scale, both in a light microscope or secondary electron microscopy (SEM). Figure 17 shows partial albitization at interfaces representing compositional changes that are clearly visible through the colour change.

Another common and interesting example of replacement is the reaction of leucite ($K[AlSi_2O_6]$) with Na-rich fluids such as seawater. Leucite occurs in mafic volcanic rocks but does not exist in rocks older than Tertiary (Deer *et al.*, 1992) because the replacement of leucite to analcime ($Na[AlSi_2O_6]$) occurs so easily. Figure 19 shows the partial replacement of leucite from Vesuvius lavas. The reaction is fast (days) at low *T* (100–150°C) and results in a pseudomorphic replacement of leucite by analcime that is highly porous (Putnis *et al.*, 2007b). This example is interesting as the reaction has a molar-volume increase but nevertheless the replacement proceeds at a compositionally sharp interface. The solubility of analcime must be lower than leucite (in the fluid at the reaction interface) allowing more leucite to dissolve into solution than the amount of analcime that precipitates. This indicates that solubility differences between parent and product are more important than simply molar volume changes in determining whether a replacement process can proceed (Pollok *et al.*, 2011). In the leucite-analcime example, interfacial strain results in reaction-induced fracturing (Jamtveit *et al.*, 2009) and this may be the case in general when a replacement has a molar volume increase. Other relevant examples were reviewed by Putnis and Fernandez-Diaz (2010).

5.3.2. Applications of dissolution–precipitation in material-synthesis and engineering processes

With knowledge of the mechanism of fluid-mediated replacement processes and how the development of porosity in a product phase can be controlled, reactions can be designed for production of materials with specific purposes, such as engineered materials from precursors that may be easier (cost and energy efficient) to manufacture or even products designed for improved drug delivery. Xia *et al.* (2009a) have shown how a hydrothermal pseudomorphic replacement route can be used to synthesize three-dimensional ordered arrays of zeolite nanocrystals. Precursor leucite crystals contain an inherent 3D ordered network of nanometer-sized lamellar twins that was preserved precisely during pseudomorphic replacement reactions in NaCl solutions, resulting in analcime nanocrystals. Furthermore the morphology of the nanocrystals is tunable by simply changing solution pH values. The replacement follows a coupled dissolution-reprecipitation mechanism where the rate of leucite dissolution equals the rate of analcime precipitation. Xia *et al.* (2009b) stated that this pseudomorphic replacement route has the potential to synthesize other ordered arrays of functional nanocrystals with controlled shape, size and crystallographic orientation. These new phases can be tailored to specific functional use, such as the manufacture of advanced materials with designed properties, especially the formation of porosity. The essential step in the control of all replacement reactions is modifying the composition of the fluid at the

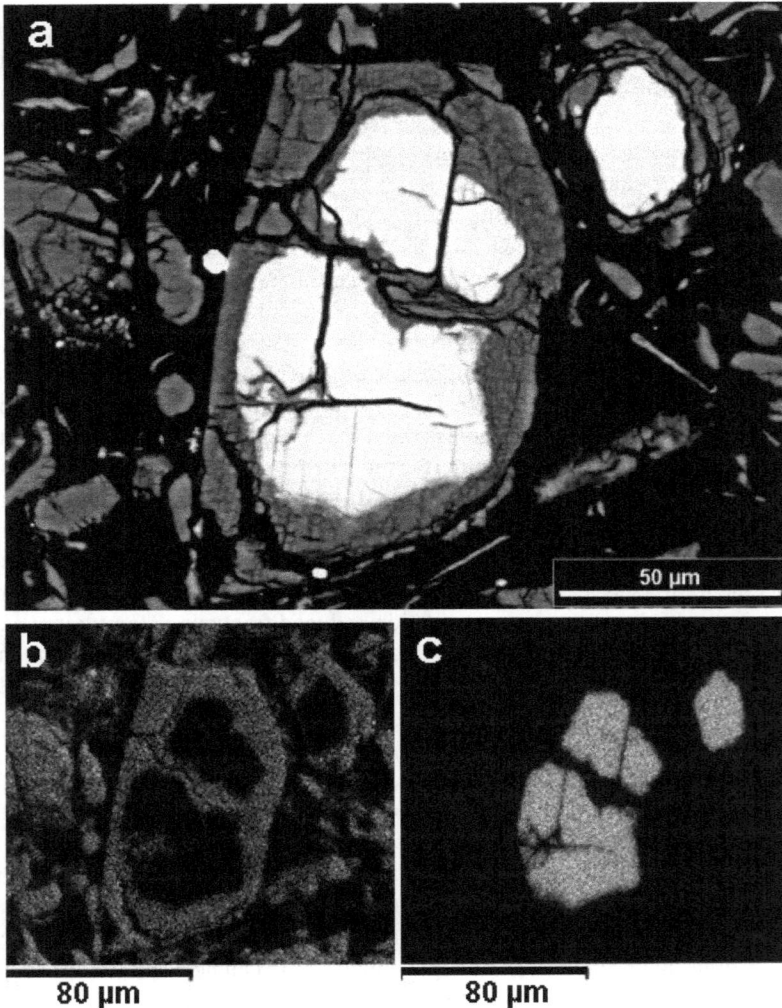

Figure 19. (a) Electron backscattered SEM image of the partial replacement of leucite (lighter centre) by analcime (darker rim). The replacement rim is porous and the interface is sharp on a nanoscale. (b) EDX map of Na and (c) K. Image from Putnis *et al.* (2007b) reprinted by permission of The Mineralogical Society of America.

mineral–fluid interface. Xia *et al.* (2009b) have also shown that an interface-coupled dissolution-precipitation reaction can be uncoupled by altering the solution pH, emphasizing the control that the solution composition has on the reaction.

Coupled dissolution-reprecipitation reactions, similar to hydrothermal mineral replacements have been presented by Brugger *et al.* (2010) showing novel routes for the synthesis of mesoporous and low thermal-stability materials. This principle has also been exploited to create porous coordination polymer (PCP) crystals by pseudomorphic replacement of a specifically shaped metal oxide (Reboul *et al.*, 2012). Other examples of compositional control and designed products include the use of apatite formed from the replacement of a carbonate such as calcite or aragonite (Kasioptas *et al.*, 2010). There is a clinical need for new bone-replacement materials that combine long implant life with compatibility and appropriate mechanical properties. Dissolution-precipitation is a process that has been proposed for the synthesis of porous biocompatible material for bone implants (Heness and Ben-Nissan, 2004).

6. Concluding remarks

The rates of fluid–mineral reactions control most geochemical natural and technological processes, including biomineralization, weathering, soil formation, the interaction of minerals with CO_2-saturated water, the remediation of contaminated waters or the durability of glasses used as containers for nuclear waste materials. This explains the urgent need to obtain a deep understanding of the mechanism of mineral growth, dissolution and replacement reactions. During the 1970s and 1980s, most investigations evaluated the rates and mechanisms of growth and dissolution using bulk, indirect methods based on measurements of the evolution of solution chemistry. With the development of high-resolution analytical and *in situ* imaging techniques, emphasis can now be focused on obtaining a new understanding of the microscopic processes that govern mineral reactivity. Studies using these new methods show that processes taking place during mineral–solution interactions are more complex than envisioned by classical approaches and theories. An increasing number of investigations are reporting crystallization processes that cannot be explained exclusively assuming the formation of critical nuclei and subsequent growth by monomer-by-monomer incorporation into the crystal lattice. In addition, 'incongruent' or 'non-stoichiometric' dissolution is essentially equal to any other mineral reequilibration process in which the dissolution of a mineral results in the fluid at the mineral surfaces becoming supersaturated with respect to a new mineral phase, which nucleates within this interfacial region. Demonstrating that these phenomena are general to all mineral-fluid systems remains a research challenge for the future, as well as the determination of the conditions under which these processes control the kinetics of mineral-solution reactions. Measurements of mineral-reactions rates under such conditions are needed in order to advance the quantitative prediction of the evolution of our environment.

Acknowledgements

The authors acknowledge support from the Spanish Government (grants MAT2012-37584-ERDF and CGL2012- 35992) and the Junta de Andalucia (research group RNM-179 and project P11-RNM-7550). E. Ruiz-Agudo also acknowledges the Spanish Ministry of Economy and Competitiveness for financial support through a Ramón y Cajal grant. C.V. Putnis acknowledges funding from the EU Marie Curie Initial Training Networks: CO2 React; FlowTrans and Minsc. Ostwald

References

Abart, R., Petrishcheva, E., Fischer, F.D. and Svoboda, J. (2009) Thermodynamic model for diffusion controlled reaction rim growth in a binary system: Application to the forsterite-enstatite-quartz system. *American Journal of Science*, **309**, 114−131.

Adamson, A.W. and Gast, A.P. (1997) *Physical Chemistry of Surfaces* (6th edition). Wiley, New York.

Addadi, L., Raz, S. and Weiner, S. (2003) Taking advantage of disorder: Amorphous calciumcarbonate and its roles in biomineralization. *Advanced Materials*, **15**, 959−970.

Aizenberg, J., Black, A.J. and Whitesides, G.M. (1999) Control of crystal nucleation by patterned self-assembled monolayers. *Nature*, **398**, 495−498.

Arvidson, R.S., Collier, M., Davis, K.J., Vinson, M.D., Amonette, J.E. and Lüttge, A. (2006) Magnesium inhibition of calcite dissolution kinetics. *Geochimica et Cosmochimica Acta*, **70**, 583−594.

Baumgartner, J.,Dey, A., Bomans, P.H.H., Le Coadou, C., Fratzl, P., Sommerdijk, N.A.J.M. and Faivre, D. (2013) Nucleation and growth of magnetite from solution. *Nature Materials*, **12**, 310−314.

Becker, R. and Doring, W. (1935) Kinetische Behandlung der Keimbildung in übersättigen Dämpfen. *Annalen der Physik*, **24**, 719−752.

Bergström, L., Sturm, E.V., Salazar-Alvarez, G. and Cölfen, H. (2015) Mesocrystals in biominerals and colloidal arrays. *Accounts of Chemical Research*, **48**, 1391−1402.

Berner, R.A. (1990) Atmospheric carbon dioxide levels over Phanerozoic time. *Science*, **21**, 1382−1386.

Bewernitz, M.A., Gebauer, D., Long, J., Cölfen, H. and Gower, L.B. (2012) A metastable precursor phase of calcium carbonate and its interactions with polyaspartate. *Faraday Discussions*, **199**, 291−312.

Bosbach, D., Hall, C. and Putnis, A. (1998) Mineral precipitation and dissolution in aqueous solution: in-situ microscopic observations on barite (001) with atomic force microscopy. *Chemical Geology*, **151**, 143−160

Brandt, F., Bosbach, D., Krawczyk-Bärsch, E., Arnold, T. and Bernhard, G. (2003) Chlorite dissolution in the acid pH-range: A combined microscopic and macroscopic approach. *Geochimica et Cosmochimica Acta*, **67**, 1451−1461.

Brantley, S.L. (2008) Kinetics of mineral dissolution. Pp. 151−210 in: *Kinetics of Water–Rock Interaction* (S.L. Brantley, J.D. Kubicki and A.F. White, editors). Springer, New York.

Brantley, S.L. and Olsen, A. (2013) Reaction kinetics of primary rock-forming minerals under ambient conditions. Pp. 73−117 in: *Surface and Ground Water, Weathering and Soils* (J.I. Drever, editor). Treatise on Geochemistry, **5**, Elsevier, New York.

Brady, P.V. (1991) The effect of silicate weathering on global temperature and atmospheric CO_2. *Journal of Geophysical Research: Solid Earth*, **96**, 18101−18106.

Brugger, J., McFadden, A., Lenehan, C. E., Etschmann, B., Xia, F., Zhao, J. and Pring, A. (2010) A novel route for the synthesis of mesoporous and low-thermal stability materials by coupled dissolution-reprecipitation reactions: mimicking hydrothermal mineral formation. *Chimia*, **64**, 693−698.

Buhmann, D. and Dreybrodt, W. (1987) Calcite dissolution kinetics in the system $H_2O-CO_2-CaCO_3$ with participation of foreign ions. *Chemical Geology*, **64**, 89−102.

Burton, W.K., Cabrera, N. and Frank, F.C. (1951) The growth of crystals and the equilibrium structure of their surfaces. *Royal Society of London, Philosophical Transactions*, **A243**, 299−358.

Busenberg, E. and Plummer, L.N. (1982) The kinetics of dissolution of dolomite in CO_2-H_2O systems at 1.5 to 65°C and 0 to 1 atm pCO_2. *American Journal of Science* **282**, 45−78.

Cartwright, J.H.E., Checa, A., Gale, J.D., Gebauer, D. and Sainz-Díaz, I. (2012) Calcium carbonate polymorphism and its role in biomineralization: How many amorphous calcium carbonates are there? *Angewandte Chemie, International Edition*, **51**, 11960−11970.

Casey, W.H., Westrich, H.R., Arnold, G.W. and Banfield, J.F. (1989a) The surface chemistry of dissolving labradorite feldspar. *Geochimica et Cosmochimica Acta*, **53**, 821−832.

Casey, W.H., Westrich, H.R., Massis, T., Banfield, J.F. and Arnold, G.W. (1989b) The surface of labradorite feldspar after acid hydrolysis. *Chemical Geology*, **78**, 205−218.

Casey, W.H., Westrich, H.R., Banfield, J.F., Ferruzzi, G. and Arnold, G.W. (1993) Leaching and reconstruction at the surfaces of dissolving chain silicate minerals. *Nature*, **366**, 253−255.

Chada, V.G.R., Hausner, D.B., Strongin, D.R., Rouff, A.A. and Reeder, R.J. (2005) Divalent Cd and Pb uptake on calcite {101 4} cleavage faces: An XPS and AFM study. *Journal of Colloid and Interface Science*, **288**, 350−360.

Chernov, A.A. (1984) *Modern Crystallography III, Crystal Growth.* Springer, Berlin,.

Chung, S., Shin, S-H., Bertozzi, C.R. and De Yoreo, J.J. (2010) Self-catalyzed growth of S layers via an amorphous-to-crystalline transition limited by folding kinetics. *Proceedings of the National Academy of Sciences, USA*, **107**, 16536−16541.

Colfen, H. and Antonietti, M. (2005) Mesocrystals: Inorganic superstructures made by highly parallel crystallization and controlled alignment. *Angewandte Chemie, International Edition*, **44**, 5576−5591.

Cölfen, H. and Antonietti, M. (2008) *Mesocrystals and Non-classical Crystallization.* Wiley, San Francisco, California, USA.

Collins, K.D. (1997) Charge density-dependent strength of hydration and biological structure. *Biophysical Journal*, **72**, 65−76.

Cubillas, P. and Higgins, S.R. (2009) *In situ* friction characteristics of Cd-rich carbonate films on calcite surfaces: implications for compositional differentiation at the nanometer scale. *Geochemical Transactions*, **10**, 7.

Daval, D., Sissmann, O., Menguy, N., Saldi, G.D., Guyot, F., Martinez, I., Corvisier, J., Garcia, B., Machouk, I., Knauss, K.G. and Hellmann, R. (2011) Influence of amorphous silica layer formation on the dissolution rate of olivine at 90°C and elevated p_{CO_2}. *Chemical Geology*, **284**, 193−209.

Daval, D., Hellmann, R., Saldi, G.D., Wirth, R. and Knauss, K.G. (2013) Linking nm-scale measurements of silicate surface reactivity to macroscopic dissolution rate laws: new insights based on diopside. *Geochimica et Cosmochimica Acta*, **107**, 121−134.

Davis, K.J., Dove, P.M. and De Yoreo, J.J. (2000) The role of Mg^{2+} as an impurity in calcite growth. *Science*, **290**, 1134−1137.

De Yoreo, J.J. and Vekilov, P.G. (2003) Principles of crystal nucleation and growth. Pp. 57−93 in: *Biomineralization* (P.M. Dove, J.J. De Yoreo and S. Weiner, editors). Reviews in Mineralogy and Geochemistry, **54**. Mineralogical Society of America and the Geochemical Society, Chantilly, Virginia, USA.

De Yoreo, J.J., Orme, C.A. and Land, T.A. (2001) Using atomic force microscopy to investigate solution crystal growth. Pp. 361−380 in: *Advances in Crystal Growth Research* (K. Sato, Y. Furukawa and K. Nakajima, editors). Elsevier Science, Amsterdam.

De Yoreo, J.J., Zepeda-Ruiz, L.A., Friddle, R.W., Qui, S.R., Wasylenki, L.E., Chernov, A.A., Gilmer G.H. and Dove, P.M. (2009) Rethinking classical crystal growth models through molecular scale insights: Consequences of kink-limited kinetics. *Crystal Growth and Design*, **9**, 5135−3144.

De Yoreo, J.J., Gilbert, P.U.P.A., Sommerdijk, N.A.J.M., Penn, R.L., Whitelam, S., Joester, D., Zhang, H., Rimer, J.D., Navrotsky, A., Banfield, J.F., Wallace, A.F., Michel, F.M., Meldrum, F.C., Cölfen, H. and Dove, P.M. (2015) Crystallization by particle attachment in synthetic, biogenic and geologic environments. *Science*, **349**, DOI: 10.1126/science.aaa6760.

Deer, W.A., Howie, R.A. and Zussman, J. (1992) *An Introduction to the Rock-Forming Minerals.* Pearson Education Limited, Harlow, Essex, UK.

Dejewska, B. (1999) The characteristics of the mixed crystals of the KBr−KCl−H₂O system at 298 K. *Crystal Research and Technology*, **34**, 975−979.

Demichelis, R., Raiteri, P., Gale, J.D., Quigley, D. and Gebauer, D. (2011) Stable prenucleation mineral clusters are liquid-like ionic polymers. *Nature Communications*, **2**, 590.

Dorvee, J.R. and Veis, A. (2013) Water in the formation of biogenic minerals: Peeling away the hydration layers. *Journal of Structural Biology*, **183**, 278−303.

Dove, P.M. and Han, N. (2007) Kinetics of mineral dissolution and growth as reciprocal microscopic surface processes across chemical driving force. Pp. 215−234 in: *Perspectives on Inorganic, Organic and Biological Crystal Growth: From Fundamentals to Applications Directions* (M. Skowronski, J.J. DeYoreo and C. Wang, editors) American Institute of Physics Conference Series, **916**.

Erdemir, D., Lee, A.Y. and Myerson, A.S. (2009) Nucleation of crystals from solution: Classical and two-step models. *Accounts of Chemical Research*, **42**, 621−629.

Etschmann, B., Brugger, J., Pearce, M.A., Ta, C., Brautigan, D., Jung, M. and Pring, A. (2014) Grain boundaries as microreactors during reactive fluid flow: experimental dolomitization of a calcite marble. *Contributions to Mineralogy and Petrology* **168**, 1–12.

Faatz, M., Gröhn, F. and Wegner, G. (2004) Amorphous calcium carbonate: Synthesis and potential intermediate in biomineralization. *Advanced Materials*, **16**, 996–1000.

Fischer, C., Arvidson, R.S. and Lüttge, A. (2012) How predictable are dissolution rates of crystalline material? *Geochimica et Cosmochimica Acta* **98**, 177–185.

Frank, F.C. (1949) The influence of dislocations on crystal growth. *Discussions of the Faraday Society*, **5**, 48–54.

Freij, S.J., Godelitsas, A. and Putnis, A. (2005) Crystal growth and dissolution processes at the calcite–water interface in the presence of zinc ions. *Journal of Crystal Growth*, **273**, 535–545.

Freij, S.J., Putnis, A. and Astilleros, J.M. (2004) Nanoscale observations of the effect of cobalt on calcite growth and dissolution. *Journal of Crystal Growth*, **267**, 288–300.

Gal, A., Habraken, W., Gur, D., Fratzl, P., Weiner, S. and Addadi, L. (2013) Calcite crystal growth by a solid-state transformation of stabilized amorphous calcium carbonate nanospheres in a hydrogel. *Angewandte Chemie, International Edition*, **52**, 4867–4870.

Gal, A., Kahil, K., Vidavsky, N., DeVol, R.T., Gilbert, P.U.P.A., Fratzl, P., Weiner, S. and Addadi, L. (2014) Particle accretion mechanism underlies biological crystal growth from an amorphous precursor phase. *Advanced Functional Materials*, **24**, 5420–5426.

Gal, A., Weiner, S. and Addadi, L. (2015) A perspective on underlying crystal growth mechanisms in biomineralization: solution mediated growth versus nanosphere particle accretion. *CrystEngComm*, **17**, 2606–2615.

García-Ruiz, J.M. (2003) Nucleation of protein crystals. *Journal of Structural Biology*, **142**, 22–31.

Gebauer, D. and Cölfen, H. (2011) Prenucleation clusters and non-classical nucleation. *Nanotoday*, **6**, 564–584.

Gebauer, D., Völkel, A. and Cölfen, H. (2008) Stable prenucleation clusters. *Science*, **322**, 1819–1822.

Gebauer, D., Cölfen, H., Verch, A. and Antonietti, M. (2009) The multiple roles of additives in CaCO$_3$ crystallization: A quantitative case study. *Advanced Materials*, **21**, 435–439.

Gebauer, D., Gunawidjaja, P.N., Ko, J.Y.P., Bacsik, Z., Aziz, B., Liu, L., Hu, Y., Bergström, L., Tai, C.-W., Edán, M. and Hedin, N. (2010) Proto-calcite and proto-vaterite in amorphous calcium carbonate. *Angewandte Chemie, International Edition*, **49**, 8889–8891.

Gebauer, D., Kellermeier, M., Gale, J.D., Bergström,L. and Cölfen, H. (2014) Pre-nucleation clusters as solute precursors in crystallization. *Chemical Society Reviews*, **43**, 2348–2371.

Geisler, T., Nagel, T., Kilburn, M.R., Janssen, A., Icenhower, J.P., Fonseca, R.O.C., Grange, M. and Nemchin, A.A. (2015) The mechanism of borosilicate glass corrosion revisited. *Geochimica et Cosmochimica Acta*, **158**, 112–129.

Gibbs, J.W. (1876) On the equilibrium of heterogeneous substances (first part). *Transactions of the Connecticut Academy of Arts and Sciences*, **3**, 108– 248.

Gibbs, J.W. (1878) On the equilibrium of heterogeneous substances (concluded). *Transactions of the Connecticut Academy of Arts and Sciences*, **16**, 343– 524.

Glynn, P.D., Reardon, E.J., Plummer, L.N. and Busenberg, E. (1990) Reaction paths and equilibrium end-points in solid solution-aqueous solution systems. *Geochimica et Cosmochimica Acta*, **54**, 267–282.

Godelitsas, A., Astilleros, J.M., Hallam, K.R., Löns, J. and Putnis, A. (2003) Microscopic and spectroscopic investigation of the calcite surface interacted with Hg(II) in aqueous solutions. *Mineralogical Magazine*, **67**, 1193–1204.

Godinho, J.R.A., Piazolo, S. and Evins, L.Z. (2012) Effect of surface orientation on dissolution rates and topography of CaF$_2$. *Geochimica et Cosmochimica Acta*, **86**, 392–403.

Gower, L.B. (2008) Biomimetic model systems for investigating the amorphous precursor pathway and its role in biomineralization. *Chemical Review*, **108**, 4551– 4627.

Gower, L.B. and Odom, D.J. (2000) Deposition of calcium carbonate firlms by a polymer-induced liquid-precursor (PILP) process. *Journal of Crystal Growth*, **210**, 719–734.

Green, E. and Lüttge, A., (2006) Incongruent dissolution of wollastonite measured with vertical scanning interferometry. *American Mineralogist* **91**, 430–434.

Grambow, B. (2006) Nuclear waste glasses – how durable? *Elements*, **2**, 357–364.

Günther, C., Becker, A., Wolf, G. and Epple, M. (2005) In vitro synthesis and structural characterization of amorphous calcium carbonate. *Zeitschrift für Anorganische Allgemeine Chemie*, **631**, 2830−2835.

Habraken, W.J.E.M., Tao, J., Brylka, L.J., Freidrich, H., Bertinetti, L., Schenk, A.S., Verch, A., Dmitrovic, V., Bomans, P.H.H., Frederik, P.M., Laven, J., van der Schoot, P., Aichmayer, B., de With, G., De Yoreo, J.J. and Sommerdijk, N.A.J.M. (2013) Ion-association complexes unite classical and non-classical theories for the biomimetic nucleation of calcium phosphate. *Nature Communications*, **4**, 1507.

Harstad, A.O. and Stipp, S.L.S. (2007) Calcite dissolution: Effects of trace cations naturally present in Iceland spar calcites. *Geochimica et Cosmochimica Acta*, **71**, 56−70.

Hay, M.B., Workman, R.K. and Manne, S. (2003) Mechanisms of metal ion sorption on calcite: Composition mapping by lateral force microscopy. *Langmuir*, **19**, 3727−3740.

Hellmann, R., Penisson, J.-M., Hervig, R.L., Thomassin, J-H., and Abrioux, M.-F. (2003) An EFTEM/HRTEM high-resolution study of the near surface of labradorite feldspar altered at acid pH: evidence for interfacial dissolution-reprecipitation. *Physics and Chemistry of Minerals* **30**, 192−197.

Hellmann, R., Wirth, R., Daval, D., Barnes, J-P., Penisson, J-M., Tisserand, D., Epicier, T., Florin, B. and Hervig, R.L. (2012) Unifying natural and laboratory chemical weathering with interfacial dissolution-reprecipitation: A study based on the nanometer-scale chemistry of fluid-silicate interfaces. *Chemical Geology*, **294−295**, 203−216.

Hellmann, R., Cotte, S., Cadel, E., Malladi, S., Karlsson, L.S., Lozano-Perez, S., Cabié, M. and Seydeux, A. (2015) Nanometre-scale evidence for interfacial dissolution−reprecipitation control of silicate glass corrosion. *Nature Materials*, **14**, 307−311.

Heness, G. and Ben-Nissan, B. (2004) Innovative bioceramics. *Materials Forum*, **27**, 104−114.

Hoffmann, U. and Stipp, S.L.S. (2001) The behaviour of Ni^{2+} on calcite surfaces. *Geochimica et Cosmochimica Acta*, **65**, 4131−4139.

Hövelmann, J., Putnis, A., Geisler, T., Schmidt, B.C. Golla-Schindler, U. (2010) The replacement of plagioclase feldspars by albite: observations from hydrothermal experiments. *Contributions to Mineralogy and Petrology* **159**, 43−59.

Ivanov, V.K., Fedorov, P.P., Baranchikov, A.Y. and Osiko, V.V. (2014) Oriented attachment of particles: 100 years of investigations of non-classical crystal growth. *Russian Chemical Reviews*, **83**, 1204−1222.

Jamtveit, B., Putnis, C.V. and Malthe-Sørenssen, A. (2009) Reaction induced fracturing during replacement processes. *Contributions to Mineralogy and Petrology* **157**, 127−133.

Jonas, L., John, T. and Putnis, A. (2013) Influence of temperature and Cl on the hydrothermal replacement of calcite by apatite and the development of porous microstructures. *American Mineralogist*, **98**, 1516−1525.

Jonas, L., John, T., King, H.E., Geisler, T. and Putnis, A. (2014) The role of grain boundaries and transient porosity in rocks as fluid pathways for reaction front propagation. *Earth and Planetary Science Letters*, **386**, 64−74.

Kamiya, N., Kagi, H., Tsunomori, F., Tsuno, H. and Notsu, K. (2004) Effect of trace lanthanum ion on dissolution and crystal growth of calcium carbonate. *Journal of Crystal Growth*, **267**, 635−645.

Kasioptas, A., Geisler, T., Putnis, C.V., Perdikouri, C. and Putnis, A. (2010) Crystal growth of apatite by replacement of an aragonite precursor. *Journal of Crystal Growth*, **312**, 2431−2440.

King, H.E., Plümper, O., Geisler, T. and Putnis, A. (2011) Experimental investigations into the silicification of olivine: Implications for the reaction mechanism and acid neutralization. *American Mineralogist* **96**, 1503−1511.

King, H.E., Satoh, H., Tsukamoto, K. and Putnis, A. (2014) Surface specific measurements of olivine dissolution by phase-shift interferometry. *American Mineralogist*, **99**, 377−386.

Klasa, J., Ruiz-Agudo, E., Wang, L.J., Putnis, C.V., Valsami-Jones, E., Menneken, M. and Putnis, A. (2013) An atomic force microscopy study of the dissolution of calcite in the presence of phosphate ions. *Geochimica et Cosmochimica Acta*, **117**, 115−128.

Kossel, W. (1927) Zur Theorie des Kristallwachstums. *Nachrichten von der Gesellschaft der Wissenschaften zu Göttingen, Mathematisch-Physikalische Klasse*, 135−143.

Kowacz, M. and Putnis, A. (2008) The effect of specific background electrolytes on water structure and solute hydration: Consequences for crystal dissolution and growth. *Geochimica et Cosmochimica Acta*, **72**, 4476−4487.

Kowacz, M., Putnis, C.V. and Putnis, A. (2009) The control of solution composition on ligand-promoted dissolution: DTPA−barite interactions. *Crystal Growth and Design*, **9**, 5266−5272.

Lam, R.S.K., Charnock, J.M., Lennie, A. and Meldrum, F.C. (2007) Synthesis-dependant structural variations in amorphous calcium carbonate. *CrystEngComm*, **9**, 1226−1236.

Larsen K., Bechgaard K. and Stipp S.L.S. (2010a) The effect of the Ca^{2+} to CO_3^{-3} activity ratio on spiral growth at the calcite $\{1\ 0-1\ 4\}$ surface. *Geochimica et Cosmochimica Acta*, **74**, 2099−2109.

Larsen, K., Bechgaard, K. and Stipp S.L.S. (2010b) Modelling spiral growth at dislocations and determination of critical step lengths from pyramid geometries on calcite $\{1\ 0\ \bar{1}\ 4\}$ surfaces. *Geochimica et Cosmochimica Acta* **74**, 558−567.

Lea, A.S., Amonette, J.E., Baer, D.R., Liang, Y. and Colton, N.G. (2001) Microscopic effects of carbonate, manganese and strontium ions on calcite dissolution. *Geochimica et Cosmochimica Acta*, **65**, 369−379.

Lea, A.S., Hurt, T.T., El-Azab, A., Amonette, J.E. and Baer, D.R. (2003) Heteroepitaxial growth of a manganese carbonate secondary nano-phase on the $(1\ 0\ \bar{1}\ 4)$ surface of calcite in solution. *Surface Science*, **524**, 63−77.

Li, D., Nielsen, M.H., Lee, J.R.I., Frandsen, C., Banfield, J.F. and De Yoreo, J.J. (2012) Direction-specific interactions control crystal growth by oriented attachment. *Science*, **336**, 1014−1018.

Li, L., Kolle, S., Weaver, J.C., Ortiz, C., Aizenberg, J. and Kolle, M. (2015) A highly conspicuous mineralized composite photonic architecture in the translucent shell of the blue-rayed limpet. *Nature Communications*, **6**, 6322.

Liang, Y., Baer, D.R., McCoy, J.M., Amonette, J.E. and LaFemina, J.P. (1996a) Dissolution kinetics at the calcite-water interface. *Geochimica et Cosmochimica Acta*, **60**, 4883−4887.

Liang Y., Baer, D.R., McCoy, J.M., Amonette, J.E. and LaFemina, J.P. (1996b) Interplay between step velocity and morphology during the dissolution of $CaCO_3$ surface. *Journal of Vacuum Science and Technology A*, **14**, 1368−1375.

Lippmann, F. (1980) Phase diagrams depicting the aqueous solubility of binary mineral systems. *Neues Jahrbuch für Mineralogie Abhandlungen*, **139**, 1−25.

Lowenstam, H.A. and Weiner, S. (1985) Transformation of amorphous calcium phosphate to crystalline dahillite in the radular teeth of chitons. *Science*, **227**, 51−53.

Lupulescu, A.I. and Rimer, J.D. (2014) *In situ* imaging of silicalite-1 surface growth reveals the mechanism of crystallization. *Science*, **344**, 1729−732

Lüttge, A., Arvidson, R.S. and Fischer, C. (2013) A stochastic treatment of crystal dissolution kinetics. *Elements*, **9**, 183−188.

Mahamid, J., Sharir, A., Addadi, L. and Weiner, S. (2008) Amorphous calcium phosphate is a major component of the forming fin bones of zebrafish: Indications for an amorphous precursor phase. *Proceedings of the National Academy of Sciences, USA*, **105**, 12748−12753.

Malkin, A.J., Kuznetsov, Y.G., Land, T.A., De Yoreo, J.J. and McPherson, A. (1995) Mechanisms of growth for protein and virus crystals. *Nature Structural Biology*, **2**, 956−959.

Meldrum, F.C. and Cölfen, H. (2008) Controlling mineral morphologies and structures in biological and synthetic systems. *Chemical Reviews*, **108**, 4332−4432.

Meldrum, F.C. and Sears, R.P. (2008) Now you see them. *Science*, **322**, 1802−1803.

Michel, F.M., MacDonald, J., Feng, J., Phillips, B.L., Ehm, L., Tarabrella, C., Parise, J.B. and Reeder, R.J. (2008) Structural characteristics of synthetic amorphous calcium carbonate. *Chemistry of Materials*, **20**, 4720−4728.

Mullin, J.W. (2001) *Crystallization* (4th edition). Butterworth, Oxford, UK.

Murphy, W.M. and Helgeson, H.D. (1989) Thermodynamics and kinetic constraints on reaction rates among minerals and aqueous solutions. IV. Retrieval of rate constants and activation parameters for the hydrolysis of pyroxene, wollastonite, olivine, andalusite, quartz, and nepheline. *American Journal of Science*, **289**, 17−101.

Mutaftschiev, B. (2001) *The Atomistic Nature of Crystal Growth*. Springer, Berlin.

Myerson, A.S. (2002) *Handbook of Industrial Crystallization* (2nd edition). Butterworth, Boston, Massachusetts, USA.

Navrotsky, A. (2004) Energetic clues to pathways to biomineralization: Precursors, clusters, and nanoparticles. *Proceedings of the National Academy of Sciences, USA*, **101**, 12096−12101.

Nebel, H., Neumann, M., Mayer, C. and Epple, M. (2008) On the structure of amorphous calcium carbonate – A detailed study by solid-state NMR spectroscopy. *Inorganic Chemistry*, **47**, 7874–7879.

Niederberger, M. and Cölfen, H. (2006) Oriented attachment and mesocrystals: Non-classical crystallization mechanisms based on nanoparticle assembly. *Physical Chemistry and Chemical Physics*, **8**, 3271–3287.

Nielsen, A.E. (1984) Electrolyte crystal growth mechanisms. *Journal of Crystal Growth*, **67**, 289–310.

Nielsen, M.H., Aloni, S. and De Yoreo, J.J. (2014) *In situ* TEM imaging of CaCO₃ nucleation reveals coexistence of direct and indirect pathways. *Science*, **345**, 1158–1162.

Nudelman, F. and Sommerdijk, N.A.J.M. (2012) Biomineralization as an inspiration for materials chemistry. *Angewandte Chemie, International Edition*, **51**, 6582–8596.

Nývlt, J. (1995) The Ostwald rule of stages. *Crystal Research and Technology*, **30**, 445–451

Oelkers, E.H. (2001) General kinetic description of multioxide silicate mineral and glass dissolution. *Geochimica et Cosmochimica Acta* **65**, 3703–3719.

Olszta, M.J., Odom, D.J., Douglas, E.P. and Gower, L.B. (2003) A new paradigm for biomineral formation: Mineralization via an amorphous liquid-phase precursor. *Connective Tissue Research*, **44**, 326–334.

Ostwald, W. (1897) Studien über die Bildung und Umwandlung fester Körper. *Zeitschrift für Physikalische Chemie*, **22**, 289–330.

Ostwald, W. (1901) *Analytische Chemie*, 3rd edition. Engelmann, Leipzig, Germany.

Pachon-Rodriguez, E.A. and Colombani, J. (2013) Pure dissolution kinetics of anhydrite and gypsum in inhibiting aqueous salt solutions. *AIChE Journal*, **59**, 1622–1626

Penn, R.L. and Banfield, J.F. (1998) Imperfect oriented attachment: Dislocation generation in defect-free nanocrystals. *Science*, **281**, 969–971.

Penn, R.L. and Soltis, J.A. (2014) Characterizing crystal growth by oriented aggregation. *CrystEngComm*, **16**, 1409–1418.

Perez-Garrido, C., Fernández-Díaz, L., Pina, C.M. and Prieto, M. (2007) *In situ* AFM observations of the interaction between calcite surfaces and Cd-bearing aqueous Solutions. *Surface Science*, **601**, 5499–5509.

Perez-Garrido, C., Astilleros, J.M., Fernandez-Diaz, L. and Prieto, M. (2009) *In situ* AFM study of the interaction between calcite {10Ī4} surface and supersaturated Mn²⁺-CO₃²⁻ aqueous solutions. *Journal of Crystal Growth*, **311**, 4730–4739

Petit, J.C., Dran, J.C., Paccagnella, A. and Della Mea, G. (1989) Structural dependence of crystalline silicate hydration during aqueous dissolution. *Earth and Planetary Science Letters*, **93**, 292–298.

Petit, J.C., Dran, J.C. and Della Mea, G. (1990) Energetic ion beam analysis in the earth sciences. *Nature*, **344**, 621–626.

Petrishcheva, E. and Abart, R. (2017) Diffusion in mineral reaction kinetics: microstructures, textures, chemical and isotopic signatures. Pp. 255–294 in: *Mineral Reaction Kinetics: Microstructures, Textures and Chemical Compositions* (R. Abart and W. Heinrich, editors). EMU Notes in Mineralogy, **16**. European Mineralogical Union and Mineralogical Society of Great Britain & Ireland, London.

Piana, S., Jones, F. and Gale, J.D. (2006) Assisted desolvation as a key kinetic step for crystal growth. *Journal of the American Chemical Society*, **128**, 13568–13574.

Pina, C.M., Becker, U., Risthaus, P., Bosbach, D. and Putnis, A. (1998) Molecular-scale mechanisms of crystal growth in barite. *Nature*, **395**, 483–486.

Pinto, A.J., Ruiz-Agudo, E., Putnis, C.V., Putnis, A., Jiménez, A. and Prieto, M. (2010) AFM study of the epitaxial growth of brushite (CaHPO₄.2H₂O) on gypsum cleavage surfaces. *American Mineralogist* **95**, 1747–1757.

Pokrovsky, O.S. and Schott, J. (2000) Forsterite surface composition in aqueous solutions: a combined potentiometric, electrokinetic, and spectroscopic approach. *Geochimica et Cosmochimica Acta*, **64**, 3299–3312.

Pokrovsky, O.S. and Schott, J. (2001) Kinetics and mechanism of dolomite dissolution in neutral to alkaline solutions revisited. *American Journal of Science* **301**, 597–626.

Pokrovsky, O.S. and Schott, J. (2002) Surface chemistry and dissolution kinetics of divalent metal carbonates. *Environmental Science & Technology*, **36**, 426–432.

Politi, Y., Arad, T., Klein, E., Weiner, S. and Addadi, L. (2004) Sea urchin spine calcite forms via a transient amorphous calcium carbonate phase. *Science* **306**, 1161–1164.

Politi, Y., Levi-Kalisman, Y., Raz, S., Wilt, F., Addadi, L. and Weiner, S. (2006) Structural characterization of the transient amorphous calcium carbonate precursor phase in sea urchin embryos. *Advanced Functional Materials*, **16**, 1289−1298.

Pollok, K., Putnis, C.V. and Putnis, A. (2011) Mineral replacement reactions in solid solution-aqueous solution systems: Volume changes, reaction paths and end points using the example of model salt systems. *American Journal of Science*, **311**, 211−236.

Pouget, E.M., Bomans, P.H.H., Goos, J.A.C.M., Frederik, P.M., de With, G. and Sommerdijk, N.A.J.M. (2009) The initial stages of template-controlled $CaCO_3$ formation revealed by cryo-TEM. *Science*, **323**, 1455−1458.

Putnis, A. (2002) Mineral replacement reactions:from macroscopic observations to microscopic mechanisms. *Mineralogical Magazine*, **66**, 689−708.

Putnis, A. (2009) Mineral replacement reactions. Pp. 87−124 in: *Thermodynamics and Kinetics of Water–Rock Interaction* (E.H. Oelkers and J. Schott, editors). Reviews in Mineralogy and Geochemistry, **70**, Mineralogical Society of America and the Geochemical Society, Chantilly, Virginia, USA.

Putnis, C.V. and Fernandez-Diaz, L. (2010) Ion partitioning in low temperature aqueous systems: from fundamentals to applications in climate, proxies and environmental geochemistry. Pp. 189−226 in: *Ion Partitioning in Ambient-temperature Aqueous Systems* (M. Prieto and H. Stoll, editors). EMU Notes in Mineralogy, **10**. European Mineralogical Union and Mineralogical Society of Great Britain & Ireland, London.

Putnis, C.V. and Mezger, K. (2004) Isotope tracing of a mineral replacement reaction: The KCl-KBr-H_2O system as a model example. *Geochimica et Cosmochimica Acta*, **66**, A618.

Putnis, A. and Putnis, C.V. (2007) The mechanism of reequilibration of solids in the presence of a fluid phase. *Journal of Solid State Chemistry*, **180**, 1783−1786.

Putnis, C.V. and Ruiz-Agudo, E. (2013) The mineral-water interface: where minerals react with the environment. *Elements*, **9**, 177−182.

Putnis, A., Junta-Rosso, J.L. and Hochella, M.F. (1995) Dissolution of barite by a chelating ligand: An atomic force microscopy study. *Geochimica et Cosmochimica Acta*, **59**, 4623−4632.

Putnis, C.V., Tsukamoto, K. and Nishimura, Y. (2005) Direct observations of pseudomorphism: compositional and textural evolution at a fluid–solid interface. *American Mineralogist*, **90**, 1909−1912.

Putnis, A., Hinrichs, R., Putnis, C.V., Golla-Schindler, U. and Collins, L.G. (2007a) Hematite in porous red-clouded feldspars. Evidence of large-scale crustal fluid−rock interaction. *Lithos*, **95**, 10−18.

Putnis, C.V., Geisler, T., Schmid-Beurmann, P., Stephan, T. and Giampaolo, C. (2007b) An experimental study of the replacement of leucite by analcime. *American Mineralogist*, **92**, 19−26.

Putnis, C.V., Renard, F., King, H.E., Montes-Hernadez, G. and Ruiz-Agudo, E. (2013) Sequestration of selenium on calcite surfaces revealed by nanoscale imaging. *Environmental Science & Technology* **47**, 13469−13476.

Putnis, C.V., Ruiz-Agudo, E. and Hövelmann, J. (2014) Coupled fluctuations in element release during dolomite dissolution. *Mineralogical Magazine*, **78**, 1−7.

Qin, L., Lu, J., Stack, L. and Wang, L. (2013) Direct imaging of nanoscale dissolution of dicalcium phosphate dihydrate by an organic ligand: concentration matters. *Environmental Science & Technology*, **47**, 13365−13374.

Rashkovich, L.N., De Yoreo, J.J., Orme, C.A. and Chernov, A.A. (2006) *In situ* atomic force microscopy of layer-by-layer crystal growth and key growth concepts. *Crystallography Reports*, **51**, 1063−1074.

Reboul, J., Furukawa, S., Horike, N., Tsotsalas, M., Hirai, K., Uehara, H., Kondo, M., Louvain, N., Sakata, O. and Kitagawa, S. (2012) Mesoscopic architectures of porous coordination polymers fabricated by pseudomorphic replication. *Nature Materials*, **11**, 717−723.

Rieger, J., Frechen, T., Cox, G., Heckmann, W., Schmidt, C. and Thieme, J. (2007) Precursor structures in the crystallization/precipitation processes of $CaCO_3$ and control of particle formation by polyelectrolytes. *Faraday Discussions*, **136**, 265−277.

Rodriguez-Navarro, C. and Benning, L. (2013) Control of crystal nucleation and growth by additives. *Elements*, **9**, 203−209.

Rodriguez-Navarro, C. and Ruiz-Agudo, E. (2013) Carbonates: An overview of recent TEM research. Pp. 337−375 in: *Minerals at the Nanoscale* (F. Nieto and K.J.T. Livi, editors.). EMU Notes on Mineralogy, **14**.

European Mineralogical Union and the Mineralogical Society of Great Britain & Ireland, London.

Rodriguez-Navarro, C., Kudlakz, K., Cizer, O. and Ruiz-Agudo, E. (2015) Formation of amorphous calcium carbonate and its transformation into mesostructured calcite. *CrystEngComm*, **17**, 58–72.

Rodriguez-Navarro, C., Burgos Cara, A., Elert, K., Putnis, C.V. and Ruiz-Agudo, E. (2016) Direct nanoscale imaging reveals the growth of calcite crystals via amorphous nanoparticles. *Crystal Growth and Design*, DOI: 10.1021/acs.cgd.5b01180.

Ruiz-Agudo, E. and Putnis, C.V. (2012) Direct observations of mineral–fluid reactions using atomic force microscopy: the specific example of calcite. *Mineralogical Magazine*, **76**, 227–253.

Ruiz-Agudo, E., Putnis, C.V., Jiménez-López, C. and Rodríguez-Navarro, C. (2009) An AFM study of calcite dissolution in concentrated saline solutions: the role of magnesium ions. *Geochimica et Cosmochimica Acta*, **73**, 3201–3217.

Ruiz-Agudo, E., Di Tommaso, D., de Leeuw, N.H., Putnis, C.V. and Putnis, A. (2010a) Interactions between organophosphate-bearing solutions and calcite surfaces: An AFM and first principles molecular dynamics study. *Crystal Growth and Design*, **10**, 3022–3035.

Ruiz-Agudo, E., Kowacz, M., Putnis, C.V. and Putnis, A. (2010b) The role of background electrolytes on the kinetics and mechanism of calcite dissolution. *Geochimica et Cosmochimica Acta*, **74**, 1256–1267.

Ruiz Agudo, E., Putnis, C.V., Rodriguez-Navarro, C. and Putnis, A. (2011a) Effect of pH on calcite growth at constant and supersaturation. *Geochimica et Cosmochimica Acta*, **75**, 284–296.

Ruiz-Agudo, E., Putnis, C.V., Wang, L.J. and Putnis, A. (2011b) Specific effects of background electrolytes on the kinetics of step propagation during calcite growth. *Geochimica et Cosmochimica Acta*, **75**, 3803–3814.

Ruiz Agudo, E., Urosevic, M., Putnis, C.V., Rodriguez Navarro, C., Cardell, C. and Putnis, A. (2011c) Ion-specific effects on the kinetics of mineral dissolution. *Chemical Geology*, **281**, 364–371.

Ruiz-Agudo, E., Putnis, C.V., Rodriguez-Navarro, C. and Putnis, A. (2012) The mechanism of leached layer formation during chemical weathering of silicate minerals. *Geology*, **40**, 947–950.

Ruiz-Agudo, E., Alvarez-Loret, P., Putnis, C.V., Rodriguez-Navarro, A.B. and Putnis, A. (2013) Influence of chemical and structural factors on the calcite-calcium oxalate transformation. *CrystEngComm*, **15**, 9968–9979.

Ruiz-Agudo, E., Putnis, C.V. and Putnis, A. (2014) Coupled dissolution and precipitation at mineral–fluid interfaces. *Chemical Geology*, **383**, 132–146.

Ruiz-Agudo, E., King, H.E., Patio-López, L.D., Putnis, C.V., Geisler, T., Rodriguez-Navarro, C. and Putnis, A. (2016) The control of silicate weathering by interface coupled dissolution-precipitation processes at the mineral-solution interface. *Geology*, **44**, 567–570.

Sathiyanarayanan, R., Alimohammadi, M., Zhou, Y. and Fichthorn, K.A. (2011) Role of solvent in the shape-controlled synthesis of anisotropic colloidal nanostructures. *Journal of Physical Chemistry C*, **115**, 18983–18990.

Schott, J., Pokrovsky, O.S., Spalla, O., Devreux, F., Gloter, A. and Mielczarski, J.A. (2012) Formation, growth and transformation of leached layers during silicate minerals dissolution: The example of wollastonite. *Geochimica et Cosmochimica Acta*, **98**, 259–281.

Schwahn, D., Ma, Y. and Cölfen, H. (2007) Mesocrystal to single crystal transformation of D,L-alanine evidenced by small angle neutron scattering. *Journal of Physical Chemistry C*, **111**, 3224–3227.

Seto, J., Ma, Y., Davis, S.A., Meldrum, F., Gourrier, A., Kim, Y.-Y., Schilde, U., Sztucki, M., Burghammer, M., Maltsev, S., Jäger, C. and Cölfen, H. (2012) Structure-property relationships of a biological mesocrystal in the adult sea urchin spine. *Proceedings of the National Academy of Sciences, USA*, **109**, 3699–3704.

Sleutel, M. and van Driessche, A.E.S. (2014) Role of clusters in nonclassical nucleation and growth of protein crystals. *Proceedings of the National Academy of Sciences, USA*, E546–E553.

Smith, M.E., Knauss, K.G. and Higgins, S.R. (2013) Effects of crystal orientation on the dissolution of calcite by chemical and microscopic analysis. *Chemical Geology* **360–361**, 10–21

Sommerdijk, N.A.J.M. and de With, G. (2008) Biomimetic CaCO$_3$ mineralization using designer molecules and interfaces. *Chemical Reviews*, **108**, 4499–4550.

Song, R.-Q., Cölfen, H., Xu, A.-W., Hartmann, J. and Antonietti, M. (2009) Polyelectrolyte-directed nanoparticle aggregation: Systematic morphogenesis of calcium carbonate by nonclassical crystallization. *ACSNano*, **3**, 1966–1978.

Song, R.-Q. and Cölfen, H. (2010) Mesocrystals – Ordered nanoparticle superstructures. *Advanced Materials*, **22**, 1303–1330.

Song, R.-Q. and Cölfen, H. (2011) Additive controlled crystallization. *CrytEngComm*, **13**, 1249–1276.

Stack, A.G. and Grantham, M.C. (2010) Growth rate of calcite steps as a function of aqueous calcium-to-carbonate ratio: independent attachment and detachment of calcium and carbonate ions. *Crystal Growth & Design*, **10**, 1409–1413.

Stack, A.G., Gale, J.D. and Raiteri, P. (2013) Virtual probes of mineral–water interfaces: the more flops the better! *Elements*, **9**, 211–216.

Stranski, I.N. (1928) Zur Theorie des Kristallwachstums. *Zeitschirft für Physicalische Chemie*, **136**, 259–278

Stranski, I.N. and Totomanov, D. (1933) Rate of formation of (crystal) nuclei and the Ostwald step rule. *Zeitschirft für Physikalische Chemie*, **163**, 399–408.

ten Wolde, P.R. and Frenkel, D. (1997) Enhancement of protein crystal nucleation by critical density fluctuations. *Science*, **277**, 1975–1978

Teng, H.H. (2004) Control by saturation state on etch pit formation during calcite dissolution. *Geochimica et Cosmochimica Acta*, **68**, 253–262.

Teng, H.H., Dove, P.M., Orme, C.A. and De Yoreo, J.J. (1998) Thermodynamics of calcite growth: Baseline for understanding biomineral formation. *Science*, **282**, 724–727.

Teng, H.H., Dove, P.M. and De Yoreo, J.J. (2000) Kinetics of calcite growth: surface processes and relationships to macroscopic rate laws. *Geochimica et Cosmochimica Acta*, **64**, 2255–2266.

Teng, H.H., Fenter, P., Cheng, L. and Sturchio, N.C. (2001) Resolving orthoclase dissolution processes with atomic force microscopy and X-ray reflectivity. *Geochimica et Cosmochimica Acta*, **65**, 3459–3474

Teng, H.H. Chen, Y. and Pauli, E. (2006) Direction specific interactions of 1,4-dicarboxcylic acid with calcite surfaces. *Journal of the American Chemical Society*, **128**, 14482–14484.

Threlfall, T. (2003) Structural and thermodynamic explanations of Ostwald's rule. *Organic Process Research and Development*, **7**, 2017–2027.

Tisserand, D. and Hellmann, R. (2008) Bringing the gap between laboratory dissolution and natural chemical weathering. *Geochimica et Cosmochimica Acta*, **72**, A948.

Towe, K.M. and Lowenstam, H.A. (1967) Ultrastructure and development of iron mineralization in the radular teeth of *Cryptochiton stelleri* (Mollusca). *Journal of Ultrastructure Research*, **17**, 1–13.

Travaille, A.M., Steijven, E.G.A., Meekes, H. and van Kempen, H. (2005) Thermodynamics of epitaxial calcite nucleation on self-assembled monolayers. *Journal of Physical Chemistry B*, **109**, 5618–5626.

Urosevic, M., Rodriguez-Navarro, C., Putnis, C.V., Cardell, C., Putnis, A. and Ruiz-Agudo, E. (2012) *In situ* nanoscale observations of the dissolution of $10\bar{1}4$ dolomite cleavage surfaces. *Geochimica et Cosmochimica Acta*, **80**, 1–13.

van Santen, R.A. (1984) The Ostwald step rule. *Journal of Physical Chemistry*, **88**, 5768–5769.

Vavouraki, A.I., Putnis, C.V., Putnis, A. and Koutsoukos, P. (2010) Crystal growth and dissolution of calcite in the presence of fluoride ions: An atomic force microscopy study. *Crystal Growth and Design*, **10**, 60–69.

Vekilov, P.G. (2004) Dense liquid precursor for the nucleation of ordered solid phases from solution. *Crystal Growth and Design*, **4**, 671– 685

Vekilov, P.G. (2010) The two-step mechanism of nucleation of crystals in solution. *Nanoscale*, **2**, 2346–2357.

Verch, A., Gebauer, D., Antonietti, M. and Cölfen, H. (2011) How to control the scaling of $CaCO_3$: a "fingerprinting technique" to classify additives. *Physical Chemistry and Chemical Physics*, **13**, 16811–16820.

Vinson, M.D. and Lüttge, A. (2005) Multiple length-scale kinetics: An integrated study of calcite dissolution rates and strontium inhibition. *American Journal of Science*, **305**, 119–146.

Vinson, M.D., Arvidson, R.S. and Lüttge, A. (2007) Kinetic inhibition of calcite (104) dissolution by aqueous manganese (II). *Journal of Crystal Growth*, **307**, 116–125.

Volmer, M. (1939) *Kinetik der Phasenbildung*. Steinkopff, Dresden, Germany.

Volmer, M. and Weber, A. (1926) Keimbildung in übersättigten Gebilden. *Zeitschrift für Phyikalische Chemie*, **119**, 277–301.

Walker. J.C.G., Hays, B. and Kasting, F. (1981) A negative feedback mechanism for the long-term stabilization of Earth's surface temperature. *Journal of Geophysical Research*, **86**, 9776–9782

Wallace, A.F., Hedges, L.O., Fernandez-Martinez, A., Raiteri, P., Gale, J.D., Waychunas, G.A., Whitelam, S., Banfield, J.F. and De Yoreo, J.J. (2013) Microscopic evidence for liquid-liquid separation in supersaturated CaCO₃ solutions. *Science*, **341**, 885–889.

Wang, L.J., Ruiz-Agudo, E., Putnis, C.V. and Putnis, A. (2011) Direct observations of the modification of calcite growth morphology by Li⁺ through selectively stabilizing energetically unfavourable faces. *ChemEngComm*, **13**, 3962–3966.

Wang, L., Li, S., Ruiz Agudo, E., Putnis, C.V. and Putnis, A. (2012a) Posner's cluster revisited: direct imaging of nucleation and growth of nanoscale calcium phosphate at the calcite-water interface. *CrystEngComm*, **14**, 6252–6256.

Wang, L.J., Ruiz-Agudo, E., Putnis, C.V., Menneken, M. and Putnis, A. (2012b) Kinetics of calcium phosphate nucleation and growth on calcite: implications for predicting the fate of dissolved phosphate species in alkaline soils. *Environmental Science & Technology*, **46**, 834–842.

Wang, L.J., Putnis, C.V., Ruiz-Agudo, E., King, H.E. and Putnis, A. (2013) Coupled dissolution and precipitation at the cerussite-phosphate solution interface: implications for immobilization of lead in soils. *Environmental Science & Technology*, 47, 13502–13510.

Weissbart, E.J. and Rimstidt, J.D. (2000) Wollastonite incongruent dissolution and leached layer formation. *Geochimica et Cosmochimica Acta*, **64**, 4007–4016.

White, A.F. and Brantley, S.L. (2003) The effect of time on the weathering rates of silicate minerals: why do weathering rates differ in the laboratory and field? *Chemical Geology*, **201**, 479–506.

Wolf, S.E., Leiterer, J., Kappl, M., Emmerling, F. and Tremel, W. (2008) Early homogeneous amorphous precursor stages of calcium carbonate and subsequent crystal growth in levitated droplets. *Journal of the American Chemical Society*, **130**, 12342–12347.

Wolf, S.L.P., Jähme, K. and Gebauer, D. (2015) Synergy of Mg²⁺ and poly(aspartic acid) in additive-controlled calcium carbonate precipitation. *CrystEngComm*, **17**, 6857–6862.

Xia, F., Brugger, J., Ngothai, Y., O'Neill, B., Chen, G. and Pring, A. (2009a) Three-dimensional ordered arrays of zeolite nanocrystals with uniform size and orientation by a pseudomorphic coupled dissolution-reprecipitation replacement route. *Crystal Growth and Design*, **9**, 4902–4906.

Xia, F., Brugger, J., Chen, G., Ngothai, Y., O'Neill, B., Putnis, A. and Pring, A. (2009b) Mechanism and kinetics of pseudomorphic mineral replacement reactions: a case study of the replacement of pentlandite by violarite. *Geochimica et Cosmochimica Acta* **73**, 1945–1969.

Xu, M. and Higgins, S.R. (2011) Effects of magnesium ions on near-equilibrium calcite dissolution: Step kinetics and morphology. *Geochimica et Cosmochimica Acta*, **75**, 719–733

Xu, X., Han, J.T. and Cho, K. (2005) Deposition of amorphous calcium carbonate hemispheres on substrates. *Langmuir*, **21**, 4801–4804.

Xu, A.-W., Ma, Y. and Cölfen, H. (2007) Biomimetic mineralization. *Journal of Materials Chemistry*, **17**, 415–449.

Yuwono, V.M., Burrows, N.D., Soltis, J.A. and Penn, R.L. (2010) Oriented aggregation: Formation and transformation of mesocrystal intermediates revealed. *Journal of the American Chemical Society*, **132**, 2163– 2165.

Zhang, H., De Yoreo, J.J. and Banfield, J.F. (2014) A unified description of attachment-based crystal growth. *ACS Nano*, **8**, 6226–6530.

Zhang, R., Hu, S., Zhang, X. and Yu, W. (2007) Dissolution kinetics of dolomite in water at elevated temperatures. *Aqueous Geochemistry*, **13**, 309–338.

Zhou, L. and O'Brien, P. (2012) Mesocrystals − Properties and applications. *Journal of Physical Chemistry Letters*, **3**, 620–628.

Metamorphic mineral reactions: Porphyroblast, corona and symplectite growth

FRED GAIDIES[1,*], R. MILKE[2], W. HEINRICH[3] and R. ABART[4]

[1] Department of Earth Sciences, Carleton University, 1125 Colonel By Drive, Ottawa, ON, K1S 5B6, Canada, e-mail: Fred.Gaidies@carleton.ca
[2] Instiute of Geological Sciences, Free University Berlin, Malteserstr. 74-100, Haus N Raum N 16 12249 Berlin, Germany, e-mail:milke@zedat.fu-berlin.de
[3] Helmholtz-Zentrum Potsdam, Deutsches GeoForschungsZentrum (GFZ) Telegrafenberg, D-14473 Potsdam, Germany, e-mail: whsati@gfz-potsdam.de
[4] University of Vienna, Department of Lithospheric Research, Althanstrasse 14, 1090 Vienna, Austria, e-mail: rainer.abart@univie.ac.at
* Corresponding author

Much of the Earth's dynamics is related to mineral reactions in the solid-state. Classically, this is referred to as metamorphic crystallization (Kretz, 1994). Based on the chemical compositions of the phases involved in a metamorphic mineral reaction, two basic reaction types may be distinguished. Reactions that involve only structural re-arrangements, while the compositions of the reactant and product phases are identical, are referred to as partitionless and 'polymorphic phase transformations'. If, in contrast, one or more reactant phases are replaced by one or more product phases with different compositions, this implies that chemical components are supplied to or removed from the reaction interfaces separating the reactants from the product phases. In the absence of advective transport via a fluid or melt, the necessary chemical mass transport can occur only by diffusion. Accordingly, this reaction type is partitioning and is referred to as 'diffusive phase transformation'. Some treatments of the kinetics of mineral reactions are based on partitionless polymorphic phase transformations and are reviewed only briefly in this chapter. However, because most metamorphic mineral reactions are partitioning diffusive phase transformations, the following discussion will focus mainly on this reaction type.

In this chapter, three types of reactions that play a key role in metamorphic crystallization are addressed. During prograde metamorphism continuous supply of aqueous fluid by dehydration reactions may facilitate relatively rapid inter-crystalline diffusion so that a state close to chemical equilibrium on the scale of mineral grains and beyond may be attained resulting in 'porphyroblastic mineral growth'. Interface-reaction controlled and diffusion-controlled growth are two end-member models in the kinetics of porphyroblastic growth and differ in terms of the spatial extent of chemical equilibration and its influence on the distribution and compositional zoning of porphyroblasts. The first section of this chapter may serve as a review of some of the key works in metamorphic petrology addressing the factors that control the abundance and size distribution of porphyroblasts and their chemical zoning patterns.

During retrograde stages of metamorphism or during metamorphic overprint of a previously largely dehydrated rock, crystallization may take place in a relatively 'dry' environment, where inter-crystalline diffusion is comparatively sluggish. In such a situation, reaction microstructures such as 'reaction bands' or 'corona structures' may develop. Typically, both the reactant and product phases are present providing evidence of incomplete reaction and indicating an overall disequilibrium situation. Chemical equilibrium may be restricted to microscopic domains along the phase boundaries or may not be attained at all. Nevertheless, important rate and time information may be obtained from the analysis of such reaction microstructures, if the processes underlying their formation are known and their rates are calibrated. The formation mechanisms of reaction bands and corona microstructures are discussed in the second section of this chapter.

Finally, the mechanisms underlying symplectite formation, another phenomenon that is typically associated with metamorphic overprint of magmatic or metamorphic rocks, will be addressed in the third section of this chapter. Symplectites are spatially highly organized, fine-grained intergrowths of two or more different phases, replacing a more coarser-grained precursor phase at a sharp reaction interface. Symplectite microstructures are characterized by a specific length scale of phase alternation, by specific lamellar or rod-shaped microstructure and compositional patterns. In this chapter different avenues for extracting petrogenetic information from symplectite microstructures are discussed.

1. Porphyroblast growth

Porphyroblasts are grains of metamorphic minerals that are characterized by a size significantly larger than that of the matrix phases. The mechanisms and rates of porphyroblastic growth have been the focus of research for more than 50 years and are still an active topic of modern metamorphic petrology (*e.g.* Hollister, 1966; Jones and Galwey, 1966; Kretz, 1966, 1969, 1973, 1974, 1993; Foster, 1981, 1999; Loomis, 1982, 1986; Cashman and Ferry, 1988; Carlson, 1989, 1991b, 2002, 2011; Miyazaki, 1991, 1996, 2015; Spear *et al.*, 1991a; Carlson *et al.*, 1995; Denison and Carlson, 1997; Daniel and Spear, 1998, 1999; Spear and Daniel, 1998, 2001; Hirsch *et al.*, 2000; Gaidies *et al.*, 2008a,b, 2011; Hirsch, 2008; Schwarz *et al.*, 2011; Ketcham and Carlson, 2012; Kelly *et al.*, 2013; Petley-Ragan *et al.*, 2016).

The scenarios developed to describe porphyroblastic growth differ in terms of the assumptions made about the spatial extent of chemical equilibration of the components involved in the respective metamorphic phase transformation and its impact on the size distribution and chemical composition of the porphyroblasts. In general, porphyroblastic growth is understood as the result of the interplay of various atomic-scale processes during the metamorphic phase transformation, including the dissolution of reactants, the attachment and detachment of atoms at the interface between the crystallizing product and its surrounding matrix, and the long-range diffusion of components between the reactant and product sites. The kinetics of porphyroblastic growth are closely related to the energy barriers associated with the interface processes and the diffusion across the bulk system. This is in stark contrast to nucleation, the process that necessarily precedes growth (see Gaidies, 2017, this volume). Whereas the critical energy barrier to nucleation is associated largely with the formation of the interface, energy barriers for growth are associated with the movement of the interface into the metastable matrix by the interplay of atomic jumps across the interface and long-range diffusion. In general, energy carriers to diffusion may be considered smaller than those for nucleation, possibly explaining the relatively large sizes but low abundances of porphyroblasts in metamorphic rocks.

Commonly, two end-member scenarios of porphyroblastic growth are distinguished: (1) In models of 'interface-reaction controlled growth', long-range diffusion is assumed to be significantly faster than the attachment and detachment processes at the propagating crystal/matrix interface so that chemical disequilibrium is expected to develop only in the vicinity of the interface. (2) 'Diffusion-controlled growth' is assumed to result from interface processes significantly faster than the diffusion across the bulk of the system. In this case, local chemical equilibrium is assumed commonly only at the interface, and gradients in the chemical potentials are expected to develop across the rock matrix as the porphyroblast grows. Because the spatial extent of chemical equilibration influences directly the size distribution and chemical composition of porphyroblasts in natural systems, methods that integrate microstructural and chemical datasets with quantitative models of metamorphic crystallization allow evaluation of the applicability of these end-member scenarios providing insight into the mechanisms and rates of porphyroblastic growth.

In the following, a brief introduction to the kinetics associated with interface reaction and diffusion-controlled crystal growth is provided, and the implications for the micro-structural and chemical evolution of porphyroblasts are discussed. Statistical methods developed to investigate the abundance and size distribution of porphyroblasts in three dimensions (3D) are reviewed and examples for the spatial disposition of porphyroblasts in contact and regional metamorphic rocks are presented. In addition, the reader is introduced to the numerical simulation of metamorphic crystallization that may help to decipher metamorphic pressure-temperature-time (*P-T-t*) paths by considering the microstructural and compositional record contained in the spatial disposition of chemically zoned garnet porphyroblasts.

1.1. Interface-reaction controlled porphyroblast growth

Crystal growth controlled by the rate of attachment and detachment processes at the crystal/matrix interface may be modelled at the atomic level where atoms continuously and independently cross the entire interface. The bulk of these attachment and detachment processes may be referred to as interface-reaction, driven by the drop in the chemical potentials of the atoms as they cross the interface. If the chemical compositions of the phases separated by the interface are identical then the phase transformation is partitionless, and the energetics of the interface-reaction will be a function of the activation energies associated with the atomic jumps across the interface and the thermodynamic driving force for the interface movement. For such a specific case, the formalism outlined by Balluffi *et al.* (2005) for interface-reaction controlled growth during solidification may be used.

According to Balluffi *et al.* (2005), the rate with which the interface moves away from the centre of a spherical stable phase β into the metastable phase α is the growth

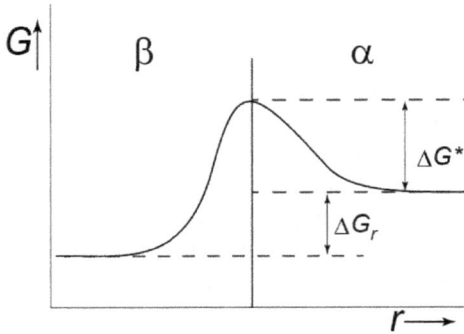

Figure 1. Chemical driving force for growth, ΔG_r, and activation energy associated with the atomic jump from α to β, ΔG^*, at the α/β interface for partitionless interface reaction-controlled growth. Modified after Christian (2002) and Balluffi et al. (2005).

rate, dr/dt, where r refers to the spatial coordinate of the interface (Fig. 1). The flux of α atoms across the interface is given by

$$N v \exp\left(\frac{-\Delta G^*}{k_B T}\right) \qquad (1)$$

where N is the total number of atoms per unit area of the interface, v is their vibrational frequency, ΔG^* is the activation energy associated with the jump across the interface, and k_B is Boltzmann's constant. Similarly, and assuming that the vibrational frequencies are identical, the jump frequency associated with the transfer of atoms of β across the interface is given by

$$N v \exp\left(\frac{-(\Delta G^* + \Delta G_r)}{k_B T}\right) \qquad (2)$$

where ΔG_r is the chemical driving force for the interface movement, acting on each atom, and is defined by the difference in the two local minima of the Gibbs energy centred on α and β (Fig. 1). It is important to note that ΔG_r increases with departure from equilibrium reflecting the diffusiveness of the interface and the energy barrier associated with the atomic jumps across it.

The net rate of transfer from α to β is the difference of (1) and (2) which, multiplied by the molecular volume of β, Ω, gives the interface-reaction controlled growth rate

$$\frac{dr}{dt} = N v \left[\exp\left(\frac{-\Delta G^*}{k_B T}\right) - \exp\left(\frac{-(\Delta G^* + \Delta G_r)}{k_B T}\right)\right] \Omega \qquad (3)$$

Rearranging yields

$$\frac{dr}{dt} = N v \left[\exp\left(\frac{-\Delta G^*}{k_B T}\right) \left[1 - \exp\left(\frac{-\Delta G_r}{k_B T}\right)\right]\right] \Omega \qquad (4)$$

which for $\Delta G_r \ll k_B T$ simplifies to

$$\frac{dr}{dt} = M \Delta G_r \qquad (5)$$

with the interface mobility, M, defined as

$$M = \frac{\Omega N_\mathrm{F}}{k_\mathrm{B} T} \exp\left(\frac{-\Delta G^*}{k_\mathrm{B} T}\right) \tag{6}$$

Note that for the more common case of partitioning phase transformations, the chemical compositions of the phases on either side of the interface are different. Hence, atoms of different components will cross the interface, each with component-specific energy barriers associated with the atomic jumps. In this case, the simple formalism outlined above cannot be used to derive at expressions for M, and a phenomenological approach may be used instead.

Expression 5 is strictly valid only for the movement of planar interfaces. A more general expression accounting also for the influence of interface curvature on the overall Gibbs energy change during crystal growth is

$$\frac{dr}{dt} = M\left(\Delta G_r - \frac{2\sigma}{R}\right) \tag{7}$$

where σ is the interfacial energy and R is the radius of interface curvature (see Petrishcheva and Abart 2017, this volume). According to equation 7, for relatively small crystals and reactions close to equilibrium, the bulk energy term may approach values similar in magnitude to the interfacial term so that the thermodynamic driving force effectively acting on the interface decreases, and the growth rate becomes size-dependent reflecting the increase in Gibbs energy with interface curvature. Only in that case, and amplified by large interfacial energies, small crystals will grow slower than larger ones for a given step in time and under the same conditions. In terms of porphyroblastic growth, compositional gradients that develop as the interface moves would be steeper in smaller crystals compared with larger but would be the same for crystals of identical size if these crystals grew simultaneously (Fig. 2), independent of their positions in the rock. Such a scenario, where crystals are smaller than a critical size so that their interface curvatures reduce the thermodynamic driving force for growth, may be referred to as 'size-dependent interface-reaction controlled growth'.

In most cases, however, interface curvature and the contribution of the interfacial energy term to the Gibbs energy of a crystal during growth may be too small so that the rate with which a porphyroblast grows will be size-independent (see expression 5). As a result, for crystals that are larger than the critical size for size-dependent interface-reaction controlled growth, porphyroblasts will grow with the same radial rate for a given step in time and P-T-X space, and identical compositional gradients will develop, independent of the sizes and positions of the porphyroblasts in the rock (Fig. 2). The interface during 'size-independent interface-reaction controlled growth' moves proportional to time according to

$$r \propto t \tag{8}$$

Note that this simple relation holds only if M and ΔG_r can be considered constant, such as for an infinitesimal step in P-T-X-t space. However, as P-T alter during metamorphism, variations in the growth rate of a porphyroblast may be expected

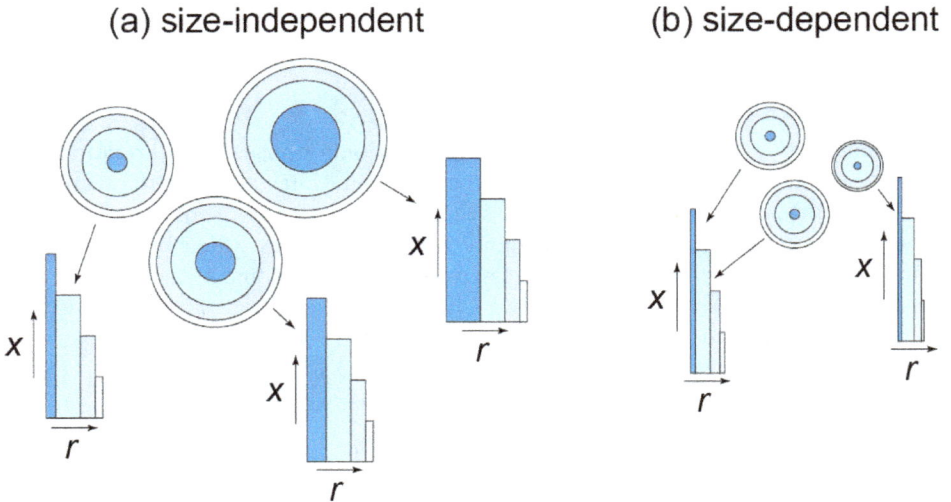

Figure 2. Two possible scenarios of interface reaction-controlled growth: (a) relatively large crystals grow with rates that change proportional to time and chemical driving force, independent of crystal size. (b) Slow growth and steep compositional gradients of small crystals due to the large interface curvature reducing the driving force for growth at any point in time. In both cases, growth rates are position-independent.

reflecting changes in the interface mobility and the chemical driving force for growth. The chemical driving force for growth relates directly to the thermodynamically effective bulk-rock composition so that chemical fractionation, commonly associated with the growth of refractory phases such as garnet or plagioclase may also influence the growth rate of a porphyroblast.

The departure from equilibrium at the interface between porphyroblast and matrix during interface reaction-controlled growth reflects the kinetics of the interface processes relative to the change of the thermodynamic driving force that acts on the interface during metamorphism. If the interface processes are fast relative to the change of the chemical driving forces for porphyroblastic growth then chemical reaction may be close to equilibrium and crystal growth may be modelled as a succession of equilibrium states. Scenarios where significant degrees of disequilibrium at propagating interfaces may be expected include porphyroblastic growth in rocks subjected to relatively fast changes in the chemical driving force for growth, and containing a fluid phase along their grain- and phase-boundary networks facilitating rapid long-range diffusion, such as during prograde metamorphism of pelites in contact aureoles.

1.2. Diffusion-controlled porphyroblast growth

If long-range diffusion of nutrients and waste products is sufficiently slow compared to their attachment and detachment at the interface as the porphyroblast grows, then gradients in the chemical potentials of the associated chemical components will develop across the rock matrix, and the composition of the matrix at the interface will be position-dependent. The rate of growth will then be controlled by the diffusion of components through the matrix surrounding the porphyroblast associated with the supply of nutrients and the removal of waste products. It is important to note that if the long-range diffusion of a component controls the rate of porphyroblastic growth but the component itself does not mix on a crystallographic site of the porphyroblast then the chemical composition of the porphyroblast may be position-independent. Assuming local equilibrium at the interface and ignoring any influence of interface curvature on the Gibbs energy of a crystal, the classic formalism by Zener (1949) may be adopted to derive an expression for the rate of diffusion-controlled spherical growth. More general expressions for diffusion-controlled growth involving solution phases and considering finite interface mobilities for different geometries are discussed in sections 2 and 3 of this chapter.

According to Zener (1949), c_i^α and c_i^β may be referred to as the equilibrium concentrations of component i at the interface of the matrix and porphyroblast, respectively (Fig. 3), and D_i may be the diffusivity of i through the matrix, so that

$$(c_i^\beta - c_i^\alpha)\mathrm{d}r = D_i \left(\frac{\partial c_i}{\partial r}\right)\mathrm{d}t \tag{9}$$

Rearrangement results in the kinetic law for diffusion-controlled spherical growth in an infinite matrix

$$\frac{dr}{dt} = \frac{D_i}{(C_i^\beta - C_i^\alpha)}\left(\frac{\partial C_i}{\partial r}\right) \tag{10}$$

assuming that differences in the molecular volumes of α and β can be ignored. Following Zener (1949), r varies with t according to

$$r = \alpha\sqrt{D_i t} \tag{11}$$

with α defined as

$$\alpha = \bar{K}\frac{\sqrt{c_i^\infty - c_i^\alpha}}{\sqrt{c_i^\beta - c_i^\infty}} \tag{12}$$

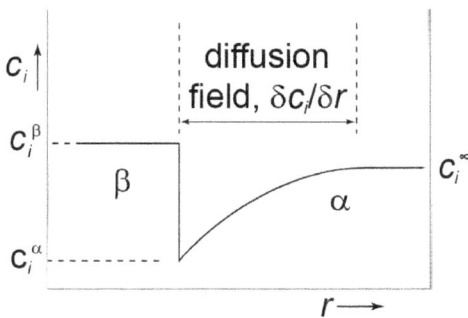

Figure 3. Diffusion-controlled partitioning growth of an isolated grain of stable β at the expense of an infinite metastable α matrix assuming local equilibrium with respect to component i at the α/β interface. Modified after Zener (1949).

and K varying between 1.4 and 2.4 as c_i^∞ varies between c_i^α and c_i^β. In other words, the interface during diffusion-controlled growth moves proportional to the square root of time whereas a linear growth law is obtained for size-independent interface reaction-controlled growth (see expression 8).

The compositional gradient in the vicinity of the porphyroblast, $\partial c_i/\partial r$, controls the rate of diffusion-controlled interface movement, and decreases with an increase of porphyroblast size. This decrease in compositional gradient is related to the width of the diffusion field surrounding the porphyroblast (Fig. 3) which is wider for larger grains compared to smaller ones. As a result, diffusion-controlled growth is size-dependent, with smaller grains growing with a greater radial growth rate than larger ones. In contrast to interface reaction-controlled growth, chemical potential gradients develop in the rock matrix during porphyroblastic growth controlled by long-range diffusion (*e.g.* Carlson, 1989, 1991; Carlson *et al.*, 1995). Both the radial growth rates and the compositional gradients that develop in the porphyroblast as it grows reflect the positions and orientations of the interfaces in the diffusion zone. Hence, initially similarly sized grains may grow with different rates according to the local diffusion field, and may develop asymmetric compositional gradients as governed by the varying availability of nutrients along their interfaces. Systematic variations in the chemical potential gradients across diffusion fields around porphyroblasts with uneven surfaces may result in the morphological instability of these surfaces. Gradients in the metastable matrix may be steeper in the vicinity of protruding elements of the porphyroblast surface compared with flat surface areas. This leads to locally increased diffusive fluxes of the nutrients towards the growth front resulting in growth rate variations along the interface, amplifying the protrusions and, hence, destabilizing flat surfaces (Mullins and Sekerka, 1964). Because morphological instability of growth fronts and dendritic growth are uncommon during porphyroblast growth in metamorphic rocks, Miyazaki (2015) questions the significance of the diffusion-controlled growth scenario in the kinetics of porphyroblast crystallization.

1.3. Statistical analysis of the spatial disposition of crystal centres and sizes

The spatial distribution of porphyroblast centres and sizes may provide insight into the kinetics of associated metamorphic mineral reactions (*e.g.* Kretz, 1974; Carlson, 1989). For example, a 'random spatial distribution' of porphyroblast sizes may be expected to develop during interface reaction-controlled growth because the rates of porphyroblast growth would be independent of the positions of the crystals in this kinetic scenario. Furthermore, if the energetically preferable nucleation sites (see Gaidies, 2017, this volume) are distributed uniformly across the rock volume then a random distribution of porphyroblast centres may also be envisaged during interface reaction-controlled kinetics. Interactions between neighbouring crystals during interface-reaction controlled growth take place only if nucleation sites are close to each other so that the crystals impinge as their interfaces move. Hence, deviations from a random spatial disposition of porphyroblast centres may be expected only if preferable nucleation sites or chemical components are distributed inhomogeneously, such as during the static metamorphism of sedimentary layering. In such a case, the number of crystal centres for a given rock volume may be significantly larger than in a random distribution and the spatial pattern may be referred to as 'clustered'.

A random spatial distribution of porphyroblast centres and sizes may also develop during diffusion-controlled porphyroblast growth but only at length scales between neighbouring crystals exceeding the effective diffusion length. For shorter length-scales, an ordered texture is expected to form during diffusion-controlled growth as a consequence of the competition for nutrients between porphyroblasts (*e.g.* Carlson, 1991). This ordering is characterized by a positive correlation between the sizes of porphyroblasts and their distances to each other caused by the interference of diffusion fields of neighbouring crystals. Whereas the decrease of porphyroblast size with a decrease of next-neighbour distance may indicate diffusion-controlled growth, the positive correlation of the distance between porphyroblasts and their abundances may reflect diffusion-controlled nucleation.

In order to study the 3D distribution of porphyroblasts statistically, X-ray computed tomography (XR-CT) (*e.g.* Ketcham and Carlson, 2001) is commonly used as it allows investigation of the relatively large rock volumes required for statistically meaningful analyses (*e.g.* Chernoff and Carlson, 1997, 1999; Petley-Ragan *et al.*, 2016). The methods developed to conduct the statistical analyses of the spatial distribution of crystals utilize microstructural information, such as the *x-y-z* coordinates of each grain centre, grain radius or semi-major axis. *Reduce3D* (Hirsch *et al.*, 2000; Hirsch, 2011) may be the most advanced software to conduct sophisticated statistical analyses of porphyroblast distributions. In the following, the theoretical background of the most useful statistics commonly applied to the study of the spatial disposition of porphyroblasts is reviewed and examples of applications to natural rocks are discussed.

1.3.1. Theoretical background

In the statistical analysis of the spatial distribution of porphyroblast centres and sizes, single-scale and scale-dependent statistics have been applied. These statistics have been used to determine whether the porphyroblasts of a given volume of rock can be characterized by a clustered, random or ordered spatial distribution (*e.g.* Kretz, 1973, 1993; Carlson, 1989; Denison and Carlson, 1997) aiding in the analysis of the kinetics of porphyroblastic growth. A fundamental assumption in the interpretation of the pattern statistics is that the spatial distribution of crystal centres and sizes is a result of nucleation and growth and is not modified by secondary processes such as resorption or relative displacement associated with deformation.

The single-scale statistics commonly used in the pattern statistics include the ordering and clustering indices (Kretz, 1966, 1969), and the impingement and isolation indices (Carlson, 1989; Hirsch, 2000). A limitation of the single-scale statistics is their inability to consider variations in the spatial distribution of porphyroblasts with distance because they reflect global averages of microstructural observations such as porphyroblast size and distances between porphyroblasts across the rock volume studied. Hence, the more sophisticated approaches inherent in scale-dependent correlation functions such as the L'-function (L'F), the pair correlation function

(PCF) and the mark correlation function (MCF′) (*e.g.* Daniel and Spear, 1999) are preferred to the single-scale measures (Carlson, 2011). These scale-dependent correlation functions allow for the quantification of textural relationships across a range of length scales instead of global averages. The most useful scale-dependent correlation functions may be the PCF and MCF′ as they can be used to quantify the 3D distribution of porphyroblast centres and sizes for any test distance across the rock volume of interest. These correlation functions will be outlined below. Additional information about other scale-dependent statistics, such as the L′-function, or the single-scale statistics may be found in Daniel and Spear (1999) and Hirsch *et al.* (2000).

   In *Reduce3D* (Hirsch *et al.*, 2000; Hirsch, 2011), both the PCF and MCF′ statistics of the natural dataset are compared with the statistics obtained through the numerical simulation of interface reaction-controlled non-clustered nucleation and spherical growth. The number density of the crystals as well as the size and shape of the sample volume and the crystal size frequency relationships used in the simulation are identical to the natural dataset. However, the numerical simulation produces a crystal array which lacks any ordering and clustering trends so that the corresponding statistics can be used as a null hypothesis case. Any deviations in the statistics of the natural crystal array compared to the null hypothesis case may then be interpreted to indicate either clustering or ordering providing insight into the mechanisms and rates of porphyroblast crystallization. The shaded regions in the diagrams produced with *Reduce3D* represent the 2σ range obtained from 100 simulations of interface reaction-controlled non-clustered spherical crystallization. Examples of these ranges are illustrated as envelopes in Figs 4 and 5.

   The artificial array of spherical crystals produced by *Reduce3D* in order to simulate non-clustered interface reaction-controlled nucleation and growth may be considered a random spatial distribution of crystal centres and sizes for distances at which these

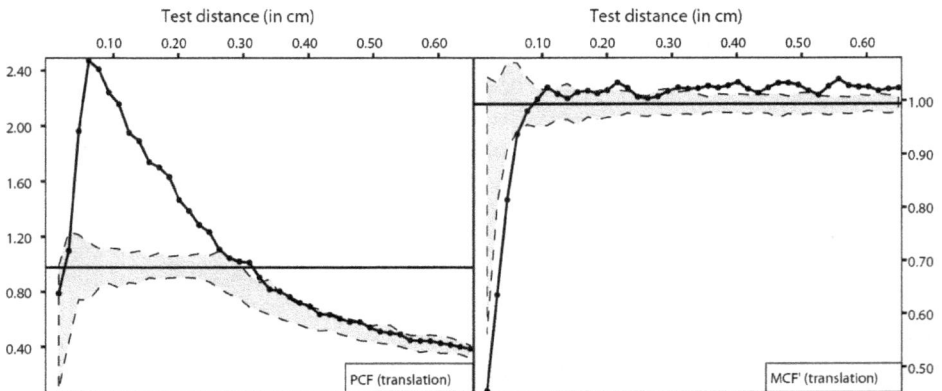

Figure 4. Results of spatial statistical correlation functions (PCF and MCF′) calculated with *Reduce3D* (Hirsch, 2011) for garnet in a metapelite from the Lesser Himalayan Belt in Sikkim, India (24-99 of Gaidies *et al.*, 2015). The blue band corresponds to theoretical results (with 2σ envelope) obtained for a random distribution of spheres. Sphere-normalized radius is used as mark.

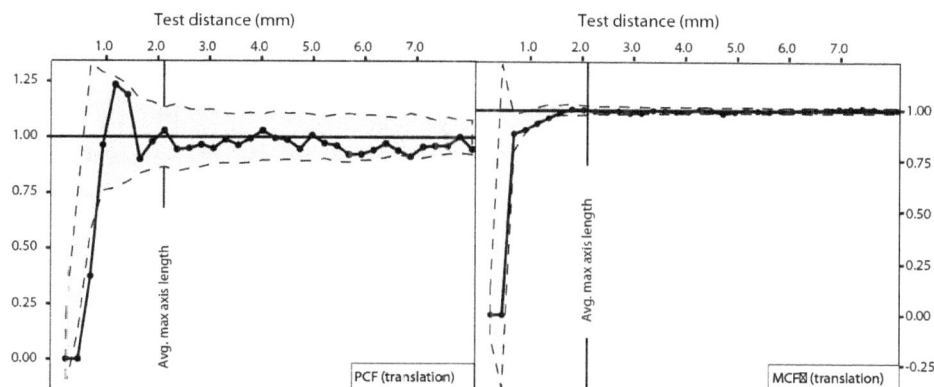

Figure 5. Statistical results of PCF and MCF′ calculations for cordierite in a hornfels from the Bugaboo contact aureole (Petley-Ragan *et al.*, 2016) using *Reduce3D* (Hirsch, 2011). The blue areas correspond to a random distribution of porphyroblasts (with 2σ envelope). Sphere-normalized radii are used for the calculation of PCF, and the semi-major axes are used for the MCF′.

crystals do not touch. However, there is always a deviation from randomness towards ordering at shorter distances even in the artificial array. The ordering with respect to the distribution of crystal centres is due to the 'volume effect' where crystals cannot nucleate inside a pre-existing crystal. This effect results in a decrease in the number of crystal centres with a decrease in test distance for scales shorter than the average diameter of the crystal population. Ordering with respect to the size of crystals at distances shorter than the average crystal size is a result of the 'Strauss hard-core process' (Strauss, 1975; Ripley, 1981; Baddeley and Turner, 2000; Illian *et al.*, 2008). This process results in an ordered pattern in a population of objects in 3D at length scales at which these objects touch each other. It is characterized by a positive correlation between the sizes of the objects and the distances between their centres because the centres of the objects can only get closer to each other if the sizes of the objects decrease. Accordingly, in a population of differently sized spheres, the Strauss hard-core process results in ordering at length scales less than the average diameter of the spheres.

The volume effect and Strauss hard-core process are considered in the numerical simulation of non-clustered interface reaction-controlled nucleation and growth by *Reduce3D* based on the assumption that crystals can be approximated as spherical objects. Accordingly, the largest crystals are placed at random positions in the model rock volume, and the remaining space is filled successively with crystals of decreasing size. This is in line with a nucleation and growth algorithm where the final sizes of the crystals reflect the time of their nucleation. Such a crystallization sequence may be expected during interface reaction-controlled growth assuming that secondary processes such as resorption or deformation did not alter crystal sizes. During the random placement of successively smaller crystals, *Reduce3D* tests whether crystal centres are positioned within the model volume occupied by previously placed crystals.

Such a placement is permitted by the software but only if the distance between the centres of the overlapping crystals is equal to or larger than the difference of their radii effectively simulating the formation of interpenetrating crystal pairs. If the distance between the centres of the overlapping crystals is smaller than the difference of their radii then the placement is not permitted and new random positions are tested until this placement criterion is fulfilled. Additional tests are run by the software to minimize the error that forms if highly interpenetrating crystal pairs are misidentified as larger single crystals in the 3D dataset.

Pair correlation function, PCF. The pair correlation function, $g(r)$, is a measure of the number of crystal centres, n, within a spherical shell volume of the sample region, $V(W)$, which increases with test distance, r, around the centre of each crystal i. It is defined as (Hirsch *et al.*, 2000)

$$g(r) = \frac{\rho(r)}{\lambda^2} \tag{13}$$

where $\rho(r)$ is given as

$$\rho(r) = \sum_{i=1}^{n} \sum_{\substack{j=1 \\ (j \neq 1)}}^{n} \frac{e_h(t)}{4\pi r^2} \tag{14}$$

with

$$t = r - \|\vec{x}_i - \vec{x}_j\| \tag{15}$$

and

$$\lambda^2 = \frac{n(n-1)}{[V(W)]^2} \tag{16}$$

$\|\vec{x}_i - \vec{x}_j\|$ is the distance between crystals i and j, and $e_h(t)$ is the Epanecnikov kernel function of bandwidth h, given by

$$e_h(t) = \frac{3}{4h}\left(1 - \frac{t^2}{h^2}\right) \text{ for } |t| \leq h \tag{17}$$

With $h = c\lambda^{-1/2}$ and c commonly ranging between 0.1 and 0.2. The Epanecnikov kernel function is 0 where $|t| > h$.

The sample region W increases with test distance to account for the change in crystal abundance as the shell volume grows. This is done around each crystal to achieve an average statistic for each test distance. The thickness of the spherical shell is determined by h allowing all crystal centres within the shell to be considered. Crystals closer to the middle of the shell are weighted more heavily than those that fall close to the outer edge of the shell. PCF values >1 for a given test distance reflect a greater abundance of crystal centres compared to a random distribution at that length scale. Such a result indicates clustering of crystal centres at the corresponding test distance, possibly indicating the uneven distribution of energetically preferable nucleation sites or chemical heterogeneities. PCF values of <1 for test distances larger than the average diameter of the crystal population result from crystal centre abundances which are smaller than expected in a random distribution indicative of an ordered pattern. Ordering at test distances shorter than the average diameter of the crystal population needs to be compared to results obtained from the numerical simulations of non-clustered interface reaction-controlled nucleation and growth. Values <1 that are positioned below the 2σ envelope may then be unrelated to the volume effect and may be interpreted to reflect ordering due to the suppression of nucleation in the diffusion fields of porphyroblasts associated with diffusion-controlled crystallization.

Mark correlation function, MCF'. The MCF' is similar to the PCF in that it also accounts for crystals present within a spherical shell around each crystal modelled using the Epanecnikov kernel function, e_κ. In addition, the MCF' allows us to consider a crystal feature, or mark, such as radius or volume.

The mark correlation function, $k(r)$, is defined as (Hirsch *et al.*, 2000)

$$k(r) = \frac{\kappa_{mm}(r)}{\kappa_{mm}(\infty)} \tag{18}$$

with

$$\kappa_{mm}(r) = \frac{\rho_{mm}(r)}{\rho(r)} \tag{19}$$

and

$$\kappa_{mm}(\infty) = \overline{m}^2 \tag{20}$$

where $\overline{m}$ is the arithmetic mean value of the mark for the entire crystal population. $\rho_{mm}(r)$ is defined as

$$\rho_{mm}(r) = \sum_{i=1}^{n} \sum_{\substack{j=1 \\ (j\neq1)}}^{n} \frac{m_i m_j e_h(t)}{4\pi r^2} \tag{21}$$

where m_i is the mark value of the *i*th crystal. Hence, the MCF' allows comparison of the arithmetic mean value of the mark for crystals separated by a certain test distance with the arithmetic mean mark value for the entire population. For example, if radius is used as the mark, and the mean radius of the crystals separated by a certain test distance is smaller than the mean radius of the population then the MCF' value will be <1. Such a statistic can be expected as a result of Strauss hard-core ordering at test distances shorter than the average diameter of the crystal population. In case the MCF' value is less than those of the 2σ envelope for that test distance then Strauss hard-core ordering is not responsible for this pattern. Instead, a reduction in porphyroblast growth rates associated with the interference of diffusion fields of neighbouring crystals may be inferred. Such a scenario may be expected if porphyroblast growth is diffusion-controlled. MCF' values larger than 1 are indicative of relatively large crystals at that test distance compared to the arithmetic mean mark value of the entire crystal population.

Strauss hard-core ordering in populations of non-spherical objects. Because crystals are non-spherical objects, Strauss hard-core ordering develops at distances equal to and less than the length of the average major axis of their best fit ellipsoids as those distances are the longest length scales at which crystals touch (Petley-Ragan *et al.*, 2016). This phenomenon develops in any population of non-spherical hard objects independent of the shape-orientation of the objects, and is not related to diffusion-controlled crystal growth. According to Petley-Ragan *et al.* (2016), there is a weakness in the correlation functions when applied to crystals if the sphere-normalized radius of these crystals is used as mark in the statistics calculations (*e.g.* Carlson, 1989; Daniel and Spear, 1999; Hirsch *et al.*, 2000; Ketcham *et al.*, 2005; Hirsch and Carlson, 2006; Hirsch, 2008). Because the length of the major axis of a non-spherical object exceeds that of the corresponding sphere-normalized diameter, Strauss hard-core ordering operates at length scales larger than the sphere-normalized diameter in a crystal population. Hence, if a sphere-normalized radius were to be used as a mark in the statistics calculations, the scale of ordering through the Strauss hard-core process would be underestimated and its characteristic ordered pattern may, incorrectly, be interpreted to be the result of diffusion-controlled growth. For this reason, Petley-Ragan *et al.* (2016) advocate the use of the semi-major axes of ellipsoids fitted around porphyroblasts as mark instead to account correctly for the Strauss hard-core process and its influence on the spatial distribution of porphyroblasts.

1.3.2. Selected examples of statistical analyses of porphyroblast distributions
Garnet is the prime mineral for the statistical analysis of the spatial distribution of porphyroblasts (*e.g.* Denison *et al.*, 1997; Daniel and Spear, 1999; Hirsch *et al.*, 2000; Ketcham *et al.*, 2005; Hirsch and Carlson, 2006; Hirsch, 2008; Gaidies *et al.*, 2015) given its key importance for the determination of metamorphic *P-T-t* trajectories (*e.g.* Spear, 1993), and because its specific attenuation properties allow it to be detected easily by XR-CT (Denison *et al.*, 1997). Other porphyroblastic phases investigated texturally using XR-CT and 3D pattern statistics include biotite (Hirsch and Carlson, 2006; Petley-Ragan *et al.*, 2016) and cordierite (Petley-Ragan *et al.*, 2016).

Figure 4 illustrates the results of PCF and MCF' calculations applied to a garnet population in a metapelite from the Barrovian garnet zone of the Lesser Himalayan Belt in Sikkim, India (sample 24-99 of Gaidies *et al.*, 2015). The total number of crystals considered for the calculations presented is 1000, and the dimensions of the prismatic sample volume investigated are ~2.5 mm × 20 mm × 20 mm. Garnet is significantly clustered at distances of <~3 mm (PCF) and is characterized by a positive correlation between size and nearest neighbour distance for test distances of <0.8 mm (MCF'). Whereas the clustering may reflect an inhomogeneous distribution of nucleation sites or nutrients for garnet crystallization at the respective length scales, the positive correlation of size and position may reflect ordering due to diffusion-controlled growth as it is developed at length scales larger than Strauss hard-core ordering. Strauss hard-core ordering is expressed in the drop to low statistical values of the null hypothesis envelope at test distances $<\sim 0.7$ mm (Fig. 4). If the ordering of garnet obtained at test distances of <0.8 mm reflects the interference of diffusion fields during growth of neighbouring crystals then the length scale of ordering may be interpreted as corresponding to the effective width of the diffusion fields. However, it is important to note that the sphere-normalized radii of the garnet porphyroblasts were used for the simulation of Strauss hard-core ordering in this sample. A possible deviation from sphericity of garnet may have increased the length scales at which Strauss hard-core ordering operated (Petley-Ragan *et al.*, 2016) and may not be considered in the statistical results obtained. In addition, resorption or syn- and post-metamorphic deformation may have modified the spatial distribution of garnet grain centres and sizes, which is not considered in the statistical analyses. The drop below values of one in the PCF for test distances $>\sim 3$ mm in both the simulated and natural array (Fig. 4), reflects test distances exceeding the length of the shortest sample dimension.

Whereas evidence for interface reaction-controlled garnet crystallization during regional metamorphism was presented by Daniel and Spear (1999), based on the statistical analysis of the 3D spatial distribution of porphyroblast centres and sizes integrated with compositional zoning data of garnet in rocks from NW Connecticut (USA), diffusion-controlled porphyroblastic growth may be an appropriate end-member scenario for the prograde crystallization of garnet in other regional metamorphic areas (*e.g.* Carlson, 1989; Denison and Carlson, 1997; Spear and Daniel, 2001). This suggests that variations in the thermal or deformational history of rocks may have a significant influence on metamorphic reaction kinetics.

Based on the analysis of the size distribution and compositional zoning of garnet in contact-metamorphic schists from the cordierite zone of a low-*P* thermal dome near Yellowknife (Canada), Kretz (1993) concluded that porphyroblastic growth was interface reaction-controlled. In a recent study, Petley-Ragan *et al.* (2016) also investigated contact metamorphic rocks. In their study, the spatial distribution of biotite and cordierite porphyroblasts in a hornfels from the contact aureole of the Bugaboo Batholith (SE British Columbia) was analysed. The PCF and MCF' statistics of the cordierite population obtained in that study are illustrated in Fig. 5 and indicate a random spatial disposition of porphyroblast centres (PCF) and sizes (MCF') across all scales investigated. Values below the null hypothesis envelope at small test distances when using the PCF are due to incorrect considerations of the Strauss hard-core process by this correlation function and are not related to diffusion-controlled crystallization (Petley-Ragan *et al.*, 2016). The statistics of the biotite population in the same rock allow for a similar interpretation (Petley-Ragan *et al.*, 2016) indicating that long-range diffusion of nutrients was relatively fast so that diffusion fields around cordierite and biotite crystals did not play a critical role in the spatial disposition of the porphyroblasts. This may indicate that interface processes were either significantly slower than long-range diffusion of nutrients, so that a departure from equilibrium developed only at the interfaces, or that both short and long-range diffusion were fast relative to the rate with which *T* changed during contact metamorphism resulting in metamorphism close to equilibrium. A final answer to these questions is largely dependent on appropriate thermodynamic descriptions of the equilibrium phase relations in metapelites at the low metamorphic *P* associated with the contact metamorphism. Only in that case can observed mineral assemblages and compositions be compared with the predicted equilibrium phase relations. However, systematic incongruities between observed and predicted phase equilibria in the low-*P* metapelites from the Bugaboo contact aureole indicate that modifications to some of the available thermodynamic models may still be required (Pattison and Debuhr, 2015).

1.4. Quantitative modelling of porphyroblastic garnet crystallization

Given its abundance in metamorphic rocks and its common chemical and isotopic zoning, garnet is arguably the most valuable porphyroblastic phase from which to infer the physicochemical environment of metamorphism (*e.g.* Spear, 1993; Ague and Carlson, 2013). Quantitative models of porphyroblast crystallization have been developed to predict the compositional zoning that develops in garnet in response to variations in the *P-T-X* conditions of metamorphism (*e.g.* Cygan and Lasaga, 1982; Loomis, 1982, 1986; Loomis and Nimick, 1982; Spear *et al.*, 1991b; Konrad-Schmolke *et al.*, 2008). The *DiffGibbs* software (Florence and Spear, 1991; Spear and Florence, 1992; Spear *et al.*, 1991b) was one of the first models designed to predict the zoning of garnet for specified *P-T-t* paths accounting for chemical fractionation associated with garnet crystallization and intragranular diffusion.

Similar to *DiffGibbs* but based on Gibbs energy minimization, the *THERIA_G* software (Gaidies *et al.*, 2008a) predicts the compositional zoning of garnet that develops during growth along a metamorphic path in *P-T-X* space considering chemical fractionation, intragranular diffusion in garnet, and diffusional fluxes between garnet and the rock matrix. However, *THERIA_G* differs from all previous models in that it allows the consideration of nucleation and growth of garnet. Hence, the compositional zoning that may develop in an entire garnet population during metamorphism can be predicted and compared with observations in natural samples allowing for a deeper understanding of porphyroblast crystallization.

Gibbs energy minimization (de Capitani and Brown, 1987) is used in *THERIA_G* at any point in *P-T-X* space to calculate the thermodynamically stable phase relations, such as the chemical compositions and volumes of all the phases present in the stable assemblages. If garnet is part of the thermodynamically stable phase assemblage then circular shells are added to all pre-existing grains according to the implications of size-independent interface reaction-controlled porphyroblastic growth (see expression 5). In other words, garnet is modelled to grow with a radial rate independent of size assuming that interface curvature does not influence the growth rate. The influences of changes to the chemical driving force for porphyroblastic growth, ΔG_r, and interface mobility, M, on garnet growth rate during metamorphism cannot be modelled because chemical equilibrium is not only considered across the rock matrix but also at the garnet/matrix interfaces. Consequently, *THERIA_G* does not account for both long-range diffusion and interface reaction, but models the evolution of a metamorphic system where chemical disequilibrium is developed only across the garnet porphyroblasts which grow with rates controlled solely by the rate at which the model system propagates through *P-T-X* space. The compositional gradients that develop in each garnet porphyroblast are constrained by the equilibrium volume and composition of garnet that forms along the *P-T-X-t* path, the number of garnet crystals per size class, and intracrystalline diffusion. Based on the assumption that bigger crystals nucleated earlier than smaller ones, *THERIA_G* nucleates garnet in pulses where the garnet density of each nucleation pulse is given by the observed number of garnet crystals per size class. A new nucleation pulse starts once garnet grows to a radius that corresponds to the size of the respective radius class. A departure from equilibrium associated with nucleation is not considered.

THERIA_G has been applied successfully to identify the metamorphic *P-T-t* paths responsible for the compositional zoning in garnet and metamorphic mineral parageneses observed in rocks from different metamorphic terrains (*e.g.* Gaidies *et al.*, 2008b,c; Hoschek, 2013; Moynihan and Pattison, 2013; Cutts *et al.*, 2014). A recent example of its application to schists of the Barrovian garnet and staurolite zones of the Lesser Himalayan Belt in Sikkim (India) is given in Fig. 6 (Gaidies *et al.*, 2015). The mineral assemblages and compositional zoning of garnet predicted with *THERIA_G* match observations remarkably well suggesting that porphyroblastic nucleation and growth close to equilibrium is a valuable model assumption to obtain high-quality *P-T* information even for the short timescales associated with the metamorphism during the Himalayan orogeny.

On the other hand, timescales associated with the metamorphism in Sikkim were too short for chemical diffusion to modify substantially garnet growth-zoning in the rocks from the garnet and staurolite zones. Research by George and Gaidies (2016) indicates that crystals from the garnet zone with a radius as small as 200 μm preserved their primary compositional zoning reflecting heating and cooling rates in excess of 100°C/Ma in accordance with independent results based on Lu-Hf garnet geochronology (Anczkiewicz *et al.*, 2014). According to George and Gaidies (2016), the major element concentrations in the rims of the entire garnet population are identical, suggestive of sample-scale equilibration with respect to these elements at the time of cessation of garnet crystallization. Furthermore, garnet crystals of equal size developed equivalent compositional zoning patterns independent of their positions in the rock, and compositional gradients gradually steepen in progressively smaller crystals. These findings may point to size-dependent porphyroblastic growth rates due to an increase of interface curvature for grains that crystallized relatively late during metamorphism, possibly indicating exceedingly small driving forces for porphyroblastic growth or relatively high interfacial energies, according to expression 7.

To conclude this section we note that both the microstructural features such as the spatial disposition and size-distribution of porphyroblasts as well as their chemical zoning, including its possible position dependence bear valuable petrogenetic information. Once nucleation has occurred, the growth of porphyroblasts may be

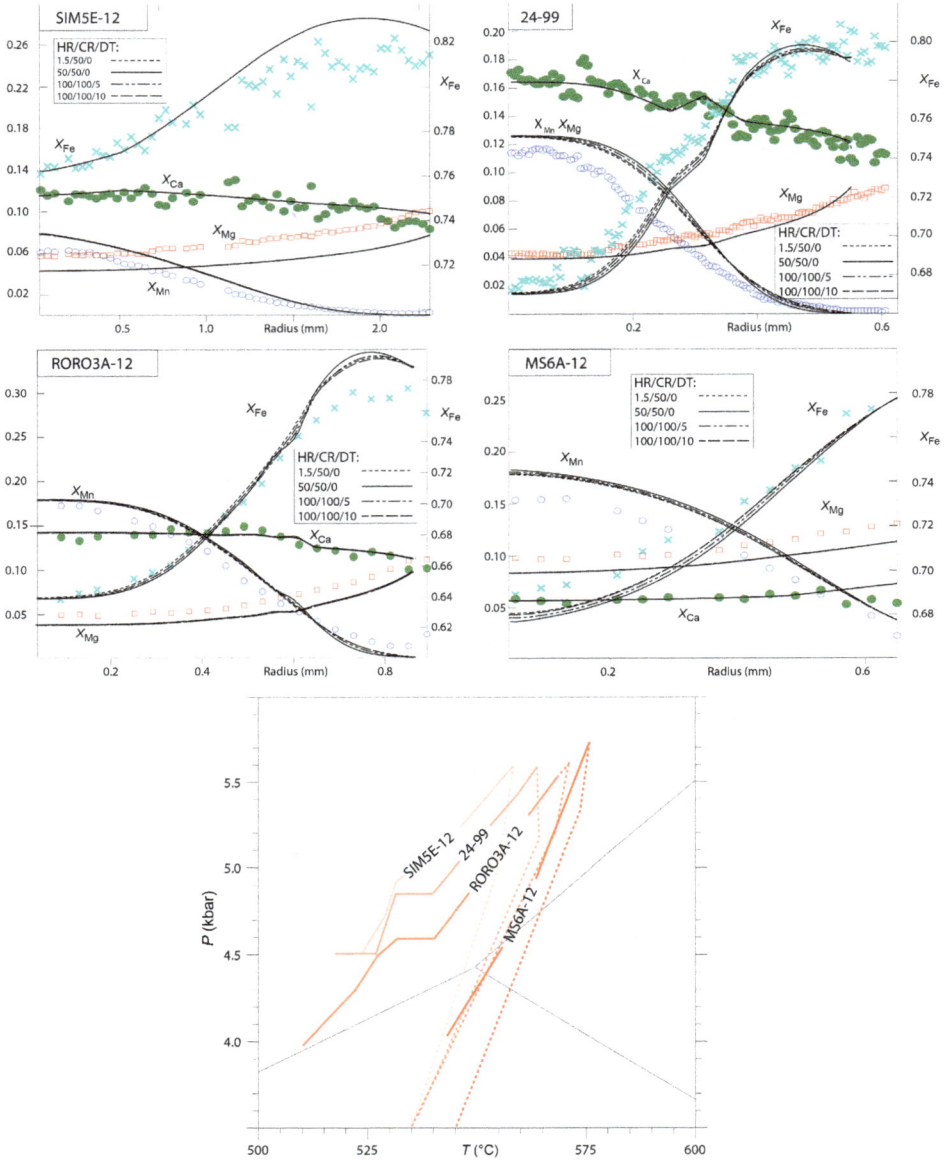

Figure 6. The observed compositional zoning of garnet (symbols) from the Lesser Himalayan Belt of Sikkim (modified after Gaidies *et al.*, 2015) *vs.* the zoning predicted with *THERIA_G* (Gaidies *et al.*, 2008a) for different heating and cooling scenarios, and the corresponding metamorphic *P-T* paths. HR = heating rate, and CR = cooling rate (both in °C/Ma), DT = dwell time (in Ma).

described by a continuum between two end-member processes represented by interface reaction-controlled and diffusion-controlled growth. Due to its refractory nature, its compositional variability and widespread occurrence, garnet is the most versatile mineral for reconstructing *P-T-t* trajectories of metamorphic rocks. The integrated analysis of the microstructural features and chemical zoning of garnet populations in metapelitic rocks from different geological environments allows reconstruction of metamorphic conditions with unprecedented accuracy. The task is demanding in terms of material characterization and data reduction. However, advances in instrumentation and increasing computational ability combined with more accurate thermodynamic and kinetic data greatly foster this integrated approach.

2. Interlayer growth – corona formation

When two phases or phase assemblages of different composition that cannot coexist stably are in contact, they may react to form a new phase or phase assemblage. When the product phase or phase assemblage takes the form of a layer or a sequence of layers along the interface between the reactant phases or phase assemblages, the process is referred to as 'interlayer growth'. In mineralogy the corresponding phenomenon is known as 'corona' (Fig. 7a), 'reaction rim' (Fig. 7b) or 'metasomatic reaction band' (Fig. 7c) (Spry, 1969; Vernon, 2004). Adopting the terminology of Joesten (1977) we refer to the individual mineralogically distinct zones in a corona or in a reaction band as the *layers*. Reaction bands or coronas may comprise a single layer or a sequence of layers. In most cases, the individual layers are polycrystalline, and they may comprise a single or of several minerals.

The stable coexistence of two phases or phase assemblages that prevailed at some point in time may be terminated by changes in pressure and/or temperature or in any other externally controlled physicochemical parameters. For example, plagioclase and

Figure. 7. (a) Transmitted light photomicrograph (plane polarized light) of garnet (pale red-grey) + quartz (transparent) coronas at the contacts between plagioclase (transparent) and clinopyroxene (green); sample from Eastern Gats, India, courtesy of M. Raith; (b) forward scatter electron image showing the polycrystalline nature of an experimentally grown enstatite reaction rim produced at 900°C and 1 GPa at forsterite (grain in the centre)–quartz (polycrystalline matrix) interfaces; (c) metasomatic reaction band comprising a tremolite-forsterite layer (inner dark green layer) and a forsterite-calcite layer (outer green layer) between an aplitic dyke (centre) and dolomite marble (host rock); image from Bergell contact aureole, N-Italy.

Fe-Mg-bearing phases such as olivine, orthopyroxene or clinopyroxene may coexist stably at granulite- or amphibolite-facies conditions or during magmatic crystal-lization, and they may become incompatible and react to form garnet-bearing coronas during eclogite facies overprinting (Joesten, 1986; Ashworth and Birdi, 1990; Johnson and Carlson, 1990; Indares and Rivers, 1995; Bethune and Davidson, 1997; Zhang and Liou, 1997; Cox and Indares, 1999; Keller *et al.*, 2004, 2006, 2008; Larikova and Zaraisky, 2009). The reverse reaction occurring during decompression of eclogites, or high-pressure granulites may produce corona microstructures of Fe-Mg phases such as orthopyroxene or clinopyroxene and plagioclase around garnet (Carlson and Johnson, 1991; Obata, 2011; Obata *et al.*, 2013; Spacek *et al.*, 2013) or plagioclase coronas around kyanite (Tajcmanova *et al.*, 2011). Alternatively, interlayer growth may occur due to the juxtaposition of incompatible rock types such as marble and chert (Joesten and Fisher, 1988; Joesten, 1991; Heinrich, 1993) or of marble and metabasic rocks (Abart *et al.*, 2001; Abart and Sperb, 2001; Fukuyama *et al.*, 2006). The reaction between two incompatible rock types is also known as bimetasomatism or diffusion metasomatism (Korzhinskii, 1959, 1970; Thompson, 1959). Typically, both the reactant and product phases are present in the microstructures produced by interlayer growth indicating that reaction did not go to completion and overall equilibrium was not attained (Carlson, 2002).

Interlayer growth requires that two basic processes occur simultaneously. Because the reactant and product phases have different compositions, interlayer growth requires 'long-range chemical mass-transfer' across the layer structure, where the term 'long range' refers to distances that are large compared to the typical interatomic spacing. In the absence of migrating melts or fluids, chemical mass-transfer can occur only by diffusion, and in what follows, 'long-range diffusion' is implied as the only mode of chemical mass-transport. At the same time, localized reactions must proceed at the 'reaction interfaces', which delimit the growing layers on either side. The overall process may be referred to as 'reactive diffusion' (Svoboda and Fischer, 2013). Both long-range diffusion and interface reaction may be rate limiting, and the coupling between the two processes determines the overall reaction kinetics.

Much of the basic understanding of interlayer growth dates back to Korzhinskii (1959, 1970) and to Thompson (1959), who applied the concept of 'local equilibrium' to derive the chemical potential gradients driving component diffusion from the mineral assemblage zoning across a metasomatic reaction band. Because then, quantitative models for reaction-band and corona formation have been presented by several authors (Fischer, 1973; Frantz and Mao, 1976; Joesten, 1977). In these models long-range diffusion is assumed as the rate-limiting process, and local equilibrium is implied at the reaction interfaces. These models, however, do not provide for deviations from local equilibrium that may arise from additional dissipative processes operating in parallel with long-range diffusion.

In the following, a more general formulation of interlayer growth is presented. To begin with, the kinematic relations linking component fluxes and interlayer growth are addressed and a rate law for diffusion-controlled interlayer growth is derived. In a second step, the 'Thermodynamic Extremal Principle' (TEP) (Onsager, 1931; Ziegler, 1961; Svoboda and Turek, 1991; Svoboda et al., 2005; Martyushev and Seleznev, 2006; Fischer et al., 2014) is applied to derive the corresponding evolution equations that account for additional potentially rate-limiting processes. The TEP is particularly well suited for considering several dissipative processes occurring in parallel, such as the simultaneous operation of long-range diffusion, sluggish interface reactions and non-ideal sources/sinks for vacancies at reaction interfaces and in the bulk. The TEP and some of its applications in materials science are discussed in detail by Svoboda et al. (2017, this volume) and only a brief introduction is given here. The TEP is then applied to diffusion-controlled interlayer growth. Finally, the effect of sluggish interface reaction and associated deviations from local equilibrium at the reaction interfaces are addressed in the frame of the TEP.

2.1. The kinematics of interlayer growth

2.1.1. Mass balance at a moving reaction interface

Many mineral reactions involve replacement of a reactant phase by a product phase at a well defined sharp 'reaction interface'. When the reactant and the product phases have different compositions, the reaction interface can move only when chemical components are supplied to or removed from the reaction interface, implying long-range chemical mass transfer. The velocity of interface motion is related to the component fluxes to and from the reaction interface.

Consider a binary system with components A and B and phases γ and β with compositions $A_{1-X_\gamma}B_{X_\gamma}$ and $A_{1-X_\beta}B_{X_\beta}$, where X_γ and X_β denote mole fractions of component B in phases γ and β, respectively, where $X_\beta > X_\gamma$ (Fig. 8a). Let a $\gamma-\beta$ assembly with unit cross-section area extend in x direction, where phase γ to the left and phase β to the right are in contact at an interface extending in the $y-z$ plane (Fig. 8b). Let both components be mobile with component fluxes J_A^γ, J_B^γ and J_A^β, J_B^β in the domains occupied by γ and β, respectively. In general, replacement of one phase by another involves a finite volume change or 'transformation strain'. As a consequence, the velocity u of the $\gamma-\beta$ interface relative to a material point in phase γ is different from the velocity w of the $\gamma-\beta$ interface relative to a material point in phase β. Mass balance across the moving $\gamma-\beta$ interface requires (Fischer and Simha, 2004)

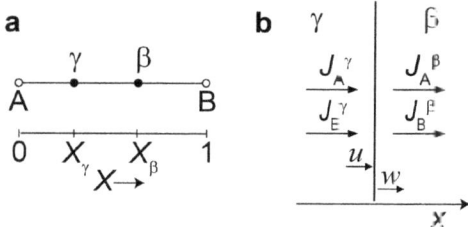

Figure 8. (a) Chemography of a binary system with components A and B and two phases γ and β; (b) one-dimensional setup with phases γ and β separated by a planar interface perpendicular to the x direction, u and w are the velocities in the x direction of the interface with respect to material points in γ and β, respectively, and J_A^γ, J_B^γ, J_A^β, J_B^β are the component fluxes in the domains occupied by the phases γ and β, respectively.

$$\frac{1 - X_\gamma}{\Omega_\gamma} w - \frac{1 - X_\beta}{\Omega_\beta} u = J_A^\beta - J_A^\gamma, \text{component A,}$$

$$\frac{X_\gamma}{\Omega_\gamma} w - \frac{X_\beta}{\Omega_\beta} u = J_B^\beta - J_B^\gamma, \text{component B} \tag{22}$$

where Ω_γ and Ω_β denote molar volumes. The expressions on the left hand sides give the rates at which components A and B are consumed or liberated at the moving $\gamma-\beta$ interface. The consumption or liberation of components must match the changes in the component fluxes across the interface (right hand side of equations 22). Given that the compositions and molar volumes of γ and β are known, the system of equations 22 provides a relation between interface velocity and component fluxes.

2.1.2. Mass balance during interlayer growth and Kirkendall effect

During interlayer growth the layer of the new phase or the sequence of layers of new phases, in general, grows from the original interface between the reactant phases or phase assemblages into both directions. The corresponding kinematic relations must account for the simultaneous motion of two or more reaction interfaces. We restrict the following discussion to single layer growth; the considerations are, however, similarly applicable to multilayer growth.

Consider the binary system A−B with phases α, β and γ with compositions $A_{1-X_\alpha}B_{X_\alpha}$, $A_{1-X_\beta}B_{X_\beta}$ and $A_{1-X\gamma}B_{X\gamma}$, where $X_\alpha < X_\gamma < X_\beta$ (Fig. 9). At $t = 0$ phases α and β extend from $x = 0$ to $x = R_0$, and from $x = R_0$ to $x = +\infty$, respectively. After a short nucleation period a layer of γ grows, and at $t > 0$ the $\alpha-\gamma$ and the $\beta-\gamma$ interfaces are at positions R_α and R_β, respectively, and the original $\alpha-\beta$ interface is at R_0*.

In single crystals of ionic substances, diffusion occurs by the motion of ions, and, due to stoichiometric and charge-balance constraints, diffusive fluxes are largely balanced so that $\Sigma J_i = 0$ (Lasaga, 1998; Glicksman, 2000; Mehrer, 2007). In metals, local charge balance constraints do not apply, and the diffusive fluxes of different components are generally not balanced. The imbalance of component fluxes is compensated by a flux of vacancies $\mathbf{J}_0$ so that $\Sigma \mathbf{J}_i = -\mathbf{J}_0$. If an ionic substance is polycrystalline such as a mineral layer produced from interlayer growth, electrically neutral components may diffuse, and the component fluxes need not necessarily be balanced. For example the electrically neutral component MgO may diffuse by the simultaneous motion of the charged species Mg^{2+} and O^{2-}, and the resulting flux of MgO is independent of any other diffusive fluxes. Interlayer growth involving decoupled fluxes of neutral components has been reported from different silicate systems (Gardes *et al.*, 2011; Joachim *et al.*, 2011, 2012). To account for the independent diffusion of components we allow for $\Sigma J_i \neq 0$ and express the relative fluxes of the A and B components as

$$\chi = \frac{J_A}{J_A - J_B} \tag{23}$$

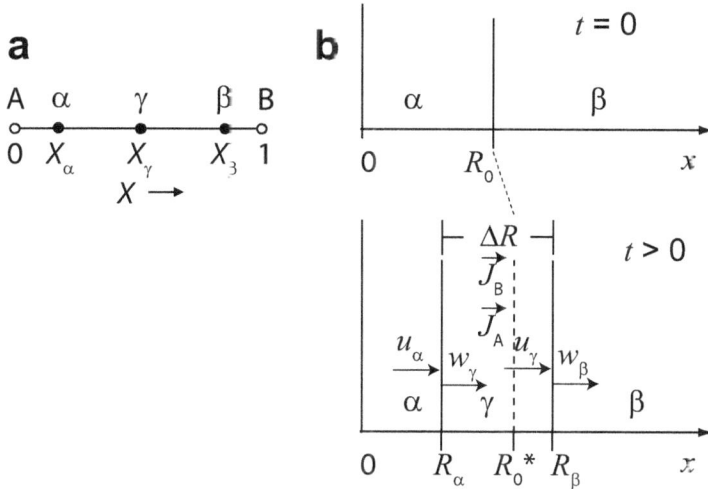

Figure 9. (a) Chemography of binary system with components A and B and phases α, β and γ with compositions $A_{1-X_\alpha}B_{X_\alpha}$, $A_{1-X_\beta}B_{X_\beta}$ and $A_{1-X_\gamma}B_{X_\gamma}$. (b) Model geometry of 1D diffusion-reaction assembly; at $t=0$ phase α occupies the region $0 \leqslant x \leqslant R_0$, and phase β occupies the region $x \geqslant R_0$; at $t>0$ a layer of phase γ forms along the α–β interface; and the α–γ and the β–γ interfaces are at R_α and R_β, respectively; the position of the original α–β interface is indicated by the dashed line labelled R_0^*; J_A and J_B denote diffusion fluxes of components A and B.

Let phase α be fixed to the global coordinate system. The velocities of the α–γ and the γ–β interfaces are $\dot{R}_\alpha$ and $\dot{R}_\beta$, respectively, in the global coordinate system. The interface reactions at the α–γ and the γ–β interfaces do not, in general, conserve volume. As a consequence, the reaction interfaces have different velocities relative to material points in the reactant and product phases. The interface velocities relative to the phases on the left and on the right of the interfaces are denoted u_α, u_γ and w_γ, w_β (Fig. 9). Given that the diffusive fluxes are restricted to the domain occupied by γ, conservation of mass at the α–γ requires (Fischer and Simha, 2004)

$$\frac{(1-X_\alpha)}{\Omega_c} u_\alpha - \frac{(1-X_\gamma)}{\Omega_\gamma} w_\gamma = -J_A^{\alpha-\gamma},$$

$$\frac{X_\alpha}{\Omega_\alpha} u_\alpha - \frac{X_\gamma}{\Omega_\gamma} w_\gamma = -J_B^{\alpha-\gamma} \tag{24}$$

and at the γ–β interface it requires

$$\frac{(1-X_\gamma)}{\Omega_\gamma} u_\gamma - \frac{(1-X_\beta)}{\Omega_\beta} w_\beta = -J_A^{\gamma-\beta},$$

$$\frac{X_\gamma}{\Omega_\gamma} u_\gamma - \frac{X_\beta}{\Omega_\beta} w_\beta = -J_B^{\gamma-\beta} \tag{25}$$

where $\Omega_\alpha, \Omega_\beta, \Omega_\gamma$ are the molar volumes, and $J_A^{\alpha-\gamma}, J_B^{\alpha-\gamma}$ and $J_A^{\gamma-\beta}, J_B^{\gamma-\beta}$ are the diffusive fluxes at the $\alpha-\gamma$ and the $\gamma-\beta$ interfaces, respectively. If the layer of γ grows only at the reaction interfaces, $J_A^{\alpha-\gamma} = J_A^{\gamma-\beta} = J_A$ and $J_B^{\alpha-\gamma} = J_B^{\gamma-\beta} = J_B$. Referring to equation 23 we note that $J_A = J\chi$ and $J_B = J(\chi - 1)$ and $J_A + J_B = J$. Combining equations 24 and 25 the different interface velocities are expressed in terms of the component fluxes

$$u_\alpha = \Omega_\alpha J \frac{(\chi - 1)(1 - X_\gamma) - \chi X_\gamma}{(1 - X_\alpha)X_\gamma - (1 - X_\gamma)X_\alpha}$$

$$w_\gamma = \Omega_\gamma J \frac{(\chi - 1)(1 - X_\alpha) - \chi X_\alpha}{(1 - X_\alpha)X_\gamma - (1 - X_\gamma)X_\alpha}$$

$$u_\gamma = \Omega_\gamma J \frac{\chi X_\beta - (\chi - 1)(1 - X_\beta)}{X_\beta(1 - X_\gamma) - (1 - X_\beta)X_\gamma}$$

$$w_\beta = \Omega_\beta J \frac{\chi X_\gamma - (\chi - 1)(1 - X_\gamma)}{X_\beta(1 - X_\gamma) - (1 - X_\beta)X_\gamma} \tag{26}$$

The rate of layer growth is given by

$$\Delta \dot{R} = u_\gamma - w_\gamma = J_A u_A + J_B u_B \tag{27}$$

where

$$u_A = \Omega_\gamma \left(\frac{X_\beta}{X_\beta(1 - X_\gamma) - (1 - X_\beta)X_\gamma} + \frac{X_\alpha}{(1 - X_\alpha)X_\gamma - (1 - X_\gamma)X_\alpha} \right)$$

$$u_B = \Omega_\gamma \left(\frac{(X_\beta - 1)}{X_\beta(1 - X_\gamma) - (1 - X_\beta)X_\gamma} + \frac{(X_\alpha - 1)}{(1 - X_\alpha)X_\gamma - (1 - X_\gamma)X_\alpha} \right) \tag{28}$$

It is seen from equation 27 that the rate of layer growth is related to the fluxes of both components, where the constant parameters u_A and u_B contain only the molar volume of the growing phase γ and the compositions of all three phases.

Recalling that α is fixed to the global coordinate system, the interface velocities in the global coordinate system are given by

$$\dot{R}_\alpha = u_\alpha, \quad \dot{R}_\beta = u_\alpha - w_\gamma + u_\gamma = \dot{R}_\alpha u \tag{29}$$

with

$$u = \frac{\Omega_\gamma}{\Omega_\alpha} \left(\frac{(1 - X_\beta)X_\alpha - (1 - X_\alpha)X_\beta}{X_\beta(1 - X_\gamma) - (1 - X_\beta)X_\gamma} \right) + 1$$

where $u < 0$ for $X_\alpha < X_\gamma < X_\beta$. The rate of layer growth expressed in terms of interface velocity reads

$$\Delta \dot{R} = \dot{R}_\alpha(u - 1) \tag{30}$$

From equation 29 together with the initial condition $R_\alpha(0) = R_\beta(0) = R_0$ the two interface positions are related by

$$R_\beta = u(R_\alpha - R_0) + R_0 \tag{31}$$

Kirkendall effect: The unbalanced flux of components gives rise to the so-called 'Kirkendall effect', which describes a net total material transfer relative to a material point in the bulk substance during the interdiffusion of two or more components. The Kirkendall effect was first described for the interdiffusion of copper and zinc in brass, where zinc diffuses faster then copper, and an overall shift of the sample mass relative to an inert marker occurs in the direction of falling zinc concentrations (Smigelskas and Kirkendall, 1947). The Kirkendall effect has also been described in the context of interlayer growth in metal systems (van Loo *et al.*, 2000) and during interlayer growth in silicate systems (Gardés *et al.*, 2011; Joachim *et al.*, 2011, 2012).

The growth rate of an interlayer depends on the fluxes of all components (equation 27), and the fluxes of the individual components cannot, in general, be determined from measured layer-growth rates alone. However, the relative rates at which the two reaction interfaces delimiting a layer on either side move, depend on the individual component fluxes (equations 26). Individual component fluxes can thus be determined, if the position of the so called 'Kirkendall plane', that is the trace of the original contact between the reactant phases, is known.

The position of the Kirknedall plane may be evident from discontinuous composition zoning (Fig. 10a) or from a microstructural or textural discontinuity (Fig. 10b) within a layer (Abart *et al.*, 2004; Götze *et al.*, 2010; Keller *et al.*, 2010; Jeřabek *et al.*, 2014). Moreover, in experiment, inert markers may be used for tracing the Kirkendall plane (Fig. 10c). Thereby an inert metal (*e.g.* platinum) is deposited on the surface of one of the reactant crystals and, provided the metal particles are not dragged along by moving grain or phase boundaries, they trace the original interface between the two reactants after the experiment (Gardés *et al.*, 2011; Joachim *et al.*, 2011, 2012).

If the compositions and molar volumes of the phases involved in layer growth are constant, the position of the Kirkendall plane is controlled only by the relative component fluxes, which are, in turn, determined by the respective component mobilities. In an inverse approach, the relative component mobilities can be determined directly from the position of the Kirkendall plane (Abart *et al.*, 2004, 2009; Svoboda *et al.*, 2006b). For example, consider an assembly comprising a crystal of periclase (MgO) to the left and a crystal of corundum (Al_2O_3) to the right, which are in contact at a vertical interface (Fig. 11). Let periclase and corundum react to form a continuous layer of spinel ($MgAl_2O_4$) along their interface according to the reaction

$$MgO + Al_2O_{3,s} = MgAl_2O_4$$

Layer growth requires that either one or both of the components MgO and Al_2O_3 are transferred across the growing layer (Fig. 11). In general, the spinel layer grows from the original periclase–corundum interface into both directions replacing periclase at the periclase–spinel reaction interface (interface I in Fig. 11) and replacing corundum at the corundum–periclase reaction interface (interface II in Fig. 11).

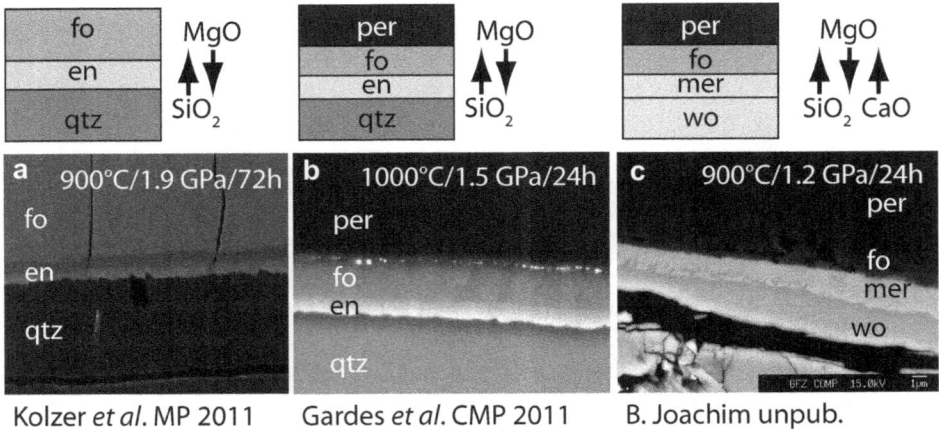

Figure 10. (a) Enstatite (en) layer forming at a forsterite (fo)−quartz (qtz) contact, different grey shades in the enstatite reflect differences in Fe content, the composition jump within the enstatite layer traces the original quartz–forsterite interface; (b) forsterite (fo)–enstatite (en) double layer forming at a periclase (per)–quartz (qtz) contact, bright spots at the periclase–forsterite interface are Pt-markers tracing the original periclase-quartz contact; image courtesy of E. Gardés; (c) forsterite (fo)−merwinite (mer) double layer at a periclase (per)−wollastonite (wo) contact; image courtesy of B. Joachim.

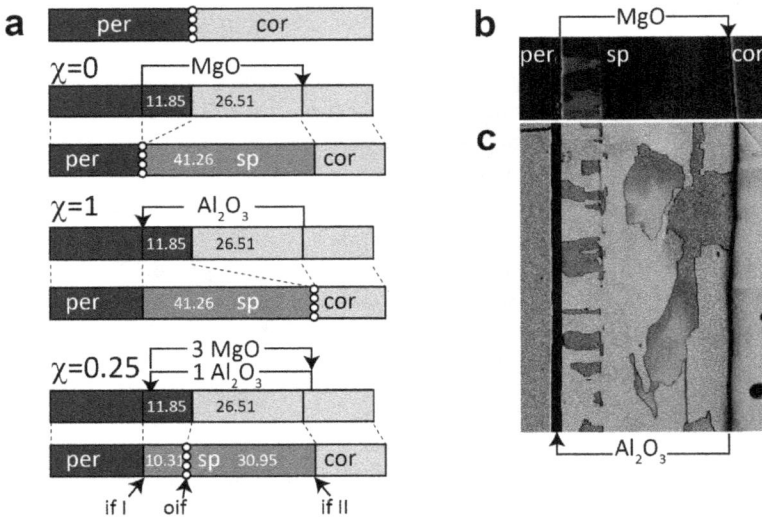

Figure 11. (a) Schematic drawing of layer growth in the system MgO−Al_2O_3; top − starting configuration with a crystal of periclase (per) and corundum (cor) in contact at a vertical interface; white circles indicate inert markers for tracing the original per–cor interface (oif). The panels below show layer growth of spinel (sp) with only MgO mobile ($\chi = 0$), only Al_2O_3 mobile ($\chi = 1$), and with transfer of both MgO and Al_2O_3 in proportions 3:1 ($\chi = 0.25$). The numbers indicate the volumes in cm^3 of phases consumed/produced per mole of mobile components transferred; (b) BSE image and (c) forward scattered electron (FSE) image of spinel rim grown at 1350°C − note the change in microstructure of spinel at the original periclase–corundum interface.

The bulk reaction may be split into two half reactions occurring at the two reaction interfaces

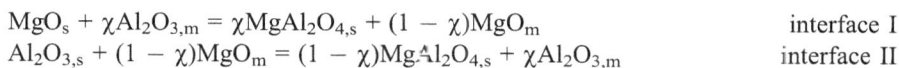

$$MgO_s + \chi Al_2O_{3,m} = \chi MgAl_2O_{4,s} + (1 - \chi)MgO_m \qquad \text{interface I}$$
$$Al_2O_{3,s} + (1 - \chi)MgO_m = (1 - \chi)MgAl_2O_{4,s} + \chi Al_2O_{3,m} \qquad \text{interface II}$$

where subscripts 's' and 'm' denote 'solid phase' and 'mobile component', respectively. Given that chemical components are neither added nor removed from the system, the two half reactions sum up to the bulk reaction. This implies that $(1 - \chi)$ moles of MgO are liberated at interface I and are transferred to interface II to form spinel by reaction with corundum. At the same time, χ moles of Al_2O_3 are liberated at interface II and are transferred to interface I, where they react with periclase to produce spinel. It can be seen from the half-reactions that the relative rates of MgO and Al_2O_3 mass transfer determine at what proportions spinel grows in the direction of periclase and corundum.

The relative interface velocities as calculated from equations 26 based on the compositions and molar volumes given in Table 1 are shown as functions of χ in Fig. 12a. All interface velocities are normalized to w_{Cor}. The interface velocities u_{Per} and w_{Cor} relative to the reactant phases are independent of χ and reflect merely the volumetric proportions at which the two reactant phases are consumed. For example, irrespective of the mass-transfer scenario, one mole of periclase corresponding to 11.85 cm^3 and one mole of corundum corresponding to 26.51 cm^3 are consumed per mole or 40.26 cm^3 of newly formed spinel (Fig. 11a). In contrast, the interface velocities w_{Sp} and u_{Sp} depend on χ. For $\chi = 0$ indicating that only MgO moves, all growth occurs at the spinel–corundum interface. Any inert markers on the original periclase-corundum interface would be located at the spinel–periclase interface. This scenario implies a large positive transformation strain at the corundum–spinel interface and a negative transformation strain at the periclase-spinel interface. For example, 26.51 cm^3 of corundum are replaced by 41.26 cm^3 of spinel at interface II, and 11.85 cm^3 of corundum are consumed while no

Table 1. Molar volumes and compositions of phases involved in layer growth as shown in Fig. 12, n_{MgO}, $n_{Al_2O_3}$, and n_{SiO_2} give the numbers of moles of component per formula unit of phase.

Phase	n_{MgO}	$n_{Al_2O_3}$	n_{SiO_2}	Ω (cm^3/mol)
Corundum	0	1	0	26.51
Periclase	1	0	0	11.85
Spinel	1	1	0	41.26
Quartz	0	0	1	23.38
Forsterite	2	0	1	44.87
Enstatite	1	0	1	31.92

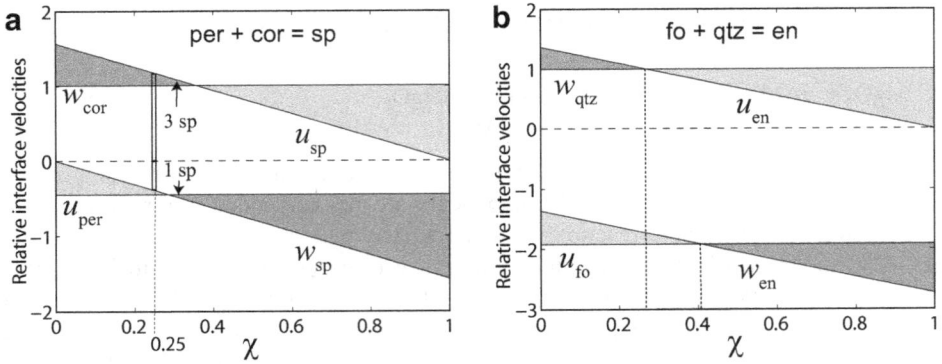

Figure 12. (a) Relative interface velocities for spinel (sp) layer growth at a periclase (per)–corundum (cor) contact in the system $MgO-Al_2O_3$; all velocities are normalized to w_{cor}, vertical bar drawn at $\chi = 0.25$ shows the relative rates at which spinel grows from the original interface into the directions of periclase and corundum; (b) relative interface velocities for enstatite (en) layer growth at a forsterite (fo)–quartz (qtz) contact in the system $MgO-SiO_2$; all velocities are normalized to w_{qtz}.

spinel is formed at interface I (Fig. 11a). In contrast, for $\chi = 1$, indicating that only Al_2O_3 is transferred between the two reaction interfaces, all growth occurs at the spinel–periclase interface, and any inert markers on the original periclase–corundum interface would be located at the spinel–corundum interface. This scenario implies a large positive transformation strain at the periclase–spinel interface and a negative transformation strain at the corundum–spinel interface. For example, 11.85 cm^3 of periclase are replaced by 40.26 cm^3 of spinel at interface I and 26.51 cm^3 of corundum are consumed while no spinel is formed at interface II.

In the general case, where $0 < \chi < 1$, the trace of the original periclase–corundum interface is located within the spinel layer at a position depending on χ, and the transformation strain will be distributed more evenly over both reaction interfaces. The triangular areas shaded in dark grey in Fig. 12 indicate domains with positive transformation strain implying that the volume of spinel formed at the respective interface is larger than the volume of the reactant phase consumed. The triangular areas shaded in light grey indicate negative volume strain. The proportions of spinel growing into the direction of corundum and periclase can be read from intersecting a vertical line drawn from w_{Sp} to u_{Sp} with the horizontal line at zero velocity, such as indicated for $\chi = 0.25$.

In an inverse approach, the relative fluxes of the two components can be determined from tracing the position of the original periclase–corundum contact within the layer of newly formed spinel. In the case illustrated in Fig. 11b the position of the original periclase–corundum interface can be discerned from a discontinuity of the microstructure within the spinel layer at about 1/4 of the rim thickness on the side of periclase. This indicates $\chi \approx 0.25$ corresponding to the transfer of 1 mole of Al_2O_3 per 3 moles of MgO. This points to interdiffusion of $2Al^{3+}/3Mg^{2+}$ in an otherwise immobile oxygen sub-lattice (Götze *et al.*, 2010; Keller *et al.*, 2010; Jeřabek *et al.*, 2014).

The relations are somewhat different for enstatite-layer growth in the system MgO–SiO$_2$. When forsterite (Mg$_2$SiO$_4$) and quartz (SiO$_2$) are put into contact at sufficiently high temperature, they react to form enstatite (MgSiO$_3$) according to the reaction

$$Mg_2SiO_{4,s} + SiO_{2,s} = 2MgSiO_{3,s}$$

Layer growth requires that either one or both of the components MgO and SiO$_2$ are transferred across the growing layer. The bulk reaction may be split into two half reactions occurring at the two reaction interfaces (Abart *et al.*, 2004)

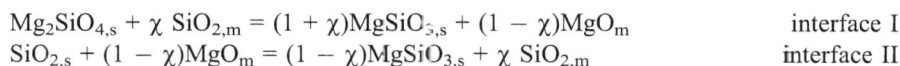

$$Mg_2SiO_{4,s} + \chi\ SiO_{2,m} = (1 + \chi)MgSiO_{3,s} + (1 - \chi)MgO_m \qquad \text{interface I}$$
$$SiO_{2,s} + (1 - \chi)MgO_m = (1 - \chi)MgSiO_{3,s} + \chi\ SiO_{2,m} \qquad \text{interface II}$$

Here $\chi = \frac{J_{SiO_2}}{J_{MgO} - J_{SiO_2}}$, and for $\chi = 0$, which indicates that only MgO and no SiO$_2$ moves, enstatite grows into both directions at the same rate, and the original forsterite–quartz interface is located in the centre of the enstatite layer (Fig. 12b). For $\chi = 1$, indicating that only SiO$_2$ and no MgO moves, all growth occurs at the forsterite–enstatite interface. The vertical dashed lines indicate constant volume replacement at the quartz–enstatite (left line at $\chi \approx 0.28$) and the forsterite-enstatite (right line at $\chi \approx 0.41$) interfaces. In this χ range the mechanical effects of growth are minimized, all other mass-transfer scenarios imply more severe transformation strains at the reaction interfaces. Milke *et al.* (2009b) presented experimental evidence for a feedback between mechanical effects and enstatite layer growth at forsterite–quartz contacts. They found different layer-growth rates when quartz grains were immersed in a forsterite matrix and when forsterite grains were immersed in a quartz matrix, which they ascribed to the different degree of mechanical-chemical feedback depending on the effective viscosity of the matrix phase (Schmid *et al.*, 2009).

The entire range of diffusive coupling, from largely independent component fluxes to charge-balanced interdiffusion has been documented for interlayer growth in silicate and oxide systems. For example, Gardés *et al* (2011) demonstrated that during layer growth in the MgO–SiO$_2$ system under 'dry' conditions at 1.5 GPa and 1100°C to 1400°C only the MgO component, supposedly represented by cooperative movement of Mg^{2+} and O^{2-}, was mobile, whereas the SiO$_2$ component remained in place. This is evident from the position of the Kirkendall plane at the periclase–forsterite contact during forsterite–enstatite double layer growth between periclase and quartz (see Pt markers in Fig. 10c). Similar findings were reported by Joachim *et al.* (2011) from layer growth in the system MgO–CaSiO$_3$, where MgO was the only mobile component under dry conditions. In contrast, charge-balanced interdiffusion of 2Al^{3+} and 3Mg^{2+} was identified as the mass-transfer scenario during the growth of magnesio-aluminate spinel (Carter, 1961; Rossi and Fulrath, 1963; Pfeiffer and Schmalzried, 1989; Götze *et al.*, 2010, 2014; Keller *et al.*, 2010) (Fig. 10b).

It must be noted that the direct link between component mobilities and the position of the Kirkendall plane is valid only for diffusion-controlled interlayer growth, where local equilibrium is maintained at the reaction interfaces. If, however, several dissipative processes such as sluggish interface reactions operate in parallel to long-range diffusion, the position of the Kirkendall plane is no longer diagnostic for the relative component mobilities (Svoboda *et al.*, 2010, 2011; Abart *et al.*, 2016).

2.2. Diffusion-controlled interlayer growth

2.2.1. Rate law for diffusion-controlled interlayer growth

Consider a binary system with components A and B and with phases α, β and γ (Fig. 13a). Let the $g–X$ surfaces of α, β and γ be so sharp convex downwards that their equilibrium compositions are essentially fixed with $X_\alpha < X_\gamma < X_\beta$ (Fig. 13b). Let the assemblage $\alpha + \beta$ be metastable with respect to the assemblages $\alpha + \gamma$ and $\beta + \gamma$. At time $t = 0$, α and β are put into contact. After a short nucleation period, a layer of γ forms (Fig. 13a). Reaction is driven by the associated lowering of Gibbs energy as indicated by $\Delta_r G$ in Fig. 13b. Following Thompson (1959) we assume that, despite the bulk system being out of equilibrium, local equilibrium prevails at the $\alpha–\gamma$ and at the $\beta–\gamma$ interfaces. At given pressure and temperature, local equilibrium between the two phases defines the chemical potentials of both components at the interface between the two phases. In a molar Gibbs energy diagram this is represented by the intersections of the common tangent to the $g–X$ surfaces of the two phases with the ordinate axes at pure A ($\mu_A^{\alpha-\gamma}, \mu_A^{\gamma-\beta}$) and B ($\mu_B^{\alpha-\gamma}, \mu_B^{\gamma-\beta}$) compositions (Fig. 13b). The differences in the chemical potentials of the A and B components between the $\gamma–\beta$ and the $\alpha–\gamma$ interfaces are related to $\Delta_r g$ by

$$\Delta\mu_A = \frac{\Delta_r g}{1 - X_\gamma} \quad \text{and} \quad \Delta\mu_B = \frac{-\Delta_r g}{X_\gamma} \tag{32}$$

In Fig. 13c the chemical potentials are shown, where a linear variation of the chemical potentials across the γ layer has been assumed. Given that component mobility is independent of position throughout the γ layer, this corresponds to constant flux. This, in turn, implies that γ grows only at the $\alpha–\gamma$ and the $\beta–\gamma$ interfaces, and no sources or sinks for components A and B exist within the γ layer. The gradients in the chemical potentials and thus the driving force for diffusion of the A and B components across the layer of γ are obtained by dividing $\Delta\mu_A$ and $\Delta\mu_B$ by the layer width

$$\nabla\mu_A = \frac{\Delta_r g}{1 - X_\gamma} \frac{1}{\Delta R}, \quad \nabla\mu_B = \frac{\Delta_r g}{X_\gamma} \frac{1}{\Delta R} \tag{33}$$

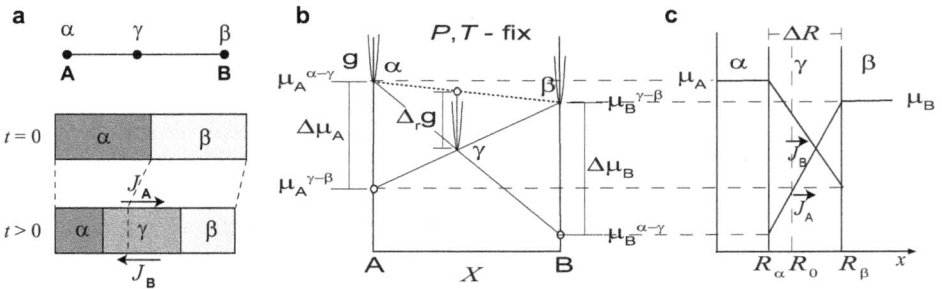

Figure 13. (a) Composition phase diagram and schematic drawing of a reaction-diffusion assembly of a binary system A–B with phases α, β and γ, where $\alpha + \beta$ react to form γ; (b) schematic molar Gibbs-energy diagram; (c) variation of chemical potentials across the diffusion-reaction assembly; R is the spatial coordinate, and ΔR is the layer width; the vertical dashed line indicates the position of the Kirkendall plane.

According to equation 27 the rate of interlayer growth is proportional to the fluxes of the A and B components, J_A and J_B, respectively, so that

$$\frac{\partial \Delta R}{\partial t} = J_A u_A + J_B u_B \tag{34}$$

where u_A and u_B are taken from equation 28. The component fluxes are related to the gradients of the respective chemical potentials (Svoboda *et al.*, 2006a) (see also Petrishcheva and Abart 2017, this volume)

$$J_i = -\sum_j L_{ij} \nabla \mu_j, \quad \text{where } i,j = A, B \tag{35}$$

and L_{ij} are the phenomenological coefficients of diffusion. Combining kinematics with the flow law yields

$$\frac{\partial \Delta R}{\partial t} = \frac{\lambda}{\Delta R}, \text{and rearrangement yields } \Delta R d\Delta R = \lambda dt \tag{36}$$

where

$$\lambda = \Delta_r g \frac{L_{AA} u_A X_\gamma + L_{AB}(u_A + u_B)(X_\gamma - 1) + L_{BB} u_B (X_\gamma - 1)}{X_\gamma (X_\gamma - 1)}$$

For a given set of kinetic parameters L_{ij} and for a given overall driving force $\Delta_r g$ the rate of layer growth decreases with layer thickness ΔR. For given layer thickness, the rate of layer growth is directly proportional to the overall driving force and increases with increasing values of the L_{ij}. Integration yields

$$(\Delta R)^2 = \lambda t \tag{37}$$

which is the parabolic growth law expected for diffusion-controlled interlayer growth in planar geometry. Parabolic growth has indeed been observed experimentally (see below), and it has been referred to as diffusion-controlled interlayer growth. Deviations from the parabolic rate law may occur when the effective component diffusivities change during layer growth or when other dissipative processes such as sluggish interface reaction or activity of non-ideal sources or sinks for vacancies are operative in parallel to long-range diffusion. Before we develop a model that accounts for several dissipative processes operating in parallel, we make a few general remarks.

2.2.2. Layer sequence

During diffusion-controlled interlayer growth in a binary system with phases of fixed composition, the number and the sequence of layers comprising the reaction band is determined exclusively by the equilibrium phase relations. If, for example, in the binary system A–B the two-phase assemblage $\alpha + \beta$ is metastable with respect to assemblages containing phase γ as shown in Fig. 13, only a single layer of γ is formed. If, however, the two-phase assemblage $\alpha + \beta$ is metastable with respect to assemblages containing either one or both of the phases γ and δ as shown in Fig. 14, two layers will

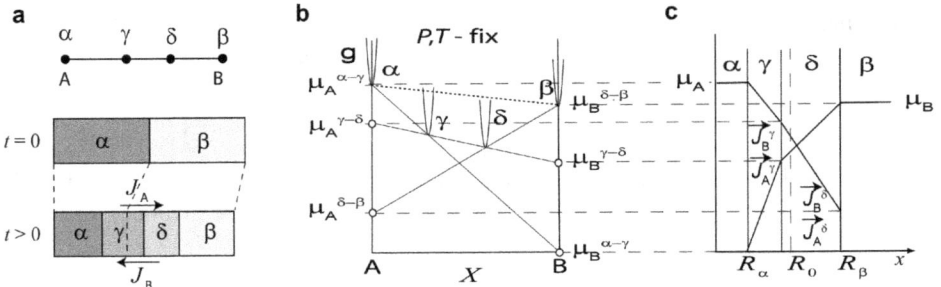

Figure 14. (a) Composition phase diagram and schematic drawing of a reaction-diffusion assembly of a binary system A−B with phases α, β, γ and δ, where α + β react to form γ and δ; (b) schematic molar Gibbs-energy diagram; (c) variation of chemical potentials across the diffusion-reaction assembly; R is the spatial coordinate; the vertical dashed line indicates the position of the Kirkendall plane.

develop, and the layer sequence will be such that the γ layer is in contact with α and the δ layer is in contact with β. In the local-equilibrium scenario, the number of layers forming is determined by the number of stable phases with their compositions between the compositions of the reactant phases, and the sequence of mineral layers is such that the compositions of the layers vary in a monotonic manner across the reaction band.

The situation is fundamentally different during diffusion-controlled interlayer growth in ternary or higher-order systems. Using the 'steady-diffusion model' of Fischer (1973) it was demonstrated in a seminal work by Joesten (1977) that the sequence, phase content and width of the individual layers is determined by the relative fluxes of the mobile components. The steady-diffusion model is based on mass balance at layer boundaries, on the assumption of local equilibrium in the bulk and at layer contacts, and on diffusion driven by chemical potential gradients. Based on the Joesten approach, relative component mobilities were determined from reaction bands and corona structures (Joesten, 1977; Frantz and Mao, 1979; Foster, 1981; Grant, 1988; Joesten and Fisher, 1988; Abart *et al.*, 2001; Larikova and Zaraisky, 2009).

2.2.3. Grain-boundary- vs. volume-diffusion

Experimentally observed layer-growth rates reflect the rates of bulk diffusive mass transfer across the growing layer or sequence of layers. They do not, however, give direct information on diffusion mechanisms and pathways. The diffusion coefficients extracted must thus be considered as 'effective diffusion coefficients', which integrate both, volume- and grain-boundary diffusion (Joesten, 1991). It was found generally in experiment that diffusion through polycrystalline layers is substantially faster than expected for diffusion in a single crystal of the same material. This suggests that grain-boundary diffusion may contribute significantly to bulk diffusion through a polycrystal (Fisler and Mackwell, 1994; Fisler *et al.*, 1997; Dohmen and Milke, 2010; Gardés and Heinrich, 2011).

Direct evidence for the significance of grain-boundary diffusion during interlayer growth was found in the system $MgO-SiO_2$, where the growth of enstatite layers between forsterite and quartz, forsterite-enstatite double layers between periclase and quartz (Gardés *et al.*, 2011), and perovskite layers between periclase and quartz (Nishi *et al.*, 2013) do not follow a parabolic rate low. TEM observations revealed pronounced grain-coarsening as the reaction proceeded. Due to the coarsening, the number of grain boundaries available for grain-boundary diffusion decreased during the course of the reaction. From this Gardés *et al.* (2011) inferred that in their experiments most of the chemical mass transfer occurred by grain-boundary diffusion. Making use of this effect, grain-boundary diffusion coefficients for the diffusion of MgO in polycrystals of forsterite, enstatite (Gardés *et al.*, 2011), and of perovskite (Nishi *et al.*, 2013) were extracted. By contrast, growth of a layer of polycrystalline akermanite between monticellite and wollastonite strictly followed a parabolic rate law, although similar grain coarsening occurred indicating that grain-boundary diffusion only played a minor role and interlayer growth was essentially controlled by volume diffusion. (Joachim *et al.*, 2011).

2.2.4. Influence of water

Mineral reactions may be enhanced substantially by the presence of even minute amounts of water (see Milke *et al.* 2017, this volume). On the other hand, nucleation and growth may be suppressed in 'dry' systems so that thermodynamic equilibrium is never attained (Harlow and Milke, 2002). The influence of minute amounts of water on reaction kinetics is difficult to quantify. Although infrared spectroscopy is a very sensitive method for the determination of water, results from infrared spectroscopy obtained from quenched samples probably do not reflect the availability of water at the *P-T* conditions of interest. The effect of trace amounts of water on the kinetics of interlayer growth was addressed in several studies in the systems $MgO-SiO_2-H_2O$ and $CaO-MgO-SiO_2-H_2O$ (Fisler *et al.*, 1997; Yund, 1997; Milke *et al.*, 2001, 2007, 2009, 2013; Gardés *et al.*, 2011, 2012; Joachim *et al.*, 2012; Milke *et al.* 2017, this volume). In the MS(H) system, growth of entstatite layers between forsterite and quartz in the presence of trace amounts of water is about four orders of magnitude faster than in a dry system. A systematic evaluation of this reaction using powder experiments with different water contents identified four different kinetic regimes (Gardés *et al.*, 2012; Milke *et al.* 2017, this volume). Intergranular transport of MgO is a function of the bulk water to solid ratio, and most of the 'wet' rim-growth experiments reported in the literature probably represent diffusion-controlled MgO transport under 'hydrous-saturated grain boundary conditions'. The jump from the virtually dry regime to conditions at 'hydrous-saturated grain boundaries' is probably very narrow (Milke *et al.*, 2013). If only a few ppm of water are present, it may concentrate at the reaction interface (Abart *et al.*, 2004). Reactions occurring by diffusion along wetted grain boundaries or fluid-filled pores may produce specific microstructures (Milke *et al.*, 2009a). By combining 3D imaging and FTIR analysis, Milke *et al.* (2013) showed that a few tens of ppm of water may have a profound influence on the resulting microstructures. Formation of porosity at the reaction front – 'active porosity' – and within the newly formed layer – 'passive porosity' – is a characteristic feature. In line with Gardés *et al.* (2012) they still found transport-controlled kinetics in their experiments.

In addition, variable amounts of water may change the relative component mobilities. For enstatite-rim growth in water-saturated conditions it has been shown that the SiO_2 component becomes mobile in addition to MgO, but is still less mobile than MgO (Abart *et al.*, 2004; Gardés *et al.*, 2012). Joachim *et al.* (2012) investigated interlayer growth in the CaO–MgO–SiO_2–H_2O system based on the overall reaction monticellite + wollastonite = diopside + merwinite. They observed systematic changes in microstructure evolution with changes in water content, which they ascribed to changes in the relative component mobility with changing water content (see Section 3 below). Finally, it was shown with TEM that water may be present as a separate phase at grain and phase boundaries even in overall water-deficient conditions (Grant *et al.*, 2014). These findings highlight the tremendous effect that minute amounts of water may have on mineral reactions. It is an indispensable pre-requisite for a sensible interpretation of reaction rims and corona structures from natural rocks that the amount of water present during reaction is known. New experimental and analytical methods are required to address this problem (see Milke *et al.* 2017, this volume).

2.3. Thermodynamic model for interlayer growth

2.3.1. The Thermodynamic Extremal Principle (TEP)

The evolution of a non-equilibrium material system is described by the rates at which its characteristic parameters such as the phase content, phase compositions or grain-size change, where these rates are functions of the characteristic parameters themselves. Constraints on the coupling among the different kinetic processes, by which system evolution occurs, can be obtained from extremal principles in the frame of the thermodynamics of irreversible processes. For example, Onsager (1931) derived the famous reciprocity relations based on maximizing the rate of entropy production associated with heat flow in an anisotropic medium. A more general variational approach, which is referred to as the 'Thermodynamic Extremal Principle' (TEP), was provided by Ziegler (1961). The TEP formulated in terms of the characteristic system parameters (Svoboda and Turek, 1991) proved to be a versatile tool for modelling irreversible processes in complex systems. The TEP is used in several studies described below, and a brief introduction is given here. For a detailed account of the approach the reader is referred to the chapter by Svoboda *et al.* (2017, this volume).

The rates of the characteristic system parameters are determined by thermodynamic fluxes $\jmath_i$. We assume system evolution close to equilibrium, so that linear kinetics applies (De Groot and Mazur, 1984). The thermodynamic fluxes $\jmath_i$ are then related to the corresponding thermodynamic forces $\mathscr{F}_k$ as (Callen, 1985)

$$\jmath_i = \sum_k L_{ik} \mathscr{F}_k \tag{38}$$

where L_{ik} is a matrix of kinetic coefficients. The dissipation Q associated with the fluxes reads (Prigogine, 1967; Callen, 1985)

$$Q = \sum_i \mathscr{F}_i \jmath_i \tag{39}$$

According to the TEP the independent thermodynamic fluxes attain the values that maximize the dissipation. In addition, in a closed system, the dissipation must be equal to the negative rate of the system Gibbs energy $\dot{G}$. The system evolution is thus governed by the constrained maximization

$$\text{Max } Q + \lambda \, (Q + \dot{G}) \tag{40}$$

where λ is a Lagrange multiplier. Noting that in the frame of linear kinetics, the dissipation function is quadratic in the fluxes (combined equations 38 and 39) this yields (Svoboda and Turek, 1991)

$$-\frac{1}{2}\sum_i \frac{\partial^2 Q}{\partial J_i \partial J_j} J_i = \frac{\partial \dot{G}}{\partial J_j} \tag{41}$$

The dissipation function and the rate of Gibbs energy are homogeneous functions of the fluxes of order 2 and 1, respectively, and $\frac{\partial^2 Q}{\partial J_i \partial J_j}$ as well as $\frac{\partial \dot{G}}{\partial J_j}$ are constants so that equation 41 represents a system of linear equations in the fluxes. Solving equations 41 for the fluxes provides a numerical scheme for describing system evolution in terms of the rates of the characteristic system parameters. The strategy is to integrate the fluxes over sufficiently small time steps and to update successively the system parameters. The TEP has been employed for describing system evolution in several non-equilibrium settings (see Svoboda *et al.*, 2017, this volume). The relation of the TEP to other extremal principles such as Prigogine's extremal principle were discussed in detail by Fischer *et al.* (2014).

2.3.2. Evolution equations for diffusion-controlled interlayer growth

Stoichiometry: We refer to the binary system A−B shown in Fig. 13a. The stoichiometric reaction equation reads

$$v_\alpha \mathbf{n}_\alpha + v_\beta \, \mathbf{n}_\beta = \mathbf{n}_\gamma \tag{42}$$

where

$$\mathbf{n}_j = \begin{pmatrix} n_A^j \\ n_B^j \end{pmatrix} = \left(n_A^j - n_B^j \right)\begin{pmatrix} 1 - X_j \\ X_j \end{pmatrix}, \quad j = \alpha, \beta, \gamma \tag{43}$$

is the composition vector of phase j, with n_A^j and n_B^j giving the numbers of moles of components A and B contained in one formula unit of phase j, and X_j is the mole fraction of component B in phase j; v_α, v_β are the stoichiometric coefficients and v_γ has been set to unity.

Kinematics: The kinematics of layer growth is described by equations 29 and 31. In the following it is assumed that diffusion through the layer of γ occurs according to ideal thermodynamic behaviour, and equation 23 may be expressed as (Abart *et al.*, 2009)

$$\chi = \frac{D_A}{D_A + D_B} \tag{44}$$

where D_A and D_B are the 'effective self-diffusion coefficients' integrating both the contributions from diffusion through the grain interiors and along the grain boundaries.

Dissipation: We assume linear kinetics (De Groot and Mazur, 1984) so that equation 38 holds. Neglecting the off-diagonal terms of the coefficient matrix and inserting into equation 39 yields an expression for the local dissipation due to diffusion

$$q = \frac{J_i^2}{L_{ii}} \tag{45}$$

The total dissipation function is obtained from integrating over the entire γ layer

$$Q_{\text{diff}} = \int_{R_\alpha}^{R_\beta} \left(\frac{J_A^2}{L_{AA}} + \frac{J_B^2}{L_{BB}} \right) A dx \tag{46}$$

where A is cross-sectional area, which is henceforth set to unity. The fluxes J_A and J_B are expressed in terms of $\dot{R}_\alpha$ using equations 23 and 26 and are inserted into equation 46. Integration yields

$$Q_{\text{diff}} = \frac{1}{\tilde{D}} R_\alpha \dot{R}_\alpha^2 \left(\frac{R_\beta}{R_\alpha} - 1 \right) \tag{47}$$

where $\tilde{D}$ is

$$\tilde{D} = -\frac{D}{R_g T} \frac{(\chi - 1)(1 - X_\gamma) - \chi X_\gamma}{\left[(1 - X_\alpha) X_\gamma - (1 - X_\gamma) X_\alpha \right]^2} \frac{\Omega_\alpha^2}{\Omega_\gamma} (1 - X_\gamma) X_\gamma \tag{48}$$

with R_g and T being the gas constant and absolute temperature, respectively; D is a combined effective diffusion coefficient accounting for diffusion of both, the A and B components. The effective tracer-diffusion coefficients of the A and B components are related to D as (see equation 44).

$$D_A = D\chi \text{ and } D_B = D(1 - \chi) \tag{49}$$

Gibbs energy: The rate of Gibbs energy change per unit area of the γ-layer is

$$\dot{G} = \Delta G_{\text{rim}} \dot{R}_\alpha \tag{50}$$

where

$$\Delta G_{\text{rim}} = -\frac{1}{\Omega_\alpha \nu_\alpha} \Delta_r G$$

and $\Delta_r G$ is the Gibbs energy of reaction (42) per mole of the product phase.

Evolution equation: Both the dissipation associated with diffusion of the A and B components as well as the free-energy change during reaction have now been expressed in terms of the single kinetic variable $\dot{R}_\alpha$. For the case of only one internal variable the TEP (equation 40) degenerates to the constraint

$$Q = -\dot{G}$$

For the problem at hand this yields

$$R_\alpha \dot{R}_\alpha \left(1 - \frac{R_0 + u(R_\alpha - R_0)}{R_\alpha} \right) = \tilde{D} \Delta G_{\text{rim}} \tag{51}$$

where equation 31 has been used to eliminate R_β.

It is convenient to introduce a reaction progress variable

$$y = 1 - \frac{R_\alpha}{R_0}$$

which varies from 0 at $t = 0$ to 1, when α has been consumed. Noting that $dR_\alpha = -R_0 dy$, equation 51 is rewritten in terms of y as

$$R_0^2[y(1 - u)]dy = \tilde{D}\Delta G_{rim}dt \tag{52}$$

Direct integration yields

$$\frac{R_0^2}{2}(1 - u)y^2 = \tilde{D}\Delta G_{rim} \tag{53}$$

Layer thickness is related to y as

$$\Delta R = R_0(1 - u)y \tag{54}$$

and the evolution equation in terms of layer thickness reads

$$(\Delta R)^2 = \tau \tag{55}$$

where normalized time τ has been introduced with

$$\tau = 2(1 - u)\tilde{D}\Delta G_{rim}t \tag{56}$$

Equation 55 represents the parabolic rate law, which is expected for diffusion-controlled layer growth in planar geometry (compare equation 37). The parabolic rate law implies that diffusion-controlled interlayer growth is particularly efficient during early stages, and that the rate of layer growth decreases with increasing layer thickness, where diffusion becomes successively more inefficient. For a given layer width, the rate of layer growth depends on the thermodynamic driving force as given by ΔG_{rim} and on the combined diffusion coefficient $\tilde{D}$. Finally, it depends on the composition jumps at the reaction interfaces as expressed by the parameter u. In an inverse approach, effective component diffusivities can be obtained from experimentally observed layer growth rates. Combining equations 55 and 56 we obtain

$$(\Delta R)^2 = 2(1 - u)\tilde{D}\Delta G_{rim}t$$

which defines a straight line through the origin in a plot of $(\Delta R)^2$ vs. t with the slope $2(1 - u)\tilde{D}\Delta G_{rim}$. Given that for a specific reaction u and ΔG_{rim} are known, a linear fit to interlayer growth data plotted in this coordinate frame provides $\tilde{D}$. Noting that $\tilde{D} = \tilde{D}(D)$ (equation 48) the individual component diffusivities can be obtained by applying equation 49.

Following the approach outlined above, corresponding evolution equations were also derived for diffusion-controlled growth of a single layer in cylindrical and spherical geometry (Abart et al., 2009). Evolution equations for diffusion-controlled multilayer growth in a binary system and planar geometry were presented by Svoboda et al. (2006b). Finally, a thermodynamic model for diffusion-controlled interlayer growth in multicomponent systems was developed by Svoboda et al. (2011).

Interlayer growth has indeed been demonstrated to follow a parabolic rate law in a number of experiments (Zaraysky et al., 1989; Liu et al., 1997). This has motivated a series of interlayer-growth experiments in simple model systems aiming to determine effective diffusion coefficients from observed layer growth rates. The systems investigated so far include the binary systems $MgO-SiO_2$ (Brady and McCallister, 1983; Fisler et al., 1997; Yund, 1997; Milke et al., 2001; Gardés et al., 2011), $CaCO_3-SiO_2$ (Milke and Heinrich, 2002), $MgO-Al_2O_3$ (Carter, 1961; Rossi and Fulrath, 1963; Watson and Price, 2001; Götze et al., 2010; Keller et al., 2010), and the ternary system $MgO-CaO-SiO_2$ (Joachim et al., 2011, 2012). For a detailed account of the experimental details of interlayer growth see Milke et al. (2017, this volume).

Although the number of systems investigated so far is rather small, some general conclusions can be drawn at least for the system $MgO-CaO-SiO_2$ ternary system. It was generally observed that under dry conditions the SiO_2 and CaO components are rather immobile, whereas MgO can diffuse readily (Gardés et al., 2011; Joachim et al., 2011). The addition of even small amounts of water may have substantial impact on component mobility and the relative mobilities among MgO, CaO and SiO_2 may change. In particular, CaO and SiO_2 become successively more mobile with increasing water content (Milke et al., 2001; Abart et al., 2004, 2012; Gardés et al., 2011, 2012; Joachim et al., 2011).

2.3.3. Effect of sluggish interface reaction

In the above derivation, diffusion of the A and B components was considered as the only dissipative process. This implies that the reaction interfaces are perfectly mobile and local equilibrium prevails at the reaction interfaces. Perfectly mobile interfaces move without a local thermodynamic driving force, and the chemical potentials of the A and B components are continuous across the interfaces. This requires that the interface reactions proceed without any dissipation. In fact, interface reactions are dissipative processes, and, as a consequence, reaction interfaces have finite mobility (Balluffi et al., 2005). To account for the potential effects of sluggish interface reaction, the dissipation associated with interface motion must be considered in addition to the dissipation due to diffusion (Dybkov, 1986; Gamsjäger, 2007; Abart and Petrishcheva, 2011). The above derivation for diffusion-controlled interlayer growth is still valid, but it needs to be extended to account for the sluggish motion of the reaction interfaces. This is done by adding a term which describes the dissipation associated with interface motion. In the following we briefly review the derivation of Abart and Petrishcheva (2011) to obtain the corresponding evolution equation.

Dissipation due to interface motion: For quantifying the dissipation associated with migration of the $\alpha-\gamma$ and the $\gamma-\beta$ interfaces we imply that in the regime of linear kinetics interface velocity is related linearly to the driving force (Christian, 2002)

$$\dot{R}_\alpha = M_\alpha F_\alpha, \quad \dot{R}_\beta = M_\beta F_\beta \tag{57}$$

where M_α and M_β are the mobilities of the $\alpha-\gamma$ and the $\gamma-\beta$ interfaces, respectively, and F_α and F_β are the corresponding driving forces. Combining equations 57 and 39 we obtain for the dissipation due to interface motion

$$Q_{if} = \frac{\dot{R}_\alpha^2}{M_\alpha} + \frac{\dot{R}_\beta^2}{M_\beta} \tag{58}$$

where Q_{if} refers to the motion of an interface segment of unit area. Using equations 29 and 31 to eliminate $\dot{R}_\beta$ and taking dissipation due to diffusion from equation 47 the total dissipation per unit area of the γ-layer reads

$$Q_{tot} = \frac{1}{\tilde{D}} R_\alpha \dot{R}_\alpha^2 \left(\frac{R_\beta}{R_\alpha} - 1 \right) + \dot{R}_\alpha^2 \left(\frac{1}{M_\alpha} + \frac{u^2}{M_\beta} \right) \tag{59}$$

Evolution equation: $\dot{R}_\alpha$ is the only kinetic variable, and the TEP (equation 40) reduces to $Q = -\dot{G}$. For the problem at hand this yields

$$R_\alpha \dot{R}_\alpha \left(1 - \frac{R_0 + u(R_\alpha - R_0)}{R_\alpha} \right) - R_\alpha \left(\frac{\tilde{D}}{M_\alpha} + \frac{u^2 \tilde{D}}{M_\beta} \right) = \tilde{D} \Delta G_{rim} \tag{60}$$

where $\dot{G}$ was taken from equation 50. In terms of the reaction progress variable y, equation 60 reads

$$[y(1 - u)R_0^2 + (a - u^2 b)R_0]dy = \tilde{D} \Delta G_{rim} dt \tag{61}$$

where

$$a = \frac{\tilde{D}}{M_\alpha} \text{ and } b = \frac{\tilde{D}}{M_\beta}$$

Integration of equation 61 yields

$$\frac{R_0^2}{2}(1 - u)y^2 + R_0(a + u^2 b)y = \tilde{D} \Delta G_{rim} t \tag{62}$$

and inserting $\frac{\Delta R}{R_0(1-u)}$ for y we obtain

$$(\Delta R)^2 + \kappa \Delta R = \tau \tag{63}$$

where τ was taken from equation 56, and $\kappa = 2(a + u^2 b)$.

Equation 63 describes mixed kinetics of layer growth with finite interface mobility. Model curves calculated from equation 63 for different values of κ are shown in Fig. 15. The parameter κ compares the efficiency of diffusion to the ease of interface motion. When interface mobilities are large compared to the diffusion coefficient, κ is small. In the case of purely diffusion-controlled layer growth $\kappa = 0$, and equation 63 reduces to the parabolic rate law of equation 55. The corresponding model curve is the straight line labelled $\kappa = 0$ in Fig. 15a. In contrast, when interface mobilities are small compared to the diffusion coefficient, κ is large, and the linear term on the left hand side dominates corresponding to a linear rate law. For a given value of κ the linear term dominates, when ΔR is small, *i.e.* during the initial stages of layer growth. With increasing layer thickness the quadratic term becomes more important, and layer growth becomes successively more parabolic. The transition from dominantly interface-reaction control to dominantly diffusion control is shown in Fig. 15b. During the initial growth stages the dissipation associated with interface motion dominates. With increasing layer thickness the dissipation associated with diffusion becomes dominant. The overall process becomes less efficient with increasing layer

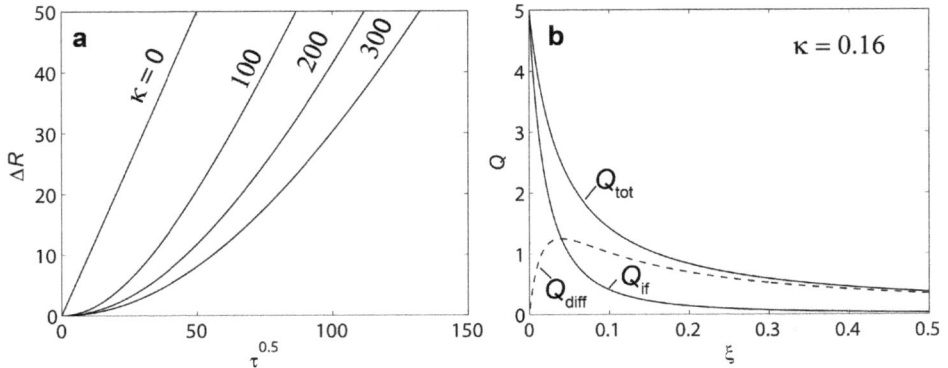

Figure 15. (a) Growth behaviour of a reaction layer in planar geometry, axes scales are in arbitrary units, κ describes the relative ease of long-range diffusion and interface reaction, κ = 0 and κ = ∞ correspond to pure diffusion-control and pure interface-reaction control, respectively – see text for further explanation; (b) contribution of diffusion (Q_{diff}) and interface motion (Q_{if}) to the total dissipation function (Q_{tot}) for a given value of κ and for reaction progress ξ ranging from 0 to 0.5.

thickness due to the increasing diffusion distance. As a consequence, the total dissipation decreases with reaction progress.

As may be expected from the above considerations, layer growth is found to follow a linear rate law only as long as the layers are thin with thicknesses on the order of several tens of nanometres (Cserhati *et al.*, 2008; Götze *et al.*, 2014). If layer thickness is of magnitude 1 μm or more, parabolic growth is generally observed. Only rarely was the transition from interface reaction-controlled to diffusion-controlled growth observed experimentally (Götze *et al.*, 2014). A gradual evolution from interface reaction-controlled to diffusion-controlled interlayer growth was suggested by Balashov and Lebedeva (1991).

2.3.4. Interlayer growth involving solution phases

When solution phases are involved in layer growth, new features including composition zoning and variations in the element partitioning at the reaction interfaces may appear. A schematic molar Gibbs energy diagram of a binary system with components A and B and three solution phases α, β and γ is shown in Fig. 16a. If the reaction interfaces are perfectly mobile, this implies local equilibrium at the reaction interfaces, and the compositions of the phases on either side of a reaction interface adhere to equilibrium element partitioning. If, however, the reaction interfaces have finite mobilities, equilibrium partitioning does not hold any more at a moving interface (Gamsjäger, 2007). Such a situation is illustrated schematically for the γ–β interface in Fig. 17. Given the γ–β interface has finite mobility, it only can propagate, when chemical potential jumps across the interface provide a local driving force for interface motion. For example, at the γ–β interface the chemical potential of component B is larger on the side of β than on the side of γ (Fig. 17). This provides a driving force for the transfer of component B from β to γ, which is required for the reaction interface to propagate

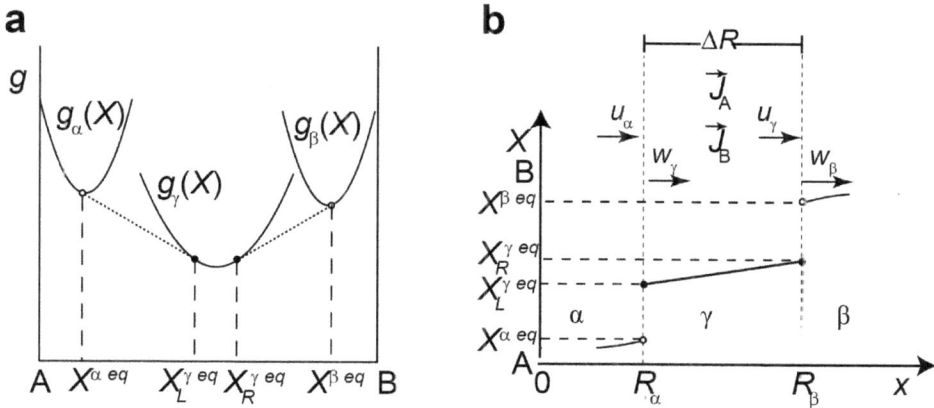

Figure 16. (a) Schematic molar Gibbs energy diagram of the binary system A−B with solution phases α, β and γ; $X^{\alpha \, eq}$, $X^{\beta \, eq}$, $X_L^{\gamma \, eq}$ and $X_R^{\gamma \, eq}$ are the equilibrium compositions of α, β and of γ at the left and right reaction interfaces, respectively, as obtained from the common tangent construction; (b) composition variation across the γ layer and composition jumps at the α−γ and γ−β interfaces during diffusion-controlled growth (local equilibrium scenario), R_α and R_β are the positions of the α−γ and γ−β interfaces, u_α, u_γ, w_γ and w_β are the interface velocities, and J_A and J_B are component fluxes.

further into the reactant phase β. With similar reasoning an opposite jump in the chemical potential of component A can be argued for. The jump in the chemical potential of component B at the γ−β interface is labelled $\Delta\mu_B^{\gamma-\beta}$ in Fig. 17b. It can be seen from the tangent construction that this chemical potential jump implies that $X^\gamma < X^{\gamma \, eq}$ and $X^\beta > X^{\beta \, eq}$ at the γ−β interface. The extent to which the element

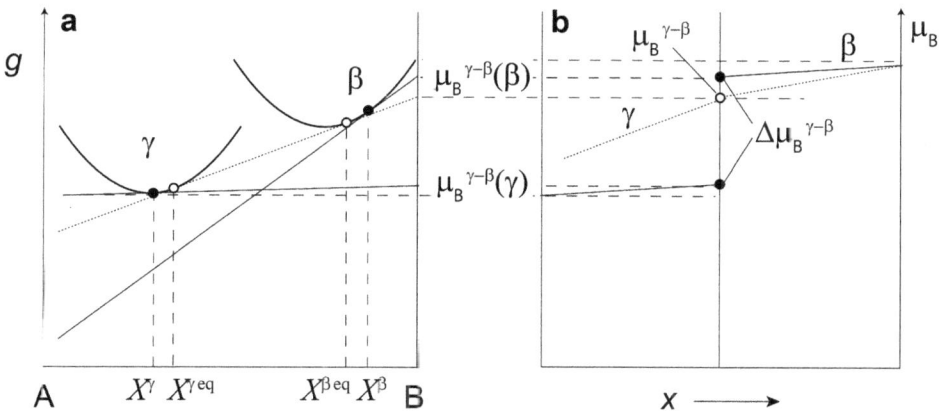

Figure 17. (a) Schematic molar Gibbs energy diagram showing the phase relations at a moving reaction interface in a binary system A−B, where solid-solution phase γ grows at the expense of solid-solution phase β; local equilibrium scenario indicated by dotted tangent and open circle symbols for tangential points; finite interface mobility scenario indicated by solid tangents and filled circle symbols for tangential points; (b) corresponding chemical potential jumps at the reaction interface.

partitioning deviates from local equilibrium is determined by the coupling among the interface reactions at both the $\gamma-\alpha$ and the $\gamma-\beta$ interfaces and long-range diffusion of the A and B components across the growing layer of γ. It is not clear, *a priori*, how these processes are coupled. Constraints on the coupling can be obtained from the TEP.

A thermodynamic model applying the TEP to layer growth with finite interface mobility and non-ideal sources and sinks for vacancies at the reaction interfaces was applied to analyse systematic deviations from equilibrium element partitioning observed during growth of magnesio-aluminate spinel at periclase-corundum interfaces (Abart *et al.*, 2016). The scenario where two phases of fixed composition react to form an intermediate solid-solution was considered. The corresponding molar Gibbs energy diagram is shown in Fig. 18a. From a set of kinematic equations relating the fluxes of the two components to interface motion and to the compositions of the intermediate solid-solution phase at the reaction interfaces, the system evolution expressed in terms of the kinetic variables was constrained to four degrees of freedom. The TEP (equations 40, 41) was then applied to solve for the kinetic variables. For a comprehensive presentation of the thermodynamic analysis the reader is referred to the original work by Abart *et al.* (2016). Here only a few salient features of the system evolution are summarized.

A systematic evolution of the compositions of the solution phase at the reaction interfaces starting with large deviations from equilibrium element partitioning during early growth and successive evolution towards local equilibrium with increasing layer thickness is predicted (Fig. 18b,c). The rate at which local equilibrium at the reaction interfaces is approached increases with increasing interface mobilities (Fig. 18c). The element partitioning at the moving reaction interfaces turned out to be a sensitive monitor for deviations from local equilibrium. In contrast, deviations from parabolic growth behaviour are comparatively subtle and can be detected only for substantial deviation from local equilibrium. Abart *et al.* (2016) showed that Mg-aluminate spinel

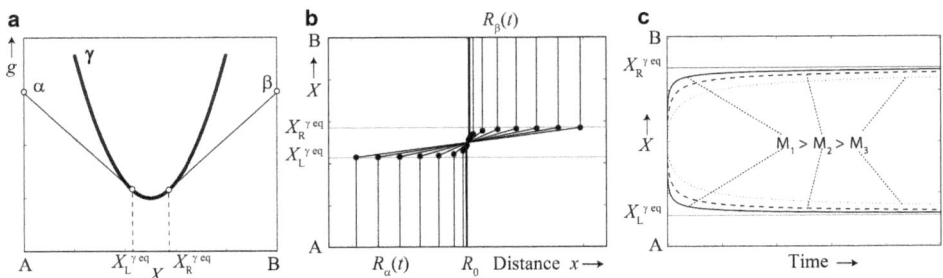

Figure 18. (a) Schematic molar Gibbs energy diagram for binary system A–B with phases α and β of fixed composition and intermediate solid-solution phase γ; $X_L^{\gamma\,eq}$ and $X_R^{\gamma\,eq}$ are the equilibrium compositions of γ at the left and right interfaces; (b) evolution of compositions across the growing layer of γ, compositions of γ at the $\alpha-\gamma$ and $\gamma-\beta$ interfaces are indicated as black dots, horizontal lines labelled $X_L^{\gamma\,eq}$ and $X_R^{\gamma\,eq}$ are the respective equilibrium compositions of γ; (c) evolution of the compositions of γ at the reaction interfaces for three different interface mobilities, modified from Abart *et al.* (2016).

grown at a temperature of 1350°C and 1 bar started to grow with an $X_{Al_2O_3}$ about 0.03 off the equilibrium values at both reaction interfaces, and after 150 h the interface compositions of the spinel approached local equilibrium partitioning to within 0.01. Although this is a relatively small effect in the system investigated by Abart *et al.* (2016), this example testifies to the systematic departure from local equilibrium at a moving reaction interface in a high-temperature environment. Given that reaction interfaces always have finite mobility, it emerges that local equilibrium can generally not be attained at a moving reaction interface. Potentially this effect may have implications for a number of applications in geochemical and petrological modelling. It remains to be seen whether the effect is big enough to be detected in different systems and, if so, how far from local equilibrium the element partitioning can be at a propagating reaction front.

Irrespective of interface mobility other interesting effects may arise from the partitioning of minor and trace elements at a moving reaction interface. For example, Milke *et al.* (2011) investigated the growth of an enstatite layer at the contact between San Carlos olivine and quartz. Whereas San Carlos olivine has a Ni content of about 0.4 wt.%, the Ni content of the orthopyroxene growing at the expense of the olivine is substantially lower. As there is no other sink for Ni available, Ni is forced to diffuse back into the reactant olivine. This produces a zone of Ni enrichment in the olivine ahead of the propagating reaction interface. A similar phenomenon was documented from partial replacement of rutile by a corona of titanite during amphibolite-facies metamorphism of a metabasite by Lucassen *et al.* (2010). There, Nb, which is present in the rutile at ~300 ppm and has a lower concentration of 100 ppm in the titanite, was forced to diffuse back into the remaining rutile ahead of the rutile-titanite reaction front. The Nb content of relict rutile turned out to be a strong function of the degree of conversion to titanite. Small grains of relict rutile showed Nb contents as high as 1200 ppm. Reaction rim or corona formation may thus have a substantial impact on trace-element contents. This effect needs to be taken into consideration in any geochemical analysis of metasomatic reaction bands or coronas (Milke *et al.*, 2011).

To conclude this section we state that, in general, layer growth follows mixed kinetics which is governed by the coupling among long-range diffusion and the processes localized at the reaction interfaces and controlling interface motion. The most important features amenable to experimentation are layer-growth rate, position of the Kirkendall plane, phase compositions and the internal microstructure and texture. All features must be taken into consideration when analysing experimental layer growth or natural corona structures in mineral and rock systems. The Thermodynamic Extremal Principle (Ziegler, 1961; Svoboda and Turek, 1991; Svoboda *et al.*, 2005; Fischer *et al.*, 2014) is a powerful concept for describing mixed kinetics Layer growth tends to be interface reaction-controlled during incipient stages and becomes successively more diffusion-controlled with increasing layer thickness. Diffusion-controlled layer growth follows a parabolic rate law, and element partitioning at reaction interfaces corresponds to equilibrium partitioning. In diffusion-controlled layer growth, relative component fluxes as obtained from the position of the Kirkendall plane, reflect the relative, effective diffusion coefficients of the mobile components. Deviations from parabolic growth may arise from changes in the effective diffusion coefficient due to coarsening of a polycrystalline material or due to changes in the availability of water. So far the potential effects of finite interface mobility and of non-ideal sources and sinks for vacancies at reaction interfaces and in the bulk have been addressed mostly in theoretical work (Deal and Grove, 1965; Schmalzried, 1974; Farrell *et al.*, 1975; Gösele and Tu, 1982; Dybkov, 1986; Abart and Petrishcheva, 2011; Svoboda and Fischer, 2013). The effect of additional dissipative processes on layer growth rates, position of the Kirkendall plane and compositions of solution phases are well understood theoretically and quantitative models accounting for these effects have been presented. At present, the experimental evidence of systematic deviations from local equilibrium at reaction interfaces or from parabolic growth are scarce (Cserhati *et al.*, 2008; Götze *et al.*, 2014; Abart *et al.*, 2016). This does not mean, however, that these effects can be ignored generally. On the contrary, given that interface reactions are dissipative processes, systematic deviations from local equilibrium partitioning at a propagating reaction front are expected generally for diffusive phase transformations. It remains to be investigated how relevant this kinetic control may be for element partitioning in reactive mineral and rock systems.

3. Cellular segregation reactions – symplectite formation

Another important phenomenon related to a specific class of diffusive phase transformations in mineral and rock systems is the formation of symplectites (Spry, 1969; Vernon, 2004). A symplectite is a fine-grained intergrowth of two or more phases replacing a more coarse-grained reactant phase at a sharp reaction interface. An example is shown in Fig. 19. Typically, the phases constituting a symplectite have lamellar or vermicular shape and alternate at close to regular intervals. The long axes of the grains are usually oriented approximately perpendicular to the reaction interface. Based on this feature, which has been referred to as the "law of normality" by Obata

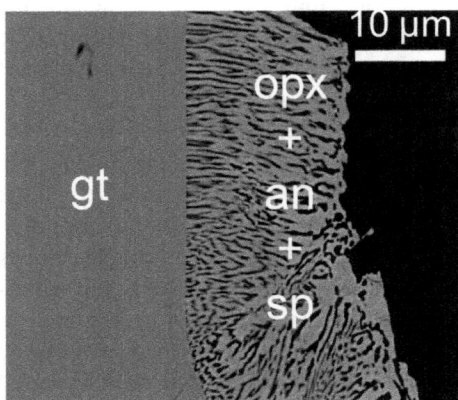

Figure 19. Example of an orthopyroxene (opx)–spinel (sp)–anorthite (an) symplectite replacing garnet from a lower crustal xenolith, Bacony volcanic field, Hungary; image courtesy of J. Degi.

(2011), the trajectories, along which the reaction interface propagated into the precursor phase, can be reconstructed from the shape orientation of the symplectite phases. Symplectite formation may occur isochemically (Degi *et al.*, 2009; Obata *et al.*, 2013), or it may involve composition change on the scale of the symplectic replacement structure (Mongkoltip and Ashworth, 1983; Nishiyama, 1983; Ashworth and Birdi, 1990; Johnson and Carlson, 1990). Typical examples are myrmekites, where K-feldspar is replaced by a quartz–plagioclase symplectite (Phillips, 1974; Abart *et al.*, 2014), pseudoleucites, where vermicular nepheline–K-feldspar intergrowths replace leucite (Gittins *et al.*, 1980), scapolite–quartz symplectites replacing K-feldspar (Harley and Santosh, 1995) and scapolite–clinopyroxene symplectites replacing garnet (Abart *et al.*, 2001). Symplectite formation may be induced by changes in pressure, temperature or fluid composition, which render the reactant phase metastable with respect to the symplectite assemblage. For example, kelyphites, which are plagioclase- and pyroxene- and/or amphibole-bearing symplectites, replace garnet during the decompression of high-pressure rocks (Messiga and Bettini, 1990; Obata, 2011; Obata and Ozawa, 2011; Obata *et al.*, 2013; Scott *et al.*, 2013). Similarly, omphacite is typically replaced by clinopyroxene–plagioclase symplectites during the decompression of eclogites (Boland and Van Roermund, 1983) or of eclogite-facies marbles (Proyer *et al.*, 2014). Symplectites have also been described from garnet peridotites, where chromian spinel-, orthopyroxene-, clinopyroxene- and olivine-bearing symplectites replace garnet during decompression (Morishita and Arai, 2003; Field, 2008; Spacek *et al.*, 2013). Formation of sapphirine–plagioclase, spinel–plagioclase or corundum plagioclase symplectites from kyanite are another phenomenon associated with decompression (Baldwin *et al.*, 2015). Symplecities may also form layers in a corona structure or in a metasomatic reaction band. In these cases their formation is driven by the supply or removal of mobile components in a chemical potential gradient (Mongkoltip and Ashworth, 1983; Nishiyama, 1983; Carlson and Johnson, 1991; Gallien *et al.*, 2013; Faryad *et al.*, 2015; Joachim *et al.*, 2012). Finally, symplectite formation may also be driven by oxidation such as the precipitation of rutile from hematite–ilmenite solid-solution (Tan *et al.*, 2015), or by dehydrogenation of H^+-bearing olivine, which may lead to the segregation of chromian spinel- and clinopyroxene-bearing symplectite from the olivine host (Khisina *et al.*, 2013).

Despite their common occurrence, relatively few studies have addressed the kinetics of symplectite formation. In materials science the corresponding phenomenon is referred to as "cellular segregation reaction" (Cahn, 1959) and, due to its practical implications, has received considerable attention. The terminology was motivated by the close to regular alternation of the product phases, which very much resembles a 'colony of cells'. Such microstructures are well known and were first investigated in steel making where the intimate lamellar intergrowth of ferrite (α-iron – bcc-structure) and cementite (iron carbide Fe_3C) is known as pearlite (Cahn and Hagel, 1962). Pearlite forms during the slow cooling of austenite (γ-iron – fcc-structure) with eutectic composition (Fe with 0.77 wt.% C) below the eutectic temperature of 727°C. Pearlite is one of the strongest materials known on Earth and is used in steel cables. More generally, cellular segregation reactions are of great interest in materials sciences and several models have been suggested that describe the underlying kinetics.

According to Cahn (1959) two kinds of cellular segregation reactions are discerned (Fig. 20). When a solid-solution phase α_0 is supersaturated with respect to phase β, it may segregate phase β. If segregation occurs at a sharp reaction front and β precipitates as regularly spaced lamellae or rods, this is referred to as 'cellular precipitation' (Fig. 20a). In this reaction, one new phase β is formed, whereas phase α only changes its composition. Irrespective of the extent of chemical segregation, precipitation of β always leads to a lowering of the Gibbs energy. The second type of cellular segregation reaction is represented by the so called 'eutectoidal reaction' $\gamma \rightarrow \alpha + \beta$, where metastable phase γ is replaced by the assemblage $\alpha + \beta$ at a sharp reaction front (Fig. 20b). In this case, the reaction products α and β must have compositions that are shifted from the original composition of γ further than to composition $X^{\alpha'}$ or $X^{\beta'}$ in Fig. 20b to ensure lowering of Gibbs energy during transformation.

Consider a binary system A-B with phases α, β, γ such as shown in Fig. 21a. Let γ be metastable with respect to the assemblage $\alpha + \beta$ so that there is a driving force for the reaction

$$\gamma \rightarrow \alpha + \beta$$

A simplified geometry of the corresponding symplectite reaction front is shown in Fig. 21b. Let the reaction front with thickness δ propagate into phase γ at a velocity u. Let the phases within the symplectite take the form of lamellae alternating at a constant interval λ.

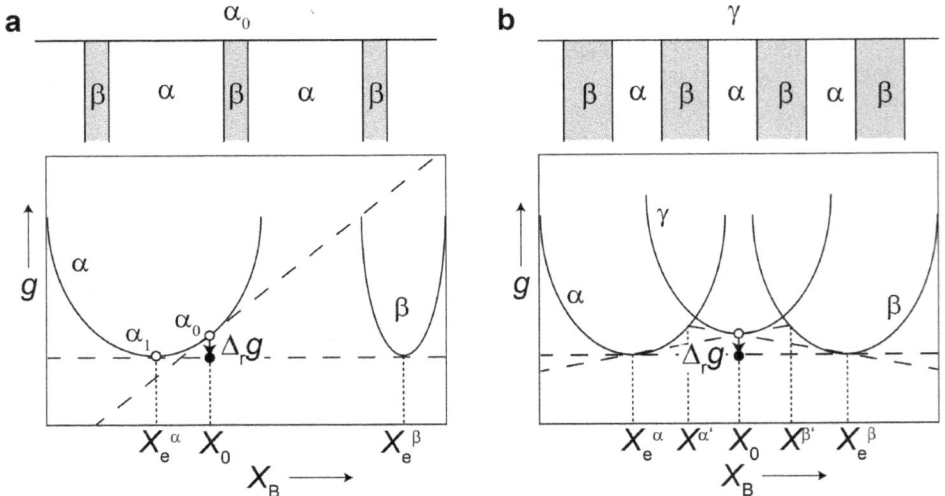

Figure 20. Schematic drawing of reaction microstructure and corresponding molar Gibbs energy diagrams for (a) cellular precipitation reaction $\alpha_0 \rightarrow \alpha_1 + \beta$; (b) eutectoidal $\gamma \rightarrow \alpha + \beta$; X_e^α and X_e^β are the equilibrium compositions of phases α and β in the product assemblage, $X^{\alpha'}$ and $X^{\beta'}$ indicate the minimum shift needed for the eutectoidal reaction to proceed, see text for further explanation.

Symplectite formation requires two processes to proceed in parallel. On the one hand, the material entering the reaction interface from the side of γ has uniform composition, whereas the material that comes out on the other side of the reaction front has segregated into two phases with different compositions. This implies that chemical components are re-distributed by diffusion within the reaction interface. On the other hand, propagation of the reaction interface into the γ phase requires that chemical bonds are broken and atomic rearrangements are made. The sum of these processes is referred to as interface reaction. Either chemical segregation by diffusion within the reaction front or interface reaction may be rate limiting. The coupling between the two processes determines the overall kinetics and microstructure evolution during cellular segregation. In the following we review briefly models of cellular segregation and discuss applications to symplectite formation. We start our considerations with reactions involving only phases of fixed composition and then turn to the more general case, where solid-solutions are involved.

3.1. Eutectoidal reaction involving phases of fixed composition

Let us begin with the system illustrated in Fig. 21a. Given that the $g-X$ curves of all phases are sharply convex downward (Fig. 22a), the equilibrium compositions of the phases may be regarded as essentially fixed with $X_\alpha < X_\gamma < X_\beta$. The reaction thus has fixed stoichiometry given by

$$n_\gamma = v_\alpha n_\alpha + v_\beta n_\beta \tag{64}$$

where n_α, n_β and n_γ are composition vectors as defined in equation 43, and the stoichiometric coefficient v_γ was set to unity. The modal proportions m_α and m_β of α and β in the symplectite are given by

$$m_\alpha = \frac{v_\alpha \Omega_\alpha}{v_\alpha \Omega_\alpha + v_\beta \Omega_\beta} \quad \text{and} \quad m_\beta = 1 - m_\alpha \tag{65}$$

where Ω_α and Ω_β are the molar volumes of the α and the β phases. In general, a finite volume change is associated with transformation. Assuming that the

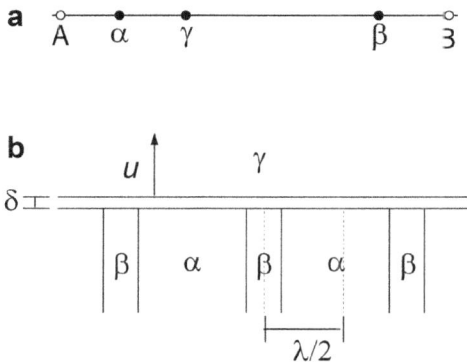

a   A α γ β 3

b u γ δ β α β α β $\lambda/2$

Figure 21. (a) Chemography of binary system A–B with phases α, β and γ; (b) schematic drawing of a symplectite reaction interface.

transformation strain is negligible in the plane of the reaction interface, any volume change is accommodated by contraction/dilation in the direction perpendicular to the reaction interface. This implies that the reaction interface has different velocities relative to phase γ and to the $\alpha + \beta$ lamellar aggregate. The interface velocities are related through

$$v = u f_\Omega$$

where u and v are the interface velocities relative to material points in γ and in the $\alpha + \beta$ aggregate, respectively (Fig. 22), and the volume factor f_Ω is given by

$$f_\Omega = \frac{v_\alpha \Omega_\alpha + v_\beta \Omega_\beta}{\Omega_\gamma}$$

It is assumed that, irrespective of the volume factor, γ and the $\alpha + \beta$ aggregate stay in contact at the reaction front. If $f_\Omega \neq 1$, this implies that material points in γ and in the $\alpha + \beta$ aggregate move with respect to one another. Reaction is driven by the associated free energy change ΔG

$$\Delta G = \Delta_r \bar{g} + \frac{2}{\lambda} f_\Omega \sigma \tag{66}$$

where

$$\Delta_r \bar{g} = \frac{\Delta_r g}{\Omega_\gamma}$$

is the free energy change of reaction per unit volume of the reacting phase γ, and σ is the interfacial energy per unit area of an $\alpha - \beta$ interface.

3.1.1. Zener-Hillert model

An early model for cellular segregation was suggested by Zener (1958) and Hillert (1972). Here it was assumed that chemical segregation occurs by diffusion within the

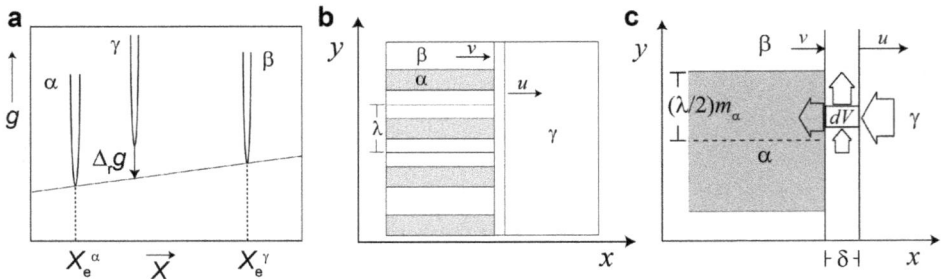

Figure 22. (a) Schematic molar Gibbs energy diagram for the binary system A–B with phases α, β and γ with sharp downwards convex g–X surfaces so that compositions are essentially fixed; (b) geometry of transforming system; (c) component fluxes into and out of the reaction interface associated with interface motion and diffusive flux along the reaction interface in the direction perpendicular to the lamellae in the $\alpha - \beta$ aggregate (y direction).

reactant phase ahead of the replacement front. Zener argued that diffusion becomes inefficient and hence overall reaction becomes slow towards large values of λ. On the other hand, the reaction rate becomes slow towards small values of λ, because the increasing volume density of interfaces and associated interfacial energy in the product assemblage successively diminishes the thermodynamic driving force (ΔG becomes less negative, see equation 66). The critical lamellar spacing λ_0 at which the driving force for reaction disappears was found from the condition

$$\Delta G = 0 = \frac{\Delta_r g}{\Omega_\gamma} + \frac{2}{\lambda_0} f_\Omega \sigma$$

Zener's model predicts the lamellar spacing λ_{max}, which maximizes the reaction rate, to be $\lambda = 2\lambda_0$. Zener's model turned out to be a poor description of cellular segregation. Experimental evidence for chemical segregation by diffusion within the reactant phase ahead of the reaction front is missing, and the predictions regarding interlamellar spacing have not been verified experimentally.

3.1.2. Thermodynamic model for eutectoidal reactions
A thermodynamic model based on the dissipation principle (Onsager, 1931; Ziegler, 1961; Svoboda and Turek, 1991; Svoboda *et al.*, 2005; Fischer *et al.*, 2014) (see also section 2.3.1 of this chapter) was employed to describe symplectite formation in mineral systems by Ashworth and Chambers (2000), Degi *et al.* (2009), and Abart *et al.* (2012). Thereby it was assumed that chemical segregation occurs by diffusion within the reaction front. A schematic sketch of the transforming system is shown in Fig. 22.

Material re-distribution within the reaction front: Due to the fact that α and β have different compositions from γ, the material that enters the reaction front from the side of γ at a particular position has different composition from the material that leaves the reaction front on the side of the symplectite. We assume that growth has reached a steady state so that component concentrations within the reaction front remain constant with time. The composition difference between incoming and outgoing material must thus be compensated by corresponding diffusive fluxes within the reaction front in the direction perpendicular to the lamellae of the $\alpha-\beta$ aggregate (y direction in Fig. 22b,c). The local mass balance for component i in a small-volume element of the reaction front reads

$$\delta \frac{dJ_i}{dy} = uS_i^\alpha \text{ for } 0 \leq y \leq \frac{\lambda}{2} m_\alpha \qquad (67)$$

$$\delta \frac{dJ_i}{dy} = uS_i^\beta \text{ for } \frac{\lambda}{2} m_\alpha \leq y \leq \frac{\lambda}{2} \qquad (68)$$

where J_i is the diffusive flux of component i in y direction within the reaction front, and S_i^α and S_i^β may be regarded as source/sink terms for component i with

$$S_i^{\alpha} = \frac{n_i^{\gamma}}{\Omega_{\gamma}} - f_{\Omega} \frac{n_i^{\alpha}}{\Omega_{\alpha}}, \quad S_i^{\beta} = \frac{n_i^{\gamma}}{\Omega_{\gamma}} - f_{\Omega} \frac{n_i^{\beta}}{\Omega_{\beta}}$$

Integration over the lamella width yields

$$J_i(y) = \frac{u}{\delta} S_i^{\alpha} y \text{ for } 0 \leq y \leq \frac{\lambda}{2} m_{\alpha} \tag{69}$$

$$J_i(y) = \frac{u}{\delta} S_i^{\beta} y \text{ for } \frac{\lambda}{2} m_{\alpha} \leq y \leq \frac{\lambda}{2} \tag{70}$$

Dissipation: Adopting the expression for the local dissipation due to diffusion from equation 45 the dissipation function for diffusion reads

$$Q_{\text{diff}} = \int\int\int \sum_i \frac{J_i^2}{L_{ii}} dV \tag{71}$$

Inserting for J_i from equations 69 and 70 and considering $2/\lambda$ lamellae per square metre of interface yields

$$Q_{\text{diff}} = \frac{u^2 \lambda^2}{12 \delta} \sum_i \frac{1}{L_{ii}} \left[(S_i^{\alpha})^2 m_{\alpha}^3 + (S_i^{\beta})^2 (1 - m_{\alpha})^3 \right] \tag{72}$$

The diffusion within the reaction front causes a resistance against interface motion. It will be convenient to express dissipation due to diffusion within the reaction interface as

$$Q_{\text{diff}} = u^2 \frac{\lambda^2}{\delta} \frac{1}{M_{\text{diff}}} \tag{73}$$

where

$$M_{\text{diff}} = \frac{12}{m_{\alpha}^3 \sum_i \frac{(S_i^{\alpha})^2}{L_{ii}} + (1 - m_{\alpha})^3 \sum_i \frac{(S_i^{\beta})^2}{L_{ii}}}$$

and

$$\frac{\lambda^2}{\delta} \frac{1}{M_{\text{diff}}}$$

may be interpreted as an interface mobility related to the necessary material re-distribution by diffusion within the reaction front (Abart *et al.*, 2012). Note that this mobility depends on λ and δ.

The dissipation due to the motion of an interface with finite intrinsic interface mobility M_{if} and with velocity u reads (see equation 58)

$$Q_f = \frac{u^2}{M_{if}} \tag{74}$$

The total dissipation is then

$$Q_{tot} = Q_{if} + Q_{diff} \tag{75}$$

Gibbs energy: The rate of Gibbs energy change associated with transformation referring to a unit cross section of the reaction interface is given by

$$\dot{G} = u\left(\Delta_r \bar{g} + \frac{2}{\lambda} f_\Omega \sigma\right) \tag{76}$$

Evolution equation: The evolution of the system has been described in terms of a single kinetic parameter u, and the Thermodynamic Extremal principle (equation 40) reduces to the constraint $Q_{tot} = -\dot{G}$, which yields

$$u = M_{tot}\Delta G \tag{77}$$

where

$$M_{tot} = 1 / \left(\frac{\lambda^2}{\delta}\frac{1}{M_{diff}} + \frac{1}{M_{if}}\right)$$

may be regarded as the bulk interface mobility, which integrates the contributions from both diffusion within the reaction front and from interface reaction. Note that equation 77 corresponds to a linear rate law indicating that for a given thermodynamic driving force and a given set of kinetic parameters and lamellar spacing, the thickness of a cellular replacement microstructure increases linearly with time. As expected, the velocity of the reaction interface increases with increasing driving force and with increasing bulk interface mobility. Due to the different dependencies of thermodynamic driving force and interface mobility on λ, σ and δ, the velocity of the reaction front shows a complex behaviour as a function of these parameters, which can be seen more clearly, when equation 77 is expanded in the form

$$u = \frac{-\Delta_r \bar{g} - \frac{2}{\lambda} f_\Omega \sigma}{\frac{\lambda^2}{\delta}\frac{1}{M_{diff}} + \frac{1}{M_{if}}} \tag{78}$$

At small values of λ the reaction front velocity decreases with decreasing λ due to the reduction of the driving force that is associated with new interfaces. At large values of λ reaction-front velocity decreases with increasing λ due to the fact that diffusion becomes successively more inefficient with increasing diffusion length. For a given set of kinetic parameters (M_{if} and M_{diff}) and interfacial energy (σ) interface velocity is maximized at a specific lamellar spacing λ_{max} (Fig. 23a). It is supposed that λ_{max} is automatically selected by the system as it maximizes the rate of free-energy dissipation. At a given degree of reaction overstepping ($\Delta_r \bar{g}$) and a fixed value

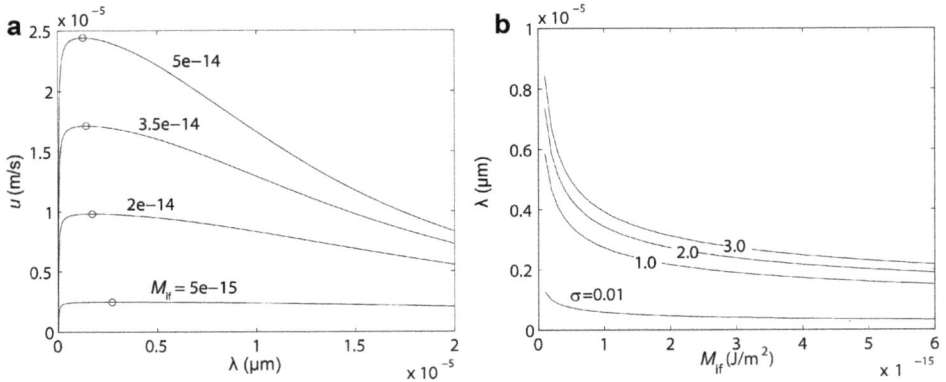

Figure 23. (a) Reaction front velocity u as a function of characteristic spacing λ for different values of M_{if}, small circles indicate u_{max}; (b) λ_{max} as a function of M_{if} for different values of σ.

of M_{diff}, λ_{max} depends on the intrinsic interface mobility M_{if} and on interface energy σ (Fig. 23b). It increases with increasing σ and with decreasing M_{if}. If intrinsic interface mobility M_{if} is high, all the driving force is available for chemical segregation by diffusion within the reaction interface, which is then rate limiting. In the diffusion-controlled regime the evolving microstructure minimizes diffusion length, and λ_{max} is relatively small. If, however, the intrinsic interface mobility M_{if} is low, and the interface reaction is rate limiting, diffusion can be effective over larger distances, and the resulting λ_{max} is relatively large, ensuring minimum possible formation of new interfaces in the symplectite. The characteristic lamellar spacing of a symplectite is thus controlled by the relative contributions of interface reaction and diffusion within the reaction front to the overall resistance to interface motion.

Apart from qualitative inferences that can be made from an analysis of equation 78, it also provides the basis for quantifying the kinetic parameters from experiment. Remmert $et\ al.$ (2017) grew synthetic symplectites in the CaO–MgO–SiO$_2$ system. To this end, monticellite with a composition of Ca$_{0.88}$Mg$_{1.12}$SiO$_4$ was treated at a pressure of 1.2 GPa and temperatures ranging from 1000°C to 1200°C and with addition of small amounts of water (0 to 0.5 wt.% of the total charge). The conditions chosen were outside the monticellite stability field, and the reactant monticellite (mtc I) decomposed forming two different types of symplectites (Fig. 24). One type is represented by a cellular aggregate of monticellite II + forsterite (sy I in Fig. 24a). Monticellite II has pure end-member composition (CaMgSiO$_4$) and is thus somewhat enriched in Ca relative to the original monticellite (Fig. 24b). The underlying reaction may be regarded as 'cellular precipitation'. The second symplectite type is represented by a lamellar aggregate of merwinite and forsterite (sy II in Fig. 24a), corresponding to a 'eutectoidal decomposition'. The product merwinite and forsterite appear to be chemically homogeneous and the forsterite-monticellite as well as the merwinite-monticellite sections of the reaction interface are curved with convex shape towards the

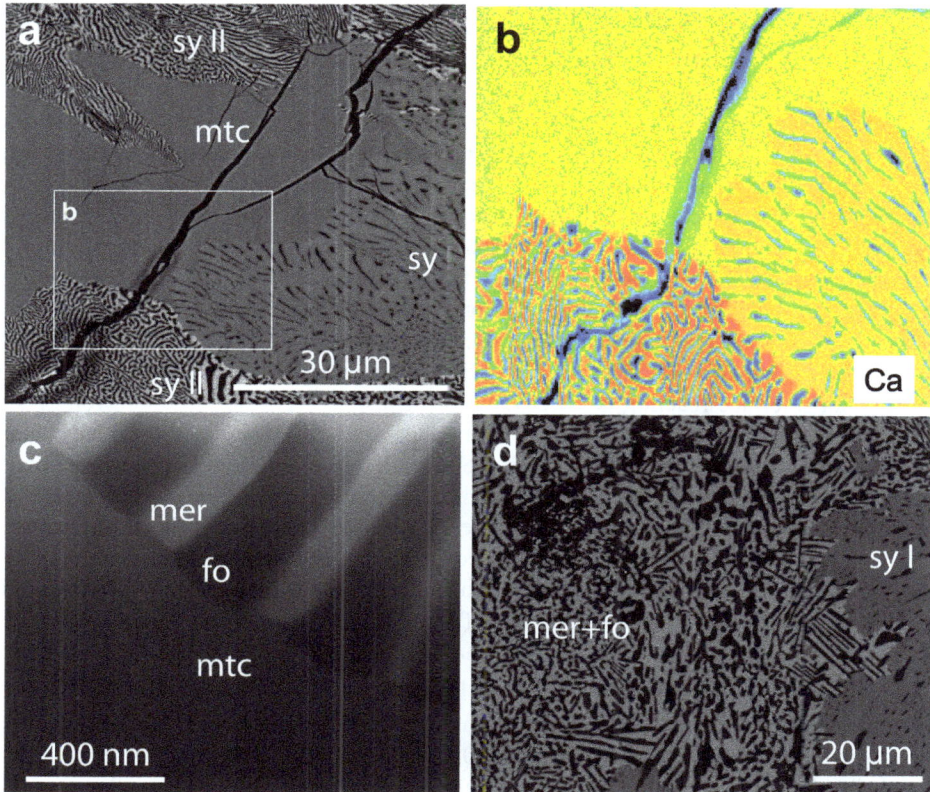

Figure 24. (a) BSE image of syI (mtc II + fo) and sy II (mer+fo) forming from mtc I (Ca$_{0.88}$Mg$_{1.12}$SiO$_4$) at 1.2 GPa, 1000°C; insert indicates location of Ca-distribution map shown in b; (b) Ca-distribution map (red – high Ca-concentration, blue – low Ca-concentration), matrix mtc II in sy I is slightly enriched in Ca relative to mtc I; high Ca-matrix in sy II is merwinite; (c) STEM dark field image of sy II reaction front; note the convex shapes of merwinite and forsterite lamellae towards mtc I; (d) mer+fo assemblage from experiment with addition of 0.5 wt.% water.

reactant monticellite (Fig. 24c) probably reflecting the relatively low mobility of the triple junctions, where all three phases are in contact. These highly organized microstructures were generated only when little water ($\leqslant 0.2$ wt.% of the total charge) was added. At higher water contents, the product phases formed a 'garben' microstructure rather than a symplectite (Fig. 24d).

Moreover, Remmert *et al.* (2017) found a substantial increase in lamellar spacing with temperature but comparatively little influence of reaction overstepping and water content. This behaviour is best explained by different activation energies of interface reaction and chemical segregation by diffusion within the reaction front. The observed increase in lamellar spacing with increasing temperature would imply that the activation energy of interface motion is lower than the activation energy of diffusion within the reaction front. Relatively low activation energy for interface motion compared to diffusion was reported from layer-growth experiments in the $MgO-Al_2O_3$ system (Götze *et al.*, 2014).

To conclude this section we note that at a given set of conditions the velocity of the reaction interface where an homogeneous precursor phase is replaced by a symplectite remains constant as long as the lamellar spacing does not change. The characteristic lamellar spacing reflects the relative contributions of the interface reaction as expressed by the intrinsic interface mobility and of the necessary component diffusion within the reaction front to the resistance against interface movement. The lamellar spacing is minimized if the diffusion within the reaction front is rate limiting, and it is comparatively large if the interface reaction is rate limiting. Experimental evidence for an increase in the characteristic lamella spacing with increasing temperature suggests that the activation energy for interface reaction is lower than for diffusion within the reaction front. As a consequence, symplectite formation is expected to be interface reaction-controlled at high temperature and likely becomes successively more diffusion-controlled towards lower temperature. In several cases, the characteristic lamellar spacing in symplectite microstructures has been observed to decrease successively towards the reaction front (Boland and Van Roermund, 1983). Based on independent constraints from mineral compositions in a myrmekite, such a pattern was interpreted as being due to symplectite growth at successively lower temperatures (Abart *et al.*, 2014).

3.2. Cellular segregation during interlayer growth

Symplectites may form layers in metasomatic reaction bands or in corona structures (Carlson and Johnson, 1991; Abart *et al.*, 2001; Yuguchi and Nishiyama, 2008; Joachim *et al.*, 2012; Gallien *et al.*, 2013). In this case, long-range diffusion across the growing layer as well as chemical segregation by diffusion within the reaction interfaces must be accounted for. Note that the chemical segregation within reaction fronts causes resistance to interface motion and implies finite interface mobility irrespective of the intrinsic interface mobility. Interlayer growth involving symplectic microstructures may thus be expected to follow similar mixed kinetics as interlayer growth involving reaction interfaces with finite intrinsic mobility (see above). In the following a thermodynamic model describing growth of a single layer with lamellar microstructure is briefly reviewed, and some implications are discussed.

Consider the ternary system A, B, C containing phases α, β, γ and ε (Fig. 25a). Let the conditions be such that the assemblage $\alpha-\beta$ (dashed tie line in Fig. 25a) is metastable with respect to the assemblage $\varepsilon-\gamma$ (solid tie line in Fig. 25a) so that a thermodynamic driving force is available for the reaction

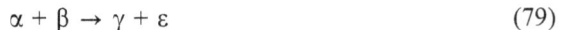

$$\alpha + \beta \rightarrow \gamma + \varepsilon \qquad (79)$$

All phases have fixed compositions with composition vectors (see equation 43) $\mathbf{n}_\alpha$, $\mathbf{n}_\beta$, $\mathbf{n}_\gamma$, $\mathbf{n}_\varepsilon$. The stoichiometric coefficients, v_j, of the phases in the corresponding reaction equation are obtained from

$$\sum_j v_j \mathbf{n}_j = 0, \quad j = \alpha,\beta,\gamma,\varepsilon \qquad (80)$$

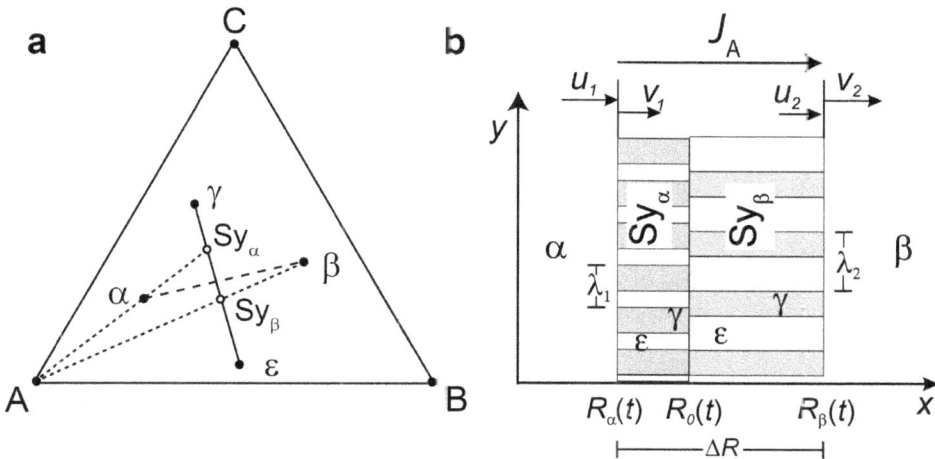

Figure 25. (a) Chemography of a ternary system A, B, C with phases α, β, ε, γ; (b) schematic drawing of the system geometry; initially phases α and β have a common interface at R_0 extending in the y–z plane; at $t > 0$ a layer comprising two symplectite domains Sy_α and Sy_β forms; the symplectites comprise γ and ε lamellae with characteristic spacing λ_1 and λ_2; the reaction interfaces propagate with velocities u and v relative to the materials on either side; the positions of the fronts are at R_α and R_β, respectively.

Let the product phases ε and γ form a layer with symplectic (lamellar) microstructure along the initial α–β contact (Fig. 25b). The ε–γ two-phase layer grows from the original α–β interface at R_0 into both directions forming the domains Sy_α and Sy_β with interlamellar spacing λ_1 and λ_2, respectively. Growth of the symplectic layer requires chemical mass transfer by diffusion over two different length scales. On the one hand, at least one component must be transferred across the growing layer. On the other hand, all three components must be re-distributed on the scale of the characteristic interlamellar spacing along the α–Sy_α and the β–Sy_β reaction interfaces. Let the component A be relatively more mobile than the components B and C. In this case, the component A is transferred across the growing layer and redistributed locally along β–Sy_β and the α–Sy_α reaction interfaces. In contrast, the less mobile components B and C are only redistributed along the reaction interfaces. For this scenario, the bulk reaction may be split into two half reactions, where we have

$$v_\alpha \mathbf{n}_\alpha + v_{\gamma 1} \mathbf{n}_\gamma + v_{\varepsilon 1} \mathbf{n}_\varepsilon + v_{A1} \mathbf{n}_A = 0 \tag{81}$$

at the α–Sy_α interface and

$$v_\beta \mathbf{n}_\beta + v_{\gamma 2} \mathbf{n}_\gamma + v_{\varepsilon 2} \mathbf{n}_\varepsilon + v_{A2} \mathbf{n}_A = 0 \tag{82}$$

at the β–Sy_β interface; $\mathbf{n}_A = (1,0,0)$ is the composition vector of the mobile component A. We assume overall closed-system behaviour so that $v_{A2} = -v_{A1}$. The dotted lines in Fig. 25a indicate compositions that are obtained from the reactant phases α and β, if only component A is added at the β–Sy_β reaction interface and removed from the α–Sy_α reaction interface. The intersections of the dotted lines with the $\varepsilon-\gamma$ tie line correspond to the compositions of the Sy_α and Sy_β domains. The corresponding modal proportions, m_α and m_β, of phase γ are given by

$$m_\alpha = \frac{v_{\gamma 1}\Omega_\gamma}{v_{\gamma 1}\Omega_\gamma + v_{\varepsilon 1}\Omega_\varepsilon} \quad \text{and} \quad m_\beta = \frac{v_{\gamma 2}\Omega_\gamma}{v_{\gamma 2}\Omega_\gamma + v_{\varepsilon 2}\Omega_\varepsilon} \tag{83}$$

for the domains Sy_α and Sy_β, respectively, where Ω_γ and Ω_ε are the molar volumes. The modal proportions of phase ε are $1 - m_\alpha$ and $1 - m_\beta$ in Sy_α and in Sy_β, respectively.

Combining the models of layer growth and cellular precipitation, Abart *et al.* (2012) obtained an explicit expression for the evolution of layer thickness ΔR

$$\frac{v_{A1}^2}{2L_{AA\perp}}(\Delta R)^2 + K\Delta R = v_\alpha(v_\gamma\Omega_\gamma + v_\varepsilon\Omega_\varepsilon)\Delta_{\text{rim}}Gt \tag{84}$$

where $L_{AA\perp}$ is the mobility of component A in the $\varepsilon-\gamma$ layer and

$$K = \left[\lambda_1^2(v_{\varepsilon 1}^2 m_\alpha^3 + \frac{\Omega_\gamma^2}{\Omega_\varepsilon^2}v_{\gamma 1}^2(1 - m_\alpha)^3) + \lambda_2^2(v_{\varepsilon 2}^2 m_\beta^3 + \frac{\Omega_\gamma^2}{\Omega_\varepsilon^2}v_{\gamma 2}^2(1 - m_\beta)^3)\right]\sum_i \frac{\Delta n_i^2}{12L_{ii}\delta}$$

where L_{ii} is the mobility of component i within the β-Sy_β and the α-Sy_α reaction interfaces ($i = A,B,C$), δ is the interface width, and

$$\Delta n_i = n_i^\gamma \frac{\Omega_\varepsilon}{\Omega_\gamma} - n_i^\varepsilon$$

Introducing

$$\bar{\kappa} = \frac{2KL_{AA\perp}}{v_{A1}^2} \quad \text{and} \quad \bar{\tau} = \frac{2L_{AA\perp}v_\alpha(v_\gamma\Omega_\gamma + v_\varepsilon\Omega_\varepsilon)\Delta_{\text{rim}}G}{v_{A1}^2}t$$

equation 84 may be written as

$$(\Delta R)^2 + \bar{\kappa}\Delta R = \bar{\tau} \tag{85}$$

which is similar to the rate law obtained for interlayer growth with finite intrinsic interface mobility (see equation 63). This indicates that similar mixed kinetic behaviour can be inferred for layer growth with symplectic internal microstructure and for layer growth involving interfaces with finite intrinsic mobility. As discussed for the latter case (see above), linear growth is expected, if interface motion is rate limiting. This is the case, when the resistance to interface motion due to chemical separation along the reaction interfaces dominates as compared to the resistance to long-range transfer of component A across the layer. Parabolic growth is expected, when long-range diffusion is rate limiting. Also, for a given set of kinetic parameters, *i.e.* for a given value of $\bar{\kappa}$, the linear term is dominant for small values of ΔR corresponding to early stages of layer growth, and the quadratic term becomes successively more important with increasing ΔR as layer growth proceeds.

It must be noted that apart from producing a single symplectic layer, reaction 79 may also produce a microstructure comprising two monomineralic layers. It can be shown, based on kinematic considerations, that the formation of a two-layer microstructure in a ternary system requires that at least two components are transferred across the layer, whereas only one component needs to be transferred across the layer if the symplectic microstructure is formed (Abart *et al.*, 2012). Hence, if a symplectic layer is formed in a ternary system, this indicates that one component is relatively more mobile than the other two. The proportions of the relatively less mobile components are then largely preserved across the β-Sy$_\beta$ and the α-Sy$_\alpha$ reaction interfaces. For example, Mongkoltip and Ashworth (1983) found that the Al/Si proportions of the precursor phase were perfectly preserved in amphibole-spinel and amphibole-anorthite symplectite forming layers at olivine plagioclase contacts. It was argued by the latter authors that, for a two-phase symplectite to form in a general open system, at least two components must have restricted diffusion ranges. In contrast, when two components are relatively more mobile than the third one, a layered microstructure will form.

Interlayers of the lamellar and multilayer microstructural types were synthesized in the CaO–MgO–SiO$_2$ system by Joachim *et al.* (2012). Some run products are shown in Fig. 26. To this end, single-crystal monticellite (CaMgSiO$_4$) and wollastonite (CaSiO$_3$) were put into contact and reacted in a piston cylinder apparatus at 1.2 GPa and 900°C for 5 to 65 h to produce merwinite (Ca$_3$MgSi$_2$O$_8$) and diopside (CaMgSi$_2$O$_6$) according to the reaction

$$2 \text{ wollastonite} + 2 \text{ monticellite} = 1 \text{ diopside} + 1 \text{ merwinite}$$

Thereby a merwinite–diopside reaction band was formed along the wollastonite–monticellite contact. For short run durations of 5 h, the reaction band took the form of a single merwinite-diopside layer with internal lamellar microstructure and with the Kirkendall plane in the centre of the layer (Fig. 26a). After 65 h a completely segregated reaction band with largely mono-phase layers showing the sequence merwinite-diopside-merwinite was formed (Fig. 26b).

The central position of the Kirkendall plane indicates that only MgO was transferred across the layer. In contrast, formation of the multilayer microstructure requires the

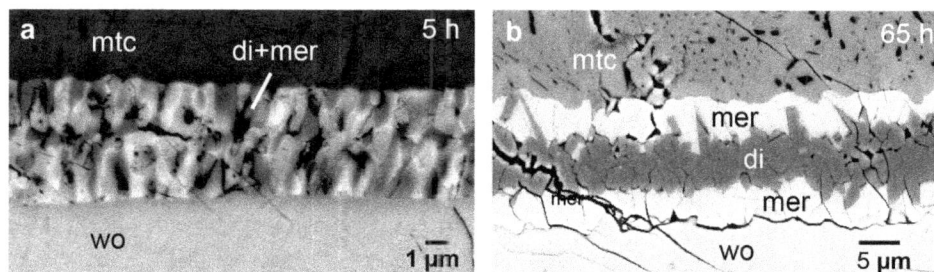

Figure 26. (a) Diopside-merwinite layer with lamellar internal microstructure, run duration 5h; (b) merwinite-diopside-merwinite multilayer reaction band, run duration 65 h, modified from Joachim *et al.* (2012).

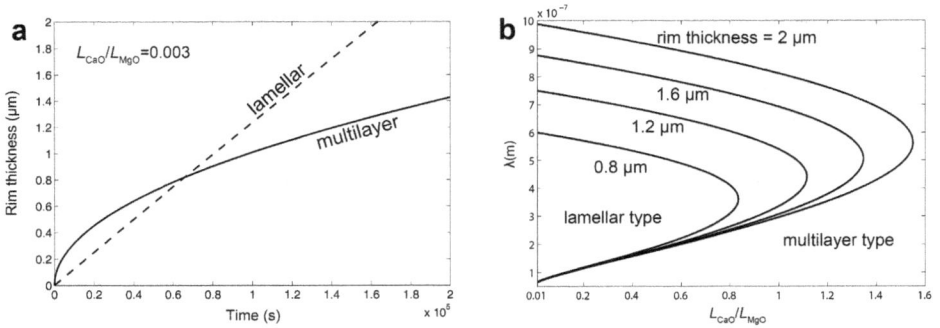

Figure 27. (a) Growth behaviour of multilayer and cellular types for a given ratio of phenomenological coefficients, where $L_{SiO2} = L_{CaO}$ is implied; (b) parameter domains where the multilayer and the cellular microstructural types are preferred for layer thicknesses ranging from 0.8 to 2 μm, modified from Joachim et al. (2012).

additional transfer of either CaO or CaO and SiO$_2$. Considering the simplest case, where, in addition to MgO, only CaO is transferred across the layer, and based on the presumption that the microstructural type that maximizes the dissipation rate is selected by the system, parameter domains can be identified, where the different microstructural types are preferred (Abart et al., 2012).

Total layer thickness is plotted vs. time in Fig. 27a for the lamellar and multilayer reaction bands and for a given L_{CaO}/L_{MgO} ratio. The material and geometrical parameters chosen are given in Table 2. The multilayer type shows parabolic growth behaviour, whereas the lamellar type shows linear growth during initial stages and parabolic growth is successively approached with increasing layer thickness. It is interesting to note that, for a given set of kinetic parameters, the growth rate is higher for the multilayer type than for the lamellar type during the initial growth stages. Towards later stages the situation reverses, and the growth rate is higher for the lamellar

Table 2. Material and system parameters used for model calculations shown in Fig. 27 (Abart et al., 2012).

V_{mer}	10.06×10^{-5}	Molar volume of merwinite [m^3/Mol]
V_{di}	6.75×10^{-5}	Molar volume of diopside [m^3/Mol]
$\Delta_r G$	-1.22×10^4	Gibbs free energy of reaction [J/Mol]
δ	2×10^{-9}	Width of reaction front [m]
λ	2×10^{-6}	Characteristic spacing of cellular intergrowth [m]
σ	2	Interfacial energy [J/m^2]
L_{MgO}	5×10^{-14}	Phenomenological coefficient of diffusion of MgO [m^3/Mol]
s	1×10^{-1}	Grain boundary area fraction [ ]

microstructural type. It is, hence, conceivable that a multilayer structure is formed during the initial stages and a symplectic microstructure is selected later. Incipient layer growth followed by symplectic intergrowth has indeed been reported from myrmekite (Abart *et al.*, 2014). Figure 27b shows the parameter domains for which the cellular and the multilayer microstructural types are preferred. The lamellar microstructure is preferred at a small L_{CaO}/L_{MgO} ratio and in a limited domain of λ values. The minimum λ value is defined by the condition $\Delta_{rim}G = 0$. Towards large λ values the material redistribution within the reaction fronts becomes successively more difficult, and the multilayer microstructural type is preferred. The domain, where the lamellar microstructure is preferred, increases with increasing rim thickness. For a given rim thickness the maximum extension towards large L_{CaO}/L_{MgO} values of the domain where the lamellar microstructure is preferred occurs at a specific λ_{max}, which maximizes the dissipation rate for a given set of kinetic parameters. λ_{max} increases slightly with increasing rim thickness suggesting that the characteristic wavelength of the lamellar microstructure tends to increase as the rim grows.

To conclude this section we remark that in systems with three or more components, interlayer growth may produce multilayer or symplectic microstructures. In a ternary system the multilayer microstructure is preferred when all three components have comparable diffusivities or two components have substantially higher diffusivities than the third one. A layer with symplectic microstructure is formed when one component has substantially higher diffusivity than the other two components. For a given overall reaction, chemical mass transfer across the layer is generally reduced if a symplectic microstructure is formed, as compared to the chemical mass transfer that is necessary for forming the multilayer type. For a given set of kinetic parameters, the multilayer type is preferred during the initial stages of interlayer growth, and the symplectic microstructure is preferred at later growth stages. A transition from the multilayer to the symplectic type is expected as interlayer growth proceeds. If the reverse transition, *i.e.* from symplectic to multilayer type is observed, this indicates that the diffusivity of one or both originally slow diffusing components was successively enhanced during the course of interlayer growth.

3.3. Cellular segregation involving solid solutions

When solid solutions are involved in a cellular segregation reaction, phase compositions become important. It was noted by Cahn (1959) that, if chemical segregation occurs only by diffusion within the reaction front of width δ, and the reaction front proceeds at velocity u, chemical segregation cannot go to completion. The product phases of a cellular segregation reaction in a binary system do not, in general, attain equilibrium compositions (X_e in Fig. 20) but have compositions that lie between X_e and X_0, where X_0 represents the composition of the precursor phase. As a consequence, only a fraction ϕ of the maximum possible thermodynamic driving force $\Delta_r\bar{g}$ is available for driving interface motion. Equation 66 from above needs to be modified as

$$\Delta_r g = \phi\Delta_r\bar{g} + \frac{2}{\lambda}f_\Omega\sigma \qquad (86)$$

Lowering the available driving force by a factor ϕ accounts for the dissipation due to diffusion within the reaction front. Cahn (1959) suggested models for cellular precipitation and for eutectoidal decomposition based on the following assumptions: (1) the reaction front is planar; (2) chemical segregation occurs by diffusion within the reaction front of thickness δ containing material with composition X_B, where X is a mole fraction; the binary inter-diffusion coefficient within the reaction front is D_B; (3) X_B varies in the direction perpendicular to the lamellae of the cellular microstructure, $X_B = X_B(y)$; (4) the system has reached a steady state, so that neither the interlamellar spacing nor the composition $X_B(y)$ changes with time; (5) local equilibrium prevails at the interfaces between the product phases.

In his derivation Cahn (1959) expressed equations 67 and 68 from above in the form

$$\delta D_b \frac{\partial^2 X_B}{\partial^2 y} = u(X_\gamma - f_\Omega X_P) \tag{87}$$

where X_B is the composition of the material within the reaction front, and X_P is the composition of a newly formed phase in the cellular microstructure. It was further assumed that the compositions in the reaction front and the compositions of the precipitates are related through

$$X_P = kX_B$$

where k is a partition coefficient. Integration of equation 87 yields

$$X_P = X_0 + A \cosh \left(\frac{ku\lambda^2}{D_B\delta} \right)^{\frac{1}{2}} \frac{y}{\lambda} \tag{88}$$

where A is an integration constant. In the following, Cahn's derivation for cellular precipitation is summarized briefly. For an equivalent derivation of the model suitable for eutectoidal decomposition, the reader is referred to the original work by Cahn (1959).

Cellular precipitation: In the case of precipitation of β from only slightly supersaturated α_0, the width of the β lamellae is small compared to the characteristic interlamellar spacing, and the integration constant is evaluated from the condition

$$X_P = X_e \text{ at } y = \lambda/2$$

where y is measured from the centre of the lamella of the depleted phase α. The composition of the depleted phase α as a function of position y is given by

$$\frac{X - X_0}{X_e - X_0} = \frac{\cosh \left(\frac{ku\lambda^2}{D_B\delta} \right)^{\frac{1}{2}} \frac{y}{\lambda}}{\cosh \left(\frac{ku\lambda^2}{D_B\delta} \right)^{\frac{1}{2}}} = \frac{\cosh \sqrt{\chi} \frac{y}{\lambda}}{\cosh \sqrt{\chi}} \tag{89}$$

where X_0 is the composition of the reactant phase α_0, and X_e is the composition of α in equilibrium with β. The important parameter

$$\chi = \frac{ku\lambda^2}{D_B\delta}$$

compares the ease of front propagation as expressed by its velocity u and the resistance to motion due to diffusion within the reaction front. Thereby it is implied that $u = M_{if}\Delta_r g$, where $\Delta_r g$ is taken from equation 86, and M_{if} is the intrinsic mobility of the reaction interface. Small values of χ correspond to scenarios where the rate of reaction-front propagation is largely controlled by the interface reaction. In this case, diffusion is effective over the entire lamellar width, and the depleted phase α has close-to-equilibrium composition right across the lamella. If, in contrast, the intrinsic interface mobility is high, diffusion within the reaction front is rate limiting, and the composition of the depleted phase α varies with position. In this case, the depleted phase α attains equilibrium compositions only at the $\alpha-\beta$ interfaces, and the composition X_0 is approached in the central parts of the α lamellae.

The composition variation across a lamella of depleted phase α is shown for different values of χ in Fig. 28. For small values of χ, the equilibrium composition is approached closely everywhere in phase α, and close to 100% of the thermodynamic driving force $\Delta_r g$ is available for driving the interface reaction at the reaction front. For increasing values of χ, equilibrium compositions are successively more restricted to the $\alpha-\beta$ interfaces at $\pm\lambda/2$, whereas the composition of the reactant phase α_0 is successively approached in the central portions of the α lamella. The internal composition zoning of

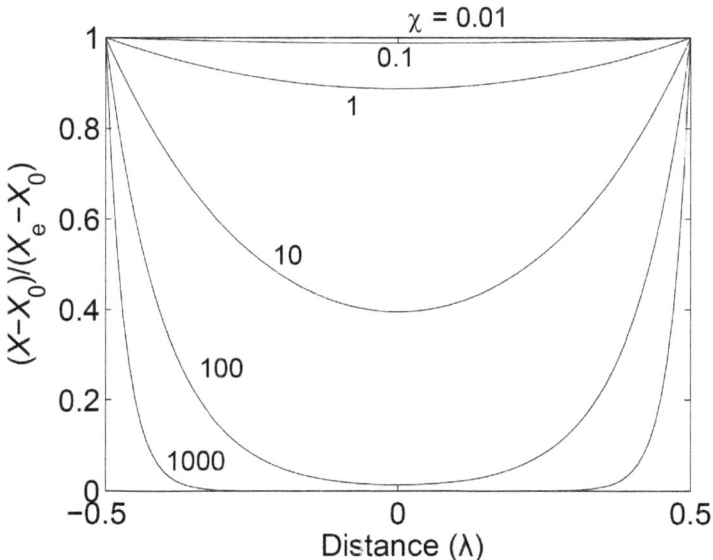

Figure 28. Composition variation across a lamella of depleted α phase for values of χ ranging from 0.01 to 1000 and expressed as fractional approach towards equilibrium composition, modified from Abart *et al.* (2014).

the lamella of depleted phase α thus bears information on the degree of departure from local equilibrium and on the fraction of the thermodynamic driving force that is locally availability at the reaction front.

Similar composition zoning as predicted from Cahn's model has been described from plagioclase feldspar forming the matrix of a myrmekite replacing perthitic alkali feldspar (Abart *et al.*, 2014). Several myrmekite generations were discerned by the latter authors with successively decreasing interlamellar spacing from the oldest portions of the myrmekite (furthest away from the reaction front) to the younger portions closer to the reaction front. The plagioclase exhibits uniform and relatively more anorthite-rich composition in the coarse-grained portions of the myrmekite, and it becomes more albite-rich, with successively greater compositional variability in the younger portions closer to the reaction front. More specifically, in the fine-grained portions of the myrmekite the plagioclase is more anorthite-rich at the interfaces with the quartz lamella than in the more central portions of the plagioclase domains, where the anorthite content approaches the one of the reactant alkali feldspar (Fig. 29). The successive decrease in characteristic grain size, the overall decrease of anorthite content, and the characteristic internal zoning of plagioclase in the latest myrmekite generation were interpreted by Abart *et al.* (2014) as an indication of myrmekite formation at decreasing temperature. The observed grain-size decrease and composition pattern indicates an increasing dominance of diffusion control over interface-reaction control with decreasing temperature. This, in turn, corroborates the view that the activation energy of the interface reaction controlling the intrinsic interface mobility is lower than the activation energy of diffusion within the reaction front.

We conclude this section by noting that symplectites are complex, but highly organized fine-grained, polyphase microstructures replacing a coarser-grained precursor minerals at a sharp reaction front. The characteristic grain size, shape, shape orientation and chemical zoning of the symplectite phases reflect the coupling between the interface reaction, by which the replacement front propagates, and diffusion within the propagating replacement front, which is necessary for the lateral chemical segregation inherent in symplectite formation. Thermodynamic models can

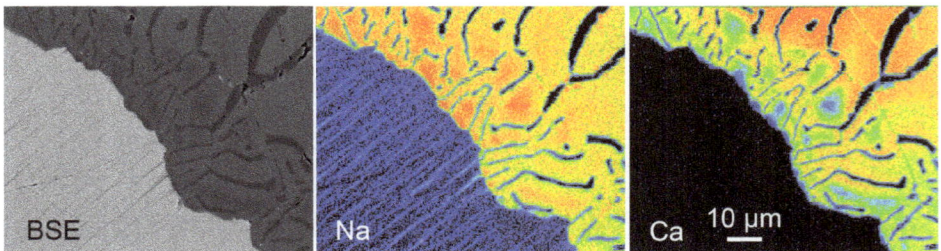

Figure 29. BSE image and Na- and Ca-element distribution maps of the youngest myrmekite generation in the Weinberg granite. Note the characteristic composition zoning with complementary Na and Ca distributions, modified from Abart *et al.* (2014).

be formulated that describe quantitatively this coupling and provide tools for extracting rate information from observed symplectite microstructures and composition patterns in an inverse approach. So far, experimental calibration of diffusion in a propagating reaction interface and of the kinetics of interface migration is largely non-existent for silicates and geologically relevant oxide systems, and the inferences drawn from this inverse approach are largely of qualitative nature. Dedicated experiments addressing these underlying processes are, however, feasible and will probably gain increasing importance so that, eventually, symplectites will be used for geo-speedometry purposes.

4. Concluding remarks

In this chapter we addressed three types of metamorphic mineral reactions: porphyroblast growth, the formation of coronas, reaction-rims or metasomatic reaction bands and symplectite formation. Typically, these reactions occur at different stages of metamorphism. Whereas porphyroblast growth is important during prograde metamorphism of rocks which have been affected previously only by diagenesis or low-grade metamorphism, corona and symplectite formation is typically associated with mineral reactions which occurred in rocks which were previously subjected to medium- or high-grade metamorphism or in magmatic rocks. All three reaction types may be classified as diffusive phase transformations, involving long-range chemical mass transport and interface reactions. It was shown that the coupling and the relative rates of chemical mass transfer and interface reactions determine the overall reaction kinetics as well as the resulting reaction microstructures and chemical patterns. During porphyroblast growth, rocks tend to be saturated with respect to a fluid phase, and chemical mass transport is fast. In contrast, corona and symplectite formation indicate sluggish chemical mass transfer. Thermodynamic models linking the underlying kinetic processes to system evolution and to reaction microstructures were discussed. The formalism presented here provides a sound basis for analysing reaction microstructures and composition patterns from experiment and from natural mineral and rock systems.

References

Abart, R. and Petrishcheva, E. (2011) Thermodynamic model for reaction rim growth: Interface reaction and diffusion control. *American Journal of Science*, **311**, 517−527.

Abart, R. and Sperb, R. (2001) Metasomatic coronas around hornblendite xenoliths in granulite facies marble, Ivrea zone, N Italy. II: Oxygen isotope patterns. *Contributions to Mineralogy and Petrology*, **141**, 494−504.

Abart, R., Schmid, R. and Harlov, D. (2001) Metasomatic coronas around hornblendite xenoliths in granulite facies marble, Ivrea zone, N-Italy: constraints on component mobility. *Contributions to Mineralogy and Petrology*, **141**, 473−492.

Abart, R., Kunze, K., Milke, R., Sperb, R. and Heinrich, W. (2004) Silicon and oxygen self-diffusion in enstatite polycrystals: the Milke et al. (2001) rim growth experiments revisited. *Contributions to Mineralogy and Petrology*, **147**, 633−646.

Abart, R., Petrishcheva, E., Fischer, F. and Svoboda, J. (2009) Thermodynamic model for diffusion controlled reaction rim growth in a binary system: Application to the forsterite-enstatite-quartz system. *American Journal of Science*, **309**, 114−131.

Abart, R., Petrishcheva, E. and Joachim, B. (2012) Thermodynamic model for growth of reaction rims with lamellar microstructure. *American Mineralogist*, **97**, 231−240.

Abart, R., Heuser, D. and Habler, G. (2014) Mechanisms of myrmekite formation: case study from the Weinsberg granite, Moldanubian zone, Upper Austria. *Contributions to Mineralogy and Petrology*, **168**, 2−14.

Abart, R., Svoboda, J., Jeřabek, P., Povoden-Karadeniz, E. and Habler, G. (2016) Kinetics of layer growth in a binary system with intermediate solid-solution and interfaces with finite mobility acting as non-ideal sources and sinks for vacancies: thermodynamic model and application to magnesio-aluminate spinel layer growth. *American Journal of Science*, **316**, 309−328.

Ague, J.J. and Carlson, W.D. (2013) Metamorphism as garnet sees it: the kinetics of nucleation and growth, equilibration, and diffusional relaxation. *Elements*, **9**, 439−445.

Anczkiewicz, R., Chakraborty, S., Dasgupta, S., Mukhopadhyay, D. and Koltonik, K (2014) Timing, duration and inversion of prograde Barrovian metamorphism constrained by high resolution Lu-Hf garnet dating: A case study from the Sikkim Himalaya, NE India. *Earth and Planetary Science Letters*, **407**, 70−81.

Ashworth, J. and Birdi, J. (1990) Diffusion modeling of coronas around olivine in an open system. *Geochimica et Cosmochimica Acta*, **54**, 2389−2401.

Ashworth, J. and Chambers, A. (2000) Symplectic reaction in olivine and the controls of intergrowth spacing in symplectites. *Journal of Petrology*, **41**, 285−304.

Baddeley, A. and Turner, R. (2000) Practical maximum pseudolikelihood for spatial point patterns. *Australian and New Zealand Journal of Statistics*, **42**, 283−322.

Balashov, V. and Lebedeva, M. (1991) Macrokinetic model of origin and development of a monomineralic bimetasomatic zone. Pp. 167−195 in: *Progress in Metamorphic and Magmatic Petrology* (L.L. Perchuk, editor). Cambridge University Press, Cambridge, UK.

Baldwin, J.A., Powell, R., White, R.W. and Stipska, P. (2015) Using calculated chemical potential relationships to account for replacement of kyanite by symplectite in high pressure granulites. *Journal of Metamorphic Geology*, **33**, 311−330.

Balluffi, R.W., Allen, S.M. and Carter, W.C. (2005) *Kinetics of Materials*. John Wiley & Sons Inc..

Bethune, K. and Davidson, A. (1997) Grenvillian metamorphism of the Sudbury diabase dyke-swarm: From protolith to two-pyroxene–garnet coronite. *The Canadian Mineralogist*, **35**, 1191−1220.

Boland, J. and Van Roermund, H. (1983) Mechanisms of exsolution in omphacites from high temperature, type b, eclogites. *Physics and Chemistry of Minerals*, **9**, 30−37.

Brady, J. and McCallister, R. (1983) Diffusion data for clinopyroxenes from homogenization and self-diffusion experiments. *American Mineralogist*, **68**, 95−105.

Cahn, J. (1959) The kinetics of cellular segregation reactions. *Acta Metallurgica*, **7**, 18−28.

Cahn, J. and Hagel, W. (1962) Theory of pearlite reaction. Pp. 131−192 in: *Decomposition of Austenite by Diffusional Processes* (V.F. Zackay and H.I. Aaronson, editors). Interscience Publishers, New York.

Callen, H. (1985) *Thermodynamics and an Introduction to Thermostatistics*, 2nd edition. John Wiley & Sons, Chichester, UK.

Carlson, W.D. (1989) The significance of intergranular diffusion to the mechanism and kinetics of porphyroblast crystallization. *Contributions to Mineralogy and Petrology*, **103**, 1−24.

Carlson, W.D. (1991) Competitive diffusion-controlled growth of porphyroblasts. *Mineralogical Magazine*, **55**, 317−330.

Carlson, W.D. (2002) Scales of disequilibrium and rates of equilibration during metamorphism. *American Mineralogist*, **87**, 185−204.

Carlson, W.D. (2011) Porphyroblast crystallization: linking processes, kinetics, and microstructures. *International Geology Review*, **53**, 406−445.

Carlson, W.D. and Johnson, C.D. (1991) Coronal reaction textures in garnet amphibolites of the Llano uplift. *American Mineralogist*, **76**, 756−772.

Carlson, W.D., Denison, C. and Ketcham, R.A. (1995) Controls on the nucleation and growth of porphyroblasts: kinetics from natural textures and numerical models. *Geological Journal*, **30**, 207−225.

Carter, R. (1961) Mechanism of solid-state reaction between magnesium oxide and aluminum oxide and between magnesium oxide and ferric oxide. *Journal of the American Ceramical Society*, **44**, 116−120.

Cashman, K.V. and Ferry, J.M. (1988) Crystal size distribution (CSD) in rocks and the kinetics and dynamics of crystallization. 3. Metamorphic crystallization. *Contributions to Mineralogy and Petrology*, **99**, 401−415.

Chernoff, C.B. and Carlson, W.D. (1997) Disequilibrium for Ca during growth of pelitic garnet. *Journal of Metamorphic Geology*, **15**, 421−438.

Chernoff, C.B. and Carlson, W.D. (1999) Trace element zoning as a record of chemical disequilibrium during garnet growth. *Geology*, **27**, 555−558.

Christian, J.W. (2002) *The Theory of Transformations in Metals and Alloys*. Elsevier, Amsterdam.

Cox, R. and Indares, A. (1999) Transformation of Fe-Ti gabbro to coronite, eclogite and amphibolite in the Baie du Nord segment, Manicouagan Imbricate Zone, eastern Grenville Province. *Journal of Metamorphic Geology*, **17**, 537−555.

Cserhati, C., Balogh, Z., Csik, A., Langer, G. and Erdelyi, Z. (2008) Linear growth kinetics of nanometric silicides in Co/amorphous-Si and Co/CoSi/amorphous-Si thin films. *Journal of Applied Physics*, **104**, Art. No. 024311.

Cutts, K.A., Stevens, G., Hoffmann, J.E., Buick, I.S., Frei, D. and Muenker, C. (2014) Paleo- to Mesoarchean polymetamorphism in the Barberton Granite-Greenstone Belt, South Africa: constraints from U-Pb monazite and Lu-Hf garnet geochronology on the tectonic processes that shaped the belt. *Geological Society of America Bulletin*, **126**, 251−270.

Cygan, R.T. and Lasaga, A.C. (1982) Crystal growth and formation of chemical zoning in garnets. *Contributions to Mineralogy and Petrology*, **79**, 187−200.

Daniel, C.G. and Spear, F.S. (1998) Three-dimensional patterns of garnet nucleation and growth. *Geology*, **26**, 503−506.

Daniel, C.G. and Spear, F.S. (1999) The clustered nucleation and growth processes of garnet in regional metamorphic rocks from north-west Connecticut, USA. *Journal of Metamorphic Geology*, **17**, 503−520.

de Capitani, C. and Brown, T.H. (1987) The computation of chemical equilibrium in complex systems containing non-ideal solutions. *Geochimica et Cosmochimica Acta*, **51**, 2639−2652.

De Groot, S. and Mazur, S. (1984) *Non-Equilibrium Thermodynamics*. Dover Publications, New York.

Deal, B. and Grove, A. (1965) General relationship for thermal oxidation of silicon. *Jounral of Applied Physics*, **36**, 3770−3778.

Degi, J., Abart, R., Török, K., Bali, E., Wirth, R. and Rhede, D. (2009) Symplectite formation during decompression induced garnet breakdown in lower crustal mafic granulite xenoliths: mechanisms and rates. *Contributions to Mineralogy and Petrology*, **159**, 293−314.

Denison, C. and Carlson, W.D. (1997) Three-dimensional quantitative textural analysis of metamorphic rocks using high-resolution computed X-ray tomography: Part II. Application to natural samples. *Journal of Metamorphic Geology*, **15**, 45−57.

Denison, C., Carlson, W.D. and Ketcham, R.A. (1997) Three-dimensional quantitative textural analysis of metamorphic rocks using high-resolution computed Xray tomography: Part I. Methods and techniques. *Journal of Metamorphic Geology*, **15**, 29−44.

Dohmen, R. and Milke, R. (2010) Diffusion in polycrystalline materials: grain boundaries, mathematical models, and experimental data, Pp. 921−970 in: *Diffusion in Minerals and Melts* (Y. Zhang and D.J. Cherniak, editors) Reviews in Mineralogy and Geochemistry, **72**. Mineralogical Society of America and the Geochemical Society, Washington, D.C.

Dybkov, V. (1986) Reaction diffusion in heterogeneous binary systems. Part 2 Growth of the chemical compound layers at the interface between two elementary substances: two compound layers. *Journal of Materials Science*, **21**, 3085−3090.

Farrell, H., Gilmer, G. and Suenaga, M. (1975) Diffusion mechanisms for growth of Nb_3Sn intermetallic layers. *Thin Solid Films*, **25**, 253−264.

Faryad, S.W., Kachlik, V., Slama, J. and Hoinkes, G. (2015) Implication of corona formation in a metatroctolite to the granulite facies overprint of HP-UHP rocks in the Moldanubian Zone (Bohemian Massif). *Journal of Metamorphic Geology*, **33**, 295−310.

Field, S. (2008) Diffusion, discontinuous precipitation, metamorphism, and metasomatism: The complex history of South African upper-mantle symplectites. *American Mineralogist*, **93**, 618−631.

Fischer, F.D. and Simha, N.K. (2004) Influence of material flux on the jump relations at a singular interface in a multicomponent solid. *Acta Mechanica*, **171**, 213–223.

Fischer, F.D., Svoboda, J. and Petryk, H. (2014) Thermodynamic extremal principles for irreversible processes in materials science. *Acta Materialia*, **67**, 1–20.

Fischer, G. (1973) Nonequilibrium thermodynamics as a model for diffusion-controlled matamorphic processes. *American Journal of Science*, **273**, 897–924.

Fisler, D. and Mackwell, S. (1994) Kinetics of diffusion-controlled growth of fayalite. *Physics and Chemistry of Minerals*, **21**, 156–165.

Fisler, D., Mackwell, S. and Petsch, S. (1997) Grain boundary diffusion in enstatite. *Physics and Chemistry of Minerals*, **24**, 264–273.

Florence, F.P. and Spear, F.S. (1991) Effects of diffusional modification of garnet growth zoning on P-T path calculations. *Contributions to Mineralogy and Petrology*, **107**, 487–500.

Foster, Jr., C.T. (1981) A thermodynamic model of mineral segregations in the lower sillimanite zone near Rangeley, Maine. *American Mineralogist*, **66**, 260–277.

Foster, Jr., C.T. (1999) Forward modeling of metamorphic textures. *The Canadian Mineralogist*, **37**, 415–429.

Frantz, J. and Mao, H. (1976) Bimetasomatism resulting from intergranular diffusion.1. Theoretical-model for monomineralic reaction zone sequences. *American Journal of Science*, **276**, 817–840.

Frantz, J. and Mao, H. (1979) Bimetasomatism resulting from intergranular diffusion. 2. Prediction of multimineralic zone sequences. *American Journal of Science*, **279**, 302–323.

Fukuyama, M., Nishiyama, T., Urata, K. and Mori, Y. (2006) Steady-diffusion modelling of a reaction zone between a metamorphosed basic dyke and a marble from Hirao-dai, Fukuoka, Japan. *Journal of Metamorphic Geology*, **24**, 153–168.

Gaidies, F. (2017) Nucleation in geological materials. Pp. 447–371 in: *Mineral Reaction Kinetics: Microstructures, Textures and Chemical Compositions* (R. Abart and W. Heinrich, editors). EMU Notes in Mineralogy, **16**. European Mineralogical Union and the Mineralogical Society of Great Britain & Ireland, London.

Gaidies, F., de Capitani, C. and Abart, R. (2008a) THERIA G: a software program to numerically model prograde garnet growth. *Contributions to Mineralogy and Petrology*, **155**, 657–671.

Gaidies, F., de Capitani, C., Abart, R. and Schuster, R. (2008b) Prograde garnet growth along complex P-T-t paths: results from numerical experiments on polyphase garnet from the Wolz Complex (Austroalpine Basement). *Contributions to Mineralogy and Petrology*, **155**, 673–688.

Gaidies, F., Krenn, E., de Capitani, C. and Abart, R. (2008c) Coupling forward modelling of garnet growth with monazite geochronology: an application to the Rappold Complex (Austroalpine crystalline basement). *Journal of Metamorphic Geology*, **26**, 775–793.

Gaidies, F., Pattison, D.R.M. and de Capitani, C. (2011) Toward a quantitative model of metamorphic nucleation and growth. *Contributions to Mineralogy and Petrology*, **162**, 975–993.

Gaidies, F., Petley-Ragan, A., Chakraborty, S., Dasgupta, S. and Jones, P. (2015) Constraining the conditions of Barrovian metamorphism in Sikkim, India: P-T-t paths of garnet crystallization in the Lesser Himalayan Belt. *Journal of Metamorphic Geology*, **33**, 23–44.

Gallien, F., Mogessie, A. and Bjerg, E. (2013) On the origin of multilayer coronas between olivine and plagioclase at the gabbro-granulite transition, Valle Fertil-la Huerta ranges, San Juan Province, Argentina. *Journal of Metamorphic Geology*, **30**, 281–302.

Gamsjäger, E. (2007) A note on the contact conditions at migrating interfaces. *Acta Materialia*, **55**, 4823–4833.

Gardés, E. and Heinrich, W. (2011) Growth of multilayered polycrystalline reaction rims in the $MgO-SiO_2$ system, part II: modelling. *Contributions to Mineralogy and Petrology*, **162**, 37–49.

Gardés, E., Wunder, B., Wirth, R. and Heinrich, W. (2011) Growth of multilayered polycrystalline reaction rims in the $MgO-SiO_2$ system, part I: experiments. *Contributions to Mineralogy and Petrology*, **161**, 1–12.

Gardés, E., Wunder, B., Marquardt, K. and Heinrich, W. (2012) The effect of water on intergranular mass transport: new insights from diffusion-controlled reaction rims in the $MgO-SiO_2$ system. *Contributions to Mineralogy and Petrology*, **164**, 1–16.

George, F. and Gaidies, F. (2016) Characterization of a garnet population from the Sikkim Himalaya: implications for the mechanisms and rates of porphyroblast crystallization. *Geophysical Research Abstracts*, **18**, EGU2016-5040.

Gittins, J., Fawcett, J., Brooks, C. and Rucklidge, J (1980) Intergrowths of nepheline, potassium feldspar and kalsilite-potassium feldspar: A re-examination of the 'pseudo-leucite problem'. *Contributions to Mineralogy and Petrology*, **73**, 119−126.

Glicksman, M.E. (2000) *Diffusion in Solids*. John Wiley & Sons, Chichester, UK.

Gösele, U. and Tu, K. (1982) Growth kinetics of planar binary diffusion couples: "thin-film case" versus "bulk cases". *Journal of Applied Physics*, **53**, 3252−3260.

Götze, L., Abart, R., Rybacki, E., Keller, L., Petrishcheva, E. and Dresen, G. (2010) Reaction rim growth in the system $MgO-Al_2O_3$ -SiO_2 under uniaxial stress. *Mineralogy and Petrology*, **99**, 2263−277.

Götze, L., Abart, R., Milke, R., Schorr, S., Zizak, I., Dohmen, R. and Wirth, R. (2014) Growth of magnesio-aluminate spinel in thin film geometry − in-situ monitoring using synchrotron X-ray diffraction and thermodynamic model. *Physics and Chemistry of Minerals*, **99**, 999−1001.

Grant, S. (1988) Diffusion-models for corona formation in metagabbros from the western-Grenville-province, Canada. *Contributions to Mineralogy and Petrology*, **98**, 49−63.

Grant, T., Milke, R. and Wunder, B. (2014) Experimental reactions between olivine and orthopyroxene with phonolite melt: implications for the origins of hydrous diopside + amphibole + phlogopite bearing metasomatic veins. *Contributions to Mineralogy and Petrology*, **168**, 1073.

Harley, S. and Santosh, M. (1995) Wollastonite at Muliyam, Kerala, southern India − a reassessment of CO_2-infiltration and charnockite formation at a classic locality. *Contributions to Mineralogy and Petrology*, **120**, 83−94.

Harlov, D. and Milke, R. (2002) Stability of corundum + quartz relative to kyanite and sillimanite at high temperature and pressure. *American Mineralogist*, **87**, 424−432.

Heinrich, W. (1993) Fluid infiltration through metachert layers at the contact aureole of the Bufa-del-Diente intrusion, northeast Mexico − implications for wollastonite formation and fluid immiscibility. *American Mineralogist*, **78**, 804−818.

Hillert, M. (1972) On theories of growth during discontinuous precipitation. *Metallurgical Transactions*, **2**, 2729−2741.

Hirsch, D.M. (2000) *Quantitative studies of porphyroblastic textures*. Dissertation, University of Texas at Austin, Texas, USA.

Hirsch, D.M. (2008) Controls on porphyroblast size along a regional metamorphic field gradient. *Contributions to Mineralogy and Petrology*, **155**, 401−415.

Hirsch, D.M. (2011) Reduce3D: A tool for three-dimensional spatial statistical analysis of crystals. *Geosphere*, **7**, 724−732.

Hirsch, D.M. and Carlson, W.D. (2006) Variations in rates of nucleation and growth of biotite porphyroblasts. *Journal of Metamorphic Geology*, **24**, 763−777.

Hirsch, D.M., Ketcham, R.A. and Carlson, W.D. (2000) An evaluation of spatial correlation functions in textural analysis of metamorphic rocks. *Geological Materials Research*, **2**, 1−42.

Hollister, L.S. (1966) Garnet zoning: an interpretation based on the Rayleigh fractionation model. *Science*, **154**, 1647−1651.

Hoschek, G. (2013) Garnet zonation in metapelitic schists from the Eclogite Zone, Tauern Window, Austria: comparison of observed and calculated profiles. *European Journal of Mineralogy*, **25**, 615−629.

Illian, J., Penttinen, P., Stoyan, H. and Stoyan, D. (2008) *Statistical Analysis and Modelling of Spatial Point Patterns*. J. Wiley & Sons, Chichester, UK.

Indares, A. and Rivers, T. (1995) Textures, metamorphic reactions and thermobarometry of eclogitized metagabbros - a proterozoic example. *European Journal of Mineralogy*, **7**, 43−56.

Jeřabek, P., Abart, R., Rybacki, E. and Habler, G. (2014) Microstructure and texture evolution during growth of magnesio-aluminate spinel at corundum−periclase interfaces under uniaxial load: the effect of loading on reaction progress. *American Journal of Science*, **314**, 1−26.

Joachim, B., Gardes, E., Abart, R. and Heinrich, W. (2011) Experimental growth of akermanite reaction rims between wollastonite and monticellite: evidence for volume diffusion control. *Contributions to Mineralogy and Petrology*, **161**, 369−399.

Joachim, B., Gardes, E., Abart, R., Velikov, B. and Heinrich, W. (2012) Experimental growth of diopside + merwinite reaction rims: The effect of water on microstructure development. *American Mineralogist*, **97**,

220–230.

Joesten, R. (1977) Evolution of mineral assemblage zoning in diffusion metamorphism. *Geochimica et Cosmochimica Acta*, **41**, 649–670.

Joesten, R. (1986) The role of magmatic reaction, diffusion and annealing in the evolution of coronitic microstructure in troctolitic gabbro from Risor, Norway. *Mineralogical Magazine*, **50**, 441–467.

Joesten, R. (1991) Local equilibrium in metasomatic processes revisited – diffusion-controlled growth of chert nodule reaction rims in dolomite. *American Mineralogist*, **76**, 743–755.

Joesten, R. and Fisher, G. (1988) Kinetics of diffusion-controlled mineral growth in the Christmas Mountains (Texas) contact aureole. *Geological Society of America Bulletin*, **100**, 714–732.

Johnson, C. and Carlson, W. (1990) The origin of olivine-plagioclase coronas in metagabbros from the Adirondack mountains, New-York. *Journal of Metamorphic Geology*, **8**, 697–717.

Jones, K.A. and Galwey, A.K. (1966) Size distribution, composition, and growth kinetics of garnet crystals in some metamorphic rocks from the west of Ireland. *Quarterly Journal of the Geological Society*, **122**, 29–44.

Keller, L., Abart, R., Stunitz, H. and De Capitani, C. (2004) Deformation, mass transfer and mineral reactions in an eclogite facies shear zone in a polymetamorphic metapelite (Monte Rosa nappe, Western Alps). *Journal of Metamorphic Geology*, **22**, 97–118.

Keller, L., Abart, R., Wirth, R., Schmid, D. and Kunze, K. (2006) Enhanced mass transfer through short-circuit diffusion: Growth of garnet reaction rims at eclogite facies conditions. *American Mineralogist*, **91**, 1024–1038.

Keller, L., Wirth, R., Rhede, D., Kunze, K. and Abart, R. (2008) Asymmetrically zoned reaction rims: assessment of grain boundary diffusivities and growth rates related to natural diffusion-controlled mineral reactions. *Journal of Metamorphic Geology*, **26**, 99–120.

Keller, L., Götze, L., Rybacki, E., Dresen, G. and Abart, R. (2010) Enhancement of solid-state reaction rates by non-hydrostatic stress effects on polycrystalline diffusion kinetics. *American Mineralogist*, **95**, 1399–1407.

Kelly, E.D., Carlson, W.D. and Ketcham, R.A. (2013) Crystallization kinetics during regional metamorphism of porphyroblastic rocks. *Journal of Metamorphic Geology*, **31**, 963–979.

Ketcham, R.A. and Carlson, W.D. (2001) Acquisition, optimization and interpretation of X-ray computed tomographic imagery: applications to the geosciences. *Computers and Geosciences*, **27**, 381–400.

Ketcham, R.A. and Carlson, W.D. (2012) Numerical simulation of diffusion-controlled nucleation and growth of porphyroblasts. *Journal of Metamorphic Geology*, **30**, 489–512.

Ketcham, R.A., Meth, C.E., Hirsch, D.M. and Carlson, W.D. (2005) Improved methods for quantitative analysis of three-dimensional porphyroblastic textures. *Geosphere*, **1**, 42–59.

Khisina, N.R., Wirth, R., Abart, R., Rhede, D. and Heinrich, W. (2013) Oriented chromite-diopside symplectic inclusions in olivine from lunar regolith delivered by "Luna-24" mission. *Geochimica et Cosmochimica Acta*, **104**, 84–98.

Konrad-Schmolke, M., Zack, T., O'Brien, P.J. and Jacob, D.E. (2008) Combined thermodynamic and rare earth element modelling of garnet growth during subduction: Examples from ultrahigh-pressure eclogite of the Western Gneiss Region, Norway. *Earth and Planetary Science Letters*, **272**, 488–498.

Korzhinskii, D. (1959) *Physicochemical Basis of the Analysis of the Paragenesis of Minerals.* Consultants Bureau.

Korzhinskii, D. (1970) *Theory of Metasomatic Zoning.* Clarendon Press, Oxford, UK.

Kretz, R. (1966) Grain size distribution for certain metamorphic minerals in relation to nucleation and growth. *Journal of Geology*, **75**, 147–173.

Kretz, R. (1969) On the spatial distribution of crystals in rocks. *Lithos*, **2**, 39–66.

Kretz, R. (1973) Kinetics of the crystallization of garnet at two localities near Yellowknife. *The Canadian Mineralogist*, **12**,1–20.

Kretz, R. (1974) Some models for the rate of crystallization of garnet in metamorphic rocks. *Lithos*, **7**, 123–131.

Kretz, R. (1993) A garnet population in Yellowknife schist, Canada. *Journal of Metamorphic Geology*, **11**, 101–120.

Kretz, R. (1994) *Metamorphic Crystallization.* Wiley, Chichester, UK.

Larikova, T.L. and Zaraisky, G.P. (2009) Experimental modelling of corona textures. *Journal of Metamorphic*

Geology, **27**, 139–151.

Lasaga, A. (1998) *Kinetic Theory in the Earth Sciences*. Princeton University Press, Princeton, New Jersey, USA.

Liu, M., Peterson, J. and Yund, R. (1997) Diffusion-controlled growth of albite and pyroxene reaction rims. *Contributions to Mineralogy and Petrology*, **126**, 217–223.

Loomis, T.P. (1982) Numerical simulation of the disequilibrium growth of garnet in chlorite-bearing aluminous pelitic rocks. *The Canadian Mineralogist*, **20**, 411–423.

Loomis, T.P. (1986) Metamorphism of metapelites – calculations of equilibrium assemblages and numerical simulations of the crystallization of garnet. *Journal of Metamorphic Geology*, **4**, 201–229.

Loomis, T.P. and Nimick, F.B. (1982) Equilibrium in Mn-Fe-Mg aluminous pelitic compositions and the equilibrium growth of garnet. *The Canadian Mineralogist*, **20**, 393–410.

Lucassen, F., Dulski, P., Abart, R., Franz, G., Rhede, D. and Romer, R.L. (2010) Redistribution of HFS elements during rutile replacement by titanite. *Contributions to Mineralogy and Petrology*, **160**, 279–295.

Martyushev, L. and Seleznev, V. (2006) Maximum entropy production principle in physics, chemistry and biology. *Physics Reports*, **426**, 1–45.

Mehrer, H. (2007) *Diffusion in Solids – Fundamentals, Methods, Materials, Diffusion-Controlled Processes*. Springer Series in Solid State Sciences, **155**. Berlin.

Messiga, B. and Bettini, E. (1990) Reactions behavior during kelyphite and symplectite formation - a case-study of mafic granulites and eclogites from the Bohemian massif. *European Journal of Mineralogy*, **2**, 125–144.

Milke, R. and Heinrich, W. (2002) Diffusion-controlled growth of wollastonite rims between quartz and calcite: Comparison between nature and experiment. *Journal of Metamorphic Geology*, **20**, 467–480.

Milke, R., Wiedenbeck, M. and Heinrich, W. (2001) Grain boundary diffusion of Si, Mg, and O in enstatite reaction rims: a SIMS study using isotopically doped reactants. *Contributions to Mineralogy and Petrology*, **142**, 15–26.

Milke, R., Dohmen, R., Becker, H.-W. and Wirth, R. (2007) Growth kinetics of enstatite reaction rims studied on nano-scale, part I: Methodology, microscopic observations and the role of water. *Contributions to Mineralogy and Petrology*, **154**, 519–533.

Milke, R., Kolzer, K., Koch-Müller, M. and Wunder, B. (2009a) Orthopyroxene rim growth between olivine and quartz at low temperatures (750–950°C) and low water concentration. *Mineralogy and Petrology*, **97**, 223–232.

Milke, R., Abart, R., Kunze, K., Koch-Mueller, M., Schmid, D. and Ulmer, P. (2009b) Matrix rheology effects on reaction rim growth I: evidence from orthopyroxene rim growth experiments. *Journal of Metamorphic Geology*, **27**, 71–82.

Milke, R., Abart, R., Keller, L. and Rhede, D. (2011) The behavior of Mg, Fe, and Ni during the replacement of olivine by orthopyroxene: experiments relevant to mantle metasomatism. *Mineralogy and Petrology*, **103**, 1–8.

Milke, R., Neusser, G., Kolzer, K. and Wunder, B. (2013) Very little water is necessary to make a dry solid silicate system wet. *Geology*, **41**, 247–250.

Milke, R., Heinrich, W., Götze, L. and Schorr, S. (2017) New avenues in experimentation on diffusion-controlled mineral reactions Pp. 5–36 in: *Mineral Reaction Kinetics: Microstructures, Textures and Chemical Compositions* (R. Abart and W. Heinrich, editors). EMU Notes in Mineralogy, **16**. European Mineralogical Union and Mineralogical Society of Great Britain & Ireland, London.

Miyazaki, K. (1991) Ostwald ripening of garnet in high P/T metamorphic rocks. *Contributions to Mineralogy and Petrology*, **108**, 118–128.

Miyazaki, K. (1996) A numerical simulation of textural evolution due to Ostwald ripening in metamorphic rocks: A case for small amount of volume of dispersed crystals. *Geochimica et Cosmochimica Acta*, **60**, 277–290.

Miyazaki, K. (2015) Diffusion-controlled growth and degree of disequilibrium of garnet porphyroblasts: is diffusion-controlled growth of porphyroblasts common? *Progress in Earth and Planetary Science*, **2**, 25.

Mongkoltip, P. and Ashworth, J. (1983) Quantitative estimation of an open-system symplectite-forming reaction – restricted diffusion of Al and Si in coronas around olivine. *Journal of Petrology*, **24**, 635–661.

Morishita, T. and Arai, S. (2003) Evolution of spinel-pyroxene symplectite in spinel-lherzolites from the

Horoman complex, Japan. *Contributions to Mineralogy and Petrology*, **144**, 509– 522.

Moynihan, D.P. and Pattison, D.R.M. (2013) An automated method for the calculation of P-T paths from garnet zoning, with application to metapelitic schist from the Kootenay Arc, British Columbia, Canada. *Journal of Metamorphic Geology*, **31**, 525–548.

Mullins, W.W. and Sekerka, R.F. (1964) Stability of a planar interface during solidification of a dilute binary alloy. *Journal of Applied Physics*, **35**, 444–451.

Nishi, M., Nishihara, Y. and Irifune, T. (2013) Growth kinetics of $MgSiO_3$ perovskite reaction rim between stishovite and periclase up to 50 GPa and its implication for grain boundary diffusivity in the lower mantle. *Earth and Planetary Science Letters*, **377**, 191–198.

Nishiyama, T. (1983) Steady diffusion model for olivine – plagioclase corona growth. *Geochimica et Cosmochimica Acta*, **47**, 283–294.

Obata, M. (2011) Kelyphite and symplectite: textural and mineralogical diversities and universality, and a new dynamic view of their structural formation, Pp. 93–122 in: *New Frontiers in Tectonic Research - General Problems, Sedimentary Basins and Island Arcs* (E.V. Sharkov, editor). In Tech Publishers, Croatia.

Obata, M. and Ozawa, K. (2011) Topotaxic relationships between spinel and pyroxene in kelyphite after garnet in mantle-derived peridotites and their implications to reaction mechanism and kinetics. *Mineralogy and Petrology*, **101**, 217–224.

Obata, M., Ozawa, K., Nakamura, K. and Miyake, A. (2013) Isochemical breakdown of garnet in orogenic garnet peridotite and its implication to reaction kinetics. *Mineralogy and Petrology*, **107**, 881–895.

Onsager, L. (1931) Reciprocal relations in irreversible processes. I. *Physical Review*, **37**, 405–426.

Pattison, D.R.M. and Debuhr, C.L. (2015) Petrology of metapelites in the Bugaboo aureole, British Columbia, Canada. *Journal of Metamorphic Geology*, **33**, 437–462.

Petley-Ragan, A., Gaidies, F. and Pattison, D.R.M. (2016) A statistical analysis of the distribution of cordierite and biotite in hornfels from the Bugaboo contact aureole: implications for the kinetics of porphyroblast crystallization. *Journal of Metamorphic Geology*, **34**, 85–101.

Petrishcheva, E. and Abart, R. (2017) Interfaces. Pp. 295–345 in: *Mineral Reaction Kinetics: Microstructures, Textures and Chemical Compositions* (R. Abart and W. Heinrich, editors). EMU Notes in Mineralogy, **16**. European Mineralogical Union and Mineralogical Society of Great Britain & Ireland, London.

Pfeiffer, T. and Schmalzried, H. (1989) Spinel formation – a detailed analysis. *Zeitschrift für Physikalische Chemie Neue Folge*, **161**, 1–17.

Phillips, E.R. (1974) Myrmekite – one hundred years later. *Lithos*, **7**, 181–194.

Prigogine, I. (1967) *Introduction to Thermodynamics of Irreversible Processes*. Interscience Publishers, New York.

Proyer, A., Rolfo, F., Castelli, D. and Compagnoni, R. (2014) Diffusion-controlled metamorphic reaction textures in an ultrahigh-pressure impure calcite marble from Dabie Shan, China. *European Journal of Mineralogy*, **26**, 25–40.

Remmert, P., Heinrich, W., Wunder, B., Morales, L., Wirth, R., Rhede, D. and Abart, R. (2017) Experimental symplectite formation in the system $CaO-MgO-SiO_2$: Influence of temperature and water content on reaction microstructures. In prep.

Ripley, B.D. (1981) *Spatial Statistics*. Wiley, New York.

Rossi, R. and Fulrath, R. (1963) Epitaxial growth of spinel by reaction in the solid state. *Journal of the American Ceramic Society*, **46**, 145–149.

Schmalzried, H. (1974) Solid-state reactions between oxides. Pp. 83–108 in: *Defects and Transport in Oxides* (M. Seltzer and R. Jaffee, editors), Plenum Press, New York.

Schmid, D.W., Abart, R., Podladchikov, Y.Y. and Milke, R. (2009) Matrix rheology effects on reaction rim growth II: coupled diffusion and creep model. *Journal of Metamorphic Geology*, **27**, 83–91.

Schwarz, J.-O., Engi, M. and Berger, A. (2011) Porphyroblast crystallization kinetics: the role of the nutrient production rate. *Journal of Metamorphic Geology*, **29**, 497–512.

Scott, J.M., Konrad-Schmolke, M., O'Brien, P.J. and Guenter, C. (2013) High-T, Low-P formation of rare olivine-bearing symplectites in Variscan eclogite. *Journal of Petrology*, **54**, 1375–1398.

Smigelskas, A. and Kirkendall, E. (1947) Zinc diffusion in alpha-brass. *Transactions of The American Institute of Mining and Metallurgical Engineers*, **171**, 130–142.

Spacek, P., Ackerman, L., Habler, G., Abart, R. and Ulrych, J. (2013) Garnet breakdown, symplectite formation and melting in basanite-hosted peridotite xenoliths from Zinst (Bavaria, Bohemian Massif). *Journal of Petrology*, **54**, 1691−1723.

Spear, F.S. (1993) *Metamorphic Phase Equilibria and Pressure-Temperature-Time Paths*. Monograph, **1**. Mineralogical Society of America, Washington, D.C.

Spear, F.S. and Daniel, C.G. (1998) Three-dimensional imaging of garnet porphyroblast sizes and chemical zoning: nucleation and growth history in the garnet zone. *Geological Materials Research*, **1**, 1−44.

Spear, F.S. and Daniel, C.G. (2001) Diffusion control of garnet growth, Harpswell Neck, Maine, USA. *Journal of Metamorphic Geology*, **19**, 179−195.

Spear, F.S. and Florence, F.P. (1992) Thermobarometry in granulites: pitfalls and new approaches. *Precambrian Research*, **55**, 209−241.

Spear, F.S., Kohn, M.J., Florence, F.P. and Menard, T. (1991a) A model for garnet and plagioclase growth in pelitic schists: implications for thermobarometry and P-T path determinations. *Journal of Metamorphic Geology*, **8**, 683−696.

Spear, F.S., Peacock, S.M., Kohn, M.J., Florence, F.P. and Menard, T. (1991b) Computer programs for petrologic P-T-t path calculations. *American Mineralogist*, **76**, 2009−2012.

Spry, A. (1969) *Metamorphic Textures*. Pergamon Press, Oxford, UK.

Strauss, D.J. (1975) A model for clustering. *Biometrika*, **62**, 467−475.

Svoboda, J. and Fischer, F. (2013) A new computational treatment of reactive diffusion in binary systems. *Computational Materials Science*, **78**, 39−46.

Svoboda, J. and Turek, I. (1991) On diffusion-controlled evolution of closed solid-state thermodynamic systems at constant temperature and pressure. *Philosophical Magazine Part B*, **64**, 749−759.

Svoboda, J., Turek, I. and Fischer, F. (2005) Application of the thermodynamic extremal principle to modeling of thermodynamic processes in material sciences. *Philosophical Magazine*, **85**, 3699−3707.

Svoboda, J., Fischer, F. and Fratzl, P. (2006a) Diffusion and creep in multi-component alloys with non-ideal sources and sinks for vacancies. *Acta Materialia*, **54**, 3043−3053.

Svoboda, J., Gasmjaeger, E., Fischer, F. and Kozeschnik, E. (2006b) Modeling of kinetics of diffusive phase transformation in binary systems with multiple stoichiometric phases. *Journal of Phase Equilibria and Diffusion*, **27**, 622−628.

Svoboda, J., Fischer, F. and Abart, R. (2010) Modeling of diffusional phase transformation in multi-component systems with stoichiometric phases. *Acta Materialia*, **58**, 2905−2911.

Svoboda, J., Fischer, F. and Abart, R. (2011) Modeling the role of sources and sinks for vacancies on the kinetics of diffusive phase transformation in binary systems with several stoichiometric phases. *Philosophical Magazine*, **92**, 67−76.

Svoboda, J., Fischer, F. and Kozeschnik, E. (2017) Thermodynamic modelling of irreversible processes. Pp. 181−214 in: *Mineral Reaction Kinetics: Microstructures, Textures and Chemical Compositions* (R. Abart and W. Heinrich, editors). EMU Notes in Mineralogy, **16**. European Mineralogical Union and Mineralogical Society of Great Britain & Ireland, London.

Tajcmanova, L., Abart, R., Neusser, G. and Rhede, D. (2011) Growth of plagioclase rims around metastable kyanite during decompression of high-pressure felsic granulites (Bohemian Massif). *Journal of Metamorphic Geology*, **29**, 1003−1018.

Tan, W., Wang, C.Y., He, H., Xing, C., Liang, X. and Dong, H. (2015) Magnetite-rutile symplectite derived from ilmenite-hematite solid solution in the Xinjie Fe-Ti oxide-bearing, mafic-ultramafic layered intrusion (SW China). *American Mineralogist*, **100**, 2348−2351.

Thompson, J.B. (1959) Local equilibrium in metasomatic processes. Pp. 427−457 in: *Researches in Geochemistry*, **1** (P.H. Abelson, editor). Wiley, New York.

van Loo, F., Pleumeerkers, M. and van Dal, M. (2000) Intrinsic diffusion and Kirkendall effect in Ni-Pd and Fe-Pd solid solutions. *Acta Materialia*, **48**, 385−396.

Vernon, R. (2004) *A Practical Guide to Rock Microstructure*. Cambridge, University Press, Cambridge, UK.

Watson, E. and Price, J. (2001) Kinetics of the reaction MgO + Al$_2$O$_3$ = MgAl$_2$O$_4$ and Al-Mg interdiffusion in spinel at 1200−2000°C and 1.0−4.0 GPa. *Geochimica et Cosmochimica Acta*, **66**, 2123−2138.

Yuguchi, T. and Nishiyama, T. (2008) The mechanism of myrmekite formation deduced from steady-diffusion modeling based on petrography: Case study of the Okueyama granitic body, Kyushu, Japan. *Lithos*, **106**, 237–260.

Yund, R. (1997) Rates of grain boundary diffusion through enstatite and forsterite reaction rims. *Contributions to Mineralogy and Petrology*, **126**, 224–236.

Zaraysky, G., Balashov, V. and Lebedeva, M. (1989) Macrokinetic model of metasomatic zonality. *Geokhimiya*, **10**, 1386–1395.

Zener, C. (1949) Theory of growth of spherical precipitates from solid solution. *Journal of Applied Physics*, **20**, 950–953.

Zener, C. (1958) Kinetics of the decomposition of austenite. *Transactions of The American Institute of Mining and Metallurgical Engineers*, **167**, 550.

Zhang, R. and Liou, J. (1997) Partial transformation of gabbro to coesite-bearing eclogite from Yangkou, the Sulu terrane, eastern China. *Journal of Metamorphic Geology*, **15**, 183–202.

Ziegler, H. (1961) Two extremal principles in irreversible thermodynamics. *Ingeneur Archive*, **30**, 410–416.

EMU Notes in Mineralogy, Vol. 16 (2017), Chapter 15, 541–585

Crystallographic orientation relationships

GERLINDE HABLER and THOMAS GRIFFITHS

Department of Lithospheric Research, University of Vienna, Althanstrasse 14, 1090 Vienna, Austria, e-mail: gerlinde.habler@univie.ac.at

Crystallographic orientation relationships (CORs) of next-neighbour crystals represent a special case of crystallographic preferred orientation (CPO), where relative crystallographic orientations of neighbour-crystals follow defined rules of misorientation systematics (COR rules). The presence/absence and nature of crystallographic orientation relationships between next-neighbour crystals can be used to infer petrogenetic information from polycrystalline materials provided that the processes of COR formation are understood and parameters that control the kinetics of COR formation can be identified.

After giving an overview on COR terminology, this chapter highlights non-genetic criteria for COR characterization, including a discussion of analytical methods that are used to constrain these criteria. The development of electron backscatter diffraction (EBSD) in scanning electron microscopy (SEM) has provided new information on CORs, which is complementary to data obtained from transmission electron microscopy (TEM) analysis. Based on these non-genetic criteria, different types of CORs are characterized.

Subsequently, physical parameters that can potentially influence COR formation are discussed. Furthermore, different scenarios and mechanisms leading to COR formation are outlined together with examples from experiments and from natural mineral and rock systems. The different boundary conditions of COR formation in various petrogenetic scenarios and the potential mechanisms that have to be taken into account when studying COR genesis are addressed.

This chapter highlights the necessity of a multi-stage investigative approach in COR studies. First, the presence/absence and nature of CORs needs to be analysed based on non-genetic criteria. In a second step the formation mechanism of the CORs under consideration must be constrained, before in a third step, petrogenetic information can potentially be inferred. Moving from the second to the third step requires understanding of the parameters controlling COR development, which is by no means complete and leaves open tasks for future COR research.

1. Terminology and occurrences

Crystallographic orientation relationships (CORs) are systematic relationships between the crystallographic orientations of next-neighbour crystal pairs sharing boundary segments. The overall distribution of next-neighbour crystal misorientations in a sample will reflect any CORs present, as well as their relative frequencies. CORs can occur between grains or lattice domains of monomineralic materials (*e.g.* twin relationships) or between crystals of different phases in polymineralic systems. CORs obey defined rules (termed 'COR rules' in this chapter) of lattice misorientation between neighbour crystals, and thus represent a special case of crystallographic preferred orientation (CPO). A CPO describes the aggregate preferred orientation of a statistically representative number of crystals, and neither requires a spatial relationship between those crystals, nor implies systematic orientation relationships between neighbour lattices. In the current definition of COR we exclude orientation relationships between grains that do not share common boundary segments, *i.e.* uncorrelated grains (random pairs of crystals) or nearest-neighbour grains (pairs of crystals located at a specific distance from each other), as well as CPOs that do not show neighbour lattices related by a systematic law (*e.g.* CPOs resulting from dynamic recrystallization that are controlled by the stress field rather than by next neighbour crystallographic interactions). Still, any group of next-neighbour crystal pairs can be studied based on the criteria employed for COR characterization.

Existing COR-related terminology is based mainly on genetic and/or structural criteria: the term epitaxy (Greek *epi* = upon; *taxis* = arrangement, order), introduced by Royer (1928), refers to the crystallographic control of an overgrowth phase by a substrate phase (Spry, 1969). Epitaxy derives from crystallization of a phase upon a pre-existing surface and therefore a given epitactic COR is associated with a particular substrate surface orientation. Epitactic (or epitaxic) crystallization denotes the process of epitaxy formation. A purely structural definition was provided by the Joint Committee on Nomenclature of the International Mineralogical Association-International Union of Crystallography, which limited epitactic relationships to "...two crystals of different species, with two-dimensional lattice control (mesh in common)..." (Bailey, 1977). Others distinguish between overgrowths of the same (homoepitaxy) or a different (heteroepitaxy) crystalline phase with respect to the substrate (Howe, 1997). In some parts of the literature, epitactic overgrowths are assumed to have a (semi)coherent and 2D periodic interface with the substrate and show a specific COR between overgrowth and substrate phase (Detavernier *et al.*, 2003; Passchier and Trouw, 2005). Other authors state that epitactic overgrowths can have "...one or more strictly defined crystallographic orientations" (Bonev, 1972).

In geosciences the definition of epitaxy is sometimes used in a purely genetic sense, implying only an overgrowth mechanism on a pre-existing crystalline surface without reference to the type of COR formed (Fettes and Desmons, 2007).

The IMA-IUCr Committee on Nomenclature also proposed the separate structural terms "monotaxy" for "mutual orientations with one-dimensional lattice control (line in common)" and "syntaxy" for "mutual orientations with three-dimensional lattice control (cell in common), usually, though not necessarily, resulting in an intergrowth." (Bailey, 1977). Syntactic phases develop specific CORs independent of the interface orientation. Contrasting with this structural definition, the term syntaxy has also been used "in a purely genetic sense ..." for "... simultaneous growth of mutually oriented crystals of two or more phases" (Bonev, 1972).

A different genetic type of COR termed topotaxy (Greek *topos* = place; *taxis* = arrangement, order) results from replacement reactions in the case where the newly crystallized phase inherited its crystallographic preferred orientation from the lattice of the precursor phase it replaced, provided that the transformation occurred within the volume of the decomposed reactant phase (Shannon and Rossi, 1964, and references therein; Bailey, 1977). Different degrees of topotaxy were distinguished based on the type of reaction (displacive *vs.* replacive), the absence/presence of symmetry changes, the amount of atomic displacement and breaking of bonds, the amount of volume change, the required activation energy for the phase transformation, and the dispersion of diffraction patterns after transformation of a single-crystal reactant (Shannon and Rossi, 1964).

Existing genetic COR terms are restricted to scenarios of overgrowth on pre-existing crystalline surfaces, intergrowths, or reactant-product crystallographic relationships developed during replacement reactions in heterophase systems, as well as to different mechanisms of twinning in homophase systems. However, COR formation can be envisaged in several other scenarios, *e.g.* coalescence of pre-existing grains/phases, exsolution, or static recrystallization (see section 4). So far no distinct genetic terminology exists for COR formation in these scenarios.

Structural terms independent of the formation mechanism have been derived from texture studies including observations of CPO development. The CPO of polycrystalline material with one particular crystallographic axis representing a single rotation axis is termed a fibre texture (Wassermann and Grewen, 1962). The rotation axis of a fibre texture can be repeated by the crystal symmetry. Furthermore, complex fibre textures develop, when more than one crystallographic axis represent rotation axes. Fibre textures may be composed of orientations rotated randomly about the fibre axis, or some orientations around the fibre axis may be preferred over others. When the fibre axis is perpendicular to the interface between substrate and overgrowth phases in thin-film materials, the lattice plane with its pole parallel to the rotation axis is parallel to this interface. When the fibre axis is inclined at <90° with respect to the interface between substrate and overgrowth, the term "axiotaxy" has been employed (Detavernier *et al.*, 2003). These authors referred axiotaxy formation to matching *d* spacing of lattice planes in the neighbour phases.

From the variety of definitions given above, it is clear that existing COR terminology is afflicted with ambiguities for several reasons:

(1) Competing definitions either regard crystallographic orientation relations as only involving systematic misorientations between crystals, or also consider the interface orientation with respect to both neighbour lattices. Lattice volume and interface information are then intermixed in one term.

(2) In geosciences and in material sciences partly different definitions are used for the same terms.

(3) Definitions based on genetic criteria are not applicable in cases of unknown genesis. Especially in the case of natural rock systems, the genesis of CORs is often unknown, rendering the usage of genetic terms difficult and potentially misleading. We therefore recommend using non-genetic criteria for COR characterization (see section 2).

In this chapter methods of COR characterization are discussed and potential controlling parameters and mechanisms underlying COR formation are addressed.

1.1. CORs of crystals sharing homophase boundary segments

Systematic crystallographic misorientations in monomineralic material are represented by twinned grains or coincidence site lattice (CSL) misorientations (Petrishcheva and Abart 2017, this volume). According to Nespolo (2015, and references therein), twinning generates multiple alternating lattice domains with a systematic misorientation due to the introduction of point group symmetry elements (mirror planes, rotation axes, inversion centres) in addition to the crystal symmetry of each homogeneous individual domain. In most cases the newly introduced twin symmetry elements coincide with symmetry elements of the respective holohedry (*i.e.* maximum point group symmetry of the crystal class), or higher symmetric classes (Grimmer and Nespolo, 2009). In contrast, coincidence site lattice (CSL) misorientations are preferred misorientations between grains that form by a combination of rotation and translation without correspondence to symmetry elements of the respective holohedry (Grimmer and Nespolo, 2009).

The twin law describes the set of all equivalent symmetry operations necessary to bring an individual lattice into coincidence with the lattice of a domain with another twin orientation. For reflection twins, which are generated by an additional mirror plane in the twin system, the composition surface (*i.e.* the interface between two twin domains) is planar, commonly (sub)perpendicular to low indexed lattice directions or (sub)parallel to low indexed lattice planes (Grimmer and Nespolo, 2009), and often coincides with the twin plane. On the other hand, rotation twins or inversion twins often show non-planar, irregular composition surfaces. COR development by twinning can occur during isochemical phase transformation (transformation twinning) or deformation (mechanical twinning) of a pre-existing homogeneous crystal, or during crystal growth (growth twinning) by single-particle nucleation and growth or by oriented attachment of coalescing homogeneous crystals (Buerger, 1945).

Grain-boundary misorientation distributions in monomineralic aggregates can also show non-random next-neighbour misorientations not arising from CORs. These develop when grains rotate out of a common orientation (*e.g.* by subgrain rotation recrystallization) or into a common orientation (*e.g.* rotation of next-neighbour grains into parallel lattice or parallel facet orientation; dynamic recrystallization induced by dislocation creep). Grain-boundary misorientations are commonly described by the lattice misorientation between neighbour grain-pairs and the frequencies of these misorientations given by the grain boundary misorientation distribution function (Engler *et al.*, 1994; Heidelbach *et al.*, 2000; Bestmann and Prior, 2003). Contrasting with CORs, the systematics of grain boundary misorientation distributions do not necessarily reflect interactions of neighbour lattices, but may be related to factors such as the orientation and anisotropy of external stresses or the deformation mechanisms effective in a system (*e.g.* Bestmann and Prior, 2003).

1.2. Crystals sharing heterophase boundary segments

Studying CORs of next-neighbour phases in polymineralic aggregates requires consideration of the differences in crystal symmetries and lattice parameters of the neighbour phases. In addition to difficulties in the comparison of lattice elements and atomic positions between different phases, boundary segments between different heterophase neighbour pairs must be distinguished (McNamara *et al.*, 2012). COR development between crystals with several different neighbour phases is thought to generate complex textures induced either by interactions between different neighbour phases or by inheritance of preferred orientations from precursor lattice(s) (Wheeler *et al.*, 2001; McNamara *et al.*, 2012; Heidelbach and Terry, 2013). The presence or absence of CORs and the nature of any CORs present have been used to determine mechanisms of phase transformations and to constrain parameters controlling reaction kinetics (Abart *et al.*, 2004; Keller *et al.*, 2010; Ageeva *et al.*, 2012; Jeřábek *et al.*, 2014).

COR investigations of heterophase materials have been applied mainly to systems with heterogeneous grain-size distribution, where a large number of grains of finer-grained phase(s) are in contact with significantly coarser-grained single crystals of another phase. In this case the orientation distribution of the fine-grained phases can be studied with respect to a single reference orientation. COR research has therefore been performed mainly using synthetic polycrystalline thin films on single-crystal substrates. In natural rocks heterogeneous grain-size distributions suitable for COR studies are present commonly in polycrystalline reaction rims around coarse-grained reactant phases (Keller *et al.*, 2006) or in inclusion-bearing porphyroblasts (Griffiths *et al.*, 2016).

2. Non-genetic criteria for characterizing next-neighbour crystal pairs

Three main approaches can be taken to investigating the systematics of CORs (the presence of which implies an interaction between the two crystals). These approaches differ mostly in terms of the degree to which relative lattice and interface characteristics are separated, and in the applied methods. Terms derived from one approach do not always map perfectly onto terms derived from another and care must be taken not to use them interchangeably.

Approach 1 is based on properties of the interface between the neighbouring lattices, including the interface geometry and structure as well as the misorientations between neighbour lattices and the interface.

Approach 2 is based on theoretically calculated matches between lattice planes or sites in neighbouring lattice volumes according to the relative orientations of the two lattices, without considering the geometry of the interface.

Approach 3 is based on the relative orientation of lattice planes/directions in neighbouring crystals neither considering the interface nor matching lattice parameters or sites.

2.1. Interface geometry and structure

A grain or phase boundary represents the contact surface between two neighbour crystals. In the case that CORs originate from mutual neighbour crystal interactions, the corresponding forces are exerted *via* the contact surface. A full understanding of CORs can thus be attained only by studying the interfaces in detail. The angular intersection of an interface with both neighbour lattices depends on the misorientation of the neighbour-pair lattices and the orientation of the interface itself, amounting to a total of five macroscopic orientational degrees of freedom (Petrishcheva and Abart, 2017, this volume).

Characterization of the interface geometry and structure for COR investigation is commonly performed by transmission electron microscopy (TEM) analysis of sections perpendicular to the interface and involves the parameters listed below.

2.1.1. Interface orientation relation with neighbour lattices (crystal morphology)
Combined microstructural and crystallographic investigations determine whether low-indexed planes of one or both neighbour crystals are parallel to interface segments, identifying crystallographic planes that represent crystal facets. Planar interface segments may have a general orientation or a special angular relationship with respect to either neighbour lattice, or coincide with low-indexed lattice planes of either neighbour crystal or both (*e.g.* Rohrer, 2011; Marquardt *et al.*, 2015; Straumal *et al.*, 2016). When an interface segment is parallel to any low-indexed plane of at least one of the neighbour crystals, the interfacial energy is at a local (or global) minimum with respect to interface orientation (see section 3.1). Determining which of the neighbour crystals has low-indexed planes parallel or vicinal to the interface segments, and which plane sets represent crystal facets, can indicate mechanisms and conditions of microstructure formation (Vernon, 2004, and references therein).

Interface orientations are investigated commonly by using TEM to look at projections of the interface plane in an electron transparent sample, and are therefore restricted to planar segments or interfaces with 2D curvature only. As interfaces can have segmented, stepped, curved or irregular geometry in 3D, only certain interface segments are accessible for investigation. Analytically accessible interface segments may not be representative of the entire contact surface of a single crystal with its neighbour(s). For example, in the case of single-crystal host-inclusion systems the phase boundary is a 3D closed surface around an inclusion. Differently oriented interface segments generate different host/inclusion lattice intersections whereas the COR is unique. Correlative studies of CORs and interface geometries in a host-inclusion system have shown that inclusion morphology does not necessarily correspond to COR systematics, which can be much more complex than the interface geometries (Proyer *et al.*, 2013). Investigations of the grain boundary character distribution (GBCD) and grain-boundary energy distribution (GBED) in polycrystal-line oxides have shown that variations in the grain-boundary plane orientation cause much higher grain boundary energy differences than variations in lattice misorientation (Rohrer, 2011, and references therein). Improvement of GBCD studies has been facilitated by advanced analytical techniques such as 3D EBSD, which allows for direct analysis of the full GBCD by combining focused ion beam (FIB) preparation and EBSD analysis (Zaefferer and Wright, 2009).

2.1.2. Coherence

A coherent interface exhibits continuity of atom rows or lattice planes across it. Full coherence in 3D can be attained only when neighbour crystals have similar symmetry, orientation and *d* spacings so that lattice misfit is small enough to be accommodated by elastic strain. Parallel orientation and continuity of a single set of lattice planes or atom rows is not sufficient to generate a 3D coherent interface (Lojkowski and Fecht, 2000). Note that Sutton and Balluffi (1995) extend the definition of coherency to include cases where continuous lattice planes belong to a reference lattice shared by two crystals at an interface, which is not necessarily the lattice of either crystal. This reference lattice is the DSC or "displacement shift complete" lattice introduced by Bollmann (1970). At an incoherent interface there is no periodic correspondence of atom rows or lattice planes across the interface. In the case that lattice misfits between two neighbour crystals are accommodated by isolated ('misfit') dislocations or some of the atom rows do not have a counterpart across the interface, coherent interface segments interchange with incoherent segments and the interface is termed semi-coherent (Howe, 1997).

A special case of coherence is so called twin coherence, which is developed when atom rows or planes are continuous though their directions change across an interface (Lojkowski and Fecht, 2000). Atomic rows or planes of both twin individuals meet with a 1:1 correspondence at the twin plane, which requires a specific misorientation of the twin individuals and a certain interface orientation (*i.e.* twin plane orientation) with respect to both twin crystals. Such 2D coherence is developed when lattices share sites or atoms only within the interface plane, and at a particular neighbour misorientation and interface orientation.

For certain combinations of interface orientations and neighbour lattice misorientations, lattice misfits at phase boundaries can be accommodated by oblique intersection of lattice planes at interfaces. Misfit accommodation by rotation (and translation) can therefore be relevant for particular lattice planes and interface orientations. The interface will intersect different lattice planes (even those that are symmetrically equivalent) at different angles, therefore only misfit between one pair of lattice planes can be accommodated this way. Relative neighbour lattice rotations can accommodate lattice misfit only in 2D interface sections and cannot establish 3D lattice coherence.

2.1.3. Atomic arrangement within the interface plane

The atomic arrangement at an interface can be investigated by aberration-corrected scanning transmission electron microscopy (STEM) at atomic scale spatial resolution. The signal of the atomic columns in a STEM high angle annular dark field (HAADF) image is proportional to the square of atomic number Z, therefore the HAADF image is also called Z-contrast image (Pennycook and Nellist, 2011). The Z-contrast images not only supply information on atomic structure, but also the precise position of atomic columns with different chemical composition from the image intensity. Besides, electron energy loss spectroscopy (EELS) or energy dispersive X-ray spectroscopy (EDX) installed in the STEM can be used to analyse the distribution. These results by STEM can reveal the atomic structure on the interface, such as the misfit dislocations, the terminating planes in each side of the interface, and the shared lattice sites among the neighbour crystals (Fig. 1). This information is crucial for inferring the bonding type across the interface plane.

Studying the atomic arrangement within the interface is limited to neighbour crystals with zone axes accessible in the given STEM sample orientation and tilt range of the microscope. The observed STEM images are the 2D projection of 3D structures along the direction of view. The atomic arrangement in the third dimension (a direction parallel to the zone axis and within the interface plane) is not accessible easily. Analysis of the full 3D interfacial atomic arrangement can be approached only by atomic-scale models calculated for a range of misorientations and interface orientations (see section 2.2; Adjaoud *et al.*, 2012).

2.1.4. Dislocation structure at interface

Grain/phase boundaries can be characterized by the types, density and arrangement of dislocations present along the interface (Petrishcheva and Abart, 2017, this volume). At a semi-coherent interface (see section 2.1.2), isolated line dislocations within the interface plane can accommodate lattice misfits between neighbouring crystals. At incoherent interfaces neighbour lattices are unstrained and single dislocations cannot be identified due to their close spacing. The degree of localization of misfit dislocations changes with the strength of bonding between the two crystals. Strong bonding leads to localization at dislocations and a semi-coherent interface, whereas weak bonding leads to poor localization of misfit and an incoherent interface (Howe, 1997). Screw dislocations in neighbour crystals crosscutting an interface generate heterogeneities and steps in the interface plane, which can act as preferred sites of nucleation and growth (Burton *et al.*, 1951).

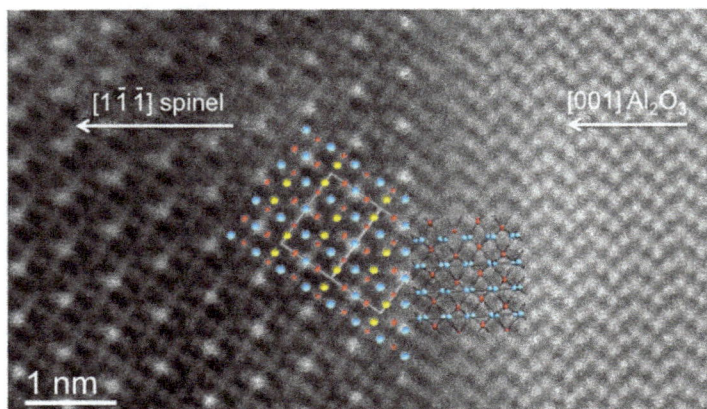

Figure 1. STEM Z-contrast image showing atomic structure on a corundum/spinel interface. The spinel and corundum are aligned to the [110] and [120] axes, respectively. The overlain atomic models show the theoretical positions of Mg (yellow), Al (blue) and O (red).

2.2. Theoretical considerations of lattice-site concordance in neighbour crystal volumes

Preferred neighbour crystal (mis)orientations have been found to correlate with concordance of lattice elements in neighbouring crystal volumes (*e.g.* Maurer, 1987). These concordances can be calculated solely from the misorientation across a boundary, whereas the interface orientation and geometry are usually disregarded in studies of this type. As the next-neighbour interactions leading to preferred misorientations are exerted *via* the interface, the theoretical models of lattice element concordances cannot exactly predict CORs. However, the CORs actually observed are probably among those estimated by this method (see section 3.1). The degree of concordance and its parameterization have been defined in various ways.

2.2.1. Coincidence site lattices and O lattices

The concepts of O lattice and coincidence-site lattice (CSL) (Petrishcheva and Abart, 2017, this volume; Sutton and Balluffi, 1995) have been employed in order to explain preferred misorientations by identification of geometric operations applied to one crystal, which lead to identical or similar 3D arrangement of specific lattice sites in the neighbour crystal. CSL considerations have mainly been applied to monomineralic systems, but were also used to investigate polymineralic CORs (Hwang *et al.*, 2015).

CSLs are described by geometric operations (combinations of rotation and translation) that do not correspond to elements of the lattice symmetry of either neighbour crystal. If these operations do correspond to elements of the lattice symmetry, they are classified as twin orientation relations.

2.2.2. Matching d spacings

Theoretical considerations are applied to test similarities and misfits in *d* spacing between certain lattice plane sets in neighbour crystals. This approach considers solely lattice elements, but disregards atomic positions within the lattice. In thin-film samples the similarity of *d* spacing has been employed to explain preferred parallel orientation of certain neighbour lattice planes and their continuity across the interface (Detavernier *et al.*, 2003). In contrast, different sample material showed the absence of COR despite excellent *d* spacing correlation (Detavernier *et al.*, 2004). Similarly, recent investigations of phase relationships in natural rocks showed that matches/misfits in neighbour *d* spacings do not necessarily correlate with COR development (Hwang *et al.*, 2015; Griffiths *et al.*, 2016).

2.3. Systematic next-neighbour misorientations

Crystallographic orientation relations, as defined in section 1 of this chapter, are characterized only by the relationship between the orientations of neighbouring crystals without considering interface properties or matching lattice sites. Using EBSD, this information can be collected much more rapidly than data about the interface itself (see section 2.1), and therefore there is great interest in attempting to use CORs as a probe of the characteristics of interfaces and also of the conditions and processes underlying COR formation.

Different notations have been employed to describe misorientations, *e.g.* misorientation axis-angle pairs, Euler angles, Rodrigues vectors and quaternions (Mason and Schuh, 2009, and references therein). In the geosciences, the misorientation axis-angle pair notation is commonly used for describing misorientations in grain/phase aggregates. An axis-angle pair describes lattice misorientations by a rotation around the misorientation axis which gives the minimum possible misorientation angle (Wheeler *et al.*, 2001). The 3D orientation and distribution of misorientation axes can be plotted with respect to the sample or crystal reference system, whereas misorientation angles are commonly used to classify GB segments according to arbitrarily defined angular value ranges. Grain-boundary misorientation distributions are biased by the crystal symmetry (Wheeler *et al.*, 2001) and therefore must be compared with measured random-pair distributions and theoretical random-pair distributions in order to infer petrogenetic information (*e.g.* Heidelbach *et al.*, 1996, 2000; Bestmann and Prior, 2003).

In contrast to the description of textures in general, characterization of CORs is based on evaluating the relationships between certain crystallographic planes/directions of crystal pairs. Systematics underlying CORs are defined by 'COR rules', which allow the assignment of neighbour pairs to groups depending on which COR rule (if any) they obey (Fig. 2). At present, COR rules are defined by the authors of each individual study and thus great care is required to avoid obscuring patterns in the data by being too restrictive (Griffiths *et al.*, 2016). The final goal of this approach is to define COR rules which have a physical meaning in order to relate COR development with environmental conditions during their formation. At the current stage of COR research, theoretical models and experimental data are not available to test whether COR rules can achieve this. At present the validity and usefulness of different (and sometimes competing) COR rules must be assessed empirically.

The choice of crystal directions/planes used to define COR rules may be arbitrary due to equivalences arising from crystal symmetry (Griffiths *et al.*, 2016). COR rules need not necessarily be based on low-indexed planes/directions, and CORs may also show deviations from a perfectly fixed relationship with one or two degrees of freedom, nonetheless repeated by crystal symmetry. Development of EBSD analysis in the last two decades has enabled statistical investigation of orientation distributions, allowing the identification of COR rules with one or more degrees of freedom based on observations of multiple neighbour-pairs in a sample.

Criteria for characterization of CORs and definition of COR rules are listed in the following subsections.

2.3.1. Number of orientational degrees of freedom

The number of degrees of freedom classifies orientation relations between neighbour phases into three different COR types based on the rule they obey, namely the specific COR type (0 degrees of freedom), the rotational statistical COR type (1 degree of freedom) and the dispersional statistical COR type (2 degrees of freedom) (Fig. 2; Griffiths *et al.*, 2016). In the literature, the terms 'crystallographic orientation

Examples of the COR types

Specific COR

Statistical CORs

Rotational (limited rotation)   Rotational (full rotation)

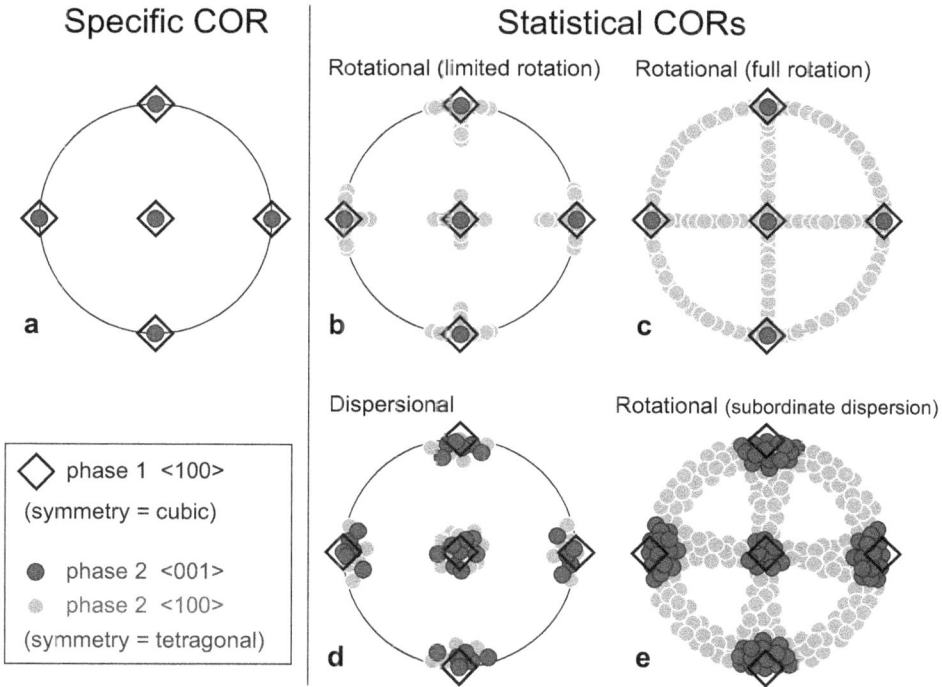

a

b

c

◇ phase 1 <100>
(symmetry = cubic)

● phase 2 <001>
● phase 2 <100>
(symmetry = tetragonal)

Dispersional   Rotational (subordinate dispersion)

d

e

Figure 2. Illustration of the different COR types defined by Griffiths *et al.* (2016). Each equal-angle upper hemisphere pole figure illustrates a different type of COR between phase 1 (cubic) and phase 2 (tetragonal). In all cases the COR is repeated fully by the cubic symmetry of phase 1, but in the following caption, one particular example of all symmetrically equivalent CORs is given. // stands for 'parallel to' and ~// stands for 'approximately parallel to'. (a) A specific COR with $[001]_{phase\ 2}//[100]_{phase\ 1}$ and $[100]_{phase\ 2}//[010]_{phase\ 1}$. (b) A rotational statistical COR with $[001]_{phase\ 2}//[100]_{phase\ 1}$, limited allowed rotation of phase 2 around its [001] axis. (c) A rotational statistical COR with $[001]_{phase\ 2}//[100]_{phase\ 1}$, full allowed rotation of phase 2 around its [001] axis. (d) A dispersional statistical COR with $[001]_{phase\ 2}~//[100]_{phase\ 1}$ and $[100]_{phase\ 2}//[010]_{phase\ 1}$. (e) A rotational statistical COR with subordinate dispersion, $[001]_{phase\ 2}~//[100]_{phase\ 1}$ with full allowed rotation of phase 2 around its [001] axis. The CORs in (b), (d) and (e) would also have precise angular ranges describing the limits of rotation or dispersion; these are omitted for this schematic example.

relationship' and 'COR' are used to refer to both the observed orientation relations and the rule defining the systematics of the orientation relations. These two are identical in the case of specific CORs, but differ in the case of statistical COR types.

In the case of a 'specific COR', the misorientation between the pair of neighbour crystals is fixed (Fig. 2a). A specific COR can always be defined by stating which lattice plane or direction in one crystal is parallel to which lattice plane or direction in the other crystal along three reference vectors. These reference vectors can be chosen arbitrarily as long as none of them is parallel to another and they do not all lie in the same plane. However, the usual practice is to choose them so that both planes or directions in a pair have low Miller indices. Two reference vectors suffice if they can be chosen such that the plane shared by them is a mirror plane in both crystal point groups. A specific COR in fact represents a number of different symmetrically equivalent orientation relationships, according to the crystal point group symmetries of the phase(s) involved (*e.g.* Angel *et al.*, 2015).

The term specific COR is already used in the literature (Hwang *et al.*, 2007; Ague and Eckert Jr., 2012) but sometimes it has not been fully separated from concepts of lattice matching or coherence at interfaces. It is important to emphasize the differences. A 3D coherent pair of lattices will always exhibit a specific COR, as all lattice planes are fixed relative to each other. However, identifying a specific COR does not necessarily imply coherence. As long as more than one set of planes in the two crystals are parallel to each other, the COR will be specific. More importantly, in the case that factors other than lattice or atom alignment at the interface (see section 3) control the specific COR between the two crystals, there is absolutely no restriction on the nature of the interface. When neighbour lattices have different crystal symmetries and/or inconsistent lattice parameters, non-random misorientation distributions can be reflected by parallel geometric lattice elements without other evidence of lattice concordance (Proyer *et al.*, 2013; Griffiths *et al.*, 2016). Two crystals may exhibit a specific COR but be separated by an incoherent interface.

In contrast to specific CORs, the 'statistical COR' types (1 or 2 degrees of freedom; Fig. 2b–e) comprise a defined range of different next-neighbour misorientations in addition to systematics caused by crystal symmetry (Griffiths *et al.*, 2016). Measurement of large numbers of orientation relationships between individual pairs of crystals in a sample is required for identification of CORs and definition of statistical COR rules. The higher the complexity of the statistical COR, the larger the dataset of measured individual pairs needed to identify the COR rule (Griffiths *et al.*, 2016). The concept of statistical COR types still presumes there is some physical restriction on the allowed next-neighbour orientation relationship, however it accommodates deviations from perfect alignment of lattice elements. The amount of deviation is specified as part of the particular statistical COR rule. The concept of statistical CORs has been developed in host-inclusion systems, where a large number of inclusions can show orientation relationships reflecting the lattice symmetry of a single host crystal. Similarly, studies of thin films (*e.g.* Detavernier *et al.*, 2003) and of interlayer growth (Keller *et al.*, 2010) have involved statistical CORs between a single-crystal substrate and polycrystalline overgrowths.

Based on the number of orientational degrees of freedom, statistical CORs can be further subdivided into two end-member types. In the case that one crystallographic plane/direction is strictly parallel to a neighbour crystallographic plane/direction, there is one orientational degree of freedom, allowing for rotation around a fixed axis in each neighbour crystal (Fig. 2b–c). The term 'rotational statistical COR' has been proposed for this end member statistical COR type (Griffiths *et al.*, 2016). This definition corresponds to the term 'monotaxy', *i.e.* the phenomenon of mutual orientation of two crystals of different species with one-dimensional lattice control (line in common) according to Bailey (1977). In the literature, the term axiotaxy has been used for rotational statistical CORs with a rotation axis at an angle <90° with respect to the interface in thin films (Detavernier *et al.*, 2003). This term implies a defined interface orientation, which is not the case for rotational statistical CORs.

Rotational statistical CORs can show either random distribution of rotations around the rotation axis or the development of several orientation maxima along the small circle distribution (Keller *et al.*, 2010; Jeřábek *et al.*, 2014). For the latter case, each symmetrically non-equivalent orientation maximum could also be treated as a separate COR, described as specific or rotational statistical COR depending on whether the maximum involves a perfect alignment or a limited degree of rotation (Fig. 3).

Limited orientation deviation from a perfectly parallel relationship between neighbour lattice elements, creating orientation clusters due to small rotations around more than one rotation axis, results in a COR type with two orientational degrees of freedom, within a stated maximum deviation angle. The term 'dispersional statistical COR' has been proposed for this end-member statistical COR type (Griffiths *et al.*, 2016). Pure dispersion leads to orientation clusters around particular crystallographic directions of the neighbour crystal (Fig. 2d).

Both end-member types of statistical CORs can occur in combination. CORs with combined rotational statistical and dispersional statistical components develop clusters of directions around the rotation axis, and associated bands of directions distributed

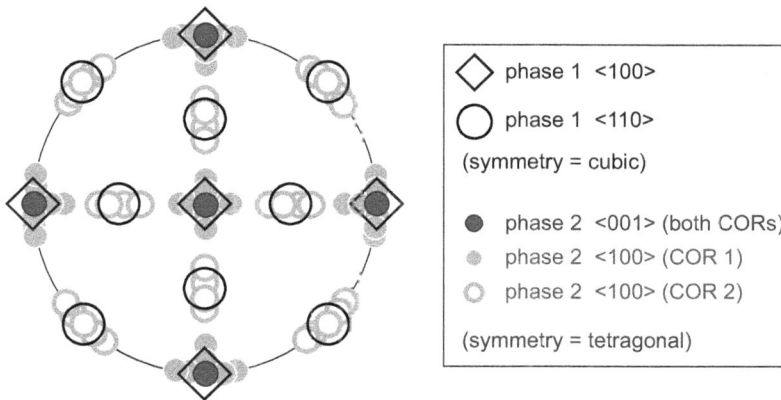

Figure 3. Illustration of two limited rotational statistical CORs between phase 1 (cubic) and phase 2 (tetragonal). The CORs are repeated fully by the cubic symmetry of phase 1, but in the following caption, one particular example of all symmetrically equivalent CORs is given. // stands for 'parallel to'. Both CORs share the fixed rotational axis $[001]_{phase\ 2}//[100]_{phase\ 1}$. However, COR 1 exhibits $[100]_{phase\ 2}//[010]_{phase\ 1} \pm \sim7.5°$ rotation and COR 2 exhibits $[100]_{phase\ 2}//[101]_{phase\ 1} \pm \sim7.5°$ rotation. Insufficient sampling of these two CORs might give the impression that this was one single complete rotational statistical COR (as in Fig. 2c).

around small circles about this dispersed rotation axis (Fig. 2e). Combinations of the two end members are named according to whether rotation or dispersion is the strongest element of the orientation distribution.

The required size of the data set for COR identification is critical especially when attempting to identify any statistical COR. Robust identification of a statistical COR requires a sufficiently large number of next-neighbour orientations, so that additional analyses will not affect the measured distribution. With small datasets, clustering of crystal directions cannot be identified robustly. For example, it might not be possible to identify whether a rotational statistical COR allows random distribution of rotations around the rotation axis (Fig. 2c), or whether rotation away from a particular preferred orientation is limited (Fig. 2b). Griffiths *et al.* (2016) recommend that in the absence of sufficient information, the most general definition of the COR rule should be given. Assuming the weakest possible restrictions on the orientation relationships avoids overinterpretation of sparse data.

Measures can be taken to 'reduce' symmetrically equivalent but different misorientations between neighbour crystals to a smaller region of misorientation space, increasing the density of measurements (*e.g.* Angel *et al.*, 2015; Griffiths *et al.*, 2016). If the full point group symmetry of both phases is taken into account every misorientation occurs only once without symmetrical equivalents and the region is called fundamental (other names include asymmetric and elementary). It should be pointed out that the symmetry of the full misorientation distribution observed for a population of crystals comprising two phases related by one statistical COR rule does not necessarily reflect crystallographic symmetry elements of either phase. When CORs are developed between phases of different crystallographic symmetries, unique orientations of the phase with lower symmetry can (but do not have to) be repeated by the crystallographic symmetry of the higher symmetric phase. Whether unique orientations of the less symmetric phase prefer one of the possible equivalents generated by the higher symmetry of the other phase or not may be diagnostic of the COR-forming process (see section 4.2.2). To avoid loss of useful information, reduction of misorientation data to the elementary region should be carried out only after inspecting the full misorientation space.

2.3.2. Angular extent of rotation and/or dispersion

For statistical CORs the angular extent of rotation and/or dispersion can be quantified. This is straightforward in the case of rotational statistical CORs. For dispersional statistical CORs, the analytical error of both EBSD measurements has to be constrained in order to distinguish true orientation dispersion from apparent orientation dispersion due to the error in orientation measurement. The error is dependent on crystal orientation, instrument type, calibrations and analytical settings, which may vary between (and for orientation within) each single dataset (Ram *et al.*, 2015). Therefore, constraining the analytical error is not straightforward. Still, dispersion by angles significantly larger than analytical error can be designated as dispersional statistical CORs without rigorous error analysis.

In order to determine cut-off values of misorientation angles for COR distinction, the percentage of orientations within an estimated misorientation angle error or precision can be determined (Griffiths *et al.*, 2016). Additionally, orientation distribution functions allow identification of combined CORs, such as a strong specific COR generating point maxima lying within a girdle belonging to a rotational statistical COR. Apart from density distributions, misorientation angle frequency histograms may also be useful to show minima at certain misorientation angles from a particular direction, marking a reasonable value for the cut-off angle separating CORs (Griffiths *et al.*, 2016).

2.3.3. Frequency of CORs

Within single datasets, several different CORs of the same or different type can be developed (Ageeva *et al.*, 2012, 2016; Hwang *et al.*, 2013, 2015; Proyer *et al.*, 2013; Xu *et al.*, 2015; Griffiths *et al.*, 2016). Identification of CORs is strongly biased by the size of the dataset of next-neighbour relative orientations (Griffiths *et al.*, 2016). Less frequent CORs may be determined only within very large datasets. Note that the strong bias of the dataset size limits the applicability of material characterization based on simple presence/absence of CORs. Measuring relative frequencies of abundant CORs in a system is one way to minimize this problem. Preliminary studies on statistical CORs indicate that predominant CORs in one sample may be subordinate in other samples containing the same neighbour phases (Proyer *et al.*, 2013; Hwang *et al.*, 2015; Griffiths *et al.*, 2016). Therefore, though only preliminary investigations are currently available, differences in relative frequency of CORs are presumed to bear petrogenetic information (*e.g.* environment conditions, COR formation mechanisms, material properties during COR formation), (Hwang *et al.*, 2015; Griffiths *et al.*, 2016).

3. Discussion of parameters with potential influence on COR formation

COR studies aim to use the observed misorientation systematics to infer information about processes and conditions of COR formation. Moreover, observed CORs are used to gain insight into the structures of poorly understood interfaces. Both aims require us to understand the controlling parameters of COR formation.

Next neighbour crystal interactions are exerted *via* the common interface. It is plausible that many CORs arise due to the preferential formation of particular interfaces between next neighbour crystals. The concept of minimization of free energy provides one way to understand which interfaces will be preferred. In this context, two contributions to the total free energy associated with an interface are the most relevant: those from the interface energy itself and those from the strain energy of an elastically distorted lattice. Most CORs are assumed to represent equilibrium orientation relationships where the system has reached at least a local energy minimum with respect to the misorientation of the neighbour crystal lattices and/or the interface orientation, for a given set of environmental conditions.

3.1. Interface energy of grain boundaries

For systems where interface energy is the most important energetic factor, it is generally assumed that the most frequent interfaces will be those with the lowest possible interface free energy. Interface free energy is dependent on the misorientation between neighbour lattices as well as on interface orientation. If two crystals meeting at an interface are free to assume any misorientation, preference for low-energy interfaces will create a systematic preferred misorientation between the two crystals, *i.e.* a COR. In order to understand what controls CORs we have to consider what controls interface energies, and more precisely, how we can predict what interfaces will have low energies.

The free energy of an interface is the sum of the free energies of the two surfaces meeting at the interface, minus the energy gained from bonding these surfaces together (Rohrer, 2011). To find low-energy interfaces we need to consider the interplay of both these variables. Crystal surfaces parallel to low-indexed planes have very low surface free energies, and therefore the lowest energy interfaces in a material tend to be parallel to low-indexed planes in at least one of the two neighbour crystals (Fig. 4; Rohrer, 2011; Marquardt *et al.*, 2015). However, many combinations of one or two low-indexed surfaces of a crystal will have similarly low surface free energies, so the actual lowest energy interface can only be found by considering the energy gain from bonding, which depends on the atomic scale structure of the interface.

Firstly, let us consider only interfaces between crystals of the same phase (grain boundaries) and ignore contributions from bonding of unlike atoms. The following trends in grain-boundary free energy with respect to varying interface geometry and structure are equally valid for grain boundaries in multi-element phases and phase boundaries, these interface types merely have additional variables on which their total free energy depends. If any of these extra variables are constant or negligible for a particular interface, then the interfacial free energy will vary with structure as described for grain boundaries of a phase containing a single element (Sutton and Balluffi, 1995).

The energy of low-angle grain boundaries is proportional to their dislocation content, which increases with increasing misorientation angle (Read and Shockley, 1950). Shallow local energy minima (cusp-shaped) are found at misorientation angles which generate periodic arrays of dislocations on a given grain-boundary plane. Which grain boundaries have the lowest energies for a given misorientation depends on the Burgers vector(s) of the dislocations involved, which vary with the orientation of the grain-boundary plane relative to the misoriented crystal lattices (Frank, 1951; Petrishcheva and Abart, 2017, this volume).

At large-angle grain boundaries the dislocation cores overlap, and the dislocation model no longer reflects the reality of the boundary at the atomic scale. However, low-energy high-angle grain boundaries are still associated with a high degree of periodicity, as this minimizes the frequency of broken bonds and maintains the coordination number of atoms at the grain boundary as close as possible to that of the bulk crystal (Howe, 1997). 'Potential' low-energy boundaries have thus been predicted

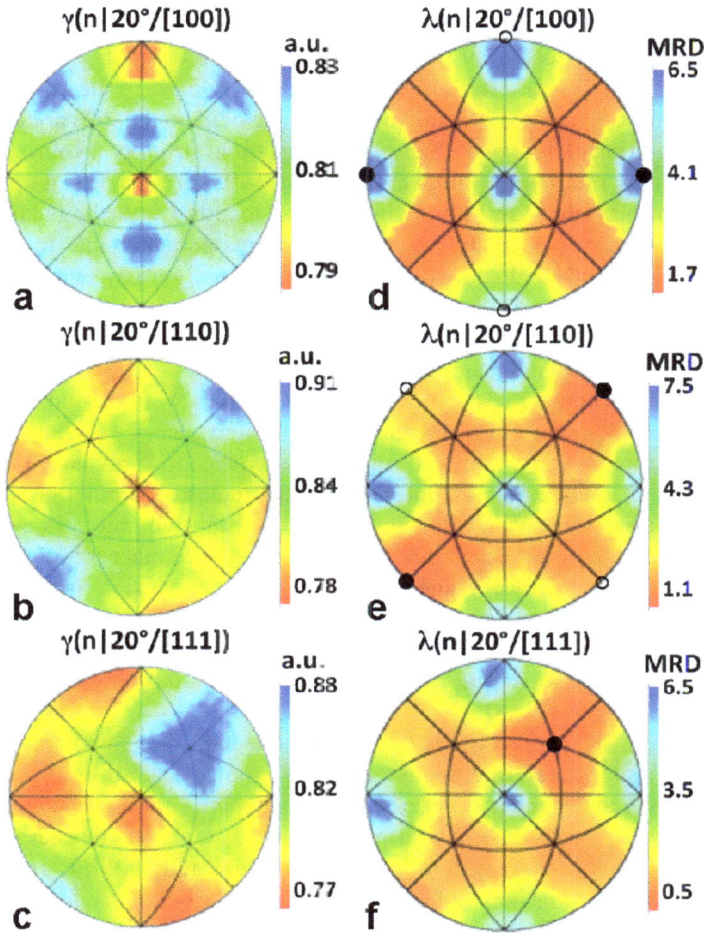

Figure 4. Stereographic pole figure plots showing how measured relative grain-boundary energies (a−c, scales in arbitrary units) and measured grain boundary frequencies (d−f, scales in multiples of the random [uniform] distribution) vary with interface orientations for a given misorientation in polycrystalline Ca-doped MgO. Low relative grain-boundary energies (warm colours in a−c) correlate with high frequencies of grain boundaries (cool colours in d−f). Each row is for a misorientation of 20° around a different axis, for rows 1, 2 and 3 the axes are [100], [110] and [111] respectively. Filled black circles in d−f indicate the interface orientation of a pure twist boundary, in figures d−e interface orientations of tilt boundaries lie on a line between the unfilled circles. Data from Saylor *et al.* (2003a; b), figure from Rohrer (2011), *Journal of Materials Science*, **46**, 5881−5895 (used under a Creative Commons Attribution Non-commercial License).

to occur at coincident site-lattice (CSL, see Petrishcheva and Abart, 2017, this volume) misorientations on planes with high densities of CSL points, and to exhibit shared small reciprocal lattice directions in the boundary plane (Sutton and Balluffi, 1987).

Which of the potential low-energy boundaries actually corresponds to a (local) energy minimum is affected by translations in 3D, known as the three microscopic degrees of freedom (DOF). Translation of one crystal relative to the other can occur in the plane of the interface or perpendicular to it. Periodic boundaries are characterized by a strong variation of energy with in-plane translation, strongly favouring translation to reach an energy minimum (Sutton and Balluffi, 1995). Translation perpendicular to the interface occurs when atoms brought too close together at the boundary repel each other, leading to an increase in the spacing of atoms perpendicular to the boundary ('boundary expansion') and an increase in boundary energy. Expansion is minimized when the mean d spacing of the lowest-index boundary-parallel planes is maximized (Sutton and Balluffi, 1995). This condition is fulfilled when boundary parallel planes are low-indexed planes in a material.

Until now the structures of the two neighbour crystals were treated as rigid, but in reality deviations from the expected geometric structure occur by shifting of individual atoms, for example to minimize grain-boundary expansion. This is called relaxation and occurs due to complex atomic interactions that require computer modelling of the individual atoms to predict (Jahn and Sun, 2017, this volume). In principle this modelling needs to be carried out for variations in all five macroscopic DOF and for translations in the interface plane. Grain-boundary expansion (relaxation) arises naturally in the simulations. In practice, for a given grain boundary some of these DOF will be fixed or have negligible effects on boundary relaxation (e.g. Adjaoud et al., 2012). Relaxation has a strong effect on which 'potential' low-energy boundaries actually have a low energy (Olmsted et al., 2009).

Now we consider the effect of atomic interactions in phases composed of more than one element, usually referred to as 'bonding' (note that all of the interactions contributing to grain-boundary energies actually involve bonding between atoms!). In ionic crystals, boundaries that juxtapose like-charged atoms will be energetically (e.g. Rohrer, 2011) unfavourable due to repulsion (Sutton and Balluffi, 1995). Periodic grain-boundary structures remain favourable, but some structures with good geometric fit will not be observed whereas others will be more frequent than expected (Chaudhari and Matthews, 1971). Ionic materials have been shown to exhibit additional low-energy cusps at pure twist boundaries.

Covalent bonding has similar effects on grain-boundary energy. The interface will relax to maximize strong covalent bonds, with additional restrictions related to preferred covalent bond angles (Sutton and Balluffi, 1995). Altogether the interactions related to covalent bonding are complex and again accessible only via atomistic modelling. However, the general rules governing relaxation remain the same: coordination should approach that in the undisturbed crystal and broken bonds are minimized (e.g. Adjaoud et al., 2012; Ghosh and Karki, 2014). Covalently bonded grain boundaries that share these features will have lower energies than expected on purely geometric grounds.

A great deal of research has looked for geometric criteria for predicting the lowest energy boundaries for a given phase, equivalent to looking for geometric criteria that can predict equilibrium CORs. Geometric criteria are often accurate for particular phases, or given certain restrictions. Unfortunately, there is no way to know *a priori* which or whether criteria will be accurate for a phase not yet studied in detail. No one set of geometric criteria has emerged which is not contradicted by some subset of experiments (Sutton and Balluffi, 1987; Rohrer, 2011). Thus, geometric criteria can, at best, narrow down the field of potential boundary-energy-driven CORs predicted for a system.

Three main factors contribute to the problems with geometric criteria, all discussed in detail by Sutton and Balluffi (1995). Firstly, relaxation has a highly complex dependence on the macroscopic DOF, and there may also be large differences in relaxation behaviour for identical boundary geometries in materials of different compositions. Secondly, the surface free energy of the boundary-parallel planes in both neighbour crystals has a strong influence on the resulting GB energy (Fig. 4). A boundary involving one or two low surface energy planes will have a lower energy than one at which only high surface energy planes meet, swamping any energy reduction due to increased periodicity and lattice matching. Thirdly, bonding between unlike atoms will favour certain geometric relationships with respect to others, and even create local energy minima where geometric constraints predict none.

A misorientation creating a CSL with low Σ has long been regarded as a geometric criterion for a low-energy boundary. On its own, this criterion turned out to be of little use (Sutton and Balluffi, 1987). For the reasons outlined above, a misorientation alone is insufficient to constrain the boundary energy, as the boundary plane is usually the greatest energy contributor (Rohrer, 2011). For the same reasons, boundary energies calculated by atomistic modelling vary widely for any given Σ value and a low Σ value does not necessarily correlate with a low boundary energy (Olmsted *et al.*, 2009).

A further development of the CSL criterion requires that a low-energy boundary should have a CSL misorientation and a boundary plane including a high density of CSL sites (Brandon *et al.*, 1964). This criterion is not universally valid because of the large influence of surface energies of boundary-parallel planes on the boundary free energy. However, this criterion is quite successful for the restricted situation of boundaries 'parallel to the same plane' in both crystals. In this case, the lowest energy boundaries/detected CORs are those which maximize the density of CSL sites in the boundary plane. This restricted criterion is successful because it removes the confounding effects of varying surface energies, and is useful when considering crystals rotating (or nucleating) on a pre-existing low free energy surface, as occurs in co-crystallization or epitactic overgrowth.

Another suggested geometric criterion is that larger mean d spacing of the lowest-indexed boundary-parallel planes correlates with lower boundary free energy. However, in general, mean d spacing correlates only weakly with boundary energy (Olmsted *et al.*, 2009), firstly again because variations in surface energies of the boundary-parallel planes outweigh the effects of changes in mean d spacing (especially in cases where the boundary is parallel to a low-indexed plane in only one neighbour crystal), and secondly because in-plane translation can generate high periodicity for certain pairs of planes, but the mean d spacing value does not indicate which ones. However, it does appear that "the lowest possible interface energies in a system are correlated with the highest possible [mean d spacings]" (Sutton and Balluffi, 1995).

A final geometric criterion is similarity of d spacing of plane sets not parallel to the interface in either of the two neighbour phases, thought to allow alignment of rows of atoms at the interface and therefore maximize coordination and minimize the frequency of broken bonds (Fecht and Gleiter, 1985; Howe, 1997). This criterion has significant overlap with others. As the boundary plane orientation is often the most important factor determining the boundary energy, the predictions of this criterion will not be correct in general, unless the criterion is restricted to a fixed initial surface. However, note that the d spacing matching criterion does not account for translation in the boundary plane, which can lead to a lower energy configuration causing certain boundaries to be more favoured than predicted even for a fixed initial surface (Sutton and Balluffi, 1987).

It follows from the above discussion that determination of the geometric and structural criteria fulfilled by a given COR cannot fully explain its existence. Most CORs will be explicable using one or more geometric criteria for formation of a low energy interface, but not all CORs predicted by geometric criteria will be developed in a sample. Of course if the most abundant COR found in a system does not conform to any geometric criteria, this indicates that a different energetic constraint was the most significant factor in COR formation. The key to further study lies in collating information about the *abundance* of CORs and their associated interfaces. These interfaces can be identified as primary candidates for subsequent computer simulation and high-resolution studies to explain their low-energy characteristics in detail and explore the energy variations at varying conditions, in order to constrain whether they reflect local or global energy minima with respect to the available degrees of freedom.

3.2. Interface energy of phase boundaries

Although the structure of phase boundaries is more complex, similar principles of rigid geometric alignment and microscopic relaxation apply (Howe, 1997). With the appropriate generalizations, the conclusions of section 3.1 are still valid, but additional factors become important that can even become the controlling influence on overall interface free energy in some cases. Chemical bonding strength is thought to be the predominant factor affecting the atomic structure and free energy of phase boundaries when a compositional change across the interface is involved.

An important complication when studying phase boundaries is the accommodation of differing crystal symmetries and/or lattice parameters. If the bonding between two phases is strong, elastic distortion can occur in order to match atomic rows 1:1 across the interface (leading to coherency). If the misfit in d spacings of the matched planes is small, elastic energy due to lattice distortion can be minimized by introducing misfit dislocations, extra planes of the unstrained phase terminating at the interface and separating coherent interface regions. The burgers vectors of these dislocations belong to one or both of the crystals. Lattice mismatch in multiple directions leads to multiple arrays of misfit dislocations. Interface energy may be increased if misfit dislocations interact, which is more likely for arrays of non-perpendicular misfit dislocations (Howe, 1997). The locations of misfit dislocations can be predicted using O-lattice theory (Petrishcheva and Abart, 2017, this volume). Zhang and Weatherly (2005) define interfaces with 1:1 correspondence of lattice planes and interfaces where the Burgers vector of misfit dislocations is a Burgers vector in either crystal as "primary" interfaces.

Interface energy increases with increasing misfit strain, whereby this increase is more pronounced at small misfit strains than at higher misfit strains (Van Der Merwe, 1963a,b). The stronger the bonding between two phases, the lower the interface energy for a given misfit, as dislocations are more localized. If there is only weak interaction between the phases, misfit dislocations are absent and the interface is incoherent. At this upper limit the energy of the interface approaches the sum of the free surface energies of the two neighbour crystals (Howe, 1997). Therefore, even weakly bonding phases will exhibit anisotropy in interface energy with interface orientation, potentially leading to preferred interface orientations and COR development.

Phase boundaries may also achieve local free energy minima for much larger lattice misfits where *d* spacing ratios differ from 1:1, equivalent to 1- or 2-dimensional CSLs. These interfaces were referred to as "secondary" by Zhang and Weatherly (2005), and consist of short periodic atomic structural units different from the arrangements in either crystal, analogous to high-angle grain boundaries. Deviations from a perfect CSL, equivalent to an exact *d*-spacing ratio such as 3:2, are accommodated by secondary misfit dislocations, which have Burgers vectors that belong to the DSC lattice, the coarsest lattice that incorporates all lattice points of both crystal lattices and thus a sublattice of the CSL (Bollmann, 1970; Sutton and Balluffi, 1995). In a secondary interface the elastic misfit strain is calculated using deviation from the undistorted DSC lattice, and is thus much less than the crystal-lattice misfit would suggest. As before, interface energy increases with increasing elastic misfit strain. Some of the CORs found by Griffiths *et al.* (2016) are consistent with low integer *d*-spacing ratios >1:1.

Phase boundaries often exhibit one or more sets of atomic-scale steps, which may or may not be associated with a partial dislocation. Each step separates a segment of interface parallel to a low-indexed plane in both phases. Regular steps accommodate inclinations of the macroscopic interface plane, which can enable a more coherent structure to emerge at the interface, reducing the density of required misfit dislocations and thus the interfacial free energy (Howe, 1997). This rearrangement of the interface to follow (at a macroscopic level) a higher indexed plane does not change the COR, suggesting that CORs involving parallel low-indexed planes are not necessarily generated by interfaces parallel to low indexed planes on the macroscopic scale.

Zhang and Weatherly (2005) have attempted to classify all possible "characteristic structures" of a phase boundary at a local energy minimum, by postulating three "optimum conditions" that low energy interfaces strive to achieve. They arrive at twelve possible structures, divided between primary and secondary interfaces and with differing numbers and combinations of misfit dislocations and steps. Of these, seven have zero orientational degrees of freedom and five have one orientational degree of freedom, suggesting that such structures would be interesting to investigate with regards to the origins of rotational statistical CORs.

3.3. Effects of impurity segregation

Impurity atoms are known to segregate preferably into interfaces (Howe, 1997). For example, Ca atoms have been observed to segregate into grain boundaries in MgO, affecting their structure and interface energy (Yan et al., 1998). Segregation to MgO grain boundaries was suggested by Verma and Karki (2010) to explain differences between experimentally observed boundaries in MgO and those calculated to have the lowest energy. Changes in boundary characteristics did not lead to a difference between expected and measured CORs in either example, but the theory does not exclude the possibility that impurity atoms can stabilize an interface with a COR not otherwise expected to form. Given the large variety of trace elements in natural rocks, this factor may prove an important influence on CORs in geosciences.

3.4. Long-range elastic strain energy

Minimization of interface energy in systems containing phases with differing lattice parameters can conflict with overall minimization of energy in a system, so that CORs may not reflect absolute (or possibly even local) minima in interface energies.

Given relatively strong bonding and significant differences in elastic properties of the neighbour phases, the lattice of the phase with lower elastic modulus can be distorted in order to accommodate elastically misfit strain and create a low-energy primary or secondary interface lacking misfit dislocations (and thus possessing long-range elastic strain). If this 'weaker' phase is considerably less volumetrically abundant (e.g. a precipitate of the order of nanometres in diameter or a layer less than a few nanometres thick) then the energy reduction from maintaining coherent interfaces outweighs the energetic cost of elastic strain (Van Der Merwe, 1963b; Sutton and Balluffi, 1995). With the growth of the smaller phase, volume increases faster than interface area for both precipitates and layers. At some point during precipitate/layer growth, elastic strain relaxation by changing the structure of the interface becomes energetically more favourable than lattice distortion. The higher energy of the new interface is balanced by the energy reduction due to elastic-strain relaxation (Sutton and Balluffi, 1995). Although in the simplest case structure changes should lead only to formation of misfit dislocations and changes in the interface geometry and/or interface orientation, this mechanism has also been suggested to explain the absence of specific CORs between precipitates and their host (Balluffi et al., 2005; Ague and Eckert Jr., 2012). This hypothesis has not yet been investigated, however.

Another possible source of volumetric elastic strain in an inclusion (whether originally a precipitate or not) could occur due to differential volume changes of the inclusion and host, either through thermal expansion, changes in hydrostatic pressure and/or phase transformation of the inclusion. It is conceivable that CORs could change in order to minimize elastic strain by rotation of an elastically anisotropic inclusion relative to the host lattice. Elastic properties can have greater rotational symmetries than crystals themselves, which suggests they might play a role in the development of rotational statistical CORs. However, this hypothesis has not yet been investigated by modelling or by experiment.

3.5. Influence of temperature and pressure on CORs

Geometric models for interfaces assume rigid crystal lattices and the majority of calculations of interface structure and energy have been carried out at 0 K and without considering the effects of pressure. As interfaces are associated with a certain entropy and excess volume expansion, it is obvious that temperature and pressure could have important effects on the relative energies of different interfaces, which could in turn lead to the preference of different CORs at different conditions.

Pressure may not have a strong effect on the CORs developed if these are primarily controlled by interface energy minimization. Simulations by Verma and Karki (2010) of grain boundaries in MgO at hydrostatic pressures between 0 and 100 GPa showed that the low-energy tilt boundaries investigated changed their boundary plane from symmetric to asymmetric, whereas the COR between the neighbour crystals was not affected. As the asymmetric boundaries still had low energies, pressure variations might affect grain-boundary plane orientations much more than CORs.

Simulations of grain boundaries in forsterite from 0 to 17 GPa showed increasing GB energies with pressure (Ghosh and Karki, 2014). Nevertheless the relative order of energetic favourability of the boundaries (and thus their CORs) was retained at high pressure. No studies have examined the effect of pressure on phase boundary free energies in geological materials so far.

Temperature increase leads to a decrease in surface and interface free energy with temperature (Howe, 1997). The 'differences' in energy between different surfaces also reduce to negligible amounts at high homologous temperature. Increasing temperature strongly reduces the energy differences between interface orientations that are due to compositional factors (*i.e.* different bonding), whereas the part of the overall interface energy generated by the structure of the interface is expected to decrease more slowly (Spanos, 1989; summarized briefly in Howe, 1997). If this conclusion can be applied generally – the calculations of Spanos (1989) are for semicoherent metallic f.c.c. interfaces – COR predictions from geometric criteria should actually be more accurate when interfaces formed at high temperatures, assuming that the effects of temperature do not alter dramatically the effects of microscopic relaxation on the final structural interface energy (which again requires further investigation).

With reduced differences between interface energies the number of local energy minima decreases (Erb and Gleiter, 1979). This reduces the number of potential interface-energy-controlled CORs. Although the effect of temperature on interface energies and thus on CORs could be large, very little information on the interrelation is currently available. The effect of temperature might be the reason that geometric criteria work as well as they do for prediction, despite the extreme simplification they represent. Further investigations comparing interface geometries and structures with associated CORs in experimental and natural samples formed at different temperatures are required.

Temperature and pressure have many other potentially important effects on CORs apart from their effect on interface energies directly. For CORs arising from precipitation or nucleation and growth, temperature and pressure determine the Gibbs free energy reduction from nucleating a new phase. If the energy reduction is extremely high, minimization of interface energy becomes relatively unimportant for the total system energy, which might lead to the development of a wider range of interfaces and CORs than would be expected at equilibrium, as hypothesized by Hwang *et al.* (2015). Rates of temperature or pressure change during COR formation determine whether CORs that develop can reach an equilibrium configuration or not. Furthermore, temperature and pressure variations also influence elastic strain (discussed in section 3.4).

4. Petrogenetic scenarios and mechanisms of COR formation

Crystallographic orientation relationships can develop in various petrogenetic scenarios and through a range of different mechanisms. The initial geometry and orientational degrees of freedom available in a system will have a large impact on the CORs which form. This section highlights the different initial settings and boundary conditions during COR formation in different petrogenetic scenarios (Fig. 5). We discuss associated mechanisms of COR formation and provide examples of COR development in geomaterials and synthetic analogues.

4.1. COR formation by overgrowth on a pre-existing crystal surface

COR formation due to overgrowth occurs when a crystalline surface with a given geometry and structure is present prior to crystallization of another grain or phase, and the overgrowth phase develops any COR to the substrate crystal (Fig. 5b). As the orientations of the substrate lattice and the interface plane are fixed, the crystal nucleating on the surface can vary by only three macroscopic degrees of freedom to minimize the energy of the interface created, as in a 'rotating crystallite' experiment, where randomly oriented small crystals deposited on a flat substrate surface rotate into low-energy orientations when the sample is heated (Sutton and Balluffi, 1987). Therefore the equilibrium COR will be a minimum for that particular interface plane, but is not necessarily the COR associated with the global minimum interface energy. Steps on the substrate surface, or dislocations intersecting it, may lead to local situations where a differently oriented interface can form and thus affect the type and relative frequency of CORs developed.

Figure 5 (facing page). Sketch of crystal-interface relations of the different scenarios of COR formation discussed in section 4 (overgrowth a–b; intergrowth c–e; phase transformation f–j; deformation, recrystallization k–n), showing the initial settings (first column) and the relationships between different precursor (white) and product (grey shaded) grains/phases (A–D), the initial interface (thick grey vertical solid line), the final reaction front (dotted grey line) and the interface(s) between the grains/phases showing a COR (black dashed line) after the COR forming process (second and third column).

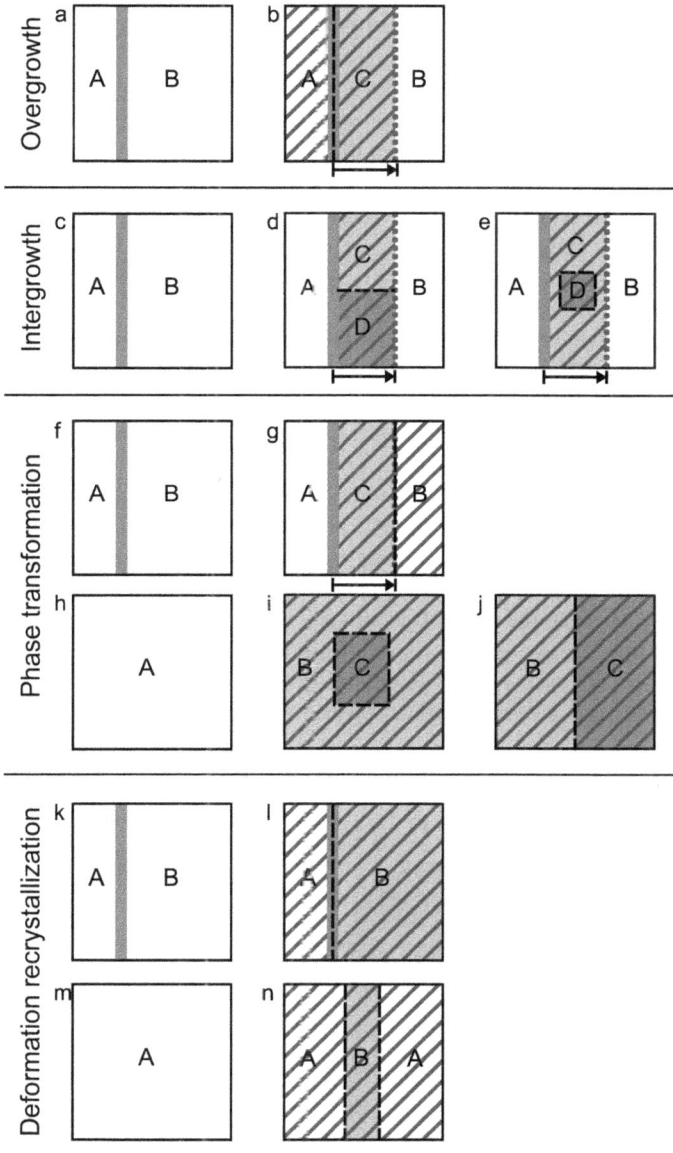

legend:

- ☐ precursor/reactant crystals
- ▨ product crystals of COR-forming process
- ⧄ crystals with mutual COR
- ▬ initial interface before COR formation
- •••••• final reaction front after COR formation
- --- interface between crystals with COR
- ⊢→ propagation direction of the reaction front

4.1.1. Nucleation and growth on a pre-existing crystal surface

COR formation by heterogeneous nucleation of a phase on a pre-existing crystalline surface and subsequent growth occurs commonly in both geological and synthetic materials. Synthetic layer growth by thin-film deposition has been applied widely to produce materials with controlled crystallographic orientation (Schlom et al., 2008). COR formation in this scenario is associated with a defined timing-relationship of next-neighbour crystallization, as substrate crystallization and formation of the substrate surface predates heterogeneous nucleation of the overgrowth phase. CORs develop at the nucleation stage of the overgrowth phase, which nucleates with a lattice orientation that minimizes the interface free energy.

At any crystal–melt, crystal–fluid or crystal–amorphous solid interface there is only one crystal present that can act as a substrate for nucleation. In these cases, after COR-forming nucleation the growth front propagates into the non-crystalline material. In contrast, at crystal–crystal interfaces both neighbour crystals could serve as nucleation sites, and selection of one of the two is required. Furthermore, crystallization of the overgrowth phase at a crystal–crystal interface requires contemporaneous decomposition of the precursor crystal across the original interface from the nucleation site in order to generate space for epitactic crystal growth. During growth of the product phase the reaction front propagates towards the interior of the precursor crystal on the opposite side of the interface with respect to the nucleation site (Fig. 6).

COR formation by overgrowth is controlled by the pre-existing substrate surface structure. Studies of crystal growth mechanisms at solid–vapour interfaces have shown that substrate lattice defects intersecting the surface represent preferred nucleation

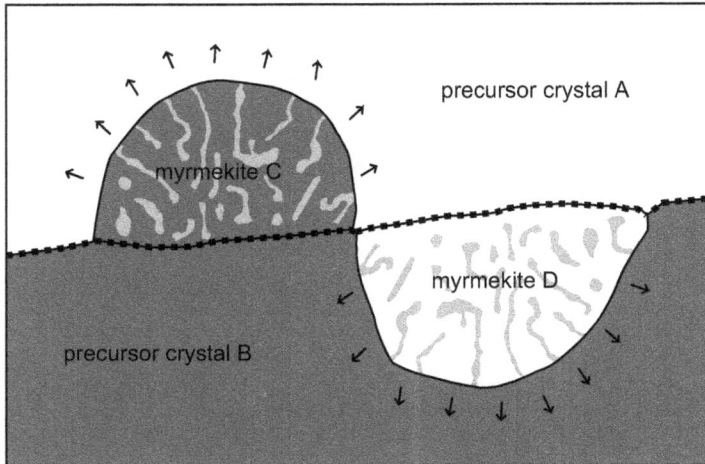

Figure 6. Sketch of a 'swapped rim' microstructure of myrmekite (Voll, 1960; Ramberg, 1962). Myrmekite feldspar has a COR with the precursor crystal on the opposite side of the initial interface (dotted line) with respect to the symplectite growth direction (arrows). Feldspar crystals with a mutual COR are indicated by identical grey shades.

sites and control the growth mechanism (Burton *et al.*, 1951). Spiral growth initiates at sites where screw dislocations impinge on the substrate surface, whereas surface steps and kinks are preferred heterogeneous nucleation sites for step-edge growth.

In natural rocks, COR formation by nucleation and overgrowth is a common phenomenon at solid–fluid interfaces, *e.g.* vein formation (Bons, 2000; Okamoto and Sekine, 2011; Bons *et al.*, 2012) or at solid–melt interfaces (*e.g.* Hammer *et al.*, 2010). COR formation at crystal–crystal interfaces has been described, for example, in symplectite forming mineral reactions including myrmekites in feldspars (Voll, 1960; Ramberg, 1962; Wirth and Voll, 1987; Abart *et al.*, 2014) or plagioclase-diopside symplectites in eclogite (Boland and Roermund, 1983). Commonly, at least one of the symplectite phases has a COR with the initially adjacent substrate crystal across the boundary from the decomposed precursor crystal, even in cases when both neighbour crystals are of the same phase prior to symplectite formation (Fig. 6). Myrmekite observed growing in opposite directions into two neighbouring precursor crystals of the same feldspar phase and showing an epitactic COR has been referred to as a 'swapped rims' microstructure (Fig. 6; Voll, 1960; Ramberg, 1962). Symplectites formed by mineral replacement commonly do not develop a COR with the replaced crystal, but instead show CORs with the initial neighbour crystal during symplectite nucleation, or with respect to other intergrown symplectite phases (Fig. 7). Within the same microstructure, different genetic types of CORs can develop at the same time.

Voll (1960) and Ramberg (1962) concluded that a myrmekite reaction front can propagate only when it represents an incoherent interface, whereas propagation of a coherent interface between the symplectite phase and the substrate crystal is prohibited by limited material transport along the interface. This theory can explain the preferred reaction front propagation direction during myrmekite growth even in monomineralic precursor material (Fig. 6), but note that it presumes that there always has to be a coherent interface between substrate crystal and overgrowth phase and an incoherent interface between the symplectite phase and the precursor crystal it replaces in order for this microstructure to form. Indeed, the natural examples mentioned above all show development of a COR between the substrate crystal and a symplectite phase with similar lattice symmetry and parameters. Myrmekite–plagioclase shows a COR with neighbouring magmatic K-feldspar or plagioclase (Voll, 1960; Ramberg, 1962). Symplectic diopsidic clinopyroxene in eclogite has a lattice orientation corresponding to the neighbour omphacite crystal adjacent to the symplectite precursor (Boland and Roermund, 1983). Therefore, the presence of a coherent interface between substrate crystal and overgrowth phase in these examples seems possible.

Within a symplectite microstructure, different symplectite phases can have different crystallographic orientation domain sizes (Obata and Ozawa, 2011). The crystallographic orientation domain sizes depend on the density of nucleation sites at the initial interface and the grain size of the neighbour crystals representing the substrate during symplectite nucleation. Crystallographic preferred orientation domains of symplectites do not necessarily correspond to shape-preferred orientation domains of the lamellar or vermicular intergrowths (Obata and Ozawa, 2011; Habler *et al.*, 2012). In case of CPO development of the precursor material, bulk crystallographic preferred orientations of symplectites can still reflect the precursor texture, even though the reactant phase has been completely replaced (Heidelbach and Terry, 2013).

4.1.2. Oriented attachment

Under certain circumstances pre-existing crystals that come into contact with a substrate surface can rotate (either *via* rigid body rotation or *via* internal defect

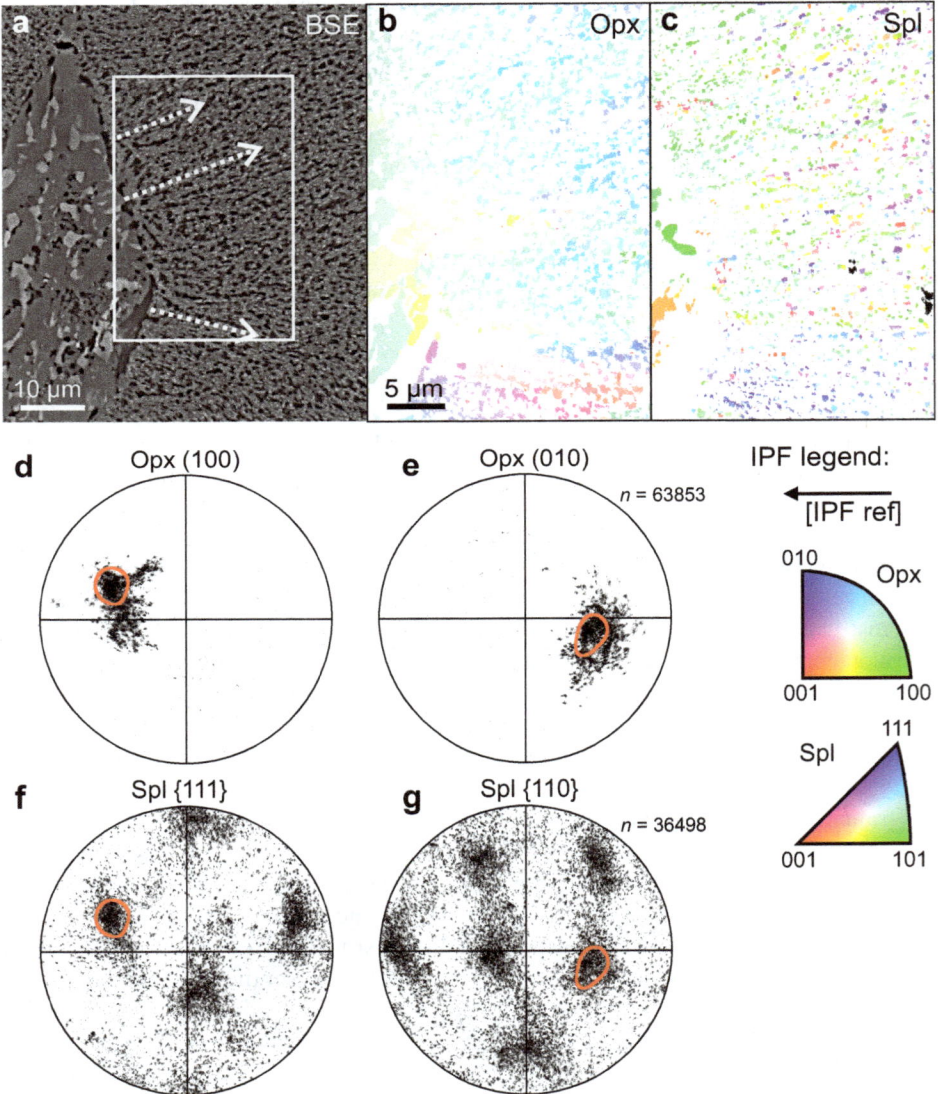

migration) in order to reduce the free energy of the system, eventually leading to the development of a COR

The free energy of a system containing two or more unattached crystals can be reduced by joining two crystal surfaces to form an interface (see this chapter section 3.1). The greatest free energy reduction is obtained by creating the interface with the strongest bonding (*i.e.* lowest interface free energy) for a given pair of surfaces. This has been demonstrated for crystalline nanoparticles of the same phase floating in a solution, which can join together to form a single perfectly aligned crystal built of multiple nanoparticles in a process called 'oriented attachment' or 'oriented aggregation' (Teng, 2013; Penn and Soltis, 2014). In this case no resulting interfaces are observed as the crystals attach with zero misorientation. Oriented attachment is not restricted to particles of the same phase, as the driving force is still creation of an interface with low free energy. Thus this mechanism can lead to COR formation in addition to being a non-classical mechanism of crystal growth (Penn and Banfield, 1999).

Slight misalignment of nanoparticles on attachment can lead to the formation of dislocations at the join (Penn and Banfield, 1998), and these dislocations may migrate through the new aggregated particle to reach its edge, leaving behind a perfectly aligned single crystal (Li *et al.*, 2012). Poorly oriented attached particles could undergo realignment after attachment due to this process; in the study of Li *et al.* (2012) alignment by whole-particle rotation before attachment was the dominant mechanism, however.

Whether oriented attachment is significant for natural mineral growth is not yet clear (Teng, 2013). The mechanism has been observed in the laboratory for a number of phases, though most seem not to be geologically relevant. All studies use aqueous solutions as a medium. Oriented attachment might be less likely in melts due to their significantly higher viscosity, which would hinder rotation of nanoparticles (Li *et al.*, 2012). However, it has been suggested that oriented attachment might be important during growth twinning (Nespolo and Ferraris, 2004).

The forces involved in oriented attachment were estimated by Li *et al.* (2012) to be small (on the order of 1×10^{-10} N) and short-ranged. In cases where both interacting

---

Figure. 7. Ultra-fine grained orthopyroxene-spinel-plagioclase symplectite in lherzolite. a) Back scattered electron image of a symplectite domain. Dotted arrows show the local propagation direction of the reaction front during the symplectite forming reaction. The white rectangle highlights the position of the EBSD inverse pole figure maps (step size 70 nm) of b) orthopyroxene and c) spinel. The IPF reference direction is given by the black arrow. Pole figures are discrete plots (upper hemisphere equal angle projection) of orthopyroxene (d–e) and spinel (f–g) single point data (n gives the number of indexed points for each phase). The red loop marks the intensity maximum (>19 times random) of the orthopyroxene orientations as derived from the density distribution calculation. Data indicate the specific COR Opx (100)//Spl (111) and Opx (010)//Spl (110) of the symplectite phases. Ultra-fine grained symplectite nucleated at the interface with coarser grained Opx-Spl-Pl domain on the left. Opx grew with preferred orientation defined at the nucleation site. Crystallographic orientation domains extend parallel to the propagation direction of the reaction front during symplectite formation.

crystals exceed nm grain sizes it therefore seems unlikely that the forces involved will be able to complete rotation into a favourable lattice alignment before the crystal faces meet.

4.2. COR formation by intergrowth

CORs can form between newly crystallizing grains or phases during contemporaneous nucleation and/or growth. In this scenario, the interface orientation and structure between the related neighbour grains/phases is generated and may evolve during co-crystallization and COR formation, as the mutual interface extends laterally at the same time as the neighbour crystals grow (Fig. 5d–e). The system is free to vary all five macroscopic degrees of freedom, the orientation of the interface plane as well as the COR of the two neighbour phases, which should allow achievement of a lower interface energy than in an overgrowth scenario of COR formation (see section 4.1).

Intergrowth-generated CORs can be identified by their potential relation with microstructural growth features. Combined analysis of COR data and microstructural growth features is required to prove an intergrowth mechanism of COR formation (*e.g.* Hwang *et al.*, 2015).

4.2.1. Contemporaneous growth of different phases

COR development between different neighbour phases can occur when growth units of two phases are able to rotate into mutual preferred orientation during crystal growth (Fig. 5d). Whereas in the overgrowth scenario the preferred crystallographic orientation is determined at the nucleation stage, COR formation by intergrowth may develop (further) during progressive growth while the interface between the neighbouring crystals is newly generated. Whereas during nucleation and initial growth the orientations of the new crystals may be influenced by the initial setting at the nucleation site, their orientations and the interface may evolve towards more favourable mutual orientation during progressive growth.

An example of COR formation by intergrowth has been reported from feldspathoid-bearing rocks (rischorrites), where orthoclase (Kfs), kalsilite (Kal) and nepheline (Ne) show specific COR development during HT metasomatism (Ageeva *et al.*, 2012). Predominantly the specific COR Kfs [010]//Kal [001] as well as Kfs [100], Kfs [001] and Kfs [101]//one of the 3 Kal [100] directions is developed. Alternatively, the specific COR Kfs [010]//one of the Kal [100] and Kfs [001]//Kal [001] has been observed. In some cases, where nepheline is decomposed to kalsilite and orthoclase by metasomatic replacement, Kal initially nucleated with Kal [001]//Ne [001] and during progressive growth rotated into Kal [001]//Kfs [010] orientation, indicating that the crystallo-graphic orientation of kalsilite was initially controlled by the pre-existing surface of nepheline at the nucleation site, whereas during progressive growth kalsilite successively approached the orientation of synchronously growing orthoclase (Ageeva *et al.*, 2012; Zaefferer and Habler, 2017, this volume).

Whereas initially the CPO of the symplectite phase(s) is controlled by the substrate lattice and surface structure at the nucleation site, the preferred orientation(s) of the symplectite phases may rotate away from an initial COR due to mutual interactions of the different phases during subsequent growth (Ageeva *et al.*, 2012; Habler *et al.*, 2012; Abart *et al.*, 2014) or due to the influence of non-hydrostatic stress.

4.2.2. Co-crystallization of phases with differing nucleation and growth rates
CORs can form when different phases crystallize contemporaneously at significantly different nucleation and growth rates. Phase(s) with relatively low nucleation rates and high growth rates will generate the coarser-grained host phase, whereas the contemporaneously crystallizing phase with high nucleation and low growth rate will form finer-grained inclusions within the host phase (Fig. 5e). COR formation in this case is expected to occur predominantly during heterogeneous nucleation of the inclusion phase upon the host phase. Whereas the COR forms during nucleation of the inclusion phase, the interface geometry can arrange to establish an energetically favourable orientation during intergrowth.

Natural examples of COR formation in such a scenario have been reported from symplectites and from host-inclusion systems in different rock types. In many symplectites one of the symplectite phases is volumetrically more prominent and coarser grained than the other(s). In this case the coarser-grained phase may partly enclose the finer-grained phase and control its preferred crystallographic orientation.

From various peridotite localities decompressional garnet breakdown to symplectitic intergrowths of orthopyroxene (Opx), clinopyroxene (Cpx), spinel (Spl), and plagioclase with mutual CORs has been described (Odashima *et al.*, 2007; Obata, 2011; Obata and Ozawa, 2011; Špaček *et al.*, 2013). In all these examples symplectitic Opx developed a specific COR with enclosed symplectitic spinel, namely Opx (100)// one of the Spl {111} and Opx (010)//one of the Spl {110} (Fig. 7). When Cpx is present, it has (100) and (010) planes parallel to those in Opx, thus showing the same specific COR with spinel. Whereas the overall crystallographic orientation of the symplectite phases can rotate progressively in the sample reference frame during growth, the mutual COR is constant even in microstructurally different symplectite domains (Odashima *et al.*, 2007; Obata and Ozawa, 2011; Habler *et al.*, 2012).

Other natural examples of COR formation by co-crystallization are mineral inclusions in single crystal hosts, which developed CORs during inclusion formation by intergrowth (Fig. 5e). Identifying an intergrowth mechanism of COR formation in host-inclusion systems requires detailed microstructural and microchemical analysis in combination with crystallographic orientation data. The intergrowth mechanism can only be proven for host-inclusion systems with high crystal symmetry of the host phase. In such cases it is possible to identify microstructural and textural inclusion features that differ between growth facets of the host and other symmetrically equivalent host planes. COR data on rutile inclusions and their garnet host combined with microstructural features of the inclusions were reported by Hwang *et al.* (2015). Due to varying inclusion microstructures with inferred growth facet of the garnet host and unequal preference of rutile needles for different symmetrically equivalent host <111> directions, the authors inferred an intergrowth mechanism for rutile inclusion formation during metamorphism of the metapelitic rocks. Different samples showed the development of different COR types, assumed to correlate with significant differences in environment conditions during inclusion formation (Hwang et al., 2015).

4.2.3. Growth twinning

During nucleation and growth or oriented attachment, growth units of monomineralic material can dock with systematically different orientation than the substrate crystal, and thus introduce twin interfaces between twin individuals having identical composition and crystal symmetry but differing lattice orientation. Twin interface orientations represent a local interface free-energy minimum with respect to general grain-boundary orientations, but still cause an increase in the twin system free energy with respect to a homogeneous single crystal (Buerger, 1945). Distinct CORs between two twin individuals are described by the twin law, which gives the symmetry operations necessary to transform the orientation of one twin individual into the orientation of the other (Grimmer and Nespolo, 2009; Nespolo, 2015, and references therein). Growth twins are characterized by straight twin interface segments, which indicate their low interface energy at a particular twin interface orientation with respect to the neighbouring twin individuals.

Growth twinning is regarded to be the consequence of "accidental departure from equilibrium" (Buerger, 1945) or "fortuitous encounters" (Vernon, 2004). Still, its common occurrence in a wide variety of natural and synthetic materials requires that growth twins represent energetically favourable configurations with respect to the total system energy reduction at least at the moment of twin nucleation.

Buerger et al. (1945) proposed the following reasons for growth twin formation: (1) High supersaturation and high chemical energy causing total system energy reduction by phase formation to outweigh energy reduction by structurally and geometric perfect arrangement; (2) high growth rates impeding interface energy minimization; (3) maximization of coordination at step edges; (4) interaction of multiple simultaneously attaching growth units; and (5) oriented attachment of already coordinated clusters.

4.3. COR formation by phase transformation

Topotactic COR formation occurs when a reaction product phase develops a COR with the replaced reactant phase during phase transformation, especially when the product phase crystallizes within the volume of the decomposed reactant phase (Shannon and Rossi, 1964). During topotactic COR formation the precursor lattice orientation is fixed, whereas both the crystallographic orientation of the product phase and the interface orientation can adjust according to the processes occurring at the propagating reaction front (Fig. 5e–i). During progressive topotactic phase transformation and propagation of the reaction front into the precursor phase, CORs and the interface (reaction front) can further develop in order to approximate favourable orientation relationships with the precursor lattice. Theoretically, topotactic systems are free to vary all five macroscopic degrees of freedom to minimize interface energy, but COR and microstructure development are highly influenced by the mechanism of atomic rearrangement at the reaction front.

4.3.1. Replacement reaction with product phase(s) inheriting (sub)lattice elements from the reactant phase

During replacement reactions CORs can develop between reactant and product phase when the latter inherits elements of the precursor phase lattice (Fig. 5f). This is the case when the reaction product takes a certain crystallographic orientation that allows minimum atomic rearrangement at the reaction front. Examples of natural and synthetic phase transformations show the preservation of the reactants' oxygen sublattice, while smaller cations change atomic positions during phase transformation (Cesare and Grobety, 1995; Li *et al.*, 2016).

Topotactic COR development has been studied with synthetic polycrystalline spinel rims that have been grown experimentally between the single-crystal reactant phases periclase and corundum (Götze *et al.*, 2010; Keller *et al.*, 2010; Jeřábek *et al.*, 2014; Gaidies *et al.*, 2017, this volume). Frequently, polycrystalline spinel rims develop two microstructural and textural layers, each of which may show topotactic CORs with respect to a different precursor phase (Fig. 8). Whereas during nucleation the spinel orientation can develop either an epitactic or a topotactic COR, progressive rim growth often leads to selective growth of topotactic spinel. In this case, spinel growing in the direction of corundum exhibits a specific COR or a rotational statistical COR with corundum, where one Spl {111}//Crn (0001). In the case that a specific COR is developed, three of the Spl {110} are also //Crn {10$\bar{1}$0}. This COR rule is valid for both Spl twin orientations, which are rotated relative each other by 60° about the Spl (111) pole//Crn (0001) (Götze *et al.*, 2010; Keller *et al.*, 2010; Jeřábek *et al.*, 2014). With increasing uniaxial load perpendicular to the initial interface between reactants, the COR type changes from a specific COR to a rotational statistical COR which allows rotation around the Spl (111)//Crn (0001) apart from the spinel twin law (Fig. 8). Consequently the grain-boundary character within the polycrystalline spinel rim changes from spinel twin boundaries to general grain boundaries, which is suggested to influence the grain-boundary diffusivity of the spinel rim (Keller *et al.*, 2010). The initial COR is rotational statistical with random rotation around the common axis and a high nucleation density at the initial reactant interface (Fig. 8). Progressive rim growth under uniaxial load leads to evolution of the COR towards preferred rotation angles of the rotational statistical COR, and the development of discrete next-neighbour misorientation maxima (Jeřábek *et al.*, 2014). Furthermore, these authors showed that grain internal misorientations decrease during progressive growth, while the grain size increases.

Spinel growing in the direction of periclase often adopts the specific COR Spl {001}//Per {001} (Fig. 8). Due to the similarities in the periclase and spinel oxygen sublattices a semi-coherent interface is developed. Still, the lattice misfit between periclase and spinel introduces misfit dislocations of extra spinel planes, which influence the kinetics of the periclase-spinel phase transformation and therefore the

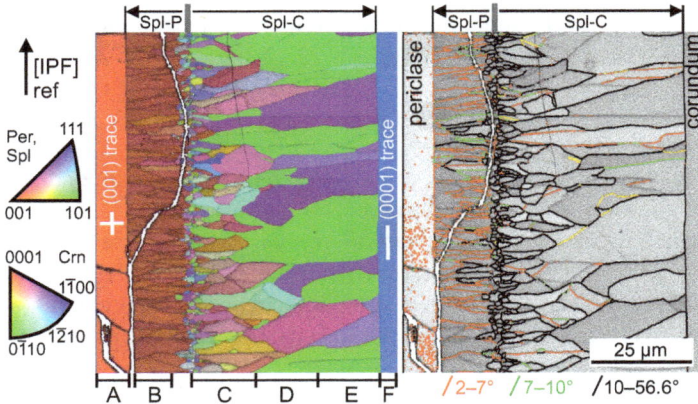

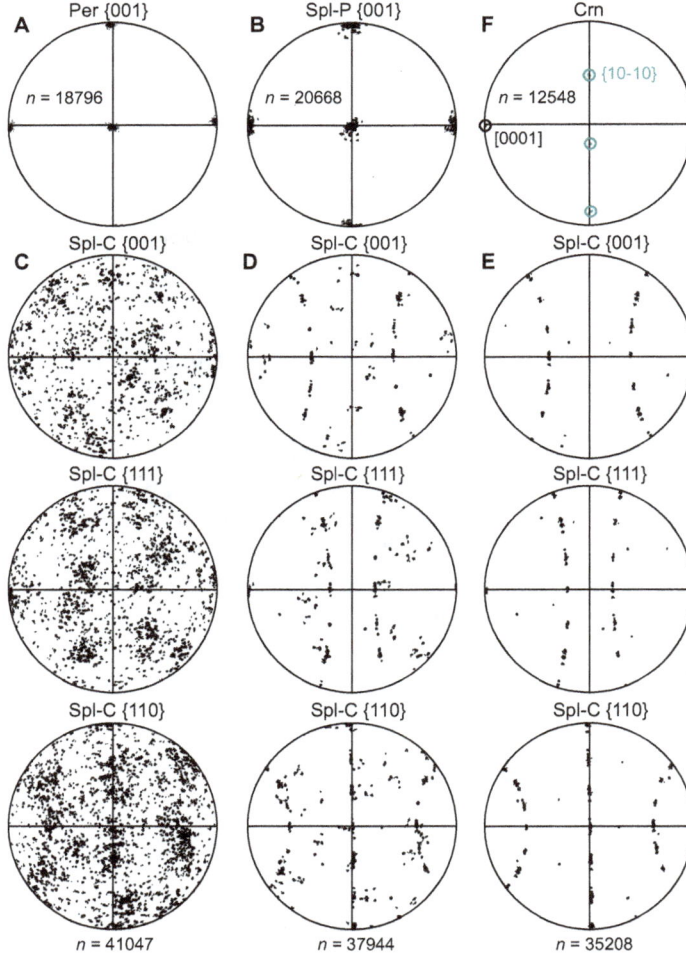

development of the nanoscale geometry of the interface (Li *et al.*, 2016). The mechanism of atomic rearrangement during phase transformation and the kinetics of associated material transport therefore contribute significantly to the interface geometry and structure (Gaidies *et al.*, 2017, this volume).

4.3.2. Intragranular precipitation and exsolution

Intragranular precipitation within or exsolution from a precursor phase is associated commonly with COR development between the newly formed phase and the precursor crystal (Fig. 5h) independent of the effective exsolution mechanism (Boudeulle, 1994). The precursor phase is present prior to the exsolution process and changes its composition while the new phase forms. In the case of a nucleation and growth mechanism, heterogeneous nucleation may occur at a particular intragranular site. The nucleus is free to adopt any crystallographic orientation and interface geometry to minimize free energy of the host-inclusion system. During subsequent growth of the product phase the reaction front, which in the case of a single crystal inclusion represents a closed interface loop, propagates into the precursor phase adjacent to the nucleation site. The precursor phase has to decompose locally, decrease in volume, or allow migration of defects in order to generate space for the newly growing phase. Whereas the host crystal orientation is fixed, the interface and the exsolution phase are free to develop orientations according to minimized system energy. A number of different interface segment orientations are required to generate a closed interface loop when the exsolution phase is entirely enclosed by a single-crystal precursor phase. Several differently oriented interface segments then contribute to the total interface energy of a single precipitate. In the case of exsolution by spinodal decomposition, the precursor lattice can be at least partially preserved and inherited by the new phase (Brown and Parsons, 1984; Abart *et al.*, 2009). Its crystallographic orientation therefore initially coincides with the pre-existing lattice of the precursor solid solution phase. When the grain size of precipitates is very small, elastic strain may accommodate lattice misfits between host and exsolution phase leading to development of a specific COR. At increasing inclusion size, the increased influence of volume strain energy may cause inclusion lattice rotation out of the initial precursor orientation (see section 3.4; Sutton and Balluffi, 1995; Balluffi *et al.*, 2005).

Figure 8 (facing page). Electron backscatter diffraction data of a polycrystalline spinel reaction rim grown between periclase (left) and corundum (right) single crystals. Inverse pole figure map (reference direction parallel to the initial interface trace) combined with EBSP quality map (upper left) and EBSP quality map with grain-boundary traces coloured according to grain-boundary misorientation angle ranges given below the image. Yellow lines show spinel twin boundary traces. Pole figures (equal angle upper-hemisphere projection) give single-point orientation data (n = number of indexed points) of periclase (A), of spinel showing COR with periclase (B), of spinel grown towards corundum (C–E) and of corundum (F). The degree of topotaxy between spinel and corundum close to the initial interface is low, but increases during progressive reaction. Selective growth favours spinel orientations with COR with corundum (rotational statistical COR type with limited rotation angle).

Natural examples of COR formation by exsolution have been described from various host-inclusion systems from a wide range of rock types. Common examples include perthite and antiperthite-lamellae in magmatic alkali-feldspar (e.g. Brown and Parsons, 1984), Fe- and/or Ti-oxide inclusions in plagioclase (Wenk et al., 2011a; Ageeva et al., 2016), in clinopyroxene (Robinson et al., 1977; Feinberg et al., 2004) or in amphibole (Mongkoltip and Ashworth, 1983), as well as pyroxene inclusions in garnet (Zhang et al., 2011).

CORs between metapelite garnet host and rutile inclusions were interpreted to have formed during inclusion precipitation post-dating host garnet crystallization (Proyer et al., 2013).

It is important to note that, whereas an exsolution mechanism in most cases will generate a COR, the presence of a COR does not necessarily imply an exsolution origin, although this assumption has been applied in previous studies (Hwang et al., 2011; Proyer et al., 2013; Xu et al., 2015). Instead, several other COR-forming mechanisms discussed in this section must also be taken into account. Detailed compositional and microstructural investigations, together with statistical information on COR frequencies and COR types are required to constrain the origin of single-crystal inclusions in host phases (Griffiths, 2015). Note that CORs do not necessarily follow the same systematics as shape-preferred orientations of acicular or platy inclusions (Proyer et al., 2013).

4.3.3. Transformation twinning

Transformation twinning occurs when isochemical phase transformation leads to reduction of the crystal symmetry, which is commonly a consequence of cooling (Buerger, 1945). Symmetry elements lost during phase transformation act as twin elements when different twin individuals replace an homogeneous precursor crystal (Fig. 5i). Transformation twin systems can convert to homogeneous single crystals during heating and transformation to the higher symmetry phase. Subsequent cooling may reproduce the same twin individuals present before heating, an effect termed "twin memory", which contrasts with the loss of the twin memory effect, termed "twin amnesia" (Frondel, 1945; Hayward and Salje, 2000; Wenk et al., 2009).

Common natural examples of transformation twins include 'microcline twinning' of K-feldspar due to transformation of monoclinic orthoclase to triclinic microcline, which leads to a combination of orthogonal twin systems according to the albite law and the pericline law (Sánchez-Munoz et al., 2012, and references therein). Another common example is Dauphiné twinning in quartz resulting from the transformation of hexagonal β-quartz to trigonal α-quartz (Frondel, 1945; Jacob and Cordier, 2010). As phase transitions leading to transformation twinning are associated with local elastic strain, they are sensitive to externally applied stress. Therefore, formation of transformation twins is influenced heavily and partly mediated by deviatoric stress (Wenk et al., 2009, and references therein).

4.4. COR formation by deformation or recrystallization

COR formation and development may post-date next-neighbour crystallization when pre-existing neighbour crystals are able to deform by mutual lattice rotation, or when intragranular deformation affects crystal domains inducing systematic lattice rotation (Fig. 5k−n).

4.4.1. Crystallite rotation

Reduction of interface energy may be established by rotation of neighbour lattices after crystals have impinged on each other (Fig. 5k). Rotating crystallite experiments involve the deposition of randomly oriented (usually ball-shaped) μm-scale crystals onto a substrate, often of the same phase. On heating, the deposited crystallites rotate and assume CORs with the substrate, usually corresponding to CSL misorientations (Sutton and Balluffi, 1987). This rotation is driven by reduction of the initially high interface energy between crystallite and substrate. Accommodation mechanisms for the rotation are diffusion of material in the 'neck' joining the crystallite to the substrate or motion of dislocations within the interface (Sutton and Balluffi, 1987). In rotating crystallite experiments the surfaces of the deposited crystallites are completely free, apart from the contact surface with the substrate. Space problems may arise if the 3D nature and 3D confinement of crystals needs to be taken into account, potentially limiting the rotation of crystals that are in contact with other crystals on all sides (Wheeler *et al.*, 2001).

Spiess *et al.* (2001) observed clusters of small, individual metamorphic garnet crystals without CPO, which in some cases had coalesced into larger grains (confirmed by compositional mapping). The aggregates had almost a single-crystal orientation, separated only by boundaries with misorientation angles <10°. Spiess *et al.* (2001) interpreted this as evidence that the garnet grains had rotated into alignment after coalescence, driven by reduction of interfacial energy. Employing a number of simplifications they modelled this process assuming it occurred due to diffusion along the interface and, taking into account the resistance of surrounding quartz, arrived at a rotation rate of 0.1−1° per several thousand years at 600°C. Rotation rate is inversely dependent on crystal size and, for a diffusion mechanism, increases with temperature. This mechanism has not been investigated further and may prove difficult to identify in natural rocks.

For larger, well faceted crystals embedded in a rheologically weak matrix, mechanical factors could lead to preferred parallel facet alignment by rigid-body rotation of crystals under externally applied stress. However, without post-attachment rotation, CORs are not expected to develop by this mechanism. In contrast, CPO formation may occur where crystallographic orientation systematics of aggregates are related to the external strain field, rather than to next-neighbour interactions.

4.4.2. Recrystallization

In some cases, COR formation and COR modification have been described in association with recrystallization phenomena. In this case, the given neighbour lattice misorientation and the orientation of the shared interface provide initial geometric constraints. Natural examples point to the additional involvement of compositional alteration.

An example of COR formation and COR alteration in natural rocks has been reported from host-inclusion systems in gabbro and plagiogranite from the Atlantic Mid Ocean Ridge (Ageeva et al., 2016). Fe-Ti oxide inclusions indicate formation and alteration of specific and rotational statistical CORs with the plagioclase host crystal. Plagioclase twinning according to the albite law partly postdates inclusion formation, and leads to inclusion lattice rotation to re-establish the COR with the newly formed hosting twin individual (Ageeva et al., 2016).

In most cases recrystallization mechanisms either lead to the adjustment of the interface geometry and orientation only, which affect the rock microstructure but not the texture, or cause preferred orientations without defined systematics in next-neighbour orientation relationships (e.g. subgrain rotation recrystallization).

4.4.3. Deformation twinning

Non-hydrostatic stress can induce deformation twinning ('mechanical twinning') at relatively low temperature and/or high strain rate, when dislocation creep is not effective. Mechanical twin systems form when potential shear planes in the homogeneous precursor parent crystal are in a particular orientation with respect to the external stress field and the yield stress parallel to the invariant plane of shear containing the shear direction in the precursor crystal is overstepped (Christian and Mahajan, 1995). Deformation twin lamellae result from small, interatomic-scale displacements parallel to the invariant plane of shear in the parent crystal and parallel to the shear direction. These displacements of certain step width generate a strained twin individual lattice orientation with a well defined specific COR with the undistorted precursor crystal domains. Commonly either a twin mirror plane parallel to the invariant plane of shear in the parent crystal (type I twin) or a two-fold rotation axis parallel to the shear direction (type II twin) are introduced as twin elements. The twin interface and/or the second undistorted plane (which is part of the strained twin individual) correspond to rational lattice planes, whereas the shear direction and/or the conjugate shear direction (within the strained twin individual) correspond to rational crystal directions (Christian and Mahajan, 1995). According to Buerger (1945) deformation twinning requires the presence of lattice planes that allow atomic rearrangement by shearing without passing a high interface energy barrier. Deformation twins therefore commonly develop twin laws different from growth twins, which do not have this energy barrier restriction. In contrast to growth and transformation twinning, deformation twinning generates lattice domains that experienced different strain paths although they have equivalent finite lattices. Whereas some precursor crystal domains retain their undistorted lattice, the strained twin lamellae underwent elastic distortion to attain the orientation of the newly formed twin individual (Christian and Mahajan, 1995).

Deformation twinning can accommodate only very limited amounts of strain, as the maximum possible strain corresponds to complete conversion of the crystal into the other twin variant. Still, deformation twinning is common in natural and synthetic materials affected by deformation at relatively low temperature and high strain rate. Examples of natural and synthetic materials include deformation twinning in calcite (*e.g.* Turner, 1953; Rybacki *et al.*, 2013), feldspars (*e.g.* Borg and Heard, 1969; Brown, 1989) and quartz (*e.g.* Wenk *et al.*, 2011b; Tochigi *et al.*, 2014). Whereas mechanical twinning of calcite and feldspar accommodates at least small amounts of crystal-plastic strain, mechanical twinning of quartz can accommodate only elastic strain. Such 'elastic twins' are caused by elastic anisotropy of the mineral phase and therefore have very close relationship with transformation twins, partly generating identical twin laws. Nevertheless, the twin systems generated by different mechanisms may show different behaviour. For example, mechanical Dauphiné twins in quartz do not show twin memory after heating, conversion to β-quartz and subsequent cooling, which contrasts with transformation Dauphiné twins (see section 4.3.4) (Wenk *et al.*, 2009).

5. Outlook

This article summarizes various aspects of systematic next-neighbour crystallographic orientation relationships (CORs) in crystalline materials. COR formation can occur in a wide range of petrogenetic scenarios and rock systems, under diverse conditions, and as a consequence of different mechanisms. CORs may develop between phases of equilibrium assemblages or between non-equilibrium phases in reactive systems. Therefore, the presence of CORs alone cannot be used as a unique tool to infer phase relations and petrogenetic processes from crystalline materials. In order to understand COR-forming mechanisms, COR data have to be combined with microstructural and microchemical data, and neighbour crystal interface characteristics including their atomic scale structure must be taken into account.

Recent COR research has concluded that statistical investigation of COR-type frequencies represents a promising way of inferring petrogenetic information from CORs. However, in order to develop this approach to the point where meaningful inferences can be made about unknown petrogenetic parameters we need to obtain comprehensive COR datasets for systems with known phase relations and mechanisms of COR formation. Further analysis of well-constrained systems along with atomistic modelling is required to identify the parameters controlling COR formation.

Whereas COR formation during overgrowth, intergrowth and phase transformations has been well known for a long time, COR formation during deformation and recrystallization is rather poorly understood due to the complex interplay of mechanical, chemical and crystallographic parameters. Investigations of crystallographic preferred orientations (CPOs) and deformation mechanisms have avoided textures showing next-neighbour interactions, as these were assumed to reflect inherited or reaction features not directly related to the material rheology. However, there is currently no reason to exclude the possibility that CORs could also form in association with certain (dynamic) recrystallization mechanisms if next-neighbour interactions contribute to texture development. Further research on potential CPO development in dynamic systems is required to prove or exclude such hypotheses.

References

Abart, R., Kunze, K., Milke, R., Sperb, R. and Heinrich, W. (2004) Silicon and oxygen self diffusion in enstatite polycrystals: the Milke et al. (2001) rim growth experiments revisited. *Contributions to Mineralogy and Petrology*, **147**, 633–646.

Abart, R., Petrishcheva, E., Wirth, R. and Rhede, D. (2009) Exsolution by spinodal decomposition II: Perthite formation during slow cooling of anatexites from Ngoronghoro, Tanzania. *American Journal of Science*, **309**, 450–475.

Abart, R., Heuser, D. and Habler, G. (2014) Mechanisms of myrmekite formation: case study from the Weinsberg granite, Moldanubian zone, Upper Austria. *Contributions to Mineralogy and Petrology*, **168**, 1–15.

Adjaoud, O., Marquardt, K. and Jahn, S. (2012) Atomic structures and energies of grain boundaries in Mg_2SiO_4 forsterite from atomistic modeling. *Physics and Chemistry of Minerals*, **39**, 749–760.

Ageeva, O., Abart, R., Habler, G., Borutzky, B. and Trubkin, N. (2012) Oriented feldspar-feldspathoid intergrowths in rocks of the Khibiny massif: genetic implications. *Mineralogy and Petrology*, **106**, 1–17.

Ageeva, O., Habler, G., Topa, D., Waitz, T., Li, C., Pertsev, A., Griffiths, T., Zhilicheva, O. and Abart, R. (2016) Plagioclase hosted Fe-Ti-oxide micro-inclusions in an oceanic gabbro-plagiogranite association from the Mid Atlantic Ridge at 13°34′ N. *American Journal of Science*, **316**, 85–109.

Ague, J.J. and Eckert Jr., J.O. (2012) Precipitation of rutile and ilmenite needles in garnet: Implications for extreme metamorphic conditions in the Acadian Orogen, U.S.A. *American Mineralogist*, **97**, 840–855.

Angel, R., Milani, S., Alvaro, M. and Nestola, F. (2015) OrientXplot: a program to analyse and display relative crystal orientations. *Journal of Applied Crystallography*, **48**, 1330–1334.

Bailey, S.W. (1977) Report of the I.M.A.-I.U.Cr. Joint Committee on Nomenclature. *American Mineralogist*, **62**, 411–415.

Balluffi, R.W., Allen, S. and Carter, W.C. (2005) *Kinetics of Materials.* John Wiley and Sons, Chichester, UK, 674 pp.

Bestmann, M. and Prior, D.J. (2003) Intragranular dynamic recrystallization in naturally deformed calcite marble: diffusion accommodated grain boundary sliding as a result of subgrain rotation recrystallization. *Journal of Structural Geology*, **25**, 1597–1613.

Boland, J.N. and Roermund, H.L.M. (1983) Mechanisms of exsolution in omphacites from high temperature, type B, eclogites. *Physics and Chemistry of Minerals*, **9**, 30–37.

Bollmann, W. (1970) *Crystal Defects and Crystalline Interfaces.* Springer, Berlin, 254 pp.

Bonev, I. (1972) On the terminology of the phenomena of mutual crystal orientation. *Acta Crystallographica Section A*, **28**, 508–512.

Bons, P. (2000) The formation of veins and their microstructures. In: *Stress, Structure and Strain: a Volume in Honour of Win D. Means* (M. Jessel and J.L. Urai, editors). *Journal of the Virtual Explorer*, DOI: 10.3809/jvirtex.2000.00007.

Bons, P.D., Elburg, M.A. and Gomez-Rivas, E. (2012) A review of the formation of tectonic veins and their microstructures. *Journal of Structural Geology*, **43**, 33–62.

Borg, I.Y. and Heard, H.C. (1969) Mechanical twinning and slip in experimentally deformed plagioclases. *Contributions to Mineralogy and Petrology*, **23**, 128–135.

Boudeulle, M. (1994) Disproportionation in mineral solid solutions: symmetry constraints on precipitate orientation and morphology. Implications for the study of oriented intergrowths. *Journal of Applied Crystallography*, **27**, 567–573.

Brandon, D.G., Ralph, B., Ranganathan, S. and Wald, M.S. (1964) A field ion microscope study of atomic configuration at grain boundaries. *Acta Metallurgica*, **12**, 813–821.

Brown, W.L. and Parsons, I. (1984) Exsolution and coarsening mechanisms and kinetics in an ordered cryptoperthite series. *Contributions to Mineralogy and Petrology*, **86**, 3–18.

Brown, W.L. (1989) Glide twinning and pseudotwinning in peristerite: twin morphology and propagation. *Contributions to Mineralogy and Petrology*, **102**, 306–312.

Buerger, M.J. (1945) The genesis of twin crystals. *American Mineralogist*, **30**, 469–482.

Burton, W.K., Cabrera, N. and Frank, F.C. (1951) The growth of crystals and the equilibrium structure of their surfaces. *Philosophical Transactions of the Royal Society of London A: Mathematical, Physical and*

Engineering Sciences, **243**, 299–358.

Cesare, B. and Grobety, B. (1995) Epitaxial replacement of kyanite by staurolite; a TEM study of the microstructures. *American Mineralogist*, **80**, 78–86.

Chaudhari, P. and Matthews, J.W. (1971) Coincidence twist boundaries between crystalline smoke particles. *Journal of Applied Physics*, **42**, 3063–3066.

Christian, J.W. and Mahajan, S. (1995) Deformation twinning. *Progress in Materials Science*, **39**, 1–157.

Detavernier, C., Ozcan, A.S., Jordan-Sweet, J., Stach, E.A., Tersoff, J., Ross, F.M. and Lavoie, C. (2003) An off-normal fibre-like texture in thin films on single-crystal substrates. *Nature*, **426**, 641–645.

Detavernier, C., Lavoie, C., Jordan-Sweet, J. and Özcan, A.S. (2004) Texture of tetragonal α-FeSi$_2$ films on Si(001). *Physical Review B*, **69**, 174106.

Engler, O., Gottstein, G., Pospiech, J. and Jura, J. (1994) Statistics, evaluation and representation of single grain orientation measurements. *Materials Science Forum*, **157–162**, 259–274.

Erb, U. and Gleiter, H. (1979) The effect of temperature on the energy and structure of grain boundaries. *Scripta Metallurgica*, **13**, 61–64.

Fecht, H.J. and Gleiter, H. (1985) A lock-in model for the atomic structure of interphase boundaries between metals and ionic crystals. *Acta Metallurgica*, **33**, 557–562.

Feinberg, J.M., Wenk, H.-R., Renne, P.R. and Scott, G.R. (2004) Epitaxial relationships of clinopyroxene-hosted magnetite determined using electron backscatter diffraction (EBSD) technique. *American Mineralogist*, **89**, 462–466.

Fettes, D. and Desmons, J. (editors) (2007) *Metamorphic Rocks: A Classification and Glossary of Terms*. Cambridge University Press, Cambridge, UK, 244 pp.

Frank, F.C. (1951) LXXXIII. Crystal dislocations – Elementary concepts and definitions. *The London, Edinburgh, and Dublin Philosophical Magazine and Journal of Science*, **42**, 809–819.

Frondel, C. (1945) Secondary Dauphiné twinning in quartz. *American Mineralogist*, **30**, 447–460.

Ghosh, D.B. and Karki, B.B. (2014) First principles simulations of the stability and structure of grain boundaries in Mg$_2$SiO$_4$ forsterite. *Physics and Chemistry of Minerals*, **41**, 163–171.

Gaidies, F., Milke, R., Heinrich, W. and Abart, R. (2017) Metamorphic mineral reactions: Porphyroblast, corona and symplectite growth. Pp. 469–540 in: *Mineral Reaction Kinetics: Microstructures, Textures and Chemical Compositions* (R. Abart and W. Heinrich, editors). EMU Notes in Mineralogy, **16**. European Mineralogical Union and the Mineralogical Society of Great Britain & Ireland, London.

Götze, L.C., Abart, R., Rybacki, E., Keller, L.M., Petrishcheva, E. and Dresen, G. (2010) Reaction rim growth in the system MgO-Al$_2$O$_3$-SiO$_2$ under uniaxial stress. *Mineralogy and Petrology*, **99**, 263–277.

Griffiths, T.A. (2015) *Solid micro-inclusions in meta-pegmatite garnet: their formation and re-equilibration due to host mineral deformation*. Ph.D. thesis, University of Vienna, Vienna, pages.

Griffiths, T.A., Habler, G. and Abart, R. (2016) Crystallographic orientation relationships in host–inclusion systems: new insights from large EBSD datasets *American Mineralogist*, **101**, 690–705.

Grimmer, H. and Nespolo, M. (2009) Geminography: the crystallography of twins. *Zeitschrift für Kristallographie - Crystalline Materials*, **221**, 28–50.

Habler, G., Spacek, P. and Abart, R. (2012) Microstructure and texture of enstatite-spinel-plagioclase symplectites: EBSD analyses of (ultra)fine-grained multi-phase aggregates. *European Microscopy Congress 2012*, Manchester, UK.

Hammer, J.E., Sharp, T.G. and Wessel, P. (2010) Heterogeneous nucleation and epitaxial crystal growth of magmatic minerals. *Geology*, **38**, 367–370.

Hayward, S.A. and Salje, E.K.H. (2000) Twin memory and twin amnesia in anorthoclase. *Mineralogical Magazine*, **64**, 195–200.

Heidelbach, F. and Terry, M.P. (2013) Inherited fabric in an omphacite symplectite: Reconstruction of plastic deformation under ultra-high pressure conditions. *Microscopy and Microanalysis*, **19**, 942–949.

Heidelbach, F., Wenk, H.R., Chen, S.R., Pospiech, J. and Wright, S.I. (1996) Orientation and misorientation characteristics of annealed, rolled and recrystallized copper. *Materials Science and Engineering: A*, **215**, 39–49.

Heidelbach, F., Kunze, K. and Wenk, H.-R. (2000) Texture analysis of a recrystallized quartzite using electron diffraction in the scanning electron microscope. *Journal of Structural Geology*, **22**, 91–104.

Howe, J.M. (1997) *Interfaces in Materials: Atomic Structure, Thermodynamics and Kinetics of Solid-Vapor, Solid-Liquid and Solid-Solid Interfaces*. John Wiley and Sons, Inc.

Hwang, S.L., Yui, T.F., Chu, H.T., Shen, P., Schertl, H.P., Zhang, R.Y. and Liou, J.G. (2007) On the origin of oriented rutile needles in garnet from UHP eclogites. *Journal of Metamorphic Geology*, **25**, 349−362.

Hwang, S.-L., Yui, T.-F., Chu, H.-T., Shen, P., Zhang, R.-Y. and Liou, J.G. (2011) An AEM study of garnet clinopyroxenite from the Sulu ultrahigh-pressure terrane: formation mechanisms of oriented ilmenite, spinel, magnetite, amphibole and garnet inclusions in clinopyroxene. *Contributions to Mineralogy and Petrology*, **161**, 901−920.

Hwang, S.L., Shen, P., Chu, H.T., Yui, T.F. and Iizuka, Y. (2013) A TEM study of the oriented orthopyroxene and forsterite inclusions in garnet from Otrøy garnet peridotite, WGR, Norway: new insights on crystallographic characteristics and growth energetics of exsolved pyroxene in relict majoritic garnet. *Journal of Metamorphic Geology*, **31**, 113−130.

Hwang, S.L., Shen, P., Chu, H.T., Yui, T.F. and Iizuka, Y. (2015) Origin of rutile needles in star garnet and implications for interpretation of inclusion textures in ultrahigh-pressure metamorphic rocks. *Journal of Metamorphic Geology*, **33**, 249−272.

Jacob, D. and Cordier, P. (2010) A precession electron diffraction study of α, β phases and Dauphiné twin in quartz. *Ultramicroscopy*, **110**, 1166−1177.

Jahn, S. and Sun, X.-S., (2017) Atomic-scale modelling of crystal defects, self-diffusion and deformation processes Pp. 215−253 in: *Mineral Reaction Kinetics: Microstructures, Textures and Chemical Compositions* (R. Abart and W. Heinrich, editors). EMU Notes in Mineralogy, **16**. European Mineralogical Union and the Mineralogical Society of Great Britain & Ireland, London.

Jeřábek, P., Abart, R., Rybacki, E. and Habler, G. (2014) Microstructure and texture evolution during growth of magnesio-aluminate spinel at corundum–periclase interfaces under uniaxial load: the effect of stress concentration on reaction progress. *American Journal of Science*, **314**, 940−965.

Keller, L.M., Abart, R., Wirth, R., Schmid, D.W. and Kunze, K. (2006) Enhanced mass transfer through short-circuit diffusion; growth of garnet reaction rims at eclogite facies conditions. *American Mineralogist*, **91**, 1024−1038.

Keller, L.M., Götze, L.C., Rybacki, E., Dresen, G. and Abart, R. (2010) Enhancement of solid-state reaction rates by non-hydrostatic stress effects on polycrystalline diffusion kinetics. *American Mineralogist*, **95**, 1399−1407.

Li, C., Griffiths, T., Pennycook, T., Mangler, C., Jeřábek, P., Meyer, J., Habler, G. and Abart, R. (2016) The structure of a propagating $MgAl_2O_4/MgO$ interface: Linked atomic- and μm-scale mechanisms of interface motion. *Philosophical Magazine and Philosophical Magazine Letters*, **96**, iss. 23.

Li, D., Nielsen, M.H., Lee, J.R.I., Frandsen, C., Banfield, J.F. and De, Y. (2012) Direction-specific interactions control crystal growth by oriented attachment. *Science* **336**, 1014−1018.

Lojkowski, W. and Fecht, H.-J. (2000) The structure of intercrystalline interfaces. *Progress in Materials Science*, **45**, 339−568.

Marquardt, K., Rohrer, G.S., Morales, L., Rybacki, E., Marquardt, H. and Lin, B. (2015) The most frequent interfaces in olivine aggregates: the GBCD and its importance for grain boundary related processes. *Contributions to Mineralogy and Petrology*, **170**, 1−17.

Mason, J.K. and Schuh, C.A. (2009) Representations of texture. Pp. 35−51 in: *Electron Backscatter Diffraction in Materials Science (Second Edition)* (A.J. Schwartz, M. Kumar and B.L. Adams, editors). Springer, Berlin.

Maurer, R. (1987) Improved technique for the determination of low energy boundaries by the rotating-sphere-on-a-plate method: Results for grain boundaries in the Cu/Ni system. *Acta Metallurgica*, **35**, 2557−2565.

McNamara, D.D., Wheeler, J., Pearce, M. and Prior, D.J. (2012) Fabrics produced mimetically during static metamorphism in retrogressed eclogites from the Zermatt-Saas zone, Western Italian Alps. *Journal of Structural Geology*, **44**, 167−178.

Mongkoltip, P. and Ashworth, J.R. (1983) Exsolution of ilmenite and rutile in hornblend. *American Mineralogist*, **68**, 143−155.

Nespolo, M. (2015) Tips and traps on crystal twinning: how to fully describe your twin. *Crystal Research and Technology*, **50**, 362−371.

Nespolo, M. and Ferraris, G. (2004) The oriented attachment mechanism in the formation of twins – a survey.

European Journal of Mineralogy, **16**, 401–406

Obata, M. (2011) Kelyphite and symplectite: Textural and mineralogical diversities and universality, and a new dynamic view of their structural formation. Pp. 94–122 in: *New Frontiers in Tectonic Research – General Problems, Sedimentary Basins and Island Arcs* (E.V. Sharkov, editor). InTech Publishers, Rijeka, Croatia.

Obata, M. and Ozawa, K. (2011) Topotaxic relationships between spinel and pyroxene in kelyphite after garnet in mantle-derived peridotites and their implications to reaction mechanism and kinetics. *Mineralogy and Petrology*, **101**, 217–224.

Odashima, N., Morishita, T., Ozawa, K., Nagahara, H., Tsuchiyama, A. and Nagashima, R. (2007) Formation and deformation mechanisms of pyroxene-spinel symplectite in an ascending mantle, the Horoman peridotite complex, Japan: An EBSD (electron backscatter diffraction) study. *Journal of Mineralogical and Petrological Sciences*, **103**, 1–15.

Okamoto, A. and Sekine, K. (2011) Textures of syntaxial quartz veins synthesized by hydrothermal experiments. *Journal of Structural Geology*, **33**, 1764–1775.

Olmsted, D.L., Foiles, S.M. and Holm, E.A. (2009) Survey of computed grain boundary properties in face-centered cubic metals: I. Grain boundary energy. *Acta Materialia*, **57**, 3694–3703.

Passchier, C.W. and Trouw, R.A.J. (2005) *Microtectonics*. Springer-Verlag, Berlin, Heidelberg, New York, 366 pp.

Penn, R.L. and Banfield, J.F. (1998) Imperfect oriented attachment: dislocation generation in defect-free nanocrystals. *Science* **281**, 969–971.

Penn, R.L. and Banfield, J.F. (1999) Morphology development and crystal growth in nanocrystalline aggregates under hydrothermal conditions: insights from titania. *Geochimica et Cosmochimica Acta*, **63**, 1549–1557.

Penn, R.L. and Soltis, J.A. (2014) Characterizing crystal growth by oriented aggregation. *CrystEngComm*, **16**, 1409–1418.

Pennycook, S.J. and Nellist, P.D. (2011) *Imaging and Analysis*. Springer-Verlag, New York. 762 pp.

Petrishcheva, E. and Abart, R. (2017) Interfaces Pp. 295–345 in: *Mineral Reaction Kinetics: Microstructures, Textures and Chemical Compositions* (R. Abart and W. Heinrich, editors). EMU Notes in Mineralogy, **16**. European Mineralogical Union and the Mineralogical Society of Great Britain & Ireland, London.

Proyer, A., Habler, G., Abart, R., Wirth, R., Krenn, K. and Hoinkes, G. (2013) TiO$_2$ exsolution from garnet by open-system precipitation: evidence from crystallographic and shape preferred orientation of rutile inclusions. *Contributions to Mineralogy and Petrology*, **166**, 211–234.

Ram, F., Zaefferer, S., Japel, T. and Raabe, D. (2015) Error analysis of the crystal orientations and disorientations obtained by the classical electron backscatter diffraction technique. *Journal of Applied Crystallography*, **48**, 797–813.

Ramberg, H. (1962) Intergranular precipitation of albite formed by unmixing of alkali feldspar. *Neues Jahrbuch für Mineralogie (Abhandlungen)*, **98**, 14–34.

Read, W.T. and Shockley, W. (1950) Dislocation models of crystal grain boundaries. *Physical Review*, **78**, 275–289.

Robinson, P., Ross, M., Nord, G.L.J., Smyth, J.R. and Jaffe, H.W. (1977) Exsolution lamellae in augite and pigeonite: fossil indicators of lattice parameters at high temperature and pressure. *American Mineralogist*, **62**, 857–873.

Rohrer, G.S. (2011) Grain boundary energy anisotropy: a review. *Journal of Materials Science*, **46**, 5881–5895.

Royer, M.L. (1928) Recherches expérimentales sur l'épitaxie ou orientation mutuelle de cristaux d'espèces différentes. *Bulletin de la Société Française de Minéralogie*, **51**, 7–159.

Rybacki, E., Evans, B., Janssen, C., Wirth, R. and Dresen, G. (2013) Influence of stress, temperature, and strain on calcite twins constrained by deformation experiments. *Tectonophysics*, **601**, 20–36.

Sánchez-Munoz, L., García-Guinea, J., Zagorsky, V.Y., Juwono, T., Modreski, P.J., Cremades, A., Van Tendeloo, G. and de Moura, O.J.M. (2012) The evolution of twin patterns in perthitic K-feldspar from granitic pegmatites. *The Canadian Mineralogist*, **50**, 989–1024.

Saylor, D.M., Morawiec, A. and Rohrer, G.S. (2003a) Distribution of grain boundaries in magnesia as a function of five macroscopic parameters. *Acta Materialia*, **51**, 3663–3674.

Saylor, D.M., Morawiec, A. and Rohrer, G.S. (2003b) The relative free energies of grain boundaries in magnesia as a function of five macroscopic parameters. *Acta Materialia*, **51**, 3675–3686.

Schlom, D.G., Chen, L.-Q., Pan, X., Schmehl, A. and Zurbuchen, M.A. (2008) A thin film approach to engineering functionality into oxides. *Journal of the American Ceramic Society*, **91**, 2429–2454.

Shannon, R.D. and Rossi, R.C. (1964) Definition of topotaxy. *Nature*, **202**, 1000–1001.

Špaček, P., Ackerman, L., Habler, G., Abart, R. and Ulrych, J. (2013) Garnet breakdown, symplectite formation and melting in basanite-hosted peridotite xenoliths from Zinst (Bavaria, Bohemian Massif). *Journal of Petrology*, **54**, 1691–1723.

Spanos, G. (1989) *The bainite and proeutectoid cementite reactions in hypereutectoid Fe-C-Mn alloys*. Ph.D. thesis, Carnegie-Mellon University, Pittsburgh, Pennsylvania, USA.

Spiess, R., Peruzzo, L., Prior, D.J. and Wheeler, J. (2001) Development of garnet porphyroblasts by multiple nucleation, coalescence and boundary misorientation-driven rotations. *Journal of Metamorphic Geology*, **19**, 269–290.

Spry, A.H. (1969) *Metamorphic Textures*. Pergamon Press, Oxford, New York, Toronto, Sydney, 358 pp.

Straumal, B.B., Kogtenkova, O.A., Gornakova, A.S., Sursaeva, V.G. and Baretzky, B. (2016) Review: grain boundary faceting–roughening phenomena. *Journal of Materials Science*, **51**, 382–404.

Sutton, A.P. and Balluffi, R.W. (1987) Overview no. 61 On geometric criteria for low interfacial energy. *Acta Metallurgica*, **35**, 2177–2201.

Sutton, A.P. and Balluffi, R.W. (1995) *Interfaces in Crystalline Materials*. Clarendon Press, Oxford, UK, 862 pp.

Teng, H.H. (2013) How ions and molecules organize to form crystals. *Elements*, **9**, 189–194.

Tochigi, E., Zepeda-Alarcon, E., Wenk, H.-R. and Minor, A.M. (2014) In situ TEM observations of plastic deformation in quartz crystals. *Physics and Chemistry of Minerals*, **41**, 757–765.

Turner, F.J. (1953) Nature and dynamic interpretation of deformation lamellae in calcite of three marbles. *American Journal of Science*, **251**, 276–298.

Van Der Merwe, J.H. (1963a) Crystal Interfaces. Part I. Semi-Infinite Crystals. *Journal of Applied Physics*, **34**, 117–122.

Van Der Merwe, J.H. (1963b) Crystal interfaces. Part II. Finite overgrowths. *Journal of Applied Physics*, **34**, 123–127.

Verma, A.K. and Karki, B.B. (2010) First-principles simulations of MgO tilt grain boundary: Structure and vacancy formation at high pressure. *American Mineralogist*, **95**, 1035–1041.

Vernon, R.H. (2004) *A Practical Guide to Rock Microstructure*. Cambridge University Press, Cambridge, UK, 594 pp.

Voll, G. (1960) New work on petrofabrics. *Liverpool and Manchester Geological Journal*, **2**, 503–567.

Wassermann, G. and Grewen, J. (1962) *Texturen metallischer Werkstoffe*. Springer, Berlin.

Wenk, H.-R., Barton, N., Bortolotti, M., Vogel, S.C., Voltolini, M., Lloyd, G.E. and Gonzalez, G.B. (2009) Dauphiné twinning and texture memory in polycrystalline quartz. Part 3: texture memory during phase transformation. *Physics and Chemistry of Minerals*, **36**, 567–583.

Wenk, H.-R., Chen, K. and Smith, R. (2011a) Morphology and microstructure of magnetite and ilmenite inclusions in plagioclase from Adirondack anorthositic gneiss. *American Mineralogist*, **96**, 1316–1324.

Wenk, H.-R., Janssen, C., Kenkmann, T. and Dresen, G. (2011b) Mechanical twinning in quartz: Shock experiments, impact, pseudotachylites and fault breccias. *Tectonophysics*, **510**, 69–79.

Wheeler, J., Prior, D., Jiang, Z., Spiess, R. and Trimby, P. (2001) The petrological significance of misorientations between grains. *Contributions to Mineralogy and Petrology*, **141**, 109–124.

Wirth, R. and Voll, G. (1987) Cellular intergrowth between quartz and sodium-rich plagioclase (myrmekite) – an analogue of discontinuous precipitation in metal alloys. *Journal of Materials Science*, **22**, 1913–1918.

Xu, H., Zhang, J., Zong, K. and Liu, L. (2015) Quartz exsolution topotaxy in clinopyroxene from the UHP eclogite of Weihai, China. *Lithos*, **226**, 17–30.

Yan, Y., Chisholm, M.F., Duscher, G., Maiti, A., Pennycook, S.J. and Pantelides, S.T. (1998) Impurity-induced structural transformation of a MgO grain boundary. *Physical Review Letters*, **81**, 3675–3678.

Zaefferer, S. and Habler, G. (2017) Scanning electron microscopy and electron backscatter diffraction. Pp. 37–95 in: *Mineral Reaction Kinetics: Microstructures, Textures and Chemical Compositions* (R. Abart and W. Heinrich, editors). EMU Notes in Mineralogy, **16**. European Mineralogical Union and the Mineralogical Society of Great Britain & Ireland, London.

Zaefferer, S. and Wright, S.I. (2009) Three-dimensional orientation microscopy by serial sectioning and EBSD-based orientation mapping in a FIB-SEM. Pp. 109–122 in: *Electron Backscatter Diffraction in Materials Science* (A.J. Schwartz, M. Kumar, B.L. Adams and D.P. Field, editors). Springer Science+Business Media, New York.

Zhang, J.F., Xu, H.J., Liu, Q., Green, H.W. and Dobrzhinetskaya, L.F. (2011) Pyroxene exsolution topotaxy in majoritic garnet from 250 to 300 km depth. *Journal of Metamorphic Geology*, **29**, 741–751.

Zhang, W.Z. and Weatherly, G.C. (2005) On the crystallography of precipitation. *Progress in Materials Science*, **50**, 181–292.

EMU Notes in Mineralogy, Vol. 16 (2017), Chapter 16, 587–615

Reaction-induced fracturing:
Chemical-mechanical feedback

OLE IVAR ULVEN[1],* and ANDERS MALTHE-SØRENSSEN[2]

*Physics of Geological Processes, Department of Physics, University of Oslo,
P.O. Box 1048 Blindern, 0316 Oslo, Norway,
e-mail: [1]o.i.ulven@fys.uio.no; [2]anders.malthe-sorenssen@fys.uio.no
* Corresponding author*

Volume-changing mineralization reactions induced by fluids play an important role in the large-scale dynamics of the Earth's crust, effectively determining *e.g.* the rates of chemical weathering on the surface of the Earth and the density of the oceanic crust through serpentinization reactions. In addition, permanent CO_2 storage through carbonation of ultramafic rock also depends on mechano-chemical processes – processes where the chemical reactions are closely coupled to fluid transport and deformation. This coupling between transport, reaction, deformation and fracture may have a first-order impact on the effective reaction rate – the coupling may significantly accelerate the overall reaction.

This chapter addresses our understanding of the chemical-mechanical feedback during reaction-induced fracturing and introduces theoretical and computer-modelling methods used to address dynamics in such coupled systems. Basic geological scenarios where such a feedback is expected to be important are introduced, and the underlying chemical processes are described. Then we provide a brief review of relevant literature that discusses reaction-induced fracturing.

We introduce a simplified one-dimensional model that can be used to understand both volume-decreasing and volume-increasing reactions, and which also produces useful quantitative predictions for the motion of the reaction front. We further address volume-changing reactions for both a pure diffusion system, which is simple to understand and describe, and for a diffusion-reaction system. Then, we introduce an analytical model for the onset of fracturing in diffusion-reaction-fracturing models, and use this theory to validate a more general numerical model, which can also predict fracture propagation beyond fracture initiation.

We introduce a theoretical description of systems with and without feedback from fracturing, and models where the effects of fluid transport in fractures can be tuned. Relevant dimensional quantities are introduced and generalized results for mechano-chemical processes are discussed in terms of the fundamental dimensionless numbers characterizing the system.

1. Introduction

Volume-changing processes occur in a variety of settings, some of which are highly relevant for geological applications. Shrinkage is observed commonly in the drying of thin films (Skjeltorp and Meakin, 1988) and in drying mud, but is less common in major geological processes. Notable exceptions are: the contraction of rock during cooling, which may lead to columnar jointing (Peck and Minakami, 1968); syneresis, which has been suggested as a mechanism for the formation of large-scale fault systems in sedimentary basins (Cartwright *et al.*, 2003); and the density increase associated with the eclogitization of crustal rocks (Austrheim, 1987; Jamtveit *et al.*, 2000). Several volume-increasing processes are important in the large-scale dynamics of the Earth's crust. Weathering breaks down rock at the surface of the Earth, and participates in the large-scale cycling of elements between crust, ocean and mantle (Martin and Meybeck, 1979; Schmidt and Poli, 2014). Serpentinization of ultramafic rock affects the density of oceanic crust, and the water in serpentine contributes significantly to the total amount of water that enters subduction zones globally (Hacker, 2008; Schmidt and Poli, 2014). In addition, carbonation of ultramafic rock might provide a safe, long-term storage solution for anthropogenically produced CO_2 (Kelemen *et al.*, 2011).

The term 'reaction-induced fracturing' most accurately denotes processes in which the volume change is caused by a chemical reaction and leads to fracture formation. Common examples are the hydration or carbonation of rock. These processes are closely coupled to fluid transport and deformation: hydration requires addition of water, and CO_2 is required for carbonation. For low-permeability rocks, the transport of the mobile reactants into the intact rock is slow, and additional reactant transport in fractures might enhance significantly the overall rate of the process compared to the situation without fractures. This coupling between transport, reaction, deformation and fracturing may have a first-order impact on the overall reaction rates. For example, this coupling might effectively determine the rates of chemical weathering on the surface of the Earth and the density of the oceanic crust through serpentinization.

In this chapter we focus on chemical processes that benefit from the enhanced reactant transport in fractures and, when relevant, links to processes less dependent on transport in fractures are included. We present our current understanding of the coupling between mechanical deformation and chemical processes during reaction-induced fracturing, and introduce theoretical and computer-modelling methods used to address the dynamics of such coupled systems. Firstly, different types of volume-changing processes with relevant field examples are introduced, the underlying mechanisms are discussed, and similarities and differences are identified. We proceed by presenting a set of numerical tools for studying reaction-induced fracturing. Further, we look into the relatively simple volume-reduction system, first with pure diffusion of mobile reactants through the solid, then with additional transport of mobile reactants in fractures. Finally we turn our attention to volume expansion, and go through the behaviour of systems with diffusion only, and with a variable effect of transport in a fracture. The main controlling parameters are discussed, and different modelling strategies are introduced, including both analytical and numerical methods.

We show throughout that fractures may provide a significant feedback in certain cases, and that the magnitude of the volume change and the sharpness of the reaction front between the initial and final material are important parameters in understanding the system.

2. Mechanisms of volume-changing processes

Reaction-induced fracturing may occur when a reaction causes changes in the stress state of the rock or changes in the material strength of the rock. Stress changes are usually related to a change in the sum of solid volume and pore volume. A dissolution process does not necessarily directly cause any large-scale stress because the volume of solid removed is usually replaced by pore space and therefore does not lead to a change in the total volume of the bulk material. However, the local stress around the pores generated may be significantly different from the original stress in the bulk material, and these changes may induce fracturing of the rock even without a change in the total volume. For example, in the formation of karst terrains, a large fraction of the initial material is dissolved, but unless the rock collapses due to gravity, a skeleton of the initial rock remains. Stylolite formation is an example of volume reduction caused by dissolution of minerals. However, pressure solution due to an externally imposed compressive stress drives the mineral dissolution, and relieves the external compressive stress instead of causing internal stress and fracturing in the material. Thus, even though many geochemical processes dissolve minerals, this does not necessarily induce fracturing. Nevertheless, external forces or pre-existing stress concentrations within the rock may cause deformation when material is removed, and ultimately lead to fracture formation. Similarly, precipitation of minerals in pores in a host rock may fill porosity and increase the solid volume, but at the same time reduce the pore space. Such a process will not necessarily cause stress changes, unless the growing minerals exert a force on the host rock. This justifies a closer look at the mechanisms behind volume change.

Volume reduction may cause stress as the material contracts. During drying of particulate matter such as mud, the capillary pressure between individual particles makes the material contract when water is removed (Coussy, 2010), which may lead to the formation of cracks (Fig. 1). Thermal contraction, the volume reduction during cooling of rock, may result in cracking and the formation of columnar joints. These two systems are expected to behave similarly, with one notable exception: transport of water in fractures might significantly enhance the drying rate, whereas the heat transport in a fracture is negligible, unless there is a significant fluid circulation in the fracture system. Another example of a shrinking system is the drying of thin films (Skjeltorp and Meakin, 1988). Thin films are, however, less relevant for bulk mineral reactions and fall outside the scope of this text. An example of volume reduction in a single crystal is related to the substitutional exchange of potassium by sodium in alkali feldspar during cation exchange with an NaCl-KCl melt (Neusser *et al.*, 2012; Scheidl *et al.*, 2014). The unit cell of alkali feldspar contracts with a shift from more K-rich towards more Na-rich compositions. This causes grain-internal stress in partially exchanged alkali feldspar, which may eventually lead to fracturing. The transport of cations into the interior of the crystal and the overall exchange reaction are accelerated by the infiltration of melt along the newly formed cracks. One geological process of significant importance that includes volume reduction is the dehydration of serpentine to form olivine during subduction. When the denser olivine is formed, pores form to maintain the total volume. However, the permeability of the serpentine is usually low,

Figure 1. Fractures formed in a layer of mud during drying. This phenomenon is commonly seen following a rainy day. Coin for scale.

and at pressures below 2–4 GPa, the volume of water released from the serpentine is higher than the volume of the pores formed (Jung *et al.*, 2004). This may lead to fracture formation caused by fluid pressure when the fluid is not allowed to escape sufficiently fast, and typical drainage patterns are often formed (Zack and John, 2007). This phenomenon is still not fully understood, and will not be discussed further in the current text.

Volume expansion may occur when a rock is subject to water or other chemical reactants, and various physical and chemical processes might lead to expansion. A characteristic example of volume increase is related to weathering, when secondary minerals precipitate in pores at the same time as some of the primary minerals are hydrated, oxidized or carbonated (Fig. 2a), thus leading to a local expansion of the material. Large-scale geochemical changes in rocks can be induced by metasomatism, which may also take place by reaction-enhanced fracturing, *e.g.* during albitization (Fig. 2b). Another example occurs during the hydration of olivine which leads to the formation of serpentine by a dissolution-reprecipitation process (Putnis, 2002). The olivine is thereby dissolved at the mineral surface, and serpentine precipitates in the vicinity of the surface being dissolved through formation of a temporary amorphous serpentine phase (Plümper *et al.*, 2012). There are two main differences between this small-scale process and large-scale weathering. Firstly, on outcrop scale, reactions may proceed gradually. Thus, a transition zone of partly reacted rock often exists between the unreacted rock and the completely reacted rock. On the other hand, dissolution-reprecipitation on the grain scale is localized essentially to a micrometre scale at the mineral surface, without an extended transition zone (Kuleci *et al.*, 2016). Secondly,

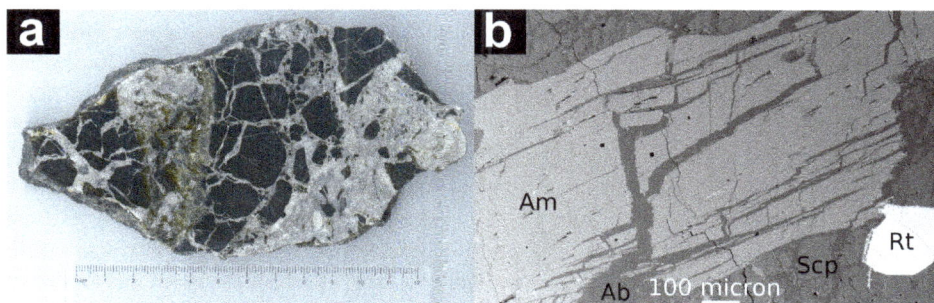

Figure 2. (a) Carbonation of a partly serpentinized peridotite (dark grey). The carbonate (light grey) is magnesite and dolomite, and the green mineral is serpentine. Sample from Feragen, Eastern Norway. (b) Amphibole (Am) broken apart by growth of albite (Ab). The surrounding scapolite (Scp) is also fractured, with albite as fracture infill. The bright grain is rutile (Rt). Sample from Kragerø, southeastern Norway. Both pictures courtesy of H. Austrheim.

during weathering, the rock is often intact through the partly reacted transition zone, whereas during dissolution-reprecipitation the dissolved and precipitated minerals are at least partly disconnected mechanically. This means that all stress components are transferred between the unreacted and reacted rock in the first case, while only compressive normal stress can be expected to be fully transferred between the dissolved and precipitated mineral. It is hard to estimate to what extent tensile stress and shear stress can be transferred in a dissolution-reprecipitation process, as little is known about the mechanical properties of the interface between the dissolved and precipitated minerals. Displacement at the interface might at least partly dissipate tensile and shear stresses and reduce the extent of fracture formation.

In most cases volume expansion depends critically on the supply of chemical reactants. One notable exception is metamictization, in which minerals with large contents of radioactive elements such as U and Th in zircon expand due to damage caused by radiation emitted by the radioactive decay of these elements (Crocombette and Ghaleb, 2001). Another example is represented by polymorphic phase transformations, such as when coesite is converted to quartz due to a pressure reduction or temperature increase (Boyd and England, 1960). This leads to volume expansion without adding any chemical reactants, as the density of quartz is lower than that of coesite. However, the presence of water has been shown to enhance the rate of transformation (Mosenfelder and Bohlen, 1997), which suggests a positive rate feedback if fractures form and enhance the supply of fluid. The precipitation of minerals can lead to armouring of surfaces or clogging of the pore space preventing further fluid transport to reactive sites. If the supply of chemical reactants is necessary for the reaction to proceed, this may stop the reaction, and potentially is a limiting factor for reaction-induced fracturing.

Whether the rock breaks or is clogged depends on a number of parameters. Firstly, when a mineral grows inside a host mineral, the response might be dissolution of the host phase, creep of the host phase or fracturing (Fletcher and Merino, 2001). The latter occurs when mineral growth is faster than dissolution and creep, and stress builds up. Additionally, the wetting behaviour of the mineral surfaces is important (Røyne and Jamtveit, 2015). Fluids are present at the interface between certain minerals, and a significant pressure is required to squeeze the fluid out. This pressure is commonly called the disjoining pressure, and has to be large enough to prevent removal of the fluid from the interface before sufficient stress builds up, and the rock fails. If the fluid is removed before the rock fails, there is no longer any fluid present to transport fresh reactant to the mineral surface. The wetting behaviour of surfaces is still not fully understood. King et al. (2010) reported formation of a dense silica layer during carbonation of olivine, and concluded that the silica layer would ultimately prevent transport of reactant to the fresh olivine and stop the reaction. This surface armouring might limit reaction-induced fracturing when low-permeability minerals form at the surface of unreacted grains. Finally, the crystallization pressure exerted by the growing mineral has to be high enough to break the rock. This depends on the supersaturation in the fluid with respect to the growing mineral. Kelemen and Hirth (2012) estimated potential crystallization pressures of several hundred MPa during carbonation or serpentinization of ultramafic rock, which would be sufficient to break any crustal rock.

Fractures due to volume expansion have been reported in a series of papers. Røyne et al. (2008) studied weathering of dolerites, and concluded that volume-expanding reactions led to both spalling of spheroidal surface layers and to large-scale hierarchical fracturing of the rock (Fig. 3a). Jamtveit et al. (2011) studied weathering of andesite, and found that growth of calcite and ferrihydrite in the largest pores led to fracturing, whereas smaller pores remained open and allowed fluid transport (Fig. 3b). Neusser et al. (2012) reported volume expansion in feldspar due to replacement of sodium cations by the larger potassium cations, thus causing the feldspar unit cell to expand. They observed both spalling of the reacted layer and domain division of the unreacted domain. Jamtveit et al. (2009) studied replacement of leucite by analcime, and reported similarly both spalling of the outer layer and domain-dividing fractures. A different system was studied by Preston and White (1934), who during heating of clay spheres, observed both spalling of the outer layer and fractures dividing the centre of the sphere.

Serpentinization provides an example of a dissolution-reprecipitation process, where the primary minerals of the ultramafic rock are dissolved, and serpentine precipitates. This includes a volume expansion of ~50% depending on the initial composition of the ultramafic rock. Jamtveit et al. (2008) reported fractures in a partly serpentinized troctolite, and presented a model in which the fractures were caused by the expansion associated with the serpentinization of olivine grains. They concluded that this provided fluid pathways into the rock even when subject to external compressive stress, and allowed the process to progress further. However, dissolution-reprecipitation partly decouples the materials mechanically, and thus, shear stress is not fully transferred between the two minerals. Plümper et al. (2012) suggested a mechanism that can explain the fractures in the olivine, with formation of etch pits on olivine surfaces, followed by precipitation of serpentine in the pits, thus causing further growth of fractures into the olivine. Dunkel and

Figure 3. (a) Spheroidal weathering of a low-porosity dolerite sill intrusion, Karoo, South Africa. Almost completely weathered saprolite is seen between remaining unaltered corestones (Røyne *et al.*, 2008). (b) Weathering of a high-porosity andesite intrusion, Argentina (Jamtveit *et al.*, 2011). The corestones are partly reacted throughout, with an extent of weathering that decreases gradually towards the corestone centres. Notice that the innermost rings of the corestones are Liesegang bands produced by the process, and not fractures. Both photos courtesy of B. Jamtveit.

Putnis (2014) observed the same pitting and cracking behaviour when replacing scolecite by tobermorite. Kuleci *et al.* (2016) reported a similar mechanism for fracture formation during hydration of periclase to form brucite.

Common to all volume-changing processes is that for fractures to form, the process has to be fast enough to avoid viscous relaxation at the current *P-T* conditions. If that is not the case, the volume change might be accommodated by viscous deformation, and prevent fracture formation. Throughout this chapter, the volume change is assumed to occur sufficiently rapidly.

3. Numerical modelling of volume-changing reactions

Models of volume-changing processes require the use of two individual solvers, one that describes the local volume change, and one that describes the mechanical deformation caused by the volume change. The volume change is usually caused by a chemical reaction, which is described by a model for reaction and fluid transport, either by diffusion or advection. The mechanical process is described by a linear elasticity-based model, in some cases with a model for fracture formation based on any type of fracture criteria from texts in structural geology. The solvers are coupled in two ways. Firstly, the volume change causes local strain in the mechanical model, and secondly, fractures in the mechanical model affect transport properties in the volume-change model. In this section, we go through some of the strategies utilized for studying such coupled processes.

3.1. Types of chemical processes

Most works on reaction-induced fracturing consider one of two general types of process. Firstly, Yakobson (1991) introduced a model for the thermal decomposition of solids, in which a solid material loses a volatile component. The simplified mechanism is A_{solid} $B_{solid} + W$, where W is a mobile component that is allowed to leave the system by some transport mechanism. The local volume of the material was assumed to decrease linearly until reaching a defined minimum volume when no W was left. The initial decomposition of the material was not treated in detail, thus further simplifying the model to a model for transport of the mobile component. Similar models have been used by a number of authors (Boeck *et al.*, 1999; Jamtveit *et al.*, 2000; Malthe-Sørenssen *et al.*, 2006), mainly for studies of volume-decreasing processes. Okamoto and Shimizu (2015) extended this type of dehydration process with a decomposition rate proportional to a reduction in the local pressure of the mobile phase.

The other type of model treats a simplified chemical reaction of the type $A_{solid} + W$ B_{solid}, in which an initial solid reacts with a mobile component and forms a final solid with different volume. The mobile component is allowed to enter the system by a transport mechanism similar to the mechanism in the first model. The rate of formation of B_{solid} is usually determined by using a second-order rate equation, and the volume change is locally a linear function of the concentration of the reaction product, from an initial volume without any B_{solid} to the final volume when all A_{solid} is converted to B_{solid}. This type of model has been used by a number of authors (Fletcher *et al.*, 2006; Jamtveit *et al.*, 2008; Røyne *et al.*, 2008; Rudge *et al.*, 2010; Ulven *et al.*, 2014a,b), mainly for studies of volume-expanding processes. Okamoto and Shimizu (2015) used a similar reaction type for studies of hydration, but computed the reaction rate as a function of the pressure of the fluid phase.

3.2. Continuum-based models for transport and deformation

Continuum-based models utilize direct solutions of the differential equations that
control linear elasticity and reaction-diffusion, often with some simplifying
assumptions. The system of equations that makes up the model is often discretized
and solved using standard numerical techniques such as the finite difference method
(FDM) or the finite element method (FEM).

The first model of this type was a one-dimensional model introduced by Yakobson
(1991), based on evaporation of a mobile component w following thermal
decomposition of a solid. The decomposition leads to an initially constant
concentration w_i of the mobile component, which is allowed to diffuse through the
intact solid in the x direction with a diffusion constant D. The mobile component can
leave the solid at the domain surface and at fracture surfaces with a rate $k(w_b - w_0)/w_i$,
where k is an evaporation constant, w_b is the concentration of the mobile component at
the surface and w_0 is concentration of the mobile component outside the considered
domain (Fig. 4). Fractures are assumed to transport the evaporated component out of
the domain instantaneously. The loss of the mobile component w is assumed to lead to a
stress according to

$$\sigma(x) = \beta E \left(1 - \frac{w(x)}{w_i} \right) \tag{1}$$

where x is the spatial dimension, σ is the normal stress component in the direction
perpendicular to x, E is Young's modulus, and β is a volume shrinkage parameter
related to the volume change $\Delta V/V$ through $\beta = \frac{\Delta V/V}{3(1-v)}$, where v is Poisson's ratio.

In this simple model, interaction between fractures is ignored, and only the
dominating stress component, which is the normal stress in the direction perpendicular
to x, is considered. A crack extending a length L beyond the front was assumed to
propagate if the stress-intensity factor at its tip exceeded a critical stress intensity
factor, K_c, where the stress intensity factor was calculated from the stress $\sigma(x)$
according to

$$K = 2 \left(\frac{L}{\pi} \right)^{1/2} \int_0^L \frac{\sigma(x)dx}{(L^2 - x^2)^{0.5}} = K_c \tag{2}$$

Notice that in the method presented by Yakobson (1991), the stress is determined
by the concentration $w(x)$, found by diffusion of a mobile component. However,
concentrations found using other reaction-diffusion models can also be applied
directly in the stress model (equation 1). This was used later by other authors. First,
Fletcher *et al.* (2006) developed a 1D model for weathering, where the extent of
weathering took the role of the concentration. The volume change was translated
directly to an inelastic strain, and the fracture criterion was based on strain energy.
Later, Rudge *et al.* (2010) developed a 1D model for serpentinization and
carbonation of ultramafic rock based on the process type $sA_{solid} + rW \rightarrow B_{solid}$,
where r and s are stoichiometric coefficients. This led to a reaction-diffusion

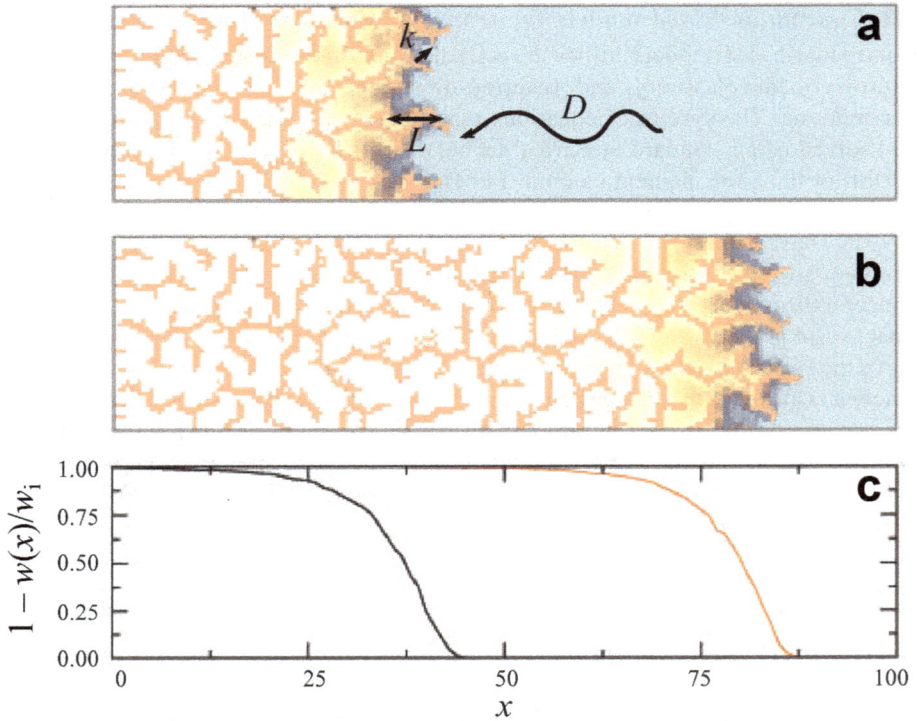

Figure 4. Illustration of the Yakobson (1991) model showing a network of cracks (red) propagating into a solid with an initial concentration w_i of a volatile component. As the reaction progresses, the concentration $w(x)$ of the volatile component sinks and the solid fractures, resulting in a fracture pattern (red) extending from the outer boundary into the solid. The volatile component diffuses with a diffusion constant D in the solid, and propagates into the fractures extending a length L from the fracture front into the solid with a transfer coefficient k. (c) Shows the concentration profiles for the two states of propagation shown in (a) and (b).

equation system given by

$$\frac{\partial w}{\partial t} = D\frac{\partial^2 w}{\partial x^2} - \frac{r\kappa}{s}\frac{w}{w_0}a \qquad (3)$$

$$\frac{\partial a}{\partial t} = -\kappa\frac{w}{w_0}a \qquad (4)$$

$$\frac{\partial b}{\partial t} = \frac{\kappa}{s}\frac{w}{w_0}a \qquad (5)$$

where a, b and w are the concentrations of the initial mineral, the reaction product and the mobile reactant, respectively, t denotes time, D is the effective diffusion constant, x is the spatial coordinate, κ is the rate constant of the chemical reaction and w_0 is the concentration of the mobile reactant outside the domain. The solution $b(x)$ was inserted as the concentration in equation 1, and the stress intensity factor was used to determine fracture formation.

One limitation of the simplified models above is that they take into account only one component of stress or strain, not the full tensor. Ulven *et al.* (2014b) developed a model that solved for the full stress tensor, using the reaction-diffusion system introduced by Rudge *et al.* (2010). In the simplest case, a cylindrical geometry with axial symmetry was assumed. That allowed recasting the reaction-diffusion model defined in equations 3−5 in one dimension for the radial direction r as

$$\frac{\partial w^*}{\partial t^*} = \frac{1}{r^*}\frac{\partial}{\partial r^*}\left(r^*\frac{\partial w^*}{\partial r^*}\right) - \frac{\Gamma}{\Theta}w^*a^* \tag{6}$$

$$\frac{\partial a^*}{\partial t^*} = -\Gamma w^*a^* \tag{7}$$

$$\frac{\partial b^*}{\partial t^*} = \Gamma w^*a^* \tag{8}$$

where $\Gamma = (R^2\kappa)/D$ and $\Theta = sw_0/\rho a_0$ are dimensionless parameters, R is the domain radius, ρ is a stoichiometric coefficient and $*$ indicates dimensionless variables. The equations of linear elasticity similarly simplify to a 1D differential equation that relates the displacement u at radial positions r to the amount of volume change:

$$\frac{\partial^2 u}{\partial r^2} + \frac{1}{r}\frac{\partial u}{\partial r} - \frac{1}{r^2}u = \frac{1+v}{(1-v)}\beta\frac{\partial b^*}{\partial r} \tag{9}$$

Notice that the right hand side term is similar to equation 1. The system of equations was solved using FDM, and stress-strain relations were then used to find the stress components. Two different stress criteria were used. The first was the Coulomb criterion, given by

$$\sigma_{CC} = [(1+\mu_i^2) - \mu_i]^2\sigma_1 - \sigma_3 - C_u > 0 \tag{10}$$

where μ_i is the internal friction coefficient, and σ_1 and σ_3 are the major and minor principal stress, respectively, and C_u is the uniaxial compressive strength. σ_{CC} is, in the following, referred to as the Coulomb stress. This determines if a shear fracture forms. The second criterion was the tensile criterion $\sigma_1 > T_u$, in which T_u is the uniaxial tensile strength, which determines whether a tensile fracture forms. This allowed insight into the actual fracture mode at the time of fracture formation. A 2D FEM was also introduced by Ulven *et al.* (2014b). In this model, the equations of linear elasticity and the reaction-diffusion equations introduced above were solved in 2D, without the simplifying assumption of axial symmetry.

The downside of these models for the full-stress tensor is that they are strictly valid only until the first fracture forms. In 1D, a fracture breaks the axial symmetry, and the

FEM is unable to describe the displacement discontinuity introduced by a fracture. Methods exist for describing fractures in FEM, but a simpler solution using a discrete element model is introduced in the next section.

3.3. Discrete element model for deformation and fracturing

The discrete element model (DEM) was first introduced in its simplest form by Cundall and Strack (1979). Fractures are introduced easily in the model, which makes DEM widely used for studying large strain and fracture formation in linear elastic solids. The solid is modelled as consisting of a collection of circular (2D) or spherical (3D) particles that fill the computational domain. The particles are connected to all surrounding particles by bonds with elastic properties. This gives a network where forces are transmitted, and it can be shown in simple cases that this model reproduces linear elasticity (Flekkøy et al., 2002). In the first version by Cundall and Strack (1979), the bonds were springs, with a force given by the deviation from an equilibrium separation. Later versions (Herrmann and Roux, 1990) have used beams that transmit both normal and shear forces between the particles. This requires tracking of particle rotation in addition to particle position, but allows more adjustment of the continuum-scale elastic properties of the material. Fractures are represented by the irreversible removal of a bond.

The particles can be distributed using two different approaches. Firstly, all particles can be given the same radius, and arranged according to a geometrical pattern. Secondly, particles of varying size can be added randomly to the computational domain, before the equilibrium arrangement of particles is found using only solid-sphere contact forces. This gives an unstructured grid, which is preferred in recent works because a grid-induced bias on fracture directions can be avoided. According to André et al. (2012), a variation of particle radius of ±20% from the average radius is sufficient to avoid ordered geometrical arrangements of particles and grid-induced fracture directions. When the equilibrium arrangement of particles is defined using one of the two approaches introduced above, bonds are inserted between particles within a given cut-off distance of each other. The equilibrium length $L_{eq,ij}$ of a bond joining particles i and j is set to the initial distance between the centres of the particles it connects. A scaling parameter $\alpha = L_{eq,ij}/(r_i + r_j)$ is introduced, where r_i and r_j are the two particle radii.

A volume change can be implemented by changing the radius r_n of particle n (Ulven et al., 2014a) according to the equation

$$r_n = r_{n,0}\left(1 + \frac{\Delta V}{V}\frac{b_n}{b_0}\right)^{1/3} \tag{11}$$

where $r_{n,0}$ is the initial particle radius, b_n is the reaction product concentration relevant to volume change in particle n, and b_0 is the concentration that corresponds to the maximum volume change. The new equilibrium length of the bond joining particles i and j after volume change is then found using the expression $L_{eq,ij} = \alpha(r_i + r_j)$. This changes the equilibrium length of the bond, and creates an elastic strain in the solid. One option for solving reaction-diffusion in DEM using equations 3–5 is presented in detail by Ulven *et al.* (2014b).

The stress tensor σ_i for each particle i can be found using the sum of outer products $\sigma_i = 0.5 V_i^{-1} \Sigma_j (F_{ij} \times d_{ij})$ (Tildesley and Allen, 1987), where V_i is the volume of particle i, F_{ij} is the force in the bond connecting particles i and j, d_{ij} is the distance vector from particle i to particle j, and $\times$ denotes the outer product of vectors.

After introducing the volume change, the equilibrium state of the material can be found using different approaches, *e.g.* over-relaxation methods (Jamtveit *et al.*, 2000; Malthe-Sørenssen *et al.*, 2006; Røyne *et al.*, 2008), conjugate gradient methods (Ulven *et al.*, 2014a,b) or direct time integration of the particle movement (Okamoto and Shimizu, 2015). Then, fractures are formed if the deformation is above some fracture criteria. The fracture criteria can be based on a critical stress or a critical strain, or be standard criteria such as the Coulomb criterion and tensile criterion defined above, or any other fracture criteria that are commonly used in structural geology.

3.4. Comparison and validation

A detailed benchmark and comparison of models was performed by Ulven *et al.* (2014b). The chosen test case was a volume-expanding process driven by fluid infiltration through effective diffusion from the boundary in circular geometry. The fluid that thus enters the domain reacts with the initial solid, and a new solid with increased volume is formed. This leads to a compressive stress near the domain boundary, where the rock is reacted, and a tensile stress in the unreacted core. See section 5.2 for more details. Three different models were used, a 1D FDM that assumed axial symmetry, 2D FEM and 2D DEM. The results are shown in Fig. 5. Note that the mean values shown for FEM and DEM are almost indistinguishable from the FDM solution, apart from a deviation near the core for FEM. This happens because the stress is averaged over a small number of elements near the core. There is a significant variation in the stress of individual DEM particles, caused by the unstructured mesh, but the mean value fits nicely with the FDM result.

4. Volume-decreasing reactions

Some of the very first quantitative works on reaction-induced fracturing focused on volume reducing processes. In this section we present the development of models, from simple one-dimensional theoretical models, to two-dimensional numerical models.

4.1. One-dimensional theoretical model: Front speed and width

Yakobson (1991) introduced a simplified model for the propagation of a one-dimensional front due to a volume-decreasing reaction. The motion for the study was the thermal decomposition of solids, where a propagating front separating the fracture from the unfractured region can be observed optically in experiments. The model was

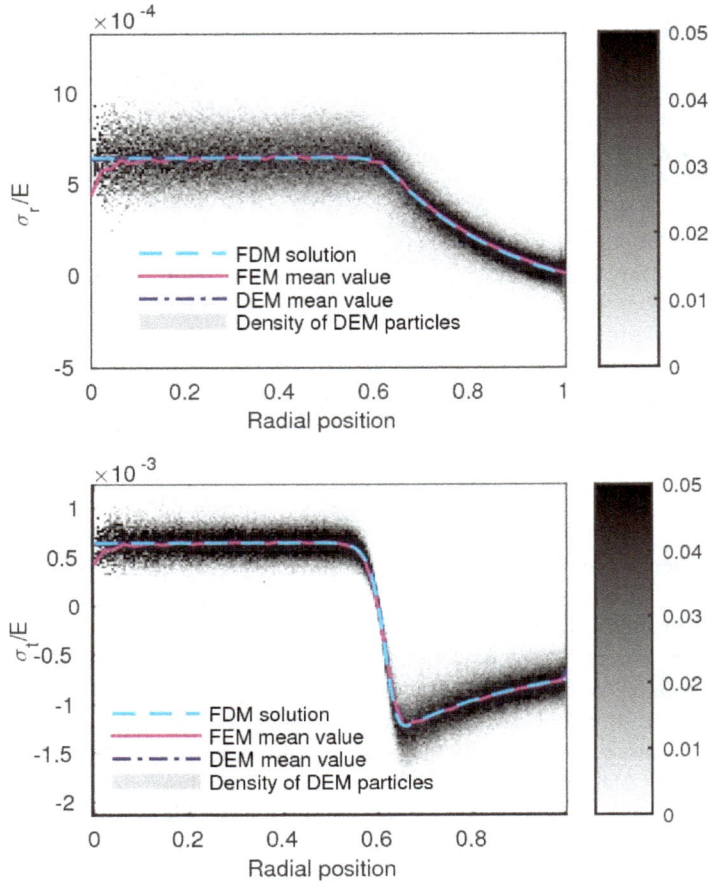

Figure 5. Comparison of normal stress in radial (σ_r, upper) and tangential (σ_t, lower) direction computed by FDM, FEM and DEM for a volume-expanding process proceeding from the boundary in a circular geometry. For FEM and DEM, a mean value is computed at each radial position. The FEM solution deviates slightly near the domain centre due to low numerical resolution, but the three plots are otherwise almost indistinguishable. The DEM density plot shows the density of individual particles with a given stress state, and we see that although most particles are close to the mean value, there is a significant deviation in the local stress. Modified from Ulven *et al.* (2014b).

simplified to one dimension and included only the diffusion of a volatile component, resulting in a local stress that depended on the local concentration according to equation 1.

The local concentration and the stress intensity factor (equation 2) were combined with a linearization of the steady-state solution for the stress profile, $\sigma(x) = 1 - (1 - w_b)\exp(-xv/D)$, to give an expression for the velocity v of the front:

$$\frac{K_c}{2E\beta}\left(1+\frac{v}{k}\right)\left(\frac{\pi v}{D}\right)^{1/2}=\sqrt{\lambda}\left(\frac{\pi}{2}-\lambda\right) \tag{12}$$

This equation includes the critical stress intensity factor, K_c, the diffusion constant, D, the surface transfer coefficient, k, and a dimensionless length, $\lambda=(v/D)L$. Yakobson assumed that the fastest fracture would dominate the front. This resulted in two limiting behaviours, depending on the ratio, Δ, of the diffusion constant and the surface transfer coefficient: $\Delta \propto (D/k)(E\beta/K_c)^2$. If the diffusion was fast, $\Delta \gg 1$, the velocity was $v = k^{2/3}D^{1/3}(E\beta/K_c)^{2/3}$ and the width of the front was $L = (DK_c/kE\beta)^{2/3}$, whereas if diffusion was slow, $\Delta \ll 1$, the velocity was $v = D(E\beta/K_c)^2$ and the width of the front was $L = (K_c/E\beta)^2$. These expressions provide quantitative estimates for the fracture spacing or the width of the front and the front velocity.

Similar results were found for an array of cracks propagating due to thermal contraction as studied by Boeck *et al.* (1999). They performed two-dimensional FEM modelling of a crack array in contact with cooler fluid, and their results for both front velocity and fracture spacing correspond to the results of Yakobson (1991).

4.2. Two-dimensional numerical model: Front geometry

The front dynamics and geometry for a volume-reducing reaction corresponding to a devolatilization reaction were studied by Malthe-Sørenssen *et al.* (2006). They introduced a discrete-element model of the reacting solid and assumed that the volatile component was immediately transported out of the system when it came in contact with a fracture connected to the free surface of the sample. This study confirmed the one-dimensional results for the velocity and the front width from Yakobson (1991), and was able to show the complicated geometry of the fracture network connecting the free surface to the reaction front. Figure 6 shows the fracture pattern for two different values of the characteristic length, demonstrating the robustness of the one-dimensional theoretical results. The simulations also demonstrated a transient development towards motion of the front with constant velocity and width, but this transient has not been studied in detail.

5. Volume-increasing reactions

Volume-expanding processes have been studied extensively only during the last decade. This section summarizes some central works.

5.1. Analytical models: Linear model

To the authors' knowledge, the first quantitative model for fracture formation during a volume-expanding process was published by Fletcher *et al.* (2006). The main focus in this work was the weathering of quartz diorite bedrock in Puerto Rico, in which oxidation of ferrous minerals was interpreted as causing a slight volume expansion and spheroidal weathering of the rock. The oxidation was assumed to be caused by a second-order chemical reaction with oxygen transported into the rock by diffusion and instantaneous transport in fractures.

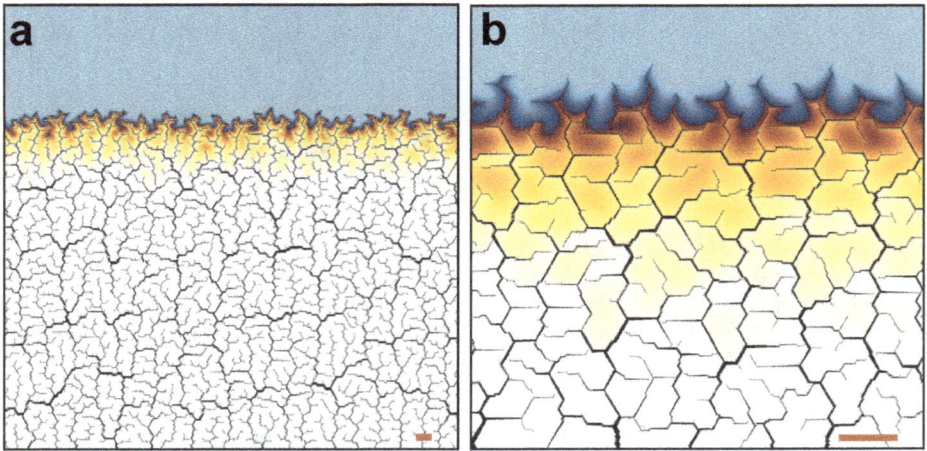

Figure 6. Snapshots from simulations from Malthe-Sørenssen *et al.* (2006) illustrating the fracture patterns and the concentration fields in two simulations with different diffusion constants, small *D* (left) and large *D* (right). The fractures started from the bottom boundary in both cases. The characteristic lengths are illustrated by the red bars.

The model was simplified to one dimension, thus describing a vertical weathering profile with only downwards transport of water with dissolved reactants. The mechanical model used in this work was highly simplified: as the rock expanded during weathering, strain energy was added. The strain energy was integrated downwards from the surface to a level where the oxygen concentration was below a given limit, and a surface-parallel fracture was assumed to form at that depth when the strain energy reached the surface energy of a fracture. It was further postulated that all of the strain energy was relieved when the fracture formed, and that water would flow rapidly into a fracture, thus giving an oxygen concentration in a fracture identical to the concentration at the surface.

Fletcher *et al.* (2006) concluded that by the mechanism they described, the weathering front will proceed at a constant rate over time: the weathering is rapid immediately after fracture formation, then slows down gradually as the transport distance increases before a new fracture forms and the process repeats itself. This produces spheroidal rindlets of constant thickness. The time scale of the computed weathering was reported to fit nicely with field observations.

Rudge *et al.* (2010) developed a similar 1D model, but applied it to reaction-induced cracking during serpentinization and carbonation of peridotite. This model was based on a second-order reaction with a mobile reactant transported into the rock by effective diffusion in intact rock and instantaneous transport in fractures. In addition, a fracture front was assumed to move at constant velocity.

While Fletcher *et al.* (2006) focused on studying parameters relevant to a specific field area, Rudge *et al.* (2010) approached a more general understanding of the speed at which fracture fronts propagate and typical crack lengths by performing a study of a wide range of parameters. By isolating the main dimensionless groups they found that the typical fracture length is, in many cases, dependent on elastic properties and volume change only, and independent of the effective diffusion constant D and reaction rate κ. For the parameter combinations they assumed relevant for most natural systems, the typical fracture length increased slightly with D, and was reduced when κ increased. The fracture front velocity in all cases increased with greater volume change, weaker material and higher D. The fracture-front velocity also increased with κ, until becoming independent of κ when the reaction rate was high enough.

Rudge *et al.* (2010) further applied their model to carbonation of peridotite. By heating the rock, increasing the partial pressure of CO_2, hydrofracturing and pressurized injection of a CO_2-bearing fluid, they estimated that natural weathering rates in the Oman ophiolite of a few tenths of a millimetre per year could be increased to several hundred metres per year, thus suggesting that accelerated carbonation of ultramafic rock is a viable option for industrial-scale CO_2 sequestration.

5.2. Analytical models: Cylindrical model

Ulven *et al.* (2014b) focused on reaction-driven expansion in a circular geometry using the system of equations defined in equations 6–8, where the dimensionless parameters Γ and Θ were introduced. Versions of Γ are otherwise often referred to as the Damköhler number II, and describe the ratio between the rate of the chemical reaction and the rate of diffusive transport into the domain. Θ simply defines the ratio between the maximum amount of the mobile reactant in a volume and the amount needed to convert all material A initially present in the same volume to the product B.

Solving the system in equations 6–8 with a constant Θ using FDM, Ulven *et al.* (2014b) showed that the sharpness of the reaction front depended on Γ (Fig. 7), but that the velocity of the front was nearly independent of Γ. For $\Theta \ll 1$, which is most often the case, it can be shown that the front velocity is proportional to Θ, and that the reaction-front shape depends on the ratio Γ/Θ.

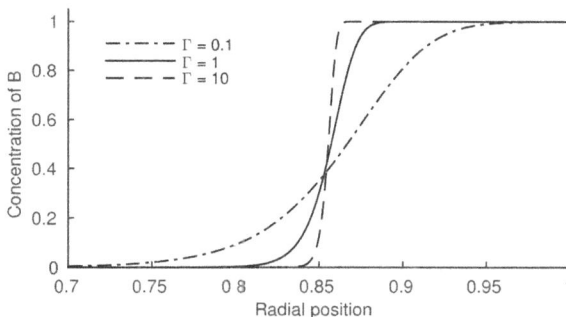

Figure. 7. Concentration of the reaction product at different radial positions in a circular geometry at a given time. The mobile reactant enters the domain from the outer boundary at radial position $r = 1$, where there is a constant concentration of the mobile reactant. The reaction front becomes sharper when Γ increases. In all cases $\Theta = 10^{-4}$. Modified from Ulven *et al.* (2014b).

The difference in reaction-front sharpness has a significant impact on the stress that is set up in the material. As seen in Fig. 8, the radial stress is the major principal stress, and has its highest value in the unreacted core of the domain. The individual stress components do not change much for different values of Γ, but the Coulomb stress increases significantly near the reaction front when Γ increases. This means that when Γ is large, there will be a tendency to initiate shear fractures near the reaction front, while a small Γ will lead instead to tensile fractures in the unreacted core of the material.

Ulven et al. (2014b) further showed that when Γ or $\Delta V/V$ is sufficiently small, no fractures will form. This is easily understood from the following: a small $\Delta V/V$ means that little strain will be caused by the volume-changing process, which obviously removes the reason for fracture formation. A very small Γ leads to a pervasive reaction in the entire domain, and a nearly uniform expansion leads to a stress below the critical strength of the material. When both Γ and $\Delta V/V$ are sufficiently large, shear fractures initiate near the reaction front. Intermediate values lead to tensile fractures in the unreacted core of the domain. These results provide a deeper insight into the fracture mode one might expect in different cases as compared to previous analytical works, even though the results only point to the type of the first fracture to initiate. Further fracture growth is studied in more detail in the next section.

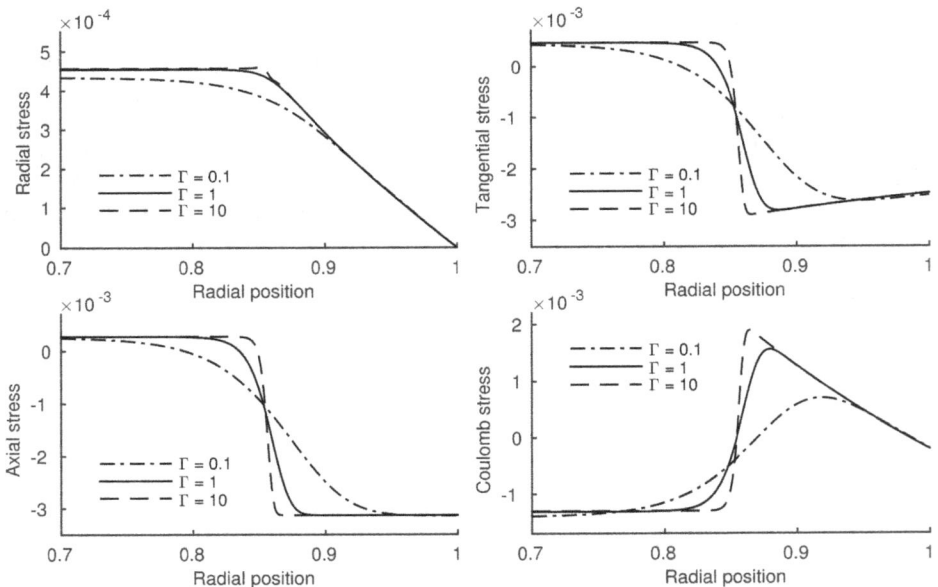

Figure 8. Radial, tangential and axial-stress components, and resulting Coulomb stress, at different radial positions, corresponding to the concentration profiles shown in Fig. 7. All stress components are shown as dimensionless fractions of the Young's modulus E. $\Delta V/V = 0.5\%$. Notice that the different stress components are similar for different Γ, apart from sharper gradients for large Γ. The Coulomb stress increases significantly when Γ increases. Modified from Ulven et al. (2014b).

5.3. Two-dimensional simulations: Diffusion-dominated case

As an extension of the previous section, Ulven *et al.* (2014b) studied the same system of reaction-diffusion using 2D DEM. Potential fluid transport in fractures was ignored, and the feedback from fracture formation was thus neglected. Eight different parameter combinations were tested (four are shown in Fig. 9), and it was observed that the fracture mode that is expected to initiate first tends to continue to dominate. When

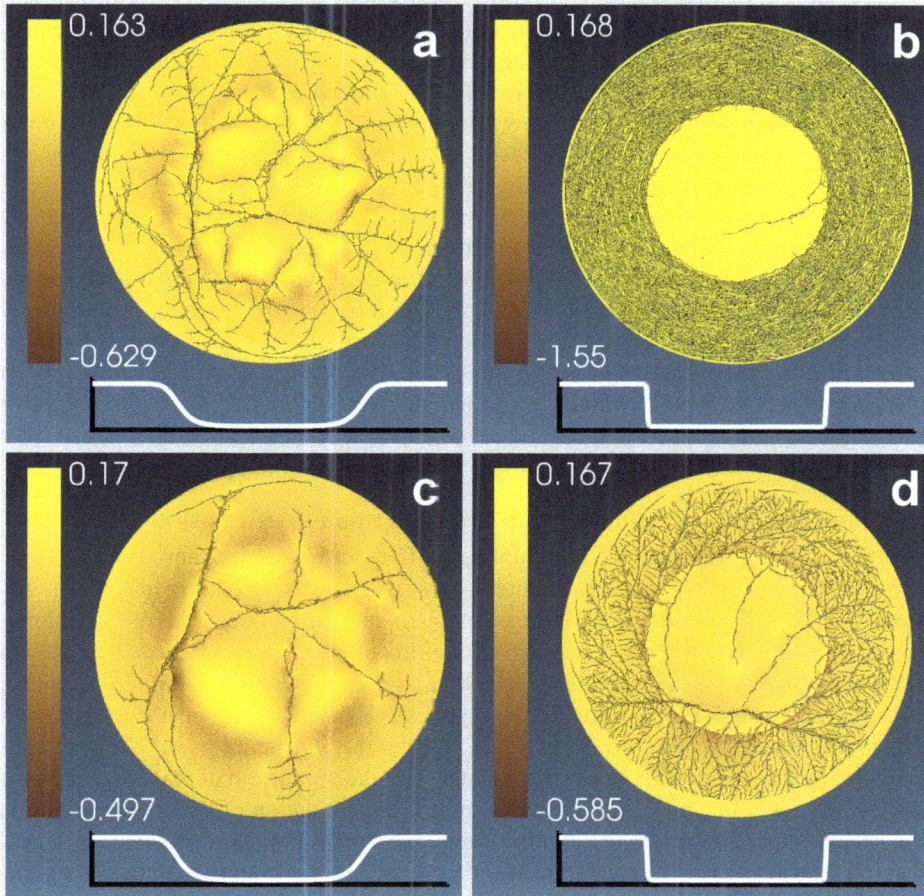

Figure 9. DEM results for four different parameter combinations. The colour scale represents the local value of the first invariant of the stress tensor divided by *E*. The white graph below shows the corresponding concentration profiles of the reaction product B for a cross-section of the circle, with completely reacted material in the rim and fresh material in the core. In all cases = 10^{-4}. (a) $\Gamma = 0.04$, $\Delta V/V$ = 0.8%. (b) $\Gamma = 100$, $\Delta V/V = 0.8\%$. (c) $\Gamma = 0.04$, $\Delta V/V = 0.3\%$. (d) $\Gamma = 100$, $\Delta V/V = 0.3\%$. Note that greater volume change leads to greater fracture density, and larger Γ leads to more fractures in the reacted material, and fewer in the unreacted core. Modified from Ulven *et al.* (2014b).

tensile fractures are initiated (Fig. 9a,c,d), fractures crosscut the unreacted core of the domain, whereas few fractures enter the unreacted core when shear fractures in the unreacted material are expected (Fig. 9b). This shows that when Γ and $\Delta V/V$ are large, spalling of reacted material is the dominant process. For lower values of Γ and $\Delta V/V$, tensile fractures will tend to subdivide domains, and create hierarchical fracture systems. Large Γ and moderate $\Delta V/V$ (Fig. 9d) lead to the expected tensile fractures, and additionally dense shear fractures in the reacted material.

This can be understood based on simple arguments. Large Γ and $\Delta V/V$ cause a sharp change in strain near the reaction front, which gives a large Coulomb stress. This causes shear failure and spalling of the reacted material, and prevents stress sufficient to form fractures in the unreacted core. Thus, the core is left mostly unfractured. When the strain changes more gradually, the reaction is allowed to proceed further without initiation of shear fractures. In this case, the stress in the unreacted core can become high enough to cause growth of tensile fractures in the radial direction.

The definition of Γ allows a deeper understanding of the system. Because Γ is proportional to the square of the domain radius, it means that large domains will tend to spall, smaller domains will tend to split, and sufficiently small fragments will be stable. This represents the size of particles that will be left after the reaction is completed. Furthermore, fast chemical reactions will tend to cause spalling, while slower reactions cause splitting or no fractures at all. Finally, the effective diffusivity of a rock is often controlled by the porosity, which means that a porous rock will tend to split and form hierarchical fractures, whereas low-porosity rock will have a higher tendency to spall. This will be studied further in the next section.

5.4. Two-dimensional simulations: Transport-dominated case

The first model including the full coupling of fracture formation caused by a volume expansion and fluid transport in fractures was presented by Jamtveit *et al.* (2008). This work focused on fracture formation in an unreactive rock matrix caused by expansion of reactive mineral grains, while the internal fracturing of the grains was ignored. Each grain consisted of several DEM particles in the simulations. One grain was initially in contact with the fluid, and fractures started growing outwards due to the expansion of this grain. The model showed that the fractures could connect the expanding grain to other grains, and initiate the volume-expanding reaction there. If the proportion of reactive grains was sufficiently high, this process was shown to lead to percolation of fluids throughout the system and expanding grains in the entire domain, possibly with some non-reacted regions that were not reached by fractures. On the other hand, a lower proportion of reactive grains led to a process that stopped because the fluid did not get access to additional reactive material, since the intact solid was assumed to be impermeable.

However, the setting in Jamtveit *et al.* (2008) is fundamentally different from a volume-expanding reaction occurring in a homogenous material. Røyne *et al.* (2008) presented the first coupled 2D model for an expanding reaction in a homogeneous block of material. Parameters were tuned to those relevant for weathering of a dolerite, in order to explain field observations. The simulations (Fig. 10) showed several features

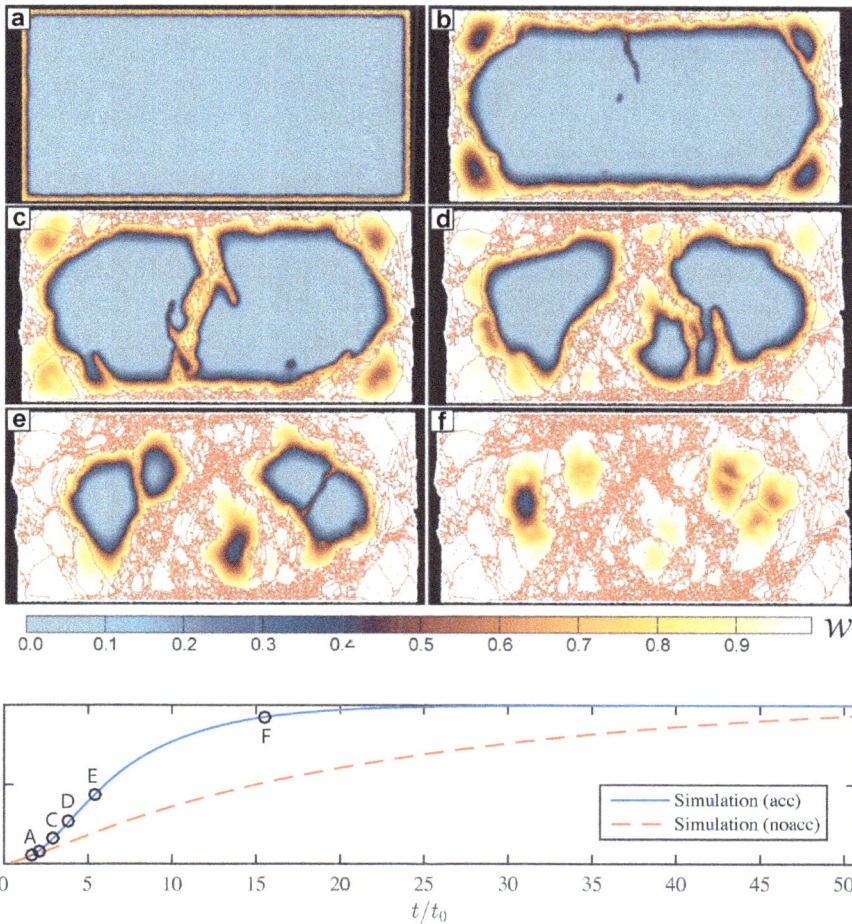

Figure 10. Snapshots from simulations from Røyne *et al.* (2008) illustrating the extent of reaction, *w*, in space in a simulation where fluid transport along fractures is instantaneous. The resulting fracture pattern clearly shows a hierarchical structure. The bottom figure illustrates the extent of reaction (V_r/V_0) as a function of time for a simulation with (acc) and a simulation without (noacc) flow in the fractures, illustrating the acceleration due to fracturing.

they also reported from field studies. Firstly, sharp corners are rounded off, and reacted fragments spall from the surface of the unreacted material that remains. Secondly, large blocks are repeatedly subdivided, thus creating a hierarchical network of fractures.

Røyne *et al.* (2008) observed that the overall material conversion proceeded faster in the simulations than would be expected from pure diffusion of reactants into the domain due to fluid transport in fractures. This led them to develop a simplified analytical model for the process. The main assumptions were that domains would subdivide after a characteristic time, and that the process followed the same kinetics from all fracture surfaces, thus increasing the total conversion rate when fractures were introduced. With appropriate parameters, this model followed closely the overall reaction progress from the DEM simulations at short times, but deviated at long times because the simple model did not include information about the finite system size or a typical size of the smallest fragments left.

This work was followed up by Ulven *et al.* (2014a), who studied the same system as in section 5.3, but with fluid transport in fractures included. This time the focus was on the weathering of rock with different initial porosity Φ. Assuming that the diffusivity of the intact rock is a linear function of porosity, large Γ corresponds to small Φ, and small Γ corresponds to large Φ. They utilized this to cast the system into a shape where the porosity was the main controlling parameter, with all other chemical and mechanical parameters held constant.

Firstly, it was observed that instantaneous fluid transport in fractures had a significant effect on fracture patterns (Fig. 11). The first fractures initiated almost identically with and without fluid transport in fractures, but afterwards the patterns

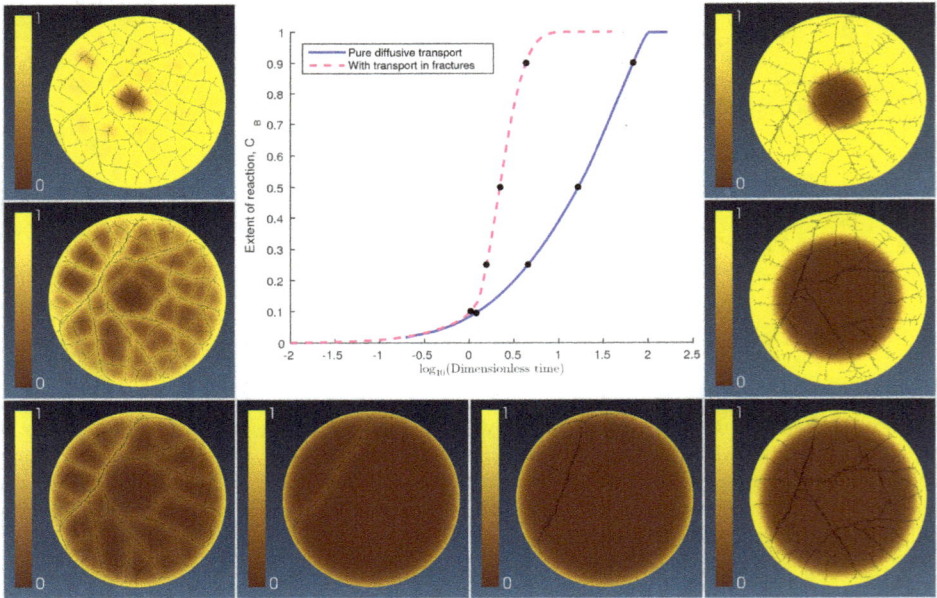

Figure 11. Contrasting fracture patterns and process development in time with (left) and without (right) fluid transport in fractures for Φ = 2.5%, Γ = 10, ΔV/V = 0.8%. The colour code shows extent of reaction, from dark brown for unreacted material to bright yellow for completely reacted material. Note that the first fractures initiate almost identically with and without transport in fractures (centre images), but that differences arise as fracture growth proceeds. Fluid transport in fractures leads to a hierarchical fracture pattern, whereas the fracture pattern without fluid transport is a combination of a hierarchical pattern and a branching network with several dead-end fractures. The central graph shows the extent of reaction as a function of a dimensionless time. Modified from Ulven *et al.* (2014a).

became different. At the same time as the pattern development branched, the overall reaction rate increased significantly with fluid transport included, as was also shown by Røyne *et al.* (2008). Secondly, different initial porosity changed the fracture pattern. The simulations (Fig. 12) showed that large Φ led to wide reaction zones along fractures and partly reacted material across quite large fragments, and fractures dominated by T-junctions creating a hierarchical pattern. Small Φ caused narrow reaction fronts separating completely reacted material from entirely fresh material, and mainly spalling of reacted material, with a smaller number of domain-cutting fractures.

It was shown that the overall reaction rate was, in all cases, quite low initially, with highest overall reaction rate for the most porous material due to the most effective transport of the mobile reactant into the rock. When fracturing commenced, the overall rate increased markedly, until the rate again levelled out when the material was almost completely reacted. A simplified model was developed to describe this behaviour. The model describes the extent of reaction C_B, $0 \leqslant C_B \leqslant 1$, as a function of time t, with $C_B = 0$ denoting an unreacted material, and $C_B = 1$ denoting a completely reacted material. Firstly, it was assumed that all fractures were connected, that the overall reaction rate was proportional to the total fracture length, and that the total fracture length was proportional to C_B. Secondly, it was assumed that the rate was proportional to the amount of fresh material remaining, $(1 - C_B)$. This leads to a differential equation

$$\frac{dC_B}{dt} = \beta C_B (1 - C_B) \tag{13}$$

where β is an undefined rate constant. The solution to this equation is

$$C_B(t) = \frac{1}{1 + e^{-\beta(t-\tau)}} \tag{14}$$

which describes sigmoidal curves, where τ is the time at which $C_B = 0.5$. When β and τ were appropriately adjusted, the sigmoidal curves were seen to fit the numerical data well. The sigmoidal curves deviate from the numerical data at very low extent of reaction, when no fractures are formed, and at high extent of reaction, when the fracture pattern is fully developed. In both cases, the assumption of fracture length proportional to extent of reaction breaks down.

It was observed that β followed a simple dependence on Φ. For low porosity, it was shown that $\beta \propto \Phi^N$, where N is an exponent that depends on the volume expansion. For sufficiently large Φ, β became independent of Φ. The latter case shows that the porosity can reach a level where even higher porosity does not affect the process because diffusive transport in pore space is rapid enough to provide all the reactant necessary to convert the material. In this case, the process is limited by the chemical reaction rate only. For low porosity, the process is heavily dependent on reactant transport. Without fractures $N = 2$. Higher volume change. and thus higher fracture density, gave lower N, down to 0.45 for a volume change of 0.8%. This shows that when fractures provide rapid fluid transport, the process becomes less dependent on fluid transport in pore space. Furthermore, it shows that fracture density is an important feedback parameter in

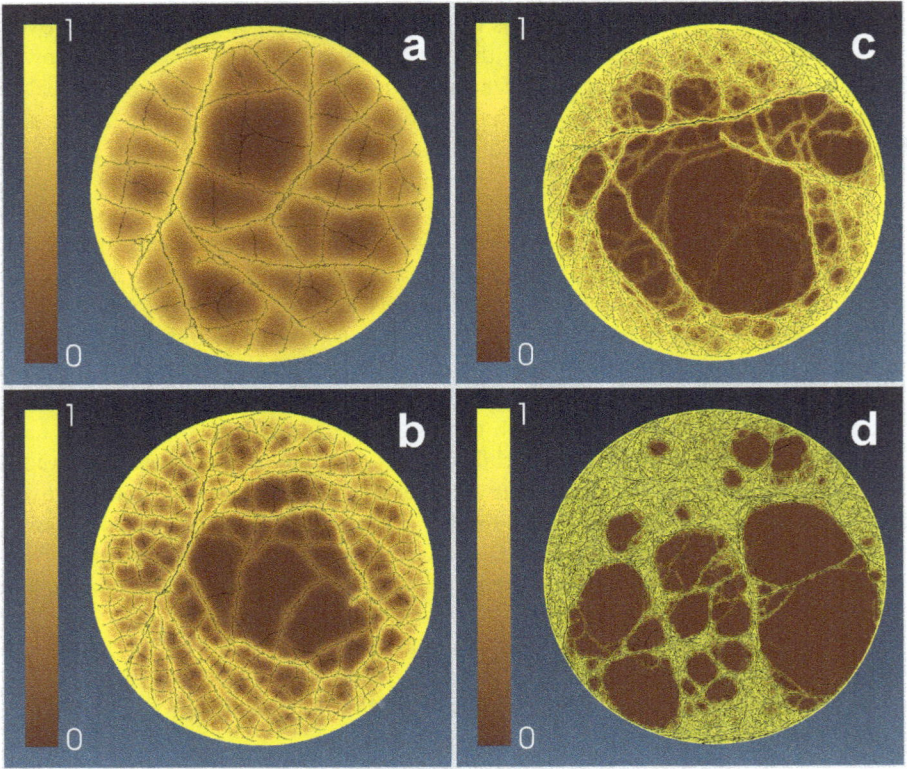

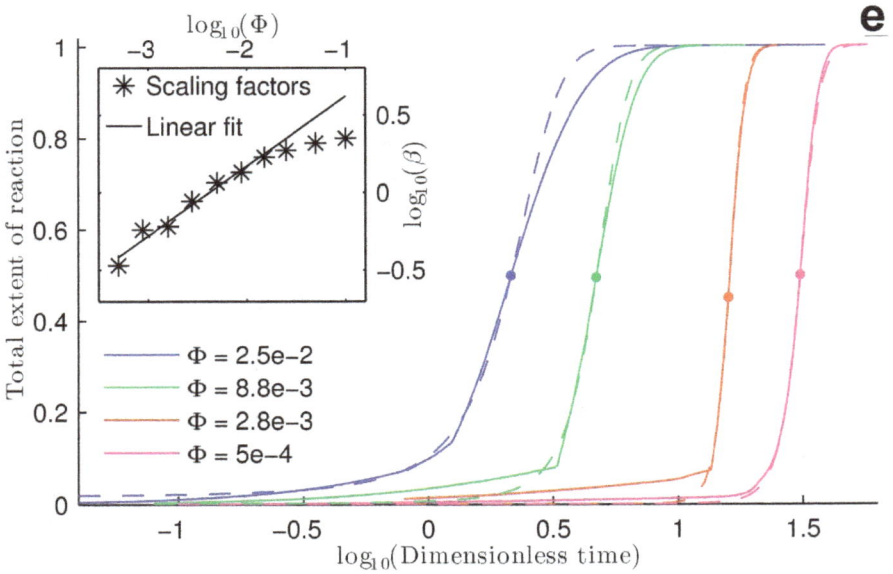

reaction-induced fracturing, and it will be difficult to determine this density in more simplified models.

Furthermore, Ulven *et al.* (2014a) showed that the domain shape had a significant effect on the process when Φ was small. The initial fracturing tended to localize at the corners with highest curvature. This caused a large variation in reaction progress for different initial shapes, even with otherwise identical rock properties and weathering reactions. Large Φ did not show the same variation in reaction progress. Thus, in a natural system with low porosity where the initial domain geometries vary significantly, a large range of local variation in weathering progress is to be expected. Completely weathered domains can be expected to be found next to intact, unweathered material. This is indeed what is often seen in areas where spheroidal weathering of low-porosity rock occurs (Fig. 3a).

If the reaction also clogs porosity and reduces permeability, the process becomes a lot more complicated. Such a system has not, to the authors' knowledge, been studied in any detail, but the rate constant $\beta \propto \Phi^N$ immediately suggests a significant rate reduction if porosity is clogged. From the definition of Γ and Θ, it follows that a reduction in porosity will cause larger Γ and smaller Θ. Based on the arguments in section 5.2, this means that the reaction front will gradually become sharper with reduced porosity, while the propagation speed of the reaction front goes down. Thus, the reaction becomes self-localizing near the domain boundary and fracture surfaces, and the overall reaction rate is significantly reduced until new fractures provide access to fresh rock and temporarily accelerates the reaction. This continuous variation in overall reaction rate makes it challenging to study this process numerically, as short time steps are necessary to achieve acceptable numerical accuracy when the overall reaction rate is high, while it would be preferable to use longer time steps when the reaction rate is low to keep the computational time at a minimum.

Okamoto and Shimizu (2015) combined a chemical-mechanical DEM with a simplified fluid-flow model, which assumed that the fluid was present in the area bounded by three or more particles from the solid model. The fluid was allowed to flow between adjacent areas if a fracture was present, and the flow between such areas was described by a Poiseuille equation for laminar flow between parallel plates. They studied a square domain under external compression, and let the material expand after reaction with a fluid that entered the domain from the boundaries. Where the previous works in this section focused on diffusive transport of reactants, and assumed an instantaneous transport in fractures, Okamoto and Shimizu (2015) ignored diffusion in the intact solid, and let the fluid be transported into the domain by pressure-driven flow

---

Figure 12 (facing page). Snapshot of reaction progress and fracture pattern for four different porosities. (a) Φ = 2.5%, (b) Φ = 0.88%, (c) Φ = 0.28%, (d) Φ = 0.05%. The colour code is as in Fig. 11. (e) The extent of reaction in time (solid line) with a fitted sigmoidal function (dashed) defined in equation 14. The dot on each solid line shows the extent of reaction corresponding to snapshots (a–d). The inset in (e) shows the scaling factor β from equation 14 for different Φ, with a line-fit demonstrating that $\beta \propto \Phi^{0.45}$ for low porosity, while β becomes independent of Φ for high porosity. Modified from Ulven *et al.* (2014a).

in fractures. This addresses two important problems: firstly, the local fluid pressure might play a role in creating fractures in a rock, and secondly, fractures under compression might not be open and allow fluid transport at all.

In the volume-expansion simulations presented by Okamoto and Shimizu (2015), the initial stages of the process had similarities with previous works (Røyne et al., 2008; Ulven et al., 2014a), with dense fractures along the domain boundary and a few fractures crosscutting the domain centre and creating a hierarchical pattern. The similarity led them to suggest that the solid volume change was more important than the details of the fluid-transport mechanism.

However, in the work by Okamoto and Shimizu (2015) the extent of reaction was limited by fluid flow in fractures. As the material expanded and became increasingly compressed along the boundary, fractures were closed and prevented fluid transport into the domain. This caused parts of the material to remain unreacted, and the process stopped before the conversion of material was complete. Thus, the more detailed treatment of fluid transport shows that the reaction can become limited by lack of the mobile reactant, which previous works ignored. Unfortunately, the full three dimensional nature of the process is somewhat more complicated, as 3D, in principle, allows fluid transport through a fracture while under compression. To the knowledge of the authors, the permeability of a fracture in this setting has not been studied in any detail.

6. Concluding remarks

We have shown that volume-changing processes can lead to fractures in many settings. In all cases, the important point is that the volume change causes a strain in the material. The magnitude of this strain (usually related to $\Delta V/V$) and the gradient of the strain (reaction front thickness) are the main controlling parameters. Sharp reaction fronts and large volume change lead to dense fracture patterns in both volume reduction and volume expansion. During volume expansion, the strain gradient controls whether the fractures formed are tensile or shear fractures. This has been shown in the present text for a volume-expanding chemical reaction, but similar situations can be expected in other settings when the strain changes sharply, e.g. rapid heating, or metamict and non-metamict material separated by a short distance.

In some cases fractures might provide a positive feedback because the fractures allow faster transport of necessary mobile components. With the assumption of instantaneous transport of reactants in fractures, it has been shown that reaction fronts during volume reduction proceed with a velocity constant in time. On the other hand, reaction fronts accelerate in time for volume expansion. This is seen in the simplified model from Ulven et al. (2014a): a sigmoidal function grows almost exponentially at early times, until the finite system size limits the overall reaction rate. This suggests that a volume-expanding reaction front under ideal conditions might move with a velocity that grows in time. Clearly the process will ultimately be limited by transport of the mobile reactant, because instantaneous transport is only a reasonable assumption when transport distances are sufficiently short.

This shows an interesting difference between volume reduction and volume expansion: volume-reducing processes with instantaneous fluid transport in fractures are stable, whereas volume-expanding processes are unstable with instantaneous fluid transport in fractures. The main difference is where the fractures form: fractures during volume reduction form in the reacted material, while volume expansion can create wedges into the fresh material, and thus provide a significantly more efficient feedback.

The acceleration of reaction fronts during volume expansion nevertheless depends on whether fractures remain open after they are formed, or whether they are closed by compressive stress. This has not been studied to any grea extent apart from the 2D work by Okamoto and Shimizu (2015). A full numerical simulation in 3D using DEM or similar particle models is necessary to provide further insight, but is at present hard to achieve due to computational cost. Furthermore, minerals may precipitate inside fractures and provide additional clogging. This is an effect that is challenging to study, and remains as a subject for further work.

Acknowledgements

Assistance from Håkon Austrheim is much appreciated. The authors thank two anonymous reviewers for their valuable comments. This work was funded by a project grant from the Norwegian Research Council.

References

André, D., Iordanoff, I., Charles, J.-L. and Néauport, J. (2012) Discrete element method to simulate continuous material by using the cohesive beam model. *Computer Methods in Applied Mechanics and Engineering*, **213**−**216**, 113−125.

Austrheim, H. (1987) Eclogitization of lower crustal granulites by fluid migration through shear zones. *Earth and Planetary Science Letters*, **81**, 221−232.

Boeck, T., Bahr, H.-A., Lampenscherf, S. and Bahr, U. (1999) Self-driven propagation of crack arrays: A stationary two-dimensional model. *Physical Review E*, **59**, 1408−1416.

Boyd, F.R. and England, J.L. (1960) The quartz-coesite transition. *Journal of Geophysical Research*, **65**, 749−756.

Cartwright, J., James, D. and Bolton, A. (2003) The genesis of polygonal fault systems: a review. Pp. 223−243 in: *Subsurface Sediment Mobilization* (P. Van Rensbergen, R.R. Hillis, A.J. Maltman, and C.K. Morley, editors). Special Publications, **216**. Geological Society, London.

Coussy, O. (2010) *Mechanics and Physics of Porous Solids*, John Wiley & Sons, Ltd, Chichester, UK.

Crocombette, J.-P. and Ghaleb, D. (2001) Molecular dynamics modeling of irradiation damage in pure and uranium-doped zircon. *Journal of Nuclear Materials*, **295**, 167−178.

Cundall, P.A. and Strack, O.D.L. (1979) A discrete numerical model for granular assemblies. *Géotechnique*, **29**, 47−65.

Dunkel, K.G. and Putnis, A. (2014) Replacement and ion exchange reactions of scolecite in a high pH aqueous solution. *European Journal of Mineralogy*, **26**, 61−69.

Flekkøy, E.G., Malthe-Sørenssen, A. and Jamtveit, B. (2002) Modeling hydrofracture. *Journal of Geophysical Research: Solid Earth*, **107**.

Fletcher, R.C. and Merino, E. (2001) Mineral growth in rocks: kinetic-rheological models of replacement, vein formation, and syntectonic crystallization. *Geochimica et Cosmochimica Acta*, **65**, 3733−3748.

Fletcher, R.C., Buss, H.L. and Brantley, S.L. (2006) A spheroidal weathering model coupling porewater chemistry to soil thicknesses during steady-state denudation. *Earth and Planetary Science Letters*, **244**, 444−457.

Hacker, B.R. (2008) H$_2$O subduction beyond arcs. *Geochemistry, Geophysics, Geosystems*, **9**, 1–24.

Herrmann, H. J. and Roux, S. (1990) *Statistical Models for the Fracture of Disordered Media*, North-Holland, Amsterdam.

Jamtveit, B., Austrheim, H. and Malthe-Sørenssen, A. (2000) Accelerated hydration of the Earth's deep crust induced by stress perturbations. *Nature*, **408**, 75–78.

Jamtveit, B., Malthe-Sørenssen, A. and Kostenko, O. (2008) Reaction enhanced permeability during retrogressive metamorphism. *Earth and Planetary Science Letters*, **267**, 620–627.

Jamtveit, B., Putnis, C. and Malthe-Sørenssen, A. (2009) Reaction induced fracturing during replacement processes. *Contributions to Mineralogy and Petrology*, **157**, 127–133.

Jamtveit, B., Kobchenko, M., Austrheim, H., Malthe-Sørenssen, A., Røyne, A. and Svensen, H. (2011) Porosity evolution and crystallization-driven fragmentation during weathering of andesite. *Journal of Geophysical Research: Solid Earth*, **116**.

Jung, H., Green II, H.W. and Dobrzhinetskaya, L.F. (2004) Intermediate-depth earthquake faulting by dehydration embrittlement with negative volume change. *Nature*, **428**, 545–549.

Kelemen, P.B. and Hirth, G. (2012) Reaction-driven cracking during retrograde metamorphism: Olivine hydration and carbonation. *Earth and Planetary Science Letters*, **345–348**, 81–89.

Kelemen, P.B., Matter, J., Streit, E.E., Rudge, J.F., Curry, W.B. and Blusztajn, J. (2011) Rates and mechanisms of mineral carbonation in peridotite: natural processes and recipies for enhanced, in situ CO$_2$ capture and storage. Pp. 545–576 in: *Annual Review of Earth and Planetary Sciences*, **39**.

King, H.E., Plümper, O. and Putnis, A. (2010) Effect of secondary phase formation on the carbonation of olivine. *Environmental Science & Technology*, **44**, 6503–6509.

Kuleci, H., Schmidt, C., Rybacki, E., Petrishcheva, E. and Abart, R. (2016) Hydration of periclase at 350°C to 620°C and 200 MPa: experimental calibration of reaction rate. *Mineralogy and Petrology*, **110**, 1–10.

Malthe-Sørenssen, A., Jamtveit, B. and Meakin, P. (2006) Fracture patterns generated by diffusion controlled volume changing reactions. *Physical Review Letters*, **96**, 245501-1–245501-4.

Martin, J.-M. and Meybeck, M. (1979) Elemental mass-balance of material carried by major world rivers. *Marine Chemistry*, **7**, 173–206.

Mosenfelder, J.L. and Bohlen, S.R. (1997) Kinetics of the coesite to quartz transformation. *Earth and Planetary Science Letters*, **153**, 133–147.

Neusser, G., Abart, R., Fischer, F.D., Harlov, D. and Norberg, N. (2012) Experimental Na/K exchange between alkali feldspar and an NaCl–KCl salt melt: chemically induced fracturing and element partitioning. *Contributions to Mineralogy and Petrology*, **164**, 341–358.

Okamoto, A. and Shimizu, H. (2015) Contrasting fracture patterns induced by volume-increasing and -decreasing reactions: Implications for the progress of metamorphic reactions. *Earth and Planetary Science Letters*, **417**, 9–18.

Peck, D.L. and Minakami, T. (1968) The formation of columnar joints in the upper part of Kilauean lava lakes, Hawaii. *Geological Society of America Bulletin*, **79**, 1151–1166.

Plümper, O., Røyne, A., Magrasó, A. and Jamtveit, B. (2012) The interface-scale mechanism of reaction-induced fracturing during serpentinization. *Geology*, **40**, 1103–1106.

Preston, F.W. and White, H.E. (1934) Observations on spalling. *Journal of the American Ceramic Society*, **17**, 137–144.

Putnis, A. (2002) Mineral replacement reactions: from macroscopic observations to microscopic mechanisms. *Mineralogical Magazine*, **66**, 689–708.

Røyne, A. and Jamtveit, B. (2015) Pore-scale controls on reaction-driven fracturing. Pp. 25–44 in *Pore-Scale Geochemical Processes* (C.I. Steefel, S. Emmanuel and L.M. Anovitz, editors) Reviews in Mineralogy and Geochemistry, **80**. Mineralogical Society of America and the Geochemical Society, Chantilly, Virginia, USA.

Røyne, A., Jamtveit, B. and Malthe-Sørenssen, A. (2008) Controls on rock weathering rates by reaction-induced hierarchical fracturing. *Earth and Planetary Science Letters*, **275**, 364–369.

Rudge, J.F., Kelemen, P.B. and Spiegelman, M. (2010) A simple model of reaction-induced cracking applied to serpentinization and carbonation of peridotite. *Earth and Planetary Science Letters*, **291**, 215–227.

Scheidl, K.S., Schaeffer, A.K., Petrishcheva, E., Habler, G., Fischer, F.D., Schreuer, J. and Abart, R. (2014)

Chemically induced fracturing in alkali feldspar. *Physics and Chemistry of Minerals*, **41**, 1−16.

Schmidt, M.W. and Poli, S. (2014) Devolatilization during subduction. Pp. 669−701 in: *Treatise on Geochemistry (Second Edition)* (H.D. Holland and K.K. Turekian, editors). Elsevier, Oxford, UK.

Skjeltorp, A.T. and Meakin, P. (1988) Fracture in microsphere monolayers studied by experiment and computer simulation. *Nature*, **335**, 424−426.

Tildesley, D. and Allen, M.P. (1987) *Computer Simulation of Liquids.* Clarendon Press, Oxford, UK.

Ulven, O.I., Jamtveit, B. and Malthe-Sørenssen, A. (2014a) Reaction-driven fracturing of porous rock. *Journal of Geophysical Research: Solid Earth*, **119**, 7473−7486.

Ulven, O.I., Storheim, H., Austrheim, H. and Malthe-Sørenssen, A. (2014b) Fracture initiation during volume increasing reactions in rocks and applications for CO_2 sequestration. *Earth and Planetary Science Letters*, **389**, 132−142.

Yakobson, B.I. (1991) Morphology and rate of fracture in chemical decomposition of solids. *Physical Review Letters*, **67**, 1590−1593.

Zack, T. and John, T. (2007) An evaluation of reactive fluid flow and trace element mobility in subducting slabs. *Chemical Geology*, **239**, 199−216.

Kinetics of stable and radiogenic isotope exchange in geological and planetary processes

JAMES A. VAN ORMAN[1] and MICHAEL J. KRAWCZYNSKI[2]

[1] Department of Earth, Environmental and Planetary Sciences, Case Western Reserve University, Cleveland, Ohio, USA, e-mail: james.vanorman@case.edu
[2] Department of Earth and Planetary Sciences, Washington University, St. Louis, Missouri, USA, e-mail: mikekraw@wustl.edu

Stable and radiogenic isotopes play a central role in the geosciences, due to their importance as geochemical tracers, thermometers, speedometers and chronometers. This chapter focuses on the importance of isotope transport and exchange in elucidating high-temperature petrogenetic processes. Stable isotopes are fractionated in several ways, including by temperature-dependent equilibrium partitioning between phases and by mass-dependent diffusion processes. The distribution of stable isotopes among a poly-phase assemblage reflects a competition between equilibrium partitioning at interfaces between phases, and diffusive transport within the mineral interiors. Both of these processes are temperature-dependent: equilibrium partitioning becomes more pronounced and diffusion of isotopes becomes more sluggish as temperature decreases. This provides the basis for a cooling speedometer, provided that the equilibrium stable isotope fractionation factors and diffusion coefficients are known. Stable isotopes can also be fractionated by chemical diffusion within a phase, due to the dependence of diffusivity on isotope mass. This diffusion-induced isotopic fractionation provides a rigorous basis for distinguishing chemical zoning profiles produced by diffusion from similar chemical zoning produced by different processes. The magnitude of the isotopic mass dependence also provides information that can help to elucidate the diffusion mechanism.

Radiogenic isotopes have a number of important uses, from tracing chemical heterogeneity in Earth's mantle to providing absolute ages for geological events. Radiogenic isotopes have long been used to provide constraints on cooling rates, based on knowledge of the diffusivity as a function of temperature. The most commonly used models assume: (1) an infinite sink for radiogenic daughters, and (2) a simple Arrhenian dependence of the diffusion coefficient. These and other simplifying assumptions are not always met, and in these situations it is necessary to model the isotope exchange process numerically. We focus here on numerical simulations of simultaneous radioactive decay and diffusive exchange in the short-lived ^{107}Pd-^{107}Ag system in iron meteorites.

1. Isotope fractionation processes at high temperature

1.1. Equilibrium fractionation

Whenever two or more phases coexist at equilibrium, elements will be partitioned so that all phases have the same chemical potential for any given component. This usually results in dissimilar concentrations of the elements in the different phases. Isotopes of the same element will behave in the same manner, and will partition themselves between phases in different proportions in order to achieve equal chemical potential. Urey (1947) was the first to develop the thermodynamics of isotopic exchange, and his original work has been expanded and advanced by innumerable authors in the succeeding decades.

Equilibrium isotope partitioning can be thought of as a special case of the analogous process of elemental partitioning. However, where element partitioning may vary with differences in the charge, size and electronic structure of the ion, the partitioning of different isotopes of the same element is based solely on mass. The substitution of one element for another often leads to an energy difference of hundreds or thousands of Joules per mole, whereas the energy difference in substituting one isotope for another may be 2–3 orders of magnitude less. The small energy difference results in much smaller differences in the abundance of isotopes. While the ratio of trace element concentrations among equilibrated phases may range over many orders of magnitude (typically 10^{-3} to 10^3), isotopic ratios typically vary from 0.990 to 1.010. Measurement of such small differences in isotope ratio requires specialized techniques, but the benefits of using isotopes as geochemical thermometers and tracers are many. For instance the miniscule volume change of the isotope exchange reaction means that pressure often has an insignificant effect on the substitution, making geothermometers based on stable isotopes more robust than thermometers based on element substitution, which are often strongly pressure-dependent.

Typically, when discussing the equilibrium partitioning of isotopes, one considers the distribution of isotopes of a single element amongst equilibrated or partially equilibrated phases. The equilibrium fractionation factor for isotopic partitioning between two phases is defined as follows:

$$\alpha_{\phi_1-\phi_2} = R_{\phi_1}/R_{\phi_2} \qquad (1)$$

where ϕ_1 and ϕ_2 refer to two distinct phases and R is the concentration ratio of two isotopes in the phase. Because α values are typically very close to 1, especially for high-temperature processes, isotopic values are reported by convention using 'delta' notation for ease of discussion and reporting of data, where δ is defined as the permil (parts per thousand) difference between the phase of interest and a standard:

$$\delta_\phi = 10^3(R_\phi/R_{std} - 1) \qquad (2)$$

The fractionation factor is related to the delta values of two phases by the expression:

$$\alpha_{\phi_1-\phi_2} = (\delta_{\phi_1} + 10^3)/(\delta_{\phi_2} + 10^3) \qquad (3)$$

The difference between delta values for two different phases is denoted with a capital delta, and as long as the fractionation factors are close to 1, the following approximation can be used:

$$\Delta_{\phi_1-\phi_2} \equiv \delta_{f_1} - \delta_{f_2} \cong 10^3 \ln\alpha_{\phi_1-\phi_2} \qquad (4)$$

The fractionation of isotopes between different phases can be understood in terms of the local vibrational properties within each of the phases. Isotopes of the same element have different masses, and this mass difference leads to a variation in the vibrational frequency of the chemical bonds formed with adjacent atoms. Heavy isotopes vibrate at a lower frequency than lighter isotopes, and this difference in vibrational properties affects the overall energy of the system under consideration. Because a system at equilibrium will be in its lowest energy state, the isotopes will distribute themselves amongst the coexisting phases to minimize the overall free energy. The frequency of bond vibration is also dependent on temperature, and thus the equilibrium isotope partitioning is also dependent on temperature. The equilibrium isotopic fractionation between two phases at high temperatures follows a function of the form:

$$\Delta_{\phi_1 - \phi_2} = A/T^2 \qquad (5)$$

where A is an empirically determined proportionality constant, and T is absolute temperature. If the equilibrium fractionation factor for a particular pair of isotopes of the same element between two phases is known at one temperature, it can be extrapolated to other temperatures using this equation. At lower temperatures, typically <200–300°C, the relationship in equation 5 breaks down, and the isotopic variation typically has the functional form of B + A/T. Note that the equilibrium fractionation factor diminishes rapidly with temperature and is typically very small at temperatures relevant to igneous processes. Reviews of the basic isotopic fractionation equations and notation can be found in Hoefs (2008) and Criss (1999).

Because isotopic exchange is often sluggish at the temperatures at which equilibrium isotope fractionation is of significant magnitude, it is often impossible to achieve an equilibrium fractionation between two phases on laboratory timescales. Fortunately, experiments on partially equilibrated phases can yield values for the equilibrium fractionation factor. Traditional methods for obtaining the equilibrium isotope fractionation factor experimentally are of two types: the partial exchange method (Northrop and Clayton, 1966), and the three-isotope method (Matsuhisha *et al.*, 1978). The partial exchange method works for all multiple-isotopic elements, but the three-isotope method is restricted to elements that have three or more isotopes that are stable on laboratory timescales. Both methods are successful at determining equilibrium fractionation factors, even at temperatures far lower than those where complete diffusive exchange and equilibrium can be obtained.

The partial exchange technique involves carrying out multiple isotopic exchange experiments between two phases under identical conditions (time and temperature), but with starting materials that differ in initial isotopic composition from each other and from each experiment. The farther away samples are from the equilibrium isotopic values initially, the greater the change in isotopic compositions will be at the end of the experiment. The approach to equilibrium isotopic exchange from two different directions allows the final 'equilibrium' values to be interpolated on plots of (ln α_{final} − ln $\alpha_{initial}$) *vs.* ln $\alpha_{initial}$ (Fig. 1a). The y intercept on these plots corresponds to the equilibrium fractionation factor for that particular temperature. The slope of the line is related to the extent of isotopic equilibration. If the experimental series is repeated for multiple temperatures, then an equation of the form of equation 5 can be determined. An example of this method applied to iron isotope fractionation between pyrrhotite and rhyolite with detailed analysis can be found in Schuessler *et al.* (2007).

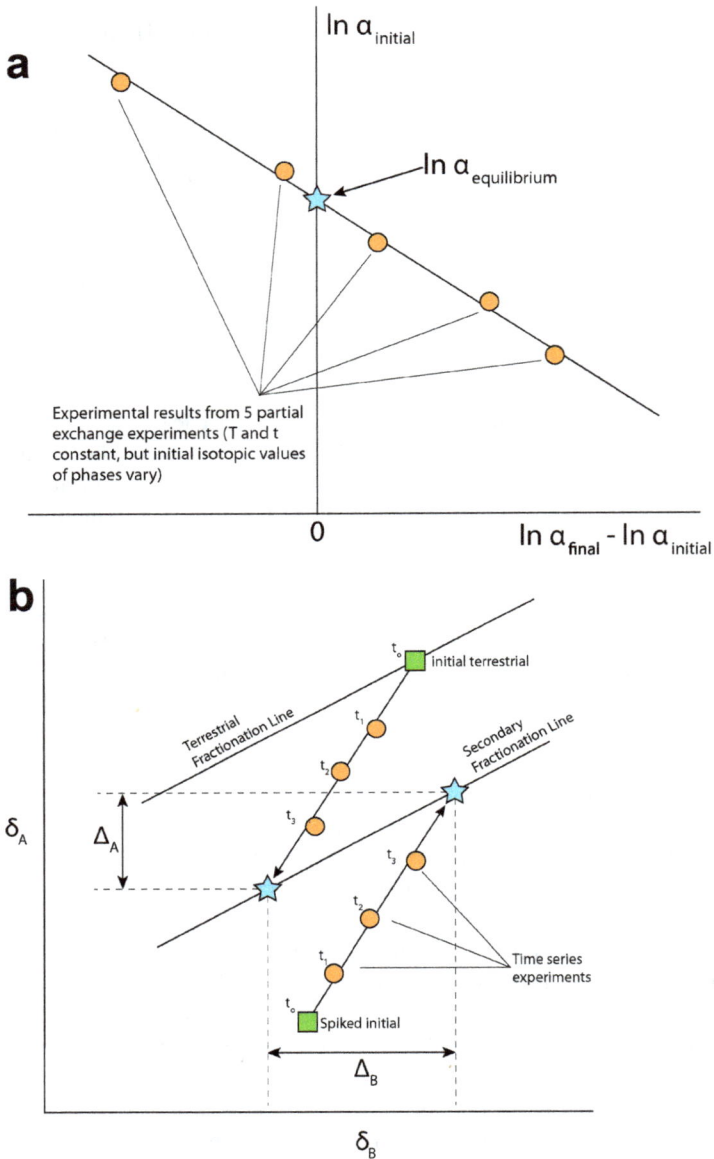

Figure 1. Schematic diagrams depicting the two most common experimental methods to determine equilibrium isotope fractionation factors. (a) The partial exchange method uses a set of two-phase equilibriation experiments conducted (in this example five experiments). Where a linear interpolation crosses the point of zero net isotope exchange corresponds to the equilibrium values. (b) The three-isotope method uses starting materials that do not lie on the same mass fractionation line. In this case experiments are run for differing lengths of time (three experiments in this case), and the linear extrapolation of these experiments will cross a secondary fractionation line where the phases are in equilibrium.

The three-isotope method uses a time series of experiments carried out under the same P-T conditions. This method relies on the fact that equilibrium isotope partitioning is solely mass dependent, and thus any equilibrium fractionation between two phases that are created from a single pool of isotopes will lie along a straight line in a plot of the delta values. The slope of this line is related to the mass differences amongst the different isotopes, and is parallel to the primary (terrestrial) mass fractionation line, which describes the isotopic compositions of natural terrestrial samples.

Equilibrium isotope partitioning data can be obtained using the three-isotope method by partially equilibrating phases that were initially formed from different isotopic pools. Starting materials for such experiments typically involve one phase that initially has normal terrestrial isotopic abundances, and a second phase synthesized with material that has an isotopic spike, thus moving it off of the terrestrial mass-dependent fractionation line. If both of the phases are then allowed to partially equilibrate they will approach equilibrium along a new mass dependent line (a secondary fractionation line) that is parallel to the terrestrial fractionation line, but offset from it (Fig. 1b). A time series of experiments is carried out, and the equilibrium values for mass-dependent fractionation can be determined by extrapolation of the isotopic values to the shared secondary fractional line. A detailed discussion of this method, and an experimental study of iron isotopes between fayalite and magnetite can be found in the paper by Shahar *et al.* (2008).

A third technique for determining the temperature dependence of equilibrium isotopic partition has been developed in the last five years using nuclear resonant inelastic X-ray scattering (NRIXS) measurements to estimate the force constants of bonds in the phases of interest (Polyakov *et al.* 2007; Dauphas *et al.* 2012). An advantage of this method is that it probes the vibrational properties that govern isotopic fractionation in each phase independently, such that no isotopic exchange among the phases is necessary to determine the equilibrium fractionation factors. Hence, it can be used even in cases where isotopic exchange is too sluggish to be observed on laboratory timescales. Previously NRIXS spectroscopy had typically been used in earth science to study sound speeds and lattice vibrational properties of minerals (Sturhahn and Jackson, 2007). This method has become viable for obtaining isotope-partitioning data due to the prevalence and accessibility of synchrotron X-ray irradiation. The NRIXS technique for determining equilibrium partitioning values for isotopes can be applied to only a narrow subset of elements that have isotopes with excited nuclear states at low energies that decay back to the ground state relatively slowly. With current technology this limits these studies to atoms that are 'Mössbauer-active' such as ^{57}Fe and ^{119}Sn. Because of its geochemical and cosmochemical relevance we will focus our discussions here on applications of NRIXS to the study of iron isotopic fractionation.

The vibrational properties of iron in condensed materials can be obtained with sufficient accuracy to estimate the magnitude of equilibrium isotopic partitioning between two phases using NRIXS spectroscopy. The NRIXS measurements provide

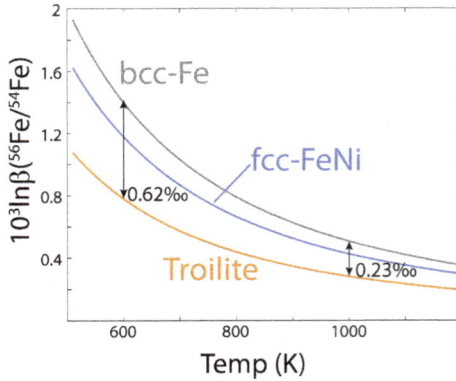

Figure 2. NRIXS reduced partition function ratios for $^{56}Fe/^{54}Fe$ *vs.* temperature calculated using the Fe force constants for the particular mineral shown. The β factors are temperature dependent and the difference between two β factors at a given temperature corresponds to the expected equilibrium isotope fractionation in permil.

information on the partial phonon density of states (PDOS) for a particular material (Sturhahn *et al.*, 1995; Sturhahn, 2004). It is partial because in this case the PDOS refers only to the vibrational properties of the iron sublattice in a material. NRIXS studies do not have to be done on crystalline material, and liquids and glasses are routinely measured. From the PDOS the force constant for the iron sublattice can be calculated, and from the force constant one can determine the reduced partition function ratio or β factors. At high temperatures the β factor for iron can be approximated by the expression (Dauphas *et al.*, 2012):

$$10^3 \ln \beta = 2904 <F>/T^2 \qquad (6)$$

where <F> is the force constant in N/m, and is related to the stiffness of iron bonds within a mineral. The equilibrium fractionation between any two phases can then be determined as the difference between the β factors for each phase:

$$\Delta^{56}Fe_{i\text{-}j} = 10^3 \ln\beta_i - 10^3 \ln\beta_j \qquad (7)$$

where *i* and *j* refer to two different phases. Figure 2 shows the β factors for several iron-bearing minerals, and the difference between the curves at each temperature represents the equilibrium fractionation at that particular temperature.

Force constants for many different iron-bearing minerals and glasses have been determined experimentally by NRIXS, including goethite, troilite, iron metal, hematite, magnetite, olivine, basalt, andesite and rhyolite (*e.g.* Sturhahn *et al.*, 1999; Alp *et al.*, 2001; Kobayashi *et al.*, 2004; Polyakov and Soultanov, 2011; Dauphas *et al.*, 2012, 2014). These data can be used to predict fractionation factors between these phases. The NRIXS technique is well suited to study minerals with very slow diffusion that are difficult to study with the partial exchange and three-isotope methods.

1.2. Kinetic fractionation

Kinetic isotope fractionation is associated with a wide range of processes. Among these, by far the most important fractionation process for minerals at relatively high temperatures is diffusion. Before turning our focus to isotope fractionation produced by chemical diffusion we will briefly discuss two other kinetic mechanisms – evaporation and Soret diffusion – that are capable of inducing substantial isotope fractionation in particular cases.

1.2.1. Evaporation

Evaporation is an important kinetic isotope fractionation process when a mineral or silicate liquid is exposed to near-vacuum conditions, or to a reducing gas in an open (or partially open) system where equilibrium between the gas and condensed phase cannot be achieved. The evaporation rate for a molecule from the mineral or liquid surface is greater when the molecule is isotopically light, and this causes evaporation residues to become progressively enriched in heavy isotopes as evaporation proceeds. High-temperature evaporation processes are particularly important in the early history of the solar system, where early condensates such as calcium aluminium inclusions (CAIs) and chondrules experienced re-heating events while exposed to the hydrogen-rich gas of the solar nebula. CAIs are often enriched in heavy isotopes of Si and Mg (*e.g.* Clayton *et al.*, 1988) and this is widely accepted as evidence that these inclusions experienced significant evaporation when heated to high temperatures (Davis and Richter, 2004). A number of experimental studies have been carried out to examine isotope fractionation in silicate liquids and minerals when they are heated and partially evaporated under vacuum (*e.g.* Davis *et al.*, 1990; Young *et al.*, 1998; Tsuchiyama *et al.*, 1999; Wang *et al.*, 1999, 2001; Richter *et al.*, 2002, 2007; Zhang *et al.*, 2014). A review of experimental observations and theoretical descriptions of isotopic fractionation by evaporation is given by Richter *et al.* (2009), and here we will mention only a few salient points regarding the basic fractionation mechanisms. Nagahara and Ozawa (2000) and Ozawa and Nagahara (2001) provide further discussion and numerical modelling of isotopic fractionation by evaporation during heating events in the early solar nebula.

At the high temperatures relevant to evaporation of silicate minerals and melts in the solar nebula ($T > 1400°C$), the kinetic fractionation factor is much larger than the equilibrium fractionation factor between the condensed phase and the gas, such that the isotopic fractionation produced by evaporation can be considered as a purely kinetic process. The kinetic isotope fractionation factor for evaporation into a vacuum, derived from the Hertz-Knudsen equation for evaporation/condensation reactions, is given by (*e.g.* Richter *et al.*, 2009):

$$\alpha_k = (\gamma_L/\gamma_H)\sqrt{m_H/m_L} \tag{8}$$

where α_k represents the instantaneous ratio of heavy (H) to light (L) isotopes in the solid (or liquid), divided by their ratio in the gas, γ is a condensation coefficient that represents the probability of condensation (or 'sticking') for a molecule striking the surface of the solid/liquid, and m is the mass of the molecule in the gas phase. Although the condensation coefficients for the two isotopes are sometimes assumed to be equal, this is often not the case, and hence the kinetic fractionation factor must be determined experimentally. Note that the mass ratio refers to that of the molecular species in the gas, which is not equivalent to the isotopic mass ratio except for monatomic gas species. For example, when SiO is the predominant silicon species in the gas phase, the masses of the ^{30}Si and ^{28}Si molecular species are 46 and 44, respectively.

Note that in equation 8 there is no explicit dependence of the fractionation factor on temperature. There can be a temperature dependence if (1) the molecular speciation in the gas changes with temperature (which would affect the molecular mass ratio) and/or (2) the condensation coefficient ratio varies with temperature. For Mg isotopic fractionation during evaporation of forsterite, no significant dependence of the kinetic fractionation factor on temperature has been found experimentally (Wang et al., 1999). On the other hand, Richter et al. (2007) found a small but systematic dependence on temperature for Mg isotopic fractionation during evaporation of a CAI-like liquid, with isotope fractionation increasing at higher temperatures. This behaviour is very different from equilibrium isotope fractionation, which diminishes rapidly with increasing temperature, as described above. In general, kinetic fractionation mechanisms have a rather weak dependence on temperature (as described further below), and as a result they tend to become more important, relative to equilibrium fractionation, at high temperatures.

1.2.2. Thermal diffusion.
A well known phenomenon in gases and liquids is the tendency of heavier isotopes to concentrate at the cold end of a temperature gradient. If the temperature gradient is held steady, the isotope concentrations evolve to a steady-state distribution along the temperature profile, on a timescale that is controlled by atomic diffusion rates within the phase. The steady-state fractionation can be characterized by the parameter δ,

$$\delta(x) = \frac{C_H(x)/C_L(x)}{C_H(x_0)/C_L(x_0)} \qquad (9)$$

where $C_H(x)$ and $C_L(x)$ are the concentrations of heavy and light isotopes at position x along the temperature profile, and x_0 is a reference position. This steady-state fractionation can be large – e.g. a temperature difference of only 50°C across molten basalt produces a 3‰ enrichment of ^{26}Mg/^{24}Mg at the cold end relative to the hot end (Richter et al., 2008).

The separation of isotopes in a thermal gradient is a natural (although not obvious) consequence of the classical kinetic theory of gases, as described, for example, by Furry et al. (1939). A first-order relation for the steady-state isotope fractionation along a temperature gradient, derived from the kinetic theory of gases, can be written as (Lacks et al., 2012):

$$\delta = -\alpha_0 \left(\frac{m_H - m_L}{m_H + m_L}\right) \left(\frac{T - T_0}{T_0}\right) \qquad (10)$$

where T_0 is the temperature for which $\delta = 0$, and α_0 is a dimensionless parameter which for hard-sphere systems has a predicted value, derived from the classical kinetic theory, of ~0.89 in the dilute gas limit. The value of α_0 has been evaluated also for dense monatomic hard-sphere fluids, and has been found to increase systematically with density, to a maximum value of ~5 at the glass transition (Kincaid *et al.*, 1987). For more complicated liquids such as silicate melts, α_0 is an empirical parameter. It has been examined theoretically *via* molecular dynamics (MD) simulations on liquid $MgSiO_3$, with absolute and relative values for Si, O and Mg in reasonable agreement with experimental data on more complex silicate liquids (Lacks *et al.*, 2012). In both experiments and MD simulations, network formers (Si, O) are found to have smaller values of α_0 than do network-modifying cations.

Note that according to equation 10 the steady-state isotopic fractionation is a function of the temperature difference but not of the temperature gradient. In other words, the steady-state fractionation does not depend on the length-scale over which the temperature changes (note, however, that the time required to achieve a steady-state distribution of isotopes along the temperature gradient does depend on the length scale). This lack of dependence of the steady-state fractionation on length-scale is confirmed by the good agreement between MD results obtained with temperature gradients of ~10^{11} K/m with experimental results obtained with gradients of ~10^4 K/m (Lacks *et al.*, 2012).

Fractionation of isotopes by thermal diffusion depends on the maintenance of a thermal gradient for a time long enough that substantial isotopic diffusion toward the steady-state distribution can occur. In gases, thermal and isotopic diffusivity are of the same order, such that a single thermal pulse can generate a gradient in temperature that persists long enough to redistribute isotopes along it. Thermal diffusion has been inferred to be responsible for fractionation of nitrogen and argon isotopes in air trapped in polar snow and ice, and this fractionation has been used to infer rapid historical changes in air temperature, on a timescale of decades (Severinghaus *et al.*, 1998; Severinghaus and Brook, 1999). For silicate melts, thermal diffusion is several orders of magnitude faster than isotopic diffusion, and special conditions are required to maintain a thermal gradient for sufficient time to produce significant isotopic fractionation. It has been hypothesized that a steady influx of magma from below to a magma chamber that is cooling from above could maintain a steady temperature gradient that leads to substantial isotopic (and chemical) fractionation in the magma (Lundstrom, 2009).

To our knowledge, no experimental or theoretical work has been done on thermally induced isotopic fractionation in crystalline solids. Isotopic fractionation by thermal diffusion in minerals is of dubious relevance to natural systems, because thermal diffusivity is many orders of magnitude faster than atomic diffusivity; maintaining a temperature gradient on the length- and time-scale of atomic diffusion would seem very difficult under natural conditions. However, isotopic fractionation by thermal diffusion is potentially significant for experimental studies of equilibrium isotopic fractionation between solids and liquids, or of the mass dependence of isotopic diffusion (discussed below). Steady-state temperature gradients are often fairly significant in high-pressure experiments, and are particularly large in diamond anvil cell experiments with laser heating (due in part to the high thermal conductivity of diamond). In experiments to determine equilibrium or diffusion-induced isotope fractionation, the results of thermal diffusion may obscure the targeted fractionation, and hence it may be critically important to understand and characterize these effects before attempting high-pressure experiments.

1.2.3. Chemical diffusion

In any phase, two isotopes of the same element generally diffuse at slightly different rates due to their difference in mass. The dependence of the diffusivity on mass can lead to significant isotopic fractionation along a chemical diffusion gradient: light isotopes, which diffuse more rapidly than their heavier counterparts, become relatively enriched at the front of a diffusion gradient, while heavy isotopes are enriched behind. As with the other kinetic fractionation mechanisms discussed above, isotope fractionation due to mass-dependent diffusivity persists at high temperatures where equilibrium fractionation becomes negligible. For this reason, and because chemical diffusion is ubiquitous in high-temperature geological processes, diffusion-induced isotope fractionation is potentially significant in a wide range of situations.

The magnitude of the mass dependence of isotope diffusivity can, however, vary greatly depending on the element and on the medium through which it diffuses. An empirical expression that is often used to describe the mass dependence of isotope diffusion in silicate melts is

$$\frac{D_L}{D_H} = \left(\frac{m_H}{m_L}\right)^{\beta} \qquad (11)$$

This expression is based on the similar expression for the mass dependence of the diffusivity in an ideal gas, for which $\beta = 1/2$. With the exception of ideal gases, β is a purely empirical parameter, and has no relation to the beta factor (reduced partition function ratio) discussed in section 1.1. In silicate melts, it has been found both in experiments and in molecular dynamics simulations that the values of β vary widely within the range $\sim 0-0.5$, and are positively correlated with the diffusivity of the ion (Watkins et al., 2011, 2014; Goel et al., 2012). Cations whose diffusive motions are not strongly bound to or inhibited by the polymerized silicate network structure have a relatively large isotopic effect (large β) compared to slower-diffusing cations like Si that are part of the network structure.

This empirical finding for silicate melts − that the isotopic mass dependence of the diffusivity increases as the diffusivity increases − does not provide reliable guidance on the isotope effect for diffusion in minerals or grain boundaries. As discussed below, the opposite trend is the normal one for ions diffusing on a particular mineral sublattice by a vacancy mechanism. Unlike in silicate melts, where the diffusion process is far more complicated, the isotope effect in minerals can be understood in terms of the classical theory for diffusion by a series of atomic jumps between sites on a crystal lattice (Vineyard, 1957; Schoen, 1958; Mullen, 1961; LeClaire, 1966). The theory and its implications for minerals were reviewed recently by Van Orman and Krawczynski (2015) and only a brief overview will be given here.

In crystals, the isotopic mass dependence of the diffusion coefficient is expressed as

$$E \equiv \left(\frac{D_L}{D_H} - 1\right) \bigg/ \left[\left(\frac{m_H}{m_L}\right)^{1/2} - 1\right] = fK \qquad (12)$$

where f is the correlation coefficient, which describes the degree to which the sequence of atomic jumps in the diffusion process deviates from a completely random walk, and K is the coupling coefficient, which essentially describes the degree to which the motion of the atom during a single jump is coupled to the motions of other nearby atoms. Note that while equation 12 takes a different form than equation 11, for isotopes with similar mass the two expressions are nearly equivalent, with $E = fK \cong 2\beta$.

Both f and K have a maximum value of unity. The correlation coefficient $f = 1$ when the diffusing atom executes a random walk, and the coupling coefficient $K = 1$ when the motion of the atom during a jump is uncoupled from that of other nearby atoms, or more precisely when the kinetic energy of the unstable vibrational mode that, when excited, leads to a jump of the atom to an adjacent unoccupied site resides completely with the migrating atom (LeClaire, 1966). The value of K has been found to vary over a relatively narrow range between ~0.5 and 1 (Allnatt and Lidiard, 1993, their table 10.2). It does not appear to vary much among different solutes diffusing in the same crystal. On the other hand, the correlation coefficient can take a wide range of values between ~0 and 1 depending on the relative frequencies of the atomic jumps involved in the diffusion process. For diffusion of a dilute impurity (trace or minor element) by a vacancy mechanism in an otherwise pure mineral, the correlation coefficient can be expressed as:

$$f = u/(u + w_2) \qquad (13)$$

where w_2 is the jump frequency of the impurity atom to an adjacent vacancy, and u is a function that contains the other atom-vacancy site exchange frequencies that are involved in diffusion of the impurity. The function u has been evaluated for only a small fraction of the mineral lattices of interest for diffusion in the geosciences, mainly simple lattices (Ghate, 1964; LeClaire, 1970). Nevertheless, even when no expression for u has been derived, equation 13 is useful for evaluating limiting cases. When the impurity jump frequency is small ($w_2 \ll u$) relative to the jump frequencies of the 'host' or solvent atoms (which means that the impurity diffuses slowly relative to the host atoms), $f \rightarrow 1$ and the isotope effect is large. In this case, the vacancy is able to randomize its position with respect to the impurity atom by migrating around and away from it between relatively infrequent impurity-vacancy site swaps. The jump sequence in this case approaches a random walk. On the other hand, when $w_2 \gg u$, $f \rightarrow 0$. In this case the jump sequence is highly correlated, because the impurity can make many back-and-forth site exchanges with the vacancy before the vacancy exchanges with other nearby atoms to change its position with respect to the impurity. Hence, an impurity that has a relatively low migration barrier (and large jump frequency) will have a small isotope effect. Note that, in contrast to melts, slowly diffusing cations in minerals are expected to have a large isotope effect.

Some other generalizations we can make about the magnitude of the isotope effect (E) in minerals are as follows.

- For impurity diffusion by a vacancy mechanism, because w_2 and u generally have different temperature dependences, f and therefore the isotope effect E are temperature dependent. In general, the isotope effect for slow-diffusing cations ($w_2 < u$) becomes somewhat smaller at higher temperatures, while the isotope effect for highly mobile cations ($w_2 > u$) increases with temperature.
- The isotope effect is a function of major-element composition, for all ions that diffuse on the same sublattice(s) on which solid solution occurs.
- The isotope effect is large ($E \approx 1$) for elements that diffuse by a simple interstitial mechanism, for which the jumps are uncorrelated ($f = 1$). However, when jumps to both interstitial and regular lattice sites are involved in diffusion (as in an interstitialcy mechanism), or when interstitials interact substantially with vacancies or other point defects, the isotope effect is reduced.
- Significant anisotropy in the isotope effect can arise (for non-cubic minerals) due to a change in diffusion mechanism with orientation or due to changes in the relative jump frequencies as a function of orientation.
- The isotope effect is generally expected to be smaller for diffusion along grain boundaries and dislocations than for diffusion in the corresponding mineral. The jumps in dislocations and grain boundaries become more correlated due both to dimensional restrictions, and to the lack of uniformity in the structure within these extended defects.

Figure 3 illustrates how isotopes are fractionated along a diffusion profile, in an effectively infinite medium with initial step function in concentration. The examples shown are for Fe^{2+} and Ni^{2+} in periclase (MgO). Both the diffusion coefficients and their isotopic mass dependence are based on the first-principles calculations reported by Crispin *et al.* (2012). In both panels of Fig. 3, the length-scale of the diffusion profiles is normalized by time, and by the nickel diffusion coefficient. The isotopic fractionation goes to zero far from the interface, where there is no diffusive flux , and is also zero at the interface where the elemental concentration remains constant (for the present case, where the diffusion coefficient is not a function of position). Light isotopes are enriched at the front of the diffusion gradient, and heavy isotopes behind. Note that this can lead to complicated effects in the special case of 'uphill diffusion', where diffusion may proceed up the concentration gradient, leading to a concentration of heavy isotopes at low elemental concentrations. For example, in experiments with ugandite-rhyolite silicate liquid diffusion couples, Watkins *et al.* (2009) found that heavy Ca isotopes were enriched on either end of the couple and depleted toward the centre, due to diffusion of Ca in both directions.

The magnitude of the isotopic fractionation produced by chemical diffusion depends on several factors. Clearly, it depends on the isotopic mass dependence of the diffusion coefficient, E; in Fig. 3, $E_{Ni} = 0.467$ and $E_{Fe} = 0.058$. It also depends on time — the isotopic fractionation profile relaxes along with the chemical diffusion profile — and on relative position along the diffusion profile, as shown in Fig. 3. The magnitude of the

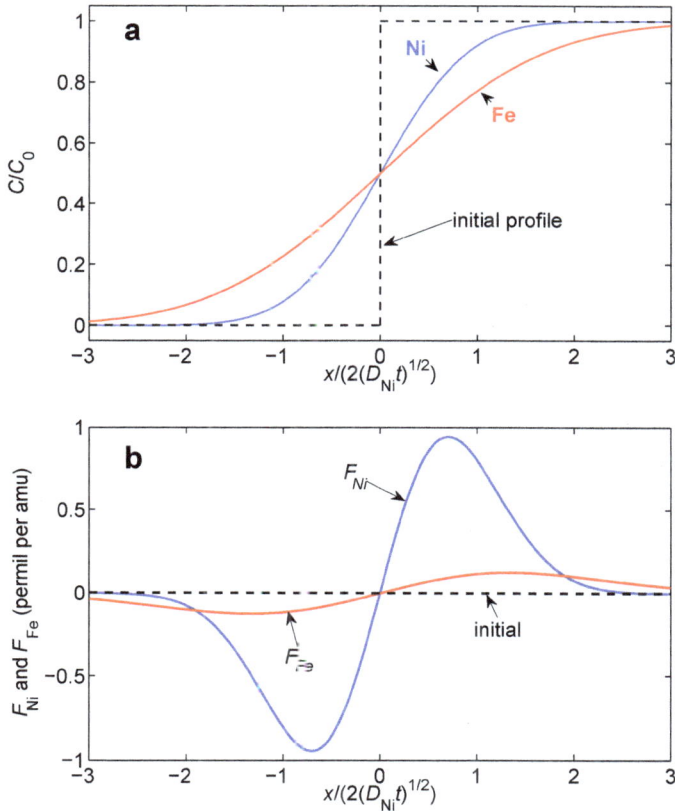

Figure 3. Theoretical diffusion (a) and associated isotopic fractionation (b) profiles for Fe^{2+} and Ni^{2+} in periclase (MgO) at 1500 K, based on first-principles calculations (after Van Orman and Krawczynski, 2015). Isotopic fractionation of Ni and Fe isotopes are represented by F_{Ni} and F_{Fe}, the ‰ per a.m.u. difference from the initial (homogeneous) isotopic composition. For example, F_{Fe} can be calculated as $\delta^{56/54}/(56-54)$, or as $\delta^{57/54}/(57-54)$.

diffusive isotopic fractionation further depends on the initial ratio of concentrations on either side of the diffusion interface, being greatest when the concentration on one side is initially zero. When there is only a small ratio of concentrations across the interface, as is typical for Si in the interdiffusion of silicate melts, for example, the magnitude of isotopic fractionation along the profile is limited (*e.g.* see fig. 9 in Richter *et al.*, 2003).

Some recent studies have attributed isotopic fractionation in natural samples to chemical diffusion, and demonstrated how this could provide important complementary information that is not provided by the chemical concentration profiles alone. In iron meteorites, Dauphas (2007) observed a relative fractionation of both Fe and Ni isotopes between kamacite and taenite, and demonstrated that this was probably produced by Fe-Ni interdiffusion during exsolution of the kamacite. Similarly, fractionation of Mg and Fe isotopes in olivine phenocrysts from a Hawaiian lava lake has been shown to be consistent with mass-dependent fractionation associated with Fe-Mg interdiffusion in the olivine (Teng *et al.*, 2011; Sio *et al.*, 2013; Oeser *et al.*, 2015).

The isotopic fractionation associated with the chemical concentration profiles gives additional information that may be important to their interpretation, and to any inferences of geological rates and timescales that are derived from them. A number of processes that do not involve diffusion are capable of producing a chemical concentration profile that is difficult to distinguish from a diffusion profile without additional information. For example, fractional crystallization from a finite well mixed melt reservoir, in the absence of any diffusion in the crystallizing phase, can produce a concentration gradient that closely mimics a diffusion profile. However, a concentration profile produced by fractional crystallization, without diffusion, will have no corresponding isotopic fractionation (other than that due to small – probably negligible at magmatic temperatures – equilibrium fractionation between the crystal and melt). A true diffusion profile, on the other hand, will have an isotopic concentration profile that corresponds in a predictable way with the chemical diffusion profile. In many cases, both diffusive and non-diffusive processes may have contributed to an observed, quenched-in concentration profile. If the mass dependence of the isotopic diffusivity is known, and the isotopic fractionation measured, the diffusive contribution to the chemical profile can be isolated.

2. Applications to cooling speedometry: Iron meteorites

The temperature dependence of isotopic fractionation (equation 5) can be used as a geothermometer. The most straightforward way to apply such a thermometer is using two totally equilibrated phases that are quenched rapidly to freeze-in their isotopic compositions. However, it is rarely the case that phases achieve an equilibrium distribution of isotopes and then preserve it perfectly upon cooling to lower temperatures. The most common situation in metamorphic materials (slowly cooled coarsely crystalline rocks), is for phases to be zoned with respect to their isotopic composition. These zoning patterns can be modelled to understand the temperature-time history of a particular rock.

In this section we will focus on the application of stable and radiogenic isotopes to determine cooling rates and cooling ages in geological and planetary samples. The methods discussed are applicable to a broad range of geological and planetary processes, but we will focus here on one relatively simple example: iron meteorites. Iron meteorites are thought to represent the solidified cores of planetesimal bodies that were among the first large bodies to form in the early solar system. Iron meteorite samples collected on Earth's surface are thought to represent more than 50 distinct planetesimals, and they provide an important record of processes in the solar nebula in the period leading up to the planet-building stage (Goldstein *et al.*, 2009). The distribution of cooling rates among the different bodies provides information on the distribution of planetesimal sizes and may provide some constraints on the collisional history of these bodies as well. Some iron meteorites thought to be derived from a single parent body (the IVA group) record a wide range of cooling rates, all relatively high, that vary systematically with their inferred radial distance from the centre of the body, and it has been inferred that this body had its silicate mantle stripped away during a hit-and-run collision (Yang *et al.*, 2007).

Our primary purpose in this section is to illustrate some of the processes that control the distribution of isotopes at high temperatures, and factors that influence the closure temperature. A number of simplifying assumptions are made in modelling these systems, some of which may not be justified for application to real systems.

2.1. Stable isotopes: Iron isotope fractionation

Cooling rates of iron meteorites are often estimated by modelling the subsolidus chemical evolution of exsolved kamacite and taenite (Yang and Goldstein, 2006). This approach, often referred to as the metallographic approach to estimating cooling rates, uses sophisticated Fe-Ni chemical interdiffusion models. It has been used successfully but is difficult to apply to meteorites with exsolution textures that are either as coarse as the hand sample size (hexahedrites), or with no clear exsolution textures due to high Ni contents (ataxites).

Here we introduce an alternative method for estimating the cooling rate of iron meteorites based on the partitioning of Fe isotopes between metal and troilite (FeS). The equilibrium partitioning of Fe isotopes is heavily dependent on temperature, with larger fractionation at lower temperatures. Slowly cooled metal–troilite pairs will preserve larger Fe isotope fractionations than rapidly cooled pairs because they are able to equilibrate diffusively to lower temperatures. Troilite is a ubiquitous phase in all Fe-meteorite groups (Buchwald, 1975; Yang and Goldstein, 2006), and Fe self-diffusion data have been determined experimentally for metal and sulfides (Condit *et al.*, 1974; Lübbehusen and Mehrer, 1990). Using the equilibrium fractionation of Fe isotopes between metal and troilite, and a numerical model for diffusion of iron isotopes, closure temperatures and cooling rates can be estimated independently from those determined by metallographic studies.

There is an observed isotopic fractionation between metal and troilite in every class of iron meteorite that has been measured. Typically the magnitude of this difference is between 0.1 and 0.8‰ $\Delta^{57}Fe_{metal-troilite}$ (Weyer *et al.*, 2005; Williams *et al.*, 2006). This fractionation is driven by the equilibrium partitioning of iron isotopes, but the magnitude of the fractionation is variable because the meteorites have various grain sizes, cooling rates and modal abundances, all of which influence their degree of reequilibration during cooling. Large fractionations generally correspond to the slowest cooling rates, because the closure temperature of the system is lower for slower cooling.

The intercrystalline closure temperature for stable isotope diffusion between two minerals is a function of cooling rate, modal abundance, grain size and diffusivity of the elements of interest. In actuality systems never fully 'close', but they will reach a temperature range below which the diffusive length scale of isotopes is so small that it has no measurable effect on the bulk isotopic partitioning. Eiler *et al.* (1993) presented a numerical finite difference model to determine the closure temperature for isotopic exchange among two or more coexisting mineral phases. This model, called the fast grain-boundary (FGB) model, assumes that all mineral grains in a system are able to maintain equilibrium compositions of isotope partitioning at their grain boundaries at all times. In other words the system is modelled based on the assumption that rapid grain-boundary transport of atoms provides for efficient chemical communication among all of the grains in the sample. In this model, the diffusive exchange of isotopes is controlled solely by volume diffusion in the minerals, not by transport through the grain-boundary network. This attribute makes the model much simpler than it would be if diffusion through the grain boundaries were treated explicitly. In that case, diffusive exchange among multiple minerals would need to be considered as a complicated three-dimensional problem. The fast grain-boundary assumption allows for simple diffusion geometries (*e.g.* spheres) to be considered instead. Here we adopt the FGB model as a matter of convenience: it is the simplest model that explicitly takes into account the diffusivity in all mineral phases, as well as mass-balance constraints, and the finite volume of each isotopic reservoir. Whether grain boundaries do indeed provide a continuously homogenizing pathway for diffusive exchange between mineral grains depends on many factors, as discussed for example in the review by Dohmen and Milke (2010). Rapid diffusion in the grain boundary relative to the minerals is a necessary but not sufficient condition. It is also necessary that the diffusive penetration along the grain boundary is large compared to the distance between mineral grains. Partitioning of elements between the minerals and the grain boundary is also important – the FGB model is more likely to apply for elements that segregate strongly to grain boundaries. Dohmen and Chakraborty (2003) explicitly considered transport through an integranular fluid in their model of diffusive exchange between two minerals, and identified several kinetic regimes among which the FGB model is one end-member.

To simulate the isotopic evolution of coexisting metal and troilite we use the FGB algorithm adapted from Eiler *et al.* (1993) by Van Orman *et al.* (2006). The equilibrium fractionation factors for $^{57/54}$Fe have been calculated using the force constant of iron in each phase determined by NRIXS spectroscopy. Iron-nickel metal initially forms as taenite (face centred cubic iron) and depending on the nickel content, unmixes upon cooling to a mixture of kamacite (body centred cubic iron) and taenite (Yang *et al.*, 1996). The force constants for iron in kamacite, taenite and troilite are 175, 148 and 100 N/m, respectively (Dauphas *et al.*, 2012; Krawczynski *et al.*, 2014). Using equation 6, the temperature dependent β factors are determined, and are plotted in Fig. 2. These β factors allow the calculation of the equilibrium fractionation factor (α) at every temperature-time step in the simulation. The β factor for the metal, which controls its partitioning of Fe isotopes with troilite, is the average value for kamacite and taenite, weighted according to their proportions in the metal.

A schematic of the fast grain-boundary isotope exchange model as applied to iron meteorites is depicted in Fig. 4. Because the grain-boundary network is assumed to remain compositionally homogeneous, with equilibrium maintained at grain surfaces,

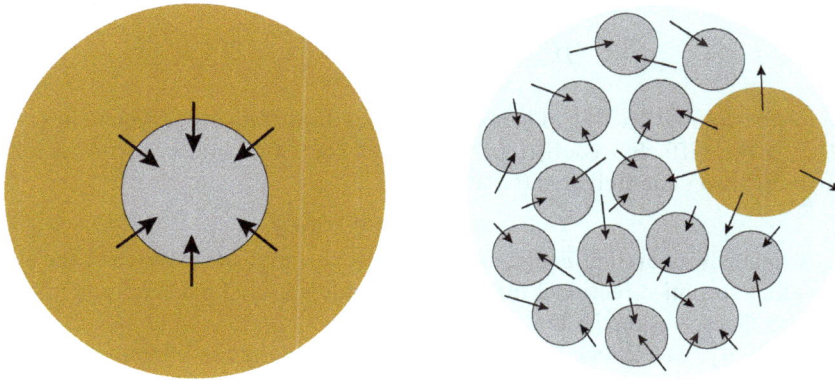

Figure 4. Iron isotope exchange models for iron meteorites. Left: Schematic of the classic Dodson model, with metal grain (centre) exchanging Fe isotopes with an infinite reservoir of troilite. Right: Fast grain boundary model in which Fe isotopes are exchanged between finite reservoirs of iron and troilite grains. The grain-boundary network (light blue) in this model is not a sink (it has effectively zero volume), but instead simply provides a medium through which the metal and troilite maintain equilibrium with each other at their surfaces. Here, the metal grains are assumed to maintain constant volume, and after exsolution begins the diffusion coefficients in the bulk metal are taken to be the weighted average of those for kamacite and taenite (see text for details).

it is not necessary to consider the spatial distribution of different mineral grains in the FGB model. Here, both troilite and metal grains are treated as spheres that maintain constant size (*i.e.* there are no moving boundaries). The spheres remain equilibrated with each other at their surfaces throughout the simulation *via* a continuously homogenizing medium which represents the fast grain-boundary network. Here, we set the volume of the grain-boundary network to be small enough that it is negligible in terms of the mass balance. In other words, the grain-boundary network is not a significant sink for Fe isotopes, as would be expected in most natural situations (unless the grain size were extremely small).

The metal, once it begins to exsolve kamacite, becomes a two-phase composite material with complicated diffusion properties. In detail, the diffusivity of Fe in the metal during exsolution will depend on the time-dependent three-dimensional distribution of the two phases within the metal, as well as the distribution of nickel within the fcc taenite phase (because the Fe diffusion coefficient is a function of composition in fcc iron-nickel alloys; *e.g.* Yunker and Van Orman, 2007). A full treatment of diffusion in the metal phase is beyond the scope of the present chapter. Instead, we simplify the problem by treating the exsolving metal as a bulk composite material with spatially uniform effective diffusivity. The exsolved metal, with Widmanstätten texture, can be considered to be a laminated composite of kamacite and taenite sheets. The effective diffusion coefficient for flow perpendicular to the sheets is (Crank, 1975)

$$D_{\text{perp}} = \left(\frac{x}{D_{\text{kam}}} + \frac{x-1}{D_{tae}}\right)^{-1} \tag{14}$$

where x is the volume fraction of kamacite, D_{kam} is the diffusion coefficient in kamacite and D_{tae} is the diffusion coefficient in taenite. For flow parallel to the sheets, the effective diffusion coefficient is (Crank, 1975)

$$D_{\text{par}} = xD_{\text{kam}} + (1-x)D_{\text{tae}} \tag{15}$$

We estimate the overall effective diffusivity in the composite metal to be the average of the values for uni-directional diffusion perpendicular and parallel to the kamacite and taenite sheets, respectively.

For simplicity, we assume here that metal and troilite are in chemical equilibrium with respect to elemental exchange and remain so throughout the simulation. Under this assumption, there is no significant diffusive fractionation of isotopes due to chemical diffusion. In reality there is likely to be some chemical exchange of iron and nickel between metal and troilite during cooling, due to the temperature dependence of the Fe-Ni exchange coefficient between metal and troilite, which could drive significant diffusive isotope fractionation during cooling. To our knowledge, diffusive fractionation of this kind has not been considered quantitatively in the context of its possible influence on stable isotope thermometry and cooling speedometry, but it could be important and should be considered in future work on this topic. There is isotopic fractionation of Ni between kamacite and taenite in iron meteorites, which has been attributed to diffusive separation during Fe-Ni interdiffusion and exsolution and modelled quantitatively (Dauphas, 2007). It is possible that such internal diffusion-driven fractionation could have an observable influence on Fe isotope fractionation between the metal and troilite, but we do not consider these effects here.

Figure 5 shows an example of an Fe isotope exchange simulation. Light isotopes prefer to be incorporated in the sulfide instead of the metal, and during cooling the equilibrium fractionation of iron isotopes between sulfide and metal becomes progressively larger. However, as diffusion slows during cooling, the metal can no longer homogenize to achieve total equilibrium, and so develops an evolving diffusion profile of stable iron isotopes (Fig. 5). Diffusion in troilite, in contrast, is fast enough that it is able to equilibrate with the surface metal throughout the entire grain. As cooling continues, the temperature eventually becomes low enough that the diffusion profiles cease to evolve and become effectively frozen in (Fig. 6). The apparent temperature is calculated at each time step by integrating the bulk isotopic composition over the entire diffusion profile for each phase, and comparing this to the equilibrium fractionation determined by the β-factor curves. The final 'frozen-in' apparent temperature represents the closure temperature for a given set of conditions.

Simulations were run at cooling rates from 1 to 10^5 °C per million years, with metal and troilite spheres that range from 0.25 to 8 mm in radius. The closure temperatures calculated for these conditions are shown in Fig. 7. Also plotted on this figure are the 'apparent' temperatures for several iron meteorites based on the bulk isotopic

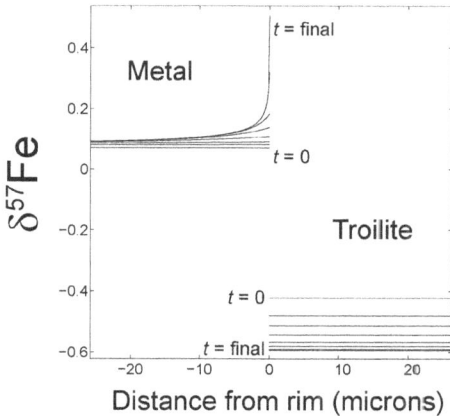

Figure 5. Isotopic compositions of coexisting bcc Fe-metal and troilite during cooling. The highest temperature is represented by the initial time t = 0, where the isotopic fractionation is smallest. At the grain interface (distance = 0), the isotopic value is maintained at equilibrium, and this boundary condition is used as a boundary condition for diffusion calculations. Diffusion is so rapid in troilite over the temperatures studied that it is able to requilibrate totally (flat lines) at each time step.

compositions of the troilite and metal phases along with the grain size of the metal in each iron meteorite (Williams *et al.*, 2006). An estimated cooling rate for a particular meteorite can then be found as the intersection between the apparent closure temperature and the metal grain size.

The closure temperatures calculated here using the fast grain-boundary model differ from those calculated using the classical Dodson (1973) model. While the Dodson (1973) model assumes diffusion from a reservoir outward to an infinite sink, troilite is actually a finite sink for the exchange of Fe isotopes, and the closure temperature depends on the fraction of troilite in the system. The difference between the Dodson (1973) closure temperature and the more realistic closure temperature calculated using the fast grain boundary model is shown in Fig. 8, as a function of the troilite fraction, cooling rate and grain size. The closure temperature decreases with the fraction of troilite; when the troilite fraction is small, even a small amount of diffusive exchange with the metal changes its isotopic composition significantly, and the system continues to evolve to lower temperatures. The difference between the Dodson and FGB model is greater at higher cooling rates and/or larger grain sizes.

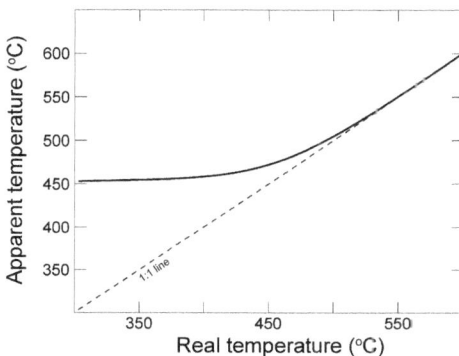

Figure 6. Calculated apparent temperature using the isotopic compositions of coexisting metal and troilite *vs.* the actual temperature of a cooling sample. After a sample reaches its closure temperature, isotopes no longer exchange at a meaningful level, locking in the temperature.

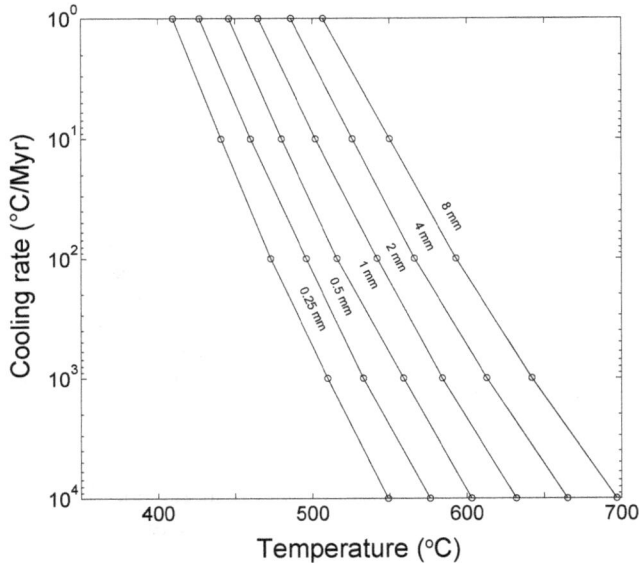

Figure. 7. Closure temperature, calculated as a function of cooling rate and contoured for grain size of the metal crystals. Each open circle represents a model run at a particular cooling rate and grain size (radius). All simulations were conducted on systems with 20 vol.% troilite. The metal phase is assumed to be pure kamacite, and thus these closure temperatures are applicable to low-Ni meteorites.

2.2. Radiogenic isotopes: The ^{107}Pd-^{107}Ag system

Variations in the abundances of radiogenic isotopes, relative to their stable, non-radiogenic counterparts, are, in most cases, controlled primarily by radioactive decay rather than the equilibrium or kinetic isotope fractionation processes discussed above. One of the primary uses of radiogenic isotopes is to provide absolute dates on geological events – and one of the critical long-standing issues is to interpret the significance of the age obtained from a particular isotopic system in a particular material. In many cases the age represents a 'cooling age' – the time at which the system became effectively closed to isotopic exchange during cooling. The classical treatment of this problem originates with Dodson (1973), who derived an analytical solution for the closure temperature for a simplified case that is nonetheless relevant to many situations and is still in wide use today. However, there are many situations where the Dodson (1973) closure temperature model, and related analytical models such as that derived by Ganguly and Tirone (1999), do not apply. In these situations it is necessary to address the closure problem numerically. Here we will address a particular example – the ^{107}Pd-^{107}Ag system in iron meteorites – to illustrate the various controls on closure temperature, and the information that can be gleaned from the system given constraints on the diffusion parameters.

Palladium-107, which undergoes beta decay to ^{107}Ag with a half-life of 6.5 million years, is now extinct but was present in the early solar system during the formation and evolution of protoplanets. The ^{107}Pd-^{107}Ag system has great potential for providing

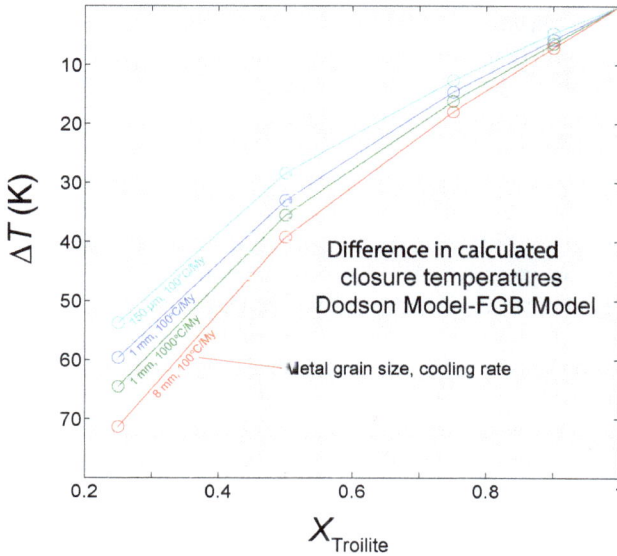

Figure 8. Difference in Fe isotope closure temperatures calculated using the Dodson (1973) model, which assumes diffusive exchange between a spherical metal grain and an infinite sink, and the fast grain boundary (FGB) model, where the sink (troilite) is finite. The FGB closure temperature decreases with the fraction of troilite; when the troilite fraction is small, even a small amount of diffusive exchange with the metal changes its isotopic composition significantly, and the system continues to evolve to lower temperatures. The difference between the Dodson and FGB model is greater at higher cooling rates and/or larger grain sizes.

detailed information on the early history of iron meteorite parent bodies, in part because: (1) the parent element, Pd, is strongly siderophile, and hence highly enriched in iron meteorites; (2) ^{107}Pd has a much shorter half-life than other siderophile radioactive isotopes such as ^{187}Re (41 billion years) and ^{190}Pt (650 billion years), and hence may provide more precise ages for early solar system events; and (3) while the metal in iron meteorites is the primary host for ^{107}Pd, there also exists a strong sink for the daughter product – ^{107}Ag partitions strongly into troilite (FeS), which is common in iron meteorites – making high-resolution dating based on metal-troilite isochrons possible.

The theoretical starting point for ^{107}Pd-^{107}Ag age considerations in iron meteorites is at the solidus of the metal–sulfide system (~990°C), where troilite crystallizes. At the solidus, palladium partitions strongly into the metal (taenite), whereas silver partitions strongly into troilite. In the case of very rapid cooling, allowing no diffusive transfer of ^{107}Ag from metal to troilite, all ^{107}Ag produced in the metal below the solidus would remain there. Hence, the metal would evolve an increasingly radiogenic Ag isotopic composition, reaching a constant value at the extinction of ^{107}Pd. Combined with the comparatively non-radiogenic Ag isotopic composition of the troilite, this provides the initial (at the solidus) value for the concentration of ^{107}Pd in the iron meteorite, as depicted in an isochron-type plot on Fig. 9. Comparison of the inferred initial $^{107}Pd/^{108}Pd$ value with that inferred from other rapidly cooled early solar objects would provide information on the time that the parent body's core solidified, relative to the timing of other events in the early solar system.

If, on the other hand, cooling is slow enough to allow the transfer of ^{107}Ag from metal to troilite below the solidus temperature, less ^{107}Ag will accumulate in the metal and the isochron age will be correspondingly younger. In this case, the inferred age would reflect not the final solidification of the core, but some later point in time along the cooling path of the parent body's core. Interpreting the age accurately requires knowledge of the diffusion properties of the minerals and simulation of the diffusive exchange process. Here we will apply to this problem essentially the same fast grain-boundary diffusion model that is described above, modified to account for the simultaneous decay of ^{107}Pd and ingrowth of ^{107}Ag in each of the phases during cooling. Jenkin *et al.* (1995) developed a numerical model that extends the fast grain-boundary model of Eiler *et al.* (1993) to radiogenic isotopes. Here, as above, we use a similar model based on the model described by Van Orman *et al.* (2006) for simultaneous decay and diffusive exchange of U-series isotopes (a process that has an additional complication: in that case the isotopes produced by decay are themselves radioactive, whereas in this case they are stable). An additional example of the model's application to the closure temperature of extinct radionuclide systems used for dating of early solar system events can be found in Kleine *et al.* (2008).

In the modelling presented here we assume, as above with respect to stable isotopes, that equilibrium partitioning of Pd and Ag is maintained at the surfaces of metal and

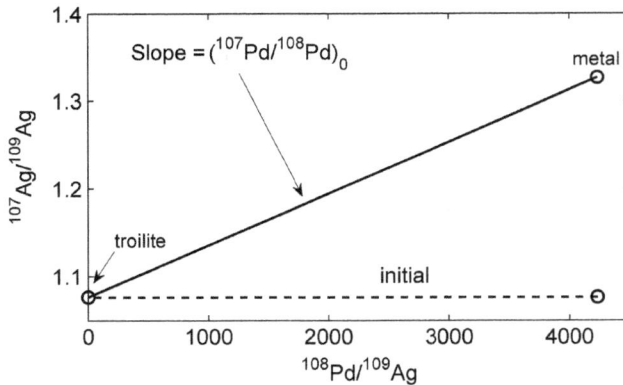

Figure 9. Schematic of an isochron-type plot for the ^{107}Pd-^{107}Ag system in iron meteorites. ^{108}Pd and ^{109}Ag are stable, non-radiogenic isotopes. Initially, at the solidus, troilite and metal have identical silver isotopic compositions (dashed horizontal line). Decay of ^{107}Pd produces ^{107}Ag in the metal. If there is no diffusive exchange of Ag isotopes with the troilite, the troilite composition remains essentially constant while the metal evolves an elevated $^{107}Ag/^{109}Ag$ ratio that stops changing when ^{107}Pd is exhausted by decay, represented by the solid line. The slope of this line is equal to the initial $^{107}Pd/^{108}Pd$ ratio. Diffusive exchange of silver isotopes between metal and troilite would rotate the solid line towards the horizontal, resulting in a lower apparent $^{107}Pd/^{108}Pd$ initial ratio, and hence a younger apparent age.

troilite. We further assume, for simplicity, that Pd and Ag partition coefficients remain constant during cooling. Because Pd partitions strongly into metal (with a metal/troilite partition coefficient of ~1700), and Ag strongly into troilite (with a metal/troilite partition coefficient of ~0.1), changes in their partition coefficients with temperature are unlikely to have a large influence on the present results. Because the partition coefficients are assumed to remain constant during the simulation, there are no net chemical fluxes, but instead only diffusion of radiogenic ^{107}Ag from metal to troilite. Hence, diffusive fractionation of Ag isotopes is not significant, and is not considered here. However, in other cases there is likely to be significant chemical diffusion of the daughter and/or parent element during cooling, which could be due to temperature-dependent partitioning among the phases, and/or to original growth zoning. For example, in the Lu-Hf system, the parent nuclide (^{176}Lu) is more mobile than the daughter, and Bloch and Ganguly (2015) showed that diffusion of the parent due to growth zoning and/or changes in the partition coefficient with temperature could have a major impact on the isochron ages. In such cases, it is also possible that diffusion-induced isotopic fractionation could be substantial — but to our knowledge this possibility has not been addressed quantitatively in any system.

As above for stable isotopes, we assume that the metal exsolves kamacite according to the equilibrium Fe-Ni phase diagram, and that diffusion in the two-phase metal can be treated in terms of an effective diffusion coefficient that is a function of the diffusion coefficients in kamacite and taenite and the relative proportions of these two phases in the metal. Silver diffusion data for kamacite are from Oikawa (1983), and Ag diffusion coefficients in taenite, as a function of Ni concentration, were estimated using a homologous temperature scaling applied to data for Ag diffusion in Ni reported in Klotsman (1989). Although no Ag diffusion data are available for troilite, cation diffusion in troilite (Condit *et al.*, 1974) is generally several orders of magnitude faster than in fcc or bcc iron alloys. We assume, therefore, that silver diffusion in troilite is sufficiently rapid to homogenize Ag isotopes in the troilite, at conditions where there is significant diffusive transfer of Ag from the metal.

Figure 10 shows the results of closure temperature simulations for the ^{107}Pd-^{107}Ag system in an iron meteorite with 5 cm metal grain radii (representing an estimate of the maximum distance of metal from a troilite surface in the Cape York meteorite). Whereas the Dodson (1973) and related closure temperature models assume that an infinite external sink exists for radiogenic daughter isotopes, troilite does not approximate an infinite sink for ^{107}Ag produced in the metal in iron meteorites. With decreasing troilite abundance, the closure temperature diminishes significantly (Fig. 10a). This behaviour can be understood in terms of the sensitivity of the troilite Ag isotope composition to the abundance of troilite. At smaller abundances, the Ag isotopic composition of the troilite is more sensitive to diffusive transfer ^{107}Ag from the metal, because there are fewer non-radiogenic silver isotopes to dilute the signature. Troilite hence continues to evolve towards more radiogenic compositions at lower temperatures, leading to a younger apparent age and lower closure temperature at smaller troilite abundances. For a troilite abundance of 5% (representing the

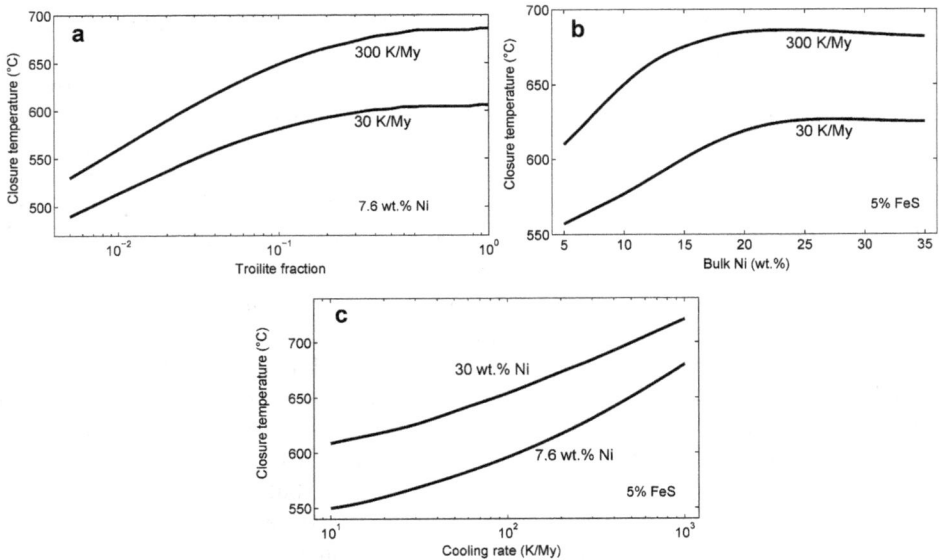

Figure 10. Closure temperature for ^{107}Pd-^{107}Ag metal-troilite isochron system in iron meteorites, as a function of: (a) the troilite abundance; (b) the bulk nickel concentration; and (c) the cooling rate. In each case the metal radius (*i.e.* maximum distance from metal to troilite) is assumed to be 5 cm, and the closure temperature refers to the simulated bulk metal-troilite isochron age.

approximate abundance in the Cape York meteorite), the closure temperature is ~50°C lower than it would be if troilite were an effectively infinite sink (corresponding to a troilite fraction approaching unity; Fig. 10a).

The dependence of the closure temperature on bulk nickel concentration (Fig. 10b) is the consequence of another factor that is not considered in standard closure temperature models. In the Dodson (1973) and other models, the diffusivity has a simple Arrhenius-type dependence on temperature. However, in iron meteorites the exsolution of kamacite leads to a strong non-Arrhenian temperature dependence of the bulk metal diffusivity over part of the cooling path. Diffusion in kamacite (bcc) is two orders of magnitude faster than in taenite (fcc), and as the temperature decreases into the two-phase kamacite-taenite field, the bulk diffusivity in the metal actually increases for a brief interval (where the kamacite proportion increases rapidly) before falling again as temperature decreases further. The smaller the bulk-nickel concentration, the larger the proportion of kamacite, and the larger the bulk diffusivity in the metal at a given temperature. Hence, the closure temperature decreases significantly with decreasing Ni abundance, below ~20–25 wt.% Ni. At higher bulk-nickel concentrations, the proportion of kamacite is too small — at temperatures where Ag diffusion is significant — to have a substantial influence on the bulk diffusivity. There is a slight decrease of the closure temperature with increasing Ni concentration above ~20–25 wt.%, because Ag diffusivity increases with Ni concentration in taenite. While the non-Arrhenian behaviour considered here is due to the exsolution of a phase with faster cation diffusivities, it should be noted that deviations from Arrhenian

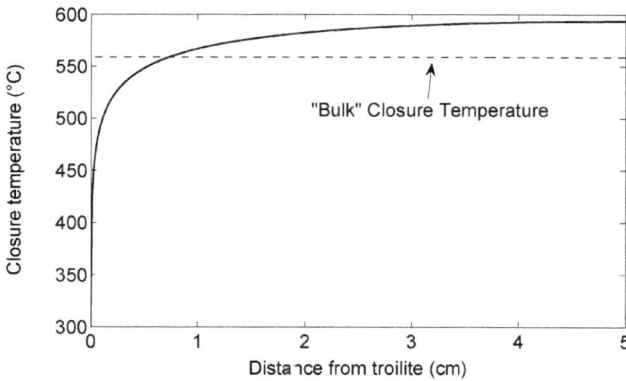

Figure 11. Example of a simulation for the ^{107}Pd-^{107}Ag metal-troilite closure temperature, as a function of distance from troilite in the metal.

behaviour can arise in other situations as well. When there is a change in diffusion mechanism — as in magnetite, for example, where the dominant cation diffusion mechanism is either an interstitialcy or a vacancy process, depending on temperature, oxygen fugacity and composition (see Van Orman and Crispin, 2010) — the diffusion coefficient does not follow a simple Arrhenius curve, and this must be considered in the closure temperature model.

The closure temperatures shown in Fig. 10 refer to the isochron ages for bulk metal and troilite. However, the isotopic composition of the metal during cooling is not homogeneous — due to diffusion, the ^{107}Ag concentration decreases near the troilite. If the metal is sampled as a function of distance from the troilite, rather than in bulk, an array of isochron ages is obtained with younger ages near the troilite and older ages further away; in other words, the closure temperature is lower near the troilite, and higher further away (Fig. 11). This change in closure temperature with distance could

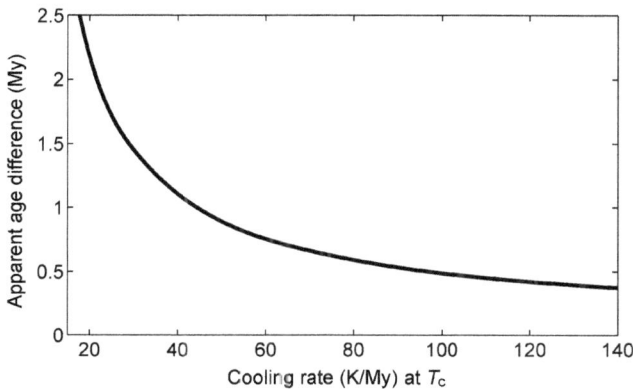

Figure 12. Difference in the time of ^{107}Pd-^{107}Ag closure in metals distant from (5 cm away) and close to troilite (0.5 cm away), as a function of the cooling rate.

be used to constrain the cooling rate of an iron meteorite. If distinct isochrons are obtained for metal close to and far from troilite, the difference in age for these two isochrons could be used to determine the cooling rate (Fig. 12).

References

Allnatt, A.R. and Lidiard, A.B. (1993) *Atomic Transport in Solids*. Cambridge University Press, Cambridge, UK.

Alp, E.E., Sturhahn, W. and Toellner, T.S. (2001) Lattice dynamics and inelastic nuclear resonant X-ray scattering. *Hyperfine Interactions*, **135**, 295–310.

Bloch, E. and Ganguly, J. (2015) ^{176}Lu-^{176}Hf geochronlogy of garnet II: numerical simulations of the development of garnet-whole rock ^{176}Lu-^{176}Hf isochrons and a new method for constraining the thermal history of metamorphic rocks. *Contributions to Mineralogy and Petrology*,**169**, 14.

Buchwald, W.F. (1975) *Handbook of Iron Meteorites: Their History, Distribution, Composition, and Structure*. Volumes 1, 2 and 3. Arizona State University, Center for Meteorite Studies, and University of California Press, Berkeley, California, USA.

Clayton, R.N., Hinton, R.W. and Davis, A.M. (1988) Isotopic variations in the rock-forming elements in meteorites. *Philosophical Transactions of the Royal Society of London*, **A325**, 483–501.

Condit, R.H., Hobbins, R.R. and Birchenall, C.E. (1974) Self-diffusion of iron and sulfur in ferrous sulfide. *Oxidation of Metals*, **8**, 409–455.

Crank, J. (1975) *The Mathematics of Diffusion*: 2nd edition. Clarendon Press, Oxford, UK.

Crispin, K.L., Saha, S., Morgan, D. and Van Orman, J.A. (2012) Diffusion of transition metals in periclase by experiment and first-principles, with implications for core-mantle equilibration during metal percolation. *Earth and Planetary Science Letters*, **357–358**, 42–53.

Criss, R.E. (1999) *Principles of Stable Isotope Distribution*. Oxford University Press, Oxford, UK.

Dauphas, N. (2007) Diffusion-driven kinetic isotope effect of Fe and Ni during formation of the Widmenstätten pattern. *Meteoritics and Planetary Science*, **42**, 1597–1613.

Dauphas, N., Roskosz, M., Alp, E.E., Golden, D.C., Sio, C.K., Tissot, F.L.H., Hu, M.Y., Zhao, J., Gao, L. and Morris, R.V. (2012) A general moment NRIXS approach to the determination of equilibrium Fe isotopic fractionation factors: application to goethite and jarosite. *Geochimica et Cosmochimica Acta*, **94**, 254–275.

Dauphas, N., Roskosz, M., Alp, E.E., Neuville, D.R., Hu, M.Y., Sio, C.K. and Cordier, C. (2014) Magma redox and structural controls on iron isotope variations in Earth's mantle and crust. *Earth and Planetary Science Letters*, **398**, 127–140.

Davis, A.M. and Richter, F.M. (2004) Condensation and evaporation of solar system materials. Pp. 407–430, in: *Meteorites, Comets, and Planets* (A.M. Davis, editor). Treatise on Geochemistry, **1** (H.D. Holland and K.K. Turekian, editors). Elsevier-Pergamon, Oxford, UK.

Davis, A.M., Hashimoto, A., Clayton, R.N. and Mayeda, T.K. (1990) Isotope mass fractionation during evaporation of forsterite (Mg_2SiO_4). *Nature*, **347**, 655–658.

Dodson, M.H. (1973) Closure temperature in cooling geochronological and petrological systems. *Contributions to Mineralogy and Petrology*, **40**, 259–274.

Dohmen, R. and Chakraborty, S. (2003) Mechanism and kinetics of element and isotopic exchange mediated by a fluid phase. *American Mineralogist*, **88**, 1251–1270.

Dohmen, R. and Milke, R. (2010) Diffusion in polycrystalline materials: grain boundaries, mathematical models, and experimental data. Pp. 921–970 in: *Diffusion in Minerals and Melts* (Y. Zhang and D.J. Cherniak, editors). Reviews in Mineralogy and Geochemistry, **72**, Mineralogical Society of America and the Geochemical Society, Chantilly, Virginia, USA.

Eiler, J.M., Valley, J.W. and Baumgartner, L.P. (1993) A new look at stable isotope thermometry. *Geochimica et Cosmochimica Acta*, **57**, 2571–2583.

Furry, W.H., Jones, R.C. and Onsager, L. (1939) On the theory of isotope separation by thermal diffusion. *Physical Review*, **55**, 1083–1095.

Ganguly, J. and Tirone, M. (1999) Diffusion closure temperature and age of a mineral with arbitrary extent of diffusion: theoretical formulation and applications. *Earth and Planetary Science Letters*, **170**, 131–140.

Ghate, P.B. (1964) Screened interaction model for impurity diffusion in zinc. *Physical Review*, **133**, A1167.

Goel, G., Zhang, L., Lacks, D.J. and Van Orman, J.A. (2012) Isotope fractionation by diffusion in silicate melts: insights from molecular dynamics simulations. *Geochimica et Cosmochimica Acta*, **93**, 205–213.

Goldstein, J.I., Scott, E.R.D. and Chabot, N.L. (2009) Iron meteorites: crystallization, thermal history, parent bodies, and origin. *Chemie der Erde-Geochemistry*, **69**, 293–325.

Hoefs, J. (2008) *Stable Isotope Geochemistry*. Springer-Verlag, Berlin.

Jenkin, G.R.T., Rogers, G., Fallick, A.E. and Farrow, C.M. (1995) Rb-Sr closure temperatures in bi-mineralic rocks: a mode effect and test for different diffusion models. *Chemical Geology*, **122**, 227–240.

Kincaid, J.M., Cohen, E.G.D. and López de Haro, M. (1987) The Enskog theory for multicomponent mixtures. IV. Thermal diffusion. *Journal of Chemical Physics*, **86**, 963–975.

Kleine, T., Touboul, M., Van Orman, J.A., Bourdon, B., Maden, C., Mezger, K. and Halliday, A. (2008) Hf-W thermochronometry: closure temperature and constraints on the accretion and cooling history of the H chondrite parent body. *Earth and Planetary Science Letters*, **270**, 106–118.

Klotsman, S.M. (1989 Diffusion and isotope effects for diffusion in transition metals. *Defects and Diffusion Forum*, **66–69**, 85–102.

Kobayashi, H., Kamimura, T., Alfé, D., Sturhahn, W., Zhao, J. and Alp, E.E. (2004) Phonon density of states and compression behavior in iron sulfide under pressure. *Physical Review Letters*, **93**, 195503.

Krawczynski, M.J., Van Orman, J.A., Dauphas, N., Alp, E.E. and Hu, M. (2014) Iron isotope fractionation between metal and troilite: a new cooling speedometer for iron meteorites. *Lunar and Planetary Science Conference Abstracts*, **45**, 2755.

Lacks, D.J., Goel, G., Bopp, C.J. IV, Van Orman, J.A., Lesher, C.E. and Lundstrom, C.C. (2012) Isotope fractionation by thermal diffusion in silicate melts. *Physical Review Letters*, **108**, 065901.

LeClaire, A.D. (1966) Some comments on the mass effect in diffusion. *Philosophical Magazine*, **14**, 1271–1284.

LeClaire, A.D. (1970) Correlation effects in diffusion in solids. Chapter 5 in: *Physical Chemistry: An Advanced Treatise, Vol. 10/Solid State* (H. Eyring, D. Handerson and W. Jost, editors). Academic Press, London.

Lübbehusen, M. and Mehrer, H. (1990) Self-diffusion in α-iron: The influence of dislocations and the effect of the magnetic phase transition. *Acta Metallurgica*, **38**, 283–292.

Lundstrom, C. (2009) Hypothesis for the origin of convergent margin granitoids and Earth's continental crust by thermal migration zone refining. *Geochimica et Cosmochimica Acta*, **73**, 5709–5729.

Matsuhisa, Y., Goldsmith, J.R. and Clayton, R.N. (1978) Mechanisms of hydrothermal crystallization of quartz at 250°C and 15 kbar. *Geochimica et Cosmochimica Acta*, **42**, 173–182.

Mullen, J. (1961) Isotope effect in intermetallic diffusion. *Physical Review*, **121**, 1649–1658.

Nagahara, H. and Ozawa, K. (2000) Isotopic fractionation as a probe of heating processes in the solar nebula. *Chemical Geology*, **169**, 45–68.

Northrop, D.A. and Clayton, R.N. (1966) Oxygen-isotope fractionations in systems containing dolomite. *Journal of Geology*, **74**, 174–196.

Oeser, M., Dohmen, R., Horn, I., Schluth, S. and Weyer, S. (2015) Processes and time scales of magmatic evolution as revealed by Fe-Mg chemical and isotopic zoning in natural olivines. *Geochimica et Cosmochimica Acta*, **154**, 130–150.

Oikwawa, H. (1983) Lattice diffusion of substitutional elements in iron and iron-base solid solutions. A critical review. *Technological Reports, Tohoku University*, **48**, 7–77.

Ozawa, K. and Nagahara, H. (2001) Chemical and isotopic fractionations by evaporation and their cosmochemical implications. *Geochimica et Cosmochimica Acta*, **65**, 2171–2199.

Polyakov, V.B., Clayton, R.N., Horita, J. and Mineev, S.D. (2007) Equilibrium iron isotope fractionation factors of minerals: reevaluation from the data of nuclear inelastic resonant X-ray scattering and Mössbauer spectroscopy. *Geochimica et Cosmochimica Acta*, **71**, 3833–3846.

Polyakov, V.B. and Soultanov, D.M. (2011) New data on equilibrium iron isotope fractionation among sulfides: constraints on mechanisms of sulfide formation in hydrothermal and igneous systems. *Geochimica et Cosmochimica Acta*, **75**, 1957–1974.

Richter F.M., Davis A.M., Ebel D.S. and Hashimoto A. (2002) Elemental and isotopic fractionation of Type B calcium-, aluminum-rich inclusions: experiments, theoretical considerations, and constraints on their thermal evolution. *Geochimica et Cosmochimica Acta*, **66**, 521–540.

Richter, F.M., Davis, A.M., DePaolo, D.J. and Watson E.B. (2003) Isotope fractionation by chemical diffusion between molten basalt and rhyolite. *Geochimica et Cosmochimica Acta*, **67**, 3905–3923.

Richter, F.M., Janney, P.E., Mendybaev, R.A., Davis, A.M. and Wadhwa, M. (2007) Elemental and isotopic fractionation of Type B CAI-like liquids by evaporation. *Geochimica et Cosmochimica Acta*, **71**, 5544–5564.

Richter, F.M., Watson, E.B., Mendybaev, R.A., Teng, F.-Z. and Janney P.E. (2008) Magnesium isotope fractionation in silicate melts by chemical and thermal diffusion. *Geochimica et Cosmochimica Acta*, **72**, 206–220.

Richter, F.M., Dauphas, N. and Teng F.-Z. (2009) Non-traditional fractionation of non-traditional isotopes: evaporation, chemical diffusion and Soret diffusion. *Chemical Geology*, **258**, 92–103.

Schoen, A.H. (1958) Correlation and the isotope effect for diffusion in crystalline solids. *Physical Review Letters*, **1**, 138–140.

Schuessler, J.A., Schoenberg, R., Behrens, H. and von Blanckenburg, F. (2007) The experimental calibration of the iron isotope fractionation factor between pyrrhotite and peralkaline rhyolitic melt. *Geochimica et Cosmochimica Acta*, **71**, 417–433.

Severinghaus, J.P. and Brook, E.J. (1999) Abrupt climate change at the end of the last glacial period inferred from trapped air in polar ice. *Science*, **286**, 930–934.

Severinghaus, J.P., Sowers, T., Brook, E.J. and Alley, R.B. (1998) Timing of abrupt climate change at the end of the Younger Dryas interval from thermally fractionated gases in polar ice. *Nature*, **391**, 141–146.

Shahar, A., Young, E.D. and Manning, C.E. (2008) Equilibrium high-temperature Fe isotope fractionation between fayalite and magnetite: an experimental calibration. *Earth and Planetary Science Letters*, **268**, 330–338.

Sio, C.K.I., Dauphas, N., Teng, F.-Z., Chaussidon, M., Helz, R.T. and Roskosz, M. (2013) Discerning crystal growth from diffusion profiles in zoned olivine by in situ Mg-Fe isotopic analyses. *Geochimica et Cosmochimica Acta*, **123**, 302–321.

Sturhahn, W. (2004) Nuclear resonant spectroscopy. *Journal of Physics: Condensed Matter*, **16**, S497.

Sturhahn, W. and Jackson, J.M. (2007) Geophysical applications of nuclear resonant spectroscopy. *Geological Society of America Specical Papers*, **421**, 157–174.

Sturhahn, W., Toellner, T.S., Alp, E.E., Zhang, X., Ando, M., Yoda, Y., Kikuta, S., Seto, M., Kimball, C.W. and Dabrowski, B. (1995) Phonon density of states measured by inelastic nuclear resonant scattering. *Physical Review Letters*, **74**, 3832.

Sturhahn, W., Alp, E.E., Quast, K.W. and Toellner, T. (1999) Lamb-Mössbauer factor and second-order Doppler shift of hematite. *Advanced Photon Source User Activity Report 1999*.

Teng, F.-Z., Dauphas, N., Helz, R.T., Gao, S. and Huang, S. (2011) Diffusion-driven magnesium and iron isotope fractionation in Hawaiian olivine. *Earth and Planetary Science Letters*, **308**, 317–324.

Tsuchiyama, A., Tachibana, S. and Takahashi, T. (1999) Evaporation of forsterite in the primordial solar nebula: rates and accompanied isotopic fractionation. *Geochimica et Cosmochimica Acta*, **63**, 2451–2466.

Urey, H.C. (1947) The thermodynamic properties of isotopic substances. *Journal of the Chemical Society (Resumed)*, 562–581.

Van Orman, J.A. and Crispin, K.L. (2010) Diffusion in oxides. Pp. 757–825 in: *Diffusion in Minerals and Melts* (Y. Zhang and D.J. Cherniak, editors). Reviews in Mineralogy and Geochemistry, **72**. Mineralogical Society of America and the Geochemical Society, Chantilly, Virginia, USA.

Van Orman, J.A. and Krawczynski, M.J. (2015) Theoretical constraints on the isotope effect for diffusion in minerals. *Geochimica et Cosmochimica Acta*, **164**, 365–381.

Van Orman, J.A., Saal, A.E., Bourdon, B. and Hauri, E.H. (2006) Diffusive fractionation of U-series radionuclides during mantle melting and shallow level melt-cumulate interaction. *Geochimica et Cosmochimica Acta*, **70**, 4797–4812.

Vineyard, G.H. (1957) Frequency factors and isotope effects in solid state rate processes. *Journal of Physics and Chemistry of Solids*, **3**, 121–127.

Wang, J., Davis, A.M., Clayton, R.N. and Hashimoto, A. (1999) Evaporation of single crystal forsterite: evaporation kinetics, magnesium isotope fractionation, and implications of mass-dependent isotopic fractionation of a diffusion-controlled reservoir. *Geochimica et Cosmochimica Acta*, **63**, 953–966

Wang, J., Davis, A.M., Clayton, R.N., Mayeda, T.K. and Hashimoto, A. (2001) Chemical and isotopic fractionation during the evaporation of the $FeO-MgO-SiO_2-CaO-Al_2O_3-TiO_2$ rare earth element melt system. *Geochimica et Cosmochimica Acta*, **65**, 479−494.

Watkins, J.M., DePaolo, D.J., Ryerson, F.J. and Peterson, B.T. (2009) Liquid composition-dependence of calcium isotope fractionation during diffusion in molten silicates. *Geochimica et Cosmochimica Acta*, **73**, 7341−7359.

Watkins, J.M., DePaolo, D.J., Ryerson, F.J. and Peterson, B.T. (2011) Influence of liquid structure on diffusive isotope separation in molten silicates and aqueous solutions. *Geochimica et Cosmochimica Acta*, **75**, 3103−3118.

Watkins, J.M., Liang, Y., Richter, F., Ryerson, F.J. and DePaolo, D.J. (2014) Diffusion of multi-isotopic chemical species in molten silicates. *Geochimica et Cosmochimica Acta*, **139**, 313−326.

Weyer, S., Anbar, A.D., Brey, G.P., Münker, C., Mezger, K. and Woodland, A.B. (2005) Iron isotope fractionation during planetary differentiation. *Earth and Planetary Science Letters*, **240**, 251−264.

Williams, H.M., Markowski, A., Quitté, G., Halliday, A.N., Teutsch, N. and Levasseur, S. (2006) Fe isotope fractionation in iron meteorites: new insights into metal-sulphide segregation and planetary accretion. *Earth and Planetary Science Letters*, **250**, 486−500.

Yang, C.W., Williams, D.B. and Goldstein J.I. (1996) A revision of the Fe-Ni phase diagram at low temperatures (<400°C). *Journal of Phase Equilibria*, **17**, 522−531.

Yang, J. and Goldstein J.I. (2006) Metallographic cooling rates of the IIIAB iron meteorites. *Geochimica et Cosmochimica Acta*, **70**, 3197−3215.

Yang, J., Goldstein, J.I. and Scott, E.R.D. (2007) Iron meteorite evidence for early formation and catastrophic disruption of protoplanets. *Nature*, **446**, 888−891.

Young, E.D., Nagahara, H., Mysen, B.O. and Audet D.M. (1998) Non-Rayleigh oxygen isotope fractionation by mineral evaporation: theory and experiments in the system SiO_2. *Geochimica et Cosmochimica Acta*, **62**, 3109−3116.

Yunker, M.L. and Van Orman, J.A. (2007) Interdiffusion of solid iron and nickel at high pressure. *Earth and Planetary Science Letters*, **254**, 203−213.

Zhang, J., Huang, S. and Davis, A.M. (2014) Calcium and titanium isotopic fractionations during evaporation. *Geochimica et Cosmochimica Acta*, **140**, 365−380.

Subject and Author index

Numbers refer to the page where a definition or explanation of, and/or a *figure* or a **table** for, a given subject is found. Author names are given with the number of the first page of the paper in which they are involved.

3D tomography, of OPX reaction rim, *15*

A

Abart, R., 1, 255, 295, 469
Analytical electron microscopy (AEM), 99
Albitization, *453*
Amorphous precursors, 433, *434*
Analcime, replacement of by leucite, *456*
Anisotropic diffusivities, 279
Anisotropy of surface energy, 324
Apparatus, crystal nucleation, 382
Aqueous solutions, reactions between minerals and, 419
Atomic force microscopy (AFM), 420, *421*, *422*, deflection image of wollastonite, *443*
Atomic interactions in a condensed phase, 296
Auger electrons, 46

B

Backscattered electrons, 42, *44*
Bicrystals, , synthesis of, 22, *23*, *26*, with symmetrical tilt-grain boundary, *341*
Bimodal distribution functions, 200
Birkhoff's idea, 268
Boltzmann's, variable, 269, equation, 276, -Matano method, 278
Born-Mayer potential for anion–cation interaction, *220*
Brachiopod, EBSD of, *70*
Bright field imaging, *112, 114, 126*

C

Cahn-Hilliard-type equation, 288, *290*
Calibration and matrix effects, 143
Cameca NanoSIMS, *137*
Capillary action, schematic, *301*
Capillary force, 297
Cathodoluminescence, 47, *49*
Cellular segregation reactions, 512, during interlayer growth, 522, involving solid solutions, 527

Chemical, composition of materials, 99, fractionation, 143, mechanical feedback, reaction-induced fracturing, 587, potentials, 287, -state imaging, 171
Chemography of a binary system, *489*, of a ternary system, *523*
Classical and non-classical nucleation theory, *431*
Classical interaction potentials, 219
Classical nucleation theory, 347
Clinopyroxene, chemistry, 387, melt partitioning, 395
Coarsening, 308, theory of, 316
Coincident site lattice model, 328
Composition-temperature diagram for the albite-orthoclase binary solid-solution, *291*
Congruent *vs.* incongruent dissolution, 439
Continuum scale, 224
Continuum-based models for transport and deformation, 595
Contrast Transfer Function (CTF), *120, 121, 122*
Cooling and decompression, 380
Cooling speedometry, 630
COR, crystallographic orientation relationships, 541, frequency of, 555, formation, 555, formation by overgrowth on a pre-existing crystal surface, 564, formation by intergrowth, 570, formation by phase transformation, 572, formation by deformation or recrystallization, 577, influence of *T* and *P* on, 563, mechanisms of formation of, 564, 570, 572, of crystals sharing heterophase boundary segments, 544, of crystals sharing homophase boundary segments, 544, types, *551*
Corona, 6, formation, 487, growth, metamorphic mineral reactions, 469
Coupled flux analysis, 361
Crystal, centres, spatial disposition of, 476, defects, atomic-scale modelling of, 226, growth, from solution, *429, rate*, effect of element partitioning on, 392, nucleation, 374, pairs, non-genetic criteria for next-neighbour, 545, symmetry or space group, 115, −interface relations, *565*, −melt boundary layer, 385

www.ingramcontent.com/pod-product-compliance
Lightning Source LLC
Chambersburg PA
CBHW080346220326
41598CB00030B/4622